Ternary Alloys Based on III–V Semiconductors

Ternary Alloys Based on III–V Semiconductors

Vasyl Tomashyk

V. Ye. Lashkaryov Institute of Semiconductor Physics of NAS of Ukraine, Kyiv

CRC Press
Taylor & Francis Group
Boca Raton London New York

CRC Press is an imprint of the
Taylor & Francis Group, an **informa** business

CRC Press
Taylor & Francis Group
6000 Broken Sound Parkway NW, Suite 300
Boca Raton, FL 33487-2742

First issued in paperback 2019

© 2018 by Taylor & Francis Group, LLC
CRC Press is an imprint of Taylor & Francis Group, an Informa business

No claim to original U.S. Government works

ISBN-13: 978-1-4987-7838-1 (hbk)
ISBN-13: 978-0-367-88976-0 (pbk)

Library of Congress Cataloging-in-Publication Data

Names: Tomashyk, Vasyl, author.
Title: Ternary alloys based on III-V semiconductors / Vasyl Tomashyk.
Description: Boca Raton, FL : CRC Press, Taylor & Francis Group, [2017] |
Includes bibliographical references and index.
Identifiers: LCCN 2016058676| ISBN 9781498778381 (hardback ; alk. paper) |
ISBN 1498778380 (hardback ; alk. paper) | ISBN 9781498778411 (e-book) |
ISBN 1498778410 (e-book)
Subjects: LCSH: Semiconductors--Materials. | Phase diagrams, Ternary. |
Ternary alloys.
Classification: LCC QC611.8.A44 T675 2017 | DDC 537.6/223--dc23
LC record available at https://lccn.loc.gov/2016058676

Visit the Taylor & Francis Web site at
http://www.taylorandfrancis.com

and the CRC Press Web site at
http://www.crcpress.com

This book is dedicated to the memory of my mother, Kateryna Tomashyk (neé Suhai, 1928–1972), and my father, Mykola Tomashyk (1923–2006).

Contents

Chapter 12 ■ Systems Based on InN

List of Figures

List of Symbols and Acronyms

$A_xB_yC_z(I)$	Low-pressure polymorph of $A_xB_yC_z$	HRTEM	High-resolution transmission electron microscopy
$A_xB_yC_z(II)$	High-pressure polymorph of $A_xB_yC_z$		
AES	Auger electron spectroscopy	L	Liquid
AFM	Atomic force microscopy	LPE	Liquid phase epitaxy
bcc	Body-centered cubic structure	MBE	Molecular beam epitaxy
Bu^t	t-butyl, C_4H_9	Me	Methyl, CH_3
c-	Cubic	mM	mmol
CALPHAD	Calculation of the phase diagram	MOCVD	Metal-organic chemical vapor deposition
CVD	Chemical vapor deposition	MOVPE	Metal-organic vapor phase epitaxy
DC	Direct current	ppm	Parts per million
DMF	Dimethylformamide	S	Solid
DSC	Differential scanning calorimetry	SEM	Scanning electron microscopy
DTA	Differential thermal analysis	T	Temperature (in K)
EDX	Energy-dispersive x-ray spectroscopy	TEM	Transmission electron microscopy
EMF	Electromotive force	THF	Tetrahydrofurane, C_4H_8O
EPMA	Electron probe microanalysis	V	Vapor
Et	Ethyl, C_2H_5	(X)	Solid solution based on X
G	Gas	XRD	X-ray diffraction
h-	Hexagonal	XPS	X-ray photoelectron spectroscopy

Preface

A SIGNIFICANT VOLUME OF SEMICONDUCTOR devices and circuits employ III–V semiconductors, the most commonly used crystal material for integrated circuits. For electronic applications, this semiconductor offers the basic advantage of higher electron mobility, which translates into higher operating speeds. In addition, devices made with III–V compounds provide lower-voltage operation for specific functions, radiation hardness (especially important for satellites and space vehicles), and semi-insulating substrates (avoiding the presence of parasitic capacitance in switching devices). Among these semiconductors, nitrides, which have recently been intensively studied, occupy a special place. The intrinsic group of superhard materials includes cubic boron nitride and ternary compounds such as B–C–N, which possess an innate hardness. The alloys formed in the In–E–Sb ternary systems are now used as Pb-free solders.

In most cases, the properties of III–V semiconductors can be modified by isovalent or heterovalent foreign impurity (E) doping, which results in forming a solid solution or creating a separate electronic level in the semiconductor lattice. Such ternary III–E–V materials enlarge and ameliorate the properties of these semiconductors and other materials based on III–V compounds.

The problem of reproducible production of the doped materials with predicted and desired properties cannot be successfully resolved without knowledge of the appropriate ternary system phase diagram, which is a "map" for technologists. Unfortunately, the main information is scattered in separate papers. This book aimed to collect and systematize all available data on the III–E–V ternary systems. It includes 577 critically compiled ternary systems based on III–V semiconductors, including literature data from 1492 papers; these data are illustrated in 505 figures. The information is divided into 15 chapters according to the number of possible combinations of B, Al, Ga, or In with N, P, As, or Sb. The chapters are structured so that first the Group III element numbers in the periodic system are presented in increasing order, that is, from B to In compounds, and then the pnictogen numbers are given in increasing order, that is, from nitrides to antimonides. The same principle is used for further description of the systems in every chapter, that is, in increasing order from the third component number in the periodic system.

Every ternary system database description contains brief information in the following order: the diagram type, the possible phase transformation and physical-chemical interaction of the components, methods of the equilibrium investigation, thermodynamic characteristics, and the method of the sample preparations. Solid and liquid phase equilibriums with vapor are also illustrated in some cases because of their importance for crystal growth from the vapor, from the melt, and by the vapor–liquid–solid technique.

The homogeneity range is of a great importance for governing the crystal defect structure. Therefore, this reference book collects all such data accessible to date. In addition, this book presents data on the baric and temperature dependences of the impurities' solubility in both the semiconductors' lattice and the liquid phase, as well as the pressure–composition relationship. As semiconductors and metal mutual solubility are usually small values, the illustrating figures present restricted concentration ranges (in mol%).

Most of the figures are presented in their original form, although some have minor corrections. If the published data essentially varies, several versions are presented in comparison. The content of system components is presented in mol% (this is not indicated in the figures). If the original phase diagram is given with mass%, this is indicated in the figures.

This book is meant for researchers at industrial and national laboratories and for university and graduate students majoring in materials science, solid-state chemistry, and engineering. It is also suitable for phase diagram researchers, inorganic chemists, and solid-state physicists.

Author

Vasyl Tomashyk is executive director and head of the department of the V. Ye. Lashkaryov Institute for Semiconductor Physics of the National Academy of the Sciences of the Ukraine. He graduated from Chernivtsi State University in Ukraine in 1972 (master of chemistry). He is a doctor of chemical sciences (1992), a professor (1999), and the author of about 600 publications in scientific journals and conference proceedings and eight books (three of them published by CRC Press), which are devoted to physical–chemical analysis, the chemistry of semiconductors, and the chemical treatment of semiconductor surfaces.

Tomashyk is a specialist at the high international level in the field of solid-state and semiconductor chemistry, including physical–chemical analysis and the technology of semiconductor materials. He was head of research topics within the international program Copernicus. He is a member of the Materials Science International Team (Stuttgart, Germany, since 1999), which prepares a series of prestigious reference books under the titles *Ternary Alloys* and *Binary Alloys*, and has published 35 articles in the Landolt–Börnstein New Series. Tomashyk is actively working with young researchers and graduate students, and under his supervision, 20 PhD theses have been prepared. For many years, he is also a professor at Ivan Franko Zhytomyr State University in Ukraine.

CHAPTER 1

Systems Based on BN

1.1 BORON–HYDROGEN–NITROGEN

Some compounds are formed in this ternary system.

H_3NBH_3. This compound melts at $104.5°C \pm 0.5°C$ (Sorokin et al. 1963), decomposes upon heating, and exhibits a first-order rotational order–disorder phase transition at 225 K (Hoon and Reynhardt 1983). The high-temperature modification crystallizes in the tetragonal structure with the lattice parameters $a = 524.0 \pm 0.5$ and $c = 502.8 \pm 0.8$ pm at 295 K (Hoon and Reynhardt 1983) ($a = 525.5$ and $c = 504.8$ pm and a calculated density of 0.74 g·cm^{-3} [Hughes 1956]; $a = 523.4$ and $c = 502.7$ pm and a calculated density of 0.74 g·cm^{-3} [Lippert and Lipscomb 1956]). The low-temperature modification crystallizes in the orthorhombic structure with the lattice parameters $a = 551.7 \pm 0.1$, $b = 474.2 \pm 0.1$, and $c = 502.0 \pm 0.1$ pm at 110 K (Hoon and Reynhardt 1983) ($a = 722$, $b = 738$, and $c = 523$ pm and calculated and experimental densities of 0.73 and 0.74 g·cm^{-3}, respectively [Sorokin et al. 1963]; $a = 542.1 \pm 0.5$, $b = 494.5 \pm 0.4$, and $c = 502.3 \pm 0.6$ pm at 200 K [Bühl et al. 1991a, 1991b]; $a = 539.5 \pm 0.2$, $b = 488.7 \pm 0.2$, and $c = 498.6 \pm 0.2$ pm at 200 K and a calculated density of 0.7799 g·cm^{-3} [Klooster et al. 1999]).

In the high-temperature tetragonal phase, the cell parameters are linear functions of temperature (Hoon and Reynhardt 1983). In the low-temperature orthorhombic phase, the length of the c axis is independent of temperature, while the a axis decreases nonlinearly as the temperature is increased. The length of the b axis increases nonlinearly with increasing temperature.

To prepare H_3NBH_3, $NaBH_4$ (1.0 M) and powdered $(NH_4)_2SO_4$ (1.0 M) were added to a 10 L three-neck round-bottom flask with indentation fitted with an overhead stirrer, a stopper, and a condenser fitted with a connecting tube (Ramachandran and Gagare 2007). The connecting tube was vented via an oil bubbler to the hood exhaust. Tetrahydrofurane (THF) (6 L) was transferred into the flask, and the contents were vigorously stirred at 40°C. Upon completion (~2 h), the reaction mixture was cooled to room temperature and filtered. The filtrate was concentrated under vacuum to obtain the title compound. All manipulations were carried out under an inert atmosphere in flame-dried glassware and cooled under N_2.

B_2H_7N. This compound was obtained only when diborane was used in excess of that required by the formula $B_2H_6 \cdot 2NH_3$ (Schlesinger et al. 1938). Its melting point is –66.5°C, and its boiling point is 76.2°C.

$B_2H_7N \cdot NH_3$. This compound, as the stable solid, was obtained from the interaction of B_2H_7N with NH_3 (Schlesinger et al. 1938). On heating to 200°C, it yields $B_3N_3H_6$.

$[(NH_3)BH_2][BH_4]$. This compound crystallizes in the tetragonal structure with the lattice parameters $a = 1072.1$ and $c = 924.1$ pm (Bowden et al. 2010). It is formed from the room temperature decomposition of NH_4BH_4, via a NH_3BH_3 intermediate.

NH_4BH_4. This compound is a white crystalline solid that is unstable above –20°C (Parry et al. 1958).

$[(NH_3)_2BH_2][BH_4]$. In a typical synthesis of this compound, $[(NH_3)_2BH_2]Cl$ (14.9 mM) and KBH_4 (14.9 mM) were placed in a 150 mL flask, which was connected to a vacuum line, and liquid NH_3 (60 mL) was condensed into the flask (Lingam et al. 2011). The solution was stirred for 90 min at 195 K. KCl was filtered from the solution, and NH_3 was removed slowly from the filtrate under vacuum to yield the title compound. It is stable at lower temperatures, slowly converting irreversibly to NH_3BN_3 as the temperature rises (263 K).

$NH_3B_3H_7$. This compound forms two crystalline modifications, a disordered, tetragonal form that is stable at about 25°C, and an ordered, monoclinic form that is stable at lower temperatures (Nordman and Reimann 1959). The lattice parameters of the high-temperature are $a = 611$ and $c = 657$ pm, and a calculated density is 0.765 g·cm⁻³. The low-temperature modification is characterized by the next lattice parameters: $a = 1040.0 \pm 1.5$, $b = 482.4 \pm 0.6$, and $c = 999.7 \pm 1.2$ and $\beta = 115.20 \pm 0.15°$ at 193 ± 10 K.

A first-order transition involving an enthalpy of transformation of 5159 kJ·mol⁻¹ and entropy increment of 17.36 J·mol⁻¹ K⁻¹ has been established at 24°C (Westrum and Levitin 1959). The heat capacity at constant pressure, entropy, enthalpy increment, and free energy function at 27°C are 159.91 J·mol⁻¹ K⁻¹, 166.94 J·mol⁻¹ K⁻¹, 28.485 kJ·mol⁻¹, and –71.96 J·mol⁻¹ K⁻¹, respectively.

$B_3N_3H_6$. This compound crystallizes in the tetragonal structure with the lattice parameters $a = 542.8 \pm 0.1$ and $c = 1627.9 \pm 0.8$ pm and a calculated density of 1.115 g·cm⁻³ at 115 K, and $a = 546.3 \pm 0.1$ and $c = 1631.5 \pm 0.4$ pm and a calculated density of 1.089 g·cm⁻³ at 160 K (Boese et al. 1994).

$(BH_2NH_2)_3$. This compound crystallizes in the orthorhombic structure with the lattice parameters $a = 438.3 \pm 0.2$, $b = 1219.3 \pm 0.2$, and $c = 1118.0 \pm 0.2$ pm at 180 ± 2 K and a calculated density of 1.317 g·cm⁻³ (Lingam et al. 2012) ($a = 440.3 \pm 0.3$, $b = 1221.0 \pm 0.7$, and $c = 1122.7 \pm 1.0$ pm at room temperature and experimental and calculated densities of 0.96 ± 0.01 and 0.95 g·cm⁻³, respectively [Corfield and Shore 1973]). To obtain it, a 100 mL THF solution of NH_3BH_2Cl (15 mM) was slowly added to a suspension of $NaNH_2$ (15 mM) in THF (Lingam et al. 2012). The reaction mixture was stirred for 2 h and allowed to stand for 6 h. NaCl was filtered from the solution, and the solvent was evaporated under a dynamic vacuum to yield the title compound. All manipulations were carried out on a high-vacuum line or in a glove box filled with high-purity N_2.

It could also be prepared by addition of HCl to borazole, followed by reduction with $NaBH_4$ in diglyme according to the equation $2(B_3N_3H_6 \cdot 3HCl) + 6NaBH_4 = 2B_3N_3H_{12} + 6NaCl + 3B_2H_6$ (Dahle and Schaeffer 1961). Although the material sublimed in vacuum readily at 100°C, the substance did not melt and no measurable

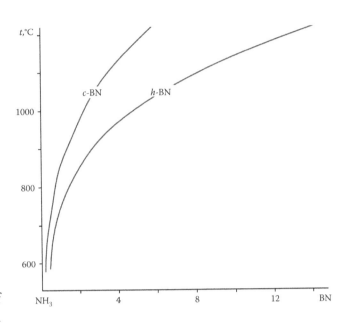

FIGURE 1.1 Values of h-BN and c-BN solubility in supercritical ammonia in the 0.065–4.2 GPa region. (From Turkevich, V.Z., *J. Phys. Condens. Matter*, 14(44), 10963, 2002. With permission.)

vapor pressure was exhibited up to 150°C. At this temperature, slow decomposition with evolution of H_2 began.

The crystals of $(BH_2NH_2)_3$ were grown from benzene solution at room temperature (Corfield and Shore 1973).

$(BH_2NH_2)_n$. Cycloborazanes of composition $(BH_2NH_2)_n$ have been synthesized by the reaction of $NaNH_2$ with $BH_2(NH_3)_2BH_4$ in liquid NH_3 (Böddeker et al. 1961). Rings with $n = 2$, 3, and 5 have been isolated. Small amounts of $(BH_2NH_2)_4$ were also formed. Cyclodiborazane is apparently thermodynamically unstable with respect to isomerization to cyclotriborazane. Cyclopentaborazane is a white microcrystalline solid, which is extremely resistant to hydrolysis. It does not sublime *in vacuo*.

$BN–NH_3$. Figure 1.1 demonstrates the results of experimental determination and thermodynamic calculations of the limiting solubility of h-BN and c-BN in supercritical ammonia (Turkevich 2002).

1.2 BORON–LITHIUM–NITROGEN

$LiB(N_3)_4$. This compound is formed in the B–Li–N ternary system. $LiBH_4$ reacts with hydrogen azide (HN_3) in etherial solution at room temperature to give a solid, white, moisture-sensitive and explosive $LiB(N_3)_4$ (Wiberg and Michaud 1954).

The possible phase transformations of BN in this ternary system have been discussed by the chemical reactions of Li_3N, h-BN, and c-BN at the conditions of 5.0 GPa and 1300°C–1500°C (Guo et al. 2015). The results of the reaction between Li_3N and h-BN show that certain B–Li–N eutectic compounds that were produced in the thermodynamically stable region of c-BN have no catalytic effect for c-BN growth. This indicates that a certain irreversible BN precipitation or dissolution process takes place in Li_3N melt. However, the reaction between Li_3N and h-BN is preferential when Li_3N, h-BN, and c-BN coexist in this system, and the regrowth of c-BN is observed.

BN–Li_3N. Two different diagrams have been proposed for this section. DeVries and Fleischer (1972) studied the BN–Li_3BN_2(I) subsystem at a pressure of 5 GPa, and eutectic was observed at 49 at% Li and 1610°C ± 15°C between Li_3BN_2(I) and c-BN. More recently, Turkevich (2002) showed that Li_3BN_2 melted incongruently around 1350°C under 5.3 GPa and observed that the temperature of the peritectic equilibrium was the lowest temperature of the c-BN synthesis in the B–Li–N system under 5.3 GPa.

This apparent contradiction between both diagrams is well explained with the hypothesis that the diagram proposed by Turkevich (2002) is a stable one, whereas that proposed by DeVries and Fleischer (1972) is a metastable. Figure 1.2 shows both stable and metastable diagrams for the BN–Li_3N section under 5.3 GPa. This

figure has been modified to take into account the true transition temperature of c-BN → h-BN given around 2950°C by Turkevich (2002). The accepted transition temperature lies around 1770°C under 5.3 GPa (Kudaka et al. 1966; DeVries and Fleischer 1972; Nakano et al. 1994b).

The **Li_3BN_2** ternary compound is formed in this system. It has a polymorphic transformation at 862°C. α-Li_3BN_2 crystallizes in the tetragonal structure with the lattice parameters $a = 464.35 \pm 0.02$ and $c = 525.92 \pm 0.05$ pm and experimental and calculated densities of 1.75 and 1.747 (1.755 [Goubeau and Anselment 1961]) g·cm⁻³, respectively (Yamane et al. 1987). β-Li_3BN_2 crystallizes in the monoclinic structure with the lattice parameters $a = 515.02 \pm 0.02$, $b = 708.24 \pm 0.02$, and $c = 679.08 \pm 0.02$ pm and $\beta = 112.956 \pm 0.002°$ and experimental and calculated densities of 1.74 and 1.737 g·cm⁻³, respectively (Yamane et al. 1986), and melts at around 916°C (Yamane et al. 1987) (at 870°C ± 5°C [Goubeau and Anselment 1961]).

α- and β-Li_3BN_2 were prepared by slow cooling of the Li_3N and BN mixtures (the Li_3N/BN molar ratio was 1.0–1.2) at 800°C and 900°C–930°C, respectively (Goubeau and Anselment 1961; Yamane et al. 1986, 1987). Single crystals of α-Li_3BN_2 were prepared from the Li_3N and BN starting mixture (the Li_3N/BN molar ratio was 1.0–1.2), which was heated at 1030°C for 20 min and cooled to 730°C at a rate of about 2°C·min⁻¹ (Yamane et al. 1987). Because of this compound's instability with respect to moisture, they were sealed in the glass capillaries. β-Li_3BN_2 crystallizes directly from the undercooled liquid at 887°C.

High-temperature oxide melt drop solution calorimetry was used to study the energetics of Li_3BN_2 formation (McHale et al. 1999). The standard enthalpy of formation is –534.5 ± 16.7 kJ·mol⁻¹. From this value and others available in the literature, the enthalpy of formation from BN and Li_3N was calculated: $\Delta H^{298}_f = -117.5 \pm 17.5$ kJ·mol⁻¹.

The p-T stability region of a high-pressure polymorph of Li_3BN_2(II) is shown in Figure 1.3 (DeVries and Fleischer 1969, 1972). The principal compound that coexists with c-BN at high pressures is a high-pressure polymorph of Li_3BN_2. It has been shown that c-BN does not form when this compound melts or decomposes. When BN is added to Li_3BN_2 in excess of the composition 53 mol% BN + 47 mol% Li_3N, then c-BN precipitates from solution.

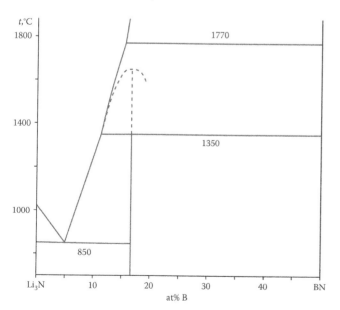

FIGURE 1.2 Phase diagram of the BN–Li_3N system at 5 GPa. (From DeVries, R.C., and Fleischer, J.F., *J. Cryst. Growth*, *13–14*, 88, 1972 and Turkevich, V.Z., *J. Phys. Condens. Matter*, *14*(44), 10963, 2002. With permission.)

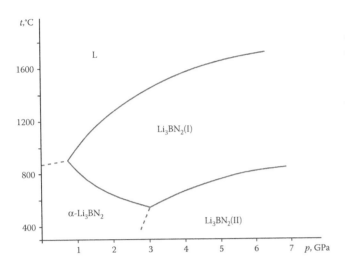

FIGURE 1.3 *p-T* diagram for Li$_3$BN$_2$. (From DeVries, R.C., and Fleischer, J.F., *J. Cryst. Growth*, *13–14*, 88, 1972. With permission.)

Another compound was synthesized from the mixtures in the compositional range of Li$_3$N/BN < 1.0, but its single phase could not be obtained (Yamane et al. 1987).

1.3 BORON–SODIUM–NITROGEN

BN–Na$_3$N. The **Na$_3$BN$_2$** ternary compound is formed in this system. It crystallizes in the monoclinic structure with the lattice parameters $a = 571.7 \pm 0.1$, $b = 793.1 \pm 0.1$, and $c = 788.3 \pm 0.1$ pm and $\beta = 111.32 \pm 0.01°$ and experimental and calculated densities of 2.10 and 2.150 g·cm^{-3}, respectively (Evers et al. 1990).

1.4 BORON–COPPER–NITROGEN

The phase equilibria in this ternary system at 900°C were investigated by means of x-ray diffraction (XRD) (Weitzer et al. 1990). No ternary compound was observed. BN coexists with Cu, and negligible mutual solubility was indicated between BN and Cu.

1.5 BORON–SILVER–NITROGEN

The phase equilibria in this ternary system at 800°C were investigated by means of XRD (Weitzer et al. 1990). No ternary compound was observed. BN coexists with Ag, and negligible mutual solubility was indicated between BN and Ag.

1.6 BORON–GOLD–NITROGEN

The phase equilibria in this ternary system at 800°C were investigated by means of XRD (Weitzer et al.

1990). No ternary compound was observed. BN coexists with Au, and negligible mutual solubility was indicated between BN and Au.

1.7 BORON–MAGNESIUM–NITROGEN

MgNB$_9$. This compound is formed in the B–Mg–N ternary system. It crystallizes in the trigonal structure with the lattice parameters $a = 549.60 \pm 0.02$ and $c = 2008.73 \pm 0.16$ pm and a calculated density of 2.570 g·cm^{-3} (Mironov et al. 2002). To obtain the title compound, a mixture of Mg and B (molar ratio of 5:1) was placed in a BN crucible. The crucible was placed in a tungsten container and subjected to a high-pressure and high-temperature reaction in a high-gas-pressure apparatus. First, Ar pressure was applied, and then the temperature was raised to 1600°C at a rate of 600°C·h^{-1}. The sample was maintained at this temperature for 1 h and then cooled from 1500°C at a rate of 60°C·h^{-1} under 100 MPa of Ar. Below 1500°C, the sample was cooled to room temperature at a rate of 600°C·h^{-1}. After crushing the crucible, the product was heated *in vacuo* at 750°C for 15 min to remove Mg.

BN–MgB$_2$. X-ray powder diffraction with synchrotron radiation was used to study the formation of *c*-BN in this system at pressures to 6.8 GPa and temperatures to 2000 K (Solozhenko et al. 1999). For the formation of *c*-BN, when it crystallizes from a BN solution in the melt of the system under study, the threshold pressure (4.5 ± 0.1 GPa) and low-temperature boundary [T (K) = 1633 – 9.2p (GPa)] are established. It was found that the position of the high-temperature boundary of the *p-T* region of the *c*-BN crystallization is specified by the nucleation of the cubic phase. The existence of the threshold pressure of *c*-BN crystallization is dictated by the strong pressure dependence of the nucleation rate. It is this dependence that is responsible for the region of the *c*-BN spontaneous crystallization being off the *h*-BN ⟷ *c*-BN equilibrium line by several hundred degrees. Below 4.4 GPa, the formation of *c*-BN has not be observed up to about 1730°C.

BN–Mg$_3$N$_2$. Figure 1.4 shows the proposal for the phase diagram of the BN–Mg$_3$BN$_3$ subsystem at 5.5 GPa (Lorenz and Orgzall 1995). Up to 1277°C, a two-phase region consisting of solid BN and **Mg$_3$BN$_3$** is observed. Starting at high pressure and ambient temperature, increasing the temperature first results in the formation of Mg$_3$BN$_3$. Enhanced Mg$_3$BN$_3$ formation proceeds above 890°C (Lorenz et al. 1995; Lorenz and Orgzall

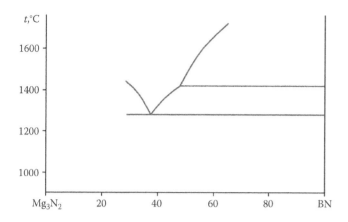

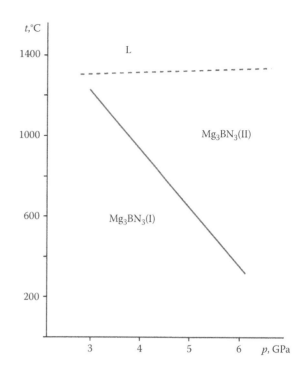

FIGURE 1.4 The phase diagram of the BN–Mg_3BN_3 system at 5.5 GPa. (From Lorenz, H., and Orgzall, I., *Diamond Relat. Mater.*, 4(8), 1046, 1995. With permission.)

FIGURE 1.5 *p-T* diagram for Mg_3BN_3. (From Nakano, S., et al., *Diamond Relat. Mater.*, 2(8), 1168, 1993. With permission.)

1995). At approximately 1277°C (1297°C ± 15°C [Pikalov et al. 1989]; 1295°C ± 7°C at 2.5 GPa [Endo et al. 1979]), the eutectic is reached. A further temperature increase leads to the formation of **$Mg_3B_2N_4$**. At 1417°C (1489°C ± 5°C at 2.5 GPa [Endo et al. 1979]), a peritectic is reached. $Mg_3B_2N_4$ is in the metastable phase, and slow pressure release results in the decomposition of this compound at 0.5 GPa and 207°C. Mg_3BN_3 melts at about 1500°C (Nakano et al. 1993) (at 1337°C ± 15°C) and decomposes at 1300°C to form *c*-BN (Nakano et al. 1992). The existence of two modifications of $Mg_3B_2N_4$ (Elyutin et al. 1981) was not confirmed (Hohlfeld 1989). The coexistence of Mg_3BN_3 and $Mg_3B_2N_4$ compounds was observed in the low-temperature *c*-BN region.

The phase relations in the BN–Mg_3N_2 system were investigated in regions of pressure from 3.0 to 8.0 GPa and temperature up to 1630°C by means of differential thermal analysis (DTA) and XRD (Gladkaya et al. 1994). It was found that the succession in formation of the intermediate compounds Mg_3BN_3(II) (high-pressure phase) and $Mg_3B_2N_4$ depends on the molar ratios of *h*-BN and Mg_3N_2 and on the *p-T* conditions. In the *p-T* region of *c*-BN growth, the system has three metastable eutectics, such as Mg_3N_2–*h*-BN, Mg_3BN_3–*h*-BN, and $Mg_3B_2N_4$–*h*-BN. It was found that eutectic temperatures are pressure dependent. The difference in the lower-temperature limits of *c*-BN growth regions is explained by *c*-BN crystallization from different eutectic melts.

The phase stability of Mg_3BN_3 was studied using the first-principles density functional calculations under isotropic hydrostatic pressures up to 50 GPa (Wang et al. 2006). This compound undergoes pressure-induced

hexagonal-to-tetragonal structural transformation at ~2 GPa. According to the data of Zhukov et al. (1994), a high-pressure phase is formed at *p* > 3 GPa and temperatures of 1030°C–1430°C, and at an increasing pressure up to 7 GPa, the transition temperature decreases up to 430°C. The phase boundary between low- and high-pressure Mg_3BN_3 was determined to be *p*(GPa) = 7.2 – 0.0035t (°C) (Figure 1.5) (Nakano et al. 1992, 1993).

The hexagonal structure of Mg_3BN_3 is characterized by the lattice parameters *a* = 354.453 ± 0.004 and *c* = 1603.536 ± 0.030 pm (Hashizume and Hiraguchi 1991; Hiraguchi et al. 1991) (*a* = 354.2 ± 0.1 and *c* = 1601 ± 1 pm and experimental and calculated densities of 2.39 ± 0.02 and 2.40 g·cm⁻³, respectively [Zhukov et al. 1994]; *a* = 351.3 and *c* = 1599 pm [calculated values] [Wang et al. 2006]).

The tetragonal structure of Mg_3BN_3 is characterized by the lattice parameters *a* = 310.7 ± 0.4 and *c* = 770 ± 1 pm and experimental and calculated densities of 2.75 ± 0.02 and 2.80 g·cm⁻³, respectively (Zhukov et al. 1994) (*a* = 306.6 and *c* = 782.9 pm [calculated values] [Wang et al. 2006]).

It should be stated that Mg_3BN_3 undergoes a series of phase transformations in the course of pressure and temperature treatment (Lorenz et al. 1992, 1993). Under the action of pressure, the structure changes from an initially hexagonal one to an orthorhombic one. Further transitions occur within this structure type if temperature is applied. This orthorhombic structure is conserved after a rapid quench, and thus is metastable under normal conditions. It is characterized by the lattice parameters $a = 309.33 \pm 0.02$, $b = 313.36 \pm 0.02$, and $c = 770.05 \pm 0.05$ pm (Hiraguchi et al. 1993).

Mg_6BN_5 is also obtained in the $BN–Mg_3N_2$ system (Nakano et al. 1993; Zhukov et al. 1998; Kulinich et al. 2000). It crystallizes in the hexagonal structure with the lattice parameters $a = 539.7 \pm 0.2$ and $c = 1058.5 \pm 0.5$ pm and experimental and calculated densities of 2.88 ± 0.05 and 2.82 g·cm^{-3}, respectively.

Mg_3BN_3(I) was synthesized using h-BN powder and Mg flakes (molar ratio ~1:3 with a slight excess of h-BN) (Hiraguchi et al. 1993). The mixture was placed in an uncapped steel capsule in a furnace filled with flowing N_2. The furnace was heated to 600°C over 2 h, held constant at this temperature for 4 h, and then cooled. Next, a screw cap was placed on the capsule and the furnace temperature was raised to 1200°C over 4 h, and then held constant at this temperature for 10 h with H_2 gas flowing through the furnace. Mg_3BN_3(I) was transformed to Mg_3BN_3(II) in a high-pressure apparatus using cubic anvils. The conversion took place at 4.0 GPa and 1200°C in 15 min.

1.8 BORON–CALCIUM–NITROGEN

The phase relations of the system Ca–B–N were investigated at a pressure of 2.5 GPa by both DTA and a quenching method (Endo et al. 1981). When BN reacted with Ca at temperatures higher than 1150°C, **$Ca_3B_2N_4$** was formed, together with small amounts of Ca_3N_2 and CaB_6.

$BN–Ca_3N_2$. The $BN–Ca_3B_2N_4$ subsystem has a eutectic relationship at 1316°C $\pm$ 6°C and 2.5 GPa (Endo et al. 1981). The synthesis of c-BN was established by using $Ca_3B_2N_4$ under the thermodynamic stable conditions of c-BN. In the system, the low-temperature limit of c-BN formation was closely related to the eutectic relationship between BN and $Ca_3B_2N_4$.

A reversible phase transition of $Ca_3B_2N_4$ to an ordered super structure was found between 215°C and 240°C (Häberlen et al. 2005) (at 265°C [Wörle et al. 1998]). α-$Ca_3B_2N_4$ (low-temperature modification) crystallizes in the orthorhombic structure with the lattice parameters $a = 1024.18 \pm 0.02$, $b = 732.43 \pm 0.02$,

and $c = 2091.60 \pm 0.04$ pm (Häberlen et al. 2005) and an experimental density of 2.68 g·cm^{-3} (Goubeau and Anselment 1961). β-$Ca_3B_2N_4$ (high-temperature modification) crystallizes in the cubic structure with the lattice parameter $a = 732.24 \pm 0.03$ pm and a calculated density of 2.5110 g·cm^{-3} (Wörle et al. 1998) ($a = 734.7 \pm 0.2$ pm [Womelsdorf and Meyer 1994]).

High-temperature oxide melt drop solution calorimetry was used to study the energetics of $Ca_3B_2N_4$ formation (McHale et al. 1999). The standard enthalpy of formation is -1062.1 ± 15.4 kJ·mol^{-1}. From this value and others available in the literature, the enthalpy of formation from BN and Ca_3N_2 was calculated: $\Delta H^{298}{}_f = -120.6 \pm 15.6$ kJ·mol^{-1}.

To prepare α-$Ca_3B_2N_4$, a stoichiometric mixture of Ca_3N_2 and BN was loaded into Cu ampoules inside a glove box (Häberlen et al. 2005). In the following procedures, the ampoules were arc welded and sealed into evacuated silica tubes. The ampoules were then placed in a furnace, heated to 950°C, and held constant for 40 h before lowering the temperature to room temperature within 6 h. Reaction products contained a white, sometimes light green, powder of α-$Ca_3B_2N_4$.

β-$Ca_3B_2N_4$ was obtained in sealed Nb ampoules from a stoichiometric mixture of Ca_3N_2 and BN (Wörle et al. 1998). The ampoules were filled with the finely ground and well-mixed starting components in a glove box with a purified Ar atmosphere. After sealing under Ar, the ampoules were placed into an Ar-filled Al_2O_3 tube and heated to 1200°C for 1 h to yield pure products. Large single crystals were prepared within 3 h at about 1630°C. The cooling rates had to be very high (>30°C·min^{-1}) to avoid the formation of α-$Ca_3B_2N_4$. Pure β-$Ca_3B_2N_4$ was in the form of a colorless single.

According to the data of Kulinich et al. (2000) and Häberlen et al. (2002a, 2002b), the **Ca_3BN_3** ternary compound is also formed in the $BN–Ca_3N_2$ system. It crystallizes in the tetragonal structure with the lattice parameters $a = 354.94 \pm 0.01$ and $c = 821.36 \pm 0.8$ pm and a calculated density of 2.513 g·cm^{-3} (Häberlen et al. 2002a, 2002b). This compound was synthesized as a light gray powder from reactions of equimolar amounts of BN and Ca_3N_2 at 1000°C.

All manipulations were performed under dry Ar atmosphere inside a glove box or in an airtight glass apparatus (Womelsdorf and Meyer 1994; Wörle et al. 1998; Häberlen et al. 2005).

Ca$_6$BN$_5$ compound is also formed in this system. It crystallizes in the orthorhombic structure with the lattice parameters $a = 821.0 \pm 0.3$, $b = 921.7 \pm 0.4$, and $c = 1056.7 \pm 0.5$ pm and experimental and calculated densities of 2.69 ± 0.02 and 2.75 g·cm^{-3}, respectively (Kulinich et al. 1995, 2000). It was obtained by the sintering of a mixture of BN and Ca$_3$N$_2$ (molar ratio 1:2) at 900°C–980°C for 2–4 h in N$_2$ atmosphere. At longstanding heating at 600°C–1400°C under a pressure of 3–6 MPa, the title compound gradually decomposes to form Ca$_3$B$_2$N$_4$ and Ca$_3$N$_2$.

1.9 BORON–STRONTIUM–NITROGEN

BN–Sr$_3$N$_2$. Sr$_3$B$_2$N$_4$ ternary compound is formed in this system. Phase stability studies revealed a phase transition between a primitive and a body-centered lattice around 820°C (Häberlen et al. 2005). α-Sr$_3$B$_2$N$_4$ (low-temperature modification) crystallizes in the cubic structure with the lattice parameter $a = 763.8 \pm 0.1$ pm (Häberlen et al. 2005) ($a = 764.56 \pm 0.03$ pm and a calculated density of 3.8 g·cm^{-3} [Womelsdorf and Meyer 1994]).

To prepare α-Sr$_3$B$_2$N$_4$, a stoichiometric mixture of Sr$_3$N$_2$ and BN was loaded into Cu ampoules inside a glove box (Häberlen et al. 2005). In the following procedures, the ampoules were arc welded and sealed into evacuated silica tubes. The ampoules were then placed in a furnace, heated to 950°C, and held constant for 40 h before lowering the temperature to room temperature within 6 h. Reaction products contained a dark brown powder of α-Sr$_3$B$_2$N$_4$. It could also be obtained by the interaction of a stoichiometric mixture of Sr$_3$N$_2$ and BN into Nb ampoules, which were sealed under vacuum in quartz and heated at 1100°C for 2 weeks (Womelsdorf and Meyer 1994).

All manipulations were performed under a dry Ar atmosphere inside a glove box or in an airtight glass apparatus (Womelsdorf and Meyer 1994; Häberlen et al. 2005).

1.10 BORON–BARIUM–NITROGEN

BN–Ba$_3$N$_2$. Ba$_3$B$_2$N$_4$ ternary compound is formed in this system (Goubeau and Anselment 1961; Reckeweg et al. 2003). It crystallizes in the orthorhombic structure with the lattice parameters $a = 424.73 \pm 0.02$, $b = 1105.60 \pm 0.04$, and $c = 1475.72 \pm 0.06$ pm and a calculated density of 4.853 g·cm^{-3} (Reckeweg et al. 2003).

To obtain the title compound, Ba$_2$N (3.45 mM) and h-BN (4.25 mM) were mixed and arc welded into a Nb container (Reckeweg et al. 2003). The container was sealed in an evacuated silica tube, which was placed in a box furnace and heated to 1030°C within 16 h. After 3 days, the furnace was switched off and allowed to cool down to room temperature. The reaction produced transparent, yellow, or sometimes light blue crystals of orthorhombic Ba$_3$B$_2$N$_4$, which are air and moisture sensitive. All manipulations of educts and products were performed in a glove box under permanently purified argon.

1.11 BORON–ZINC–NITROGEN

The phase equilibria in this ternary system at 250°C were investigated by means of XRD (Weitzer et al. 1990). No ternary compound was observed. BN coexists with Zn$_3$N$_2$ and Zn, and negligible mutual solubility was indicated between BN and Zn.

1.12 BORON–CADMIUM–NITROGEN

The phase equilibria in this ternary system at 250°C were investigated by means of XRD (Weitzer et al. 1990). No ternary compound was observed. BN coexists with Cd, and negligible mutual solubility was indicated between BN and Cd.

1.13 BORON–ALUMINUM–NITROGEN

According to the data of Rizzol et al. (2002), this ternary system at 1500°C can be divided into four triangles (Al–AlB$_{12}$–AlN, AlB$_{12}$–AlN–BN, AlB$_{12}$–B–BN, and AlN–BN–N). Phase equilibria in the B–Al–N system at 900°C under 100 kPa of argon (in the absence of external nitrogen) have been established from XRD (Rogl and Schuster 1992) and are given in Figure 1.6. They are characterized by the formation of a stable equilibrium (BN) + (AlN) + AlB$_2$. Lattice parameters suggest a mutual solubility of AlN and BN of less than ~4 mol% with no significant changes in solubility between 900°C and 1600°C. No ternary compounds were observed (Rogl and Schuster 1992; Wen 1993; Novikov et al. 1997). There is practically no nitrogen solubility in the binary aluminum borides. Some isothermal sections of the B–Al–N ternary system have been calculated using the program Thermo-Calc (Wen 1993). According to these calculations, AlN is in thermodynamic equilibrium with Al$_2$B$_3$, AlB$_2$, and AlB$_{12}$ and BN forms tie-line with AlB$_{12}$ at 230°C. Increasing

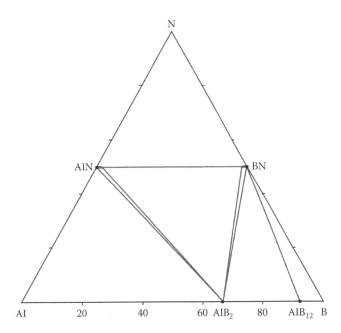

FIGURE 1.6 Isothermal section of the B–Al–N system at 900°C under 0.1 MPa of argon (in the absence of external nitrogen). (From Rogl, P., and Schuster, J.C., in *Phase Diagrams of Ternary Boron Nitride and Silicon Nitride Systems*, American Society for Metals International, Materials Park, OH, 1992, 3–5. With permission.)

the temperature to 730°C leads to disappearing of the AlN–Al$_2$B$_3$ tie-line. At 1230°C, in this ternary system, there is one two-phase region (AlN + L) and four three-phase regions (BN + AlN + N, BN + AlN + AlB$_{12}$, BN + AlB$_{12}$ + B, and AlN + AlB$_{12}$ + L). Increasing the temperature to 1730°C leads to enlarge the AlN + L two-phase region and decreasing the AlN + AlB$_{12}$ + L three-phase region. At 2060°C, a new two-phase region BN + L appears and instead of three-phase regions BN + AlN + AlB$_{12}$ and AlN + AlB$_{12}$ + L two other three-phase regions BN + AlN + L and BN + AlB$_{12}$ + L appear. A large region of the gas phase appears at 2330°C, which leads to the formation of four two-phase (AlN + L, L + G, BN + G, and BN + L) and two three-phase (BN + L + G and AlN + L + G) regions. At 3430°C, the gas phase region enlarges and occupies the part of the ternary system near Al–N binary system and the large two-phase region L + G exists.

BN–Al. At the sintering of *c*-BN with Al at 1230°C–1430°C, AlN is formed (Novikov et al. 1997).

BN–AlN. A hypothetical phase diagram of this quasi-binary system was constructed in the investigation at 8–9 GPa and 2000°C–2500°C (Bartnitskaya et al. 1980). Homogeneous solid solutions were

obtained at the simultaneous nitration in the NH$_3$ flow of BN and Al powder. It is worth noting that the phase diagram presented there is in disagreement with the phase rule, and the large solubility ranges disagree with the information about the mutual solubility of BN and AlN. Thermodynamic calculations based on a regular solution model and experimental results have predicted an unstable region of mixing to occur in the BN–AlN system (Lyutaya and Bartnitskaya 1973b; Takayama et al. 2001). The interaction parameter (138.5 kJ·mol^{-1}) that was used in the calculations has been analytically obtained by the valence force field model, modified for wurtzite structures (Takayama et al. 2001). According to the interaction parameter, the value of the critical temperature was found to be 8060°C. The phase diagram of the BN–AlN system, including the spinodal and binodal curves, was also calculated using the generalized quasi-chemical approximation and *ab initio* total energy method (Teles et al. 2002b). From these calculations, the critical temperature appears to be very high, approximately 9230°C; this results in a very large miscibility gap (Scolfaro 2002; Teles et al. 2002b). The phase diagram, as far as obtained experimentally, verifies that there is spinodal decomposition for the B$_x$Al$_{1-x}$N alloys in the interval $0.037 < x < 0.949$ at 1030°C. According to the data of Kumar et al. (2015), the calculated *T-x* phase diagram shows a broad miscibility gap with a high critical temperature equal to 2790°C.

The calculated structural and electronic properties of the B$_x$Al$_{1-x}$N solid solutions show that E_g has a strong nonlinear dependence on x (Riane et al. 2008). For $x > 0.71$, these materials have a phase transition from a direct-gap semiconductor to an indirect-gap semiconductor (the direct gap was found in the range of $0.07 < x < 0.83$ [Djoudi et al. 2012]). First-principles calculations indicated that at $x = 0.04$ and $x = 0.84$, these solid solutions have a phase transition from a direct-gap semiconductor to an indirect-gap semiconductor (Kumar et al. 2015).

Self-consistent calculations of total energies and the electronic structure based on the scalar relativistic full-potential "linearized augmented plane-wave" method were carried out (Riane et al. 2008). The lattice parameter as a function of the B composition has a positive deviation from Vegard's law. From the calculated results, it was found that the bowing parameter of the lattice constant for B$_x$Al$_{1-x}$N solid solutions is −30.6 ± 2.1 pm (Djoudi et al. 2012). The calculated concentration dependence of the lattice parameter could be expressed by the next equation: a (pm) = 459.46 + 91.3x + 3.7x^2 (Kumar et al. 2015). According to the data of Teles et al. (2002b), the composition dependence of the lattice constant and the bulk modulus are linear. A nonlinear increasing with x of the bulk modulus within the $B_xAl_{1-x}N$ solid solutions was observed (Riane et al. 2008).

Layers of $B_xAl_{1-x}N$ solid solutions with x up to 0.01 were grown on sapphire by metal-organic vapor phase epitaxy (MOVPE) at 1050°C using BEt_3, $AlMe_3$, and NH_3 as precursors (Polyakov et al. 1997b). These layers crystallize in the wurtzite-type structure. Gupta et al. (1999) noted that $B_xAl_{1-x}N$ solid solution with B composition up to 6 at% was successfully grown on sapphire substrates. From the calculated spinodal isotherms, the maximum B solubility having a uniform single-phase solid solution is 2.8 at% at 1000°C (Wei and Edgar 2000).

1.14 BORON–GALLIUM–NITROGEN

BN–GaN. Thermodynamic calculations based on a regular solution model and experimental results have predicted an unstable region of mixing to occur in this system (Lyutaya and Bartnitskaya 1973a; Takayama et al. 2001). The interaction parameter (189.1 kJ·mol⁻¹) that was used in the calculations has been analytically obtained by the valence force field model, modified for wurtzite structures (Takayama et al. 2001). According to the interaction parameter, the value of the critical temperature was found to be 11130°C. The phase diagram of the BN–GaN system, including the spinodal and binodal curves, was calculated using the generalized quasi-chemical approximation and *ab initio* total energy method (Teles et al. 2002b). From these calculations, the critical temperature appears to be very high, approximately 9230°C; this results in a very large miscibility gap (Scolfaro 2002; Teles et al. 2002b). The phase diagram, as far as obtained experimentally, verifies that there is spinodal decomposition for the $B_xGa_{1-x}N$ alloys in the interval 0.028 < x < 0.995 at 730°C. From the calculated spinodal isotherms, the maximum B solubility having a uniform single-phase solid solution is 1.8 at% at 1000°C (Wei and Edgar 2000). In MOVPE growth of $B_xGa_{1-x}N$, phase separation occurred at a B concentration of more than 1.5 at%, as determined from XRD and Auger measurements. Layers of $B_xGa_{1-x}N$ solid solutions with x up to 0.007 were grown on sapphire by MOVPE in the temperature range from 450°C to 1000°C using BEt_3, $GaMe_3$, and NH_3 as precursors (Polyakov et al. 1997a). These layers crystallize in the wurtzite-type structure. The band gap varies from 3.4 eV for x = 0 to 3.63 eV for x = 0.05. Boron can also be substitutionally incorporated into GaN in a concentration of at least 7 at% by room temperature ion implantation. $B_xGa_{1-x}N$ solid solutions with a B concentration of up to 2 at% were successfully grown on sapphire substrates (Gupta et al. 1999). Wurtzite solid solutions with x up to 0.0456 were grown by molecular beam epitaxy (MBE) (Vezin et al. 1997).

For x > 0.75, a phase transition from direct to indirect gap for $B_xGa_{1-x}N$ occurs at the band gap energy of 5.1 eV (Djoudi et al. 2012).

From the calculated results, it was found that the bowing parameter of the lattice constant for the $B_xGa_{1-x}N$ solid solutions is −42.0 ± 4.1 pm (Djoudi et al. 2012). According to the data of Teles et al. (2002b), the composition dependence of the lattice constant and the bulk modulus are linear.

1.15 BORON–INDIUM–NITROGEN

The phase equilibria in this ternary system at 120°C were investigated by means of XRD (Weitzer et al. 1990). No ternary compound was observed. BN coexists with InN and In, and negligible mutual solubility was indicated between BN and In.

BN–InN. Thermodynamic calculations based on a regular solution model and experimental results have predicted an unstable region of mixing to occur in this system (Takayama et al. 2001). The interaction parameter (389.9 kJ·mol⁻¹) that was used in the calculations has been analytically obtained by the valence force field model, modified for wurtzite structures. According to the interaction parameter, the value of the critical temperature was found to be 23230°C.

1.16 BORON–THALLIUM–NITROGEN

The phase equilibria in this ternary system at 200°C were investigated by means of XRD (Weitzer et al. 1990). No ternary compound was observed. BN coexists with TlN and Tl, and negligible mutual solubility was indicated between BN and Tl.

1.17 BORON–SCANDIUM–NITROGEN

The thermodynamic phase equilibria in the B–Sc–N ternary system at 1400°C and 0.1 MPa of Ar have been established from the data of XRD (Figure 1.7) (Klesnar and Rogl 1990). Ternary phase equilibria reveal the absence of ternary compounds. There is no compatibility between BN and Sc. Mutual solubility under select experimental conditions was found to be rather restricted. At 1400°C, BN is in thermodynamic equilibrium with ScB_2 and ScB_{12}, and ScN coexists with ScB_2.

The samples were annealed at 1400°C for 100 h under 0.1 MPa of Ar.

1.18 BORON–YTTRIUM–NITROGEN

The thermodynamic phase equilibria in the B–Y–N ternary system at 1400°C and 0.1 MPa of Ar have been established from the data of XRD (Figure 1.8) (Klesnar and Rogl 1990). Ternary phase equilibria reveal the absence of ternary compounds. There is no compatibility between BN and Y. Mutual solubility under select experimental conditions was found to be rather restricted. At 1400°C, BN is in thermodynamic equilibrium with YB_4, YB_6, YB_{12}, YB_{66}, and YN, and YN coexists with YB_2 and YB_4.

The samples were annealed at 1400°C for 100 h under 0.1 MPa of Ar.

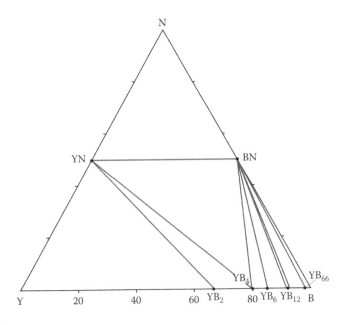

FIGURE 1.8 Isothermal section of the B–Y–N system at 1400°C under 0.1 MPa of argon. (From Klesnar, H.P., and Rogl, P., *High Temp. High Pressure*, 22(4), 453, 1990. With permission.)

1.19 BORON–LANTHANUM–NITROGEN

The isothermal sections of the B–La–N ternary system at 1400°C and 1650°C is given in Figure 1.9 (Rogl 2001). $La_3B_2N_4$, $La_3B_3N_6$ ($LaBN_2$), $La_4B_2N_5$ [$La_4(B_2N_4)N$], $La_5B_2N_6$ [$La_5(B_2N_4)N_2$], $La_5B_3N_8$, $La_5B_4N_9$, $La_6B_4N_{10}$, and $La_{15}B_8N_{25}$ ternary compounds are formed in this system.

$La_3B_2N_4$ crystallizes in the orthorhombic structure with the lattice parameters $a = 362.94 \pm 0.03$, $b = 641.25 \pm 0.06$, and $c = 1097.20 \pm 0.08$ and a calculated density of 6.43 g·cm⁻³ (Reckeweg and Meyer 1999c) ($a = 363.42 \pm 0.02$, $b = 641.41 \pm 0.05$, and $c = 1097.42 \pm 0.12$ and a calculated density of 6.42 g·cm⁻³ [Rogl et al. 1990]). For single crystal growth of this compound, a typical reaction batch consisted of Ln (1.8 mM), h-BN (2 mM), and $CaCl_2$ (0.5 mM) as a reactive melt (Reckeweg and Meyer 1999c). A little more than the stoichiometric amount of Ln powder was used. The educts were finely ground in a mortar and filled into Ta ampoules. The reaction tube was placed into a chamber furnace, which was heated at 25°C·h⁻¹ up to 1230°C. After 4 days, the temperature was slowly reduced to room temperature at a rate of 10°C·h⁻¹. Crystals of $La_3B_2N_4$ were obtained as bunches of black thin needles and as thin rectangular plates. This compound could also be prepared by the interaction of BN, LaN, and Ln at 1230°C for 4 days. Samples heat treated at 1800°C and

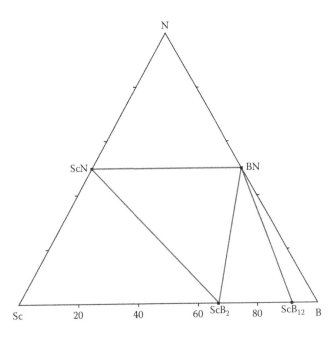

FIGURE 1.7 Isothermal section of the B–Sc–N system at 1400°C under 0.1 MPa of argon. (From Klesnar, H.P., and Rogl, P., *High Temp. High Pressure*, 22(4), 453, 1990. With permission.)

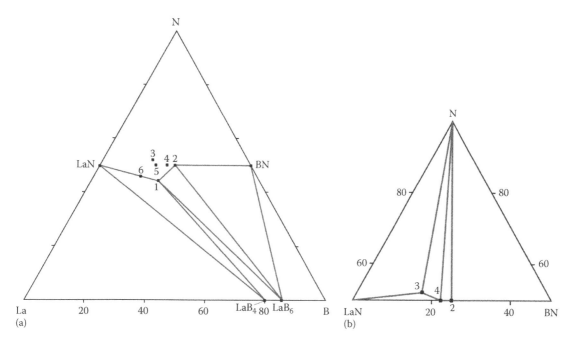

FIGURE 1.9   Isothermal sections of the B–La–N system at (a) 1400°C under 0.1 MPa of argon and (b) 1650°C under 0.1 MPa of nitrogen: 1 = $La_3B_2N_4$; 2 = $LaBN_2$; 3 = $La_{15}B_8N_{25}$; 4 = $La_5B_4N_9$; 5 = $La_5B_3N_8$; 6 = $La_5B_2N_6$. (From Rogl, P., *Int. J. Inorg. Mater.*, 3(3), 201, 2001. With permission.)

furnace cooled proved the stability of $La_3B_2N_4$ up to this temperature (Rogl 2001).

According to the data of Jing et al. (2002), exothermic solid-state metathesis reactions provide an effective route for the synthesis of lanthanum nitridoborates. An ignition of the metathesis reaction between $LaCl_3$ and Li_3BN_2 was obtained at 560°C. Single crystal growth was obtained at slightly higher temperatures by the aid of a LiCl flux, which is formed during the reaction. Modified metathesis reactions with additional Li_3N may be used to adjust certain N contents for desired compositions. Additional Li was used to synthesize metal-rich nitridoborates. The preparative route could be used not only for the synthesis of $La_3B_2N_4$, but also for the preparation of $La_3B_3N_6$ ($LaBN_2$), $La_4B_2N_5$ [$La_4(B_2N_4)N$], $La_5B_2N_6$ [$La_5(B_2N_4)N_2$], and $La_6B_4N_{10}$.

$La_3B_3N_6$ ($LaBN_2$) crystallizes in the triclinic structure with the lattice parameters $a = 661.3 \pm 0.1$, $b = 687.0 \pm 0.1$, and $c = 779.8 \pm 0.1$ pm and $\alpha = 106.06 \pm 0.01°$, $\beta = 90.55 \pm 0.01°$, and $\gamma = 115.63 \pm 0.01°$ (Reckeweg and Meyer 1999a, 1999b). It was synthesized by reaction of LaN with *h*-BN in a $CaCl_2$ flux (molar ratio 1:5:1). Arc-welded Ta ampoules were used as reaction containers. The ampoules were sealed in an evacuated silica tube to shield them from moisture and air. The title compound was obtained as colorless transparent single crystals.

$La_4B_2N_5$ [$La_4(B_2N_4)N$] crystallizes in the monoclinic structure with the lattice parameters $a = 1260.4 \pm 0.1$, $b = 366.15 \pm 0.03$, and $c = 919.8 \pm 0.1$ pm and $\beta = 129.727 \pm 0.006°$ (Jing et al. 2001b). It was prepared by reductive salt metathesis reaction, $4LaCl_3 + 2Li_3BN_2 + Li_3N + 3Li = La_4B_2N_5$ [$La_4(B_2N_4)N$] $+ 12LiCl$, at 950°C over 3 days or from LaN and *h*-BN with $CaCl_2$ as a flux at 1200°C over 3 days.

$La_5B_2N_6$ [$La_5(B_2N_4)N_2$] crystallizes in the monoclinic structure with the lattice parameters $a = 1259.5 \pm 0.2$, $b = 368.53 \pm 0.04$, and $c = 909.4 \pm 0.2$ pm and $\beta = 106.03 \pm 0.02°$ and a calculated density of 6.55 g·cm^{-3} (Jing and Meyer 2000). It was synthesized by solid-state reactions in fused Ta containers from Li_3BN_2, $LaCl_3$, Li_3N, and La at 950°C.

$La_5B_4N_9$ crystallizes in the orthorhombic structure with the lattice parameters $a = 988.25 \pm 0.05$, $b = 1263.48 \pm 0.07$, and $c = 770.33 \pm 0.04$ at 20°C and a calculated density of 6.00 g·cm^{-3}, and $a = 978.8 \pm 0.5$, $b = 1260.7 \pm 0.8$, and $c = 767.8 \pm 0.3$ at 210 K (Reckeweg and Meyer 1999c). Its single crystals have been obtained as a by-product while attempting to synthesize the crystals of $La_3B_2N_4$. Crystals appeared as black, irregular plates. They were also obtained when a transition metal was employed as a "boron trap" according to the following procedure: La powder (1 mM), Mn or Ni powder

(1 mM), and h-BN (2 mM) were finely ground with $CaCl_2$ (0.5 mM) and filled into Ta ampoules. All other reaction parameters remained the same as in the case of obtaining $La_3B_2N_4$.

$La_6B_4N_{10}$ crystallizes in the monoclinic structure with the lattice parameters $a = 971.89 \pm 0.06$, $b = 1479.41 \pm 0.09$, and $c = 762.32 \pm 0.04$ pm and $\beta = 90.005 \pm 0.009°$ and a calculated density of 6.162 g·cm^{-3} (Jing 2001a). It was synthesized by solid-state reactions at high temperatures. The starting materials were weighed in a glove box under Ar, intimately mixed, pressed into tablets, and introduced into Ta ampoules. These were then sealed in a welding device under Ar and melted under vacuum in a quartz ampoule. The mixture of h-BN and LaN (molar ratio 3:2) was heated using $CaCl_2$ as a flux for 3 days at 1230°C. The title compound could also be prepared from the reaction of $LaCl_3$, Li_3BN_2 or $Ca_3(BN_2)_2$, and Li_3N or Ca_3N_2 at 950°C for 3 days.

$La_{15}B_8N_{25}$ is only stable under nitrogen (Rogl 2001).

1.20 BORON–CERIUM–NITROGEN

The isothermal sections of the B–Ce–N ternary system at 1400°C and 1800°C is given in Figure 1.10 (Rogl 2001). $Ce_3B_2N_4$, $Ce_3B_3N_6$ ($CeBN_2$), $Ce_4B_2N_5$ [$Ce_4(B_2N_4)N$], and $Ce_{15}B_8N_{25}$ ternary compounds are formed in this system.

$Ce_3B_2N_4$ crystallizes in the orthorhombic structure with the lattice parameters $a = 356.20 \pm 0.03$, $b = 631.90 \pm 0.06$, and $c = 1071.91 \pm 0.08$ and a calculated density of 6.86 g·cm^{-3} (Reckeweg and Meyer 1999c) ($a = 356.53 \pm 0.05$, $b = 631.60 \pm 0.21$, and $c = 1071.31 \pm 0.50$ and a calculated density of 6.86 g·cm^{-3} [Rogl et al. 1990]). For single crystal growth of this compound, a typical reaction batch consisted of Ce (1.8 mM), h-BN (2 mM), and $CaCl_2$ (0.5 mM) as a reactive melt (Reckeweg and Meyer 1999c). A little more than the stoichiometric amount of Ce powder was used. The educts were finely ground in a mortar and filled into Ta ampoules. The reaction tube was placed into a chamber furnace, which was heated at 25°C·h^{-1} up to 1230°C. After 4 days, the temperature was slowly reduced to room temperature at a rate of 10°C·h^{-1}. Crystals of $Ce_3B_2N_4$ were obtained as bunches of black thin needles and as thin rectangular plates. This compound could also be prepared by the interaction of BN, CeN, and Ce at 1230°C for 4 days. Samples heat treated at 1800°C and furnace cooled proved the stability of $Ce_3B_2N_4$ up to this temperature (Rogl 2001).

$Ce_3B_3N_6$ ($CeBN_2$) crystallizes in the triclinic structure with the lattice parameters $a = 658.3 \pm 0.4$, $b = 677.5 \pm 0.5$, and $c = 773.0 \pm 0.5$ pm and $\alpha = 106.05 \pm 0.08°$, $\beta = 90.73 \pm 0.08°$, and $\gamma = 115.52 \pm 0.08°$ (Reckeweg and Meyer 1999a, 1999b).

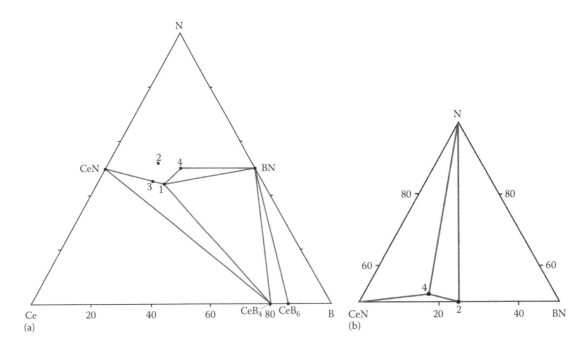

FIGURE 1.10 Isothermal sections of the B–Ce–N system at (a) 1400°C under 0.1 MPa of argon and (b) 1800°C under 0.1 MPa of nitrogen: 1 = $Ce_3B_2N_4$; 2 = $CeBN_2$; 3 = $Ce_4B_2N_5$; 4 = $Ce_{15}B_8N_{25}$. (From Rogl, P., *Int. J. Inorg. Mater.*, 3(3), 201, 2001. With permission.)

It was synthesized by reaction of CeN with h-BN in a CaCl$_2$ flux (molar ratio 1:5:1). Arc-welded Ta ampoules were used as reaction containers. The ampoules were sealed in an evacuated silica tube to shield them from moisture and air. The title compound was obtained as black single crystals with defined faces but irregular shape.

Ce$_4$B$_2$N$_5$ [Ce$_4$(B$_2$N$_4$)N] crystallizes in the monoclinic structure with the lattice parameters $a = 1238.2 \pm 0.1$, $b = 357.32 \pm 0.03$, and $c = 905.21 \pm 0.07$ pm and $\beta = 129.700 \pm 0.001°$ and a calculated density of 7.03 g·cm^{-3} (Jing et al. 2001b). It was prepared by reductive salt metathesis reaction, $4CeCl_3 + 2Li_3BN_2 + Li_3N + 3Li = Ce_4B_2N_5$ [Ce$_4$(B$_2$N$_4$)N] + 12LiCl, at 950°C over 3 days or from CeN and h-BN with CaCl$_2$ as a flux at 1200°C over 3 days.

Ce$_{15}$B$_8$N$_{25}$ crystallizes in the rhombohedral structure with the lattice parameters $a = 1094.6 \pm 0.3$ pm and $\alpha = 82.96 \pm 0.04°$ and a calculated density of 6.56 g·cm^{-3} (Gaudé et al. 1985). The title compound was obtained by the interaction of BN and CeN at 1750°C. It is only stable under nitrogen (Rogl 2001).

1.21 BORON–PRASEODYMIUM–NITROGEN

Pr$_3$B$_2$N$_4$ and Pr$_3$B$_3$N$_6$ (PrBN$_2$) ternary compounds are formed in this system.

Pr$_3$B$_2$N$_4$ crystallizes in the orthorhombic structure with the lattice parameters $a = 353.46 \pm 0.04$, $b = 630.04 \pm 0.13$, and $c = 1079.04 \pm 0.23$ pm and a calculated density of 6.92 g·cm^{-3} (Reckeweg and Meyer 1999c) ($a = 353.99 \pm 0.05$, $b = 631.04 \pm 0.04$, and $c = 1079.47 \pm 0.13$ pm and a calculated density of 6.89 g·cm^{-3} [Rogl et al. 1990]). For single crystal growth of this compound, a typical reaction batch consisted of Pr (1.8 mM), h-BN (2 mM), and CaCl$_2$ (0.5 mM) as a reactive melt (Reckeweg and Meyer 1999c). A little more than the stoichiometric amount of Pr powder was used. The educts were finely ground in a mortar and filled into Ta ampoules. The reaction tube was placed into a chamber furnace, which was heated at 25°C·h^{-1} up to 1230°C. After 4 days, the temperature was slowly reduced to room temperature at a rate of 10°C·h^{-1}. Crystals of Pr$_3$B$_2$N$_4$ were obtained as bunches of black thin needles and as thin rectangular plates. Samples heat treated at 1800°C and furnace cooled proved the stability of Pr$_3$B$_2$N$_4$ up to this temperature (Rogl 2001).

Pr$_3$B$_3$N$_6$ (PrBN$_2$) crystallizes in the trigonal structure with the lattice parameters $a = 1211.95 \pm 0.09$ and $c = 701.53 \pm 0.07$ pm and a calculated density of 6.020 g·cm^{-3} (Orth and Schnick 1999) (in a rhombohedral structure, with the lattice parameters $a = 1211.44 \pm 0.13$ and $c = 701.26 \pm 0.15$ pm [in a hexagonal setting] and a calculated density of 6.03 g·cm^{-3} [Rogl and Klesnar 1992]). It was prepared by reacting compacted h-BN powder and Pr filling (Rogl and Klesnar 1992). After a first reaction for several hours at 1800°C, the samples were heat treated for another 20 h at 1400°C before cooling to room temperature. Due to the sensitivity of the sample to moisture, handling of the specimens was performed in an Ar-filled glove box. Single crystals of the title compound were obtained by the reaction of Pr and BN$_x$(NH)$_y$(NH$_2$)$_z$ in a NaCl melt under N$_2$ atmosphere in a high-frequency furnace at 1250°C (Orth and Schnick 1999).

1.22 BORON–NEODYMIUM–NITROGEN

The phase relations and phase stabilities have been derived for the B–Nd–N ternary system at 1400°C and above by means of XRD (Klesnar and Rogl 1992). Under the experimental conditions, Nd$_3$B$_2$N$_4$ and Nd$_3$B$_3$N$_6$ (NdBN$_2$) ternary compounds are formed in this system (Figure 1.11). Phase equilibria at 1400°C under 0.1 MPa of Ar are mainly characterized by the incompatibility of Nd with BN. Each of the ternary compounds is found to be in two-phase equilibrium with h-BN; Nd$_3$B$_2$N$_4$ is

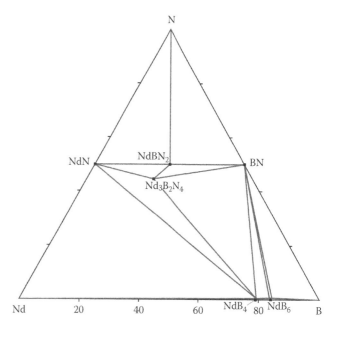

FIGURE 1.11 Isothermal section of the B–Nd–N system at 1400°C under 0.1 MPa of argon. (From Klesnar, H., and Rogl, P., *J. Am. Ceram. Soc.*, 75(10), 2825, 1992. With permission.)

observed only at temperatures below 1800°C and under 0.1 MPa of Ar.

The samples were annealed at 1400°C for 96 h (Klesnar and Rogl 1992).

Nd₃B₂N₄ crystallizes in the orthorhombic structure with the lattice parameters $a = 351.52 \pm 0.04$, $b = 627.01 \pm 0.15$, and $c = 1075.59 \pm 0.23$ and a calculated density of 7.15 g·cm⁻³ (Reckeweg and Meyer 1999c) ($a = 352.00 \pm 0.11$, $b = 627.75 \pm 0.24$, and $c = 1075.58 \pm 0.42$ and a calculated density of 7.13 g·cm⁻³ [Rogl et al. 1990; Klesnar and Rogl 1992]). For single crystal growth of this compound, a typical reaction batch consisted of Nd (1.8 mM), h-BN (2 mM), and CaCl₂ (0.5 mM) as a reactive melt (Reckeweg and Meyer 1999c). A little more than the stoichiometric amount of Nd powder was used. The educts were finely ground in a mortar and filled into Ta ampoules. The reaction tube was placed into a chamber furnace, which was heated at 25°C·h⁻¹ up to 1230°C. After 4 days, the temperature was slowly reduced to room temperature at a rate of 10°C·h⁻¹. Crystals of Nd₃B₂N₄ were obtained as bunches of black thin needles and as thin rectangular plates. Samples slowly cooled from 1800°C contained only small amounts of Nd₃B₂N₄, whereas those annealed at 1400°C and below immediately revealed the Nd₃B₂N₄ phase as the major constituent (Rogl 2001).

Nd₃B₃N₆ (NdBN₂) crystallizes in the trigonal structure with the lattice parameters $a = 1207.28 \pm 0.15$ and 1207.40 ± 0.18, and $c = 694.19 \pm 0.31$ and 694.54 ± 0.30 pm (for two samples) (Klesnar and Rogl 1992) (in a rhombohedral structure, with the lattice parameters $a = 1207.4 \pm 0.2$ and $c = 694.5 \pm 0.3$ pm [in a hexagonal setting] and a calculated density of 6.24 g·cm⁻³ [Rogl and Klesnar 1992]; in a monoclinic structure, with the lattice parameters $a = 695.0 \pm 0.3$, $b = 1207.7 \pm 0.4$, and $c = 585.6 \pm 0.2$ pm and $\beta = 99°32$ and experimental and calculated densities of 5.93 and 6.27 g·cm⁻³, respectively [Gaudé 1983]). It was prepared by reacting compacted h-BN powder and Nd filling (Rogl and Klesnar 1992). After a first reaction for several hours at 1800°C, the samples were heat treated for another 20 h at 1400°C before cooling to room temperature. This compound could also be prepared by heating a mixture of BN and NdN (molar ratio 1:1) at 1550°C for about 20 h (Gaudé 1983). Due to the sensitivity of the sample to moisture, handling of the specimens was performed in an Ar-filled glove box.

1.23 BORON–SAMARIUM–NITROGEN

The phase relations and phase stabilities have been derived for the B–Sm–N ternary system at 1400°C and above by means of XRD (Klesnar and Rogl 1992). Under the experimental conditions, Sm₃B₃N₆ (SmBN₂) ternary compounds are formed in this system (Figure 1.12). Phase equilibria at 1400°C under 0.1 MPa of Ar are mainly characterized by the incompatibility of Sm with BN. Sm₃B₃N₆ (SmBN₂) is found to be stable only at temperatures above 1400°C and under a partial pressure of 0.1 MPa of nitrogen.

The samples were annealed at 1400°C for 96 h (Klesnar and Rogl 1992).

Sm₃B₃N₆ (SmBN₂) crystallizes in the trigonal structure with the lattice parameters $a = 1196.84 \pm 0.23$, 1197.19 ± 0.30, and 1195.18 ± 0.69 and $c = 684.93 \pm 0.28$, 684.96 ± 0.23, and 682.44 ± 0.64 pm (for three samples) (Klesnar and Rogl 1992) (in a rhombohedral structure, with the lattice parameters $a = 1197.2 \pm 0.3$ and $c = 685.0 \pm 0.2$ pm [in a hexagonal setting] and a calculated density of 6.65 g·cm⁻³ [Rogl and Klesnar 1992]; in a monoclinic structure, with the lattice parameters $a = 689.7 \pm 0.4$, $b = 1192.8 \pm 0.5$, and $c = 578.3 \pm 0.2$ pm and $\beta = 99°84$ and experimental and calculated densities of 6.52 and 6.70 g·cm⁻³, respectively [Gaudé 1983]). It was prepared by reacting compacted h-BN powder and Sm filling (Rogl and

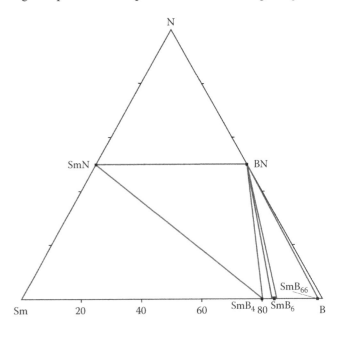

FIGURE 1.12 Isothermal section of the B–Sm–N system at 1400°C under 0.1 MPa of argon. (From Klesnar, H., and Rogl, P., *J. Am. Ceram. Soc.*, 75(10), 2825, 1992. With permission.)

Klesnar 1992). After a first reaction for several hours at 1800°C, the samples were heat treated for another 20 h at 1400°C before cooling to room temperature. This compound could also be prepared by heating a mixture of BN and SmN (molar ratio 1:1) at 1550°C for about 20 h (Gaudé 1983). Due to the sensitivity of the sample to moisture, handling of the specimens was performed in an Ar-filled glove box.

1.24 BORON–EUROPIUM–NITROGEN

Two ternary compounds, $Eu_3B_2N_4$ and $Eu_3B_3N_6$, are formed in the B–Eu–N system. $Eu_3B_2N_4$ crystallizes in the cubic structure with the lattice parameter $a = 762.4 \pm 0.1$ pm (Carrillo-Cabrera et al. 2001). The title compound could be synthesized from binaries BN and EuN (molar ratio 3:2) or from a stoichiometric mixture of BN, EuN, and Eu in Nb ampoules (Carrillo-Cabrera et al. 2001; Somer et al. 2004). After sealing, the samples were heated to 1200°C, followed by annealing at 1100°C for 12 h and rapid cooling to room temperature. It was obtained as a dark red to black crystalline product, very sensitive to air and moisture.

$Eu_3B_3N_6$ crystallizes in the trigonal structure with the lattice parameters $a = 1193.70 \pm 0.04$ and $c = 680.73 \pm 0.04$ and a calculated density of 6.79 g·cm⁻³ (Aydemir et al. 2016). Black crystals of this compound were prepared by oxidation of $Eu_3B_2N_4$ with Br_2 at 800°C.

1.25 BORON–GADOLINIUM–NITROGEN

The phase relations and phase stabilities have been derived for the B–Gd–N ternary system at 1400°C and above by means of XRD (Klesnar and Rogl 1992). Under the experimental conditions, $Gd_3B_3N_6$ ($GdBN_2$) ternary compounds are formed in this system (Figure 1.13). Phase equilibria at 1400°C under 0.1 MPa of Ar are mainly characterized by the incompatibility of Gd with BN. $Gd_3B_3N_6$ ($GdBN_2$) is found to be stable only at temperatures above 1400°C and under a partial pressure of 0.1 MPa of nitrogen.

The samples were annealed at 1400°C for 96 h (Klesnar and Rogl 1992).

$Gd_3B_3N_6$ ($GdBN_2$) crystallizes in the trigonal structure with the lattice parameters $a = 1193.17 \pm 0.46$ and $c = 673.59 \pm 0.38$ (Klesnar and Rogl 1992) (in a rhombohedral structure, with the lattice parameters $a = 1193.2 \pm 0.5$ and $c = 673.6 \pm 0.4$ pm [in a hexagonal setting] and a calculated density of 7.06 g·cm⁻³ [Rogl and Klesnar 1992]). It was prepared by reacting

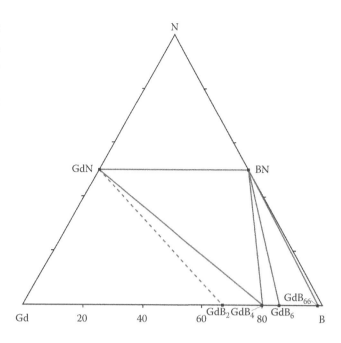

FIGURE 1.13 Isothermal section of the B–Gd–N system at 1400°C under 0.1 MPa of argon. (From Klesnar, H., and Rogl, P., *J. Am. Ceram. Soc.*, 75(10), 2825, 1992. With permission.)

compacted *h*-BN powder and Gd filling (Rogl and Klesnar 1992). After a first reaction for several hours at 1800°C, the samples were heat treated for another 20 h at 1400°C before cooling to room temperature. Due to the sensitivity of the sample to moisture, handling of the specimens was performed in an Ar-filled glove box.

1.26 BORON–TERBIUM–NITROGEN

The thermodynamic phase equilibria in the B–Tb–N ternary system at 1400°C and 0.1 MPa of Ar have been established from the data of XRD, and they are the same as in the B–Y–N ternary system (Figure 1.8) (Klesnar and Rogl 1990). Ternary phase equilibria reveal the absence of ternary compounds. There is no compatibility between BN and Tb. Mutual solubility under select experimental conditions was found to be rather restricted. At 1400°C, BN is in thermodynamic equilibrium with TbB_4, TbB_6, TbB_{12}, TbB_{66}, and TbN, and TbN coexists with TbB_2 and TbB_4.

The samples were annealed at 1400°C for 100 h under 0.1 MPa of Ar.

1.27 BORON–DYSPROSIUM–NITROGEN

The thermodynamic phase equilibria in the B–Dy–N ternary system at 1400°C and 0.1 MPa of Ar have been established from the data of XRD, and they are the same as in the B–Y–N ternary system (Figure 1.8) (Klesnar and Rogl 1990). Ternary phase equilibria reveal the absence of ternary compounds. There is no compatibility between BN and Dy. Mutual solubility under select experimental conditions was found to be rather restricted. At 1400°C, BN is in thermodynamic equilibrium with DyB_4, DyB_6, DyB_{12}, DyB_{66}, and DyN, and DyN coexists with DyB_2 and DyB_4.

The samples were annealed at 1400°C for 100 h under 0.1 MPa of Ar.

1.28 BORON–HOLMIUM–NITROGEN

The thermodynamic phase equilibria in the B–Ho–N ternary system at 1400°C and 0.1 MPa of Ar have been established from the data of XRD (Figure 1.14) (Klesnar and Rogl 1990). Ternary phase equilibria reveal the absence of ternary compounds. There is no compatibility between BN and Ho. Mutual solubility under select experimental conditions was found to be rather restricted. At 1400°C, BN is in thermodynamic equilibrium with HoB_4, HoB_{12}, HoB_{66}, and HoN, and HoN coexists with HoB_2 and HoB_4.

The samples were annealed at 1400°C for 100 h under 0.1 MPa of Ar.

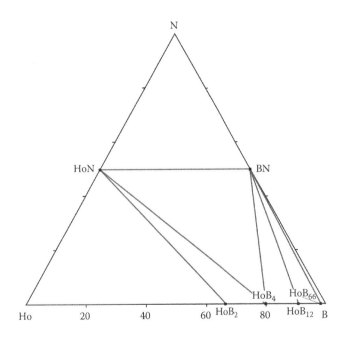

FIGURE 1.14 Isothermal section of the B–Ho–N system at 1400°C under 0.1 MPa of argon. (From Klesnar, H.P., and Rogl, P., *High Temp. High Pressure*, 22(4), 453, 1990. With permission.)

1.29 BORON–ERBIUM–NITROGEN

The thermodynamic phase equilibria in the B–Er–N ternary system at 1400°C and 0.1 MPa of Ar have been established from the data of XRD, and they are the same as in the B–Ho–N ternary system (Figure 1.14) (Klesnar and Rogl 1990). Ternary phase equilibria reveal the absence of ternary compounds. There is no compatibility between BN and Er. Mutual solubility under select experimental conditions was found to be rather restricted. At 1400°C, BN is in thermodynamic equilibrium with ErB_4, ErB_{12}, ErB_{66}, and ErN, and ErN coexists with ErB_2 and ErB_4.

The samples were annealed at 1400°C for 100 h under 0.1 MPa of Ar.

1.30 BORON–THULIUM–NITROGEN

The thermodynamic phase equilibria in the B–Tm–N ternary system at 1400°C and 0.1 MPa of Ar have been established from the data of XRD, and they are the same as in the B–Ho–N ternary system (Figure 1.14) (Klesnar and Rogl 1990). Ternary phase equilibria reveal the absence of ternary compounds. There is no compatibility between BN and Tm. Mutual solubility under select experimental conditions was found to be rather restricted. At 1400°C, BN is in thermodynamic equilibrium with TmB_4, TmB_{12}, TmB_{66}, and TmN, and TmN coexists with TmB_2 and TmB_4.

The samples were annealed at 1400°C for 100 h under 0.1 MPa of Ar.

1.31 BORON–LUTETIUM–NITROGEN

The thermodynamic phase equilibria in the B–Lu–N ternary system at 1400°C and 0.1 MPa of Ar have been established from the data of XRD, and they are the same as in the B–Ho–N ternary system (Figure 1.14) (Klesnar and Rogl 1990). Ternary phase equilibria reveal the absence of ternary compounds. There is no compatibility between BN and Lu. Mutual solubility under select experimental conditions was found to be rather restricted. At 1400°C, BN is in thermodynamic equilibrium with LuB_4, LuB_{12}, LuB_{66}, and LuN, and LuN coexists with LuB_2 and LuB_4.

The samples were annealed at 1400°C for 100 h under 0.1 MPa of Ar.

1.32 BORON–URANIUM–NITROGEN

The phase equilibria as determined by XRD on samples annealed at 1400°C for 100 h under 0.1 MPa of Ar are characterized by the formation of a ternary compound

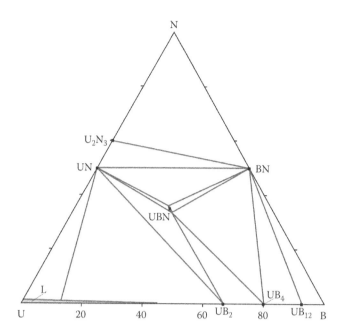

FIGURE 1.15 Isothermal section of the B–U–N system at 1400°C under 0.1 MPa of argon. (From Klesnar, H., and Rogl, P., *Boron-rich solids. AIP Conference Proc.*, **231**, 414, 1991. With permission.)

UBN (Figure 1.15) (Klesnar and Rogl 1991). It crystallizes in the orthorhombic structure with the lattice parameters $a = 358.51 \pm 0.05$, $b = 1182.73 \pm 0.53$, and $c = 332.54 \pm 0.05$ pm and a calculated density of 12.39 g·cm^{-3}. No compatibility was obtained between BN and U or between the uranium nitrides and B. Practically no nitrogen (<0.5 at% N) is soluble in the uranium borides. Similarly, solid solubility of U in BN is considered to be rather limited. Solid solubility of B in uranium nitrides is also negligible. UBN was found to be unstable under a high vacuum of 10^{-4} Pa.

1.33 BORON–CARBON–NITROGEN

Some isothermal sections of the B–C–N ternary system have been calculated using the program Thermo-Calc and including the formation of the BCN compound (Wen 1993). These sections are shown in without the BCN compound. According to the data of Ruys and Sorrell (1991), a BN–C join is stable up to ~1900°C, and above this temperature there is a B$_4$C–N$_2$ tie-line. Geometrically, a B$_4$C–BN tie-line seems likely.

In this ternary system, mainly materials of the sections BN–B$_4$C and *h*-BN–graphite were investigated. The results that have been reported by different authors are rather contradictory, and currently, there is no unambiguous answer to the question of whether the synthesis products are ternary compounds, solid solutions of carbon in *c*-BN, or a mechanical mixture of diamond and *c*-BN (Solozhenko et al. 1997). The evidence for the formation of the ternary compounds in the B–C–N system is hardly exhaustive and is open to doubt (Lundström and Andreev 1996). It is evident that substantial difficulties are encountered in the identification and characterization of phases consisting of light elements with nearly the same atomic numbers arranged in similar honeycomb networks. Thus, it is an arduous task to distinguish between a substitutional ternary network and a nanometric-scale mixture of graphite and *h*-BN, both turbostratically distorted. Samples prepared in different ways contain phases that differ in structure, which is in agreement with the fact that they are highly defective, nonequilibrium phases.

Thermodynamically stable ternary phases do not exist in the B–C–N equilibrium phase diagram (Ruys and Sorrell 1991; He et al. 2001; Huang et al. 2001; Azevedo 2006; Pan et al. 2006) (just Nicolich et al. [2001] indicated that the graphite-like BC$_2$N phase is thermodynamically stable). However, it is anticipated that some metastable phases, including solid solutions and compounds, should exist under definite extreme conditions. According to Grigor'ev et al. (2005), the ternary compounds in the B–C–N system are thermally unstable and, at the sintering temperature of 1900°C, are converted into BN and B$_4$C phases.

Noting the structural relationships between phases of carbon and boron carbide with BN and boron subnitride, their mutual solubilities have been investigated using a combination of first-principles total energies supplemented with statistical mechanics to address finite temperatures (Zhang et al. 2016). The solid-state phase part of the diagram of the B–C–N system was predicted (Figure 1.16). Owing to the large energy costs of substitution, it was found that the mutual solubilities of the ultrahard materials diamond and *c*-BN are negligible, and the same for the quasi-two-dimensional materials graphite and *h*-BN. In contrast, a continuous range of solubility connecting boron carbide to boron subnitride at elevated temperatures was determined. An electron-precise ternary compound **B$_{13}$CN** consisting of B$_{12}$ icosahedra with NBC chains was found to be stable at all temperatures up to melting. It exhibits an order–disorder transition in the orientation of NBC chains at approximately 230°C. At elevated temperatures, B$_{13}$CN joins into a solid solution that covers a triangular area, within the ternary phase diagram extending to binary

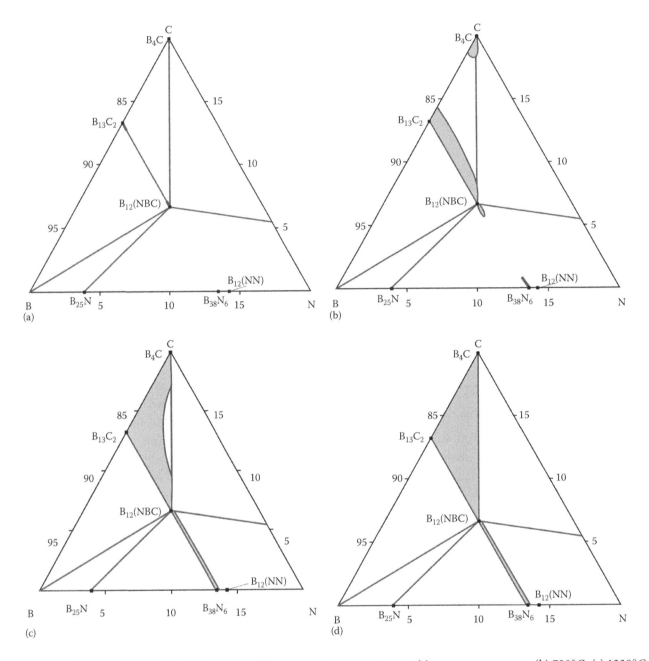

FIGURE 1.16 Solubility ranges in the B–corner of the B–C–N ternary system at (a) room temperature, (b) 730°C, (c) 1230°C, and (d) 2230°C. (From Zhang, H., et al., *Phys. Rev. B*, *93*(14), 144107_1, 2016. With permission.)

boron carbide, and along a line extending to boron subnitride.

In the literature, there are the data about obtaining some compounds in the B–C–N ternary system with compositions that do not belong to the BN–C section. $BC_{0.47}N_{0.85}$ crystallizes in the cubic structure with the lattice parameter $a = 364.5$ pm and a calculated density of 3.356 g·cm^{-3} (Filonenko et al. 2008). This compound was obtained from the mixture of C_3N_4 and B at approximately 1500°C and 8 GPa. BCN_2 crystallizes in the tetragonal structure with the calculated lattice

parameters $a = 357.85$ and $c = 701.78$ pm and a calculated density of 3.816 g·cm^{-3} (Basalaev 2016). Crystal lattice parameters were obtained from the first-principles calculations.

BC_3N_3 or $B(CN)_3$ crystallizes in the tetragonal structure with the lattice parameters $a = 821.4$ and $c = 1383.2$ pm (Williams et al. 2000). To obtain it, a sample of pure $B(CN)_3$·$CNSiMe_3$ was heated in a Pyrex tube at 250°C for 1.5 h under dynamic vacuum. A nonvolatile brown solid BC_3N_3 formed near the bottom of the tube, and modest amounts of starting material sublimed to the top of the tube. This compound is thermally stable up to about 450°C. Reactions were performed under purified N_2 using standard Schlenk and dry box techniques.

B_2CN is characterized by some polymorphic modifications. The *first polymorph* crystallizes in the hexagonal structure with the lattice parameters $a = 442$ and $c = 670$ pm for the samples obtained using an Fe catalyst and $a = 238$ and $c = 335$ pm for the samples obtained using a Ni catalyst (He et al. 2002). It was synthesized at approximately 5.5 GPa and 1600°C for 15 min using various catalysts (Ni or Fe). The recovered samples were boiled in a solution of H_2SO_4 and HNO_3 to remove the catalyst and clinging graphite.

The *second polymorph* could crystallize in the tetragonal structure with the calculated lattice parameters $a = 363.70$ and $c = 720.46$ pm and a calculated density of 3.372 g·cm^{-3} (Basalaev 2016) ($a = 354.4$ and $c = 389.3$ pm and bulk modulus 332.1 GPa [He et al. 2004]; $a = 252.6$ and $c = 397.3$ pm [Li et al. 2012]).

The *third polymorph* could crystallize in the trigonal structure with the lattice parameters $a = 257.4$ and $c = 444.2$ pm or $a = c = 464.0$ pm (calculated values for various space groups) (Li et al. 2012).

The *fourth polymorph* crystallizes in the orthorhombic structure with the lattice parameters $a = 477.6$, $b = 458.5$, and $c = 362.9$ pm (He et al. 2001, 2002) ($a = 286.1$, $b = 254.7$, and $c = 450.4$ pm [calculated values] [Wu et al. 2007]; $a = 256.0$ and 251.0, $b = 250.7$ and 251.0, and $c = 791.0$ and 789.8 pm [calculated values for various space groups] [Li et al. 2012]). It was synthesized at approximately 5.5 GPa and 1600°C for 15 min using $Ca_3B_2N_4$ as the catalyst (He et al. 2001, 2002). The recovered samples were boiled in a solution of H_2SO_4 and HNO_3 to remove the catalyst and clinging graphite.

B_6CN_3. $B_xC_yN_z$ layers have been prepared by laser-driven synthesis in a gas atmosphere containing C_2H_4, NH_3, and B_2H_6, where the starting composition ratios could vary in a large range (C/B = 0–4, C/N = 0–8, N/B = 0.5–2) (Morjan et al. 1999). Due to carbon loss and nitrogen content preservation, the chemical composition of the layers tended to a "focal point" close to a composition B_6CN_3 on the tie-line BN–"B_3C" in the ternary composition diagram. The obtained layers exhibited a turbostratic or modified turbostratic structure.

According to the first-principles calculations, BC_xN_y monolayers behave as a semiconductor or metal with band gap energy ranging from 0 to 2.45 eV, depending on the atomic arrangement (Azevedo 2006). The phase stability for the monolayer B–N–C ternary system was examined by Monte Carlo simulations and the cluster expansion technique based on the first-principles calculations (Yuge 2009). It was shown that lattice vibration significantly enhances solubility limits l-BNC: within the framework of harmonic approximation, complete miscibility is achieved at around 3230°C, which is below the melting lines between h-BN and graphite. l-BNC has no stable intermediate phase and undergoes phase separation into l-BN and grapheme.

BN–B_4C. Lattice parameter measurements indicate a slight solubility of B_4C in BN, but no solubility of BN in B_4C for samples prepared at 2250°C (Ruh et al. 1992) (the solubility of B_4C in BN at 1900°C–2800°C and pressures above 0.5 GPa is equal to 7–20 mol% [Sirota et al. 1977]). No ternary compounds were found in the B_4C–BN system.

BN–B_4C composites were fabricated by vacuum hot pressing at 1900°C–2250°C (Ruh et al. 1992)

BN–C. The section between graphite and BN is shown in Figure 1.17 (Kasper 1996). No mutual solubility between graphite and c-BN was found by Huang et al. (2001). Gago et al. (2002) indicated that the solubility limit of graphite in the h-BN structure is ~15 at%. Under equilibrium conditions, the BN solubility in diamond is quite limited (Langenhorst and Solozhenko 2002).

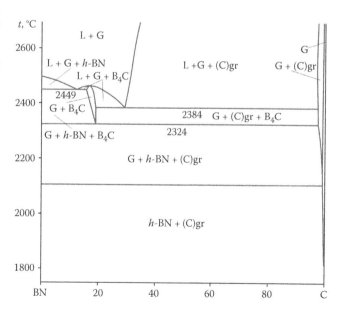

FIGURE 1.17 Phase relations in the BN–C system. (From Kasper, B., Thesis, Max-Planck-Institut, Stuttgart, 1–225, 1996. With permission.)

Synthesis of several samples across the cubic $(BN)_{1-x}C_x$ solid solutions ($x = 0.3–0.33$, 0.5, and 0.6) at pressures in excess of 30 GPa and temperatures above 1230°C indicates that they are isostructural with diamond and c-BN (Knittle et al. 1995). Measurement of the lattice parameters of the samples quenched to ambient conditions shows that the solid solution between C (diamond) and c-BN is nonideal, with unit cell volumes up to 1% larger than predicted based on ideal mixing (Vegard's law). The lattice parameter is equal to 361.6 ± 0.3 pm for $x = 0.3–0.33$, 360.2 ± 0.3 pm for $x = 0.5$, and 359.6 ± 0.3 pm for $x = 0.6$ The isothermal bulk modulus of $(BN)_{0.67}C_{0.33}$ has been measured by XRD through a diamond cell to more than 100 GPa at room temperature. The bulk modulus is equal to 355 ± 19 GPa.

According to the data of Badzian (1981), mixed crystals of diamond and c-BN were obtained by the direct phase transformation at a pressure of 14.0 GPa and temperature of about 3030°C of mixed crystals of graphite and h-BN. The crystals, which can be described by the formula $(BN)_xC_{1-x}$ ($0 < x < 1$), possess a sphalerite structure. h-BN–C obtained by chemical vapor deposition (CVD) was used as a starting material. The ternary phase was formed by reaction between BCl_3, CCl_3, N_2, and H_2, which were passed over the surface of a graphite rod placed in a quartz tube. The rod was heated by a high-frequency coil up to 1930°C. The XRD analysis described this polycrystalline material as mixed crystals of h-BN and graphite (solid solutions).

Electronic structure calculations allow us to predict that hypothetical ordered mixed crystals of c-BN and diamond representative of the expected short-range order in the alloys show a very pronounced band gap bowing (Lambrecht and Segall 1993). The miscibility phase diagram of this system was estimated using a simple quasi-binary "regular solution" model for the short-range order and energies of formation of the representative ordered compounds calculated from first principles. This indicates that there is only very limited mutual solubility in the solid state.

An amorphous BN–C phase was synthesized by ball milling of a mixture of h-BN and graphite with a nominal composition of $(BN)_{0.5}C_{0.5}$ (Huang et al. 2001).

The phase stability of c-BNC with a composition range of $(BN)_{1-x}(C_2)_x$ ($0 \leq x \leq 1$) was examined by the combination of the cluster expansion technique and Monte Carlo simulation based on first-principles calculations (Yuge et al. 2008). It was shown that the highest critical temperature of the miscibility gap (4230°C) is higher than the melting point of both the diamond and c-BN, which indicates that complete miscibility cannot be achieved. The phase stabilities of $(BN)_{1-x}(C_2)_x$ alloys were also investigated by *ab initio* pseudopotential calculations (Zheng et al. 2002). It was found that complete miscibility is possible. The critical temperature of the miscibility gap is 3273°C, which is lower than the melting temperature of $(BN)_{0.5}(C_2)_{0.5}$ alloy.

Some borocarbonitrides of the general formula BC_xN have been obtained in the BN–C system.

BCN crystallizes in the hexagonal structure with the lattice parameters $a = 250.56 \pm 0.05$, 250.57 ± 0.01, and 250.88 ± 0.02 and $c = 671.51 \pm 0.09$, 676.49 ± 0.01, and 680.80 ± 0.05 and calculated densities of 2.257, 2.237, and 2.221 g·cm^{-3} for three different samples (Filonenko et al. 2014). This compound is a degenerate semiconductor. Its powder particles were synthesized from boron–melamine mixtures. These mixtures were prepared by grinding three times in an agate mortar under acetone. The boron content of the mixtures was varied from 25 to 40 mass%. The mixtures thus obtained were pressed into pellets, which were then placed in a lithographic limestone cell. Synthesis was performed in thoroid chambers at pressures from 3.5 to 8.0 GPa and temperatures in the range of 1000°C–1500°C. After the pressure was raised and stabilized, the working volume of the cell was heated at a rate of about 50°C·s^{-1}. The holding time of the highest temperature was varied from 10 to 300 s. The temperature difference across the samples during holding was no greater than 100°C.

BC$_{1.6}$N was obtained as a result of a high-temperature gas phase reaction of liquid BBr_3 with a mixture of ethylene and NH_3 (Kumar et al. 2011). This compound possesses a graphene-like structure.

BC$_2$N is characterized by some polymorphic modifications. The *first polymorph* crystallizes in the cubic structure with the lattice parameter $a = 359.9$ pm (Komatsu 2004) ($a = 361.7$ pm and a calculated density of 3.501 g·cm^{-3} [Widany et al. 1998]; 364.2 ± 0.2 pm [Solozhenko and Novikov 2001, Solozhenko et al. 2001]; 360.0 ± 1.0 and 359.3 ± 0.4 pm for two different samples [Langenhorst and Solozhenko 2002]; $a = 359.5 \pm 7$ pm [Zhao et al. 2002]). It may be the first of a new family of ternary B–C–N superhard materials with hardness and elastic modulus higher than those of c-BN (Solozhenko and Novikov 2001). Young's modulus and modulus of shearing of c-BC$_2$N are equal to 980 ± 40 GPa and 447 ± 18

GPa, respectively. The bulk modulus of this polymorph is equal to 282 ± 15 GPa (Solozhenko et al. 2001).

Cubic BC_2N was synthesized from graphite-like BC_2N at pressures above 18 GPa and temperatures higher than 1930°C (Solozhenko and Novikov 2001, Solozhenko et al. 2001) or using shock syntheses (Komatsu 2004). Nanostructured superhard material bulks have been synthesized at high pressures and high temperatures (Zhao et al. 2002). The starting material for synthesis was a mixture of graphite and h-BN at a 2:1 molar ratio, rendered completely amorphous following 34 h of ball milling in a WC vial. The amorphous form was compressed to 20 GPa and heated to 1930°C for 5 min using a multianvil press. The final product was well-sintered chunks of millimeter size with a grain size of approximately 5 nm and was translucent and yellowish in color. Langenhorst and Solozhenko (2002) used a nanopowder of turbostratic graphite-like $(BN)_{0.48}C_{0.52}$ solid solution as the starting material for the synthesis of c-BC_2N. The precursor was synthesized by simultaneous nitridation of H_3BO_3 and carbonization of saccharose in molten urea, followed by annealing in N_2 at 1500°C. c-BC_2N was synthesized by direct solid-state transition of h-BC_2N at 30 GPa and approximately 3230°C.

A graphite relative of composition BC_2N has been subjected to high-pressure (7.7 GPa) and high-temperature (2000°C–2400°C) conditions to explore the possibility for the formation of a cubic phase via direct transformation (Nakano et al. 1994a). Several cubic phases with a diamond-like structure were confirmed in the products above 2150°C by XRD. The products obtained at 2150°C–2300°C consisted of c-BN, a "diamond" (containing minor amounts of B and N), and a cubic B–C–N substance. At 2400°C, however, the cubic products tended to segregate into two major phases assigned as c-BN and diamond. It could be concluded that not a cubic B–C–N compound, but a mixture of c-BN and diamond, exists as the thermodynamically stable phases in the ternary system under the conditions employed.

The structural form of c-BC_2N was studied with the *ab initio* pseudopotential density functional method for all the possible c-BC_2N structures, starting from an eight-atom zinc blende–structured cubic unit cell (Sun et al. 2001; Pan et al. 2006). The c-BC_2N structures are expected to be metastable. The hardest structures of c-BC_2N studied have bulk and shear moduli comparable to or slightly higher than those of c-BN.

A non-self-consistent, *ab initio*–based tight-binding molecular dynamics method has been used to investigate the stability and structural and electronic properties of c-BC_2N (Widany et al. 1998).

The *second polymorph* of BC_2N crystallizes in the hexagonal structure with the lattice parameters $a = 248.20 \pm 0.07$ and $c = 662.0 \pm 0.3$ pm (Nicolich et al. 2001) ($a = 247 \pm 2$ and $c = 727 \pm 5$ pm [Solozhenko et al. 2001]; $a = 243$ and 240 and $c = 700$ and 682 pm for two different samples [Komatsu 2004]; $a = 244$ and $c = 340$ pm for the material prepared at 850°C [Sasaki et al. 1993]). h-BC_2N is a small-band-gap semiconductor (Kouvetakis et al. 1989).

To prepare h-BC_2N, the vapors of BCl_3 (~8 kPa) and MeCN (~2.7 kPa) were carried separately by He and mixed in the reaction chamber inside a furnace at 800°C–1000°C (Kouvetakis et al. 1989; Sasaki et al. 1993). The former was controlled by diluting BCl_3 gas (0.1 MPa). The latter was set by bubbling He through liquid MeCN cooled with ice. The overall flow rate was regulated at 100 mL·min^{-1}. The 12 h interaction resulted in a black monolithic deposit of 10–50 μm thickness. High-quality h-BC_2N was also bulk synthesized by pyrolysis of the polymers between BCl_3 and acetonitrile, acrylonitrile, malononitrile, tetracyanoethylene, or polyacrylonitrile at 1500°C (Komatsu 2004). It was also obtained under high-temperature (1200°C–1500°C) and high-pressure (3–5 GPa) conditions (Nicolich et al. 2001).

The structural property of the h-BC_2N was investigated by means of a semiclassical method (Nozaki and Itoh 1996). In order to examine the dependence of the structural stability of monolayers on atomic arrangements, cohesive energies of some polymorphic structures having various-sized unit cells were estimated based on the chemical bond energies. From calculations, it has been found that the stable structure of h-BC_2N is formed so as to increase the number of both C–C and B–N bonds.

The *third polymorph* of BC_2N crystallizes in the tetragonal structure with the lattice parameters $a = 361.18$ and $c = 714.56$ pm and a calculated density of 3.468 g·cm^{-3} (Basalaev 2016) ($a = 361.3$ and $c = 714.7$ pm, a calculated density of 3.478 g·cm^{-3}, an energy gap of about 3.3 eV, and bulk modulus 364.7 GPa [Sun et al. 2006]; $a = 356.5$ and $c = 716.8$ pm, a calculated density of 3.561 g·cm^{-3}, an energy gap of about 2.7 eV, and bulk modulus 402.7 GPa [Zhou et al. 2007]). The most likely structure of this polymorph was predicted using first-principles calculation (Sun et al. 2006; Zhou et al. 2007). This is a more stable metastable phase with respect to other structures.

Theoretical characterization of the *fourth polymorph* of BC_2N was performed using methods based on the density functional theory in its local density approximation (LDA) (Mattesini and Matar 2001). The two orthorhombic crystals with the lattice parameters $a = 355.36$ and 252.80, $b = 359.86$ and 250.24, and $c = 355.28$ and 358.71 pm and bulk moduli 459.41 and 408.95 GPa, respectively, were found to be metastable with respect to the starting materials (c-BN and diamond).

The thermodynamic properties of BC_2N with an orthorhombic structure have been obtained from first-principles theory and the quasi-harmonic Debye model (Zhou et al. 2014). From first-principles calculations, the lattice constants a, b, and c and bulk moduli B_0 and B' of BC_2N are 355.3, 360.1, and 355.7 pm, and 393 GPa and 3.72, respectively. From the quasi-harmonic Debye model, the isochoric heat capacity, Debye temperature, and expansion coefficient have been studied at different temperatures and pressures for BC_2N. At normal temperature and pressure, $c_v = 23.15$ kJ·mol^{-1} K^{-1}. The heat capacity of BC_2N increases with the temperature but decreases with the pressure. In the low-temperature region, the decrease of c_v is very slow. In the medium-temperature region, the decrease of c_v becomes steep. In the high-temperature region, the decrease of c_v becomes slow. Debye temperature decreases with the temperature. The value of the thermal expansion coefficient along the c axis is more than that along the a and b axes.

$BC_{2.5}N$ crystallizes in the cubic structure with the lattice parameter $a = 360.5 \pm 0.1$ pm (Komatsu et al. 1996). It was synthesized by shock compression of h-$BC_{2.5}N$. The material was a polycrystal composed of nanocrystals of 5–20 nm in size.

BC_3N is characterized by two polymorphic modifications. The *first polymorph* crystallizes in the cubic structure with the lattice parameter $a = 358.6$ pm ($a = 360.3 \pm 1.0$, 361.1 ± 0.6, and 358.2 ± 1.0 for three different samples [Langenhorst and Solozhenko 2002]), and the *second polymorph* crystallizes in the hexagonal structure with the lattice parameters $a = 241$ and $c = 682$ pm (Komatsu 2004). High-quality h-BC_3N was bulk synthesized by pyrolysis of the polymers between BCl_3 and acetonitrile, acrylonitrile, malononitrile, tetracyanoethylene, or polyacrylonitrile at 1500°C (Komatsu 2004). It could also be obtained upon treatment of

polyacrylonitrile with BCl_3 at 400°C or acrylonitrile with BCl_3 at 1000°C (Kawaguchi and Kawashima 1993). Electron diffraction data and XRD indicated a turbostratic structure for this compound. Using h-BC_3N, c-BC_3N was shock synthesized (Komatsu 2004). Langenhorst and Solozhenko (2002) used a nanopowder of turbostratic graphite-like $(BN)_{0.48}C_{0.52}$ solid solution as the starting material for the synthesis of c-BC_3N. The precursor was synthesized by simultaneous nitridation of H_3BO_3 and carbonization of saccharose in molten urea, followed by annealing in N_2 at 1500°C. c-BC_3N was synthesized by direct solid-state transition of h-BC_3N at 20 GPa and approximately 1830°C–2030°C.

BC_4N is also characterized by two polymorphic modifications. The *first polymorph* crystallizes in the cubic structure with the lattice parameter $a = 358.6 \pm 0.9$ pm (Zhao et al. 2002) ($a = 357.6 \pm 0.2$ pm [Onodera et al. 2001]). Composites made up of B, C, and nitrogen have been studied using a combination of CVD and high-pressure techniques (Onodera et al. 2001). A diamond-type solid solution was obtained after $(BN)_{0.2}C_{0.8}$ sample was treated at 10 GPa and 1600°C or at 8 GPa and 1700°C, whereas the $(BN)_{0.97}C_{0.03}$ and $(BN)_{0.5}C_{0.5}$ samples exhibited decomposition into C and BN under all p-T conditions. Nanostructured superhard material bulks have been synthesized in the same way as c-BC_2N using a mixture of graphite and h-BN at a 4:1 molar ratio (Zhao et al. 2002).

The *second polymorph* crystallizes in the hexagonal structure with the lattice parameters $a = 247 \pm 2$ and $c = 1090 \pm 20$ pm (Solozhenko et al. 1997, 2001). A zero-pressure bulk modulus is equal to $B_0 = 18.1 \pm 0.2$ GPa, and its pressure derivative is 6.6 ± 0.1. Thermodynamic analysis has shown that this polymorph decomposition is a nonequilibrium process and the phase itself is metastable (Solozhenko et al. 1997). It has been shown that the decomposition process at ambient pressure starts at 1780°C, and as the pressure increases up to 6.6 GPa, the temperature of decomposition decreases to 800°C.

BC_5N is also characterized by two polymorphic modifications. The *first polymorph* crystallizes in the cubic structure with the lattice parameter $a = 358.1$ pm, and the *second polymorph* crystallizes in the hexagonal structure with the lattice parameters $a = 242$ and $c = 676$ pm (Komatsu 2004). High-quality h-BC_5N was bulk synthesized by pyrolysis of the polymers between BCl_3 and acetonitrile, acrylonitrile, malononitrile, tetracyanoethylene, or polyacrylonitrile at 1500°C. Using h-BC_5N, c-BC_5N was shock synthesized.

BC_8N crystallizes in the cubic structure with the lattice parameter $a = 357.2 \pm 1.0$ pm (Langenhorst and Solozhenko 2002). A nanopowder of turbostratic graphite-like $(BN)_{0.48}C_{0.52}$ solid solution was used as the starting material for the synthesis of c-BC_8N. The precursor was synthesized by simultaneous nitridation of H_3BO_3 and carbonization of saccharose in molten urea, followed by annealing in N_2 at 1500°C. c-BC_8N was obtained by direct solid-state transition of h-BC_8N at 30 GPa and approximately 2730°C.

$B_3C_{10}N_3$ crystallizes in the monoclinic structure with the lattice parameters $a = 510.7$, $b = 505.3$, and $c = 888.3$ pm and $\beta = 126.12°$, a calculated density of 3.501 g·cm^{-3}, an energy gap of 1.65 eV, and bulk modulus 398 GPa (Li et al. 2009). The diamond-like $B_3C_{10}N_3$ model was built using the monolayer $B_3C_{10}N_3$ having the lowest energy, and calculated with the density functional theory. The diamond-like $B_3C_{10}N_3$ is a semiconductor with an energy gap of about 1.65 eV. It might be synthesized from the layered $B_3C_{10}N_3$ at about 25.3 GPa.

$B_7C_6N_7$, as a structural relative of graphite, has been synthesized from the interaction of BCl_3, acetone, and NH_3 at 400°C–700°C (Kaner et al. 1987). A hexagonal layer structure type was indicated by XRD and electron diffraction data. This material is a semiconductor.

1.34 BORON–SILICON–NITROGEN

The isothermal sections at 1000°C and 1900°C have been calculated using the program Thermo-Calc (Figure 1.18a and b) (Kasper 1996).

BN–Si. A solid solution of Si in c-BN is formed as a result of the sintering of c-BN and Si at 1750°C and 7.7 GPa for 60 s (Gromyko et al. 1990).

The vapor composition of this system has been investigated through the Knudsen effusion method and mass spectroscopy of the vapor components (Aleshko-Ozhevskaya et al. 1993). It was established that the vapor consists of nitrogen, Si, Si_2, and Si_2N. The concentration of Si_2N in the vapor phase at 1410°C–1610°C is within the interval from 0.8 to 1.7 at%.

BN–Si_3N_4. A solid solution of Si_3N_4 in c-BN is formed as a result of the sintering of c-BN and Si_3N_4 at 1750°C and 7.7 GPa for 60 s (Gromyko et al. 1990).

A short-ranged classical force field was used for the modeling of the structural parameters of hypothetical $Si_3B_3N_7$ polymorphs (Marian et al. 2000). Thirteen polymorphs were optimized, and the results for the two most stable conformations are as follows. Both of them crystallize in the hexagonal structure with the lattice parameters $a = 749.5$ and $c = 537.2$ pm and bulk modulus 170.1 GPa and $a = 745.5$ and $c = 524.1$ pm and bulk modulus 290.9 GPa, respectively.

Amorphous $Si_3B_3N_7$ was synthesized from (trichlorosilyl)aminodichloroborane (Hagenmayer et al. 1999). The structural model for this ternary nitride may be described as follows. Almost trigonal planar BN_3 units and slightly distorted SiN_4 tetrahedra bridged via common nitrogen atoms form a three-dimensional random network in which Si and B atoms are homogeneously distributed. The thermodynamic properties of the $Si_3B_3N_7$ ceramic material that has been synthesized as an amorphous compound have been investigated using Monte Carlo simulations (Hannemann et al. 2008). The condensed amorphous phase remains solid up to about 1480°C. A hypothetical crystalline phase was found to be stable up to 2230°C. Up to about 2730°C–3730°C, the liquid state with a small admixture of gas phase molecules was found. This coexistence region continues up to about 7730°C–8730°C. A critical point in the liquid–gas region exists at about 8230°C and 1.3 GPa.

All calculated structures of hypothetical $Si_3B_3N_7$ polymorphs are semiconductors with wide band gaps (4–6 eV) (Kroll and Hoffmann 1998). The calculated densities of such structures are within the interval from 2.79 to 2.88 g·cm^{-3}. $Si_3B_3N_7$ is not thermodynamically stable at low temperatures.

1.35 BORON–TIN–NITROGEN

The phase equilibria in this ternary system at 200°C were investigated by means of XRD (Weitzer et al. 1990). No ternary compound was observed. BN coexists with Sn, and negligible mutual solubility was indicated between BN and Sn.

1.36 BORON–LEAD–NITROGEN

The phase equilibria in this ternary system at 200°C were investigated by means of XRD (Weitzer et al. 1990). No ternary compound was observed. BN coexists

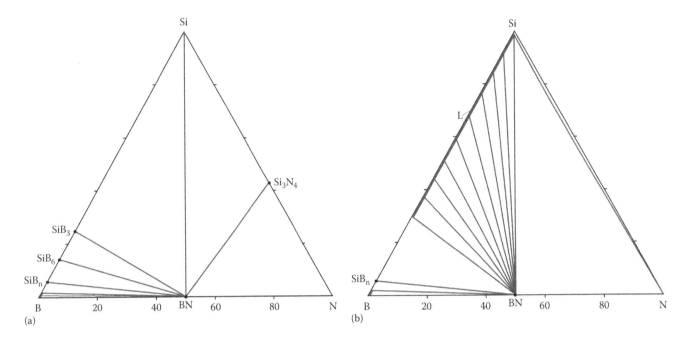

FIGURE 1.18   Calculated isothermal sections of the B–Si–N ternary system at (a) 1000°C and (b) 1900°C under 0.1 MPa of nitrogen. (From Kasper, B., Thesis, Max-Planck-Institut, Stuttgart, 1–225, 1996. With permission.)

with Pb, and negligible mutual solubility was indicated between BN and Pb.

1.37 BORON–TITANIUM–NITROGEN

The isothermal section of the B–Ti–N ternary system at 1500°C is given in Figure 1.19 (Rogl 2001). No ternary compounds are formed in this system.

TiB$_2$, TiB, and TiN are formed as a result of the BN interaction with Ti (Bondarenko and Khalepa 1977; Bondarenko et al. 1978). Generalizing the results of the DTA, XRD, and chemical analysis, it is possible to conclude that in the system c-BN–Ti in the temperature range 947°C–1104°C, the reverse phase transition of boron nitride from the cubic to the hexagonal structure occurs, after which borides and nitrides of titanium are formed (Korzun et al. 2009). On the basis of the fact that the temperature of the initial reverse transition c-BN → h-BN in the system c-BN–Ti decreases considerably, it is possible to conclude that titanium catalyzes the boron nitride phase transition from the cubic to the hexagonal structure

1.38 BORON–ZIRCONIUM–NITROGEN

The isothermal section of the B–Zr–N ternary system at 1500°C is given in Figure 1.20 (Rudy and Benesovsky 1961). No ternary compounds were found in this system

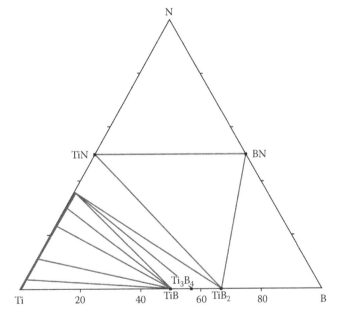

FIGURE 1.19   Isothermal section of the B–Ti–N system at 1500°C. (From Rogl, P., *Int. J. Inorg. Mater.*, 3(3), 201, 2001. With permission.)

(Nowotny et al. 1960; Rudy and Benesovsky 1961; Ruys and Sorrell 1991). According to the thermodynamic simulations, ZrB$_2$, ZrB, and ZrN are formed as a result of the BN interaction with Zr (Bondarenko and Khalepa 1977).

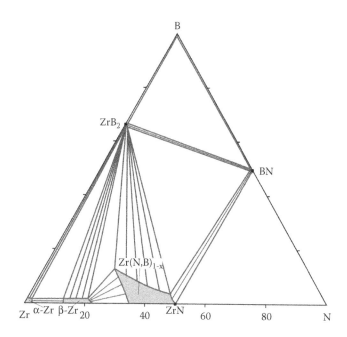

FIGURE 1.20 Isothermal section of the B–Zr–N system at 1500°C. (From Rudy, E., and Benesovsky, F., *Monatsch. Chem.*, *92*(2), 415, 1961. With permission.)

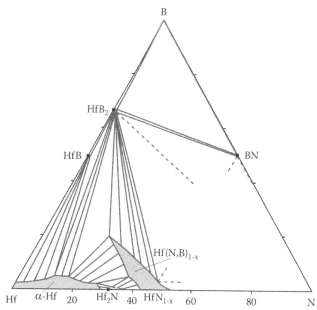

FIGURE 1.21 Isothermal section of the B–Hf–N system at 1500°C. (From Rudy, E., and Benesovsky, F., *Monatsch. Chem.*, *92*(2), 415, 1961. With permission.)

1.39 BORON–HAFNIUM–NITROGEN

The isothermal section of the B–Hf–N ternary system at 1500°C is given in Figure 1.21 (Rudy and Benesovsky 1961). According to the thermodynamic simulations, HfB_2, HfB, and HfN are formed as a result of the BN interaction with Hf (Bondarenko and Khalepa 1977).

1.40 BORON–PHOSPHORUS–NITROGEN

BN–BP. The structural and electronic properties of BN_xP_{1-x} solid solutions have been investigated by means of first-principles density functional total energy using the all-electron full-potential linear augmented plane-wave method (Hassan El Haj and Akbarzadeh 2005; Mohammad and Katircioğlu 2009b; Benkraouda and Amrane 2013). A strong deviation of the lattice constants from Vegard's law and the bulk modulus from linear concentration dependence was observed. The calculated phase diagram, including the spinodal and bimodal curves, indicates that the critical temperature is equal to 15.3 K (Hassan El Haj and Akbarzadeh 2005). BN_xP_{1-x} solid solutions are direct-gap materials when the nitrogen concentration exceeds 12.5 at% (Mohammad and Katircioğlu 2009b). The band gap bowing parameter has been calculated to be 3.055 and 4.784 eV for the ranges

of $0 < x < 0.125$ and $0.125 < x < 1.0$, respectively. The overall band gap bowing parameter has been calculated to be 7.946 eV in the range $0 < x < 1.0$. According to the data of Benkraouda and Amrane (2013), upon increasing x, the energy gap decreases and then increases again, having a minimum value at $x \approx 0.35$.

1.41 BORON–ARSENIC–NITROGEN

BN–BAs. First-principles calculations have been used to investigate the structural and electronic properties of BN_xAs_{1-x} solid solutions using the electron full-potential linearized plane-wave method within the density functional theory (Hassan El Haj and Akbarzadeh 2005). A strong deviation of the lattice constants from Vegard's law and the bulk modulus from linear concentration dependence were observed. The calculated phase diagram, including the spinodal and binodal curves, indicates that the critical temperature is equal to 21.7 K.

1.42 BORON–ANTIMONY–NITROGEN

The phase equilibria in this ternary system at 575°C were investigated by means of XRD (Weitzer et al. 1990). No ternary compound was observed. BN coexists with Sb, and negligible mutual solubility was indicated between BN and Sb.

1.43 BORON–BISMUTH–NITROGEN

The phase equilibria in this ternary system at 200°C were investigated by means of XRD (Weitzer et al. 1990). No ternary compound was observed. BN coexists with Bi, and negligible mutual solubility was indicated between BN and Bi.

1.44 BORON–VANADIUM–NITROGEN

According to the thermodynamic simulations, vanadium boride and vanadium nitride are formed as a result of the BN interaction with V (Bondarenko and Khalepa 1977).

1.45 BORON–NIOBIUM–NITROGEN

The thermodynamic phase equilibria in this ternary system at 1200°C under 0.1 MPa of Ar have been established from XRD (Figure 1.22) (Rogl et al. 1988a). Three ternary compounds were found to exist, of which Nb_2BN_{1-x} ($x \approx 1$) is the most stable. There is no compatibility between BN and Nb (Bondarenko and Khalepa 1977; Rogl et al. 1988a, 1988b). Practically no B dissolves in the niobium nitrides, and no nitrogen in the niobium borides. Nb_2BN_{1-x} crystallizes in the orthorhombic structure with the lattice parameters $a = 317.12 \pm 0.03$, $b = 1785.04 \pm 0.63$, and $c = 311.45 \pm 0.01$ pm and a calculated density of 7.85 g·cm^{-3} (Rogl et al. 1988a, 1988b).

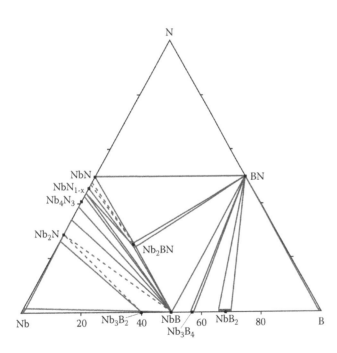

FIGURE 1.22 Isothermal section of the B–Nb–N system at 1200°C under 0.1 MPa of argon. (From Rogl, P., et al., *J. Mater. Sci. Lett.*, 7(11), 1229, 1988a. With permission.)

Two more ternary phases have been observed in the region bound by BN, NbN, and Nb_2BN_{1-x}.

To obtain Nb_2BN_{1-x}, powders of NbN_{1-x}, Nb, and B were first compacted and reacted under an atmosphere of Ar at 1200°C for 36 h and then finally heat treated at 1400°C for 200 h (Rogl et al. 1988a).

1.46 BORON–TANTALUM–NITROGEN

From thermodynamic calculations, it follows that Ta reacts with *c*-BN, forming tantalum boride and tantalum nitride (Bondarenko and Khalepa 1977; Benko et al. 2004). XRD determined the presence of TaB and nonstoichiometric $Ta_6N_{2.57}$ phases in this system (Benko et al. 2004). Transmission electron microscopy (TEM) observation exhibited a system showing the formation of TaB, TaB_2, and TaN at the BN/Ta interface.

1.47 BORON–OXYGEN–NITROGEN

The **B_3NO_3** ternary compound could exist in this system. Calculations of its optimized geometries, relative stabilities, and their equilibrium properties were performed within the density functional theory (Okeke and Lowthe 2008). The computational approach uses projector augmented wave potentials and plane-wave formalism with an LDA and a generalized gradient approximation (GGA). A cubic spinel-type structure is considered to be derived from a reaction of the form $BN + B_2O_3 = B_3NO_3$. The calculated lattice parameter of B_3NO_3 is 684.2 pm (LDA) or 694.7 pm (GGA), and the calculated energy gap is 4.60 eV (LDA) or 4.43 eV (GGA).

1.48 BORON–CHROMIUM–NITROGEN

According to the thermodynamic simulations and XRD, chromium borides and chromium nitride are formed as a result of the BN interaction with Cr (Bondarenko and Khalepa 1977; Bondarenko et al. 1978). It was shown that the products of the chromium borides nitriding in a dry NH_3 stream are BN and chromium nitride, which is stable in streaming NH_3 at the reaction temperature (Kiessling and Liu 1951). The stability of chromium borides in NH_3 increases with increasing B content. No ternary compounds are formed in this system.

1.49 BORON–MOLYBDENUM–NITROGEN

According to the thermodynamic simulations and XRD, molybdenum boride and molybdenum nitride are formed as a result of the BN interaction with Mo (Bondarenko and Khalepa 1977; Bondarenko et al. 1978).

1.50 BORON–TUNGSTEN–NITROGEN

At 1200°C and under 0.1 MPa of Ar, boron nitride is in equilibrium with W_2B, α-WB, W_2B_5, and $W_{1-x}B_3$, and W_2B is in equilibrium with nitrogen (Rogl 2001). Upon increasing the temperature up to 1400°C, instead of BN–α-WB equilibrium, α-WB is in equilibrium with nitrogen, while at a rather low temperature (<1000°C), tungsten and lower tungsten borides are in equilibrium with BN. At slightly higher temperatures and in the absence of external N_2, an increasing instability of W and lower tungsten borides was observed with respect to BN.

It was shown that the products of the tungsten borides nitriding in a dry NH_3 stream are BN and tungsten nitride, which is stable in streaming NH_3 at the reaction temperature (Kiessling and Liu 1951). Tungsten borides begin to be attacked by NH_3 between 800°C and 900°C. No ternary compounds are formed in this system.

1.51 BORON–FLUORINE–NITROGEN

$(BF_2N_3)_3$ ternary compound is formed in this system. To obtain it, BF_3 was dissolved in CH_2Cl_2 (25 mL) at –196°C with Me_3SiN_3 (molar ratio 1:1) (Wiberg et al. 1972). During thawing, the reaction starts at the melting point of the solvent. The solvent was removed with Me_3SiF. It remains colorless and easily sublimable $(BF_2N_3)_3$.

1.52 BORON–CHLORINE–NITROGEN

$(BCl_2N_3)_3$ and $(BCl_2N)_3$ ternary compounds are formed in this system. $(BCl_2N_3)_3$ melts at 67°C (Paetzold 1963; Paetzold et al. 1965) and crystallizes in the monoclinic structure with the lattice parameters $a = 887.4 \pm 0.3$, $b = 1449.4 \pm 0.4$, and $c = 1053.8 \pm 0.8$ pm and $\beta = 99.74 \pm 0.01°$ and a calculated density of 1.84 g·cm^{-3} (Müller 1971). To obtain this compound, BCl_3 was dissolved in CH_2Cl_2 (25 mL) at –196°C with Me_3SiN_3 (molar ratio 1:1) (Wiberg et al. 1972). During thawing, the reaction starts at –70°C. The solvent was removed with Me_3SiCl. It remains colorless, sublimable at 60°C under high vacuum $(BCl_2N_3)_3$. It could also be prepared by the interaction of BCl_3 and Cl_3N at a temperature below 0°C (Paetzold et al. 1965) or from BCl_3 and LiN_3 in CH_2Cl_2 (Paetzold 1963). At 200°C, N_2 is split off and $(BCl_2N)_3$ is formed.

1.53 BORON–BROMINE–NITROGEN

$(BBr_2N_3)_3$ ternary compound is formed in this system. It is volatile and explosive and melts at 94.5°C (Paetzold et al. 1965). To obtain it, BBr_3 was dissolved in CH_2Cl_2 (25 mL) at –196°C with Me_3SiN_3 (molar ratio 1:1) (Wiberg et al. 1972). During thawing, the reaction starts at 0°C. The solvent was removed with Me_3SiBr contaminated with BBr_3 and Br_2. It remains light yellow, sublimable at 90°C under high vacuum $(BBr_2N_3)_3$. It could also be prepared at the interaction of BBr_3 and Br_3N (Paetzold et al. 1965).

1.54 BORON–IODINE–NITROGEN

BI_2N_3 ternary compound is formed in this system. It was prepared as follows (Dehnicke and Krüger 1978). BI_3 (4.22 g) was dissolved in CH_2Cl_2 (20 mL) and treated at –20°C with a solution of IN_3 (1.82 g) in CH_2Cl_2 (90 mL). Iodine precipitation takes place; the mixture was thawed and stirred for 2 h at room temperature. The precipitate, which corresponded approximately to the composition BI_2N_3, was filtered under dry N_2, washed with small amounts of CH_2Cl_2, and dried under vacuum. The experiments require the exclusion of moisture.

1.55 BORON–IRON–NITROGEN

A brief review of the earlier experimental data for the B–Fe–N system was carried out by Raghavan (1987). An isothermal section for the BN–B–Fe region at 900°C is shown in Figure 1.23 (Smid and Rogl 1986; Raghavan 1993b). The iron borides Fe_2B and FeB form tie-lines with BN. No ternary compounds were found in this system. The samples were annealed at 900°C for 3 days in an Ar or N_2 atmosphere. The phase equilibria were studied by XRD.

The associated mixture model has been applied to describe equilibria in the Fe-rich corner of the B–Fe–N ternary system (Yaghmaee and Kaptay 2003). It has been found that 37% of boron is complexed into BN molecules (associates) at 1550°C, when the nitrogen content of the melt is above 80 ppm and the boron content of the melt is below 60 ppm. The phase boundary between liquid Fe and N_2 gas, on the one hand, and between liquid Fe and solid BN, on the other hand, has been determined. It was determined that the solubility

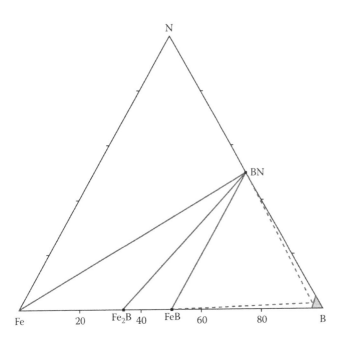

FIGURE 1.23 Isothermal section of the B–Fe–N system at 900°C under 0.1 MPa of argon. (From Smid, I., and Rogl, P., *Inst. Phys. Conf. Ser.*, (75), 249, 1986. With permission.)

of N_2 gas increases with increasing concentration of B in the liquid Fe.

It was shown that the products of the Fe_2B and FeB nitriding in a dry NH_3 stream are BN and iron nitride, which is stable in streaming NH_3 at the reaction temperature (Kiessling and Liu 1951). The stability of iron borides in NH_3 increases with increasing B content.

According to the data of Vishnevskiy et al. (1978), the interaction of the carbon steel with c-BN, h-BN, and w-BN is absent at 1130°C and Fe_2B is formed at the interaction at 1230°C–1330°C. The interaction of c-BN with this steel is the most active.

The solubility of boron and nitrogen in Fe melts in equilibrium with solid BN at 1600°C, and various N_2 pressures were determined by Schenck and Steinmetz (1968).

Solubilities of nitrogen in the boron-containing γ-Fe alloys were determined by Fountain and Chipman (1962) at 950°C–1150°C in the range 0.001–0.9 mass% B. It was demonstrated that the solubility of nitrogen decreases with increasing boron content. The solubility data indicated that B decreases the solubility of nitrogen in liquid Fe (Evans and Pehlke 1964). Within the limits of experimental accuracy, the data indicated that

Sieverts' law was observed below the BN solubility limit at all B concentrations studied.

It was estimated by Fountain and Chipman (1962) that the activity coefficient of nitrogen in the solid B–Fe–N alloys is decreased by B addition. According to the data of Evans and Pehlke (1964), the activity coefficient of nitrogen in liquid B–Fe–N alloys increases with increasing B content in the investigated range 0–7 mass% B. The activity coefficients of boron and nitrogen in the B–Fe–N melts were calculated by Isayev and Morozov (1964) from experimental data on B and N solubility in the liquid phase, and it was shown that both activity coefficients increase with increasing boron content. The solid BN films are formed on the Fe–B melt surfaces upon nitrogen contact within certain conditions.

It was shown that the synthesis of c-BN with the Fe additions at stationary heating leads to the formation of polycrystals containing FeB and Fe_2B, and at pulse heating, the impurity of γ-Fe is included (Vasil'yev et al. 1982). The samples with γ-Fe inclusions have higher cutting properties than the samples with Fe borides. Therefore, to obtain c-BN polycrystals from h-BN with Fe addition with high mechanical properties, it is necessary to use pulse heating, which excludes Fe boride formation.

The interaction of h-BN with carbonyl iron at a non-catalytic synthesis leads to the formation of the magnetically ordered phase ε (Fe_3N) and amorphous B–Fe, which decomposes at elevated temperatures to metallic Fe precipitates (Fedotova et al. 2003). In samples prepared by catalytic synthesis, the dominating stable fraction of Fe is paramagnetic Fe^{3+} in low-symmetry surroundings. These results may give some hints for the degradation in the mechanical properties of c-BN cutting edges.

The process and resulting products of the solid-state reaction between Fe and h-BN were studied by ball milling a mixture with an Fe-to-h-BN volume ratio of 1:12.5 and/or heat treating under normal or high pressure (Tao et al. 2004). A small amount of Fe_4N was observed in the mixture annealed at 500°C, and a little inclusion of γ-Fe(N) was obtained at 900°C. All Fe atoms in the mixture reacted with nitrogen in the BN to form a single ε (Fe_xN) phase when the mixture was milled for 60 h or annealed for 1 h under

4 GPa at 420°C–530°C. However, no B–Fe phases were observed.

1.56 BORON–COBALT–NITROGEN

The isothermal section for the BN–B–Co region at 900°C is shown in Figure 1.24 (Smid and Rogl 1986). The cobalt borides and Co form tie-lines with BN (Bondarenko et al. 1980; Smid and Rogl 1986). No ternary compounds were found in this system. The samples were annealed at 900°C for 3 days in an Ar or

N$_2$ atmosphere. The phase equilibria were studied by XRD (Smid and Rogl 1986).

1.57 BORON–NICKEL–NITROGEN

The isothermal section for the BN–B–Ni region at 900°C is shown in Figure 1.25 (Smid and Rogl 1986). The nickel borides form tie-lines with BN. No ternary compounds were found in this system. The samples were annealed at 900°C for 3 days in an Ar or N$_2$ atmosphere. The phase equilibria were studied by XRD.

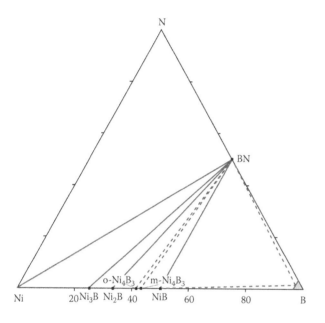

FIGURE 1.24 Isothermal section of the B–Co–N system at 900°C under 0.1 MPa of argon. (From Smid, I., and Rogl, P., *Inst. Phys. Conf. Ser.*, (75), 249, 1986. With permission.)

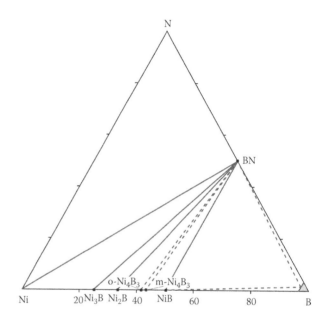

FIGURE 1.25 Isothermal section of the B–Ni–N system at 900°C under 0.1 MPa of argon. (From Smid, I., and Rogl, P., *Inst. Phys. Conf. Ser.*, (75), 249, 1986. With permission.)

CHAPTER 2

Systems Based on BP

2.1 BORON–SODIUM–PHOSPHORUS

Na₃BP₂ ternary compound is formed in this system. It crystallizes in the monoclinic structure with the lattice parameters $a = 699.5 \pm 0.2$, $b = 927.9 \pm 0.1$, and $c = 915.9 \pm 0.2$ pm and $\beta = 111.03 \pm 0.01°$, a calculated density of 1.696 g·cm⁻³, and an energy gap of 1.9 eV (Somer et al. 1990g, 2000b). The title compound was synthesized as bright yellow prisms from the elements or from BP and Na₃P in an evacuated Nb or steel ampoule at 730°C.

2.2 BORON–POTASSIUM–PHOSPHORUS

K₃BP₂ ternary compound is formed in this system. It crystallizes in the monoclinic structure with the lattice parameters $a = 936.2 \pm 0.2$, $b = 889.4 \pm 0.2$, and $c = 901.3 \pm 0.2$ pm and $\beta = 110.99 \pm 0.02°$ and a calculated density of 1.802 g·cm⁻³ (von Schnering et al. 1990a, 1990b; Somer et al. 1990b, 2000b). The title compound was synthesized as bright yellow crystals from the pure elements or from BP, K, and P. The starting components (K/B/P molar ratio 3:1:2) were heated in sealed Nb crucibles (protective tube—quartz) for 4 h at 730°C and cooled within 40 h to room temperature. Distillation of K excess was carried out in high vacuum at 252°C.

2.3 BORON–RUBIDIUM–PHOSPHORUS

Rb₃BP₂ ternary compound is formed in this system. It crystallizes in the monoclinic structure with the lattice parameters $a = 953.3 \pm 0.1$, $b = 922.9 \pm 0.1$, and $c = 941.8 \pm 0.1$ pm and $\beta = 110.53 \pm 0.01°$, a calculated density of 2.439 g·cm⁻³, and an energy gap of 1.9 eV (Somer et al. 1995a, 2000b). The title compound was prepared as bright yellow prisms from a mixture of BP, Rb₄P₆, and

Rb (molar ratio 1:0.17:4.5) in sealed steel ampoules at 730°C. The excess Rb was removed in high vacuum at 202°C–252°C.

2.4 BORON–CESIUM–PHOSPHORUS

Cs₃BP₂ ternary compound is formed in this system. It crystallizes in the monoclinic structure with the lattice parameters $a = 983.4 \pm 0.2$, $b = 967.4 \pm 0.2$, and $c = 985.9 \pm 0.2$ pm and $\beta = 109.77 \pm 0.02°$ and a calculated density of 3.548 g·cm⁻³ (Somer et al. 1990f, 2000b). The title compound could be prepared as bright yellow prisms from a stoichiometric mixture of elements or from BP, Cs₄P₆, and Cs in an evacuated and sealed Nb ampoule at 630°C–730°C. Distillation of Cs excess was carried out in high vacuum at 202°C.

2.5 BORON–COPPER–PHOSPHORUS

BP–Cu. The solubility of BP in the Cu melt is negligible (Goryunova et al. 1964).

BP–Cu₃P. The Cu₃P melt dissolves about 1 mol% BP at 1200°C (Baranov et al. 1967).

2.6 BORON–SILVER–PHOSPHORUS

BP–Ag. The solubility of BP in the Ag melt is negligible (Goryunova et al. 1964).

2.7 BORON–ZINC–PHOSPHORUS

BP–Zn. Zinc does not wet the boron phosphide and BP is not dissolved in the Zn melt (Goryunova et al. 1964).

2.8 BORON–ALUMINUM–PHOSPHORUS

BP–Al. At the heating of Al melt with BP, aluminum phosphide is formed (Goryunova et al. 1964).

2.9 BORON–GALLIUM–PHOSPHORUS

The boron concentration in the GaP ingots grown by the Czochralski process under B_2O_3 flux has been found in the range of $3 \cdot 10^{16}$ to $3 \cdot 10^{18}$ cm^{-1}, and boron incorporation was seen to be suppressed by the presence of oxygen and nitrogen (Lightowlers 1972).

BP–Ga. At the heating of Ga melt with BP, gallium phosphide is formed (Goryunova et al. 1964).

2.10 BORON–INDIUM–PHOSPHORUS

BP–In. At the heating of In melt with BP, indium phosphide is formed (Goryunova et al. 1964).

2.11 BORON–THALLIUM–PHOSPHORUS

BP–Tl. Thallium does not wet the boron phosphide, and BP is not dissolved in the Tl melt (Goryunova et al. 1964).

2.12 BORON–SILICON–PHOSPHORUS

BP–Si. The solubility of BP in Si is equal to 4 mol% (Grigor'eva et al. 1968).

2.13 BORON–GERMANIUM–PHOSPHORUS

BP–Ge. The solubility of BP in the Ge melt is negligible (Goryunova et al. 1964).

2.14 BORON–TIN–PHOSPHORUS

BP–Sn. The solubility of BP in the Sn melt is negligible (Goryunova et al. 1964).

2.15 BORON–LEAD–PHOSPHORUS

BP–Pb. Lead does not wet the boron phosphide, and BP is not dissolved in the Pb melt (Goryunova et al. 1964).

2.16 BORON–ARSENIC–PHOSPHORUS

The structural and electronic properties of BP_xAs_{1-x} alloys have been investigated by the full-potential linearized augmented plane-wave method based on the density functional theory (Hassan El Haj and Akbarzadeh 2005; Mohammad and Katircioğlu 2009a). The equilibrium lattice constants closely follow Vegard's law. A very small downward bowing has been found in the range of $0 < x < 0.3125$, with the bowing parameter of 1.34 pm. The bowing of the lattice constant is again very small, but upward in the ranges of $0.3125 < x < 0.5$, $0.5625 < x < 0.6875$, and $0.875 < x < 0$, with bowing parameters of −1.3, −1.2, and 3.8 pm, respectively (Mohammad

and Katircioğlu 2009a). The calculated phase diagram, including the spinodal and bimodal curves, indicates that the critical temperature is equal to 0.6 K (Hassan El Haj and Akbarzadeh 2005).

The bowing of the bulk moduli with respect to the bowing of those of the linear concentration dependence rule was found to be not smooth but downward totally, with a total bowing parameter of 5.376 GPa (Mohammad and Katircioğlu 2009a). In the ranges of $0.0 < x < 0.1250$, $0.1250 < x < 0.3750$, and $0.4375 < x < 1.0$, the bowing parameters were calculated to be 4.006, 4.293, and 5.881 GPa, respectively. BP_xAs_{1-x} alloys are indirect-gap semiconductor materials when the concentrations of phosphorus are in the range of $0.0625 < x < 0.1875$. They also have indirect band gap characteristics for the x values of 0.3125, 0.6875, 0.8125, and 0.9375. The band gap bowing parameter was calculated to be 0.565 eV in the range of $0.0 < x < 1.0$.

2.17 BORON–ANTIMONY–PHOSPHORUS

BP–Sb. Antimony does not wet the boron phosphide, and BP is not dissolved in the Sb melt (Goryunova et al. 1964).

2.18 BORON–BISMUTH–PHOSPHORUS

BP–Bi. Bismuth does not wet the boron phosphide, and BP is not dissolved in the Bi melt (Goryunova et al. 1964).

2.19 BORON–OXYGEN–PHOSPHORUS

BPO_4 ternary compound is formed in this system. It is characterized by two polymorphic modifications. The *first polymorph* crystallizes in the tetragonal structure with the lattice parameters $a = 434.04 \pm 0.01$ and $c = 665.02 \pm 0.02$ pm and a calculated density of 2.804 g·cm^{-3} for the sample prepared by the chemical transport, and $a = 433.79 \pm 0.01$ and $c = 665.13 \pm 0.03$ pm and a calculated density of 2.807 g·cm^{-3} for the sample prepared by the hydrothermal growth (Schmidt et al. 2004) ($a = 434.2$ and $c = 664.5$ pm [Baykal et al. 2001]; $a = 433.4 \pm 0.8$ and $c = 663.6 \pm 0.8$ pm [Schulze 1933]). The tetragonal structure of this compound is stable up to about 830°C and sublimate above this temperature (Kosten and Arnold 1980). According to the data of Hummel and Kupinski (1950), BPO_4 begins to vaporize in the neighborhood of 1450°C and will disappear completely at 1462°C if held for 1 h at this temperature. It does not decompose into the constituent oxides, B_2O_3 and P_2O_5, although there is no direct proof that this is the case.

The *high-pressure polymorph* crystallizes in the trigonal structure with the lattice parameters $a = 447.0 \pm 0.5$ and $c = 992.6 \pm 1.0$ pm and a calculated density of 3.069 $g \cdot cm^{-3}$ (Dachille and Glasser 1959; Dachille and Roy 1959) ($a = 494$ and $c = 554$ pm and an experimental density of 2.24 $g \cdot cm^{-3}$ [Shafer et al. 1956]). According to the data of Mackenzie et al. (1959), this polymorph crystallizes in the hexagonal structure with the lattice parameters $a = 775 \pm 2$ and $c = 995 \pm 2$ pm and experimental and calculated densities of 3.05 ± 0.05 and 3.05 $g \cdot cm^{-3}$, respectively. Dachille and Glasser (1959) noted that the material prepared by Mackenzie et al. (1959) was identical to the earlier material and was in fact quartz form, and that the unit cells assigned by them are incorrect.

Some different routes were employed for the BPO_4 synthesis. For microwave-assisted synthesis, a dry and ground solid mixture of $(NH_4)_2B_4O_7 \cdot 4H_2O$ and H_3PO_4 (molar ratio 1:4) was placed in a Teflon beaker and was treated in a conventional microwave system for a short period (3 min) (Baykal et al. 2001). This experiment was also performed for the mixture of B_2O_3 and P_2O_5. The time required for this reaction is about 5 min.

The synthesis of BPO_4 was carried out by a hydrothermal method starting from mixtures H_3BO_3 and P_2O_5 in a 6:1 molar ratio (Baykal et al. 2001). The reactants were mixed in 10 mL distilled water and treated under stirring with 25 mL of concentrated HNO_3 until the components dissolved completely. The clear solution was heated without boiling and concentrated to 15 mL and then transferred to a Teflon-coated steel autoclave (degree of filling 65%) and heated at 160°C for 2 days. The crystalline products were filtered off *in vacuo*, washed with distilled water, and dried at 60°C.

For the solid-state synthesis of BPO_4, the reaction was performed at 1000°C in a porcelain crucible (Baykal et al. 2001). The mixture was heated to 300°C at a rate of 2°C·min^{-1}. Then the temperature was gradually increased to 1000°C and the sample was kept at this temperature for 7 h.

It was also prepared by containing the BPO_4 polycrystalline samples in graphite capsules where they were subjected to pressures of 8.6 GPa at 1000°C–1200°C for about 10 min (Mackenzie et al. 1959).

Single crystals of BPO_4 were grown by chemical transport reactions with PCl_5 using a gradient from 800°C to 700°C, as well as with solvothermal syntheses in the temperature range between 140°C and 250°C using H_2O, EtOH or 2-propanol as a polar protic solvent (Schmidt et al. 2004).

2.20 BORON–SULFUR–PHOSPHORUS

BPS_4 ternary compound is formed in this system. It is characterized by two polymorphic modifications. **α-BPS_4** crystallizes in the orthorhombic structure with the lattice parameters $a = 560 \pm 6$, $b = 525 \pm 6$, and $c = 904 \pm 6$ pm and experimental and calculated densities of 1.98 and 2.12 $g \cdot cm^{-3}$, respectively (Weiss and Schäfer 1963). **β-BPS_4** crystallizes in the monoclinic structure with the lattice parameters $a = 1038$, $b = 605$, and $c = 669$ pm and $\beta = 75°$.

BPS_4 could be obtained by the interaction of B, P, and S, or BP and S, or B_2S_3, P, and S at temperatures above 450°C in the atmosphere without moisture and oxygen (Weiss and Schäfer 1963). Depending on the operating temperature, it could be prepared in two modifications or the mixture of them. α-BPS_4 forms at temperatures between 450°C and 500°C, and β-BPS_4 between 650°C and 700°C. α-BPS_4 forms hygroscopic, extremely sensitive, crystalline aggregates, which completely decompose on the air in a few minutes. β-BPS_4 forms brown, flaky crystals, which are much more stable in air than α-BPS_4. α-BPS_4 could be converted by annealing at 650°C in β-BPS_4. In vacuum at temperatures above 750°C, both modifications decompose.

2.21 BORON–SELENIUM–PHOSPHORUS

BP–B_2Se_3. According to the data of XRD, the solubility of B_2Se_3 in 3BP is not less than 50 mol% (the system is regarded as 3BP–B_2Se_3) (Goryunova and Radautsan 1964).

2.22 BORON–CHROMIUM–PHOSPHORUS

Cr_3P dissolves boron and **$Cr_3B_{0.2}P_{0.8}$** crystallizes in the tetragonal structure with the lattice parameters $a = 913.4$ and $c = 454.5$ pm (Rundqvist 1962).

2.23 BORON–IODINE–PHOSPHORUS

It was shown that in the equilibrium conditions at 1075°C–1195°C, the reaction $2BP(s) + BI_3(g) \longleftrightarrow 3BI(g) + P_2(g)$ takes place (Grinberg et al. 1966). The temperature dependence of the equilibrium constant, enthalpy, and entropy of this reaction have been determined.

2.24 BORON–MANGANESE–PHOSPHORUS

Mn₅B₂P ternary compound is formed in this system. It crystallizes in the tetragonal structure with the lattice parameters $a = 554$ and $c = 1049$ pm (Rundqvist 1962). Mn₃P dissolves boron and Mn₃B₀.₅P₀.₅ also crystallizes in the tetragonal structure with the lattice parameters $a = 912.1$ and $c = 453.0$ pm.

2.25 BORON–IRON–PHOSPHORUS

An isothermal section of the B–Fe–P ternary system at 1000°C is given in Figure 2.1 (Rundqvist 1962). The investigation was restricted to the range BP–FeB–Fe–FeP₂. This system contains **Fe₅B₂P** ternary compounds and **Fe₃BₓP₁₋ₓ** solid solution. Fe₅B₂P crystallizes in the tetragonal structure with the lattice parameters $a = 548.2$ and $c = 1033.2$ pm. Fe₃BₓP₁₋ₓ also crystallizes in the tetragonal structure with the lattice parameters $a = 881.2$ and $c = 437.5$ pm for $x = 0.63$. At 1000°C, the solid solutions based on Fe₃P reach the composition of Fe₃B₀.₃₀P₀.₇₀, and for $x = 0.30$–0.48, the two-phase region exists.

2.26 BORON–COBALT–PHOSPHORUS

This system contains **Co₅B₂P** ternary compounds and **Co₃BₓP₁₋ₓ** solid solution. Co₅B₂P crystallizes in the tetragonal structure with the lattice parameters $a = 542$ and $c = 1020$ pm (Rundqvist 1962). At 950°C,

the Co₃BₓP₁₋ₓ solid solutions reach the composition of Co₃B₀.₅₀P₀.₅₀, and for $x = 0.5$–0.7, the two-phase region ($\varepsilon + \varepsilon_1$) exists. They also crystallize in the tetragonal structure with the lattice parameters from $a = 878.0$ and $c = 433.6$ pm to $a = 875.2$ and $c = 432.6$ pm for ε, and from $a = 868.6$ and $c = 430.3$ pm to $a = 867.3$ and $c = 429.7$ pm for ε_1. Co₂P and CoP dissolve some boron as their lattice parameters change a little.

2.27 BORON–NICKEL–PHOSPHORUS

An isothermal section of this system at 800°C is shown in Figure 2.2 (Mikhalenko et al. 1984). **NiB₃P₂** ternary compound is formed in the B–Ni–P system. It crystallizes in the tetragonal structure with the lattice parameters $a = 670 \pm 3$ and $c = 3840 \pm 20$ pm. Solid solutions based on Ni₃P and Ni₂P exist in this system (according to the data of Rundqvist [1962], Ni₃P does not dissolve any appreciable amounts of boron). The homogeneity region based on Ni₃P reaches the composition of Ni₃B₀.₃P₀.₇, and the homogeneity region based on Ni₂P reaches the composition of Ni₂B₀.₆P₀.₄. NiB₃P₂ is in thermodynamic equilibrium with Ni₂BₓP₁₋ₓ solid solution and BP.

Stearns and Greene (1965) suggested that BP–Ni₂P, Ni₂B–Ni₂P, and Ni₂B–B₁₂P₂ thermodynamic equilibria exist in the B–Ni–P ternary system. It would also be indicated upon the existence of the ternary compound in the region near Ni₂P.

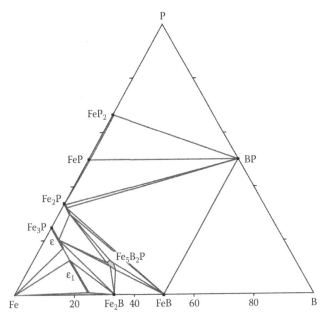

FIGURE 2.1 Isothermal section of the B–Fe–P system at 1000°C. (From Rundqvist, S., *Acta Chem. Scand.*, 16(1), 1, 1962. With permission.)

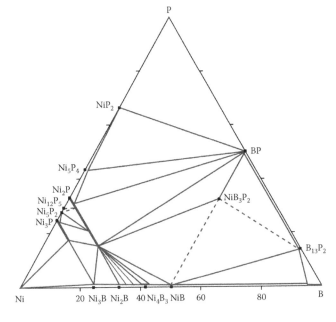

FIGURE 2.2 Isothermal section of the B–Ni–P system at 800°C. (From Mikhalenko, S.I., et al., *Izv. AN SSSR Neorgan. Mater.*, 20(10), 1746, 1984. With permission.)

CHAPTER 3

Systems Based on BAs

3.1 BORON–POTASSIUM–ARSENIC

K$_3$BAs$_2$ ternary compound is formed in this system. It crystallizes in the monoclinic structure with the lattice parameters $a = 960.9 \pm 0.8$, $b = 910.9 \pm 0.7$, and $c = 919.4 \pm 0.6$ pm and $\beta = 111.68 \pm 0.03°$ and a calculated density of 2.468 g·cm^{-3} (von Schnering et al. 1990a, 1990b; Somer et al. 1990a) ($a = 960.2 \pm 0.1$, $b = 909.4 \pm 0.2$, and $c = 921.2 \pm 0.2$ pm and $\beta = 111.41 \pm 0.01°$ and a calculated density of 2.464 g·cm^{-3} [Somer et al. 2000b]). The title compound was synthesized as yellow-red prismatic crystals from the pure elements or from BAs, K, and As. The starting components (K/B/As molar ratio 3:1:2) were heated in sealed Nb crucibles (protective tube—quartz) for 4 h at 730°C and cooled within 40 h to room temperature (von Schnering et al. 1990a, 1990b; Somer et al. 2000b). Distillation of K excess was carried out in high vacuum at 252°C.

3.2 BORON–RUBIDIUM–ARSENIC

Rb$_3$BAs$_2$ ternary compound is formed in this system. It crystallizes in the monoclinic structure with the lattice parameters $a = 977.2 \pm 0.3$, $b = 944.3 \pm 0.2$, and $c = 964.9 \pm 0.4$ pm and $\beta = 111.08 \pm 0.03°$ and a calculated density of 3.334 g·cm^{-3} (Somer et al. 1995a, 2000b). The title compound was prepared as yellow-red prismatic crystals from a mixture of BAs, Rb$_4$As$_6$, and Rb (molar ratio 1:0.17:4.5) in sealed steel ampoules at 730°C. The excess Rb was removed in high vacuum at 252°C.

3.3 BORON–CESIUM–ARSENIC

Cs$_3$BAs$_2$ ternary compound is formed in this system. It crystallizes in the monoclinic structure with the lattice

parameters $a = 1006.9 \pm 1.0$, $b = 986.6 \pm 0.9$, and $c = 1017.1 \pm 0.8$ pm and $\beta = 110.62 \pm 0.07°$, a calculated density of 3.929 g·cm^{-3}, and an energy gap of 1.8 eV (Somer et al. 1990e, 2000b). The title compound could be prepared as bright yellow prisms from a stoichiometric mixture of elements or from BAs, Cs$_4$As$_6$, and Cs in an evacuated and sealed Nb ampoule at 630°C–730°C. Distillation of Cs excess was carried out in high vacuum at 202°C.

3.4 BORON–ALUMINUM–ARSENIC

BAs–AlAs. B$_x$Al$_{1-x}$As solid solution has been grown by low-pressure metal-organic vapor phase epitaxy (MOVPE) on GaAs substrates using the precursors BEt$_3$, AlMe$_3$, and AsH$_3$ (Gottschalch et al. 2003). The maximum boron concentrations that could be incorporated in AlAs were about 4 at%. In the range from 550°C to 600°C, the temperature dependence of boron incorporation is rather small. For AlAs, the boron incorporation is drastically reduced for temperatures above 600°C. For temperatures below 550°C, the boron incorporation is reduced due to surface-kinetic effects.

3.5 BORON–GALLIUM–ARSENIC

BAs–GaAs. The calculated spinodal curve indicates that the critical temperature for this system is equal to 5476°C at 50 mol% GaAs (Dumont and Monteil 2006). Thus, under normal growth conditions at thermodynamic equilibrium, for instance, at 640°C, it was predicted that the homogeneous solid solution formation could reach approximately 4 at% B. This result was confirmed experimentally. At low temperatures (520°C–600°C), a maximum B composition of 2–4 at% can be achieved in both low-pressure and atmospheric pressure metal-organic chemical vapor deposition (MOCVD) reactors using GaEt$_3$ or GaMe$_3$ at growth rates of 2–5 μm·h^{-1} if sufficient AsH$_3$ flow is supplied (Geisz et al. 2000, 2001a). Extremely high As/(B + Ga) ratios appear to be necessary near the solubility limit. But at higher temperatures (650°C), it is possible to achieve only 0.5–1.3 at% B. The boron incorporation using constant growth flow conditions is highly temperature dependent. Using BEt$_3$ together with GaEt$_3$, a higher boron composition (x = 0.04–0.07) has been achieved (Geisz et al. 2001b).

The layers of the B$_x$Ga$_{1-x}$As solid solutions were obtained only within a narrow range of growth conditions (Geisz et al. 2000).

B$_x$Ga$_{1-x}$As solid solutions have also been grown by low-pressure MOVPE on GaAs substrates using the precursors BEt$_3$, GaMe$_3$, and AsH$_3$ (Gottschalch et al. 2003). The maximum B concentrations that could be incorporated in GaAs were about 4 at%. In the range from 550°C to 600°C, the temperature dependence of B incorporation is rather small. For GaAs, the B incorporation is drastically reduced for temperatures above 600°C. For temperatures below 550°C, the B incorporation is reduced due to surface-kinetic effects.

Solid solutions with B composition up to x = 0.06 have been grown on GaAs substrates (Dumont et al. 2003). The epitaxial layers were grown by MOVPE using GaEt$_3$ or GaMe$_3$, B$_2$H$_6$, and AsH$_3$. A growth temperature of circa 570°C–600°C may be optimal for the growth of these solid solutions.

Solid solutions up x = 0.01 (x = 0.078 [Groenert et al. 2004]) have been grown by molecular beam epitaxy (MBE) (Gupta et al. 2000). The layers of 0.9 μm thicknesses were grown at various temperatures ranging from 455°C to 620°C, while the boron cell temperature was varied between 1700°C and 1900°C.

The band gap of B$_x$Ga$_{1-x}$As increases by only 4–8 meV per 1 at% B at x increases up to 2–4 (Geisz et al. 2000). According to the data of Ku (1966), variation of the energy gap from 1.36 to 1.27 eV was observed within the solid solution compositions.

Thermodynamic parameters have been derived and calculated for the B$_x$Ga$_{1-x}$As solid solutions (Dumont and Monteil 2006). The Gibbs free energy change was calculated for the gas phase reaction of these solid solutions.

3.6 BORON–OXYGEN–ARSENIC

BAsO$_4$ ternary compound is formed in this system. It is characterized by two polymorphic modifications. The *first polymorph* crystallizes in the tetragonal structure with the lattice parameters a = 445.9 ± 0.6 and c = 679.6 ± 0.6 pm and a calculated density of 3.642 g·cm^{-3} (Schulze 1933; Shafer et al. 1956). The tetragonal structure of this compound is stable up to about 830°C and sublimate above this temperature (Kosten and Arnold 1980).

The *high-pressure polymorph* crystallizes in the trigonal structure with the lattice parameters a = 456.2 ± 0.5 and c = 1033 ± 1 pm and a calculated density of 4.004 g·cm^{-3} (Dachille and Glasser 1959; Dachille and Roy 1959). According to the data of Mackenzie et al. (1959), this polymorph crystallizes in the hexagonal structure with the lattice parameters a = 791 ± 2 and c = 1032 ± 2 pm and experimental and calculated densities of 3.9 ± 0.2 and 3.98 g·cm^{-3}, respectively. Dachille and Glasser (1959) noted that the material prepared by Mackenzie et al. (1959) was identical to the earlier material and was in fact quartz form, and that the unit cells assigned by them are incorrect.

BAsO$_4$ was prepared by mixing together the solutions of H$_3$BO$_3$ and H$_3$AsO$_4$ or As$_2$O$_5$ and evaporating to dryness (Shafer et al. 1956; Mackenzie et al. 1959). The high-pressure polymorph was prepared by containing the BAsO$_4$ polycrystalline samples in graphite capsules where they were subjected to pressures of 8.6 GPa at 1000°C–1200°C for about 10 min (Mackenzie et al. 1959).

CHAPTER 4

Systems Based on AlN

4.1 ALUMINUM–LITHIUM–NITROGEN

AlN–Li$_3$N. Li$_3$AlN$_2$ ternary compound is formed in this system. It crystallizes in the cubic structure with the lattice parameter $a = 942.7$ pm ($a = 946.1 \pm 0.3$ pm and an experimental density of 2.33 g·cm^{-3} [Juza and Hund 1946, 1948; Goel and Cahoon 1989]; $a = 947.0$ pm [Yamane et al. 1985]) and an energy gap of 4.4 eV (Kushida et al. 2004; Kuriyama et al. 2005, 2007).

The title compound is a pure ionic conductor (Yamane et al. 1985) and hydrolyzes easily (Juza and Hund 1946, 1948).

A single phase of Li$_3$AlN$_2$ was obtained by heating Li$_3$N and Al in the N$_2$ atmosphere at 730°C, or by interaction of Li$_3$Al and N$_2$, or by interaction of Li$_3$Al and AlN (molar ratio Li$_3$Al/AlN = 1.2–1.5) above 750°C in the N$_2$ flow for 1 h (Juza and Hund 1946, 1948; Yamane et al. 1985). It was also synthesized by direct reaction between Li$_3$N and Al (molar ratio 1:1) when the reaction was performed in the Ta crucible under a N$_2$ pressure of 93.3 kPa after the evacuation to 0.13 Pa (Kushida et al. 2004; Kuriyama et al. 2005). The typical reaction temperature and time were 750°C and 5 h, respectively.

High-temperature oxide melt drop solution calorimetry was used to study the energetics of Li$_3$AlN$_2$ formation (McHale et al. 1999). The standard enthalpy of formation is -567.8 ± 12.4 kJ·mol^{-1}. From this value and others available in the literature, the enthalpy of formation from BN and Li$_3$N was calculated: $\Delta H^{298}_f = -90.6 \pm 12.7$ kJ·mol^{-1}.

4.2 ALUMINUM–BERYLLIUM–NITROGEN

AlN–Be$_3$N$_2$. When powder mixtures of AlN and α-Be$_3$N$_2$ were hot pressed, phase transformation from α-Be$_3$N$_2$ to β-Be$_3$N$_2$ occurred (Schneider et al. 1980). The transformation started at 1600°C and was completed to 1760°C. The solubility of β-Be$_3$N$_2$ in AlN at 1760°C is equal to 2.5 mol%.

4.3 ALUMINUM–MAGNESIUM–NITROGEN

AlN–Mg$_3$N$_2$. According to the data of x-ray diffraction (XRD), five compounds are formed in this system, which crystallizes in the hexagonal structure (Zhukov et al. 1996). The lattice parameters for these compounds are as follows: $a = 336.8 \pm 0.1$ and $c = 2550 \pm 2$ pm and experimental and calculated densities of 2.80 and 2.82 g·cm^{-3}, respectively, for **Mg$_3$AlN$_3$**; $a = 328.7 \pm 0.2$ and $c = 1075 \pm 1$ pm and experimental and calculated densities of 3.05 and 3.02 g·cm^{-3}, respectively, for **Mg$_3$Al$_2$N$_4$**; $a = 325.7 \pm 0.1$ and $c = 2636 \pm 1$ pm and experimental and calculated densities of 3.05 and 3.07 g·cm^{-3}, respectively, for **Mg$_3$Al$_3$N$_5$**; $a = 323.9 \pm 0.1$ and $c = 1560 \pm 1$ pm and experimental and calculated densities of 3.10 g·cm^{-3} for **Mg$_3$Al$_4$N$_6$**; and $a = 323.1 \pm 0.1$ and $c = 3622 \pm 4$ pm and experimental and calculated densities of 3.10 g·cm^{-3} for **Mg$_3$Al$_5$N$_7$**.

According to the data of XRD, AlN dissolves up to 2 mol% Mg$_3$N$_2$ (Lesunova and Burmakin 2005).

The samples were annealed at 1200°C for 100 h (Zhukov et al. 1996).

4.4 ALUMINUM–CALCIUM–NITROGEN

According to the data of Beznosikov (2003), **Ca$_3$AlN** compound, which crystallizes in the cubic structure of perovskite type, could form in the Al–Ca–N ternary system.

AlN–Ca₃N₂. $Ca_3Al_2N_4$ and **$Ca_6Al_2N_6$** ternary compounds are formed in this system. $Ca_3Al_2N_4$ exists in two polymorphic modifications, and both crystallize in the monoclinic structure (Ludwig et al. 2000). They are characterized by the following lattice parameters: a = 957.2 ± 0.3, b = 580.2 ± 0.3, and c = 956.3 ± 0.5 pm and β = 116.62 ± 0.03° and a calculated density of 3.097 g·cm⁻³ for the *first polymorph* and a = 1060.6 ± 0.2, b = 826.0 ± 0.2, and c = 551.7 ± 0.1 pm and β = 92.1 ± 0.1° and a calculated density of 3.166 g·cm⁻³ for the *second polymorph*. Green transparent single crystals of the first polymorph were obtained from reactions of mixtures of the representative metals with N_2 above temperatures of 1000°C. It could also be formed from mixtures of Ca_3N_2 and AlN by heating to temperatures between 1000°C and 1100°C in welded Nb tubes sealed within quartz tubes. The second polymorph was formed as a by-product of a reaction of Ca with Al_2O_3 under N_2 at 930°C in the form of colorless crystals. Both phases are moisture sensitive, forming NH_3.

$Ca_6Al_2N_6$ also crystallizes in the monoclinic structure, with the lattice parameters a = 693.7 ± 0.3, b = 614.9 ± 0.3, and c = 987.1 ± 0.5 pm and β = 94.01 ± 0.03° and a calculated density of 2.993 g·cm⁻³ (Ludwig et al. 1999). For its preparation, mixtures of Ca with Al (molar ratio 5:1) were heated under static N_2 at ambient pressure to 1000°C, reacted at this temperature for 10 h, and then allowed to cool down rapidly at a maximum rate of 150°C·h⁻¹. Moisture-sensitive, slightly transparent crystals of the title compound were obtained.

4.5 ALUMINUM–STRONTIUM–NITROGEN

AlN–Sr₃N₂. $Sr_3Al_2N_4$ ternary compound is formed in this system. It crystallizes in the orthorhombic structure with the lattice parameters a = 590.1 ± 0.3, b = 1000.5 ± 0.5, and c = 958.0 ± 0.4 pm and a calculated density of 4.38 g·cm⁻³ (Blase et al. 1994). It was prepared as colorless crystals from the mixture of Sr and Al (molar ratio 3:1) at 1000°C under 0.1 MPa of N_2 for 24 h. The rate of heating and cooling was 50°C·h⁻¹. The title compound is sensitive to hydrolysis.

4.6 ALUMINUM–BARIUM–NITROGEN

AlN–Ba₃N₂. $Ba_3Al_2N_4$ ternary compound is formed in this system. It crystallizes in the orthorhombic structure with the lattice parameters a = 617.9 ± 0.2, b = 1005.2 ± 0.4, and c = 1023.0 ± 0.4 pm and a calculated density of

5.457 g·cm⁻³ (Ludwig et al. 1999). Single crystals of this compound were obtained from mixtures of Ba and Al (molar ratio 3:1). The mixture was first premolten under Ar at 1000°C. In a second step, it was heated under static N_2 at ambient pressure to 1000°C, reacted at this temperature for 10 h, and then allowed to cool down rapidly at a maximum rate of 150°C·h⁻¹. Moisture-sensitive, colorless transparent crystals were obtained.

4.7 ALUMINUM–GALLIUM–NITROGEN

AlN–GaN. Thermodynamic calculations based on a regular solution model and experimental results have predicted that an unstable region of mixing occurs in the AlN–GaN system (Takayama et al. 2000, 2001). The interaction parameter (3 kJ·mol⁻¹) that was used in the calculations has been analytically obtained by the valence force field model, modified for wurtzite structures. According to the interaction parameter, the value of the critical temperature was found to be 181 K.

The thermodynamic properties of the Al–Ga–N₂ system under high N_2 pressure up to 1 GPa and 1800°C were investigated by Belousov et al. (2010a). The stability of $Al_xGa_{1-x}N$ with a particular composition x strongly depends on the pressure and temperature. Using the thermodynamic functions, the T-x phase diagram (Figure 4.1) for the AlN–GaN system was determined. For 0.1 MPa, the equilibrium isobar of

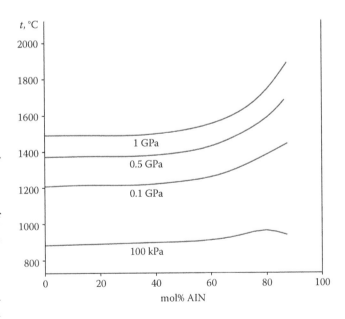

FIGURE 4.1 *T-x* phase diagram of $Al_xGa_{1-x}N$ at various nitrogen pressures. (From Belousov, A., et al., *J. Cryst. Growth*, *312*(18), 2579, 2010. With permission.)

the temperature seems to depend linearly on composition. However, with increase in pressure, the isobars have a tendency to bend to higher temperatures, particularly for higher Al content, $x > 0.6$. Accordingly, the $Al_xGa_{1-x}N$ solid solutions are more stable at higher pressures. For $x = 0.5$, the equilibrium temperature at 0.1 GPa is 1230°C. An increase in the pressure up to 0.5 or 1 GPa will lead to a higher stability, and the equilibrium temperature has been calculated to be 1392°C and 1513°C, respectively.

$Al_xGa_{1-x}N$ solid solutions are direct-gap semiconductors over the entire range of Al content (Djoudi et al. 2012).

According to the experimental data, while the lattice parameters essentially follow Vegard's law, minor, yet significant, deviations were observed: the lattice parameter a tends to be smaller and c slightly larger (Belousov et al. 2010b).

First-principles calculations by means of the full-potential augmented plane-wave method using the local density approximation (LDA) were carried out for the structural and electronic properties of the $Al_xGa_{1-x}N$ alloys in the wurtzite structure (Dridi et al. 2003). The lattice constants a and c were found to change linearly. The calculated band gap variation for these alloys exhibits a downward bowing of 0.71 eV. From the calculated results, it was found that the bowing parameter of the lattice constant for $Al_xGa_{1-x}N$ solid solutions is 1.7 ± 0.1 pm (Djoudi et al. 2012).

The crystal growth of $Al_xGa_{1-x}N$ solid solutions ($0.22 < x < 0.91$) has been carried out from the solution of Ga melt under high N_2 pressure (up to 1 GPa) and at high temperature (up to 1780°C) (Belousov et al. 2009, 2010b). Aluminum and prereacted polycrystalline $Al_yGa_{1-y}N$ or AlN powder were used as a source of Al and N. The best results were achieved using prereacted polycrystalline $Al_yGa_{1-y}N$ and/or AlN. The crystals grew spontaneously on the wall and bottom of the crucible. The p-T equilibrium phase diagram of the $Al_xGa_{1-x}N(s) + Al/Ga(l) + N_2(g)$ system for the various Al contents in solid phase was determined.

$Al_xGa_{1-x}N$ bulk single crystals at x up to 0.27 have also been grown by a solution growth technique in a cubic anvil cell with a solid pressure medium (Geiser et al. 2005). GaN powder pellets as starting materials serve as a nitrogen source for crystallization in a Ga/Al alloy. The growth process was carried out at 1750°C and 3 GPa for up to 3 days.

4.8 ALUMINUM–INDIUM–NITROGEN

AlN–InN. Thermodynamic calculations based on a regular solution model and experimental results have predicted that an unstable region of mixing occurs in this system (Takayama et al. 2000, 2001). The interaction parameter (56.5 kJ·mol⁻¹) that was used in the calculations for the wurtzite structure has been analytically obtained by the valence force field model, modified for wurtzite structures. According to the interaction parameter, the value of the critical temperature was found to be 3130°C. The critical temperature for cubic $Al_xIn_{1-x}N$ alloys was calculated to be approximately 1210°C (Scolfaro 2002) (2080°C [Ferhat and Bechstedt 2002]). The first-principles calculations show that the lattice parameter of the $Al_xIn_{1-x}N$ alloys with a sphalerite structure has a very small deviation from Vegard's law and even exhibits linear composition dependence (Ferhat and Bechstedt 2002; Liou and Liu 2007; Wang et al. 2010). Direct and indirect bowing parameters of 4.731 ± 0.794 eV and 0.462 ± 0.285 eV, respectively, for the energy gap were obtained, and there is a direct–indirect crossover near $x = 0.817$. The bulk modulus is monotonically increased with an increase of the aluminum composition, and a deviation parameter of the bulk modulus of 10.34 ± 9.37 GPa was obtained. On the contrary, the pressure derivative of the bulk modulus is monotonically decreased with an increase of x.

First-principles calculations by means of the full-potential augmented plane-wave method using the LDA were carried out for the structural and electronic properties of the $Al_xIn_{1-x}N$ alloys in the wurtzite structure (Teles et al. 2002a; Dridi et al. 2003). The lattice constant c was found to change linearly, while the lattice parameter a exhibits an upward bowing. According to the data of Liou et al. (2005), the Vegard's law deviation parameter for such alloys is 6.3 ± 1.4 pm for the a lattice constant and -16 ± 1.5 pm for the c constant. The calculated band gap variation for these alloys exhibits a downward bowing of 4.09 eV (Dridi et al. 2003). Liou et al. (2005) indicated that the band gap bowing parameter is 3.668 ± 0.147 eV with the lattice constants obtained by means of the minimized equilibrium energy, and 3.457 ± 0.152 eV with the lattice constants derived from Vegard's law.

The first-principles evolutionary techniques have been performed on the Al–In–N ternary system in order to obtain its stable structure at atmospheric pressure

(Chang et al. 2016). Besides identifying orthorhombic **Al₄In₂N₆** ($a = 556.8$, $b = 555.1$, and $c = 514.6$ pm) as the thermodynamically stable one, the explorations also uncovered five metastable ternary structures, specifically **AlIn₇N₈** and **Al₅InN₆** (both trigonal structures, $a = 688.9$ and 545.0 and $c = 533.9$ and 505.4 pm, respectively), and **Al₂In₄N₆**, **Al₃In₃N₆**, and **Al₆In₂N₈** (all monoclinic structures, $a = 580.8$, 568.0, and 550.4; $b = 580.8$, 526.8, and 508.8; and $c = 630.0$, 568.0, and 638.1 pm). It can be seen that the lattice constant a is decreasing monotonically with an increase in Al composition. In contrast, dependence of the c lattice constant on Al composition is quite irregular. The nonlinear interpolation equation of composition dependence energy band gap E_g for Al$_x$In$_{1-x}$N composites is given by the following equation: $E_g(x) = 1.235x^2 + 4.806x + 0.298$.

Epitaxial layers of Al$_x$In$_{1-x}$N have been grown on sapphire substrates by microwave-excited metal-organic vapor phase epitaxy (MOVPE) (Guo et al. 1995). At a growth temperature of 600°C, single crystal layers with x up to 0.14 were obtained.

4.9 ALUMINUM–THALLIUM–NITROGEN

AlN–TlN. *Ab initio* calculations have been performed to study structural, electronic, and optical properties of the Al$_x$Tl$_{1-x}$N alloys (Dantas et al. 2008; Shi et al. 2012). It was shown that the lattice constants obey Vegard's law and the band gap bowing coefficients show very strong composition dependence.

4.10 ALUMINUM–SCANDIUM–NITROGEN

An isothermal section of this system at 1000°C under 0.1 MPa of argon is shown in Figure 4.2 (Schuster and Bauer 1985; Schuster et al. 1985). **Sc₃AlN$_{1-x}$** ternary compound is formed in the Al–Sc–N system. It crystallizes in the cubic structure with the lattice parameter $a = 439.6–443.5$ pm (Schuster and Bauer 1985; Schuster et al. 1985) ($a = 440$ pm and an elastic modulus of 249 GPa for $x = 0$ [Höglund et al. 2008]). At 900°C, this compound is in equilibrium with ScN, (Sc), Sc₂Al, and ScAl. AlN does not coexist with Sc but does so with ScN, ScAl₂, and ScAl₃.

The samples were annealed at 1000°C for 500–2000 h (Schuster and Bauer 1985; Schuster et al. 1985).

4.11 ALUMINUM–YTTRIUM–NITROGEN

An isothermal section of this system at 1000°C and 600°C under 0.1 MPa of argon is shown in Figures 4.3 and 4.4

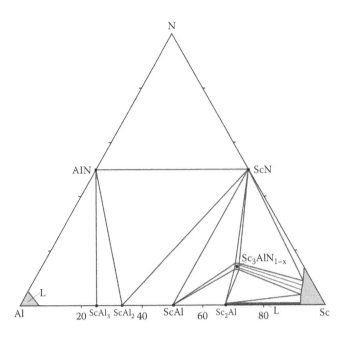

FIGURE 4.2 Isothermal section of the Al–Sc–N system at 1000°C under 0.1 MPa of argon. (From Schuster, J.C., et al., *Rev. Chim. Miner.*, 22, 546, 1985; Schuster, J.C., and Bauer, J., *J. Less Common Metals*, 109(2), 345, 1985. With permission.)

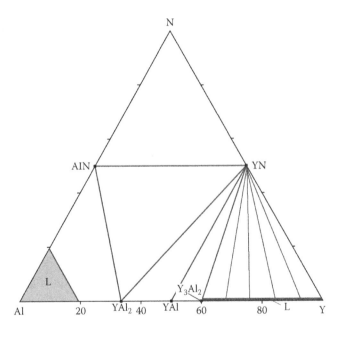

FIGURE 4.3 Isothermal section of the Al–Y–N system at 1000°C under 0.1 MPa of argon. (From Schuster, J.C., et al., *Rev. Chim. Miner.*, 22, 546, 1985; Schuster, J.C., and Bauer, J., *J. Less Common Metals*, 109(2), 345, 1985. With permission.)

(Schuster and Bauer 1985; Schuster et al. 1985; Schweitz and Mohney 2001). No ternary phases exist at temperatures between 600°C and 900°C. At 1200°C, the three-phase field AlN + YAl₂ + YN dominates the phase

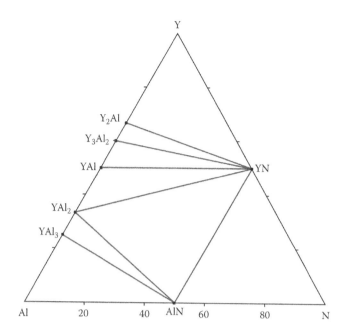

FIGURE 4.4 Calculated isothermal section of the Al–Y–N system at 600°C under 0.1 MPa of N₂. (From Schweitz, K.O., and Mohney, S.E., *J. Electron. Mater.*, *30*(3), 176, 2001. With permission.)

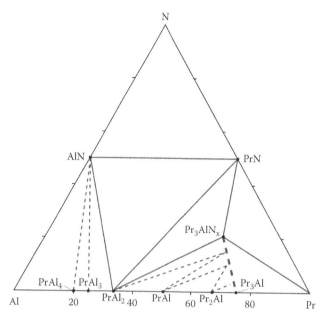

FIGURE 4.5 Isothermal section of the Al–Pr–N system at 600°C under 0.1 MPa of argon. (From Schuster, J.C., et al., *Rev. Chim. Miner.*, *22*, 546, 1985; Schuster, J.C., *J. Less Common Metals*, *105*(2), 327, 1985. With permission.)

diagram since no other phases of this system are solid at this temperature.

The samples were annealed at 1000°C for 500–2000 h.

AlN–YN. No ternary compound occurs in this system (Sun et al. 1991).

4.12 ALUMINUM–LANTHANUM–NITROGEN

An isothermal section of this system at 1000°C under 0.1 MPa of argon is the same as for the Al–Sc–N system (Figure 4.2) (Schuster et al. 1985). **La₃AlN₁₋ₓ** ternary compound is formed in the Al–La–N system (Schuster et al. 1985; Schuster 1985). It crystallizes in the cubic structure with the lattice parameter $a = 507.0$ pm. At 900°C. this compound is in equilibrium with LaN, (La), La₂Al, and LaAl. AlN does not coexist with La but does so with LaN, LaAl₂, and LaAl₃.

4.13 ALUMINUM–CERIUM–NITROGEN

An isothermal section of this system at 600°C under 0.1 MPa of argon is the same as for the Al–Pr–N system (Figure 4.5) (Schuster et al. 1985; Schuster 1985). **Ce₃AlN₁₋ₓ** ternary compound is formed in the Al–Ce–N system. It crystallizes in the cubic structure with the lattice parameter $a = 500.8$ pm.

4.14 ALUMINUM–PRASEODYMIUM–NITROGEN

An isothermal section of this system at 600°C under 0.1 MPa of argon is shown in Figure 4.5 (Schuster et al. 1985). **Pr₃AlN₁₋ₓ** ternary compound is formed in the Al–Pr–N system (Schuster et al. 1985; Schuster 1985). It crystallizes in the cubic structure with the lattice parameter $a = 497.7$ pm.

4.15 ALUMINUM–NEODYMIUM–NITROGEN

An isothermal section of this system at 1000°C under 0.1 MPa of argon is the same as for the Al–Sc–N system (Figure 4.2) (Schuster et al. 1985). **Nd₃AlN₁₋ₓ** ternary compound is formed in the Al–Nd–N system (Schuster et al. 1985; Schuster 1985). It crystallizes in the cubic structure with the lattice parameter $a = 493.9$ pm (Schuster et al. 1985; Schuster 1985) ($a = 491.0$ pm [Haschke et al. 1967]).

4.16 ALUMINUM–SAMARIUM–NITROGEN

An isothermal section of this system at 1000°C under 0.1 MPa of argon is the same as for the Al–Sc–N system (Figure 4.2) (Schuster et al. 1985). **Sm₃AlN₁₋ₓ** ternary compound is formed in the Al–Sm–N system (Schuster et al. 1985; Schuster 1985). It crystallizes in the cubic structure with the lattice parameter $a = 486.2$ pm.

4.17 ALUMINUM–GADOLINIUM–NITROGEN

An isothermal section of this system at 1000°C under 0.1 MPa of argon is the same as for the Al–Y–N system (Figure 4.3) (Schuster et al. 1985). No ternary compounds were observed in the Al–Gd–N system (Schuster et al. 1985; Schuster 1985).

4.18 ALUMINUM–TERBIUM–NITROGEN

An isothermal section of this system at 1000°C under 0.1 MPa of argon is the same as for the Al–Y–N system (Figure 4.3) (Schuster et al. 1985). No ternary compounds were observed in the Al–Tb–N system (Schuster et al. 1985; Schuster 1985).

4.19 ALUMINUM–DYSPROSIUM–NITROGEN

An isothermal section of this system at 1000°C under 0.1 MPa of argon is the same as for the Al–Y–N system (Figure 4.3) (Schuster et al. 1985). No ternary compounds were observed in the Al–Dy–N system (Schuster et al. 1985; Schuster 1985).

4.20 ALUMINUM–HOLMIUM–NITROGEN

An isothermal section of this system at 1000°C under 0.1 MPa of argon is the same as for the Al–Y–N system (Figure 4.3) (Schuster et al. 1985). No ternary compounds were observed in the Al–Ho–N system (Schuster et al. 1985; Schuster 1985).

4.21 ALUMINUM–ERBIUM–NITROGEN

An isothermal section of this system at 1000°C under 0.1 MPa of argon is the same as for the Al–Y–N system (Figure 4.3) (Schuster et al. 1985). No ternary compounds were observed in the Al–Er–N system (Schuster et al. 1985; Schuster 1985).

4.22 ALUMINUM–THULIUM–NITROGEN

An isothermal section of this system at 1000°C under 0.1 MPa of argon is the same as for the Al–Y–N system (Figure 4.3) (Schuster et al. 1985). No ternary compounds were observed in the Al–Tm–N system (Schuster et al. 1985; Schuster 1985).

4.23 ALUMINUM–YTTERBIUM–NITROGEN

An isothermal section of this system at 1000°C under 0.1 MPa of argon is the same as for the Al–Y–N system (Figure 4.3) (Schuster et al. 1985). No ternary compounds were observed in the Al–Yb–N system (Schuster et al. 1985; Schuster 1985).

4.24 ALUMINUM–LUTETIUM–NITROGEN

An isothermal section of this system at 1000°C under 0.1 MPa of argon is the same as for the Al–Y–N system (Figure 4.3) (Schuster et al. 1985). No ternary compounds were observed in the Al–Lu–N system (Schuster et al. 1985; Schuster 1985).

4.25 ALUMINUM–CARBON–NITROGEN

Some isothermal sections of the Al–C–N ternary system have been calculated using the program Thermo-Calc and including the formation of the Al_5C_3N compound (Wen 1993). According to these calculations, AlN is in thermodynamic equilibrium with Al_5C_3N and C and Al_5C_3N forms tie-line with Al_4C_3 at 530°C. Increasing the temperature to 1730°C and then to 2130°C leads to appearing and enlarging of the AlN + L two-phase region and to decreasing of AlN + L + Al_5C_3N and Al_5C_3N + L + Al_4C_3 three-phase regions. At 2330°C, in this ternary system, there is two two-phase region (AlN + L and C + G) and two three-phase regions (AlN + L + C and AlN + C + G). Increasing the temperature to 2430°C leads to appearing the L + G two-phase region near the Al–N binary system and C + L + G three-phase region and increasing the C + G two-phase region near the C–N binary system. An isothermal section at 1800°C was also constructed by Pietzka and Schuster (1996). None of the solid phases have a measurable homogeneity range (Wen 1993; Pietzka and Schuster 1996; Raghavan 2006b).

According to the data of Wittig and Bille (1951), **Al(CN)₃** ternary compound could be obtained in this system by the interaction of AlH_3 with HCN.

AlN–Al₄C₃. The phase diagram of this system was calculated by excluding the gas phase (Figure 4.6a) (Qiu and Metselaar 1997). The eutectic temperature is 2180°C. Only one ternary compound, **Al₅C₃N**, is formed in this system. It melts incongruently at 2253°C (Qiu and Metselaar 1997) (at 2185°C ± 15°C [Pietzka and Schuster 1996; Raghavan 2006a]) and crystallizes in the hexagonal structure with the lattice parameters $a = 327.6$ and $c = 215.8$ pm (Pietzka and Schuster 1996) ($a = 328.1$ and $c = 216.7$ pm [Jeffrey and Wu 1966; Raghavan 2006a; Xu et al. 2011]). Upon annealing Al_5C_3N under 0.1 MPa of pure H_2 at 1500°C for 8 h, no decomposition of this compound and the formation of other phases occurred (Pietzka and Schuster 1996). Other possible compounds have been ignored

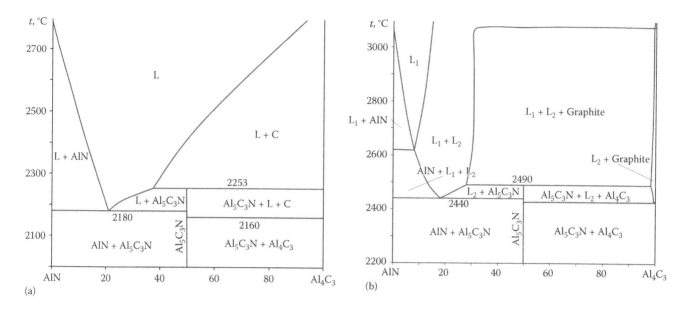

FIGURE 4.6 Two versions of the calculated phase diagram of the AlN–Al_4C_3 system. (Panel a from Qiu, C., and Metselaar, R., *J. Am. Ceram. Soc.*, 80(8), 2013, 1997; panel b from Pavlyuchkov, D., et al., *J. Phase Equilibria Diffus.*, 33(5), 357, 2013.)

in the evaluation of the AlN–Al_4C_3 system (Qiu and Metselaar 1997).

The new version of the AlN–Al_4C_3 phase diagram was calculated by Pavlyuchkov et al. (2012) (Figure 4.6b). The comparison of the obtained phase diagram with that presented in Figure 4.6a shows an essential difference: the calculations exhibit the miscibility gap in the liquid phase.

According to *ab initio* calculations, Al_5C_3N is a semiconductor with an energy gap of 0.81 eV (Xu et al. 2011). The calculated bulk and Young's moduli are 201 and 292 GPa, respectively. The bulk modulus at zero pressure (B_0) and its derivative (B') were estimated as 195 GPa and 3.84, respectively.

According to the data of Jeffrey and Wu (1963, 1966), **$Al_6C_3N_2$, $Al_7C_3N_3$, $Al_8C_3N_4$, $Al_9C_3N_5$,** and **$Al_{10}C_3N_6$** ternary compounds are formed in the Al–C–N ternary system (the last two are hypothetical compounds). Stoichiometrically and structurally, these aluminum carbonitrides, including Al_5C_3N, can be expressed as members of a homologous series $(AlN)_n Al_4C_3$ where n = 1 to 4. All these compounds crystallize in the hexagonal or trigonal structure with the lattice parameters a = 324.8 and c = 400.3 pm and experimental and calculated densities of 3.04 and 3.076 g·cm⁻³, respectively, for $Al_6C_3N_2$; a = 322.6 and c = 317.0 pm and experimental and calculated densities of 3.046 and 3.102 g·cm⁻³, respectively, for $Al_7C_3N_3$; a = 321.1 and c = 550.8 pm and

experimental and calculated densities of 3.05 and 3.133 g·cm⁻³, respectively, for $Al_8C_3N_4$; a = 319.7 and c = 416.9 pm and experimental and calculated densities of 3.06 and 3.141 g·cm⁻³, respectively, for $Al_9C_3N_5$; and a = 318.6 and c = 700.0 pm and experimental and calculated densities of 3.07 and 3.157 g·cm⁻³, respectively, for $Al_{10}C_3N_6$. The compounds $Al_6C_3N_2$, $Al_7C_3N_3$, and $Al_8C_3N_4$ were prepared by heating AlN in a carbon crucible with a N_2 environment at about 2000°C and formed on the walls of the vessel.

The existence of all these aluminum carbonitrides, except Al_5C_3N, remains controversial (Qiu and Metselaar 1997), and it has been found that they are impurity stabilized (Pietzka and Schuster 1996; Raghavan 2006a).

4.26 ALUMINUM–SILICON–NITROGEN

The phase relations in the whole Al–Si–N ternary system were calculated by applying an ideal solution model to the liquid phase and the gas pressure of 0.1 MPa (Hillert and Jonsson 1992b). Isothermal sections at 600°C, 1300°C, 1800°C, 1900°C, and 2400°C have been constructed, and two of them, at 600°C and 2400°C, are shown in Figure 4.7a and b. An isothermal section at 600°C was also constructed by Weitzer et al. (1990). No ternary compound was observed. At 600°C, AlN is in thermodynamic equilibrium with Si_3N_4, Si, and the liquid in the Al–Si system. Heating up to 1300°C leads to an increase of the AlN + L two-phase

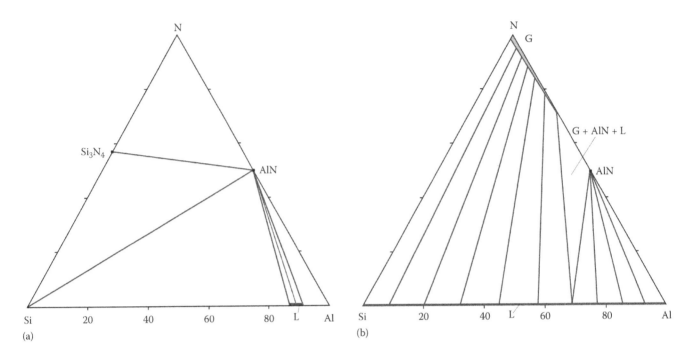

FIGURE 4.7   Calculated isothermal sections of the Al–Si–N ternary system at (a) 600°C and (b) 2400°C. (From Hillert, M., and Jonsson, S., *CALPHAD*, *16*(2), 199, 1992. With permission.)

field and a decrease of the AlN + Si₃N₄ + L three-phase field. At 1800°C, the AlN + Si₃N₄ + L three-phase field disappeared, and at 1900°C, the thermodynamic equilibrium between AlN and Si₃N₄ also disappeared and the three-phase field AlN + L + G enlarged. At 2400°C, two two-phase regions (AlN + L and L + G) and one three-phase region (AlN + L + G) exist in the isothermal section.

Kasu et al. (2001) synthesized **Al₁₋ₓSiₓN** solid solutions up to $x = 0.12$ by MOVPE. From the thermodynamic point of view, it is very likely that this solid solution has to be considered as metastable supersaturated, although the crystal quality is very perfect. The temperature during preparation (900°C) may be far too low to enable equilibration. The solid solution was characterized as substitutional; one Si atom replaces one Al atom. The same authors (Taniyasu et al. 2001) indicated that these solid solutions crystallize in the hexagonal structure. The a and c axis constants of the strain-free $Al_{1-x}Si_xN$ solid solutions grown on the AlN layers linearly decrease with the Si content as $a = 311.13 - 14.12x$ and $c = 498.14 - 22.99x$ (pm) at x up to 0.08.

Al–Si–N films were sputtered in an Ar + N₂ mixture from a composed target (a Si plate fixed by an Al ring) (Musil et al. 2008). It was found that while the films with a low (≤10 at%) Si content are crystalline, those

with a high (≥25 at%) Si content are amorphous when sputtered at the substrate temperature of 500°C. Both groups of the films exhibit a high hardness, 21 and 25 GPa, respectively. The hardness of the amorphous films does not vary with increasing annealing temperature up to 1000°C even after 4 h.

Al–Si–N films were also deposited by reactive direct current (DC) magnetron cosputtering of Al and Si targets in an Ar + N₂ atmosphere at 200°C and 500°C sample temperatures (Pélisson et al. 2007; Pélisson-Schecker et al. 2014). The chemical composition was varied from pure AlN to Al–Si–N alloys with 23 at% Si. The films were crystalline with the *h*-AlN structure up to 12–16 at% Si, as was found from XRD and transmission electron microscopy (TEM). A shift of XRD peaks indicates a substitutional incorporation of Si in *h*-AlN lattice up to a solubility limit identified at 6 at% Si. By further increasing the Si content, a nanocomposite Al₀.₈₈Si₀.₁₂N exists with Si₃N₄. Hardness measurements show a diffuse hardness maximum exceeding 30 GPa around 10 at% Si. The crystalline phase may be stable up to 900°C.

The influence of the growth conditions on the Si incorporation in AlN films grown by plasma-assisted molecular beam epitaxy (MBE) has been studied by Hermann et al. (2005). It was shown that nitrogen-rich growth conditions allow controlled incorporation of Si

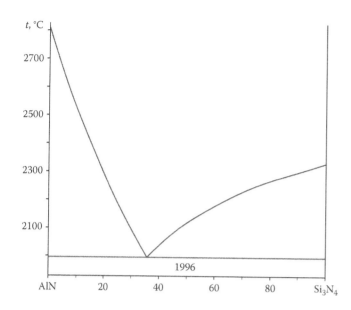

FIGURE 4.8 Calculated phase diagram of the AlN–Si_3N_4 system. (From Hillert, M., and Jonsson, S., *CALPHAD*, *16*(2), 199, 1992. With permission.)

up to a concentration of $5.2 \cdot 10^{21}$ cm^{-3}. The structural and morphological properties determined by XRD and AFM were not affected up to a Si concentration of $1.2 \cdot 10^{21}$ cm^{-3}.

AlN–Si_3N_4. The phase diagram of this system is a eutectic type (Figure 4.8) (Hillert and Jonsson 1992b). It was predicted that a eutectic liquid could form at 1996°C and 64.3 mol% AlN under a pressure of at least 0.67 MPa. According to the data of Kaufman (1979), **$Al_5Si_3N_9$** ternary compound that melts incongruently at 2230°C is formed in this system. Schneider et al. (1980), however, assumed this phase to be unstable in the Al–N–Si ternary system, as it needs some oxygen to be stabilized. Other investigations regarding quaternary systems with AlN–Si_3N_4 as a boundary system (Land et al. 1978; Huang et al. 1983, 1986; Fukuhara 1988; Weitzer et al. 1990) do not mention this phase and assume AlN to be in equilibrium with Si_3N_4.

In the reaction of AlN with Si_3N_4 at 1745°C–1775°C, silicon nitride is partially decomposed to form free silicon (Khoruzhaya et al. 2002).

4.27 ALUMINUM–GERMANIUM–NITROGEN

Coatings of the Al–Ge–N system were synthesized using reactive DC magnetron sputtering and analyzed using XRD, x-ray photoelectron spectroscopy (XPS), and both scanning electron microscopy (SEM) and TEM (Lewin et al. 2012). Ternary alloys with Ge contents from 5 to 28 at% were produced and found to be either a solid solution phase of nanocrystalline **$Al_{1-x}Ge_xN$**, or nanocomposites of this phase and an amorphous Ge_3N_{4-y} phase. Ternary coatings were found to have hardness between 18 and 24 GPa.

4.28 ALUMINUM–TIN–NITROGEN

Coatings consisting of Al, Sn, and N have been deposited using cosputtering from Al and Sn targets in a reactive atmosphere containing N_2 (Lewin and Patscheider 2016). The coatings were analyzed using XRD, XPS, and SEM. It was shown that the coatings consist of a single-phase solid solution based on wurtzite **$Al_{1-x}Sn_xN_y$**, with x varying between 0 and 0.5, and y close to unity. The attained material is metastable with respect to decomposition into AlN, Sn, and N_2. The material is hard at room temperature, with nanoindentation values of 17–24 GPa. Coatings on silica substrates are transparent and yellow to red-brown in color.

4.29 ALUMINUM–TITANIUM–NITROGEN

The equilibria in the Ti-rich part of the ternary system have been determined by van Thyne and Kessler (1954) for 0–10 mass% Al and 0–1 mass% N. This study applied micrograph analysis and XRD of samples annealed at 600°C to 1250°C for 576 to 6 h. These samples were prepared from high-purity arc molten alloys. The obtained results are given as vertical sections for constant N content.

There are three ternary compounds in this system. **Ti_2AlN** (τ_1) is apparently the most stable ternary phase, which belongs to the group of H phases (Nowotny et al. 1964) and was observed to exist at temperatures as low as 850°C and up to 1600°C (Ivchenko and Kosolapova 1976; Wu et al. 1995). This phase is a line compound, and at 1300°C, it is homogeneous at the composition $Ti_2AlN_{0.82}$ (Wu et al. 1995; Pietzka and Schuster 1996). Upon heating single-phase Ti_2AlN_{1-x}, no liquid phase occurs up to 1950°C. According to the data of Pang et al. (2010), this compound was susceptible to decomposition above 1500°C through sublimation of high-vapor-pressure Al, resulting in a porous surface layer of TiN_x ($0.5 \le x \le 0.75$) being formed.

Ti_2AlN crystallizes in the hexagonal structure with the lattice parameters $a = 298.69$ and $c = 1362.2$ pm for $Ti_2AlN_{0.82}$ (Khan et al. 2004b; Raghavan 2006c) ($a = 299.4$ and $c = 1361$ pm and a calculated density of 4.30 g·cm^{-3} [Jeitschko et al. 1963]; $a = 298.9$ and $c = 1361.5$ pm and a calculated density of 4.29 g·cm^{-3} [Ivchenko and Kosolapova 1977]; a = 299.12 and c = 1362.1 pm [Schuster et al. 1985]; a = 304 and c = 1369 pm [Kaufman et al. 1986]; a = 299.5 and c = 1361 pm [Wu et al. 1995]; a = 299.11 ± 0.02 and c = 1361.9 ± 0.1 pm [Pietzka and Schuster 1996; Hug et al. 2005]; a = 300 and c = 1370 pm [Farber et al. 1999]; a = 299.1 and c = 1361.1 pm [calculated values] [Ivanovskii et al. 2000]; a = 298.9 ± 0.2 and c = 1361.4 ± 0.5 pm and a calculated density of 4.31 g·cm⁻³ at 25°C [Barsoum 2000; Barsoum et al. 2000a; Gamarnik et al. 2000]; a = 299.9 and c = 1365.0 pm at 400°C and a = 300.9 and c = 1365.0 pm at 800°C [Barsoum et al. 2000a]; a = 298.6 ± 0.3 and c = 1360 ± 2 [Manoun et al. 2006]; a = 299.029 ± 0.002 and 302.138 ± 0.003 and c = 1361.34 ± 0.02 and 1374.07 ± 0.03 at 100°C and 1100°C, respectively [Lane et al. 2011]). The thermal expansion coefficients in the a and c directions are, respectively, (10.3 ± 0.2)·10⁻⁶ and (9.3 ± 0.2)·10⁻⁶ K⁻¹ (Lane et al. 2011).

The lattice parameters of τ_1 were measured as a function of pressure up to circa 50 GPa (Manoun et al. 2006). No phase transformations were observed. The compressibility along the c axis is larger than that along the a axis. The bulk modulus of this compound is 169 ± 3 GPa.

τ_1 has been prepared by hot pressing a powder mixture of TiN, Al, and Ti (molar ratio 1:1:1) at 850°C for 200 h in sealed quartz ampoules (Jeitschko et al. 1963). According to the data of Pang et al. (2010), TiN, Al, and Ti were initially mixed in EtOH for 24 h, and then hot pressed in an Ar atmosphere at a heating rate of 50°C·min⁻¹ until the temperature of 1400°C was reached, and held at this temperature for 2 h. The pressure used during hot pressing was 30 MPa. The title compound could also be fabricated by reactive sintering of AlN and Ti for 16 h under a vacuum of 10⁻³ Pa (Gamarnik et al. 2000) or by heating such mixtures at 1400°C for 48 h under a pressure of circa 40 MPa (Barsoum et al. 2000a). The samples of τ_1 were also made by hot pressing AlN and Ti powders, which were stoichiometrically weighed, ball milled for 12 h, and dried in vacuum for 12 h at 150°C (Lane et al. 2011). The powder mixtures were then poured and wrapped in graphite foil, placed in a graphite die, and heated under vacuum at 10°C·min⁻¹ in a graphite-heated hot press up to 1400°C and held for 8 h. During heating, a load corresponding to a stress of circa 45 MPa was applied when the temperature reached 500°C and was maintained throughout the run. Fully dense polycrystalline samples of τ_1 were processed by mixing elemental or binary powders to the desired stoichiometry, followed by cold pressing and then hot isostatic pressing in sealed evacuated containers at temperatures ranging from 1275°C to 1600°C and pressures of up to 110 MPa for up to 24 h (Farber et al. 1999).

Crystals of τ_1 were grown using a modified Bridgman method and subsequently annealed at 1100°C for 60 h, followed by cooling to room temperature at a rate of 65°C·h⁻¹ (Wu et al. 1995).

Ti₃AlN (Ti₃AlN₀.₅₆) (τ_2) melts incongruently at 1590°C ± 10°C (Pietzka and Schuster 1996). It crystallizes in the cubic structure with the lattice parameter a = 411.27 ± 0.17 pm for Ti₃AlN₀.₅₆ (Han et al. 2004; Khan et al. 2004b; Raghavan 2006c) (a = 411.2 pm [Schuster et al. 1985]; a = 411.70 ± 0.07 pm for Ti₃AlN₀.₅₆ [Pietzka and Schuster 1996]; a = 411 pm, a calculated density of 4.406 g·cm⁻³, and bulk modulus B = 194.62 GPa [Kanchana 2009]; a = 409 pm, a calculated density of 4.52 g·cm⁻³, and bulk modulus B = 173 GPa [Wang et al. 2014]). Pietzka and Schuster (1996) reported a low-temperature modification (<1200°C) of Ti₃AlN₁₋ₓ, where the cubic structure transforms to the orthorhombic structure with the lattice parameters $a \approx$ 306.5, $b \approx$ 1074.8, and $c \approx$ 845.5 pm.

There is some controversy regarding the composition of the third compound in the Al–Ti–N ternary system. Barsoum et al. (1999) and Procopio et al. (2000b) concluded that the stoichiometry of this compound is neither Ti₃Al₂N₂ (Schuster and Bauer 1984b; Schuster et al. 1985; Pietzka and Schuster 1996; Barsoum and Schuster 1998) nor Ti₃Al₁₋ₓN₂ (Lee and Petuskey 1997, 1998), but it is **Ti₄AlN₃₋δ** (τ_3, 0 < δ < 0.1) (Finkel et al. 2000). This compound is stable only in a narrow temperature range between 1250°C and 1400°C (Lee and Petuskey 1997) (it is not stable at 1200°C or below and decomposes at 1335°C ± 12°C, but no liquid phase was observed to occur upon heating this compound up to 1900°C [Pietzka and Schuster 1996]; it is stable up to 1500°C in Ar, but decomposes in air to form TiN and Ti₂AlN at circa 1400°C [Lee and Petuskey 1997; Procopio et al. 2000a]). It crystallizes in the hexagonal structure with the lattice parameters a = 298.75 and c = 2335.0 pm (Schuster et al. 1985; Raghavan 2006c) (a = 298.84–299.16 and c = 2339.3–2341.3 pm [Pietzka and Schuster 1996]; a = 299.0 and c = 2338.54 pm [Lee and Petuskey 1997]; a = 300 and c = 2330 pm [Barsoum et al. 1999]; a = 298.80 ± 0.02 and c = 2337.2 ± 0.2 pm and a calculated density of 4.76 g·cm⁻³ [Barsoum 2000; Rawn et al. 2000]; a = 299.05 ± 0.01, 299.96 ± 0.02, 300.45 ±

0.02, 301.01 ± 0.02, and 302.22 ± 0.02 and c = 2338.0 ± 0.1, 2344.7 ± 0.1, 2348.1 ± 0.2, 2352.2 ± 0.2, and 2360.8 ± 0.2 pm at 25°C, 360°C, 570°C, 735°C, and 1094°C, respectively [Barsoum et al. 2000b]; experimental and calculated densities of 4.58 and 4.77 g·cm⁻³, respectively [Procopio et al. 2000a; Rawn et al. 2000; Manoun et al. 2005]). The thermal expansion coefficients along the a and c axes are, respectively, (9.6 ± 0.1)·10⁻⁶ and (8.8 ± 0.1)·10⁻⁶ K⁻¹ (Barsoum et al. 2000b).

The heat capacity of the $Ti_4AlN_{3-\delta}$ compound was measured between 2 and 10 K using a standard adiabatic calorimeter in a liquid helium cryostat (Ho et al. 1999). It was determined that $c_p = 0.00812T + 0.033·10^{-3}T^3$ J·mol⁻¹·K⁻¹ and the characteristic Debye temperature equals 506°C.

Fully dense polycrystalline samples of Ti_4AlN_3 were processed by mixing AlN, TiN, and TiH_2 with the desired stoichiometry (Barsoum et al. 1999,2000b; Ho et al. 1999; Procopio et al. 2000a; Rawn et al. 2000; Manoun et al. 2005). The mixed powders were cold pressed at circa 200 MPa, dehydrided by heating in a vacuum at 900°C for 9 h, and sealed in evacuated borosilicate tubes. These were hot isostatically pressed at 1275°C for 24 h under a pressure of circa 70 MPa. To complete the reaction, the isostatically pressed samples were further annealed at 1325°C for 168°C. Because the annealing step resulted in the formation of voids, the samples were hot isostatically pressed one more time without a can at 1275°C for 10 h, under an Ar pressure of 110 MPa.

AlN–TiN, AlN–TiAl₃, and TiAl₃–TiN are stable tie-lines in this system at low temperatures (Beyers et al. 1984; Bhansali et al. 1990). The existence of the AlN–TiN tie-line implies that a tie-line cannot exist between nitrogen and $TiAl_2$. The most important tie plane is the TiN–AlN–TiAl₃ one (Caron et al. 1996). This plane limits the formation of a lot of other equilibria involving the AlN phase. Two other important tie planes of this diagram are the Ti–TiN–Ti₃Al and TiAl₃–AlN–Al planes.

A number of experimental and computed isothermal sections have been determined for this system: at 1000°C and 1300°C (Schuster and Bauer 1984b; Schuster et al. 1985); at 1300°C (Holleck 1988; Pietzka and Schuster 1996); at 900°C (Durlu et al. 1997); at 1000°C (Schmid-Fetzer and Zeng 1997); at 900°C, 1000°C, 1200°C, 1300°C, 1400°C, 1580°C, 1600°C, 1900°C, and 2500°C (Chen and Sundman 1998); at 700°C and 900°C (Anderbouhr et al. 1999); at 1325°C (Procopio et al. 2000b); at 1000°C (Han et al. 2004; Khan et al. 2004b);

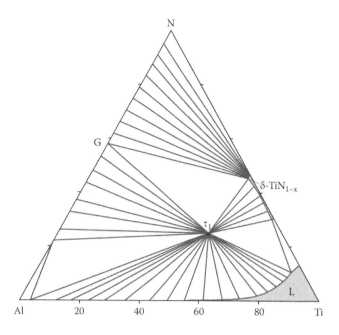

FIGURE 4.9 Isothermal section of the Al–Ti–N system at 2500°C. (From Chen, G., and Sundman, B., *J. Phase Equilibria*, 19(2), 146, 1998. With permission.)

and at 600°C, 900°C, 1000°C, and 1300°C (Gao et al. 2008). Some of these sections are included in the review of Raghavan (2006c).

The computed isothermal section at 2500°C is shown in Figure 4.9 (Chen and Sundman 1998; Raghavan 2006c). It must be noted that there no is experimental point to validate this section. Only Ti_2AlN is stable at this temperature (and at 1600°C also). Ti_2AlN and Ti_3AlN are stable in the computed sections at 1580°C and 1400°C, and at 1300°C all three ternary compounds are stable. The isothermal section at 1325°C is given in Figure 4.10 (Procopio et al. 2000b; Raghavan 2006c). It was obtained from samples prepared by hot isostatic pressing with a subsequent anneal at 1325°C for 168°C. All three ternary compounds exist at this temperature. The location of Ti_4AlN_{3-x} is significantly different from that of $Ti_3Al_2N_2$ in the section at 1300°C (Schuster and Bauer 1984b).

An isothermal section at 1000°C (Figure 4.11) was constructed using annealed for 670 h under Ar atmosphere alloys (Han et al. 2004; Khan et al. 2004b). Diffusion couple experiments with an AlN/Ti pair and nitriding from a gaseous phase were also adopted. The phase equilibria were studied with optical microscopy and SEM, XRD, and electron probe microanalysis (EPMA). The three-phase equilibrium between AlN, TiAl₃, and Ti_2AlN appears to be established. There are

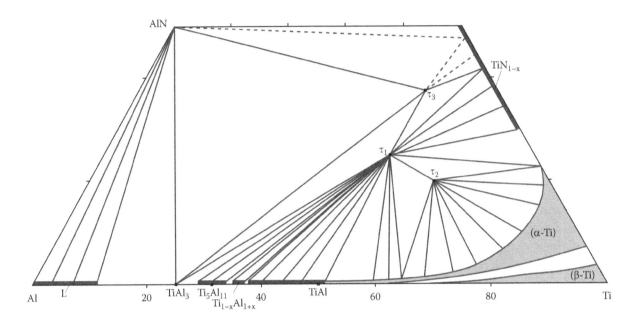

FIGURE 4.10 Isothermal section of the Al–Ti–N system at 1325°C. (From Procopio, A.T., et al., *Metal. Mater. Trans. A, 31*(2), 373, 2000. With permission.)

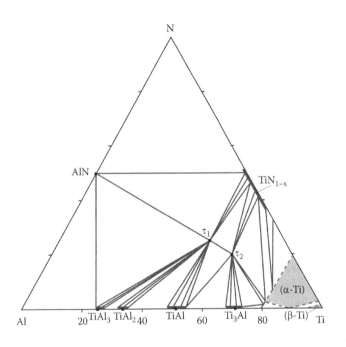

FIGURE 4.11 Isothermal section of the Al–Ti–N system at 1000°C. (From Han, Y.S., et al., *J. Phase Equilibria Diffus., 25*(5), 427, 2004; Khan, Yu.S., et al., *Dokl. RAN, 396*(6), 788, 2004. With permission.)

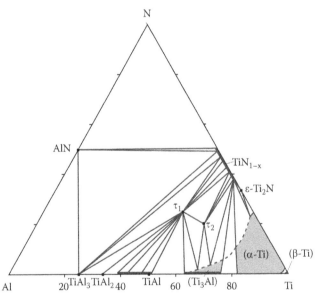

FIGURE 4.12 Isothermal section of the Al–Ti–N system at 900°C. (From Durlu, N., *Z. Metallkde, 8*(5), 390, 1997. With permission.)

some differences between Han et al. (2004) and Khan et al. (2004), and between Chen and Sundman (1998) and Schuster and Bauer (1984) and Schuster et al. (1985) in the triangulation of the isothermal section at this temperature.

An isothermal section at 900°C is given in Figure 4.12 (Durlu et al. 1997). Comparing Figures 4.11 and 4.12, a transition reaction $AlN + Ti_2AlN \longleftrightarrow TiAl_3 + TiN_{1-x}$ is expected to occur between 1000°C and 900°C.

According to the calculations, the TiN phase was predicted to be in equilibrium with both AlN and N_2 at

600°C and 1000°C under a N_2 pressure of less than 0.1 MPa (Schweitz and Mohney 2001).

The isothermal sections of the metastable diagrams were also established at 400°C and 700°C, which are the deposition temperatures that are used for the two competitive chemical vapor deposition (CVD) processes (Anderbouhr et al. 1999). Equilibrium between the metastable $Al_xTi_{1-x}N$ phase and the $TiAl_3$ phase was obtained. This equilibrium is valid in the 20°C–1220°C temperature range. Therefore, $TiAl_3$ could be deposited simultaneously with the expected $Al_xTi_{1-x}N$ solid solution. All the equilibria in which TiN and AlN are not involved are similar to those simulated for the stable phase diagram.

AlN–TiN. The phase diagram of this system has been calculated using the model of regular solutions for the solid phases and that of an ideal solution for the liquid phase (Figure 4.13) (Holleck 1988, 1989). The eutectic temperature and eutectic composition depend experimentally on the dimensions of both AlN and TiN particles (Andrievskiy and Anisimova 1991) because of the possible formation of $Al_xTi_{1-x}N$ metastable solid solutions. Hard coatings prepared by the cathodic arc ion plating method allow formation of a cubic solid solution $Al_xTi_{1-x}N$ ($0 < x < 0.7$) and a wurtzite-type solid solution $Al_xTi_{1-x}N$ ($0.8 < x < 1$) (Ikeda and Satoh 1991; Tanaka et al. 1992). The existing experimental results and thermodynamic calculations lead to the so-called vapor deposition phase

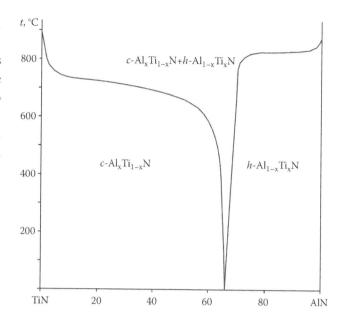

FIGURE 4.14 Metastable phase diagram of the AlN–TiN system. (From Spencer, P.J., *CALPHAD*, 25(2), 163, 2001. With permission.)

diagram (Figure 4.14) (Spencer 2001). Because of differences in chemical composition, the sputtered $Al_xTi_{1-x}N$ coatings show colors changing from metallic silver for low-nitrogen coatings to very dark blue for layers with high nitrogen contents (Jehn et al. 1986); Inamura et al. (1987) indicates that these solid solutions in the composition range of $0.13 \leq x \leq 0.58$ were greenish brown in color.

Mixtures of AlN and TiN containing from 30 to 90 mass% AlN did not show evidence of a reaction between these two materials (Kuzenkova et al. 1976). Only phases based on the initial TiN and AlN were found by XRD in the products of reaction of TiN with AlN (Khoruzhaya et al. 2002). $Al_xTi_{1-x}N$ metastable solid solutions can be obtained using reactive magnetron sputtering (Knotek et al. 1986). The lattice parameter of the $Al_xTi_{1-x}N$ films linearly decreases with increasing Al content.

Ab initio calculations were employed to demonstrate a strong pressure dependence of the maximum AlN fraction preserving the cubic phase in the $Al_xTi_{1-x}N$ hard coating before transforming to the wurtzite structure (Holec et al. 2010). Under compression of 4 GPa, an increase of circa 10 mol% AlN was obtained. Critical analysis of the results obtained using *ab initio* simulation showed that the calculated "crossover" composition is about $x_c \approx 0.68 \pm 0.03$, where the cubic structure is more stable for $x < x_c$ and the hexagonal structure is more stable for $x > x_c$ (Zhang and Veprek 2007). This agrees very well with experimental data.

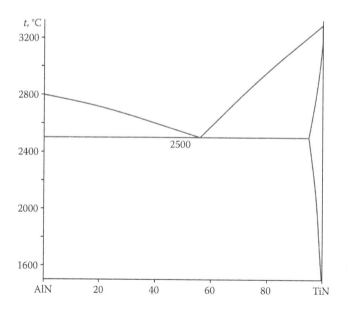

FIGURE 4.13 Calculated phase diagram of the AlN–TiN system. (From Holleck, H., *Surf. Coat. Technol.*, 36(1–2), 151, 1988. With permission.)

$Al_xTi_{1-x}N$ metastable solid solutions ($0 \leq x < 0.7$) can be obtained in the AlN–TiN quasi-binary system using the cathodic arc plasma depositing process (Beensh-Marchwicka et al. 1981) or reactive DC and radio-frequency magnetron sputtering (Jehn et al. 1986; Inamura et al. 1987). The amorphous $Al_xTi_{1-x}N$ films can be obtained when the N_2 content in the Ar + N_2 atmosphere is greater than 20% (Beensh-Marchwicka et al. 1981).

Based on the first-principles method of calculations, the miscibility region exists in the AlN–TiN system (Mayrhofer et al. 2006, 2007; Alling et al. 2007; Zhang and Veprek 2007). Figure 4.15 shows the T-x phase diagram for this system with calculated binodal and spinodal curves (Mayrhofer et al. 2006, 2007). At room temperature, the chemical binodal and spinodal include compositions in the ranges of $x \approx 0.00$–1.00 and 0.20–0.99, respectively. The binodal and spinodal meet at the critical temperature and composition of ~4100°C and $x \sim 0.69$, respectively.

A computed stability diagram of the AlN–TiN system at 1000°C is shown in Figure 4.16, where the partial pressure of N_2 is plotted against the mole fraction $x_{Ti}/(x_{Ti} + x_{Al})$ (Schmid-Fetzer and Zeng 1997; Raghavan 2006c). At the left end, N_2 remains dissolved in liquid Al at low pressures. As the pressure increases, AlN becomes stable. At the right end, N_2 remains dissolved in (α-Ti) initially. As the N_2 pressure increases, Ti_2N and TiN_{1-x} progressively became stable. The formation of Ti_3AlN

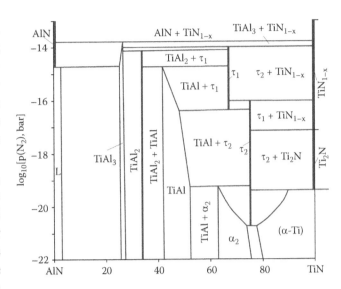

FIGURE 4.16 Computed stability diagram of the AlN–TiN system at 1000°C. (From Schmid-Fetzer, R., and Zeng, K., *Metal. Mater. Trans. A*, 28(9), 1949, 1997. With permission.)

is very sluggish, and in real-time process applications, this ternary compound may not form at all. To simulate this metastable condition, Schmid-Fetzer and Zeng (1997) computed additional stability diagram at 1000°C by suppressing the presence of Ti_3AlN.

Nitriding of the intermetallic $TiAl_3$ in nitrogen and ammonium flow was studied by Pshenichnaya et al. (1983) in a temperature range of 600°C to 1200°C. This work states that Al and Ti are nitrated simultaneously, which results in the formation of a heterogeneous mixture of practically not interacting binary nitrides.

4.30 ALUMINUM–ZIRCONIUM–NITROGEN

Two compounds are formed in this ternary system. **Zr_3AlN** (τ_1) melts congruently at 1507° ± 20°C (Schuster et al. 1983). An orthorhombic unit cell having twice the volume of the monoclinic unit cell ($a = 598.8$, $b = 896.6$, and $c = 336.7$ pm and $\beta = 116.39°$) used previously to characterize this compound (Schuster et al. 1983, 1985) was found. The lattice parameters are $a = 336.90 \pm 0.05$, $b = 1149.8 \pm 0.2$ nm, and $c = 898.25 \pm 0.15$ pm (Schuster 1986). According to the calculation by the pseudopotential plane-wave method based on the density functional theory, this compound could have once more polymorphic modification, which crystallizes in the cubic structure with the lattice parameter $a = 443$ pm, a calculated density of 6.07 g·cm⁻³, and bulk modulus $B = 153$ GPa (Wang et al. 2014).

Zr_5Al_3 is a low-temperature phase that is stabilized by a small amount (ca. 1 at%) of nitrogen to higher

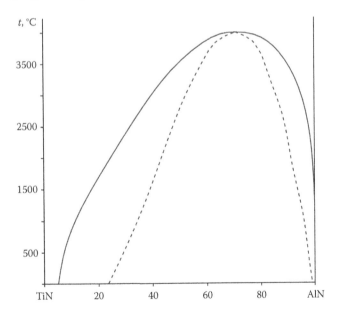

FIGURE 4.15 Calculated temperature-composition phase diagram for c-$Ti_{1-x}Al_xN$ showing the binodal (solid line) and spinodal (dashed line). (From Mayrhofer, P.H., et al., *Appl. Phys. Lett.*, 88(7), 071922, 2006. With permission.)

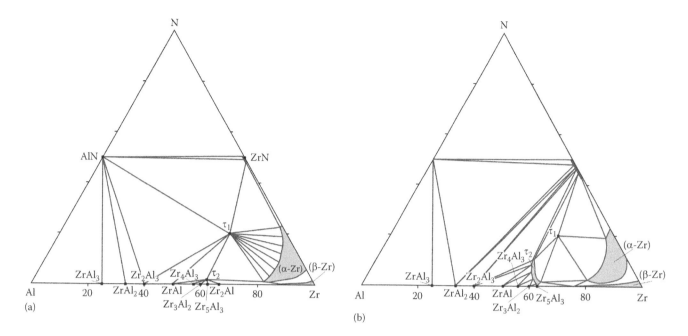

FIGURE 4.17 Calculated isothermal sections of the Al–Zr–N ternary system at (a) 1000°C and (b) 1300°C. (From Schuster, J.C., et al., *J. Nucl. Mater.*, 116(2–3), 131, 1983. With permission.)

temperatures and dissolves up to 10 at% of nitrogen at 1200°C to form **Zr₅Al₃N₁₋ₓ** (τ_2) ternary compound, which melts incongruently (Schuster et al. 1983). It crystallizes in the hexagonal structure with the lattice parameters $a = 824.0$ and $c = 569.4$ pm (Schuster et al. 1985) ($a = 817.0$ and $c = 565.54$ pm [Schuster et al. 1983]).

Isothermal sections of this system at 1000° and 1300°C were constructed (Schuster et al. 1983,1985; Holleck 1988; Khan et al. 2004a). Isothermal sections at 1000°C and 1300°C are given in Figure 4.17 (Schuster et al. 1983, 1985). At 1000°C, AlN coexists with ZrAl₃, ZrAl₂, Zr₂Al₃, Zr₃AlN, and ZrN, and at 1200°C, it is in equilibrium with ZrAl₃, ZrAl₂, and ZrN. Zr₃AlN does not exhibit a detectable homogeneity range and is in equilibrium with AlN, Zr₂Al₃, ZrAl, Zr₅Al₃N₁₋ₓ, (α-Zr), and ZrN at 1000°C and Zr₅Al₃N₁₋ₓ, (α-Zr), and ZrN at 1200°C. (α-Zr) has an extended solubility for Al up to the melting region when stabilized by about 2 at% of nitrogen. No other ternary solubility was observed.

A liquidus surface projection of the Zr-rich corner of the Al–Zr–N ternary system is shown in Figure 4.18 (Schuster et al. 1983).

According to the data of Khan et al. (2004a), AlN is not in equilibrium with Zr₃AlN at 1000°C.

Schuster et al. (1983) used two methods of sample preparations for obtaining the ternary alloys:

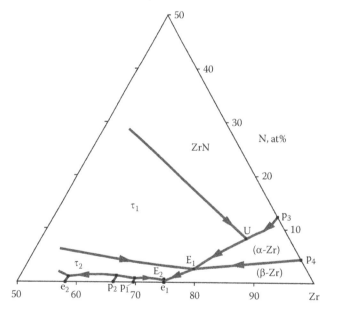

FIGURE 4.18 Liquidus surface projection of the Zr-rich corner of the Al–Zr–N ternary system. (From Schuster, J.C., et al., *J. Nucl. Mater.*, 116(2–3), 131, 1983. With permission.)

1. Ingots of binary Al–Zr alloys were heated in an induction furnace and nitrided with pure N₂. The alloys were arc melted and annealed for homogenization for 70 h at 1200°C and for 400 h at 1000°C,

2. Mixtures of powders of Zr, AlZr alloys, AlN, and ZrN were cold pressed, sintered under purified

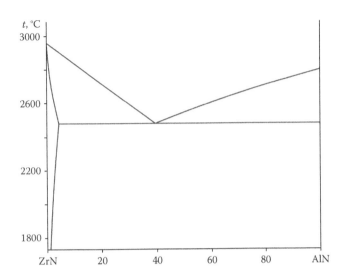

FIGURE 4.19 Calculated phase diagram of the AlN–ZrN system. (From Holleck, H., *Surf. Coat. Technol.*, 36(1–2), 151, 1988. With permission.)

Ar, cooled to room temperature, crushed, pressed again, and sintered for a second time. The sintering times were twice 50 h at 1200°C and twice 400 h at 1000°C, respectively. BN and Mo were used as crucible materials for both methods.

Khan et al. (2004a) constructed the isothermal section at 1000°C using annealed for 670 h under Ar atmosphere alloys. Diffusion couple experiments with an AlN/Zr pair and nitriding from a gaseous phase were also adopted. The phase equilibria were studied with optical microscopy and SEM, XRD, and EPMA.

AlN–ZrN. The phase diagram of this system has been calculated using the model of regular solutions for the solid phases and that of an ideal solution for the liquid phase (Figure 4.19) (Holleck 1988).

4.31 ALUMINUM–HAFNIUM–NITROGEN

Two compounds, **Hf₃AlN** (τ_1) and **Hf₅Al₃N$_{1-x}$** (τ_2), are formed in this ternary system. An orthorhombic unit cell having twice the volume of the monoclinic unit cell ($a = 590.11$, $b = 886.46$, and $c = 331.85$ pm and $\beta = 106.33°$) used previously to characterize Hf₃AlN (Schuster and Bauer 1984a; Schuster et al. 1985) was found. The lattice parameters are $a = 331.85 \pm 0.05$, $b = 1132.61 \pm 0.02$ nm, and $c = 886.46 \pm 0.12$ pm (Schuster 1986). According to the calculation by the pseudopotential plane-wave method based on the density functional theory, this compound could have once more polymorphic modification, which crystallizes in the

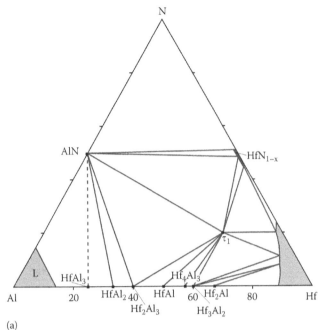

(a)

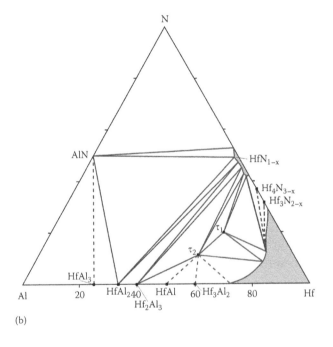

(b)

FIGURE 4.20 Calculated isothermal sections of the Al–Hf–N ternary system at (a) 1000°C and (b) 1400°C. (From Schuster, J.C., and Bauer, J., *J. Nucl. Mater.*, 120(2–3), 133, 1984. With permission.)

cubic structure with the lattice parameter $a = 444$ pm, a calculated density of 11.00 g·cm⁻³, and bulk modulus $B = 174$ GPa (Wang et al. 2014).

Hf₅Al₃N$_{1-x}$ crystallizes in the hexagonal structure with the lattice parameters $a = 806.2$ and $c = 560.3$ pm (Schuster and Bauer 1984a; Schuster et al. 1985)

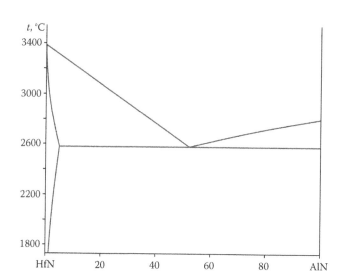

FIGURE 4.21 Calculated phase diagram of the AlN–HfN system. (From Holleck, H., *Surf. Coat. Technol.*, 36(1–2), 151, 1988. With permission.)

Isothermal sections of this system at 1000° and 1400°C were constructed and are given in Figure 4.20 (Schuster and Bauer 1984a; Schuster et al. 1985; Holleck 1988). At 1000°C, Hf_3AlN coexists with HfN_{1-x}, AlN, Hf_2Al_3, HfAl, Hf_4Al_3, Hf_3Al_2, and (α-Hf). At 1400°C, a second ternary compound $Hf_5Al_3N_{1-x}$ occurs. AlN in contact with Hf or Hf–Al-rich alloys decomposes at this temperature into HfN_{1-x} and $HfAl_2$. No ternary solubility was observed further to the extended solid solutions of α-Hf.

AlN–HfN. The phase diagram of this system has been calculated using the model of regular solutions for the solid phases and that of an ideal solution for the liquid phase (Figure 4.21) (Holleck 1988).

4.32 ALUMINUM–PHOSPHOR–NITROGEN

AlN–AlP. The electronic structure of zinc blende $AlN_{1-x}P_x$ alloy has been calculated from first principles (Winiarski et al. 2013). Structural optimization has been performed within the framework of LDA and the band gaps calculated with the modified Becke–Jonson method. Two approaches have been examined: the virtual crystal approximation and the supercell-based calculations. The composition dependence of the lattice parameter obtained from the supercell-based calculations obeys Vegard's law, whereas the volume optimization in the virtual crystal approximation leads to an anomalous bowing of the lattice constant. A strong correlation between the band gaps and the structural parameter in the virtual crystal approximation method has been observed. On the other hand, in the supercell-based calculations method,

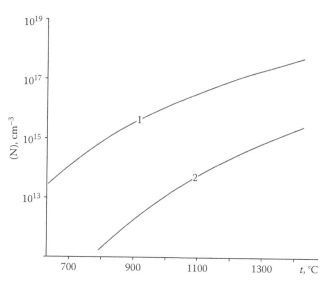

FIGURE 4.22 Calculated temperature dependence of nitrogen solubility in (1) AlP and (2) AlAs. (From Ho, I.H., and Stringfellow, G.B., *Mater. Res. Soc. Symp. Proc.*, 449, 871, 1996. With permission.)

the supercell size and atoms' arrangement appear to have a great influence on the computed band gaps.

The calculated temperature dependence of nitrogen solubility in AlP is shown in Figure 4.22 (Ho and Stringfellow 1996a). The calculations indicate that the solubility of nitrogen increases significantly with increasing temperature.

4.33 ALUMINUM–ARSENIC–NITROGEN

The calculated temperature dependence of nitrogen solubility in AlAs is shown in Figure 4.22 (Ho and Stringfellow 1996a). The calculations indicate that the solubility of nitrogen increases significantly with increasing temperature.

4.34 ALUMINUM–VANADIUM–NITROGEN

Isothermal sections of this system at 1000°C and 1300°C under 0.1 MPa of Ar were constructed using calculation and experimental investigations (Schuster and Nowotny 1985b; Schuster et al. 1985; Du et al. 1998). No ternary phases were observed within the temperature range between 1000°C and 1500°C. At 1000°C, AlN is in equilibrium with VAl_3, V_5Al_8, (V), and V_2N_y (Figure 4.23) (Du et al. 1998). No solubility of the respective third elements was observed in AlN, VN_{1-x}, VAl_3, or V_5Al_8. Three-phase equilibrium AlN + (V) + V_2N_y is the true phase equilibrium state at both 1000°C and 1300°C. The phase assemblage AlN + (V) + VN_{1-x} is probably an intermediate state, which could appear during an

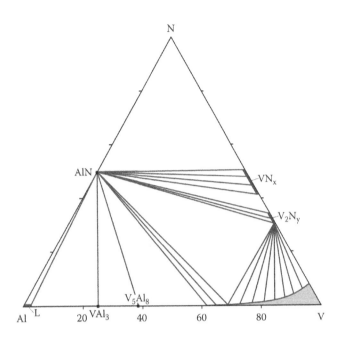

FIGURE 4.23 Calculated isothermal section of the Al–V–N ternary system at 1000°C. (From Du, Y., et al., *CALPHAD*, *22*(1), 43, 1998. With permission.)

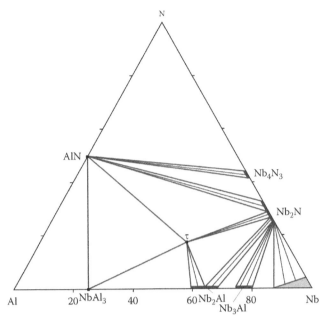

FIGURE 4.24 Isothermal section of the Al–Nb–N ternary system at 1500°C. (From Schuster, J.C., and Nowotny, H., *Z. Metallkde*, *76*(11), 728, 1985. With permission.)

approach to the true equilibrium state. According to the calculations, the VN phase was predicted to be in equilibrium with both AlN and N_2 at 600°C under a N_2 pressure of less than 0.1 MPa (Schweitz and Mohney 2001).

AlN–V. The phase relations in this system were calculated by Du et al. (1998).

4.35 ALUMINUM–NIOBIUM–NITROGEN

An isothermal section of this system at 1500°C under 0.1 MPa of Ar (Figure 4.24) was constructed by Schuster and Nowotny (1985a) and Schuster et al. (1985). **Nb_3Al_2N** (τ) ternary compound, which melts incongruently, is formed in the Al–Nb–N ternary system. It crystallizes in the cubic structure with the lattice parameter $a = 703.14$–703.87 pm (Schuster and Nowotny 1985a; Schuster et al. 1985) ($a = 703.4$ pm and a calculated density of 6.61 g·cm^{-3} [Jeitschko et al. 1965]). At 1500°C, this compound coexists with AlN, NbAl$_3$, Nb$_2$Al, and Nb$_2$N, thus dominating the phase fields of this ternary system. AlN coexists with NbAl$_3$, Nb$_3$Al$_2$N, and all Nb nitrides but not with Nb. Solubility of Nb in AlN is negligible. In none of the binary compounds was solubility of the respective third component observed.

At 1000°C, AlN is in thermodynamic equilibrium with Nb$_4$N$_3$, Nb$_2$N, NbN, Hb$_3$Al$_2$N, and NbAl$_3$; Hb$_3$Al$_2$N coexists with Nb$_2$N, (Nb$_2$Al), and NbAl$_3$, and Nb$_2$N is in equilibrium with (Nb), (Nb$_3$Al), and (Nb$_2$Al) (Schuster and Nowotny 1985a; Schuster et al. 1985).

According to the calculations, the NbN phase was predicted to be in equilibrium with both AlN and N_2 at 600°C under a N_2 pressure of less than 0.1 MPa (Schweitz and Mohney 2001).

Hb$_3$Al$_2$N was obtained by the annealing of a mixture of Nb, NbN, and Al for a long time (Jeitschko et al. 1965). Formation energy for the title compound was found to be represented by the relation ΔG (kJ·M^{-1}) = $-547.24 + 0.1584T$ between 1000°C and 1500°C under 0.1 MPa inert gas pressure, where T is the temperature in kelvin (Schuster and Nowotny 1985a).

4.36 ALUMINUM–TANTALUM–NITROGEN

Isothermal sections of this ternary system were calculated at 1000°C, 1250°C (at this temperature, the isothermal section was also constructed experimentally by Schuster and Nowotny [1985a] and Schuster et al. [1985]), 1500°C, and 1750°C (Du et al. 1998). No ternary compound is observed in the temperature range between 1000°C and 1500°C. According to the calculation at 1000°C, the tie-lines exist from AlN to all solid phases (TaAl$_3$, Ta$_2$Al$_3$, Ta$_5$Al$_7$, TaAl, Ta$_2$Al, Ta$_2$N$_y$, and TaN) except for (Ta) (Figure 4.25a). At 1250°C and 1300°C, the tie-lines also exist from AlN to all solid phases

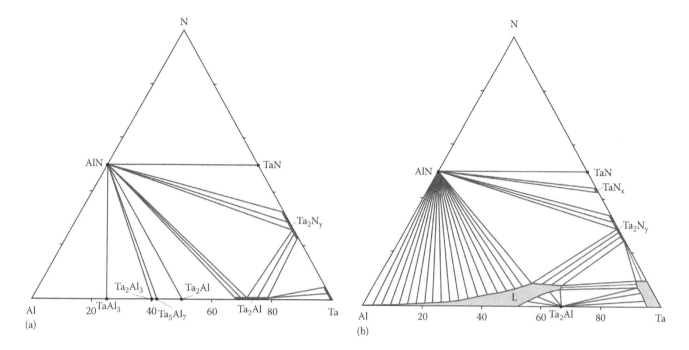

FIGURE 4.25 Calculated isothermal sections of the Al–Ta–N ternary system at (a) 1000°C and (b) 1750°C. (From Du, Y., et al., *CALPHAD*, *22*(1), 43, 1998. With permission.)

(TaAl$_3$, Ta$_{31}$Al$_{69}$, Ta$_5$Al$_7$, TaAl, Ta$_2$Al, Ta$_2$N$_y$, and TaN) except for (Ta). No ternary solubilities were observed. At 1500°C, the liquid phase is in thermodynamic equilibria with neighboring solid phases. There is one very narrow two-phase (L + TaAl$_3$) equilibrium between two three-phase equilibria, AlN + TaAl$_3$ + L and TaAl$_3$ + Ta$_{31}$Al$_{69}$ + L. An isothermal section at 1750°C is above the four-phase reaction L + (Ta) = Ta$_2$Al + Ta$_2$N$_y$ (Figure 4.25b) (Du et al. 1998).

According to the calculations, the TaN phase was predicted to be in equilibrium with both AlN and N$_2$ at 600°C under a N$_2$ pressure of less than 0.1 MPa (Schweitz and Mohney 2001).

AlN–Ta. The phase relations in this system were calculated by Du et al. (1998).

4.37 ALUMINUM–OXYGEN–NITROGEN

Isothermal sections of the Al–O–N ternary system at 2500°C, 2400°C, 2300°C, 2200°C, and 2100°C were obtained with an ideal solution description of the liquid, based on the ionic two-sublattice model (Hillert and Jonsson 1992c). Potential phase diagrams were also calculated at a series of temperatures. Isothermal sections at 2400°C, 2300°C, and 2200°C are given in Figure 4.26.

An isothermal section of this ternary system from Caron et al. (1996) is consistent with that reported by

Bhansali et al. (1990), and its range of validity can be extended to temperatures between 450°C and 627°C. The tie plane including AlN, Al, and Al$_2$O$_3$ is the only one in the solid state that could possibly exist.

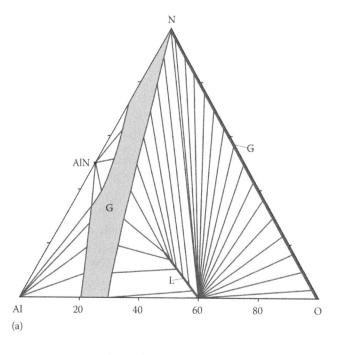

FIGURE 4.26 Isothermal sections of the Al–O–N ternary system at (a) 2400°C.

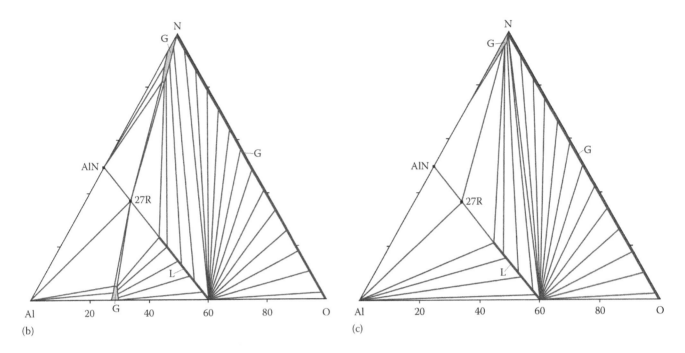

(b)

(c)

FIGURE 4.26 (Continued) (b) 2300°C, and (c) 2200°C. (From Hillert, M., and Jonsson, S., *Z. Metallkde*, *83*(10), 714, 1992. With permission.)

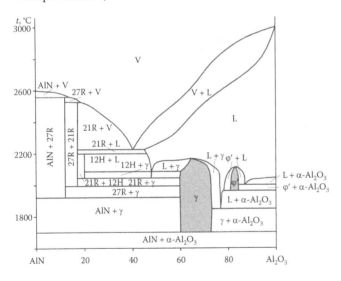

FIGURE 4.27 Phase diagram of the AlN–Al$_2$O$_3$ system. (From McCauley, J.W., et al., *J. Eur. Ceram. Soc.*, *29*(2), 223, 2009. With permission.)

AlN–Al$_2$O$_3$. The first phase diagram of this system was constructed by Lejus (1962, 1964) and included in the review of Collongues et al. (1967), indicating a γ-phase centered at about 25 mol% AlN. McCauley and Corbin (1979) published a new diagram in the region of the γ-phase, showing the composition centered at about 35.7 mol% AlN. This was followed by a more complete phase equilibrium diagram (Figure 4.27) (McCauley and Corbin 1979; McCauley et al. 2009), determined

experimentally. The solid line at about 1700°C represents the temperature at which equilibrium could not be reached conveniently. It is important to point out some of the key aspects of this diagram: since AlN sublimes at these conditions, whereas alumina melts, the diagram contains solid/vapor (S/V), liquid/vapor (L/V), and L/S equilibrium. Three L/S eutectics, one V/S eutectic, and several polytypoid phases are also present—there are others that have been identified, but it is not clear where their stability regions are located. McCauley et al. (2009) suggested that the γ-phase melts incongruently; however, in subsequent experiments and analysis, it was concluded that for the conditions of these experiments (i.e., about 0.1 MPa of flowing N$_2$) the γ-phase melted congruently, but this is probably a function of the total overpressure and the pressure media (gas composition).

Almost in parallel with the experimental determination of the phase equilibrium diagram of the AlN–Al$_2$O$_3$ system, many have attempted to calculate the γ-phase stability region and the full system using available thermodynamic data and experimental identification of the various phases. Kaufman (1979) calculated the first comprehensive diagram of this system. Other calculated diagrams have also been constructed (Dörner et al. 1981; Hillert and Jonsson 1992c; Dumitrescu and Sundman 1995; Qiu and Metselaar 1997; Tabary and Servant 1998; Medraj et al. 2003; Mao and Selleby 2007). Tabary and

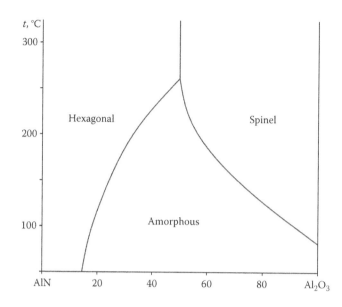

FIGURE 4.28 Calculated metastable phase regions in the AlN–Al₂O₃ system. (From Spencer, P.J., *CALPHAD*, 25(2), 163, 2001. With permission.)

Servant (1998) also calculated L/V equilibrium, which is in general agreement with the proposed experimental phase diagram.

Calculated metastable phase regions in the AlN–Al₂O₃ system are given in Figure 4.28 (Spencer 2001).

The addition of AlN to fused Al_2O_3 yielded δ-Al_2O_3 (Long and Foster 1961). About 3 mass% AlN was adequate to convert all Al_2O_3 to the δ-phase at 2050°C. Repeated fusion of this phase continued to yield δ-Al_2O_3 as long as the nitrogen content was not reduced to a very low value. The δ-phase will transform to α-Al_2O_3 upon removal of the nitrogen. The formation of ζ-Al_2O_3 with the addition of larger amounts of AlN to fused Al_2O_3 completes the high-temperature crystal phases between AlN and Al_2O_3. There appears to be no other intermediate phase inasmuch as a mixture of ζ-Al_2O_3 and AlN was observed when the nitrogen content exceeded the 3.9 mass% required for ζ-Al_2O_3. The ζ-phase, like the δ-phase, will revert to the α-phase upon removal of the nitrogen by oxidation.

Some compounds exist in this system. **Al₃NO₃** ternary compound can be produced in a reductive atmosphere above 1650°C (Yamaguchi and Yanagida 1959). It crystallizes in the cubic structure with the lattice parameter $a = 794.0$ pm and a calculated density of 3.78 g·cm⁻³. Calculations of its optimized geometries, relative stabilities, and their equilibrium properties were performed within the density functional theory (Okeke

and Lowthe 2008). The computational approach uses projector augmented wave potentials and plane-wave formalism with an LDA and a generalized gradient approximation (GGA). A cubic spinel-type structure is considered to be derived from a reaction of the form AlN + Al_2O_3 = Al_3NO_3. The calculated lattice parameter of Al_3NO_3 is 793.7 pm (LDA) or 805.7 pm (GGA), and calculated energy gap is 4.02 eV (LDA) or 3.70 eV (GGA).

According to the data of Tabary et al. (2000a), **Al₄N₂O₃** (8H) and **Al₅N₃O₃** (15R) are formed at the interaction of AlN and Al_2O_3.

Al₆N₄O₃ (12H) was observed in the specimens hot pressed higher than 1900°C in the composition range of AlN higher than 50 mol% (Sakai 1978). It was prepared in coexistence with the γ-phase within the range 80–85 mol% AlN (Bassoul et al. 1976). $Al_6N_4O_3$ crystallizes in the hexagonal structure with the lattice parameters $a = 304.1$ and $c = 3326.6$ pm (Tabary and Servant 1999) ($a = 304.3$ and $c = 3327.6$ pm [Sakai 1978]).

Al₇N₅O₃ (21R) crystallizes in the trigonal structure with the lattice parameters $a = 305.06 \pm 0.01$ and $c = 5721.6 \pm 0.01$ pm and a calculated density of 3.316 g·cm⁻³ (Asaka et al. 2013b) ($a = 304.95$ and $c = 5737.4$ pm [Tabary and Servant 1999]; $a = 304.5$ and $c = 5720.4$ pm and a calculated density of 3.31 g·cm⁻³ [Sakai 1978]). To prepare the title compound, the well-mixed AlN and Al_2O_3 (molar ratio 5:1) were heated under a N_2 pressure of 1.0 MPa at 2000°C for 2 h, followed by cooling to ambient temperature by cutting the furnace power (Asaka et al. 2013b). It was also prepared the same way as $Al_6N_4O_3$ (12H) was obtained (Sakai 1978).

Al₈N₆O₃ (16H) was prepared the same way as $Al_6N_4O_3$ (12H) was obtained. It crystallizes in the hexagonal structure with the lattice parameters $a = 305.9$ and $c = 4307.2$ pm (Sakai 1978). The phase stability of this compound remains equivocal (Asaka et al. 2013b).

Al₉N₇O₃ (27R) crystallizes in the trigonal structure with the lattice parameters $a = 306.56 \pm 0.02$ and $c = 7200.8 \pm 0.3$ pm and a calculated density of 3.306 g·cm⁻³ (Asaka et al. 2013a) ($a = 308.0$ and $c = 7224$ pm [Tabary and Servant 1999]; $a = 306.0$ and $c = 7217.1$ pm [Sakai 1978]). According to the data of Bartram and Slack (1979), the title compound crystallizes in the rhombohedral structure with the lattice parameters $a = 308 \pm 1$ and $c = 7224 \pm 5$ pm (in a hexagonal setting) and a calculated density of 3.26 g·cm⁻³. To obtain it, AlN and

Al$_2$O$_3$ (molar ratio 7:1) were mixed and heated under a N$_2$ pressure of 1.0 MPa at 2050°C for 2 h, followed by cooling to ambient temperature by cutting the furnace power (Asaka et al. 2013a). It could also be prepared the same way as Al$_6$N$_4$O$_3$ (12H) was obtained (Sakai 1978; Cannard et al. 1991) or grown by sublimation (Bartram and Slack 1979).

Al$_{10}$N$_8$O$_3$ (20H) crystallizes in the hexagonal structure with the lattice parameters a = 307.082 ± 0.005 and c = 5294.47 ± 0.08 pm and a calculated density of 3.302 g·cm^{-3} (Banno et al. 2015) (a = 309 ± 1 and c = 5310 ± 5 pm and a calculated density of 3.33 g·cm^{-3} [Bartram and Slack 1979]). To prepare it, the mixture of AlN and Al$_2$O$_3$ (molar ratio 8:1) was heated in the hot isostatic press at 2400°C for 2 h under a N$_2$ pressure of 178 MPa, and then cooled to ambient temperature by cutting the furnace power (Banno et al. 2015). The obtained polycrystalline sample was composed of not only 20H, but also AlN, 27R, 21R, and 12H.

The crystals of the title compound have been grown by sublimation (Bartram and Slack 1979).

The phase stability of this compound remains equivocal (Asaka et al. 2013b).

Al$_{12}$N$_{10}$O$_3$ (2Hδ) was prepared the same way as Al$_6$N$_4$O$_3$ (12H) was obtained. It crystallizes in the hexagonal structure with the lattice parameters a = 307.7 and c = 526.8 pm (Sakai 1978).

According to the data of Corbin (1989), **Al$_{16}$N$_{14}$O$_3$** (32H) compound is formed in this system.

Al$_{23}$N$_5$O$_{27}$ (γ-phase) melts incongruently at circa 2050°C (McCauley and Corbin 1979). It is characterized by the homogeneity region and crystallizes in the cubic structure with the lattice parameter a = 794.3 ± 0.6 pm (Tyurin et al. 2015) (a = 792.7 pm at 16 mol% AlN and 795.0 at 33 mol% AlN [Lejus 1962; Collongues et al. 1967; Goursat et al. 1976]; a = 793.8 pm at 30 mol% AlN and 795.1 at 37.5 mol% AlN [McCauley and Corbin 1979]; a = 792.3 pm at 9.8 mol% AlN and 795.0 at 28.6 mol% AlN [Sakai 1978]; a varies with composition from 795.1 pm for the N-rich phase to 793.8 pm for the O-rich phase [Corbin 1989]; a = 794.0 ± 0.5 pm [Dravid et al. 1990]; a = 794.17 pm [Labbe et al. 1992]; a = 793.2 pm at 19 mol% AlN and 795.3 at 34 mol% AlN [Willems et al. 1992a]; a = 795.3, 794.4, and 793.8 pm for the γ-phase prepared from mixtures containing 32.5, 27, and 22.5 mol.% AlN [Willems et al. 1993]; a = 795 pm [Tabary and Servant 1999]).

A homogeneity region of the γ-phase is from 20 to 33 mol% AlN (Tabary and Servant 1999) (from 16 to 33

mol% AlN at 1700°C and from 16 to 50 mol% AlN for products crystallized from a 2000°C melt [Lejus 1964; Land et al. 1978]; from Al$_{2.74}$O$_{3.78}$N$_{0.22}$ to Al$_{2.84}$O$_{3.48}$N$_{0.52}$ at 1700°C [Goursat et al. 1976]; from circa 27 to 40 mol% AlN at 1975°C [McCauley and Corbin 1979]; from 19 to 34 mol% AlN at 1850°C [Willems et al. 1992a]).

The formation of the γ-phase from the reactions of AlN and Al$_2$O$_3$ was observed at temperatures higher than 1700°C (Sakai 1978; Cannard et al. 1991). This phase is stable only within a small region of O$_2$ and N$_2$ partial pressures, and is not stable below 1640 ± 10°C (Willems et al. 1992a, 1992b). It has been observed to form at lower temperatures when gaseous precursors are used (Corbin 1989). The preparation of the γ-phase has been carried out by the solid-state reaction of AlN and Al$_2$O$_3$ (molar ratio 1:4) heated in an Ar atmosphere at 1850°C for 1 h (Labbe et al. 1992).

The temperature-dependent heat capacity has been determined by adiabatic calorimetry in the range 0–340 K (Tyurin et al. 2015). Smoothed heat capacity data have been used to calculate the thermodynamic functions: c^0_p (298.15 K) = 891.0 ± 1.8 J·mol^{-1} K^{-1}, $S^0_{298.15}$ = 614.4 ± 1.2 J·mol^{-1} K^{-1}, $\Delta H^0_{298.15}$ = 117.3 ± 0.2 kJ·mol^{-1}, and reduced Gibbs energy 221.0 ± 1.8 J·mol^{-1} K^{-1}.

δ-Spinel (**Al$_{19}$NO$_{27}$**) and φ′-spinel (**Al$_{22}$N$_2$O$_{30}$**) are also formed in the AlN–Al$_2$O$_3$ system (Corbin 1989; Tabary and Servant 1999; Tabary et al. 2000b). The formation temperature of these phases was measured equal to 1925°C for φ′-spinel and 1980°C for δ-spinel (Tabary et al. 2000b).

4.38 ALUMINUM–CHROMIUM–NITROGEN

An isothermal section of the Al–Cr–N ternary system at 1000°C under 0.1 MPa of Ar is shown in Figure 4.29 (Schuster and Nowotny 1985b; Schuster et al. 1985). No ternary compound is observed between 1000°C and 1500°C. According to the data of Beznosikov (2003), **Cr$_3$AlN** compound, which crystallizes in the cubic structure of perovskite type and could have structural phase transformations, could form in this system.

At 1000°C, solid-state equilibria were found between AlN and (Cr), γ$_2$-Cr$_5$Al$_8$, γ$_3$-Cr$_4$Al$_9$, and ε-CrAl$_4$ (Schuster and Nowotny 1985b; Schuster et al. 1985). At 1200°C, the phase equilibria remain unchanged except that ε-CrAl$_4$ is already liquid and the solubility within the Al–Cr system varies somewhat. According to the calculations, the CrN phase was predicted to be in equilibrium with both

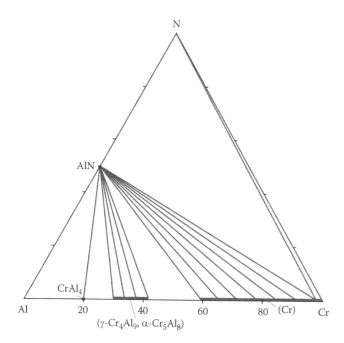

FIGURE 4.29 Isothermal section of the Al–Cr–N system at 1000°C. (From Schuster, J.C., and Nowotny, H., *J. Mater. Sci.*, *20*(8), 2787, 1985. With permission.)

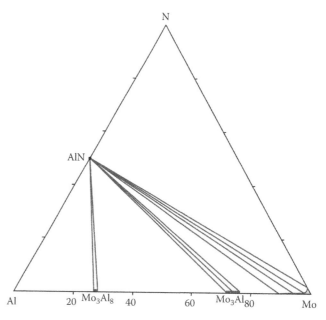

FIGURE 4.30 Isothermal section of the Al–Mo–N system at 1500°C. (From Schuster, J.C., and Nowotny, H., *J. Mater. Sci.*, *20*(8), 2787, 1985. With permission.)

AlN and N_2 at 600°C under a N_2 pressure of less than 0.1 MPa (Schweitz and Mohney 2001).

The investigation of the $Al_{1-x}Cr_xN$ coatings prepared by the reactive cathodic arc process revealed that it is possible to keep the face-centered cubic structure stable until at least 71 at% Al, where the face-centered cubic AlN phase is dominant (Reiter et al. 2005). For a higher Al content, a hexagonal closest-packed structure was observed.

Ab initio calculations were employed to demonstrate a strong pressure dependence of the maximum AlN fraction preserving the cubic phase in the $Al_xCr_{1-x}N$ hard coating before transforming to the wurtzite structure (Holec et al. 2010). Under compression of 4 GPa an increase by circa 10 mol%, AlN was obtained.

4.39 ALUMINUM–MOLYBDENUM–NITROGEN

An isothermal section of the Al–Mo–N ternary system at 1000°C under 0.1 MPa of Ar is shown in Figure 4.30 (Schuster and Nowotny 1985b; Schuster et al. 1985). No ternary compound is observed between 1000°C and 1500°C. At both temperatures, the three-phase fields AlN + Mo_3Al + Mo_3Al_8 and AlN + (Mo) + Mo_3Al exist. Neither the Mo-containing constituents

nor AlN shows solubility for the respective third elements.

According to the calculations, Mo was predicted to be in equilibrium with both AlN and N_2 at 600°C under a N_2 pressure of less than 0.1 MPa (Schweitz and Mohney 2001).

4.40 ALUMINUM–TUNGSTEN–NITROGEN

Three isothermal sections of the Al–W–N ternary system were constructed at 1000°C, 1100°C, and 1500°C (Schuster and Nowotny 1985b; Schuster et al. 1985). No ternary compound is observed in this temperature range. The three-phase field AlN + (W) + WAl_4 dominates the phase diagram at all temperatures investigated. The solubility of the respective third elements in (W), WAl_4, or AlN is virtually negligible as concluded from lattice parameter measurements.

According to the calculations, W was predicted to be in equilibrium with both AlN and N_2 at 600°C under a N_2 pressure of less than 0.1 MPa (Schweitz and Mohney 2001).

4.41 ALUMINUM–RHENIUM–NITROGEN

According to the calculations, Re was predicted to be in equilibrium with both AlN and N_2 at 600°C under a N_2 pressure of less than 0.1 MPa (Schweitz and Mohney 2001).

4.42 ALUMINUM–CHLORINE–NITROGEN

AlN₃Cl₂ and **AlCl(N₃)₂** compounds are formed in the Al–Cl–N ternary system (Wiberg et al. 1972). To obtain AlN₃Cl₂, AlCl₃ (12.4 mM) was reacted with Me₃SiN₃ (12.4 mM) in CH₂Cl₂ (25 mL) at room temperature for 5 h. The solvent was removed with Me₃SiCl. Residual colorless title compound was recrystallized from CH₂Cl₂. Interaction of AlCl₃ (9.8 mM) with Me₃SiN₃ (19.5 mM) in CH₂Cl₂ (25 mL) leads to the formation of AlCl(N₃)₂. It crystallizes upon freezing of the solution.

4.43 ALUMINUM–BROMINE–NITROGEN

AlN₃Br₂ ternary compound is formed in the Al–Br–N system. To obtain it, AlN₃I₂ (1.20 g) was introduced into a 50 mL flask, which was connected with a container with Br₂ (4 mL) (Dehnicke and Krüger 1978). The standing under the autogeneous vapor pressure Br₂ reacts slowly via the gas phase with AlN₃I₂ for 5 h. Then the rest of the bromine was added to the reaction mixture, which was stirred for 12 h to achieve complete substance conversion. The residue was washed several times with small amounts of benzene and filtered, and after drying under vacuum, AlN₃Br₂ was obtained. The experiments require the exclusion of moisture.

4.44 ALUMINUM–IODINE–NITROGEN

AlN₃I₂ ternary compound is formed in this system. It was prepared as follows (Dehnicke and Krüger 1978).

AlI₃ (2.79 g) was suspended in benzene (30 mL) and added slowly at room temperature to a solution of IN₃ (1.15 g) in benzene (40 mL). The iodine excretion takes place. The mixture was stirred for 2 h, filtered, and evacuated under vacuum. The residue was washed several times with small amounts of benzene and filtered, and after drying under vacuum, the title compound was obtained. The experiments require the exclusion of moisture.

4.45 ALUMINUM–MANGANESE–NITROGEN

Isothermal sections of the Al–Mn–N ternary system at 800°C and 1000°C under 0.1 MPa of Ar are shown in Figure 4.31 (Schuster and Nowotny 1985b; Schuster et al. 1985). No ternary compound is observed up to 1000°C. According to the data of Beznosikov (2003), **Mn₃AlN** compound, which crystallizes in the cubic structure of perovskite type and could have structural phase transformations, could form in this system.

At 800°C, AlN was found coexisting with MnAl₄, Mn₄Al₁₁, γ₂-Mn₅Al₈, (β-Mn), (α-Mn), and ζ-Mn₂N (Schuster and Nowotny 1985b). No solubility of the respective third element was detected in AlN or any of the Al–Mn phases, but (γ-Mn) dissolves at least 10 at% Al. At 100°C, only (β-Mn) was observed in solid equilibrium with AlN. The liquid region in the Al-rich corner increases considerably and participates in the phase equilibria to a large extent.

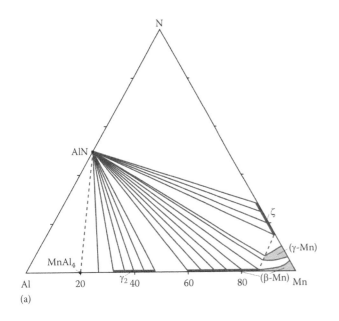

(a)

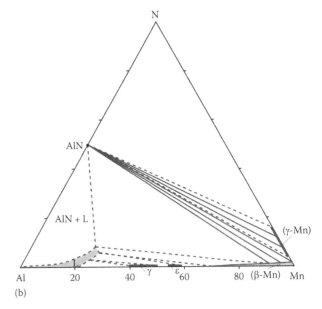

(b)

FIGURE 4.31 Isothermal sections of the Al–Mn–N system at (a) 800°C and (b) 1000°C. (From Schuster, J.C., and Nowotny, H., *J. Mater. Sci.*, 20(8), 2787, 1985. With permission.)

According to the calculations, the Mn_4N phase was predicted to be in equilibrium with both AlN and N_2 at 600°C under a N_2 pressure of less than 0.1 MPa (Schweitz and Mohney 2001).

4.46 ALUMINUM–RHENIUM–NITROGEN

An isothermal section of the Al–Re–N ternary system was constructed at 1500°C under 0.1 MPa of Ar (Schuster and Nowotny 1985b; Schuster et al. 1985). No ternary compound is observed between 1000°C and 1500°C. At 1000°C, AlN coexists with all binary rhenium aluminides ($ReAl_4$, $ReAl_3$, ReAl, and Re_2Al) and with Re without showing signs of mutual solubility. The only binary phases besides AlN that are stable at 1500°C are $ReAl_3$ and Re_2Al. The only solid-state three-phase fields existing at that temperature are therefore AlN + $ReAl_3$ + Re_2Al and AlN + Re + Re_2Al.

4.47 ALUMINUM–IRON–NITROGEN

A liquidus surface of the Al–Fe–N ternary system in the region up to 50 at% N is shown in Figure 4.32, and the solubility of N_2 (0.1 MPa) in liquid Al–Fe alloys at 1630°C is given in Figure 4.33 (Hillert and Jonsson 1992a; Raghavan 1993a). The stability of solid AlN is very high, and its liquidus surface thus comes very close to the Fe–Al and Fe–N sides. The liquidus surfaces of the

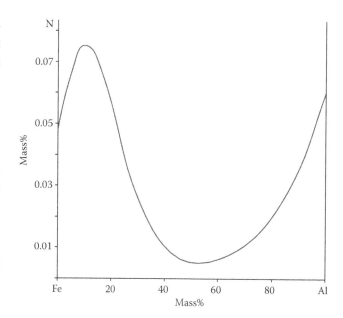

FIGURE 4.33 Solubility of N_2 (0.1 MPa) in liquid Al–Fe alloys at 1630°C. (From Hillert, M., and Jonsson, S., *Metall Trans. A*, 23(11), 3141, 1992. With permission.)

binary Fe–Al and Fe–N phases cannot be seen, but the invariant four-phase equilibria were calculated.

Fe_3AlN ternary compound, which crystallizes in the cubic structure with the lattice parameter $a = 379.67 \pm 0.03$ pm and a calculated density of 6.33 g·cm^{-3}, is formed in this system (Houben et al. 2009). It was synthesized using two-step ammonolysis. The decomposition of the title compound starts at about 540°C and is completed above 600°C.

The associated mixture model has been applied to describe equilibria in the Fe-rich corner of the Al–Fe–N ternary system (Yaghmaee et al. 2000; Yaghmaee and Kaptay 2003). It has been found that 90% of nitrogen is complexed into AlN molecules (associates) at 1600°C, when the molar ratio Al/N is higher than 2.5. The phase boundary between liquid Fe and N_2 gas, on the one hand, and between liquid Fe and the solid AlN phase, on the other hand, has been determined). It was shown that the solubility of N_2 gas increases with an increasing concentration of Al in the liquid Fe.

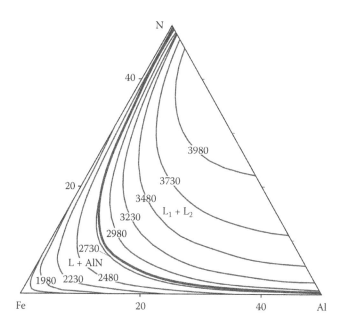

FIGURE 4.32 Liquidus surface of the Al–Fe–N system in the region up to 50 at% N. (From Hillert, M., and Jonsson, S., *Metall Trans. A*, 23(11), 3141, 1992. With permission.)

4.48 ALUMINUM–COBALT–NITROGEN

In the Al–Co–N ternary system, Co was revealed to be in equilibrium with both AlN and N_2 (0.1 MPa) at 600°C (Schweitz and Mohney 2001).

According to the data of Beznosikov (2003), **Co$_3$AlN** compound, which crystallizes in the cubic structure of perovskite type, could form in the Al–Co–N ternary system.

4.49 ALUMINUM–NICKEL–NITROGEN

In the Al–Ni–N ternary system, Ni was revealed to be in equilibrium with both AlN and N$_2$ (0.1 MPa) at 600°C (Schweitz and Mohney 2001).

According to the data of Beznosikov (2003), **Ni$_3$AlN** compound, which crystallizes in the cubic structure of perovskite type, could form in the Al–Ni–N ternary system. It is possible that this compound, which crystallizes in the cubic structure with the lattice parameter $a = 380$–388 pm, was prepared by Stadelmaier and Yun (1961).

4.50 ALUMINUM–RUTHENIUM–NITROGEN

In the Al–Ru–N ternary system, Ru was revealed to be in equilibrium with both AlN and N$_2$ (0.1 MPa) at 600°C (Schweitz and Mohney 2001).

4.51 ALUMINUM–RHODIUM–NITROGEN

In the Al–Rh–N ternary system, Rh was revealed to be in equilibrium with both AlN and N$_2$ (0.1 MPa) at 600°C (Schweitz and Mohney 2001).

4.52 ALUMINUM–PALLADIUM–NITROGEN

In the Al–Pd–N ternary system, Pd$_2$Al gives rise to the tie-line between this phase and N$_2$ gas (Figure 4.34) (Schweitz and Mohney 2001).

4.53 ALUMINUM–OSMIUM–NITROGEN

In the Al–Os–N ternary system, Os was revealed to be in equilibrium with both AlN and N$_2$ (0.1 MPa) at 600°C (Schweitz and Mohney 2001).

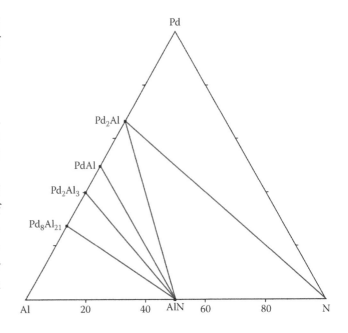

FIGURE 4.34 Calculated isothermal section of the Al–Pd–N system at 600°C under 0.1 MPa of N$_2$. (From Schweitz, K.O., and Mohney, S.E., *J. Electron. Mater.*, *30*(3), 176, 2001. With permission.)

4.54 ALUMINUM–IRIDIUM–NITROGEN

In the Al–Ir–N ternary system, Ir was revealed to be in equilibrium with both AlN and N$_2$ (0.1 MPa) at 600°C (Schweitz and Mohney 2001).

4.55 ALUMINUM–PLATINUM–NITROGEN

In the Al–Pt–N ternary system, Pt$_3$Al gives rise to the tie-line between this phase and N$_2$ gas (Schweitz and Mohney 2001).

CHAPTER 5

Systems Based on AlP

5.1 ALUMINUM–LITHIUM–PHOSPHORUS

AlP–Li₃P. Li₃AlP₂ ternary compound is formed in this system. It crystallizes in the orthorhombic structure with the lattice parameters $a = 1147.1$, $b = 1161.0$, and $c = 1173.1$ pm and experimental and calculated densities of 1.90 and 1.866 g·cm⁻³, respectively (Juza and Schulz 1952; Goel and Cahoon 1989) ($a = 1143$, $b = 1193$, and $c = 1203$ pm and an energy gap of ca. 2.75 eV [Kuriyama et al. 2008]). It is considered doubtful that this compound is in equilibrium with Al.

Li₃AlP₂ was prepared by interaction of Li₃Al and P, or Li₃P and AlP, or Li₃P, Al, and P (Juza and Schulz 1952). It was also obtained by direct reaction of Li, Al, and P (Kuriyama et al. 2008). These elements with a molar ratio of 3:1:2 were inserted in a Ta crucible, adding a few atomic percent of Li to compensate for their evaporation losses. The charged crucible was sealed under vacuum in a quartz ampoule. The ampoule set was heated to 800°C at a rate of 323°C·h⁻¹, kept for 100 h, and cooled to room temperature at the same rate. The obtained crystals were yellow in color and very sensitive to oxidization and hydroxides in air.

5.2 ALUMINUM–SODIUM–PHOSPHORUS

AlP–Na₃P. Na₃AlP₂ ternary compound is formed in this system. It crystallizes in the orthorhombic structure with the lattice parameters $a = 1317.6 \pm 0.1$, $b = 676.4 \pm 0.1$, and $c = 606.5 \pm 0.1$ pm and a calculated density of 1.94 g·cm⁻³ (Ohse et al. 1993) ($a = 677.3$, $b = 1319.4$, and $c = 607.7$ pm [Somer et al. 1995b]).

Na₃AlP₂ was prepared from a stoichiometric mixture of elements or from Na, NaP, and AlP at 730°C (Ohse et al. 1993). It was also obtained from the stoichiometric mixture of Na₃P, Al, and P at 765°C (Somer et al. 1995b).

5.3 ALUMINUM–POTASSIUM–PHOSPHORUS

AlP–K₃P. K₁₂(AlP₂)₂(Al₂P₄) or **K₃AlP₂** ternary compound is formed in this system. It crystallizes in the triclinic structure with the lattice parameters $a = 887.1 \pm 0.3$, $b = 1187.9 \pm 0.4$, and $c = 1528.0 \pm 0.5$ pm and $\alpha = 72.47 \pm 0.01°$, $\beta = 73.35 \pm 0.01°$, and $\gamma = 71.62 \pm 0.01°$ (Somer et al. 1990j). It was synthesized from a stoichiometric mixture of elements or from KP, Al, and K in an evacuated and sealed Nb ampoule at 680°C.

5.4 ALUMINUM–CESIUM–PHOSPHORUS

AlP–Cs₃P. Cs₆Al₂P₄ ternary compound, which crystallizes in the monoclinic structure with the lattice parameters $a = 1123.3 \pm 0.3$, $b = 864.1 \pm 0.2$, and $c = 1898.6 \pm 0.5$ pm and $\beta = 100.056 \pm 0.001°$, is formed in this system (Somer et al. 1990k). It was prepared from the stoichiometric mixture of the elements or from Cs, Al, and Cs₄P₆ at 680°C in an evacuated and sealed Nb ampoule.

5.5 ALUMINUM–CALCIUM–PHOSPHORUS

AlP–Ca₃P₂. Ca₃Al₂P₄ ternary compound, which crystallizes in the monoclinic structure with the lattice parameters $a = 1262.36 \pm 0.11$, $b = 983.58 \pm 0.08$, and $c = 644.69 \pm 0.06$ pm and $\beta = 90.454 \pm 0.001°$ and a calculated density of 2.474 g·cm⁻³ at 200 ± 2 K, is formed in this system (He et al. 2012). It was obtained as orange crystals from the mixture of the elements with the stoichiometric ratio Ca/P of 3:4 and a 25-fold excess of Al, which was used as a flux. The reaction was equilibrated at 960°C for 20 h, and then cooled to 750°C at a rate of 5°C·h⁻¹. The molten Al was decanted at this temperature. All synthetic and postsynthetic manipulations were performed inside an argon-filled glove box or under vacuum.

5.6 ALUMINUM–STRONTIUM–PHOSPHORUS

AlP–Sr₃P₂. Sr₃Al₂P₄ ternary compound, which crystallizes in the monoclinic structure with the lattice parameters $a = 1314.6 \pm 0.1$, $b = 1017.6 \pm 0.1$, and $c = 665.7 \pm 0.1$ pm and $\beta = 91.13 \pm 0.01°$, is formed in this system (Somer et al. 1998d).

It was formed as a by-product of the NaSr₂AlP₃ synthesis (Somer et al. 1998d). The pure title compound was prepared from a stoichiometric mixture of AlP and SrP (molar ratio 2:3) in an alumina crucible enclosed in a steel ampoule at 1102°C.

5.7 ALUMINUM–BARIUM–PHOSPHORUS

AlP–Ba₃P₂. Three compounds, **Ba₃AlP₃**, **Ba₃Al₂P₄**, and **Ba₃Al₃P₅**, are formed in this system. Ba₃AlP₃ crystallizes in the orthorhombic structure with the lattice parameters $a = 1935.72 \pm 0.14$, $b = 674.33 \pm 0.05$, and $c = 1297.07 \pm 0.10$ pm at 200 ± 2 K (Stoyko et al. 2015). For obtaining this compound, starting materials were Ba pieces, Al granules, and red P powder. Mixtures of the elements were loaded into alumina crucibles covered on the top with quartz wool and placed within fused silica tubes, which were evacuated and sealed. The tubes were heated to 960°C at a rate of 60°C·h⁻¹, held at that temperature for 20 h (samples with Al flux) or 40 h (samples with Pb flux), and cooled to 750°C·h⁻¹ at a rate of 5°C·h⁻¹ (for samples with Al flux) or to 500°C·h⁻¹ at a rate of 30°C·h⁻¹ (for samples with Pb flux). The excess of metal flux was removed at 750°C·h⁻¹ (in samples with Al flux) or at 500°C·h⁻¹ (in samples with Pb flux) by using a centrifuge. All reagents and products were handled within an argon-filled glove box with controlled oxygen and a moisture level below 1 ppm.

Ba₃Al₂P₄ also crystallizes in the orthorhombic structure with the lattice parameters $a = 725.40 \pm 0.05$, $b = 1154.16 \pm 0.08$, and $c = 1157.70 \pm 0.08$ pm at 200 ± 2 K, a calculated density of 4.042 g·cm⁻³, and an energy gap of 1.6 eV (He et al. 2012). It was obtained as red crystals from the mixture of the elements with the stoichiometric ratio Ba/P of 3:4 and a 25-fold excess of Al, which was used as a flux (He et al. 2012). The reaction was equilibrated at 960°C for 20 h, and then cooled to 750°C at a rate of 5°C·h⁻¹. The molten Al was decanted at this temperature. All synthetic and postsynthetic manipulations

were performed inside an argon-filled glove box or under vacuum.

Ba₃Al₃P₅ crystallizes in the rhombohedral structure with the lattice parameters $a = 1458.86 \pm 0.09$ and $c = 2899.0 \pm 0.3$ pm (in a hexagonal setting), a calculated density of 3.624 g·cm⁻³ at 200 ± 2 K, and an energy gap of 1.9 eV (He et al. 2013). It was prepared by the interaction of Ba, Al, and P (molar ratio 1:20:2). The mixture was heated at 960°C for 20 h, and then cooled at a rate of 10°C·h⁻¹ to 750°C, at which temperature the excess Al flux was removed. All manipulations of the starting materials and the reaction products were conducted inside an Ar-filled glove box or under vacuum.

5.8 ALUMINUM–ZINC–PHOSPHORUS

The isothermal section of the Al–Zn–P ternary system at 450°C was investigated using scanning electron microscopy (SEM) coupled with energy-dispersive x-ray spectroscopy (EDX), and x-ray diffraction (XRD) (Figure 5.1) (Zhu et al. 2012). AlP is in equilibrium with Zn_liq., α-Al, Zn₃P₂, and ZnP₂. Phosphorus solubility in liquid Zn and in α-Al is limited. AlP dissolves about 1 at% Zn at 450°C. Zn_liq. + Zn₃P₂ + AlP, Zn_liq. + α-Al + AlP, and Zn₃P₂ + ZnP₂ + AlP three-phase equilibria were established. No ternary phase was detected in the experiments.

The mixtures were heated to 1000°C into a corundum crucible, which was sealed in an evacuated quartz

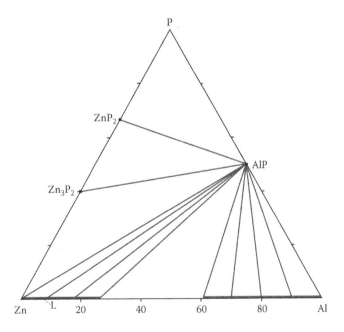

FIGURE 5.1 Isothermal section of the Al–Zn–P system at 450°C. (From Zhu, Y., et al., *J. Alloys Compd.*, 515, 161, 2012. With permission.)

tube and kept at this temperature for 3 days, followed by quenching in H_2O (Zhu et al. 2012). The quenched samples were annealed at 450°C for 60 days. The treatment was completed with rapid water quenching.

The isothermal section of the Al–Zn–P ternary system at 450°C was also determined experimentally and assessed by Tu et al. (2009) and included in the review of Raghavan (2010c). In the Zn-rich corner of this system, there exist a single-phase region (liquid Zn) and a two-phase region ($Zn_{liq.}$ + Zn_3P_2). But the most stable compound in this system is AlP; therefore, the phase equilibria given by Tu et al. (2009) are not correct.

5.9 ALUMINUM–GALLIUM–PHOSPHORUS

The calculated liquidus surface of the Al–Ga–P ternary system is shown in Figure 5.2 (Khuber 1975). Liquidus isotherms at 800°C-1100°C and corresponding solidus isotherms at 800°C–1000°C in the Ga-rich region have also been determined by Ilegems and Panish (1973) and Tanaka et al. (1982). The calculated liquidus and solidus isotherms were obtained with a simplified thermodynamic model for the liquid and solid solutions (Panish and Ilegems 1972; Ilegems and Panish 1973). Liquidus isotherms have also been obtained by the saturation technique. Tanaka et al. (1982) determined experimentally and calculated on the model of the quasi-regular solution the liquidus and solidus isotherms in the low-temperature region (Figures 5.3 and 5.4).

At a temperature higher than 400°C, GaP reacted with Al, forming AlP (Bonapace and Kahn 1983).

AlP–GaP. The calculated phase diagram of this system is given in Figure 5.5 (Panish and Ilegems 1972). The lattice parameter of the $Al_xGa_{1-x}P$ solid solutions follows Vegard's law (Bessolov et al. 1982).

5.10 ALUMINUM–INDIUM–PHOSPHORUS

The calculated liquidus surface in the In corner of the Al–In–P ternary system is shown in Figure 5.6 (Panish and Ilegems 1972). Liquidus isotherms were also calculated in the Al–In region of this system.

AlP–InP. The calculated phase diagram of this system is given in Figure 5.7 (Panish and Ilegems 1972). The electronic structure and the electronic properties of $Al_xIn_{1-x}P$ solid solutions have been studied from first-principles calculations (Ameri et al. 2011). It was found that the lattice parameter follows Vegard's law, the bulk modulus varies significantly with the composition, and the electronic band structure has a nonlinear dependence on the composition.

5.11 ALUMINUM–EUROPIUM–PHOSPHORUS

$Eu_3Al_2P_4$ and **$Eu_{14}AlP_{11}$** ternary compounds are formed in the Al–Eu–P system. $Eu_3Al_2P_4$ crystallizes in the monoclinic structure with the lattice parameters $a = 1303.04 \pm 0.17$, $b = 1009.74 \pm 0.13$, and $c = 658.57 \pm 0.08$ pm and $\beta = 90.607 \pm 0.002°$ and a calculated density of 4.858 g·cm^{-3} at

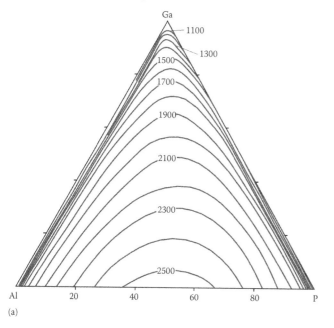

(a)

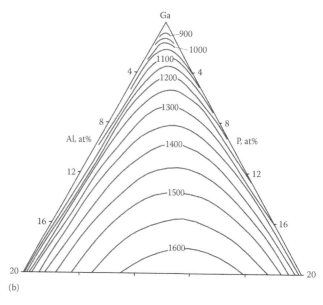

(b)

FIGURE 5.2 Calculated liquidus surface of the (a) Al–Ga–P ternary system and (b) Ga-rich corner of this system. (From Khuber, D.V., in *Protsesy Rosta i Sinteza Poluprovodn. Kristallov i Plenok*, Part 2, Nauka Publish., Novosibirsk, 1975, 212–218. With permission.)

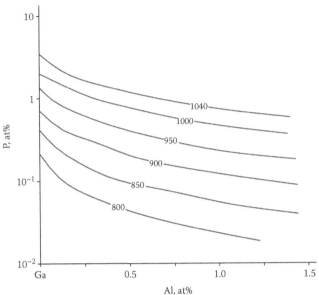

FIGURE 5.3 Calculated liquidus isotherms in the Ga corner of the Al–Ga–P ternary system. (From Tanaka, A., et al., *J. Cryst. Growth*, 60(1), 120, 1982. With permission.)

200 ± 2 K (He et al. 2012). It was obtained as dark brown crystals from the mixture of the elements with the stoichiometric ratio Eu/P of 3:4 and a 25-fold excess of Al, which was used as a flux. The reaction was equilibrated at 960°C for 20 h, and then cooled to 750°C at a rate of 5°C·h⁻¹. The molten Al was decanted at this temperature.

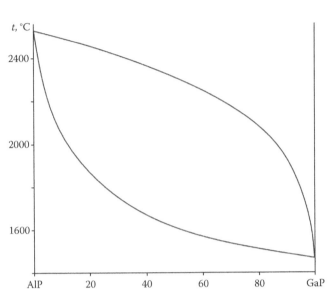

FIGURE 5.5 Phase diagram of the AlP–GaP system. (From Panish, M.B., and Ilegems, M., *Progr. Solid State Chem.*, 7, 39, 1972. With permission.)

All synthetic and postsynthetic manipulations were performed inside an argon-filled glove box or under vacuum.

$Eu_{14}AlP_{11}$ crystallizes in the tetragonal structure with the lattice parameters $a = 1519.3 \pm 0.1$ and $c = 2134.3 \pm 0.2$ pm (Somer et al. 1998c). It was prepared as a by-product from the reaction of EuP with Al (molar ratio 3:1) in a sealed Nb ampoule at 1100°C.

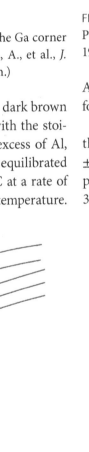

FIGURE 5.4 Calculated solidus isotherms in the Ga corner of the Al–Ga–P ternary system. (From Tanaka, A., et al., *J. Cryst. Growth*, 60(1), 120, 1982. With permission.)

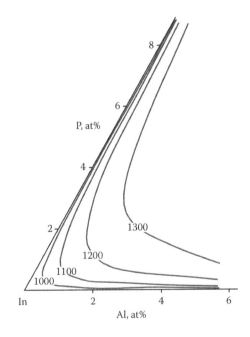

FIGURE 5.6 Calculated liquidus isotherms in the In corner of the Al–In–P ternary system. (From Panish, M.B., and Ilegems, M., *Progr. Solid State Chem.*, 7, 39, 1972. With permission.)

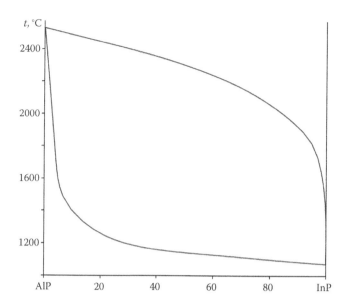

FIGURE 5.7 Calculated phase diagram of the AlP–InP system. (From Panish, M.B., and Ilegems, M., *Progr. Solid State Chem.*, 7, 39, 1972. With permission.)

5.12 ALUMINUM–SILICON–PHOSPHORUS

Isotherms of the liquidus surface of the AlP–Al–Si subsystem of the Al–Si–P ternary system were calculated using the theory of the regular solutions (Figure 5.8) (Kuznetsov and Rotenberg 1971). The surface of AlP primary crystallization occupies the biggest part of the concentration triangle. Calculated isotherms of the liquidus surface of the Al–Si–P system from the Al corner are shown in Figure 5.9.

The region of the solid solution based on Si in the Al–Si–P system at 700°C–1200°C was constructed according to the data of metallography and measurements of microhardness (Figure 5.10) (Abrikosov et al. 1962). The maximum joint solubility of Al and P in Si is observed in the AlP–Si section. The ingots were annealed at 600°C, 700°C, 800°C, 900°C, 1000°C, and 1200°C for 850, 700, 700, 500, 500, and 200 h, respectively.

AlP–Si. According to the calculations, a phase diagram of this system is of the eutectic type (Kuznetsov and Rotenberg 1971). The eutectic crystallizes at 1402°C and contains 96.3 at% Si.

5.13 ALUMINUM–GERMANIUM–
PHOSPHORUS

The region of the solid solution based on Ge in the Al–Ge–P system at 600°C–800°C was constructed according to the data of metallography and measurements of microhardness (Figure 5.11) (Abrikosov et al. 1962). The maximum joint solubility of Al and P in Ge is observed

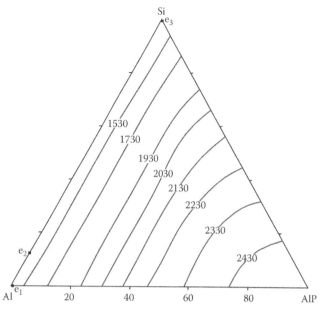

FIGURE 5.8 Calculated liquidus surface of the AlP–Al–Si subsystem. (From Kuznetsov, G.M., and Rotenberg, V.A., *Izv. AN SSSR Neorgan. Mater.*, 7(6), 943, 1971. With permission.)

in the AlP–Ge section. The ingots were annealed at 500°C, 600°C, 700°C, 800°C, 850°C, and 900°C for 1000, 700, 500, 400, 320, and 320 h, respectively.

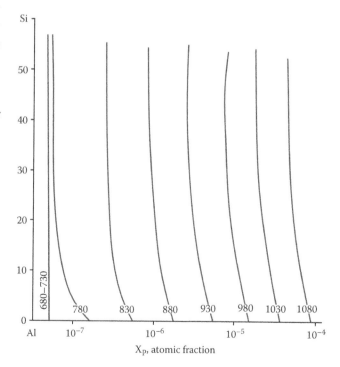

FIGURE 5.9 Calculated isotherms of the liquidus surface of the Al–Si–P system from the Al corner. (From Kuznetsov, G.M., and Rotenberg, V.A., *Izv. AN SSSR Neorgan. Mater.*, 7(6), 943, 1971. With permission.)

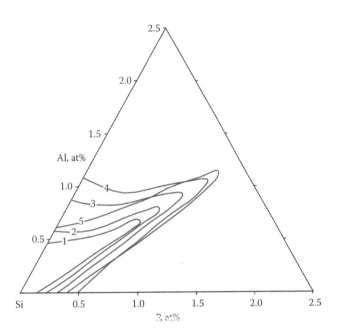

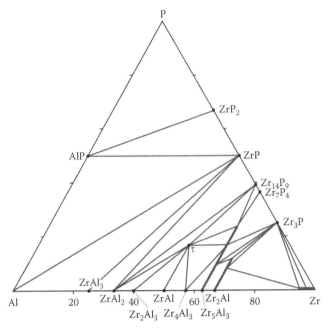

FIGURE 5.10 Solid solubility of Al and P in Si in the Al–Si–P ternary system at (1) 700°C, (2) 800°C, (3) 900°C, (4) 1000°C, and (5) 1200°C. (From Abrikosov, N.Kh., et al., *Zhurn. Neorgan. Khimii*, 7(4), 831, 1962. With permission.)

FIGURE 5.12 Isothermal section of the Al–Zr–P system at 800°C. (From Lomnitskaya, Ya.F., and Doskoch, A.M., *Neorgan. Mater.*, 29(6), 870, 1993. With permission.)

5.14 ALUMINUM–ZIRCONIUM–PHOSPHORUS

An isothermal section of the Al–Zr–P ternary system at 800°C was constructed using XRD and is shown in Figure 5.12 (Lomnitskaya and Doskoch 1993). The compound with the approximate composition Zr_3Al_2P (τ)

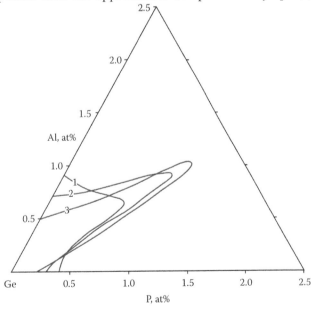

FIGURE 5.11 Solid solubility of Al and P in Ge in the Al–Ge–P ternary system at (1) 600°C, (2) 700°C, and (3) 800°C. (From Abrikosov, N.Kh., et al., *Zhurn. Neorgan. Khimii*, 7(4), 831, 1962. With permission.)

and unknown structure was determined in this system. Solid solutions are formed on the base of Zr_5Al_3 and Zr_2Al compounds with limiting compositions $Zr_5Al_{1.1}P_{1.9}$ and $Zr_2Al_{0.7}P_{0.3}$, respectively. The samples with more than 75 at% Al were annealed at 600°C, and those containing less than 75 at% Al were annealed at 800°C for 800 h.

5.15 ALUMINUM–ARSENIC–PHOSPHORUS

The equilibria between the liquid and $AlAs_xP_{1-x}$ were calculated by Panish and Ilegems (1972) on the basis of the regular solution theory (these data are included in the review of Schmid-Fetzer [1989a]) and by Ishida et al. (1989b). The estimated enthalpy data for mixing in the $AlAs_xP_{1-x}$ solid solutions indicate small positive values comparable to those obtained from a calculation based on Van Vechten's theory (Stringfellow 1972a). The calculated liquidus surface for the Al–As–P system is given in Figure 5.13 (Ishida et al. 1989b).

AlP–AlAs. The calculated phase diagram of this system is shown in Figure 5.14 (Ishida et al. 1989b).

5.16 ALUMINUM–ANTIMONY–PHOSPHORUS

The calculated liquidus surface of the Al–Sb–P ternary system (Figure 5.15) shows the entire composition range to be governed by the solidification of the high melting AlP compound (Ishida et al. 1989b).

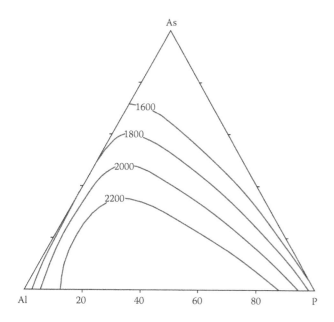

FIGURE 5.13 Calculated liquidus surface of the Al–As–P system. (From Ishida, K., et al., *J. Cryst. Growth*, 98(1–2), 140, 1989. With permission.)

AlP–AlSb. The calculated phase diagram of this system is shown in Figure 5.16 (Ishida et al. 1989b).

5.17 ALUMINUM–NIOBIUM–PHOSPHORUS

An isothermal section of the Al–Nb–P ternary system at 800°C was constructed using XRD and is shown in Figure 5.17 (Lomnitskaya and Doskoch 1993). No ternary compounds or solid solutions based on the binary compounds were found. The authors indicated the existence of Nb_8P_5 and Nb_5P_3 binary compounds, but the composition of these compounds is practically the same. Therefore, only the Nb_8P_5 binary compound is indicated on the isothermal section. The samples with more than 75 at% Al were annealed at 600°C, and those containing less than 75 at% Al were annealed at 800°C for 800 h.

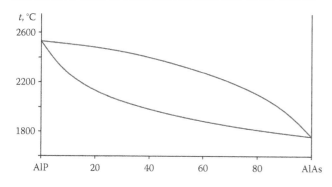

FIGURE 5.14 Calculated phase diagram of the AlP–AlAs system. (From Ishida, K., et al., *J. Cryst. Growth*, 98(1–2), 140, 1989. With permission.)

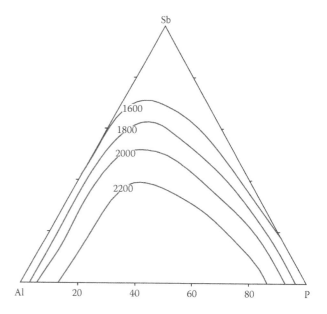

FIGURE 5.15 Calculated liquidus surface of the Al–Sb–P system. (From Ishida, K., et al., *J. Cryst. Growth*, 98(1–2), 140, 1989. With permission.)

5.18 ALUMINUM–OXYGEN–PHOSPHORUS

Some compounds are formed in the Al–O–P ternary system. **AlPO₄ (Al₂O₃·P₂O₅)** melts at 1850°C ± 50°C (at a temperature higher than 1600°C [Beck 1949]) and crystallizes in several polymorphic modifications (Beck 1949; Shafer and Roy 1957; Flörke 1967; Kosten and Arnold 1980; Graetsch 2001). According to the data of Huttenlocher (1935), this compound is stable up to 1460°C. The phase transitions take place at 215°C ± 2°C, at 705°C ± 7°C,

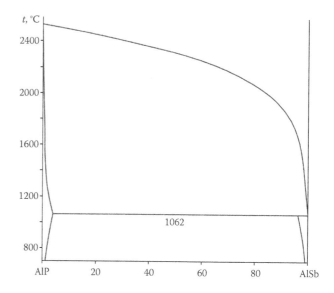

FIGURE 5.16 Calculated phase diagram of the AlP–AlSb system. (From Ishida, K., et al., *J. Cryst. Growth*, 98(1–2), 140, 1989. With permission.)

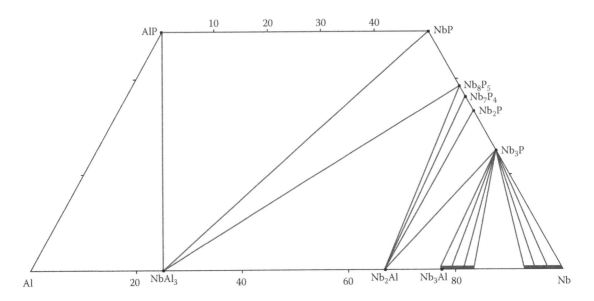

FIGURE 5.17 Isothermal section of the Al–Nb–P system at 800°C. (From Lomnitskaya, Ya.F., and Doskoch A.M., *Neorgan. Mater.*, *29*(6), 870, 1993. With permission.)

and at 1047°C ± 8°C (Shafer and Roy 1957). According to the data of Muraoka and Kihara (1997), the polymorph transition also takes place at 583°C.

Upon cooling from its hexagonal high-temperature modification, $AlPO_4$ successively transforms to several displacively distorted forms (Graetsch 2001, 2002). $AlPO_4$ tridymite prepared by annealing noncrystalline $AlPO_4$ at 950°C for 1 day was transformed to hexagonal high-temperature tridymite by heating to 300°C–320°C with a hot-air jet directed perpendicular to the sample and then cooling to the desired temperatures. A gradual phase transition from hexagonal to monoclinic symmetry was observed near 300°C. Below 200°C, the appearance of weak satellite reflections indicated the formation of incommensurate modulation.

The α-β transition in the cristobalite phase of $AlPO_4$ at about 200°C has been studied using XRD (Leadbetter and Wright 1976). The transition shows marked hysteresis but also occurs over a range of about 20°C, in which α and β phases coexist. The lattice parameters show a smooth and slow variation with temperature for each phase and a discrete difference between phases corresponding to volume differences of about 4%, showing that the transition is of the first order. The phase coexistence in the bulk sample occurs because different crystallites undergo the transition at different temperatures. The high-temperature cristobalite phase of $AlPO_4$ is metastable at low temperatures and undergoes

a first-order but reversible transition at around 200°C (Hatch et al. 1994).

Careful measurements of the specific heat, c_p, of $AlPO_4$ near the α → β phase transition show that the intermediate phase is realized in a temperature range of about 1°C (Durand et al. 1983). An intermediate phase exists on both heating and cooling. Neutron scattering experiments have shown that this intermediate phase is modulated (Bachheimer et al. 1984).

The *first modification* crystallizes in the trigonal structure with the lattice parameters $a = 494.38 \pm 0.03$ to 503.87 ± 0.7 and $c = 1094.98 \pm 0.08$ to 1106.1 ± 0.2 pm at 25°C and 585°C, respectively (Muraoka and Kihara 1997) ($a = 493 \pm 1$ and $c = 1094 \pm 2$ pm and a calculated density of 2.56 g·cm^{-3} [Huttenlocher 1935; Strunz 1941]; $a = 494.3 \pm 0.2$ and $c = 1094.8 \pm 0.4$ pm and the experimental and calculated densities of 2.66 and 2.62 g·cm^{-3}, respectively, for mineral berlinite [Gallagher and Gerards 1963]; $a = 494.291 \pm 0.016$ and $c = 1094.761 \pm 0.046$ pm and an experimental density of 2.64 g·cm^{-3} [Schwarzenbach 1966]; $a = 493.9 \pm 0.1$ and $c = 1093.2 \pm 0.5$ pm and a calculated density of 2.630 g·cm^{-3} [Wise and Loh 1976]; $a = 494.23 \pm 0.03$ and $c = 1094.46 \pm 0.13$ pm [Thong and Schwarzenbach 1979]; $a = 494.2 \pm 0.5$ and $c = 1095 \pm 1$ pm [Kosten and Arnold 1980]; $a = 492.7 \pm 0.1$ and $c = 1091.8 \pm 0.2$ pm at –100°C, $a = 493.7 \pm 0.2$ and $c = 1092.6 \pm 0.3$ pm at 20°C, and $a = 495.0 \pm 0.1$ and $c = 1094.8 \pm 0.3$ pm at 100°C [Goiffon et al. 1986]; $a =$

494.1 ± 0.2 and c = 1094.0 ± 0.3 at ambient conditions and a = 460.5 ± 0.3 and c = 1055.8 ± 0.4 at 8.51 ± 0.7 GPa [Sowa et al. 1990]; a = 494.14 and c = 1094.7 at normalized volume V/V_0 = 1 and a = 451.91 and c = 1047.1 at V/V_0 = 0.80; a and c decrease linearly with decreasing V/V_0 [Christie and Chelikowsky 1998]; a = 494 ± 4 and c = 1087 ± 1 pm for mineral berlinite [Onac and White 2003]; a = 494.58 ± 0.10 and c = 1095.26 ± 0.20 pm for mineral berlinite [Onac and Effenberger 2007]).

The *second modification* crystallizes in the hexagonal structure with the lattice parameters a = 509.76 ± 0.03 and c = 834.41 ± 0.04 pm at 320°C and a calculated density of 2.157 g·cm⁻³ (Graetsch 2001) (a = 504.0 ± 0.1 and c = 1106.1 ± 0.3 pm at 598°C–640°C [Muraoka and Kihara 1997]; a = 502.9 ± 0.5 and c = 1105 ± 2 pm at 650°C [Kosten and Arnold 1980]).

The *third modification* crystallizes in the cubic structure with the lattice parameter a = 709.9 ± 0.3 (Ng and Calvo 1977) (a = 719.51 pm at 205 ± 3°C and a = 721.03 pm at 1000°C [Wright and Leadbetter 1975]).

The *fourth modification* crystallizes in the orthorhombic structure (pseudotetragonal cell) with the lattice parameters a = 709.037 ± 0.012, b = 708.395 ± 0.012, and c = 699.430 ± 0.009 pm (Graetsch 2003) (a = b = 709.9 ± 0.3 and c = 700.6 ± 0.3 pm and a calculated density of 2.28 g·cm⁻³ [Mooney 1956b]; a = b = 707.8 ± 0.4 and c = 699.1 ± 0.3 pm [Ng and Calvo 1977]). The lattice parameters are also given for 75°C, 100°C, 125°C, 150°C, 175°C, and 200°C (Graetsch 2003). They change from 709.04 ± 0.02 to 711.49 ± 0.03 for a, from 708.40 ± 0.02 to 711.06 ± 0.03 for b, and from 699.43 ± 0.01 to 703.89 ± 0.02 for c.

The *fifth modification* crystallizes in the monoclinic structure with the lattice parameters a = 3738.63 ± 0.16, b = 504.55 ± 0.02, and c = 2622.17 ± 0.10 pm and β = 117.8164 ± 0.0013° and a calculated density of 2.222 ± 0.001 g·cm⁻³ (Graetsch 2000) (a = 508.03 ± 0.02 and 508.00 ± 0.02, b = 507.03 ± 0.02 and 507.48 ± 0.02, and c = 829.92 ± 0.03 and 830.09 ± 0.03 pm and β = 119.6037 ± 0.0003° and 119.6253 ± 0.0002° and a calculated density of 2.179 and 2.177 ± 0.001 g·cm⁻³ at 190°C and 200°C, respectively [Graetsch 2002]). Monoclinic white powders of $AlPO_4$ were prepared by firing at 950°C for 1 day in a quartz glass capillary filled with a powder of the triclinic modifications.

The *sixth modification* crystallizes in the triclinic structure with the lattice parameters a = 1001.32 ± 0.04, b = 1736.35 ± 0.06, and c = 8249.6 ± 0.6 pm and α = 90.006 ± 0.008°, β = 90.026 ± 0.004°, and γ = 89.983 ± 0.005° and a calculated density of 2.259 ± 0.001 g·cm⁻³ (Graetsch 2000). A triclinic white powder of $AlPO_4$ was prepared by firing noncrystalline $AlPO_4$ at 1000°C for 1 day with subsequent grinding in an agate mortar.

High-pressure XRD experiments on berlinite (α-$AlPO_4$) showed that it transforms to a crystalline orthorhombic phase beyond 13 GPa (Sharma et al. 2000). The persistence of the diffraction pattern up to 40 GPa does not confirm the conclusions of high-pressure amorphization of $AlPO_4$ around 12–18 GPa. The so-called memory glass effect observed earlier (Kruger and Jeanloz 1990; Gillet et al. 1995) may in fact be due to the reversibility of hexagonal ⟷ orthorhombic phase transformation.

Cohen and Klement (1973) determined the high-low temperature inversion in berlinite to 0.6 GPa by DTA in hydrostatic apparatus. From near 584°C at 0.1 MPa, the transition temperature rises linearly with pressure at the rate of 260°C ± 5°C for 1 GPa. According to the data of Dachille and Roy (1959), $AlPO_4$ transforms to a new form near 5.5 GPa at 450°C.

$AlPO_4$ was obtained by precipitation of ammonium alum by ammonium phosphate after the addition of sodium acetate or by reaction of a concentrated Na aluminate solution in concentrated H_3PO_4 (Huttenlocher 1935). It was also prepared by dissolving Al in an excess of 85% H_3PO_4 (Stone et al. 1956). When the reaction was complete, the excess acid was washed from the precipitate with acetone and H_2O. The precipitate was dried at room temperature. The product, a fine white powder, was $AlPO_4 \cdot 2H_2O$. It is also possible to synthesize the title compound in a sealed gold tube at a H_2O pressure of 50 MPa to 0.3 GPa in a standard, externally heated, cold-seal pressure vessel (Wise and Loh 1976). The starting materials was $Al(OH)_3$ or Al_2O_3 with $AlPO_4$. The crystals of $AlPO_4$ were produced from Al_2O_3 and concentrated H_3PO_4 at 260°C and 5 MPa (Kosten and Arnold 1980). They could be also grown by the hydrothermal method, in a medium of H_3PO_4 (Durand et al. 1983). Good-quality cubic $AlPO_4$ was prepared by heating precipitated anhydrous compound at 1300°C (Wright and Leadbetter 1975). Crystals of $AlPO_4$ were grown in the melt by using V_2O_5 as a flux (Ng and Calvo 1977). About 1 mol% of polycrystalline $AlPO_4$ was mixed with V_2O_5 in a Pt crucible and heated to 1300°C. The melt was cooled at a rate of 6°C·h^{-1} to 700°C and quenched to room temperature. Some colorless crystals in the form of hexagonal plates were found in the recrystallized matrix of V_2O_5.

The bulk compressibility of $AlPO_4$ is 36 GPa (Sowa et al. 1990).

$Al_2O_3 \cdot 3P_2O_5$ [$Al(PO_3)_3$] crystallizes in the cubic structure with the lattice parameter $a = 1363$ pm and the experimental and calculated densities of 2.71 and 2.73 g·cm^{-3}, respectively (Hendricks and Wyckoff 1927; Pauling and Sherman 1937; Stone et al. 1956). It loses some P_2O_5 at elevated temperatures (Stone et al. 1956), and its melting point was estimated to be 1490°C ± 5°C. Although relatively stable at temperatures below 700°C, $Al(PO_3)_3$ decomposes rapidly at temperatures above 1100°C.

$Al(PO_3)_3$ was prepared by adding Al_2O_3 to an excess of fused HPO_3 or by dissolving $Al(OH)_3$ in an excess of aqueous HPO_3, after which the solution was evaporated and the residue was fused for several hours (Hendricks and Wyckoff 1927). It could be also prepared by dissolving $Al_2O_3 \cdot 3H_2O$ in excess of H_3PO_4 with the next evaporation the solution to dryness and volatilization of the P_2O_5 excess at 600°C (Stone et al. 1956). The crystalline product was washed with H_2O and dried at 110°C.

An amorphous aluminum phosphate approximating the composition $2Al_2O_3 \cdot P_2O_5$ ($Al_4P_2O_{11}$) was precipitated from Na pyrophosphate and $AlCl_3$ in aqueous solution (Stone et al. 1956). The precipitate was washed free of chlorides and dried at room temperature to yield a white powder that contained 35.8% total H_2O.

5.19 ALUMINUM–SULFUR–PHOSPHORUS

AlPS$_4$ and **Al$_4$(P$_2$S$_6$)$_3$** compounds are formed in the Al–S–P ternary system. $AlPS_4$ exists in two polymorphic modifications (Kuhn et al. 2014). α-$AlPS_4$ crystallizes in the tetragonal structure with the lattice parameters $a = 565.72 \pm 0.09$ and $c = 922.0 \pm 0.2$ pm, and β-$AlPS_4$ crystallizes in the orthorhombic structure with the lattice parameters $a = 566.0 \pm 0.2$, $b = 575.9 \pm 0.2$, and $c = 918.9 \pm 0.2$ pm (Kuhn et al. 2014) ($a = 561 \pm 6$, $b = 567 \pm 6$ and $c = 905 \pm 6$ pm and a calculated density of 1.99 g·cm^{-3} [Weiss and Schäfer 1960]).

α-$AlPS_4$ was obtained when the mixture of AlP with 10% excess of S was heated to 650°C (50°C·h^{-1}), held at that temperature for 30 days, and then cooled down to room temperature (50°C·h^{-1}). Well-shaped colorless transparent platelets of α-$AlPS_4$ were formed at the cold

end of the silica tube in the natural gradient of the tube furnace. The crystals are sensitive against moisture and have to be stored and handled under inert conditions (Kuhn et al. 2014).

β-$AlPS_4$ was prepared when the mixture of AlP with 10% excess of S was heated to 750°C (50°C·h^{-1}), held at that temperature for 30 days, and then cooled down to room temperature (50°C·h^{-1}) (Kuhn et al. 2014). Long, colorless, transparent, needle-shaped crystals were formed in the natural gradient of the furnace. The needle-shaped crystallites were very fragile and split easily into thinner fibers under mechanical stress. β-$AlPS_4$ is much more sensitive against moisture than α-$AlPS_4$. All preparations and manipulations were carried out in an argon atmosphere. This modification decomposes in the air with formation of H_2S (Weiss and Schäfer 1960).

$Al_4(P_2S_6)_3$ crystallizes in the monoclinic structure with the lattice parameters $a = 1758.4 \pm 0.3$, $b = 1015.6 \pm 0.2$, and $c = 669.8 \pm 0.1$ pm and β = 106.93 ± 0.01° (Kuhn et al. 2014). To obtain this compound, stoichiometric amounts of Al, P, and S were first treated in a high-energy ball mill for 6 h at 500 rpm in an Ar atmosphere. For the high-pressure synthesis, the resulting black precursor powder was tightly precompacted in a gold crucible. While being heated to temperatures of 750°C for 10 h, a pressure of 5 GPa was applied to the crucible. Intergrown and easily cleavable platelets of $Al_4(P_2S_6)_3$ were obtained as the only phase. $Al_4(P_2S_6)_3$ is fairly stable toward air and moisture; that is, decomposition, when exposed to air, is observed only after several days.

5.20 ALUMINUM–MOLYBDENUM– PHOSPHORUS

An isothermal section of the Al–Mo–P ternary system at 800°C was constructed using XRD and is shown in Figure 5.18 (Lomnitskaya and Kuz'ma 1996). The **MoAlP** (τ) compound, which crystallizes in the hexagonal structure with the lattice parameters $a = 475.72 \pm 0.02$ and $c = 663.89 \pm 0.04$ pm ($a = 476$ and $c = 664$ pm [Boller and Parthé 1963]), is formed in this system. Solid solution based on Mo_3P reaches the composition $Mo_3Al_{0.12}P_{0.88}$. The samples rich in Al were annealed at 600°C; the rest were annealed at 800°C for 800 h.

To obtain MoAlP, the well-mixed metal powders were pressed into pellets and induction melted in BN crucibles under an Ar atmosphere (Boller and Parthé 1963). The alloys thus formed were homogenized in sealed-off evacuated silica tubes at 1000°C for 24 h.

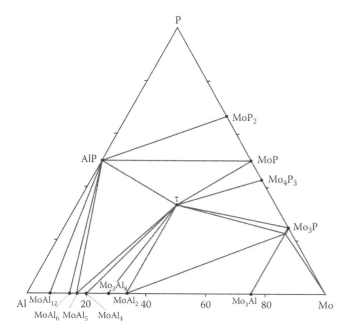

FIGURE 5.18 Isothermal section of the Al–Mo–P system at 800°C. (From Lomnitskaya, Ya.F., and Kuz'ma, Yu.B., *Poroshk. Metallurgiya*, (5–6), 70, 1996. With permission.)

5.21 ALUMINUM–IRON–PHOSPHORUS

A thermodynamic description of this ternary system was optimized and some isothermal and vertical sections were calculated by Miettinen et al. (2015). The calculated liquidus surface of the Al–Fe–P system is given in Figure 5.19. The next invariant reactions exist on the liquidus surface: E_1 (1207°C), L $\longleftrightarrow$ AlP + Fe$_2$P + FeP;

U_1 (1141°C), L + Al$_5$Fe$_2$ $\longleftrightarrow$ AlP + Al$_{13}$Fe$_4$; U_2 (1132°C), L + Al$_5$Fe$_2$ $\longleftrightarrow$ AlP + Al$_2$Fe; U_3 (1120°C), L + Al$_2$Fe $\longleftrightarrow$ AlP + Al$_5$Fe$_4$; U_4 (1112°C), L + Al$_5$Fe$_4$ $\longleftrightarrow$ AlP + (Al, Fe, P); U_5 (1033°C), L + Fe$_3$P $\longleftrightarrow$ (Al, Fe, P) + Fe$_2$P; and E_2 (996°C), L $\longleftrightarrow$ AlP + Fe$_2$P + (Al, Fe, P). The liquidus surface of the AlP–Al–Fe–Fe$_2$P subsystem was also constructed by Vogel and Klose (1952). These data agree with those of Kaneko et al. (1965a).

An isothermal section of the Al–Fe–P ternary system at 450°C in the P learn region is shown in Figure 5.20 (Wu et al. 2012). The maximum solubility of Fe in AlP and P in FeAl is 1.8 and 2.7 at%, respectively. No phosphorus can be detected in other Al–Fe binary compounds. Fe$_3$P and Fe$_2$P were found to dissolve 2.1 and 1.7 at% Al. No ternary phase was found.

The solubility of P in (α-Fe) containing Al was investigated by Kaneko et al. (1965b). It was determined that solubility of P in (α-Fe), containing 1 at% Al, increases linearly from about 0.7 at% at 650°C to about 4 at% at 1050°C and decreases from 4.4 at% in pure (α-Fe) to 2.1 at% at 7.9 at% Al at 1000°C.

Amorphous phase formation with good ductility has been found in the Al–Fe–P ternary system by using a melt-spinning technique (Inoue et al. 1983). Formation of a completely amorphous phase was achieved for a wide range of compositions, as shown in Figure 5.21 (0–18 at% Al and 13–21 at% P).

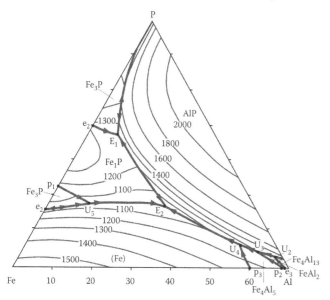

FIGURE 5.19 Calculated liquidus surface of the Al–Fe–P system. (From Miettinen, J., et al., *J. Phase Equilibria Diffus.*, **36**(4), 317, 2015. With permission.)

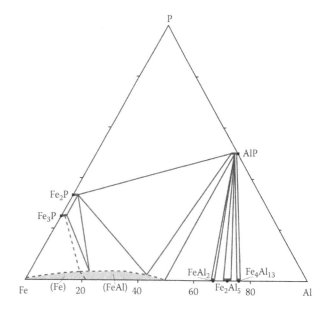

FIGURE 5.20 Isothermal section of the Al–Fe–P system at 450°C at low phosphorus content. (From Wu, C., et al., *CALPHAD: Computer Coupling of Phase Diagrams and Thermochemistry*, **38**(1), 1, 2012. With permission.)

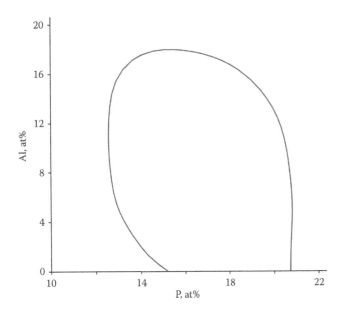

FIGURE 5.21 Glass-forming region in the Al–Fe–P system. (From Inoue, A., et al., *J. Mater. Sci.*, *18*(3) 753, 1983. With permission.)

Aluminum was found to increase the activity of P in liquid Fe at 1600°C, the interaction coefficients being $\varepsilon^{Al}_p = 4.6 \pm 0.7$ and $e^{Al}_p = 0.037$ (Yamada and Kato 1979, 1983) (according to the thermodynamic calculation, the value of ε^{Al}_p is equal to 8.78 [Ding et al. 1993]).

Treatments of a pure Al melt at 800°C with the equivalent of 200–800 mass ppm P by means of an Al–Fe–P addition prior to casting result in a distribution of polyhedral AlP particles of medium size, ~6–9 mkm, with some tendency of cluster at the higher levels of addition (Kyffin et al. 2001).

The mixture of Al, Fe, and P was made in evacuated quartz tubes by heating to 1100°C for 2 days (Wu et al. 2012). A final annealing was given at 450°C for 60 days, followed by quenching in water. The phase equilibria were studied by optical microscope, SEM with EDX, and XRD.

Critical assessments of the Al–Fe–P ternary system have been published by Raghavan (1989, 2013).

AlP–Fe₂P. This section belongs to the simple eutectic quasi-binary systems with the eutectic at 1225°C (Vogel and Klose 1952).

AlP–Fe₂Al₅. This section also belongs to the simple eutectic quasi-binary systems (Vogel and Klose 1952).

5.22 ALUMINUM–NICKEL–PHOSPHORUS

Amorphous phase formation with good ductility has been found in the Al–Ni–P ternary system by using a melt-spinning technique (Inoue et al. 1983). Formation of a completely amorphous phase was achieved for a wide range of compositions, as shown in Figure 5.22 (0–6 at% Al and 15–23 at% P).

5.23 ALUMINUM–PALLADIUM–PHOSPHORUS

Pd₂Al₀.₅P₀.₅ compound is formed in the Al–Pd–P ternary system. It crystallizes in the hexagonal structure with the lattice parameters $a = 662.6 \pm 0.1$ and $c = 335.5 \pm 0.1$ pm (Ellner et al. 1985). The title compound was prepared from Pd, Al, and P in evacuated quartz ampoules under an Ar atmosphere (40 kPa). The homogenization was carried out for heat treatment of alloys at 600°C for 1 (powder mixture) or 2 (mixture of globules) days.

5.24 ALUMINUM–PLATINUM–PHOSPHORUS

Pt₅AlP ternary compound, which crystallizes in the tetragonal structure with the lattice parameters $a = 389.6$ and $c = 681.8$ pm, is formed in the Al–Pt–P system (El-Boragy and Schubert 1970).

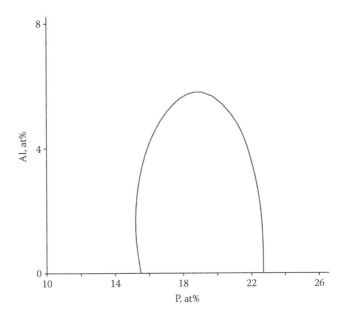

FIGURE 5.22 Glass-forming region in the Al–Ni–P system. (From Inoue, A., et al., *J. Mater. Sci.*, *18*(3) 753, 1983. With permission.)

CHAPTER 6

Systems Based on AlAs

6.1 ALUMINUM–LITHIUM–ARSENIC

AlAs–Li$_3$As. Li$_3$AlAs$_2$ ternary compound is formed in this system. It crystallizes in the orthorhombic structure with the lattice parameters a = 1186.5, b = 1198.1, and c = 1211.4 pm and experimental and calculated densities of 2.99 and 3.048 g·cm^{-3}, respectively (Juza and Schulz 1952; Goel and Cahoon 1989). It is considered doubtful that this compound is in equilibrium with Al.

Li$_3$AlAs$_2$ was prepared by interaction of Li$_3$Al and As or Li$_3$As and AlAs (Juza and Schulz 1952).

6.2 ALUMINUM–SODIUM–ARSENIC

Na$_3$AlAs$_2$ and **Na$_2$Al$_2$As$_3$** ternary compounds are formed in this system. Na$_3$AlAs$_2$ crystallizes in the orthorhombic structure with the lattice parameters a = 1360.4 ± 0.5, b = 689.5 ± 0.3, and c = 622.7 ± 0.2 pm and a calculated density of 2.80 g·cm^{-3} (Cordier and Ochmann 1988). It was prepared by the heating of the Al + 2As mixture with 10% excess of Na in an Ar atmosphere up to 1030°C at a rate of 125°C·h^{-1}, kept at that temperature for 2 h, and cooled at a rate of 10°C·h^{-1}.

Na$_2$Al$_2$As$_3$ crystallizes in the monoclinic structure with the lattice parameters a = 1311.4 ± 0.3, b = 671.0 ± 0.2, and c = 1444.8 ± 0.4 pm and β = 90.0 ± 0.2° (Cordier and Ochmann 1991j). It was synthesized from the elements at 1030°C.

6.3 ALUMINUM–POTASSIUM–ARSENIC

AlAs–K$_3$As. K$_{12}$(AlAs$_2$)$_2$(Al$_2$As$_4$) or K$_3$AlAs$_2$ ternary compound is formed in this system. It crystallizes in the triclinic structure with the lattice parameters a = 906.2 ± 0.2, b = 1216.4 ± 0.2, and c = 1557.0 ± 0.2 pm, and

α = 72.40 ± 0.01°, β = 73.05 ± 0.01°, and γ = 71.63 ± 0.01° (von Schnering et al. 1990d). It was synthesized from a stoichiometric mixture of elements or from KAs, Al, and K in an evacuated and sealed Nb ampoule at 680°C.

6.4 ALUMINUM–CESIUM–ARSENIC

Cs$_3$AlAs$_2$ ternary compound, which crystallizes in the monoclinic structure with the lattice parameters a = 1145.8 ± 0.2, b = 883.1 ± 0.2, and c = 1945.3 ± 0.4 pm and β = 99.68, is formed in this system (Somer et al. 1990i). It was prepared from the stoichiometric mixture of the elements or from AlAs, CsAs, and Cs at 630°C in an evacuated and sealed Nb ampoule.

6.5 ALUMINUM–COPPER–ARSENIC

The isothermal section of the Al–Cu–As ternary system at room temperature was approximately calculated by Klingbeil and Schmid-Fetzer (1989) and at 600°C determined experimentally (Figure 6.1) (Klingbeil and Schmid-Fetzer 1994). At this temperature, AlAs is in thermodynamic equilibrium with all binary compounds in the Al–Cu system except Cu$_3$Al. It also coexists with (Cu$_5$As$_2$) and two liquid regions. There is also a Cu$_3$Al–Cu$_3$As tie-line in this system.

6.6 ALUMINUM–SILVER–ARSENIC

The isothermal section of the Al–Ag–As ternary system at room temperature was approximately calculated (Klingbeil and Schmid-Fetzer 1989). The calculations are based on the following approximations: ternary phases and solid solubility are disregarded and the Gibbs energy of formation of binary compounds is mostly estimated by the enthalpy of formation calculated from Miedema's

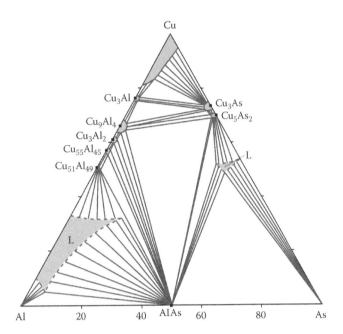

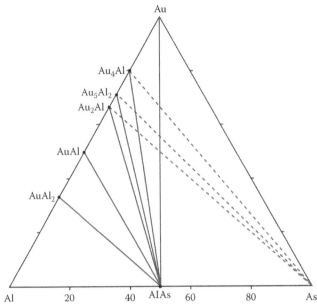

FIGURE 6.1 Isothermal section of the Al–Cu–As ternary system at 600°C. (From Klingbeil, J., *CALPHAD*, *18*(4), 429, 1994. With permission.)

FIGURE 6.2 Calculated isothermal section of the Al–Au–As ternary system at room temperature. (From Klingbeil, J., and Schmid-Fetzer, R., *13*(4), 367, 1989. With permission.)

model. At room temperature, AlAs is in thermodynamic equilibrium with Ag, Ag_3Al, and Ag_2Al.

6.7 ALUMINUM–GOLD–ARSENIC

The calculated isothermal section of the Al–Au–As ternary system at room temperature is shown in Figure 6.2 (Klingbeil and Schmid-Fetzer 1989). The calculations are based on the following approximations: ternary phases and solid solubility are disregarded and the Gibbs energy of formation of binary compounds is mostly estimated by the enthalpy of formation calculated from Miedema's model. At this temperature, AlAs is in thermodynamic equilibrium with all binary compounds in the Al–Au system. The dotted lines indicate possible alternative tie-lines.

6.8 ALUMINUM–CALCIUM–ARSENIC

AlAs–Ca₃As₂. Ca_3AlAs_3 ternary compound crystallizes in the orthorhombic structure with the lattice parameters $a = 1221.2 \pm 0.3$, $b = 420.1 \pm 0.2$, and $c = 1343.4 \pm 0.4$ pm and a calculated density of 4.584 g·cm^{-3} (Cordier and Schäfer 1981a, 1981b; Cordier et al. 1982).

Stoichiometric amounts of the elements were heated under Ar at 930°C in corundum crucibles contained in quartz bombs (Cordier and Schäfer 1981a, 1981b). After cooling, the reaction product was homogenized,

reheated to 930°C, and annealed for 24 h at 630°C. The title compound is relatively stable, but shows distinct signs of decomposition on exposure to moist atmosphere for a few days.

6.9 ALUMINUM–STRONTIUM–ARSENIC

Two ternary compounds, **Sr₃AlAs₃** and **Sr₃Al₂As₄**, are formed in the Al–Sr–As system.

Sr_3AlAs_3 crystallizes in the orthorhombic structure with the lattice parameters $a = 1914.9 \pm 0.2$, $b = 656.52 \pm 0.08$, and $c = 1268.71 \pm 0.15$ pm and a calculated density of 4.29 g·cm^{-3} at 200 K (Stoyko et al. 2015). For obtaining this compound, starting materials were Sr pieces, Al granules, and As lumps. Mixtures of the elements were loaded into alumina crucibles covered on the top with quartz wool and placed within fused silica tubes, which were evacuated and sealed. The tubes were heated to 960°C at a rate of 60°C·h^{-1}, held at that temperature for 20 h (samples with Al flux) or 40 h (samples with Pb flux), and cooled to 750°C·h^{-1} at a rate of 5°C·h^{-1} (for samples with Al flux) or to 500°C·h^{-1} at a rate of 30°C·h^{-1} (for samples with Pb flux). The excess of metal flux was removed at 750°C·h^{-1} (in samples with Al flux) or at 500°C·h^{-1} (in samples with Pb flux) by using a centrifuge. All reagents and products were handled within an argon-filled glove box with controlled oxygen and a moisture level below 1 ppm.

$Sr_3Al_2As_4$ crystallizes in the monoclinic structure with the lattice parameters $a = 1350.17 \pm 0.16$, $b = 1045.65 \pm 0.13$, and $c = 681.09 \pm 0.08$ pm and $\beta = 90.457 \pm 0.002°$ at 200 ± 2 K, a calculated density of 4.259 g·cm⁻³, and an energy gap of 1.6 eV (He et al. 2012). It was obtained as red crystals from the mixture of the elements with the stoichiometric ratio Sr/As of 3:4 and a 25-fold excess of Al, which was used as a flux. The reaction was equilibrated at 960°C for 20 h, and then cooled to 750°C at a rate of 5°C·h⁻¹. The molten Al was decanted at this temperature. All synthetic and postsynthetic manipulations were performed inside an argon-filled glove box or under vacuum.

6.10 ALUMINUM–BARIUM–ARSENIC

Three ternary compounds, **Ba₃AlAs₃**, **Ba₃Al₂As₄**, and **Ba₃Al₃As₅**, are formed in the Al–Ba–As system.

Ba_3AlAs_3 crystallizes in the orthorhombic structure with the lattice parameters $a = 1985.4 \pm 0.3$, $b = 686.36 \pm 0.09$, and $c = 1328.49 \pm 0.17$ pm at 200 ± 2 K, a calculated density of 4.87 g·cm⁻³, and an energy gap of 0.57 eV (Stoyko et al. 2015). For obtaining this compound, starting materials were Ba pieces, Al granules, and As lumps. The next procedure is the same as for $Sr_3Al_2As_4$ preparation.

$Ba_3Al_2As_4$ also crystallizes in the orthorhombic structure with the lattice parameters $a = 742.5 \pm 0.2$, $b = 1178.4 \pm 0.3$, and $c = 1184.2 \pm 0.3$ pm at 200 ± 2 K, a calculated density of 4.909 g·cm⁻³, and an energy gap of 1.3 eV (He et al. 2012). It was obtained as black crystals from the mixture of the elements with the stoichiometric ratio Ba/As of 3:4 and a 25-fold excess of Al, which was used as a flux (He et al. 2012). The reaction was equilibrated at 960°C for 20 h, and then cooled to 750°C at a rate of 5°C·h⁻¹. The molten Al was decanted at this temperature. All synthetic and postsynthetic manipulations were performed inside an argon-filled glove box or under vacuum.

$Ba_3Al_3As_5$ crystallizes in the rhombohedral structure with the lattice parameters $a = 1497.27 \pm 0.13$ and $c = 2968.9 \pm 0.4$ pm (in a hexagonal setting) at 200 ± 2 K, a calculated density of 4.499 g·cm⁻³, and an energy gap of 1.6 eV (He et al. 2013). It was prepared by the interaction of Ba, Al, and As (molar ratio 1:20:2). The mixture was heated at 960°C for 20 h, and then cooled at a rate of 10°C·h⁻¹ to 750°C, at which temperature the excess Al flux was removed. All manipulations of the starting materials and the reaction products were conducted inside an Ar-filled glove box or under vacuum.

6.11 ALUMINUM–GALLIUM–ARSENIC

The liquidus temperatures in the Ga-rich region of the Al–Ga–As ternary system have been determined using differential thermal analysis (DTA) by Panish and Sumski (1969), Panish and Ilegems (1972), Leonhardt et al. (1974), and Khuber (1975). Liquidus equations have been derived from experimental data by Muszyński and Ryabcev (1975, 1976) using a simplex lattice method in order to interpolate the liquidus curves over the entire composition range. The full liquidus surface has also been calculated by Ishida et al. (1989b). The experimental phase diagram of this system has been assessed comprehensively and critically by Li et al. (2001), and the resulting liquidus surface of the Al–Ga–As ternary system is shown in Figure 6.3. The calculated liquidus surface of the Ga-rich corner of the Al–Ga–As ternary system is given in Figure 6.4 (Khuber 1975). In Figures 6.5 and 6.6, thermodynamically calculated liquidus and solidus isotherms are presented for the Ga-rich corner (Ansara and Dutartre 1984; Ansara 1992; Li et al. 2001). These diagrams are the result of a critical evaluation of the experimental data.

AlAs–GaAs. A continuous series of solid solutions exists in this system, ranging from AlAs to GaAs. The liquidus temperatures in the composition range from 0 to 20 mol% AlAs and were determined using a gravimetric method (Mirtskhulava 1971; Mirtskhulava and Sakvarelidze 1977), and the solidus temperatures were measured by Foster et al. (1972a) using DTA. The phase

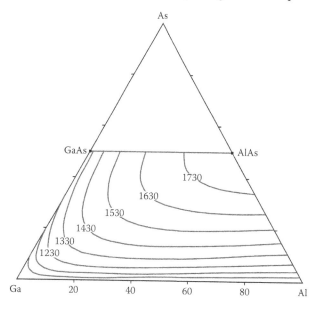

FIGURE 6.3 Liquidus surface of the Al–Ga–As ternary system. (From Li, Ch., et al., *J. Phase Equilibria*, 22(1), 26, 2001. With permission.)

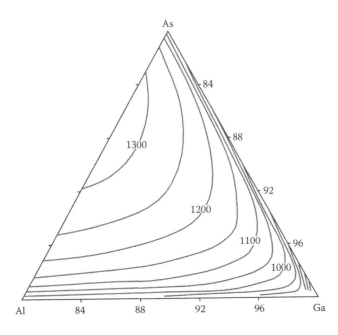

FIGURE 6.4 Calculated liquidus surface of the Ga-rich corner of the Al–Ga–As ternary system. (From Khuber, D.V., in *Protsesy Rosta i Sinteza Poluprovodn. Kristallov i Plenok*, Part 2, Nauka Publish., Novosibirsk, 1975, 212–218. With permission.)

diagram of the AlAs–GaAs system was also calculated using thermodynamic data of binary compounds and various solution models (Panish and Ilegems 1972; Bublik and Leikin 1978; Ishida et al. 1989b; Ansara and Dutartre 1984; Ansara 1992; Li et al. 2001). The calculated phase diagram (Figure 6.7) (Li et al. 2001) is in excellent agreement with the existent experimental data.

First-principles calculations were performed to study the structural, electronic, optical, and thermodynamic

properties of $Al_xGa_{1-x}As$ ternary alloys using the full-potential linearized augmented plane-wave plus local orbitals method within the density functional theory. The investigation on the effect of composition on lattice constant, bulk modulus, band gap, effective mass, and refractive index for these ternary alloys shows almost nonlinear dependence on the composition (El Haj Hassan et al. 2010). A regular solution model was used to investigate the thermodynamic stability of the alloys, which mainly indicates a phase miscibility gap with a critical temperature of 280°C and a critical composition of 46 mol% AlAs (64 K and 46 mol% AlAs [Wei et al. 1990]). The concentration dependence of the lattice parameters has a positive deviation from Vegard's law (Bublik et al. 1974).

The thermodynamic calculation on nonstoichiometric factor δ at the boundary of the homogeneity region in $Al_xGa_{1-x}As$ solid solutions is given in Figure 6.8 (Ivashchenko et al. 1990). These results may speak only about qualitative tendency. It is seen that increasing x up to circa 0.8 leads to reduction of δ at low temperature (<1080°C).

The three-phase vapor–liquid–solid equilibrium in this quasi-binary system, for AlAs contents between 0 and 15 mol%, has been determined by Mirtskhulava (1971) and Mirtskhulava and Sakvarelidze (1977) by means of a gravimetric method.

A first-principles calculation of a temperature-composition epitaxial phase diagram of the AlAs–GaAs system has been performed, and calculated values of the equilibrium molar volumes, the bulk moduli,

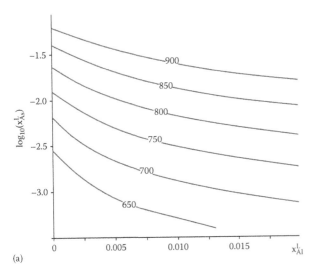

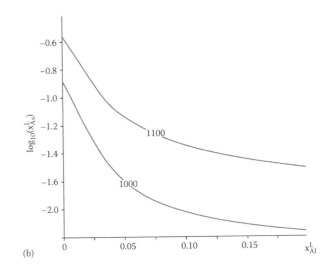

FIGURE 6.5 Calculated liquidus isotherms (a) at 650°C–900°C and (b) at 1000°C–1100°C in the Ga-rich corner of the Al–Ga–As ternary system. (From Ansara, I., et al., *J. Phase Equilibria*, 13(6), 624, 1992. With permission.)

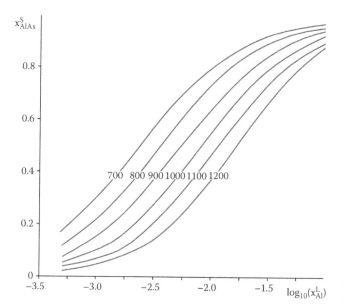

FIGURE 6.6 Calculated solidus in the Ga-rich corner of the Al–Ga–As ternary system. (From Ansara, I., et al., *J. Phase Equilibria*, *13*(6), 624, 1992. With permission.)

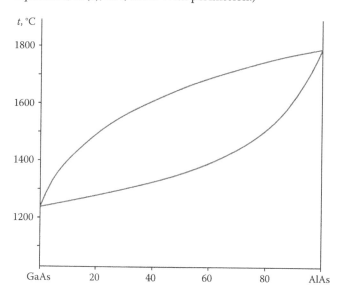

FIGURE 6.7 Phase diagram of the AlAs–GaAs system. (From Li, Ch., et al., *J. Phase Equilibria*, *22*(1), 26, 2001. With permission.)

and pressure derivatives for AlGaAs$_2$, AlGa$_3$As$_4$, and Al$_3$GaAs$_4$ compositions were obtained (Wei et al. 1990).

Al$_x$Ga$_{1-x}$As solid solutions could be obtained by the chemical transport reactions using HCl as the transport agent (Kabutov et al. 1980). These solid solutions are formed by annealing the Al/GaAs interface and are only on the order of a few atomic layers thick (Chambers 1989). There was no evidence of any excess As at the interface either before or after annealing.

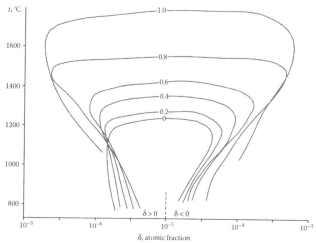

FIGURE 6.8 Calculated behavior of the nonstoichiometric factor δ at the boundary of the homogeneous region in the Al$_x$Ga$_{1-x}$As ternary solid solution. (From Ivashchenko, A.I., et al., *Cryst. Res. Technol.*, *25*(6), 661, 1990. With permission.)

6.12 ALUMINUM–INDIUM–ARSENIC

The calculated liquidus surface of the Al–In–As ternary system is shown in Figure 6.9 (Ishida et al. 1989b). The part of this liquidus surface near the Al–In binary system was also calculated by Panish and Ilegems (1972).

AlAs–InAs. The phase diagram of the AlAs–InAs system is shown in Figure 6.10 (Ishida et al. 1989b). The liquidus of this system was determined experimentally up to 7 mol% AlAs (Matyas 1977), and the solidus was determined for the

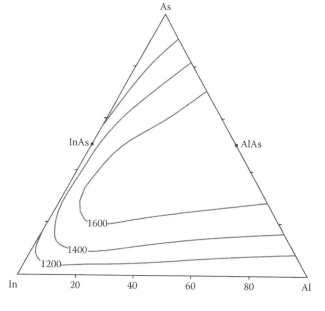

FIGURE 6.9 Calculated liquidus surface of the Al–In–As ternary system. (From Ishida, K., et al., *J. Cryst. Growth*, *98*(1–2), 140, 1989. With permission.)

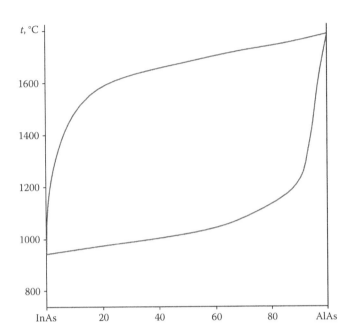

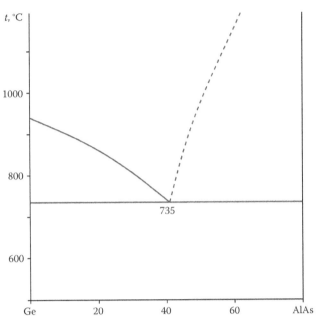

FIGURE 6.10 Calculated phase diagram of the AlAs–InAs system. (From Ishida, K., et al., *J. Cryst. Growth*, *98*(1–2), 140, 1989. With permission.)

FIGURE 6.11 Phase diagram of the AlAs–Ge system. (From Glazov, V.M., et al., *Izv. AN SSSR Neorgan. Mater.*, *2*(9), 1692, 1966. With permission.)

entire system by Foster and Scardefield (1971). A continuous series of solid solutions exists in this system at higher temperatures, but below 280°C, a miscibility gap exists in the AlAs–InAs system (Matyas 1977). The phase diagram was also calculated using thermodynamic data of binary compounds and various solution models (Osamura et al. 1972b; Panish and Ilegems 1972; Ishida et al. 1989b).

6.13 ALUMINUM–EUROPIUM–ARSENIC

Eu₃Al₂As₄ ternary compound, which crystallizes in the monoclinic structure with the lattice parameters $a = 1340.4 \pm 0.2$, $b = 1039.35 \pm 0.17$, and $c = 675.02 \pm 0.11$ pm and $\beta = 90.023 \pm 0.002°$ and a calculated density of 5.718 g·cm⁻³ at 200 ± 2 K, is formed in this system (He et al. 2012).

It was obtained as black crystals from the mixture of the elements with the stoichiometric ratio Eu/As of 3:4 and a 25-fold excess of Al, which was used as a flux (He et al. 2012). The reaction was equilibrated at 960°C for 20 h, and then cooled to 750°C at a rate of 5°C·h⁻¹. The molten Al was decanted at this temperature. All synthetic and postsynthetic manipulations were performed inside an argon-filled glove box or under vacuum.

6.14 ALUMINUM–GERMANIUM–ARSENIC

AlAs–Ge. The phase diagram of this system is a eutectic type (Figure 6.11) (Glazov et al. 1966). The eutectic crystallizes at 735°C and contains 44 mol% AlAs.

6.15 ALUMINUM–TITANIUM–ARSENIC

The isothermal section of the Al–Ti–As ternary system at room temperature was approximately calculated by Klingbeil and Schmid-Fetzer (1989) (Figure 6.12). The calculations are based on the following approximations: ternary phases and solid solubility are disregarded and

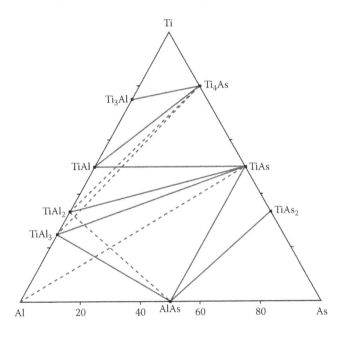

FIGURE 6.12 Calculated isothermal section of the Al–Ti–As ternary system at room temperature. (From Klingbeil, J., and Schmid-Fetzer, R., *CALPHAD*, *13*(4), 367, 1989. With permission.)

the Gibbs energy of formation of binary compounds is mostly estimated by the enthalpy of formation calculated from Miedema's model. The dotted lines indicate possible alternative tie-lines.

6.16 ALUMINUM–ZIRCONIUM–ARSENIC

The isothermal section of the Al–Zr–As ternary system at room temperature was approximately calculated by Klingbeil and Schmid-Fetzer (1989) (Figure 6.13). The calculations are based on the following approximations: ternary phases and solid solubility are disregarded and the Gibbs energy of formation of binary compounds is mostly estimated by the enthalpy of formation calculated from Miedema's model. The dotted lines indicate possible alternative tie-lines.

6.17 ALUMINUM–HAFNIUM–ARSENIC

The isothermal section of the Al–Hf–As ternary system at room temperature was approximately calculated by Klingbeil and Schmid-Fetzer (1989) (Figure 6.14). The calculations are based on the following approximations: ternary phases and solid solubility are disregarded and the Gibbs energy of formation of binary compounds is mostly estimated by the enthalpy of formation calculated from Miedema's model. The dotted lines indicate possible alternative tie-lines.

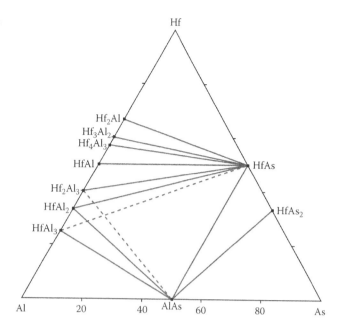

FIGURE 6.14 Calculated isothermal section of the Al–Hf–As ternary system at room temperature. (From Klingbeil, J., and Schmid-Fetzer, R., *CALPHAD*, *13*(4), 367, 1989. With permission.)

6.18 ALUMINUM–ANTIMONY–ARSENIC

The liquidus surface of the Al–Sb–As ternary system, calculated using the parameters of Ishida et al. (1989b), is shown in Figure 6.15. The liquidus surface of the AlAs$_x$Sb$_{1-x}$ solid solution covers almost the entire ternary system, and the liquidus surfaces of (Al) and

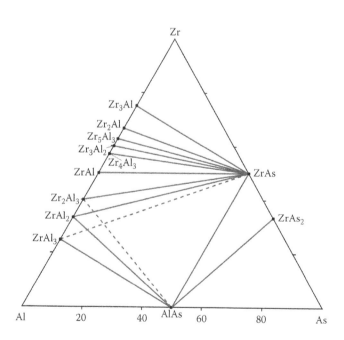

FIGURE 6.13 Calculated isothermal section of the Al–Zr–As ternary system at room temperature. (From Klingbeil, J., and Schmid-Fetzer, R., *CALPHAD*, *13*(4), 367, 1989. With permission.)

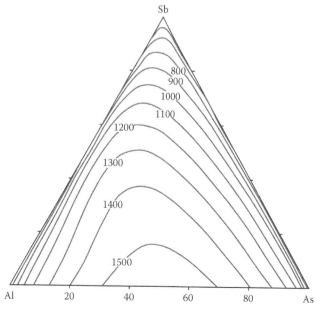

FIGURE 6.15 Calculated liquidus surface of the Al–Sb–As ternary system. (From Ishida, K., et al., *J. Cryst. Growth*, *98*(1–2), 140, 1989. With permission.)

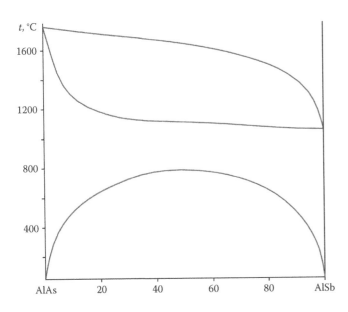

FIGURE 6.16 Calculated phase diagram of the AlAs–AlSb system. (From Ishida, K., et al., *J. Cryst. Growth*, 98(1–2), 140, 1989. With permission.)

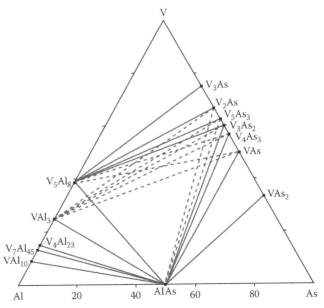

FIGURE 6.17 Calculated isothermal section of the Al–V–As ternary system at room temperature. (From Klingbeil, J., and Schmid-Fetzer, R., *CALPHAD*, 13(4), 367, 1989. With permission.)

As_xSb_{1-x} solid solution degenerate into the Al corner and the As–Sb binary edge, respectively.

AlAs–AlSb. A calculated phase diagram of the AlAs–AlSb system is shown in Figure 6.16 (Ishida et al. 1989b). According to the data of Schmid-Fetzer (1989b), a complete solid solubility in this system could be expected at and above 600°C with either value in a regular solution approximation.

First-principles calculations were performed to study the structural, electronic, optical, and thermodynamic properties of $AlSb_xAs_{1-x}$, ternary alloys using the full-potential linearized augmented plane-wave plus local orbitals method within the density functional theory. The investigation on the effect of composition on lattice constant, bulk modulus, band gap, effective mass, and refractive index for these ternary alloys shows almost nonlinear dependence on the composition (El Haj Hassan et al. 2010). A regular solution model was used to investigate the thermodynamic stability of the alloys, which mainly indicates a phase miscibility gap with the critical temperature 1420°C (691°C [Stringfellow 1982a]) and critical composition 63 mol% AlAs.

6.19 ALUMINUM–VANADIUM–ARSENIC

The isothermal section of the Al–V–As ternary system at room temperature was approximately calculated by Klingbeil and Schmid-Fetzer (1989) (Figure 6.17). The

calculations are based on the following approximations: ternary phases and solid solubility are disregarded and the Gibbs energy of formation of binary compounds is mostly estimated by the enthalpy of formation calculated from Miedema's model. The dotted lines indicate possible alternative tie-lines.

6.20 ALUMINUM–NIOBIUM–ARSENIC

The isothermal section of the Al–Nb–As ternary system at room temperature was approximately calculated by Klingbeil and Schmid-Fetzer (1989) (Figure 6.18). The calculations are based on the following approximations: ternary phases and solid solubility are disregarded and the Gibbs energy of formation of binary compounds is mostly estimated by the enthalpy of formation calculated from Miedema's model. The dotted lines indicate possible alternative tie-lines.

6.21 ALUMINUM–TANTALUM–ARSENIC

The isothermal section of the Al–Ta–As ternary system at room temperature was approximately calculated by Klingbeil and Schmid-Fetzer (1989) (Figure 6.19). The calculations are based on the following approximations: ternary phases and solid solubility are disregarded and the Gibbs energy of formation of binary compounds is mostly estimated by the enthalpy of formation calculated from Miedema's model. The dotted lines indicate possible alternative tie-lines.

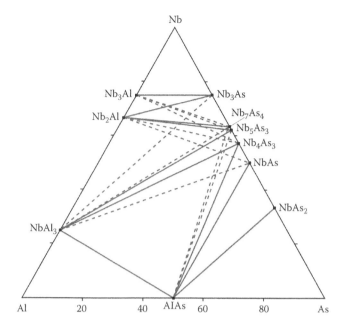

FIGURE 6.18 Calculated isothermal section of the Al–Nb–As ternary system at room temperature. (From Klingbeil, J., and Schmid-Fetzer, R., *CALPHAD*, *13*(4), 367, 1989. With permission.)

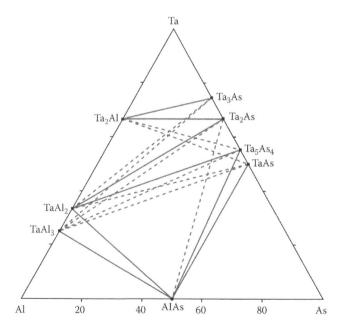

FIGURE 6.19 Calculated isothermal section of the Al–Ta–As ternary system at room temperature. (From Klingbeil, J., and Schmid-Fetzer, R., *CALPHAD*, *13*(4), 367, 1989. With permission.)

6.22 ALUMINUM–OXYGEN–ARSENIC

AlAsO₄ compound is formed in the Al–O–As ternary system. It is stable in air upon heating up to 1400°C and crystallizes in the trigonal structure with the lattice parameters $a = 503.1 \pm 0.1$ and $c = 1122.6 \pm 0.6$ pm and experimental and calculated densities of 3.32 ± 0.01 and 3.34 g·cm⁻³, respectively (for mineral alarsite) (Semenova et al. 1994; Jambor et al. 1995) ($a = 503$ and $c = 1122$ pm and a calculated density of 3.34 g·cm⁻³ [Machatschki 1936]; $a = 503.0 \pm 0.4$ and $c = 1123 \pm 2$ pm [Kosten and Arnold 1980]; $a = 502.2 \pm 0.1$ and $c = 1121.5 \pm 0.4$ pm at 173 K, $a = 502.7 \pm 0.1$ and $c = 1121.3 \pm 0.3$ pm at 20°C, and $a = 502.8 \pm 0.1$ and $c = 1120.7 \pm 0.2$ pm at 100°C [Goiffon et al. 1986]; $a = 503.2 \pm 0.1$, 506.7 ± 0.1, and 511 ± 1, and $c = 1123.0 \pm 0.1$, 1125.8 ± 0.1, and 1130 ± 2 pm and a calculated density of 3.36, 3.30, and 3.23 g·cm⁻³ at 20°C, 500°C, and 750°, respectively [Baumgartner et al. 1989]; $a = 502.8 \pm 0.4$ and $c = 1122.8 \pm 0.3$ pm at ambient conditions and a decrease to $a = 480.0 \pm 0.3$ and $c = 1100.1 \pm 0.3$ pm at 4.87 ± 0.07 GPa [Sowa 1991]). The pressure dependence of the lattice constants can be expressed as a (pm) $= (503 \pm 1) - (6 \pm 1)p + (0.2 \pm 0.3)p^2$ and c (pm) $= (1123 \pm 1) - (6 \pm 1)p + (0.3 \pm 0.2)p^2$, with p in gigapascals (Sowa 1991).

Dachille and Roy (1959) did not find new forms of AlAsO₄ in the region extending from ambient conditions up to approximately 600°C and 6 GPa. A tetragonal polymorph of the title compound with the lattice parameters $a = 435.9$ and $c = 281.5$ pm and a calculated density of 5.15 g·cm⁻³ has been synthesized at 9 GPa and 900°C (Young et al. 1963). AlAsO₄ prepared under 0.1 GPa at 800°C for 4 h starting either from a homogenized mixture of Al₂O₃ and As₂O₃ under a pressure of dry O₂ or from a homogenized mixture of Al₂O₃ and As₂O₅ under a pressure of a mixture of dry O₂/N₂ (volume ratio 1:9) crystallizes in the hexagonal structure with the lattice parameters $a = 503.7 \pm 0.2$ and $c = 1122.6 \pm 0.5$ pm (Matar et al. 1990) ($a = 503.0 \pm 0.5$ and $c = 561.2 \pm 0.6$ pm [Machatschki 1935]). This compound prepared at 4–7 GPa crystallizes in the orthorhombic structure with the lattice parameters $a = 554.0$, $b = 825.1$, and $c = 602.4$ pm, and prepared at 9 GPa and approximately 900°C crystallizes in the tetragonal structure with the lattice parameters $a = 435.9 \pm 0.3$ and $c = 563.7 \pm 0.5$ pm (Matar et al. 1990).

AlAsO$_4$ was also obtained as a gelatinous precipitate from aluminum alum by the using of potassium diarsenate (Sharan 1959). The excess of H$_2$SO$_4$ produced in the reaction was neutralized by NaCH$_3$COO. The compound is amorphous at low temperatures and becomes crystalline upon heating it to about 850°C. It crystallizes in the orthorhombic structure with the lattice parameters $a = 1090$, $b = 896$, and $c = 822$ pm. The compound heated to about 850°C has proven to be unstable. AlAsO$_4$ also has another orthorhombic modification with the lattice parameters $a = 1040$, $b = 873$, and $c = 740$ pm.

According to the data of Stanley (1956), an $\alpha \rightarrow \beta$ inversion of AlAsO$_4$ takes place at 571°C. The bulk modulus of AlAsO$_4$ is equal to 36 GPa (Sowa 1991).

AlAsO$_4$ was produced from Al$_2$O$_3$ and concentrated H$_3$AsO$_4$ at 260°C and 5 MPa (Kosten and Arnold 1980). Solutions for growing the crystals of this compound were prepared by heating to 200°C sufficient As$_2$O$_5$, H$_2$O, and AlAsO$_4$ to make a 33 N solution containing 1.2 M AlAsO$_4$ (Stanley 1956). This preliminary heating period of 16 h was accomplished in a sealed autoclave. The solution obtained was clear and viscous. Further heating of this solution in a sealed autoclave at 235°C–240°C for an 18 h period produced small crystals of AlAsO$_4$.

6.23 ALUMINUM–CHROMIUM–ARSENIC

The isothermal section of the Al–Cr–As ternary system at room temperature was approximately calculated by Klingbeil and Schmid-Fetzer (1989) (Figure 6.20). The calculations are based on the following approximations: ternary phases and solid solubility are disregarded and the Gibbs energy of formation of binary compounds is mostly estimated by the enthalpy of formation calculated from Miedema's model. The dotted lines indicate possible alternative tie-lines.

6.24 ALUMINUM–MOLYBDENUM–ARSENIC

The isothermal section of the Al–Mo–As ternary system at room temperature was approximately calculated by Klingbeil and Schmid-Fetzer (1989) (Figure 6.21). The calculations are based on the following approximations: ternary phases and solid solubilities are disregarded and the Gibbs energy of formation of binary compounds is mostly estimated by the enthalpy of formation calculated from Miedema's model. The dotted lines indicate possible alternative tie-lines.

Al$_2$Mo$_5$As$_4$ ternary compound, which is a semiconductor and crystallizes in the tetragonal structure with

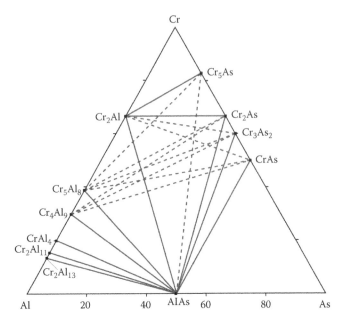

FIGURE 6.20 Calculated isothermal section of the Al–Cr–As ternary system at room temperature. (From Klingbeil, J., and Schmid-Fetzer, R., *CALPHAD*, 13(4), 367, 1989. With permission.)

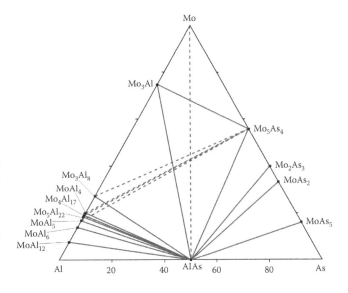

FIGURE 6.21 Calculated isothermal section of the Al–Mo–As ternary system at room temperature. (From Klingbeil, J., and Schmid-Fetzer, R., *CALPHAD*, 13(4), 367, 1989. With permission.)

the lattice parameters $a = 964.3 \pm 0.2$ and $c = 328.3 \pm 0.3$ pm, was prepared by reaction of the elements in the required stoichiometry in evacuated sealed silica ampoules at 1000°C–1050°C for about 15 days with one grinding in between (Nanjundaswamy and Gopalakrishnan 1988). Dronkowski et al. (1991) noted that is necessary to reject the existence of this compound. Together with the findings that the lattice constants of Mo_5As_4 obtained in the absence or presence of Al are identical within the standard deviation and that no characteristic x-ray emission of Al has been found, it was concluded that Mo_5As_4 does not incorporate a significant amount of Al.

6.25 ALUMINUM–TUNGSTEN–ARSENIC

The isothermal section of the Al–W–As ternary system at room temperature was calculated by Klingbeil and Schmid-Fetzer (1989) and experimentally determined at 600°C by Klingbeil and Schmid-Fetzer (1994). At both temperatures, AlAs is in equilibrium with all binary compounds in the Al–Cu system except Cu_3Al. It also coexists with WAl_{12}, WAl_5, (WAl_4), W, WAs_2, and possibly W_2As_3.

6.26 ALUMINUM–MANGANESE–ARSENIC

The isothermal section of the Al–Mn–As ternary system at room temperature was approximately calculated by Klingbeil and Schmid-Fetzer (1989) (Figure 6.22). The calculations are based on the following approximations: ternary phases and solid solubility are disregarded and the Gibbs energy of formation of binary compounds is mostly estimated by the enthalpy of formation calculated

from Miedema's model. The dotted lines indicate possible alternative tie-lines.

6.27 ALUMINUM–RHENIUM–ARSENIC

The isothermal section of the Al–Re–As ternary system at room temperature was approximately calculated by Klingbeil and Schmid-Fetzer (1989). The calculations are based on the following approximations: ternary phases and solid solubility are disregarded and the Gibbs energy of formation of binary compounds is mostly estimated by the enthalpy of formation calculated from Miedema's model. AlAs is in equilibrium with all binary compounds in the Al–Re and Re_3As_7.

6.28 ALUMINUM–IRON–ARSENIC

The isothermal section of the Al–Fe–As ternary system at room temperature was approximately calculated by Klingbeil and Schmid-Fetzer (1989) (Figure 6.23). The calculations are based on the following approximations: ternary phases and solid solubility are disregarded and the Gibbs energy of formation of binary compounds is mostly estimated by the enthalpy of formation calculated from Miedema's model. The dotted lines indicate possible alternative tie-lines.

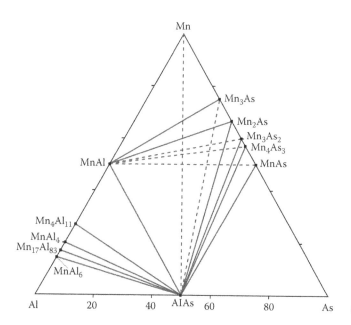

FIGURE 6.22 Calculated isothermal section of the Al–Mn–As ternary system at room temperature. (From Klingbeil, J., and Schmid-Fetzer, R., *CALPHAD*, *13*(4), 367, 1989. With permission.)

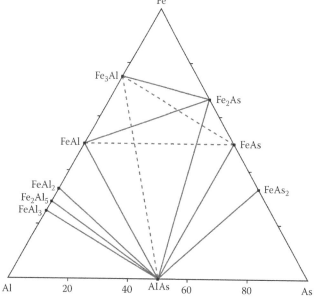

FIGURE 6.23 Calculated isothermal section of the Al–Fe–As ternary system at room temperature. (From Klingbeil, J., and Schmid-Fetzer, R., *CALPHAD*, *13*(4), 367, 1989. With permission.)

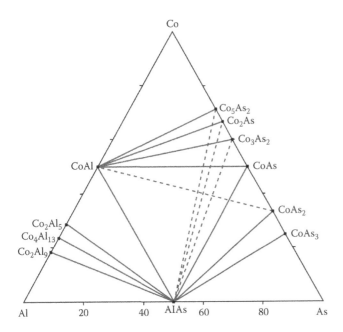

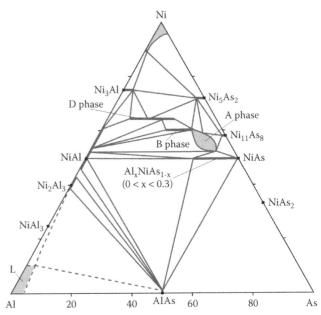

FIGURE 6.24   Calculated isothermal section of the Al–Co–As ternary system at room temperature. (From Klingbeil, J., and Schmid-Fetzer, R., *CALPHAD*, 13(4), 367, 1989. With permission.)

FIGURE 6.25   Calculated isothermal section of the Al–Ni–As ternary system at 800°C. (From Députier, S., et al., *J. Alloys Compd.*, 217(1), 13, 1995. With permission.)

6.29 ALUMINUM–COBALT–ARSENIC

The isothermal section of the Al–Co–As ternary system at room temperature was approximately calculated by Klingbeil and Schmid-Fetzer (1989) (Figure 6.24). The calculations are based on the following approximations: ternary phases and solid solubility are disregarded and the Gibbs energy of formation of binary compounds is mostly estimated by the enthalpy of formation calculated from Miedema's model. The dotted lines indicate possible alternative tie-lines.

6.30 ALUMINUM–NICKEL–ARSENIC

Klingbeil and Schmid-Fetzer (1989) did not include any ternary phases in their computation of the phase equilibria at 25°C. Three ternary phases, which crystallize in the hexagonal structure and are all structurally derived from the NiAs type, were found in samples annealed at 800°C and quenched in H_2O (Figure 6.25) (Députier et al. 1995a; Raghavan 2006a). Two of the phases denoted **A** and **D** have a fully disordered structure with lattice parameters comparable to those of NiAs, whereas the **B** phase is a superstructure with $a = a'\sqrt{3}$ and $c = 3c'$, where a' and c' are the parameters of the NiAs subcell. The **A** phase has the compositional range of $Al_{0.35-0.60}Ni_{2.2-3.0}As_{1.40-1.65}$ with $a = 368.5-383.6$ and $c = 503.5-511.5$ pm. The **B** phase has the composition range of $Al_{0.50-0.85}Ni_{3.0}As_{1.50-1.15}$ with $a = 666.1-668.7$

and $c = 1535.1-1531.8$ pm. The **D** phase has the composition range of $Al_{0.8-1.5}Ni_{3.5}As_{0.5-1.2}$ with $a = 393.7-397.3$ and $c = 508.9-502.7$ pm. The Al–Ni and Ni–As binary compounds show very limited solubility for the third component, with the exception of NiAs, which dissolves Al up to composition $Al_{0.3}NiAs_{0.7}$. The three ternary phases and Ni do not form tie-lines with AlAs. The phases AlNi, Al_3Ni_2, and NiAs form tie-lines with AlAs. The ingots were annealed at 800°C for 20 days and investigated using x-ray diffraction (XRD), electron probe microanalysis (EPMA), and scanning electron microscopy (SEM).

According to the data of Députier et al. (1995b), NiAl is the "key" compound around which the Ni/AlAs interaction progresses. The mixture of the binaries NiAl + NiAs is the final stage of the Ni/AlAs interaction.

6.31 ALUMINUM–RUTHENIUM–ARSENIC

The isothermal section of the Al–Ru–As ternary system at room temperature was approximately calculated by Klingbeil and Schmid-Fetzer (1989) (Figure 6.26). The calculations are based on the following approximations: ternary phases and solid solubility are disregarded and the Gibbs energy of formation of binary compounds is mostly estimated by the enthalpy of formation calculated from Miedema's model.

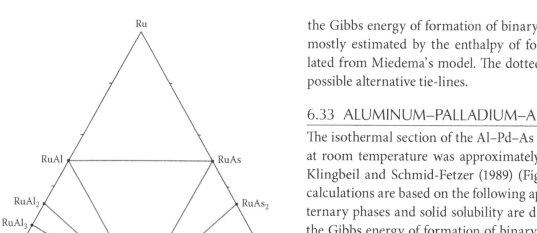

FIGURE 6.26 Calculated isothermal section of the Al–Ru–As ternary system at room temperature. (From Klingbeil, J., and Schmid-Fetzer, R., *CALPHAD*, *13*(4), 367, 1989. With permission.)

6.32 ALUMINUM–RHODIUM–ARSENIC

The isothermal section of the Al–Rh–As ternary system at room temperature was approximately calculated by Klingbeil and Schmid-Fetzer (1989) (Figure 6.27). The calculations are based on the following approximations: ternary phases and solid solubility are disregarded and

the Gibbs energy of formation of binary compounds is mostly estimated by the enthalpy of formation calculated from Miedema's model. The dotted lines indicate possible alternative tie-lines.

6.33 ALUMINUM–PALLADIUM–ARSENIC

The isothermal section of the Al–Pd–As ternary system at room temperature was approximately calculated by Klingbeil and Schmid-Fetzer (1989) (Figure 6.28). The calculations are based on the following approximations: ternary phases and solid solubility are disregarded and the Gibbs energy of formation of binary compounds is mostly estimated by the enthalpy of formation calculated from Miedema's model. The dotted lines indicate possible alternative tie-lines.

6.34 ALUMINUM–OSMIUM–ARSENIC

The isothermal section of the Al–Os–As ternary system at room temperature was approximately calculated by Klingbeil and Schmid-Fetzer (1989) (Figure 6.29). The calculations are based on the following approximations: ternary phases and solid solubility are disregarded and the Gibbs energy of formation of binary compounds is mostly estimated by the enthalpy of formation calculated from Miedema's model. The dotted line indicates a possible alternative tie-line.

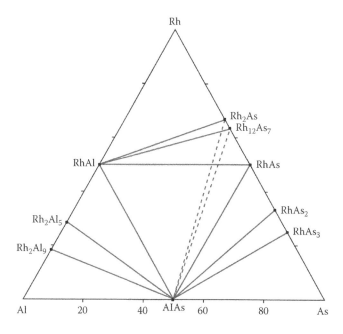

FIGURE 6.27 Calculated isothermal section of the Al–Rh–As ternary system at room temperature. (From Klingbeil, J., and Schmid-Fetzer, R., *CALPHAD*, *13*(4), 367, 1989. With permission.)

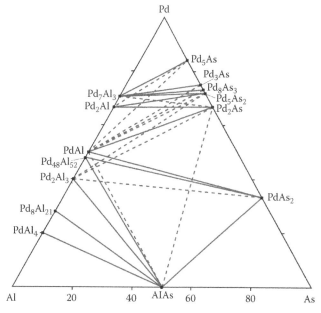

FIGURE 6.28 Calculated isothermal section of the Al–Pd–As ternary system at room temperature. (From Klingbeil, J., and Schmid-Fetzer, R., *CALPHAD*, *13*(4), 367, 1989. With permission.)

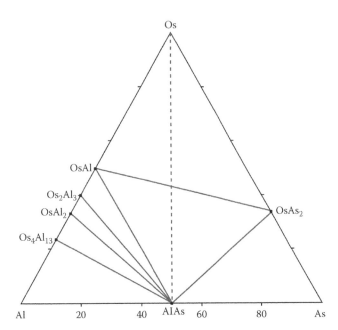

FIGURE 6.29 Calculated isothermal section of the Al–Os–As ternary system at room temperature. (From Klingbeil, J., and Schmid-Fetzer, R., *CALPHAD*, 13(4), 367, 1989. With permission.)

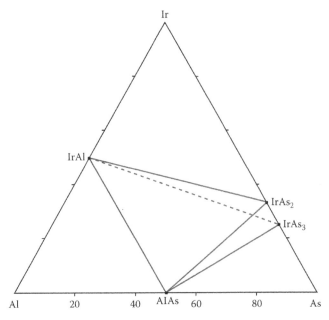

FIGURE 6.30 Calculated isothermal section of the Al–Ir–As ternary system at room temperature. (From Klingbeil, J., and Schmid-Fetzer, R., *CALPHAD*, 13(4), 367, 1989. With permission.)

6.35 ALUMINUM–IRIDIUM–ARSENIC

The isothermal section of the Al–Ir–As ternary system at room temperature was approximately calculated by Klingbeil and Schmid-Fetzer (1989) (Figure 6.30). The calculations are based on the following approximations: ternary phases and solid solubility are disregarded and the Gibbs energy of formation of binary compounds is mostly estimated by the enthalpy of formation calculated from Miedema's model. The dotted line indicates a possible alternative tie-line.

6.36 ALUMINUM–PLATINUM–ARSENIC

The isothermal section of the Al–Pt–As ternary system at room temperature was approximately calculated by Klingbeil and Schmid-Fetzer (1989) (Figure 6.31). The calculations are based on the following approximations: ternary phases and solid solubility are disregarded and the Gibbs energy of formation of binary compounds is mostly estimated by the enthalpy of formation calculated from Miedema's model. The dotted line indicates a possible alternative tie-line.

According to the data of El-Boragy and Schubert (1970), **Pt₅AlAs** ternary compound, which crystallizes in the tetragonal structure with the lattice parameters $a = 389.5$ and $c = 693.4$ pm, is formed in this system.

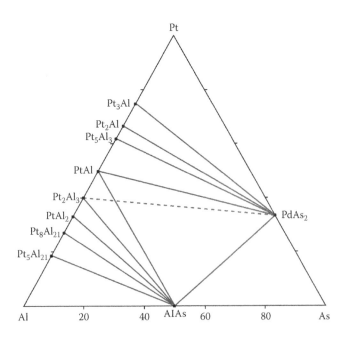

FIGURE 6.31 Calculated isothermal section of the Al–Pt–As ternary system at room temperature. (From Klingbeil, J., and Schmid-Fetzer, R., *CALPHAD*, 13(4), 367, 1989. With permission.)

CHAPTER 7

Systems Based on AlSb

7.1 ALUMINUM–LITHIUM–ANTIMONY

No intermediate ternary phase was found in this system (Luedecke et al. 1983; Goel and Cahoon 1989), and attempts to prepare the ternary **Li₃AlSb₂** were unsuccessful (Juza and Schulz 1952).

7.2 ALUMINUM–SODIUM–ANTIMONY

Na₇Al₂Sb₅ ternary compound, which crystallizes in the monoclinic structure with the lattice parameters $a = 819.0 \pm 0.2$, $b = 1359.0 \pm 0.3$, and $c = 772.0 \pm 0.2$ pm and $\beta = 118.0 \pm 0.1°$ and a calculated density of 3.60 g·cm⁻³, is formed in this system (Cordier et al. 1984a). It was prepared by the melting of a stoichiometric mixture of elements at 830°C under an Ar atmosphere.

7.3 ALUMINUM–POTASSIUM–ANTIMONY

KAlSb₄ ternary compound, which crystallizes in the orthorhombic structure with the lattice parameters $a = 1036.6 \pm 0.3$, $b = 422.0 \pm 0.2$, and $c = 1786.5 \pm 0.5$ pm, is formed in this system (Cordier and Ochmann 1991m). It was prepared by the melting of a stoichiometric mixture of elements at 650°C.

7.4 ALUMINUM–RUBIDIUM–ANTIMONY

Rb₆AlSb₃ ternary compound, which crystallizes in the monoclinic structure with the lattice parameters $a = 1062.4 \pm 0.3$, $b = 626.0 \pm 0.3$, and $c = 1237.7 \pm 0.4$ pm and $\beta = 100.7 \pm 0.1°$, is formed in this system (Blase et al. 1992b). It was prepared by the melting of a stoichiometric mixture of elements at 600°C.

7.5 ALUMINUM–CESIUM–ANTIMONY

Cs₆AlSb₃ ternary compound, which crystallizes in the monoclinic structure with the lattice parameters $a = 1084.5 \pm 0.2$, $b = 650.7 \pm 0.2$, and $c = 1270.7 \pm 0.4$ pm and $\beta = 100.95 \pm 0.02°$ (von Schnering et al. 1990c; Blase et al. 1991a, 1991b), is formed in this system. It was synthesized from the stoichiometric mixture of elements or from AlSb, CsSb, and Cs in an evacuated and sealed Nb ampoule at 630°C–680°C.

7.6 ALUMINUM–COPPER–ANTIMONY

The liquidus projection, invariant equilibria, three vertical sections, and isothermal sections at 400°C and 200°C were predicted using the thermodynamically CALPHAD method (Minić et al. 2013). In addition, phase transition temperatures of the selected samples with compositions along calculated isopleths were experimentally determined using differential thermal analysis (DTA). Experimentally obtained phase transition temperatures were compared with the results of thermodynamic calculation. Predicted isothermal sections were compared with the results of the scanning electron microscopy–energy-dispersive x-ray spectroscopy (SEM-EDX) analysis. In both cases, a good agreement between theoretical calculations and experimental results was obtained.

On the calculated liquidus surface of the Al–Cu–Sb ternary system (Figure 7.1), there are 10 fields of primary crystallizations of the next phases: (Cu), (Sb), (Al), AlSb, Al₂Cu, AlCu, (Cu₃Sb, Cu₃Al), Cu₂Sb, Al₅Cu₈, and AlCu₂. Three of them, (Al), AlCu, and Al₂Cu, are very small (Minić et al. 2013). The AlSb primary crystallization field is the biggest among the others. The next six invariant reactions exist on the liquidus surface: L + AlCu₂ ⟷ (Cu₃Sb, Cu₃Al) + AlSb (U_1, 725°C), L + AlCu₂ ⟷ AlSb + AlCu (U_2, 626°C), L + AlCu ⟷ AlSb

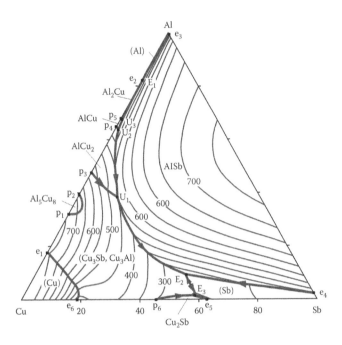

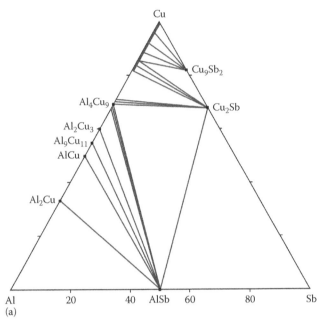

FIGURE 7.1 Calculated liquidus surface of the Al–Cu–Sb ternary system. (From Minić, D., et al., *J. Alloys Compd.*, *555*, 347, 2013. With permission.)

+ Al$_2$Cu (U_3, 593°C), L $\longleftrightarrow$ AlSb + Al$_2$Cu + (Al) (E_1, 547°C), L $\longleftrightarrow$ (Cu$_3$Sb, Cu$_3$Al) + (Sb) + AlSb (E_2, 534°C), and L $\longleftrightarrow$ (Cu$_3$Sb,Cu$_3$Al) + Cu$_2$Sb + (Sb) (E_3, 522°C).

Isothermal sections of this ternary system at 200°C and 400°C are given in Figure 7.2 (Minić et al. 2013). Thirteen different regions can be observed at 200°C, among which nine are three-phase regions and four are two-phase regions (Figure 7.2a). Twenty-six different regions exist at 400°C (Figure 7.2b), out of which 2 are single-phase regions, 13 are two-phase regions, and 11 are three-phase regions.

AlSb–Cu. According to the calculation of Minić et al. (2013), the fields of primary crystallization of AlSb, (Cu$_3$Sb,Cu$_3$Al), and Cu exist in this section.

7.7 ALUMINUM–MAGNESIUM–ANTIMONY

The first experimental data concerning the Al–Mg–Sb ternary system were obtained by Guertler and Bergmann (1933). The ternary phase equilibria in this system were first calculated without introducing any ternary parameters (Balakumar and Medraj 2005; Raghavan 2007). Later, the validity of these calculations was questioned (Paliwal and Jung 2010; Raghavan 2011). The computed liquidus surface of the Al–Mg–Sb system is shown in Figure 7.3 (Paliwal and Jung 2010; Raghavan

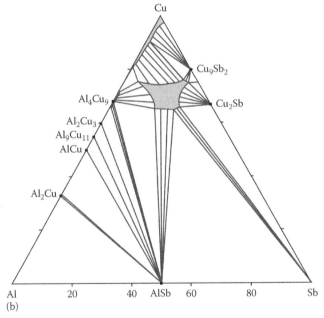

FIGURE 7.2 Calculated isothermal sections of the Al–Cu–Sb ternary system at (a) 200°C and (b) 400°C. (From Minić, D., et al., *J. Alloys Compd.*, *555*, 347, 2013. With permission.)

2011). The invariant reactions on the liquidus surface are L$_1$ + β-Mg$_3$Sb$_2$ $\longleftrightarrow$ L$_2$ + α-Mg$_3$Sb$_2$ (U_1, 928°C), L$_1$ + β-Mg$_3$Sb$_2$ $\longleftrightarrow$ L$_2$ + α-Mg$_3$Sb$_2$ (U_2, 910°C), L$_1$ $\longleftrightarrow$ L$_2$ + α-Mg$_3$Sb$_2$ + AlSb (M, 877°C), L $\longleftrightarrow$ (Al) + α-Mg$_3$Sb$_2$ + AlSb (E_1, 656°C), L $\longleftrightarrow$ (Sb) + α-Mg$_3$Sb$_2$ + AlSb (E_2, 579°C), L $\longleftrightarrow$ (Al) + α-Mg$_3$Sb$_2$ + β-AlMg (E_3, 451.3°C), L $\longleftrightarrow$ α-Mg$_3$Sb$_2$ + β-AlMg + γ-Al$_{12}$Mg$_{17}$ (E_4, 451.05°C), and (Mg) + α-Mg$_3$Sb$_2$ + γ-Al$_{12}$Mg$_{17}$ (E_5, 439°C). The

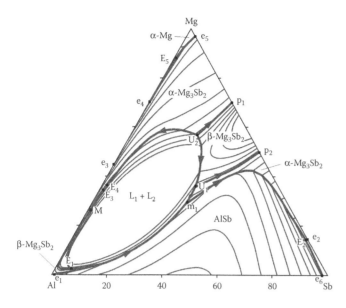

FIGURE 7.3 Calculated liquidus surface of the Al–Mg–Sb ternary system. (From Paliwal, M., and Jung, I.-H., *CALPHAD*, *34*(1), 51, 2010. With permission.)

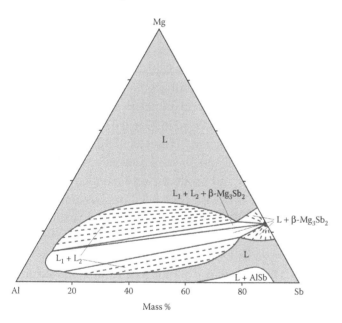

FIGURE 7.4 Calculated liquid miscibility gap in the Al–Mg–Sb ternary system at 1000°C. (From Paliwal, M., and Jung, I.-H., *CALPHAD*, *34*(1), 51, 2010. With permission.)

ternary eutectics reactions E_1 through E_5, except E_2, all lie very close to the Al–Mg binary system.

In Figure 7.4, the computed liquid miscibility gap at 1000°C is shown (Paliwal and Jung 2010; Raghavan 2011).

AlSb–Mg$_3$Sb$_2$. This system is a quasi-binary section of the Al–Mg–Sb ternary system (Guertler and Bergmann 1933). The calculated phase diagram of this

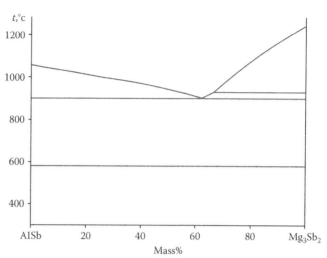

FIGURE 7.5 Calculated phase diagram of the AlSb–Mg$_3$Sb$_2$ system. (From Paliwal, M., and Jung, I.-H., *CALPHAD*, *34*(1), 51, 2010. With permission.)

system is given in Figure 7.5 (Paliwal and Jung 2010; Raghavan 2011).

7.8 ALUMINUM–CALCIUM–ANTIMONY

Some ternary compounds are formed in this system. **Ca$_3$AlSb$_3$** crystallizes in the orthorhombic structure with the lattice parameters $a = 1283.5 \pm 0.5$, $b = 448.9 \pm 0.2$, and $c = 1428.2 \pm 0.2$ pm and a calculated density of 4.14 g·cm^{-3} (Cordier et al. 1984b). It was prepared by the melting a stoichiometric mixture of elements in a corundum crucible in the Ar flowing at 1000°C.

According to the optimization using the density functional theory, **Ca$_5$Al$_2$Sb$_6$** crystallizes in the orthorhombic structure with the lattice parameters $a = 1212.43$, $b = 1409.29$, and $c = 451.04$ pm (Yan and Wang 2011) ($a = 1376.0$, $b = 1178.5$, and $c = 437.2$ pm [Yan et al. 2014]), a calculated density of 4.31 g·cm^{-3}, and an energy gap of 0.65 eV (Zevalkink et al. 2012) (0.5 eV [Toberer et al. 2010]; 0.35 eV [Yan et al. 2014]; 0.53 eV [Ye et al. 2015]).

Ca$_5$Al$_2$Sb$_6$ samples were prepared by ball milling elemental reagents, followed by hot pressing (Toberer et al. 2010; Zevalkink et al. 2012). Starting with Ca dendrites, Al shot, and Sb lumps, the elements were loaded into stainless steel vials with stainless steel balls in an Ar-filled glove box. The mixtures were drying ball milled for 1 h, and the fine powder that resulted from ball milling was hot pressed in high-density graphite dies. A maximum temperature of 700°C for 2 h in Ar was used during hot pressing, followed by a 2 h stress-free annealing at 600°C and a 3 h cooldown under vacuum.

Ca₁₁AlSb₉ crystallizes also in the orthorhombic structure with the lattice parameters $a = 1183.2 \pm 0.3$, $b = 1250.5 \pm 0.2$, and $c = 1667.4 \pm 0.4$ pm (Young and Kauzlarich 1995). This compound was prepared by adding stoichiometric amounts of the elements in a Nb tube that was sealed on one end. This tube was then sealed in a quartz ampoule under vacuum. Heating the reactants at a rate of $60°C·h^{-1}$ to 850°C for 2 weeks and subsequently cooling at the same rate to room temperature provided the highest yield of the title compound, which is air sensitive.

Ca₁₄AlSb₁₁ crystallizes in the tetragonal structure with the lattice parameters $a = 1667.6 \pm 0.5$ and $c = 2242.3 \pm 0.7$ pm and the same experimental and calculated densities of 4.11 g·cm⁻³ (Cordier et al. 1984c) ($a = 1667.2 \pm 0.6$ and $c = 2443 \pm 1$ pm [Brock et al. 1993]). This compound was synthesized by mixing stoichiometric amounts of the elements (Cordier et al. 1984c; Brock et al. 1993). The reactants and the products were handled in a N₂-filled dry box. The mixture of the elements was placed into a Nb tube, which was arc welded closed under an atmosphere of Ar. The filled Nb tube was then sealed in a quartz ampoule under vacuum. The reaction vessel was placed in a furnace and heated to 600°C ($8°C·min^{-1}$), followed by heating to 1250°C ($20°C·h^{-1}$). It was maintained at that temperature for 24 h and first cooled to 600°C ($20°C·h^{-1}$) and then rapidly to room temperature ($8°C·min^{-1}$). The resulting air-sensitive product was made of a mixture of powder, chunks of a silver material, and a small amount of elongated diamond-shaped or needle crystals.

Ca₁₄AlSb₁₁ is an intrinsic semiconductor with an energy gap of 0.0143 eV (Brock et al. 1993).

7.9 ALUMINUM–STRONTIUM–ANTIMONY

Three compounds are formed in this system. **Sr₅Al₂Sb₆** crystallizes in the orthorhombic structure with the lattice parameters $a = 1212.4 \pm 0.4$, $b = 1034.1 \pm 0.4$, and $c = 1340.9 \pm 0.5$ pm and a calculated density of 4.83 g·cm⁻³ (Cordier and Stelter 1988) ($a = 1192.70$, $b = 1022.46$, and $c = 1310.86$ pm [calculated values] [Ye et al. 2015]). This compound is a semiconductor with an indirect band gap of 0.78 eV (Ye et al. 2015).

The structure of Sr₅Al₂Sb₆ was optimized with the Vienna *ab initio* simulation package, which is based on the density functional theory (Ye et al. 2015).

Sr₆Al₂Sb₆ also crystallizes in the orthorhombic structure with the lattice parameters $a = 2041.0 \pm 0.9$, $b = 690.0 \pm 0.3$, and $c = 1350.1 \pm 0.6$ pm and experimental and calculated densities of 4.68 and 4.58 g·cm⁻³, respectively (Cordier et al. 1984d). It was prepared by the melting of a stoichiometric mixture of elements at 1200°C for 30 min. The title compound is moisture sensitive.

Sr₁₄AlSb₁₁ crystallizes in the tetragonal structure with the lattice parameters $a = 1754.2 \pm 0.6$ and $c = 2333 \pm 1$ pm at room temperature and $a = 1749.3 \pm 0.4$ and $c = 2348.0 \pm 0.8$ pm and a calculated density of 4.82 g·cm⁻³ at 130 K (Brock et al. 1993). Sr₁₄AlSb₁₁ is an intrinsic semiconductor with an energy gap of 0.0667 eV.

Sr₁₄AlSb₁₁ was synthesized the same way as Ca₁₄AlSb₁₁ was prepared.

7.10 ALUMINUM–BARIUM–ANTIMONY

Three compounds are formed in this system. **Ba₃AlSb₃** crystallizes in the orthorhombic structure with the lattice parameters $a = 2113.3 \pm 1.0$, $b = 719.4 \pm 0.5$, and $c = 1406.9 \pm 0.8$ pm and a calculated density of 4.998 g·cm⁻³ (Cordier et al. 1982). It was prepared by melting a stoichiometric mixture of elements in a corundum crucible in the Ar flowing at 500°C–930°C.

Ba₇Al₄Sb₉ also crystallizes in the orthorhombic structure with the lattice parameters $a = 1080.20 \pm 0.11$, $b = 1805.75 \pm 0.18$, and $c = 716.77 \pm 0.07$ pm and a calculated density of 5.14 g·cm⁻³ (He et al. 2016). To obtain this compound, Ba, Al, and Sb were used as starting materials. All reagents and products were handled within an Ar-filled glove box with controlled oxygen and a moisture level below 1 ppm. Single crystals were first isolated from reactions attempted to extend the $AETr_2Pn_2$ and $AE_3Tr_2Pn_4$ series (AE = Ca, Sr, Ba; Tr = Al, Ga, In; and Pn = P, As, Sb) through use of the Al or Pb fluxes. Mixtures of the elements were loaded into alumina crucibles covered on the top with quartz wool and placed within fused silica tubes, which were then evacuated and flame sealed. The tubes were heated to 960°C at a rate of $60°C·h^{-1}$, held at that temperature for 20 h (samples with Al flux) or 40 h (samples with Pb flux), and cooled to 750°C at a rate of $5°C·h^{-1}$ (for samples with Al flux) or to 500°C at a rate of $30°C·h^{-1}$ (for samples with Pb flux). The molten metal fluxes were removed at 750°C (Al flux) or at 500°C (Pb flux) by centrifugation.

Ba₁₄AlSb₁₁ crystallizes in the tetragonal structure with the lattice parameters $a = 1836.0 \pm 0.8$ and $c = 2415 \pm 2$ pm at room temperature and $a = 1829.3 \pm 0.2$ and $c = 2422.2 \pm 0.9$ pm and a calculated density of 5.40 g·cm⁻³ at 130 K

(Brock et al. 1993). $Ba_{14}AlSb_{11}$ is an intrinsic semiconductor with an energy gap of 0.4814 eV.

$Ba_{14}AlSb_{11}$ was synthesized the same way as $Ca_{14}AlSb_{11}$ was prepared.

7.11 ALUMINUM–ZINC–ANTIMONY

The first experimental data about the liquidus surface of the Al–Zn–Sb system were obtained by Köster (1942), who investigated the AlSb–Al–Zn–Zn_3Sb_2 part of this system through DTA and metallography. It was shown that AlSb primarily crystallizes in almost the entire investigated region. The calculated liquidus surface of this ternary system was determined by Klančnik and Medved (2011b). The next six invariant reactions exist on the liquidus surface: L $\longleftrightarrow$ β-Zn_3Sb_2 + γ-Zn_4Sb_3 + AlSb (E_1, 559°C), L + γ-Zn_4Sb_3 $\longleftrightarrow$ ZnSb + AlSb (U_1, 538°C), L $\longleftrightarrow$ ZnSb + (Sb) + AlSb (E_2, 509°C), L + β-Zn_3Sb_2 + AlSb $\longleftrightarrow$ α-Zn_3Sb_2 (P_1, 447°C), L $\longleftrightarrow$ (Zn) + α-Zn_3Sb_2 + AlSb (E_3, 410°C), and L $\longleftrightarrow$ (Al) + (Zn) + AlSb (E_4, 381°C).

Some isothermal sections of this ternary system have been constructed. No ternary compound has been found. At 100°C, the Al–Zn–Sb ternary system is divided into four quasi-ternary subsystems: AlSb–Al–Zn, AlSb–Zn–α-Zn_4Sb_3, AlSb–ZnSb–α-Zn_4Sb_3, and AlSb–ZnSb–Sb (Figure 7.6) (Klančnik and Medved 2013).

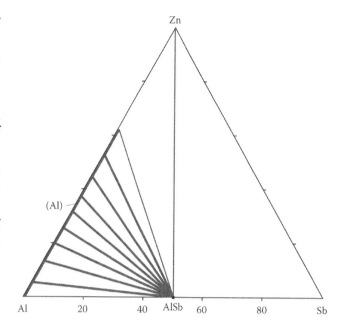

FIGURE 7.7 Isothermal section of the Al–Zn–Sb ternary system at 374°C. (From Klančnik, G., and Medved, J., *J. Min. Metall. Sect. B Metall.*, 47(2), 179, 2011. With permission.)

The isothermal section of the Al–Zn–Sb ternary system at 374°C is given in Figure 7.7 (Klančnik and Medved 2011a). AlSb has a great influence on the constitution of the ternary phase diagram. A small solubility of Zn in AlSb was determined. The monotectoid reaction β-Al $\longleftrightarrow$ α-Al + η-Zn was determined to be in average at 286°C at the heating and at 261°C at the cooling. The solubility of Al in η-Zn was determined to be 4.56 at%. This confirms a small extension of the η-Zn region from the binary system into the ternary system.

Based on the microstructures of the alloys and the phase compositions obtained using the SEM-EDX technique, phase equilibria states available in the Al–Zn–Sb ternary system at 450°C were determined (Figure 7.8) (Zhu et al. 2009; Raghavan 2012). The ingots were annealed at this temperature for 30 days. There are five three-phase regions and six two-phase regions in the system. AlSb can equilibrate with all phases in the system, while the Zn-rich liquid phase can equilibrate with the Zn_3Sb_2, AlSb, and α-Al. The Zn solubility in AlSb is 2.1 at%, and the Al solubility in Zn_4Sb_3 is not higher than 3.3 at%. The Al solubility is up to 0.5 at% in Zn_3Sb_2 and is negligible in ZnSb.

The isothermal section of the Al–Zn–Sb system at 600°C constructed using optical microscopy, SEM with

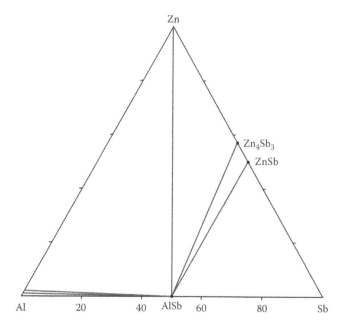

FIGURE 7.6 Calculated isothermal section of the Al–Zn–Sb ternary system at 100°C. (From Klančnik, G., and Medved, J., *Comput. Mater. Sci.*, 66, 14, 2013. With permission.)

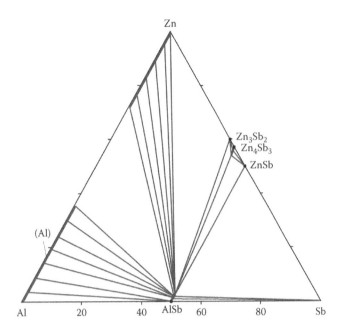

FIGURE 7.8 Isothermal section of the Al–Zn–Sb ternary system at 450°C. (From Zhu, Z., and Su, X., *J. Phase Equilibria Diffus.*, *30*(6), 595, 2009. With permission.)

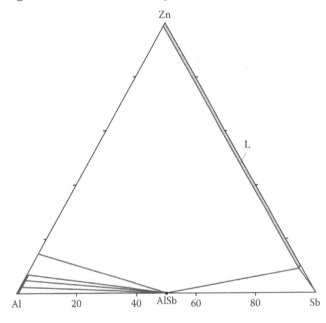

FIGURE 7.9 Isothermal section of the Al–Zn–Sb ternary system at 600°C. (From Su, X., et al., *Int. J. Mater. Res.*, *102*(3), 241, 2011. With permission.)

EDX, and XRD is given in Figure 7.9 (Su et al. 2011; Raghavan 2012). The samples were annealed at this temperature for 30 days. AlSb dissolves up to 0.6 at% Zn.

The calculated isothermal section at 800°C is shown in Figure 7.10 (Klančnik and Medved 2013). The position of the two-phase region AlSb + L is asymmetric and pushed into the Zn–Sb binary system.

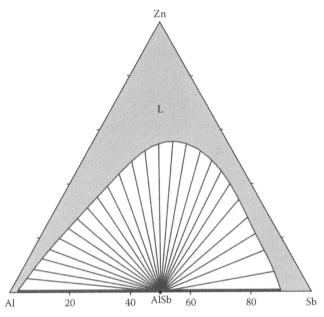

FIGURE 7.10 Calculated isothermal section of the Al–Zn–Sb ternary system at 800°C. (From Klančnik G., and Medved, J., *Comput. Mater. Sci.*, *66*, 14, 2013. With permission.)

The Al–Sb–Zn ternary phase diagram was calculated from thermodynamic Gibbs energies using the CALPHAD method, and experimentally investigated using DTA, differential scanning calorimetry (DSC), and a Calvet-type microcalorimeter (Zhu et al. 2009; Su et al. 2011; Klančnik and Medved 2013). SEM equipped with an EDX, and XRD were used to confirm the predicted phase diagram. Good agreement was found between the experimental results and the predictions.

AlSb–Zn. This system is a quasi-binary section of the Al–Zn–Sb ternary system (Köster 1942). A part of the AlSb–Zn phase diagram from the Zn side is shown in Figure 7.11 (Klančnik and Medved 2013). Eutectic contains 0.4 mol% AlSb and crystallizes at 405.5°C. At the growing of AlSb single crystals in the [111] and [$\bar{1}\bar{1}\bar{1}$] directions, the effective coefficient of zinc distribution is 0.45 ± 0.06 and 1.2 ± 0.14, respectively (Strel'nikova and Mirgalovskaya 1965a).

7.12 ALUMINUM–CADMIUM–ANTIMONY

The liquidus surface of the Al–Cd–Sb ternary system is given in Figure 7.12 (Belotskiy et al. 1985). The field of the AlSb primary crystallization occupies the biggest part of the liquidus surface. This system is divided for three subsystems, AlSb–CdSb–Sb, AlSb–CdSb–Cd, and AlSb–Al–Cd, in which the ternary eutectic crystallizes

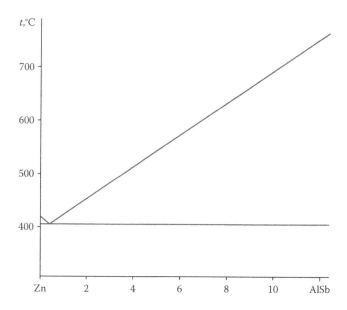

FIGURE 7.11 Part of the phase diagram of the AlSb–Zn system. (From Klančnik, G., and Medved, J., *Comput. Mater. Sci.*, *66*, 14, 2013. With permission.)

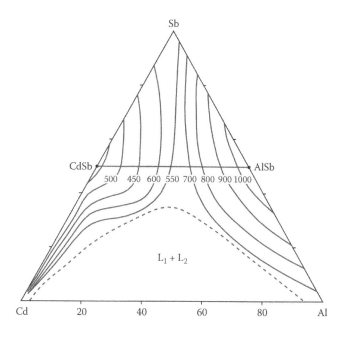

FIGURE 7.12 Liquidus surface of the Al–Cd–Sb ternary system. (From Belotskiy, D.P., et al., *Izv. AN SSSR Neorgan. Mater.*, *21*(7), 1093, 1985. With permission.)

at 400°C, 548°C, and 250°C with a Cd/Sb/Al molar ratio of 42:56:2, 92:6:2, and 98:1:1, respectively.

AlSb–Cd. This system is a quasi-binary section of the Al–Cd–Sb ternary system (Figure 7.13) (Belotskiy et al. 1985). There is an immiscibility gap within the interval from 30 to 95 mol% Cd with the monotectic temperature 630°C.

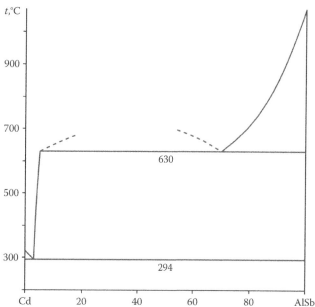

FIGURE 7.13 Phase diagram of the AlSb–Cd system. (From Belotskiy, D.P., et al., *Izv. AN SSSR Neorgan. Mater.*, *21*(7), 1093, 1985. With permission.)

AlSb–CdSb. This system is also a quasi-binary section of the Al–Cd–Sb ternary system (Belotskiy et al. 1985). The eutectic contains 1 mol% AlSb and crystallizes at 317°C.

This ternary system was investigated through DTA, metallography, and measurement of microhardness (Belotskiy et al. 1985). The ingots were annealed at 450°C for 240 h and then at 350°C–250°C for 400 h.

7.13 ALUMINUM–GALLIUM–ANTIMONY

Liquidus temperatures in the Ga-rich region of the Al–Ga–Sb ternary system have been determined using DTA by Panish and Ilegems (1972), Pelevin and Chupakhina (1975), and Osamura et al. (1979). Liquidus equations have been derived from experimental data by Muszyński and Ryabcev (1975, 1976) using a simplex lattice method in order to interpolate the liquidus curves over the entire composition range. The full liquidus surface has been determined experimentally by Köster and Thoma (1955) and calculated on the basis of various solution models by Joullié et al. (1979), Osamura et al. (1979), Lendvay (1984), Ishida et al. (1989b), and Sharma and Srivastava (1992b).

The liquidus and the corresponding solidus isotherms in the Ga-rich corner of the Al–Ga–Sb phase diagram have been determined at 600°C by a saturation technique (Bedair 1975). The mass loss technique was used to determine the liquidus data at 500°C, 518°C, and 536°C (van Nguyen Mau et al. 1976). Solidus data were measured by electron probe microanalysis (EPMA) on epitaxial layers. The calculations were based on a quasi-regular model for the three binary liquids and on an ideal solution for the ternary solid. The dependence of Sb solubility on the Al concentration in the liquid phase at 403°C, 452°C, and 500°C has been established, and the dependence of the AlSb concentration in the $Al_xGa_{1-x}Sb$ solid phase on the composition of the liquid phase has been determined at 452°C by Yordanova and Tret'yakov (1982). Using the chemical constants equilibrium method, the phase equilibrium in the region of liquid phase composition near the Ga-rich corner has been calculated. Ohshima et al. (1985) calculated the solidus and liquidus isotherms in a wide temperature range from 300°C to 600°C for the Ga-rich region. The liquidus isotherms in the temperature range 600°C–640°C at the Sb-rich corner for the region of 0 to 0.25 at% in the liquid were determined by the seed solution technique (Kuwatsuka et al. 1989).

The liquidus surface of the Al–Ga–Sb ternary system given in Figure 7.14 was calculated by Li et al. (1999). It is shown that the field of the $Al_xGa_{1-x}Sb$ solid solutions' primary crystallization covers almost the entire system.

Liquidus isotherms at 400°C, 450°C, 500°C, and 550°C (Figure 7.15) and solidus isotherms at 500°C and

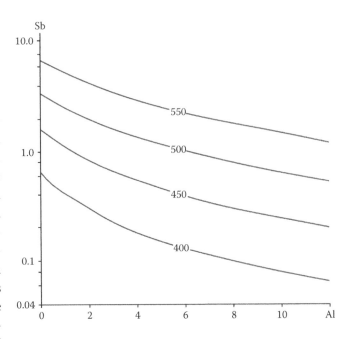

FIGURE 7.15 Liquidus isotherms at 400°C, 450°C, 500°C, and 550°C of the Al–Ga–Sb ternary system in the Ga-rich region. (From Cheng, K.Y., and Pearson, G.L., *J. Electrochem. Soc.*, 124(5), 753, 1977. Reproduced with permission of The Electrochemical Society.)

550°C (Figure 7.16) were determined experimentally (Cheng and Pearson 1977). A saturation technique was used to determine the liquidus data. Under the assumption that the ternary liquid is a simple solution and that AlSb–GaSb solid forms an ideal solution, excellent agreement was obtained between the calculated liquidus and solidus isotherms and the experiment.

Isothermal sections of this ternary system have been constructed at 600°C, 700°C, 750°C, 800°C, 900°C, and 1000°C (Kaufman et al. 1981; Sharma and Srivastava 1992b; Li et al. 1999). An isothermal section at 600°C is given in Figure 7.17 (Li et al. 1999).

The Al–Ga–Sb alloys have been studied by the electromotive force (EMF) method between 350°C and 1030°C, and partial molar functions of Al and some equilibrium temperatures were determined (Girard et al. 1988). The obtained results are in satisfactory agreement with the calculation of Kaufman et al. (1981). The mixing enthalpy of the Al–Ga–Sb alloys was investigated in the temperature range of 863°C–954°C using the drop method into a Calvet microcalorimeter (Girard et al. 1987). Because of the volatility of Sb, the Sb-rich part of the ternary composition could not be studied, but an extrapolation from the experimental results allowed proposing the

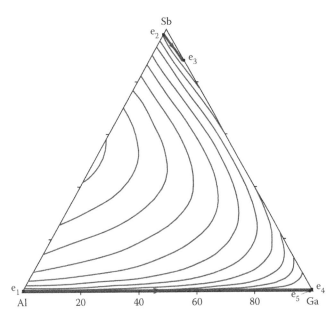

FIGURE 7.14 Calculated liquidus surface of the Al–Ga–Sb ternary system. (From Li, J.-B., et al., *J. Phase Equilibria*, 20(3), 316, 1999. With permission.)

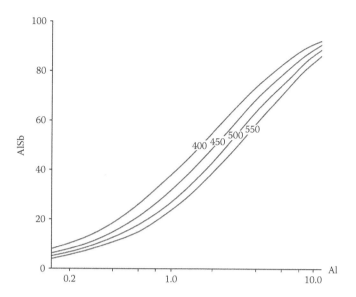

FIGURE 7.16 Solidus isotherms at 400°C, 450°C, 500°C, and 550°C of the Al–Ga–Sb ternary system in the Ga-rich region. (From Cheng, K.Y., and Pearson, G.L., *J. Electrochem. Soc.*, *124*(5), 753, 1977. Reproduced with permission of The Electrochemical Society.)

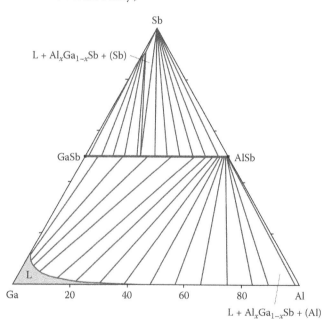

FIGURE 7.17 Calculated isothermal section of the Al–Ga–Sb ternary system at 600°C. (From Li, J.-B., et al., *J. Phase Equilibria*, *20*(3), 316, 1999. With permission.)

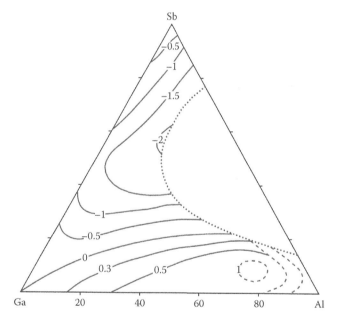

FIGURE 7.18 Isoenthalpic lines for Al–Ga–Sb system (the values are given in kJ·M^{-1}). (From Girard, C., et al., *J. Less Common Metals*, *128*(1–2), 101, 1987. With permission.)

Al–Ga–Sb isoenthalpic curves (Figure 7.18). According to the data of Gerdes and Predel (1979b), the maximum deviation of the mixing enthalpy value from those linearly interpolated between the mixing enthalpies of the AlSb and GaSb melts is $\delta(\Delta H)^{max}_{exp} = +562$ J·M^{-1}, and the concentration dependence of $\delta(\Delta H)$ is not in agreement with the values predicted by the regular solution model.

AlSb–GaSb. In the first study of the liquidus surface of this quasi-binary system by thermal analysis and metallography (Köster and Thoma 1955), it was concluded that the compounds AlSb and GaSb do not exhibit mutual solid solubilities. This was corrected later by the same group (Köster and Ulrich 1958b) and shown to be due to the low diffusivities in the AlSb–GaSb system, where homogenization was not completed after annealing for 1000 h at 500°C or 100 h at 700°C in pressed powder samples. The slow solid-state diffusion was also detected by Woolley and Smith (1958a). A complete solid solubility was also suggested by Gorshkov and Goryunova (1958), Goryunova and Radautsan (1958a), and Miller et al. (1960).

The phase diagram of the AlSb–GaSb system was calculated on the basis of a simple solution model (Joullié et al. 1979), or of a regular solution model (Osamura et al. 1972b; Aulombard and Joullie 1979), or of quasi-sub-subregular solution model (Sharma and Srivastava 1992b), or considering the existence of associated complexes in the liquid state (Osamura et al. 1979). It was also calculated by Steininger (1970), Panish and Ilegems (1972), Stringfellow (1972), Kaufman et al. (1981), Lendvay (1984), Ishida et al. (1989b), and Li et al. (1999) and is shown in Figure 7.19. Vegard's law was found to be at least roughly applicable for the $Al_xGa_{1-x}Sb$ solid solution (Miller et al. 1960).

First-principles calculations were performed to study the structural, electronic, optical, and thermodynamic properties of $Al_xGa_{1-x}Sb$, ternary alloys using the full-potential linearized augmented plane-wave plus local orbitals method within the density functional theory. The investigation on the effect of composition on lattice constant, bulk modulus, band gap, effective mass, and refractive index for these ternary alloys shows almost nonlinear dependence on the composition (El Haj Hassan et al. 2010). A regular solution model was used to investigate the thermodynamic stability of the alloys, which mainly indicates a phase miscibility gap with a critical temperature of 805°C and a critical composition of 58 mol% AlSb.

By extremely rapid cooling of the liquid alloys, solid solutions in the AlSb–GaSb were obtained (Bucher et al. 1986). In order to attain the high cooling rates (10^7–10^8 °C·s^{-1}), a shock wave tube has been used. Bulk $Al_xGa_{1-x}Sb$ ingots have been grown by a vertical Bridgman method (Aulombard and Joullie 1979).

7.14 ALUMINUM–INDIUM–ANTIMONY

The liquidus surface of the Al–In–Sb ternary system was constructed by Köster and Thoma (1955), Ishida et al. (1988b, 1989b), and Sharma and Srivastava (1992a) and is given in Figure 7.20. Isothermal sections at 700°C, 800°C, and 900°C are shown in Figure 7.21 (Ishida et al. 1988b). It should be noted that the three-phase region composed of the solid solution based on AlSb and two molten phases is formed in this system at temperatures below about 750°C.

AlSb–InSb. In the first study of this quasi-binary system by thermal analysis and metallography (Köster and Thoma 1955), the solidus data were misinterpreted in terms of a degenerated eutectic. This misinterpretation was due to the very slow solid-state diffusion in

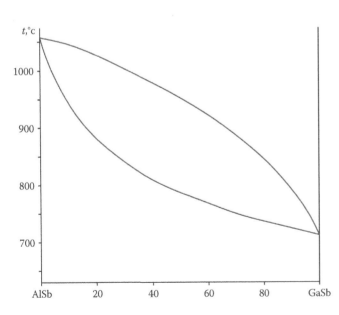

FIGURE 7.19 Calculated phase diagram of the AlSb–GaSb system. (From Li, J.-B., et al., *J. Phase Equilibria*, 20(3), 316, 1999. With permission.)

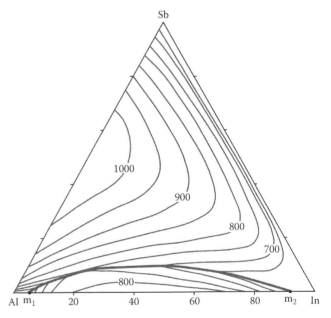

FIGURE 7.20 Calculated liquidus surface of the Al–In–Sb ternary system. (From Ishida, K., et al., *J. Less Common Metals*, 143(1–2), 279, 1988. With permission.)

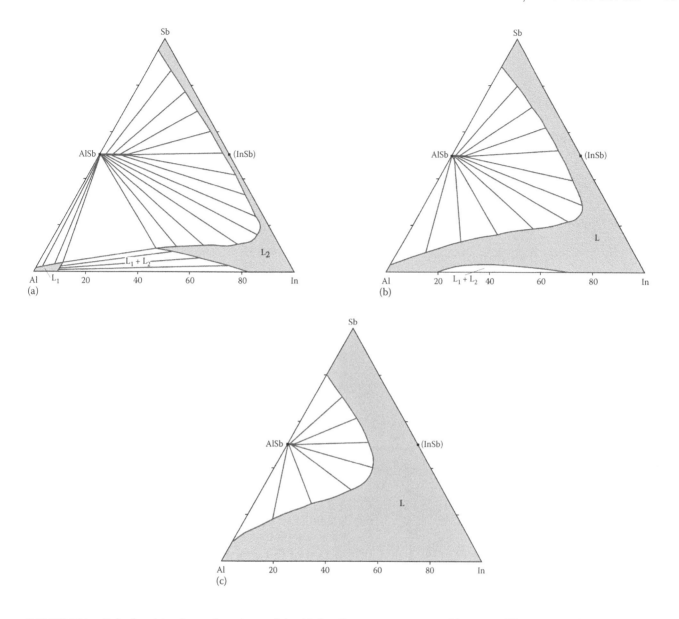

FIGURE 7.21 Calculated isothermal sections of the Al–In–Sb ternary system at (a) 700°C, (b) 800°C, and (c) 900°C. (From Ishida, K., et al., *J. Less Common Metals*, *143*(1–2), 279, 1988. With permission.)

the $Al_xIn_{1-x}Sb$ phase and was corrected in a follow-up work (Köster and Ulrich 1958b), where the development of a homogenous solid solution was found by XRD on samples annealed at 500°C up to 1000 h. These data were confirmed by Woolley and Smith (1958a), who also detected the slow solid-state diffusion in this system.

A nearly linear variation of lattice parameters with the composition of the $Al_xIn_{1-x}Sb$ solid solutions was found by Baranov and Goryunova (1960). In a later paper (Baranov et al. 1965), the same authors used two samples in the AlSb–InSb section to illustrate that due to nonequilibrium conditions, DTA measurements on cooling should not be used for determination of solidus

temperatures. The calculated phase diagram of the AlSb–InSb system, given in Figure 7.22 (Ishida et al. 1988b, 1989b), is in good agreement with experimental data (Köster and Thoma 1955; Gorshkov and Goryunova 1958; Woolley and Smith 1958a; Baranov et al. 1965) and calculations (Steininger 1970; Osamura et al. 1972b; Panish and Ilegems 1972; Stringfellow 1972a; Brebrick and Panlener 1974; Sharma and Srivastava 1992a).

The mixing enthalpy of liquid AlSb–InSb alloys has been determined with the aid of a high-temperature calorimeter (Gerdes and Predel 1979b). The maximum deviation of the mixing enthalpy value from those linearly interpolated between the mixing enthalpies of the AlSb

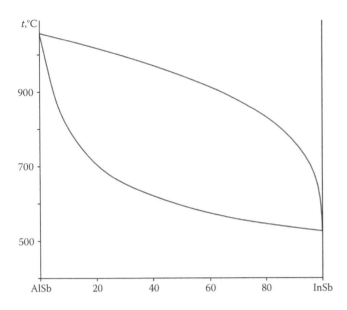

FIGURE 7.22 Calculated phase diagram of the AlSb–InSb system. (From Ishida, K., et al., *J. Less Common Metals*, *143*(1–2), 279, 1988. With permission.)

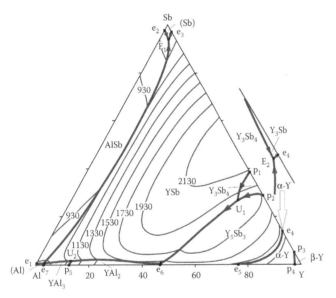

FIGURE 7.23 Calculated liquidus surface of the Al–Y–Sb ternary system in an Al-rich region. (From Zhang, L.G., et al., *J. Alloys Compd.*, *475*(1–2), 233, 2009. With permission.)

and GaSb melts is $\delta(\Delta H)^{max}_{exp} = -417$ J·M^{-1}, and the concentration dependence of $\delta(\Delta H)$ is not in agreement with the values predicted by the regular solution model.

7.15 ALUMINUM–YTTRIUM–ANTIMONY

The calculated liquidus surface of the Al–Y–Sb ternary system is given in Figure 7.23 (Zhang et al. 2009; Raghavan 2010e). No experimental data are available for comparison with the computed liquidus projection.

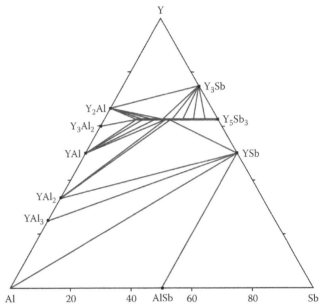

FIGURE 7.24 Calculated isothermal section of the Al–Y–Sb ternary system at 500°C. (From Zhang, L.G., et al., *J. Alloys Compd.*, *475*(1–2), 233, 2009. With permission.)

The isothermal sections at 500°C and 527°C were computed by Zhang et al. (2009). The computed section at 527°C agrees with that determined experimentally (Zeng and Wang 2003; Raghavan 2008). Y$_3$Sb$_5$ dissolves up to 27 at% Al at constant yttrium content. The third component solubility in the other binary phases is negligible. No ternary compounds were found. The calculated isothermal section of this system at 500°C is shown in Figure 7.24 (Zhang et al. 2009; Raghavan 2010e). The triangulations of Al + α-YAl$_3$ + YSb, Al + AlSb + YSb, and AlSb + YSb + Sb agree with those found by Muravieva et al. (1971), who studied the system up to 33.3 at% Y.

7.16 ALUMINUM–LANTHANUM–ANTIMONY

The isothermal section of the Al–La–Sb ternary system at 500°C was constructed between 0 and 50 at% Sb (Figure 7.25) (Chen et al. 2010; Raghavan 2010b). There are no ternary phases and the solubility of the third component in the binary compounds is negligible. The samples were annealed at 500°C for 30 days and quenched in liquid N$_2$. The phase equilibria were studied with XRD, and SEM equipped with an EDX analyzer.

7.17 ALUMINUM–CERIUM–ANTIMONY

AlSb + CeSb + Al and Al + CeSb + Al$_4$Ce ternary equilibria were determined in the Al–Ce–Sb system (Zarechnyuk and Tsygulya 1966). This indicates that AlSb is in

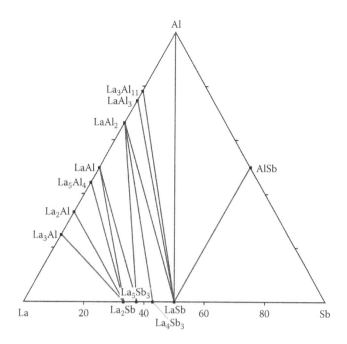

FIGURE 7.25 Isothermal section of the Al–La–Sb ternary system at 500°C. (From Chen, Y., et al., *J. Alloys Compd.*, 492(1–2), 208, 2010. With permission.)

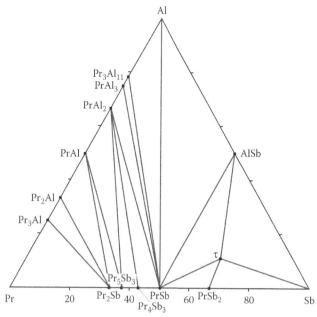

FIGURE 7.26 Isothermal section of the Al–Pr–Sb ternary system at 500°C. (From Zeng, L., et al., *J. Alloys Compd.*, 450(1–2), 252, 2008. With permission.)

thermodynamic equilibrium with CeSb and CeSb is in equilibrium with Al and Al$_4$Ce. The samples were annealed at 500°C for 150 h and investigated through XRD.

7.18 ALUMINUM–PRASEODYMIUM–ANTIMONY

The isothermal section of the Al–Pr–Sb ternary system at 500°C over the whole concentration region has been investigated by XRD and SEM (Figure 7.26) (Zeng et al. 2008). A ternary compound of an approximately **Al$_{11}$Pr$_{24}$Sb$_{65}$** (τ) composition has been found. The mutual solid solubilities of Al in Pr–Sb and Sb in Pr–Al compounds are negligible. The samples with more than 50 at% Sb were annealed at 600°C for 25 days, and the samples with less than 50 at% Sb were annealed at 730°C for 25 days. Subsequently, they were cooled to 500°C at a rate of 10°C·h^{-1} and kept at this temperature for 10 days and then quenched in liquid N$_2$.

7.19 ALUMINUM–NEODYMIUM–ANTIMONY

The isothermal section of the Al–Nd–Sb ternary system at 500°C has been determined with the use of XRD, and SEM with EDX (Wei et al. 2007). **~Al$_6$Nd$_{28}$Sb$_{66}$** (**~AlNd$_{4.67}$Sb$_{11}$**) ternary compound, which crystallizes in the orthorhombic structure with the lattice parameters $a = 1519.1 \pm 0.1$, $b = 1991.4 \pm 0.3$, and $c = 437.7 \pm 0.1$ pm, was found in this system.

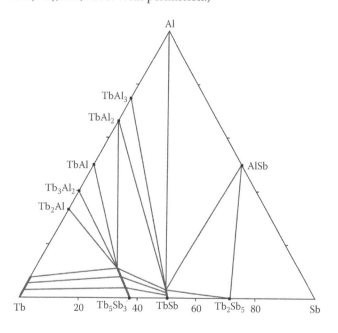

FIGURE 7.27 Isothermal section of the Al–Tb–Sb ternary system at 500°C. (From Feng Q., et al., *J. Phase Equilibria Diffus.*, 37(5), 564, 2016. With permission.)

7.20 ALUMINUM–TERBIUM–ANTIMONY

The isothermal section of the Al–Tb–Sb ternary system at 500°C has been investigated by means of XRD, and SEM with EDX and is shown in Figure 7.27 (Feng et al. 2016). No ternary compound was found. Tb$_4$Sb$_3$ is unstable at this temperature. The solubility of Al in Tb, Tb_5Sb_3, and TbSb was determined to be 2, 11, and 3 at%, respectively.

The compacted Sb-rich alloys with more than 50 at% Sb and Al-rich alloys with more than 70 at% Al were homogenized at 500°C for 960 h in a vacuum tube. The other samples were homogenized at 600°C for 600 h and then cooled at a rate of 2°C·h^{-1} to 500°C and kept at this temperature for 480 h. Finally, all the alloys were quenched in liquid nitrogen.

7.21 ALUMINUM–DYSPROSIUM–ANTIMONY

The isothermal section of the Al–Dy–Sb ternary system at 500°C is given in Figure 7.28 (Zeng et al. 2009; Raghavan 2010a). No ternary compound was found, and the solubility of the third component in the binary compounds was found to be less than 1 at%. It was found that α-Dy_4Sb_3 is not stable at 500°C. This section consists of 12 single-phase regions, 21 two-phase regions, and 10 three-phase regions. The alloys were annealing at 500°C for 480–960 h and then quenched in liquid N_2. The phase equilibria were studied with optical and scanning electron metallography and XRD.

7.22 ALUMINUM–YTTERBIUM–ANTIMONY

Two ternary compounds, **$Yb_5Al_2Sb_6$** and **$Yb_{11}AlSb_9$**, are formed in the Al–Yb–Sb system. $Yb_5Al_2Sb_6$ is stable up to

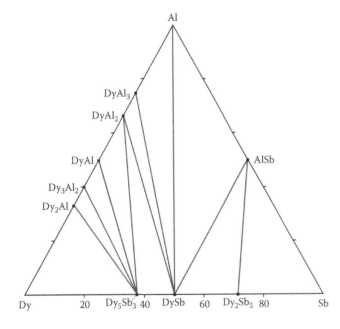

FIGURE 7.28 Isothermal section of the Al–Dy–Sb ternary system at 500°C. (From Zeng, L., et al., *J. Alloys Compd.*, *479*(1–2), 173, 2009. With permission.)

1100°C (Todorov et al. 2009) and crystallizes in the orthorhombic structure with the lattice parameters $a = 732.34 \pm 0.07$, $b = 2287.7 \pm 0.2$, and $c = 440.99 \pm 0.04$ pm and a calculated density of 7.47 g·cm^{-3} (Aydemir et al. 2015) ($a = 732.1 \pm 0.2$, $b = 2287.8 \pm 0.4$, and $c = 440.61 \pm 0.04$ pm [Fornasini and Manfrinetti 2009]; $a = 729.71 \pm 0.15$, $b = 2278.0 \pm 0.5$, and $c = 441.15 \pm 0.09$ pm, a calculated density of 7.471 g·cm^{-3}, and an energy gap of 0.3–0.4 eV [Todorov et al. 2009]).

Various methods were used to prepare $Yb_5Al_2Sb_6$. In a typical direct combination reaction, Yb (3.32 mM), Al (1.33 mM), Sb (4.00 mM), and Ge (0.333 mM) were mixed in a Nb ampoule that is sealed by arc welding under Ar (Todorov et al. 2009). The Nb ampoule was then placed in a fused silica tube and flame sealed under a vacuum. The assembly was heated at 950°C for 2 days and cooled to room temperature at a rate of 1°C·min^{-1}.

Arc melting under an argon atmosphere is another method used to produce this compound by using YbSb (2.5 mM), Sb (0.5 mM), Al (1 mM), and Ge (0.5 mM) (Todorov et al. 2009).

The title compound could be also obtained if stoichiometric amounts of Yb, Al, and Sb were sealed in Ta crucibles and melted in an induction furnace, raising the temperature up to about 1310°C, while shaking (Fornasini and Manfrinetti 2009). The alloys were then annealed at 900°C for 7 days. $Yb_5Al_2Sb_6$ is very brittle and air sensitive.

It was also synthesized by ball milling, followed directly by hot pressing (Aydemir et al. 2015). Stoichiometric amounts of small-cut Yb ingot, Sb shot, and Al shot were loaded under Ar into stainless steel vials with stainless steel balls. The contents were ball milled for 1 h. The resulting fine powder was hot pressed under Ar using a maximum pressure and temperature of 45 MPa and 550°C, respectively, for 1 h. The samples were cooled down to room temperature slowly under no load.

$Yb_{11}AlSb_9$ also crystallizes in the orthorhombic structure with the lattice parameters $a = 1176 \pm 2$, $b = 1239 \pm 2$ and $c = 1668 \pm 2$ pm (Magnavita et al. 2016). This compound behaves as a metal above circa 100 K and as a small gap semiconductor at a low temperature. Its single crystals were grown from the reaction of Yb, Al, and Sb with Sn as flux (molar ratio 11:9:9:70). The elements were mixed inside alumina crucibles, and then sealed in evacuated quartz ampoules. The ampoules were heated from room temperature to 1100°C in 8 h, soaked at 1100°C for 5 h, and then cooled at 2°C·h⁻¹ to 700°C. At that point, the ampoules were removed from the furnace and the excess flux was separated by centrifugation.

7.23 ALUMINUM–SILICON–ANTIMONY

The computed liquidus projection of the Al–Si–Sb ternary system is shown in Figure 7.29 (Hansen and Loper 2000; Raghavan 2010d). The univariant line starting at the eutectic maximum on the AlSb–Si (e_1) join descends to the ternary eutectic points on either side. At the Al end, the reaction E_2 occurs at 576°C: L ⟷ (Al) + AlSb + (Si). At the Sb end, E_1 occurs at 623°C: L ⟷ (Sb) + AlSb + (Si). Hansen and Loper (2000) used the thermodynamic descriptions of the binary system from literature with minor modifications. No ternary interaction parameters were introduced.

AlSb–Si. The computed quasi-binary section along the AlSb–Si join is given in Figure 7.30 (Hansen and Loper 2000; Raghavan 2010d). The computed eutectic is at 18.2 at% Si and 994°C, whereas the data of Glazov and Lyu Chzhen'-yuan' (1962) give the eutectic at 30 at% Si and 1014°C.

7.24 ALUMINUM–GERMANIUM–ANTIMONY

The solubility isotherms of Al and Sb in germanium at 600°C–900°C are shown in Figure 7.31 (Glazov

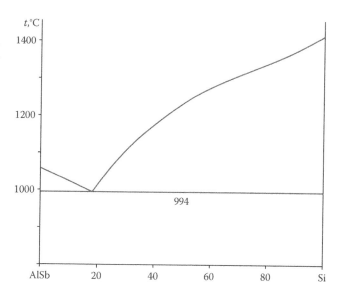

FIGURE 7.30 Calculated phase diagram of the AlSb–Si system. (From Hansen, S.C., and Loper Jr., C.R., *CALPHAD*, *24*(3), 339, 2000. With permission.)

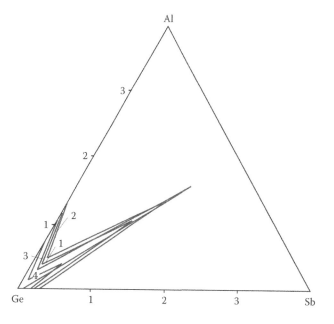

FIGURE 7.31 Solubility isotherms of Al and Sb in germanium at (1) 600°C, (2) 700°C, (3) 800°C, and (4) 900°C. (From Akopyan, R.A., et al., *Zhurn. Neorgan. Khimii*, *32*(5), 1219, 1987. With permission.)

et al. 1962; Akopyan et al. 1987). It was shown that Al addition increases the solubility of Sb, which reaches a maximum value on the AlSb–Ge section. The samples were annealed at 600°C, 700°C, 800°C, and 900°C for 1100, 900, 750, and 700 h, respectively, and investigated through metallography and measurement of microhardness.

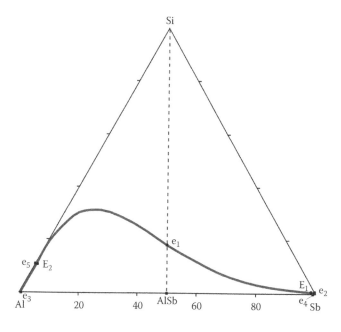

FIGURE 7.29 Calculated liquidus projection of the Al–Si–Sb ternary system. (From Hansen, S.C., and Loper Jr., C.R., *CALPHAD*, *24*(3), 339, 2000. With permission.)

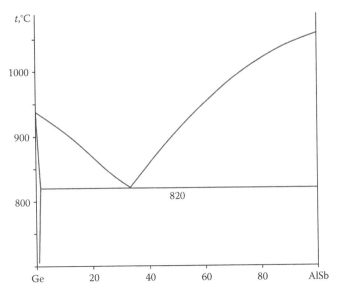

FIGURE 7.32 Phase diagram of the AlSb–Ge system. (From Glazov, V.M., and Lyu, Chzhen'-yuan', *Zhurn. Neorgan. Khimii*, 7(3), 582–589, 1962. With permission.)

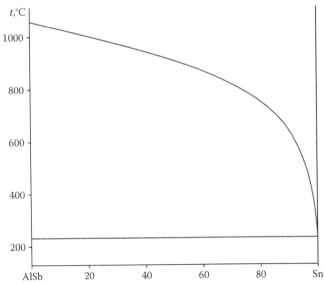

FIGURE 7.33 Phase diagram of the AlSb–Sn system. (From Gerdes, F., and Predel, B., *J. Less Common Metals*, 79(2), 281, 1981. With permission.)

AlSb–Ge. The phase diagram of this system is a eutectic type (Figure 7.32) (Glazov and Lyu Chzhen'-yuan' 1962). The eutectic crystallizes 820°C and contains 33 mol% AlSb. The solubility of AlSb in Ge is equal to 1.20, 1.80, 2.70, 3.00, and 0.60 mass% at 400°C, 500°C, 725°C, 820°C, and 900°C, respectively (Glazov et al. 1959, 1962). The samples were annealed at 600°C, 700°C, and 725°C for 300 h and investigated through metallography and measurement of microhardness.

As a result of ultrafast crystallization (10^7°C·s) of the AlSb–Ge melts, a two-phase structure was fixed according to the equilibrium phase diagram (Glazov et al. 1977a).

7.25 ALUMINUM–TIN–ANTIMONY

AlSb–Sn. The phase diagram of this system constructed using DTA is shown in Figure 7.33 (Gerdes and Predel 1981a). The calculated value of the Sn maximum solubility in AlSb at 902°C is equal to 0.014 at% (Arbenina et al. 1992).

Using a high-temperature calorimeter, the change $\delta(\Delta H)$ in the enthalpy on mixing the molten AlSb with Sn in the entire concentration range was determined (Gerdes and Predel 1981c). It was shown that the concentration dependence of the $\delta(\Delta H)$ values are positive over the whole concentration range, which is in close accordance with the regular solution behavior.

7.26 ALUMINUM–LEAD–ANTIMONY

AlSb–Pb. Along this quasi-binary section, the change enthalpy $\delta(\Delta H)$ on mixing liquid Pb with the molten AlSb was determined using a high-temperature calorimeter, and positive $\delta(\Delta H)$ values were found (Gerdes and Predel 1981b).

7.27 ALUMINUM–TITANIUM–ANTIMONY

The partial isothermal sections of the Al–Ti–Sb system at 1100°C and 1300°C are given in Figure 7.34 (Kimura et al. 1997; Raghavan 2005). **AlTi$_5$Sb$_2$** (τ) ternary compound was found to be stable up to 1500°C, with a homogeneity range of 11.0%–17.1 at% Al, 19.2%–25.5 at% Sb, and 63.0%–65.4 at% Ti. This compound forms tie-lines with Ti$_3$Sb, (β-Ti), (α-Ti), and TiAl (γ) at 1300°C and with Ti$_3$Sb, (β-Ti), (Ti$_3$Al) (α$_2$), and TiAl (γ) at 1100°C. It crystallizes in the tetragonal structure with the lattice parameters a = 1044.7–1048.4 and c = 523.9–527.5 pm (Kimura et al. 1997; Raghavan 2005) (a = 1068.9 ± 0.1 and c = 536.7 ± 0.1 pm [Kozlov and Pavlyuk 2003]; a = 1049.99 and c = 521.39 pm and enthalpy of formation −49.1 kJ·M^{-1} [calculated values] [Colinet and Tedenac 2015]).

The phase equilibria in this ternary system were studied by XRD, DTA, and EPMA (Kimura et al. 1997). The samples were annealed at 1100°C for 1 week or at 1300°C for 16 h. AlTi$_5$Sb$_2$ was prepared by arc melting of a mixture of Al, Ti, and Sb in a stoichiometric ratio

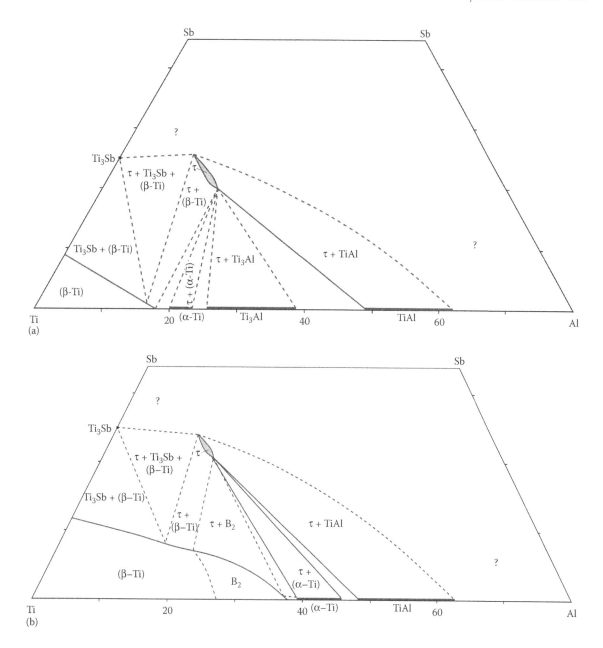

FIGURE 7.34 Partial isothermal sections of the Al–Ti–Sb system at (a) 1100°C and (b) 1300°C. (From Kimura, T., et al., *Nippon Kinzoku Gakkaishi*, 61(5), 385, 1997. With permission.)

in an Ar atmosphere, annealed at 400°C for 720 h, and quenched in cold water (Kozlov and Pavlyuk 2003).

7.28 ALUMINUM–BISMUTH–ANTIMONY

AlSb–Bi. The enthalpy change due to the mixing of molten AlSb with liquid Bi has been determined by using a high-temperature calorimeter (Gerdes and Predel 1979a). Melting of a quasi-binary system yields a maximum value of enthalpy change of +3200 J·mol⁻¹ at 1077°C. The calculated result is consistent within an order of magnitude with experimental data.

7.29 ALUMINUM–SULFUR–ANTIMONY

Miscibility gaps occupy a significant part of the Al–S–Sb ternary system (Strel'nikova and Mirgalovskaya 1965b).

AlSb–AlS. This system is a non-quasi-binary section of the Al–S–Sb ternary system (Strel'nikova and Mirgalovskaya 1965b). It crosses the fields of primary crystallization of AlSb, Al, and Al_2S_3 (AlS). The solubility of AlS in AlSb is negligible.

AlSb–S. This system is a non-quasi-binary section of the Al–S–Sb ternary system (Strel'nikova and Mirgalovskaya 1965b).

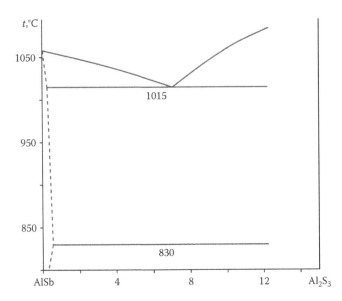

FIGURE 7.35 Part of the phase diagram of the AlSb–Al$_2$S$_3$ system. (From Strel'nikova, I.A., and Mirgalovskaya, M.S., *Izv. AN SSSR Neorgan. Mater.*, 1(1), 96, 1965. With permission.)

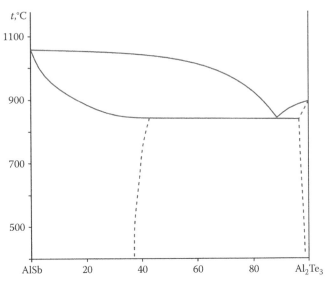

FIGURE 7.36 Phase diagram of the AlSb–Al$_2$Te$_3$ system. (From Mirgalovskaya, M.S., and Skudnova, E.V., *Zhurn. Neorgan. Khimii*, 4(5), 1113, 1959. With permission.)

AlSb–Al$_2$S$_3$. The phase diagram of the AlSb–Al$_2$As$_3$ system is a eutectic type with a monotectic at 7 mol% Al$_2$S$_3$ and 1015°C (Figure 7.35) (Strel'nikova and Mirgalovskaya 1965b). The eutectic crystallizes at 830°C. The solubility of Al$_2$S$_3$ in AlSb is negligible.

AlSb–Sb$_2$S$_3$. This system is a non-quasi-binary section of the Al–S–Sb ternary system (Strel'nikova and Mirgalovskaya 1965b). It crosses the fields of primary crystallization of AlSb, Sb, and Sb$_2$S$_3$. There is an immiscibility gap in this section.

The samples of the Al–S–Sb ternary system were annealed at 750°C for 6 weeks (Strel'nikova and Mirgalovskaya 1965b). The system was investigated by XRD, DTA, metallography, and measurement of microhardness.

7.30 ALUMINUM–TELLURIUM–ANTIMONY

AlSb–Al$_2$Te$_3$. The phase diagram of this system is shown in Figure 7.36 (Mirgalovskaya and Skudnova 1959). The solubility of Al$_2$Te$_3$ in AlSb at 800°C reaches 45 mass%. The system was investigated through DTA, XRD, metallography, and measurement of microhardness and density.

7.31 ALUMINUM–IODINE–ANTIMONY

Al$_2$Sb$_2$I$_{12}$ ternary compound, which melts at 150°C and crystallizes in the monoclinic structure with the lattice parameters $a = 1644.0 \pm 0.3$, $b = 1318.3 \pm 0.2$, and $c =$

728.9 ± 0.1 pm and β = 119.33 ± 0.01° and a calculated density of 4.39 g·cm^{-3}, is formed in the Al–I–Sb system (Pohl 1983). It was prepared by reaction of stoichiometric amounts of AlI$_3$ and SbI$_3$ in CS$_2$ at –20°C.

7.32 ALUMINUM–MANGANESE–ANTIMONY

Al$_x$Mn$_{1.1}$Sb$_{1-x}$ ($0 < x \leq 0.2$) solid solutions were prepared in the Al–Mn–Sb system (Budzyński et al. 2013). They crystallize in the hexagonal structure with the lattice parameters $a = 417.7$ and $c = 573.4$ pm for $x = 0.1$. To prepare these solid solutions, a powder mixture of the elements was sealed in a silica ampoule under vacuum of 0.1 Pa. In the first step, a homogeneous mixture of starting powders was slowly heated over a period of 24 h to fusion temperature (900°C–950°C). Next, the mixture was held at these temperatures for 4 h and cooled to 840°C–860°C. The mixture was annealed at this temperature for 24 h and then quenched in water.

7.33 ALUMINUM–IRON–ANTIMONY

Isothermal sections of the Al–Fe–Sb ternary system at 450°C, 680°C, and 800°C (Figure 7.37) have been determined experimentally by means of optical microscopy, SEM with EDX, and XRD (Hou et al. 2013, 2014). No ternary compound has been found in this system. Experimental results indicate that Sb cannot dissolve in FeAl$_2$, Fe$_2$Al$_5$, and FeAl$_3$, while the maximum solubility of Al in FeSb$_2$ is 1.3 at%. The maximum solubility of

Fe in AlSb at 450°C, 680°C, and 800°C reaches 0.5, 1.0, and 1.2 at%, respectively, and the maximum solubility of Al in FeSb at these temperatures is, respectively, about 3.2, 3.7, and 3.8 at%.

There are seven three-phase regions in the Al–Fe–Sb system at 450°C (Figure 7.37a) (Hou et al. 2013). AlSb can equilibrate with all phases except FeSb, while $FeSb_2$ can equilibrate with FeSb, FeAl, AlSb, and α-Sb. Seven three-phase regions have been at 680°C, and six three-phase regions exist in the 800°C isothermal section (Figure 7.37b and c) (Hou et al. 2014).

The samples were annealed at 450°C and 680°C for 30 days and at 800°C for 25 days (Hou et al. 2013,

2014). The treatment was completed with rapid water quenching.

$AlFe_4Sb_{12}$ metastable compound, which crystallizes in the cubic structure with the lattice parameter $a = 917.7 \pm 0.1$ pm, could be obtained in the Al–Fe–Sb ternary system (Sellinschegg et al. 1998). This compound forms at 177°C and decomposes at 494°C. Modulated elemental reactants were used to form amorphous ternary reaction intermediates of the desired compositions. The amorphous reaction intermediate crystallizes exothermically below 200°C, forming kinetically stable "filled" skutterudite. This metastable compound can only be prepared by controlling the reaction intermediates, avoiding the

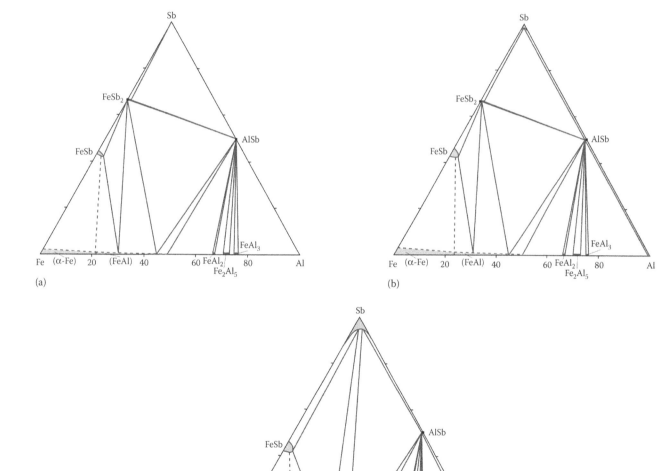

FIGURE 7.37 Isothermal sections of the Al–Fe–Sb ternary system at (a) 450°C, (b) 680°C, and (c) 800°C. (From Hou, T., and Zhu, Z., *J. Phase Equilibria Diffus.*, 34(3), 188, 2013; Hou, T., et al., *J. Alloys Compd.*, 586, 295, 2014. With permission.)

formation of more thermodynamically stable binary compounds.

7.34 ALUMINUM–NICKEL–ANTIMONY

Ni₃AlSb ternary compound, which melts at temperatures higher than 1150°C and crystallizes in the hexagonal structure with the lattice parameters $a = 403$ and $c = 513$ pm, is formed in the Al–Ni–Sb system (Jan and Chang 1991). It is quite likely that this phase merely represents a specific composition of extensive solid solutions in this ternary system. The solid solutions would extend from NiSb toward the nonexistent phase "Ni₃Al₂" and would terminate somewhere in the ternary region. To prepare Ni₃AlSb, AlSb and pieces of Ni were sealed in evacuated quartz ampoules, annealed at 700°C for 1 day, and gradually heated to 1200°C over a period of 2 h. They were subsequently cooled to 700°C and annealed at that temperature for at least 2 weeks. The samples were then quenched in ice water.

CHAPTER 8

Systems Based on GaN

8.1 GALLIUM–HYDROGEN–NITROGEN

$[H_2GaNH_2]_3$ and $[D_2GaND_2]_3$ ternary compounds are formed in this system. $[H_2GaNH_2]_3$ crystallizes in the monoclinic structure with the lattice parameters $a = 576.15 \pm 0.05$, $b = 850.79 \pm 0.07$, and $c = 808.48 \pm 0.06$ pm and $\beta = 110.843 \pm 0.002°$ at 106 ± 2 K and a calculated density of 2.361 g·cm^{-3} (Campbell et al. 1998). The second compound also crystallizes in the monoclinic structure with the lattice parameters $a = 578.93 \pm 0.04$, $b = 856.35 \pm 0.04$, and $c = 816.17 \pm 0.06$ pm and $\beta = 111.038 \pm 0.006°$ at 106 ± 2 K.

To prepare $[H_2GaNH_2]_3$, trimethylamine gallane (4 mM) was charged to a 100 mL Schlenk flask, which was cooled to –196°C, under vacuum, and approximately 1 mL (45 mM) of liquid NH_3 was condensed on top of the gallane (Campbell et al. 1998). The flask was then filled with N_2, and the mixture was slowly warmed to –78°C, and then ultimately to room temperature. Excess NH_3 was purged with a stream of N_2. A white solid of the title compound was isolated. A melting point was not observed up to 150°C. The same product could also be obtained by direct reaction of gaseous NH_3 over solid $(GaH_3)NMe_3$ at room temperature.

For obtaining $[D_2GaND_2]_3$, two equivalents of ND_3 were condensed onto $(GaD_3)NMe_3$ (22.8 mM) cooled to liquid nitrogen temperature. The mixture was then slowly warmed to –78°C and eventually room temperature over 2 h. The reaction mixture bubbled very slowly at circa –25°C and turned cloudy at circa –10°C. A white solid of $[D_2GaND_2]_3$ was isolated.

Air and moisture were excluded from all reactions through the use of standard Schlenk techniques or a N_2-filled glove box (Campbell et al. 1998).

8.2 GALLIUM–LITHIUM–NITROGEN

An isothermal section of the Ga–Li–N ternary system at 800°C under 0.1 MPa (Figure 8.1) was constructed and assessed by using the CALPHAD method (Wang et al. 2004). At 800°C, there are a liquid phase region, a two-phase region with L and GaN, and a L + GaN + Li$_3$GaN$_2$ three-phase region. The isothermal section suggests that the richer the Li in the molten solution, the higher the concentration of nitrogen in the solution.

GaN–Li$_3$N. Li$_3$GaN$_2$ ternary compound is formed in this system. It crystallizes in the cubic structure with the lattice parameter $a = 960.5$ pm ($a = 959.4 \pm 0.3$ pm and

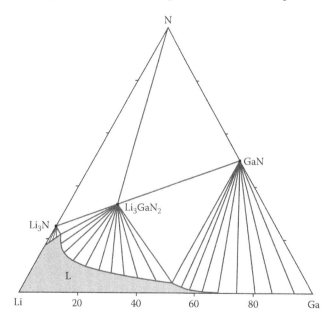

FIGURE 8.1 Calculated isothermal section of the Ga–Li–N ternary system at 800°C under 0.1 MPa. (From Wang, W.J., et al., *J. Cryst. Growth*, 264(1–3), 13, 2004. With permission.)

an experimental density of 3.35 g·cm⁻³ [Juza and Hund 1946, 1948]) and an energy gap of 4.15 eV (Kuriyama et al. 2007). The title compound was prepared by interaction of Li_3Ga with N_2 at 600°C (Juza and Hund 1946, 1948). It hydrolyzes easily.

8.3 GALLIUM–SODIUM–NITROGEN

The calculated isothermal sections of the Ga–Na–N ternary system under the pressure of 0.1 MPa at 600°C and 750°C are shown in Figure 8.2 (Wang et al. 2007). The ternary liquid immiscibility occurs in the system. From the isothermal sections, there are a liquid phase, three two-phase fields of GaN + L, and two three-phase fields of $GaN + L_1 + L_2$. The nitrogen concentration to dissolve in the liquid phase reaches a maximum of about 2 mol% at 750°C.

There is little difference between isothermal sections at 750°C under 0.1 MPa and under 5 MPa. However, it should be pointed out that the pressure has a large effect on the solubility of gaseous N_2 in Ga–Na melts (Wang et al. 2007).

8.4 GALLIUM–COPPER–NITROGEN

The subsolidus phase relations in the Ga–Cu–N ternary system were studied by x-ray diffraction (XRD) (Figure 8.3) (Zhang et al. 2007a). No ternary compound was found. There are a total of eight triangles in the diagram. Two of them are two-phase regions. The other six triangles are three-phase regions.

The samples were annealed at 700°C for 5 days, at 350°C for 11 days, and at 200°C for 40 days.

8.5 GALLIUM–SILVER–NITROGEN

The phase relations in the Ga–Ag–N ternary system at room temperature were determined by XRD (Figure 8.4) (Zhang et al. 2007b). No ternary compound was observed, and no solubility of Ag in GaN was found. Ag has not reacted with GaN. There are a total of seven triangles in the diagram. Two of them are two-phase regions: a region containing (α-Ag) and GaN and a region containing (Ag_2Ga) and GaN. The other five triangles are three-phase regions.

The samples were annealed at 800°C for 5 days and at 500°C for 7 days (Zhang et al. 2007b).

8.6 GALLIUM–MAGNESIUM–NITROGEN

$GaN–Mg_3N_2$. Mg_3GaN_3 ternary compound is formed in this system. It is a semiconductor with a direct band

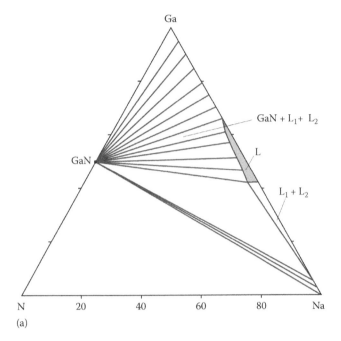

(a)

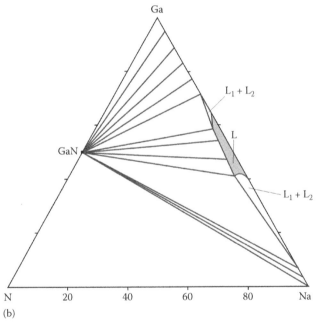

(b)

FIGURE 8.2 Calculated isothermal sections of the Ga–Na–N ternary system at (a) 600°C and (b) 750°C under 0.1 MPa. (From Wang, J., et al., *J. Cryst. Growth*, *307*(1), 59, 2007. With permission.)

gap of 3.0 eV (Hintze et al. 2013). The title compound in the form of green powder was obtained from the mixture containing Mg_3N_2 and GaN (molar ratio 1:1), which was heated at 930°C for 48 h in the quartz tubes sealed under vacuum (Verdier et al. 1970). It is moisture sensitive and loses nitrogen in the form of NH_3.

Light yellow single crystals of Mg_3GaN_3 were synthesized from NaN_3, Mg, Ga, and sodium flux in welded shut Ta ampoules, which were placed into quartz tubing under vacuum and then heated in a furnace up to 760°C (Hintze et al. 2013). The sodium flux was removed from the reaction products by sublimation at 320°C under vacuum.

8.7 GALLIUM–CALCIUM–NITROGEN

The isothermal sections of the Ga–Ca–N ternary system at 800°C and 900°C (Figure 8.5) were predicted from the corresponding binary systems by the CALPHAD method (Wang et al. 2008). It was assumed that **CaGaN** was the only stable ternary nitride. It crystallizes in the tetragonal structure with the lattice parameters $a =$

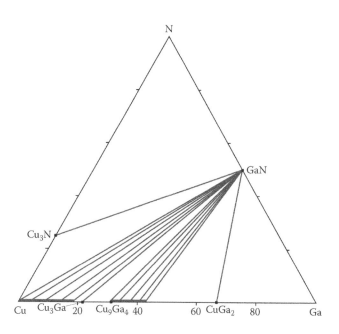

FIGURE 8.3 Phase relations in the Ga–Cu–N ternary system at room temperature. (From Zhang, Y., et al., *J. Alloys Compd.*, 438(1–2), 158, 2007. With permission.)

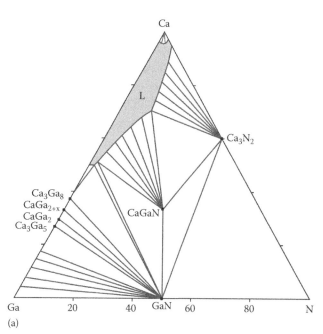

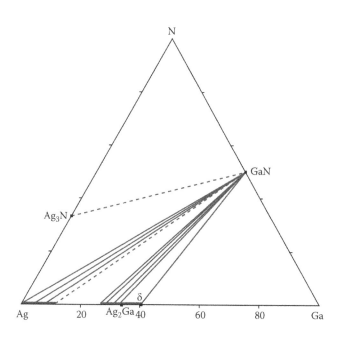

FIGURE 8.4 Phase relations in the Ga–Ag–N ternary system at room temperature. (From Zhang, Y., et al., *J. Alloys Compd.*, 429(1–2), 184, 2007. With permission.)

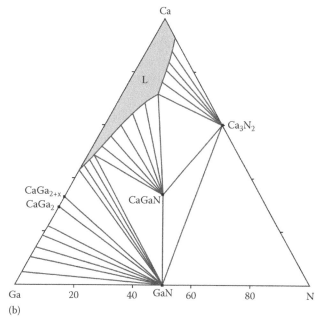

FIGURE 8.5 Calculated isothermal sections of the Ga–Ca–N ternary system at (a) 800°C and (b) 900°C under 0.2 MPa of nitrogen. (From Wang, G., et al., *Scr. Mater.*, 58(4), 319, 2008. With permission.)

357.0 ± 0.1 and $c = 755.8 ± 0.2$ pm and experimental and calculated densities of 4.29 and 4.268 g·cm^{-3}, respectively (Verdier et al. 1973, 1974). Its single crystals were prepared by the interaction of Ca_3N_2 and GaN or Ca_3N_2 and Ga in a N_2 atmosphere at 800°C–1000°C. This compound decomposes in vacuum forming $CaGa_{1.92}$ and $CaGa_{2.07}$ and is stable in a N_2 atmosphere up to 1000°C (Verdier et al. 1973).

According to the literature data, three more ternary compounds, **$Ca_5Ga_2N_4$**, **$Ca_7Ga_{1.33 ± 0.01}N_4$**, and **$Ca_{18.5}Ga_8N_7$**, are formed in this ternary system (Cordier 1988; Höhn et al. 2004, 2006; Kirchner et al. 2005). $Ca_5Ga_2N_4$ crystallizes in the orthorhombic structure with the lattice parameters $a = 487.3 ± 0.2$, $b = 1110.5 ± 0.3$, and $c = 1421.7 ± 0.4$ pm and a calculated density of 3.42 g·cm^{-3} (Cordier 1988). It was prepared from Ca and Ga in nitrogen medium at 1150°C.

$Ca_7Ga_{1.33 ± 0.01}N_4$ also crystallizes in the orthorhombic structure with the lattice parameters $a = 1147.4 ± 0.1$, $b = 1208.0 ± 0.1$, and $c = 363.6 ± 0.1$ pm and a calculated density of 2.84 g·cm^{-3} (Höhn et al. 2006). Single-phase powder samples of this compound were synthesized by the reaction of appropriate mixtures of Ca_3N_2 and Ga in a Ta crucible at 850°C for 48 h under an Ar atmosphere. Repeated reaction with intermediate regrinding and reforming of pellets was employed. Small amounts of needlelike single crystals were prepared by using a small excess of Ca in the reaction mixture. The title compound readily hydrolyzes when exposure to air.

GaN crystals up to 1.5 mm were grown from the Ga–Ca–N system at 900°C under 0.2 MPa of nitrogen pressure (Wang et al. 2008).

GaN–Ca_3N_2. $Ca_3Ga_2N_4$ and Ca_6GaN_5 ternary compounds are formed in this system. $Ca_3Ga_2N_4$ exists in two polymorphic modifications. α-$Ca_3Ga_2N_4$ crystallizes in the monoclinic structure with the lattice parameters $a = 1069.01 ± 0.11$, $b = 836.55 ± 0.07$, and $c = 557.01 ± 0.04$ pm and $β = 91.194 ± 0.006°$ and a calculated density of 4.211 ± 0.001 g·cm^{-3} (Clarke and DiSalvo 1997). The starting materials for obtaining α-$Ca_3Ga_2N_4$ were NaN_3, Na, Ca, and Ga. The mixture of NaN_3 (80 mg), Na (50 mg), Ca (10 mg), and Ga (15 mg) (Na/Ca/Ga/N molar ratio of ca. 15:1.2:1:17) was loaded in the Nb tube and sealed under 0.1 MPa of clean Ar in an arc furnace, taking care that the NaN_3 was not allowed to reach its decomposition. The Nb tube was then sealed inside a quartz tube under vacuum at elevated temperatures and placed upright in a muffle furnace. The temperature was

raised to 760°C over 15 h, maintained at that temperature for 24 h, and then lowered linearly to 100°C over 200 h, whereupon the furnace was turned off. The sample was washed anaerobically with anhydrous liquid NH_3 to remove Na and any excess Ca, and the sample was then dried for several hours under vacuum. Colorless crystals of α-$Ca_3Ga_2N_4$ were obtained. All materials were handled in a dry box in which the Ar atmosphere was constantly circulated.

β-$Ca_3Ga_2N_4$ crystallizes in the tetragonal structure with the lattice parameters $a = 1121.0 ± 0.1$ and $c = 1591.4 ± 0.2$ and a calculated density of 4.195 ± 0.001 g·cm^{-3} (Clarke and DiSalvo 1998). To obtain almost colorless transparent crystals of this modification, the following reactants were placed in the Nb tube: NaN_3, Na, Ca, Ge, and Ga, so that the molar ratio of Na/Ca/Ge/Ga/N was about 20:4:1:1:19. The next procedure was analogous to that for obtaining α-$Ca_3Ga_2N_4$.

Ca_6GaN_5 crystallizes in the hexagonal structure with the lattice parameters $a = 627.7 ± 0.3$ and $c = 1219.8 ± 0.3$ pm (Cordier et al. 1990). It was prepared by the reaction of Ca and Ga (molar ratio 1:1) with nitrogen at 850°C.

8.8 GALLIUM–STRONTIUM–NITROGEN

Sr_6GaN_6 and **Sr_6Ga_5N** ternary compounds are formed in this system. Sr_6GaN_6 crystallizes in the hexagonal structure with the lattice parameters $a = 666.67 ± 0.06$ and $c = 1299.99 ± 0.17$ and a calculated density of 4.417 g·cm^{-3} at 173 ± 2 K (Park et al. 2003a). Sr_6GaN_6 was synthesized in containers made out of Nb tubing. Under argon in a dry box, the Nb container was loaded with NaN_3, Na, Ga, Sr, and Li (Na/Ga/Sr/Li molar ratio of 18:1:4:1). The sealed Nb container was put into a quartz tube, which was then sealed under vacuum and placed in a muffle furnace. The temperature was raised to 760°C over 15 h, maintained at that temperature for 48 h, and lowered linearly to 200°C over 200 h. Thereafter, the furnace was turned off. Sodium was separated from the product by sublimation at 300°C under dynamic vacuum.

Sr_6Ga_5N crystallizes in the trigonal structure with the lattice parameters $a = 758.0 ± 0.3$ and $c = 4041.3 ± 0.1$ and a calculated density of 4.40 g·cm^{-3} (Cordier et al. 1995a, 1995b). To obtain it, Sr and Ga (molar ratio 8:1) were heated in a sintered corundum crucible under an Ar atmosphere to 1100°C, cooled to room temperature, ground in an inert gas atmosphere, and finally allowed to react with N_2 at 1050°C in a corundum crucible (stationary N_2 atmosphere of 0.1 MPa) and cooled to room temperature within 10 h.

GaN–Sr_3N_2. Sr_3GaN_3, $Sr_3Ga_2N_4$, and $Sr_3Ga_3N_5$ ternary compounds are formed in this system. Sr_3GaN_3 crystallizes in the hexagonal structure with the lattice parameters a = 758.4 ± 0.2 and c = 541.0 ± 0.3 and a calculated density of 4.616 g·cm⁻³ at 235 ± 2 K (Park et al. 2003a). It was obtained simultaneously with Sr_6GaN_6.

$Sr_3Ga_2N_4$ crystallizes in the orthorhombic structure with the lattice parameters a = 595.92 ± 0.06, b = 1027.53 ± 0.08, and c = 955.95 ± 0.09 pm and a calculated density of 5.204 ± 0.001 g·cm⁻³, and $Sr_3Ga_3N_5$ crystallizes in the triclinic structure with the lattice parameters a = 593.58 ± 0.06, b = 723.83 ± 0.08, and c = 868.53 ± 0.12 pm and α = 108.332 ± 0.010°, β = 103.383 ± 0.009°, and γ = 95.326 ± 0.008°, a calculated density of 5.322 ± 0.001 g·cm⁻³ (Clarke and DiSalvo 1997), and an energy gap of 1.53 eV (Hintze et al. 2012). Both compounds may be prepared from the same reaction mixture, and one can adjust their ratio by appropriate adjustment of the Sr/Ga ratio (Clarke and DiSalvo 1997). However, it has not yet proved possible to synthesize either material exclusive of the other. In a typical reaction that produced both of these phases in approximately equal abundance, the following amounts of reactants were placed in the Nb tube: NaN_3 (81 mg), Na (54 mg), Sr (20 mg), and Ga (11 mg), so that the molar ratio Na/Sr/Ga/N was about 20:1.4:1:22. The next procedure was analogous to that for obtaining α-$Ca_3Ga_2N_4$. Yellow crystals of $Sr_3Ga_2N_4$ and orange yellow crystals of $Sr_3Ga_3N_5$ were obtained.

8.9 GALLIUM–BARIUM–NITROGEN

Three ternary compounds were obtained in this system. **$Ba_3Ga_2N_4$** crystallizes in the orthorhombic structure with the lattice parameters a = 620.10 ± 0.12, b = 1051.1 ± 0.2, and c = 1007.0 ± 0.2 pm and a calculated density of 6.148 g·cm⁻³ (Yamane and DiSalvo 1996). To prepare this compound, Ba (1.0 mM), Ga (0.5 mM), NaN_3 (1.2 mM), and Na (2.4 mM) were sealed in a Nb tube using an arc furnace under 0.1 MPa of Ar. The welded Nb tube was then sealed in an evacuated quartz tube. The starting materials were heated to 750°C over 7.5 h, held for 1 h at that temperature, and then cooled at a rate of 6.4°C·h⁻¹. The sample in the Nb tube was washed with liquid NH_3 to remove the Na flux. Transparent yellow lamellar crystals and black granular single crystals of the title compound were obtained from the residual product.

All manipulations were carried out in an Ar-filled glove box.

$Ba_3Ga_3N_5$ crystallizes in the monoclinic structure with the lattice parameters a = 1680.1 ± 0.3, b = 833.01 ± 0.02, and c = 1162.3 ± 0.2 pm and β = 109.92 ± 0.03°, a calculated density of 6.004 g·cm⁻³, and an energy gap of 1.46 eV (Hintze et al. 2012). Synthesis of the title compound was carried out in a Ta ampule. All manipulations were done under an Ar atmosphere in a recirculated glove box. Single crystals were obtained from a reaction of NaN_3 (0.312 mM), Ba (0.060 mM), Sr (0.005 mM), Mg (0.062 mM), and Ga (0.252 mM) in Na flux (1.992 mM). The filled Ta ampule was weld shut under an Ar atmosphere and placed into quartz tubing. The reaction mixture was then heated in a furnace (50°C·h⁻¹) to 760°C, maintained at that temperature for 48 h, and then cooled to 200°C at a rate of 3.4°C·h⁻¹. Subsequently, the furnace was turned off. Sodium was separated from the product by sublimation at 320°C under vacuum (0.1 Pa) for 18 h.

Ba_6Ga_5N crystallizes in the trigonal structure with the lattice parameters a = 790.5 ± 0.3 and c = 4196.5 ± 0.7 and a calculated density of 5.21 g·cm⁻³ (Cordier et al. 1995a, 1995b). To obtain it, Ba and Ga (molar ratio 5:1) were heated to 1000°C in a corundum crucible, followed by reaction of the powdered product with N_2 at 1100°C. After a reaction period of 30 min, the melt was quickly cooled by removing the reaction tube from the oven.

8.10 GALLIUM–INDIUM–NITROGEN

GaN–InN. Thermodynamic calculations based on a regular solution model and experimental results have predicted an unstable region of mixing to occur in this system (Ho and Stringfellow 1996b; Saito and Arakawa 1999; Takayama et al. 2000, 2001; Caetano et al. 2006). The interaction parameter (32.7 kJ·mol⁻¹) that was used in the calculations for the wurtzite structure has been analytically obtained by the valence force field model, modified for wurtzite structures (Takayama et al. 2000, 2001). According to this interaction parameter, the value of the critical temperature was 1697°C (1417°C and critical composition of 39 mol% InN [Saito and Arakawa 1999]; 1566°C [Caetano et al. 2006]). At 800°C, $Ga_{1-x}In_xN$ solid solution is immiscible in the range of 0.04 ≤ x ≤ 0.88 (Saito and Arakawa 1999). The critical temperature for cubic $Al_xIn_{1-x}N$ alloys was calculated to be 1022°C (Caetano et al. 2006) (1130°C [Ferhat and Bechstedt 2002]; approximately 1020°C

[Scolfaro 2002]; 1250°C [Ho and Stringfellow 1996a, 1996b]). A small deviation from Vegard's law was found for the lattice parameter variation in cubic alloys (Ferhat and Bechstedt 2002). At 800°C, the solubility of InN in GaN was calculated to be less than 6 mol% (Figure 8.6) (Ho and Stringfellow 1996a, 1996b).

The phase separation in thick (~2 mkm) $Ga_{1-x}In_xN$ alloys (x = 0.2–0.4) grown at 570°C–750°C on AlN/Si(111), α-Al_2O_3(0001), and GaN/α-Al_2O_3(0001) substrates has been studied by Yamamoto et al. (2014, 2015). It was shown that the phase separation occurs when the $Ga_{1-x}In_xN$ film exceeds a critical value. The critical thickness for phase separation is markedly increased with decreasing temperature. It is around 0.2 mkm for a film grown at 750°C, while it is more than 1 mkm for that grown at 570°C. When the thickness of the epitaxial film grown at 650°C exceeds ~1 mkm, GaN-rich solid solutions (hexagonal structure) and metallic In appear. The InN composition ($x \approx 0.03$) in the obtained films is in agreement with the solubility of In in GaN at 650°C. The metallic In contains a small amount (~0.03 at%) of Ga. These results clearly show that these epitaxial films are phase separated into GaN-rich and InN-rich solid solutions. The latter was changed into metallic In–Ga owing to its thermal instability at 650°C. No substrate dependencies were found in the critical thickness.

In $Ga_xIn_{1-x}N$ epitaxial films grown by metal-organic vapor phase epitaxy (MOVPE), phase separation was experimentally found to be at an In composition of more than 20 at% (Matsuoka 1998). Phase separation in thick $Ga_xIn_{1-x}N$ films grown by metal-organic chemical vapor deposition (MOCVD) from 690°C to 780°C was also reported by El-Masry et al. (1998). A single phase was obtained for the as-grown films with <28 at% In. However, it is necessary to note that vapor phase growth techniques are nonequilibrium and can allow the growth of alloys with compositions exceeding the solubility limits.

First-principles calculations by means of the full-potential augmented plane-wave method using the local density approximation were carried out for the structural and electronic properties of the $Al_xGa_{1-x}N$ alloys in the wurtzite structure (Dridi et al. 2003). The lattice constant c was found to change linearly, while the lattice parameter a exhibited an upward bowing. The calculated band gap variation for these alloys exhibits a downward bowing of 1.7 eV.

Thermodynamic, structural, and electronic properties of wurtzite $Ga_{1-x}In_xN$ alloys are studied by combining first-principles total energy calculations with the generalized quasi-chemical approach, and compared with previous results for the zinc blende structure (Caetano et al. 2006). It was observed that the results for the wurtzite structure are not significantly different from the ones obtained for the zinc blende structure. The calculated phase diagram of the alloy shows a broad and asymmetric miscibility gap, as in the zinc blende case, with a similar range for the growth temperatures, although with a higher critical temperature. A value of 1.44 eV for the gap bowing parameter was found.

Band gap measurements have been performed on strained $Ga_{1-x}In_xN$ epilayers with x = 0.12 (McCluskey et al. 1998, 2003). According to the first-principles calculations for the band gap as a function of alloy composition, the bowing is strongly composition dependent. At x = 0.125, the calculated bowing parameter is 3.5 eV, in good agreement with the experimental value (3.8 eV for x = 0.1).

By the use of the electron plasma technique, it has been found that the thin films of $Ga_{1-x}In_xN$ solid solutions can be synthesized over the entire composition region (Osamura et al. 1975). The composition dependence of the energy gap was found to deviate downward from linearity. From the result of the annealing treatment for the

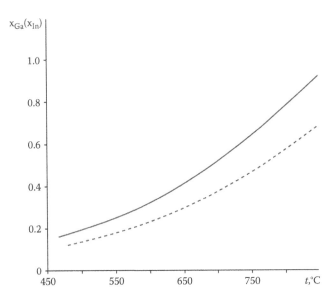

FIGURE 8.6 Calculated solubility of GaN in InN (solid line) and InN in GaN (dashed line). (From Ho, I.H., and Stringfellow, G.B., *Mater. Res. Soc. Symp. Proc.*, 449, 871, 1996. With permission.)

solid solution alloys and the theoretical consideration, it was pointed out that this quasi-binary system has a solid-phase miscibility gap in chemical equilibrium. $Ga_{1-x}In_xN$ layers were also grown by MOCVD on GaN or sapphire substrates (McCluskey et al. 1998). These solid solutions could be also synthesized by pyrolysis of molecular precursors $[Ga(NMe_2)_3]_2$ and $[In(NMe_2)_3]_2$ in a NH_3 atmosphere at 630°C (Kinski et al. 2005a). The obtained solid solutions crystallizing in the wurtzite-type structure contain 10–43 mol% GaN and follow Vegard's law. Cubic $Ga_{1-x}In_xN$ solid solutions were grown on GaN/MgO substrates in a plasma-assisted molecular beam epitaxy (MBE) system (Compeán García et al. 2015). In the case of low and high In concentrations, it is easy to obtain epitaxial layers with cubic structure (sphalerite type), but at around $x = 0.5$ the obtained layers showed hexagonal inclusions. The lattice parameter changes linearly with composition.

8.11 GALLIUM–THALLIUM–NITROGEN

Ab initio self-consistent calculation based on the full-potential linear augmented plane-wave method has been used in order to study the $Ga_{1-x}Tl_xN$ alloys (Saidi-Houat et al. 2009). Vegard's law cannot be applied to the variation of the lattice constant. The optical band gap bowing was found to be strong and composition dependent.

8.12 GALLIUM–SCANDIUM–NITROGEN

No ternary compound is formed in the Ga–Sc–N system (Mohney and Lin 1996). In this system, GaN is in thermodynamic equilibrium with ScN.

8.13 GALLIUM–YTTRIUM–NITROGEN

No ternary compound is formed in the Ga–Y–N system (Mohney and Lin 1996). In this system, GaN is in thermodynamic equilibrium with YN.

8.14 GALLIUM–LANTHANUM–NITROGEN

The calculated isothermal section of the Ga–La–N ternary system at room temperature under 0.1 MPa is shown in Figure 8.7 (Mohney and Lin 1996). This section is simplified because the ternary phases and the ranges of homogeneity of the binary phases have been neglected. In the Ga–La–N ternary system, a tie-line exists between GaN and LaN.

GaN–LaN. La_2GaN_3 ternary compound, which crystallizes in the monoclinic structure with the lattice parameters $a = 567.09 \pm 0.05$, $b = 1094.5 \pm 0.1$, and $c = 1198.6 \pm 0.1$ pm and $\beta = 93.591 \pm 0.005°$ at 167 ± 2 K

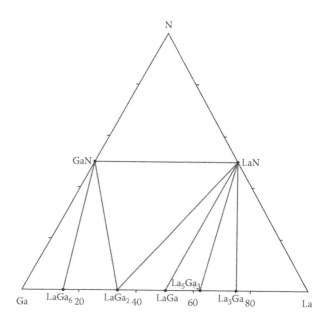

FIGURE 8.7 Calculated isothermal section of the Ga–La–N ternary system at room temperature under 0.1 MPa. (From Mohney, S.E., and Lin, X., *J. Electron. Mater.*, 25(5), 811, 1996. With permission.)

and a calculated density of 6.970 g·cm⁻³, is formed in this system (Cheviré and DiSalvo 2008). Na, Ga, La, and NaN_3 were placed into a Nb tube in the atomic ratio of $Na/Ga/La/N_2 = 6:4:1:3$. The Nb container was sealed under Ar in an arc furnace and then itself sealed under vacuum in a fused silica tube. The silica tube was then placed into a muffle furnace, heated up to 900°C in 15 h, and held at that temperature for 36 h. Then the furnace was allowed to cool to room temperature over 100 h. Following this heating sequence, the Nb tube was opened and unreacted sodium was removed by evaporation from the products by heating the Nb tube to 350°C under a pressure of circa 0.1 Pa for 8 h. Due to the air sensitivity of the reagents, all manipulations were carried out in an Ar-filled glove box.

8.15 GALLIUM–CERIUM–NITROGEN

According to the data of Beznosikov (2003), **Ce_3GaN** compound, which crystallizes in the cubic structure of perovskite type, could form in the Ga–Ce–N ternary system.

8.16 GALLIUM–PRASEODYMIUM–NITROGEN

According to the data of Beznosikov (2003), **Pr_3GaN** compound, which crystallizes in the cubic structure of perovskite type could, form in the Ga–Pr–N ternary system.

8.17 GALLIUM–NEODYMIUM–NITROGEN

Nd_3AlN ternary compound, which crystallizes in the cubic structure with the lattice parameter $a = 506.3$ pm, is formed in the Ga–Nd–N system (Haschke et al. 1967).

8.18 GALLIUM–SAMARIUM–NITROGEN

According to the data of Beznosikov (2003), Sm_3GaN compound, which crystallizes in the cubic structure of perovskite type, could form in the Ga–Sm–N ternary system.

8.19 GALLIUM–CARBON–NITROGEN

$Ga(CN)_3$ ternary compound, which crystallizes in the cubic structure with the lattice parameter $a = 529.5 \pm 0.2$ pm, is formed in the Ga–C–N system (Kouvetakis et al. 1996; Brousseau et al. 1997; Williams et al. 1997).

To obtain this compound, $GaCl_3$ (11 mM) was placed into a Schlenk flask with a magnetic stir bar, and Me_3SiCN (6 mL, 45 mM) was added dropwise (Brousseau et al. 1997). The mixture was stirred at room temperature for 6 h, and then at 60°C for another 60 h, during which time the mixture gradually darkened to deep brown. The temperature was then raised to 75°C for 4 h to complete the reaction. The reaction flask was pumped to remove Me_3SiCl and excess Me_3SiCN. The light brown solid product was washed with dry hexane and then pumped for several hours. The resulting product was a white solid $Ga(CN)_3$.

In a similar manner, a suspension of GaN_3Cl_2 (18 mM) in 10 mL of hexane was treated dropwise with $SiMe_3CN$ (7.4 mL, 55 mM) (Kouvetakis et al. 1996; Brousseau et al. 1997). Initially, the mixture becomes clear and the GaN_3Cl_2 completely dissolves. The solution is then stirred for 12 h, over which time a colorless precipitate is formed. The solid is isolated by filtration and washed several times with hexane to yield the title compound. Reactions were performed under prepurified N_2 with standard Schlenk and dry box techniques.

8.20 GALLIUM–TITANIUM–NITROGEN

Ti_2GaN ternary compound is formed in the Ga–Ti–N system. It crystallizes in the hexagonal structure with the lattice parameters $a = 300.3 \pm 0.2$ and $c = 1333 \pm 2$ pm (Manoun et al. 2010) ($a = 300.4$ and $c = 1330$ pm and a calculated density of 5.73 g·cm^{-3} [Jeitschko et al. 1964a]; $a = 296.1$ and $c = 1302.1$ pm [local density approximation] and $a = 301.8$ and $c = 1331.8$ pm [generalized

gradient approximation] [Bouhemadou 2009]; $a = 300$ and $c = 1330$ pm and a calculated density of 5.75 g·cm^{-3} [Barsoum 2000]; $a = 301$ and $c = 1332$ pm [Ivanovskii et al. 2000]; $a = 301.8$ and $c = 1331.8$ pm [Yang et al. 2013]). The calculation of the lattice parameters shows that the c axis is always stiffer than the a axis (Yang et al. 2013). The elastic constant investigations demonstrated that Ti_2GaN is stable over a wide pressure range of 0–1000 GPa, with the only exception of 350–600 GPa, owing to the elastic softening. Up to a pressure of about 50 GPa, no phase transformations were observed (Manoun et al. 2010).

The bulk modulus of Ti_2GaN was calculated to be 189 $\pm$ 4 GPa and its pressure derivative 3.5 $\pm$ 0.3 (Manoun et al. 2010) ($B_0 = 157$ GPa and $B' = 4.35$ in the pressure range of 0–50 GPa [Yang et al. 2013]; $B_0 = 182$ GPa and $B = 3.88$ [local density approximation] and $B_0 = 157$ GPa and $B' = 3.92$ [generalized gradient approximation] [Bouhemadou 2009]).

The starting powders for the fabrication of Ti_2GaN were Ti and GaN (Jeitschko et al. 1964a; Manoun et al. 2010). The powders were ball milled for 2 h, poured into a graphite mold and hot pressed. The hot press was heated from room temperature to 675°C at 500°C·h^{-1} and held at that temperature for 1.5 h, after which the temperature was again ramped at the same heating rate to 1150°C and held at this temperature for 6 h, before furnace cooling. A load that corresponded to a stress of 45 MPa was applied when the temperature reached 675°C and held at that temperature for the duration of the run.

According to the data of Beznosikov (2003), Ti_3GaN compound, which crystallizes in the cubic structure of perovskite type, could form in this ternary system.

8.21 GALLIUM–ZIRCONIUM–NITROGEN

The calculated isothermal section of the Ga–Zr–N ternary system at room temperature under 0.1 MPa is shown in Figure 8.8 (Mohney and Lin 1996). This section is simplified because the ternary phases and the ranges of homogeneity of the binary phases have been neglected. In this system, GaN is in thermodynamic equilibrium with ZrN.

8.22 GALLIUM–HAFNIUM–NITROGEN

No ternary compound is formed in the Ga–Hf–N system (Mohney and Lin 1996). In this system, GaN is in thermodynamic equilibrium with HfN.

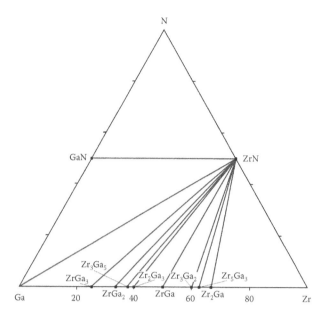

FIGURE 8.8 Calculated isothermal section of the Ga–Zr–N ternary system at room temperature under 0.1 MPa. (From Mohney, S.E., and Lin, X., *J. Electron. Mater.*, 25(5), 811, 1996. With permission.)

8.23 GALLIUM–PHOSPHOR–NITROGEN

The calculated temperature dependence of nitrogen solubility in AlP is shown in Figure 8.9 (Ho and Stringfellow 1996a; Stringfellow 1972b). The calculations indicate that the solubility of nitrogen increases significantly with increasing temperature. At 1470°C, the nitrogen

solubility in GaP reaches $(4 \pm 1) \cdot 10^{17}$ cm^{-3} (Il'yin et al. 1980). The nitrogen gas used to pressurize the crystal puller upon the growing of GaP results in nitrogen incorporation in the range of $(1–2) \cdot 10^{17}$ cm^{-3} (Lightowlers 1972). Temperature dependence of the nitrogen solubility in GaP is given in Figure 8.10 (Il'yin and Saenko 1980). Within the temperature region from 730°C to 1470°C, this solubility could be expressed by the following equation: $[N]^{max}$ (cm^{-3}) = $2.5 \cdot 10^{22} \cdot \exp(10,760/T - 17.22)$, where T is the temperature in kelvin.

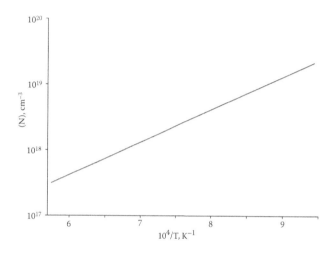

FIGURE 8.10 Temperature dependence of the nitrogen solubility in GaP. (From Il'yin, Yu.L., and Saenko, I.V., *Izv. AN SSSR Neorgan. Materialy*, 16(3), 376, 1980. With permission.)

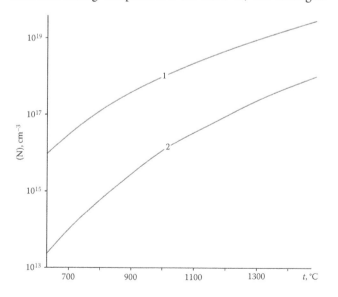

FIGURE 8.9 Calculated temperature dependence of nitrogen solubility in (1) GaP and (2) GaAs. (From Ho, I.H., and Stringfellow, G.B., *Mater. Res. Soc. Symp. Proc.*, 449, 871, 1996. With permission.)

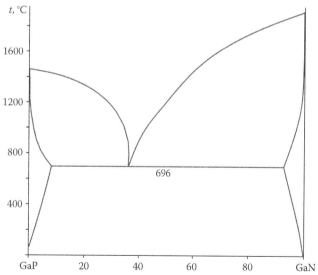

FIGURE 8.11 Calculated phase diagram of the GaN–GaP system. (From Yang, X., et al., *J. Appl. Phys.*, 77(11), 5553, 1995. With permission.)

GaN–GaP. The phase diagram of this system is a eutectic type (Figure 8.11) (Yang et al. 1995). The eutectic crystallizes at about 680°C. Il'yin and Saenko (1980) indicated that the eutectic in this system is degenerated, contains 0.54 mol% GaN, and crystallizes at a temperature 2.1°C lower than the melting temperature of GaP.

According to the data of Stringfellow (1982b), a miscibility gap with the critical temperature of 8318°C exists in the GaN–GaP system. The miscibility gap in this system ranges from 0.00003 to 99.99997 at% N at 700°C (Miyoshi et al. 1993).

GaN_xP_{1-x} solid solutions ($x = 0.02$–0.07 and $x \approx 0.99$) have been formed by ion implantation of Ga^+ and N^+ into GaP (Yang et al. 1995). By ion implantation with a heavy dose of N^+ into GaP, solid solution containing 2 at% of nitrogen was also produced (Z. Lin et al. 1988). The films of these solid solutions were epitaxially deposited on sapphire (0001) substrates by the vapor phase reaction of the Ga–NH_3–PCl_3–N_2 system (Igarashi and Okada 1985). The concentration of P increased with decreasing temperature, and the maximum concentration (4 at% P) was obtained at 1050°C. GaN_xP_{1-x} solid solutions were also grown on GaP substrates by MOVPE using $GaMe_3$, PH_3, and dimethylhidrazine as the Ga, P, and N sources, respectively (Miyoshi et al. 1993). Metastable alloys ($x \leq 0.04$) have been obtained at the growth temperatures of 630°C–700°C. The nitrogen incorporation increases with decreasing growth temperature. It is also increases with increasing growth rate.

It was shown that it is possible to incorporate up to 0.08 at% N in GaP with chemical beam epitaxy using tertiarybutylphosphine and NH_3 as precursors (Li et al. 1996). The nitrogen incorporation strongly depends on the growing surface. When the incorporation rate of P is higher than that of Ga, a 0.08 at% N composition of GaN_xP_{1-x} is achieved using coinjected tertiarybutylphosphine and NH_3 at a cracker temperature of 800°C. On the other hand, no nitrogen incorporation was observed when samples were grown under insufficient phosphor supply. Solid solutions containing up to 16 at% N have been obtained at the incorporation of nitrogen into GaP using a radio-frequency nitrogen plasma source (Bi and Tu 1997). The nitrogen composition increases slowly with increasing the N_2 flow rate, and then saturates. Lowering the growth temperature is effective in incorporating more nitrogen in GaP.

Growth of GaN_xP_{1-x} single crystals has been carried out under high pressure of N_2 (1–2 GPa) (Karpiński

et al. 1985). Increasing the crystallization temperature causes an increase of the nitrogen concentration in GaP. A maximum concentration of $3 \cdot 10^{19}$ cm^{-3} was obtained for the GaP melt. This value seems to be natural for the solid solubility of nitrogen in GaP for a given pressure.

8.24 GALLIUM–ARSENIC–NITROGEN

The calculated temperature dependence of nitrogen solubility in GaAs is shown in Figure 8.9 (Ho and Stringfellow 1996a). The calculations indicate that the solubility of nitrogen increases significantly with increasing temperature.

GaN–GaAs. The layers of GaN_xAs_{1-x} ($0 < x < 0.016$) have been grown by MOCVD at very low pressure (25 Pa) (Weyers and Sato 1993). The nitrogen source NH_3 has been decomposed in a remote microwave, and uncracked $GaEt_3$ and AsH_3 were used. The nitrogen uptake into the layers shows a strong dependence on the growth temperature. The competition for the Group V lattice sites leads to a reduction for the nitrogen content at higher AsH_3 fluxes. GaN_xAs_{1-x} solid solutions with a nitrogen content up to 1.5 at% were obtained on GaAs wafers exploring the gas source MBE, under which a nitrogen radical was used as the nitrogen source (Kondow et al. 1994). Solid solution films have also been successfully grown using active nitrogen supplied by a radio-frequency-activated plasma source (Foxon et al. 1995). A significant amount of nitrogen (≈ 20 at%) was incorporated in such films.

The composition dependence of the band gap of thin layers of GaN_xAs_{1-x} ($0 \leq x \leq 0.05$) on GaAs substrates was studied by optical transmission measurements and high-resolution XRD (Tisch et al. 2002). The bowing parameter reaches 40 eV for very low nitrogen incorporations ($x \approx 0.001$), and strongly decreases with increasing nitrogen molar fraction. According to the estimations, the bowing parameter would reach a constant value of 7.5 eV for $x \geq 0.08$. This bowing would not be sufficient to close the band gap for higher nitrogen incorporations.

8.25 GALLIUM–BISMUTH–NITROGEN

GaN–BiN. The structural and electronic properties of **$GaBi_xN_{1-x}$** alloys were calculated by using the density functional theory based on the full-potential linearized augmented plane-wave method (Mbarki and Rebey 2012). The calculated lattice parameter shows an increase by increasing the composition of Bi, while a significant deviation from Vegard's law was observed. $GaBi_xN_{1-x}$ remains a semiconductor until $x = 0.5$, and it becomes a semimetal for $x \geq 0.625$.

8.26 GALLIUM–VANADIUM–NITROGEN

Three ternary compounds, $\mathbf{V_2GaN}$, $\mathbf{V_3GaN}$, and $\mathbf{V_3Ga_2N}$, are formed in the Ga–V–N system (Jeitschko et al. 1964d, 1965; Boller 1971; Barsoum 2000; Bouhemadou 2009). V_2GaN crystallizes in the hexagonal structure with the lattice parameters $a = 300$ and $c = 1330$ pm and a calculated density of 5.94 g·cm^{-3} (Barsoum 2000) ($a = 287.2$ and $c = 1246.1$ pm and bulk modulus $B = 200$ GPa and its pressure derivative $B' = 4.08$ [local density approximation] and $a = 293.4$ and $c = 1273.2$ pm and bulk modulus $B = 160$ GPa and its pressure derivative $B' = 3.74$ [generalized gradient approximation] [Bouhemadou 2009]).

V_3GaN crystallizes in the orthorhombic structure with the lattice parameters $a = 295.0$, $b = 1030$, and $c = 793.1$ pm (Boller 1971).

V_3Ga_2N crystallizes in the cubic structure with the lattice parameter $a = 662.0$ and a calculated density of 7.01 g·cm^{-3} (Jeitschko et al. 1965). It was obtained by annealing the mixture of V, VN, and Ga at 750°C–850°C for 700 h.

8.27 GALLIUM–NIOBIUM–NITROGEN

No ternary compound is formed in the Ga–Nb–N system (Mohney and Lin 1996). In this system, GaN is in thermodynamic equilibrium with NbN. Earlier data indicated the existence of the $\mathbf{Nb_5Ga_3N_x}$ ternary compound in this system (hexagonal structure; $a = 769.6$ and $c = 531.0$ pm [Jeitschko et al. 1964c]).

8.28 GALLIUM–TANTALUM–NITROGEN

No ternary compound is formed in the Ga–Ta–N system (Mohney and Lin 1996). In this system, GaN is in thermodynamic equilibrium with TaN. Earlier data indicated the existence of the $\mathbf{Ta_5Ga_3N_x}$ ternary compound in this system (hexagonal structure; $a = 764.6$ and $c = 527.4$ pm [Jeitschko et al. 1964c]).

8.29 GALLIUM–OXYGEN–NITROGEN

The adsorption of NO at room temperature leads to the formation of an amorphous film of GaO_xN_y on a CoGa(001) surface (Schmitz et al. 1999). After annealing to 552°C, a well-ordered GaO_xN_y film was formed. The atomic ratio between the amount of oxygen and nitrogen on the surface was estimated to be O/N $\approx$ 2:1. The band gap of this film was determined to be 4.1 ± 0.2 eV. The adsorption of NO and the formation of GaO_xN_y on CoGa(001) was studied by means of high-resolution electron energy loss spectroscopy, scanning tunneling microscopy, Auger electron spectroscopy, and low-energy electron diffraction.

$\mathbf{GaN–Ga_2O_3}$. Some ternary compounds are formed in this system. $\mathbf{Ga_3O_3N}$ exists in two polymorphic modifications. The first one is thermodynamically stable above 890°C (Kroll 2005) and crystallizes in the cubic structure with the lattice parameter $a = 826.4 \pm 0.1$ pm for the composition $Ga_{2.81}O_{3.57}N_{0.43}$ (Kinski et al. 2005b) ($a = 820 \pm 7$ pm for the composition $Ga_{2.8}O_{3.5}N_{0.5}$ [Puchinger et al. 2002]; $a = 822.8$ pm and an energy gap of 2.16 eV for the composition $Ga_{2.8}O_{3.5}N_{0.5}$ [Lowther et al. 2004]; $a = 816.0$ pm at 10.5 GPa, $a = 828.1 \pm 0.2$ pm for the sample synthesized at 5 GPa and 1700°C, and an energy gap of around 4 eV [Soignard et al. 2005]; $a = 826.7$ pm and an energy gap of 1.72 eV [local density approximation] and $a = 842.5$ pm and an energy gap of 1.37 eV [generalized gradient approximation] [Okeke and Lowthe 2008]).

Bulk samples of Ga_3O_3N were synthesized using high-pressure, high-temperature techniques from GaN + Ga_2O_3 mixtures (Soignard et al. 2005). Theoretical study showed that the most stable structure for this compound corresponds to a rhombohedral distortion of the ideal spinel structure. The formation of Ga_3O_3N is endothermic at ambient pressure and low temperature, and the optimal synthesis pressure was predicted to lie close to the $\alpha \rightarrow \beta$ phase transition in Ga_2O_3 (around 6.6 GPa).

$Ga_{2.81}O_{3.57}N_{0.43}$ was crystallized under high-pressure, high-temperature conditions in a spinel-type structure from a prestructured gallium oxynitride ceramic, which was obtained from the single-source molecular precursor $[Ga(OBu^t)_2NMe_2]_2$ by thermal treatment in an ammonia atmosphere (Kinski et al. 2005b). The optimized precursor-derived gallium oxynitride ceramic remains nanocrystalline up to 600°C and can be transformed at 7 GPa and 1100°C into the crystalline phase with the same composition. All reactions and operations were carried out under inert conditions with rigorous exclusion of oxygen and moisture.

Puchinger et al. (2002) described this modification found in GaN thin films that were produced from a dimethylamide base liquid precursor. At an intermediate growth stage, a cubic phase with an elemental composition near $Ga_{2.8}O_{3.5}N_{0.5}$ was observed using high-resolution transmission electron microscopy (HRTEM).

The second modification of Ga_3O_3N is thermodynamically stable above 1213°C and crystallizes in the rhombohedral structure with the lattice parameters a = 592.12 pm and $α$ = 59.26° (Kroll et al. 2005; Martin et al. 2009). This modification did not form unless it was stabilized at elevated temperature by substantial entropy contributions.

First-principles calculations were performed to investigate the structure of spinel-type gallium oxynitrides ($Ga_{22}O_{30}N_2$ and $Ga_{23}O_{27}N_2$), as well as their formation from and relative stability to the binary phases Ga_2O_3 and GaN. Enthalpies of formation indicate that at ambient pressure, these ternary compounds are unfavorable relative to a phase assemblage of monoclinic β-Ga_2O_3 and wurtzite GaN. It was found that at 630°C, $Ga_{22}O_{30}N_2$ and $Ga_{23}O_{27}N_2$ are not stable thermodynamically with respect to the binaries. The temperatures above which these compounds will form are 1183°C and 1100°C (above 1080°C and 910°C [Kroll 2005]), respectively (Kroll et al. 2005; Martin et al. 2009).

Pyrolysis of $GaCl_3·NH_3$ in the presence of the H_2O vapor leads to the formation of oxynitride phases with the structure of GaN and amorphous gallium oxynitrides, the composition of which lies in the GaN–Ga_2O_3 section (Grekov et al. 1979).

8.30 GALLIUM–CHROMIUM–NITROGEN

Isothermal sections of the Ga–Cr–N ternary system at 700°C and 800°C are shown in Figures 8.12 and 8.13 (Mohney et al. 1997; Gröbner et al. 1999). At 800°C, CrN and liquid Ga are the only condensed phases in thermodynamic equilibrium with GaN (Mohney et al. 1997). Ternary phases may exhibit a detectable range of homogeneity. The samples were annealed at 700°C for different times up to 162 h (Gröbner et al. 1999) and at 800°C for 7–10 days (Mohney et al. 1997). For the ternary calculations, the binary data sets were combined and extrapolated into the ternary system (Gröbner et al. 1999). Calculations were performed, including the gas phase at 0.1 MPa of total gas pressure.

Isothermal sections at 670°C, 740°C, 800°C, and 1000°C were also constructed by Farber and Barsoum (1999), but there is some discrepancy between these sections and the results of Mohney et al. (1997) and

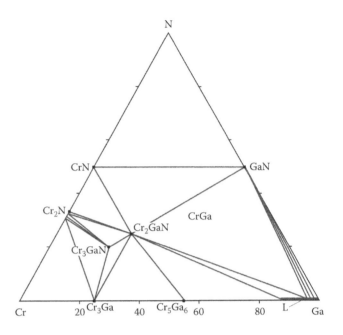

FIGURE 8.12   Calculated isothermal section of the Ga–Cr–N ternary system at 700°C under 0.1 MPa. (From Gröbner, J., et al., *J. Phase Equilibria*, 20(6), 615, 1999. With permission.)

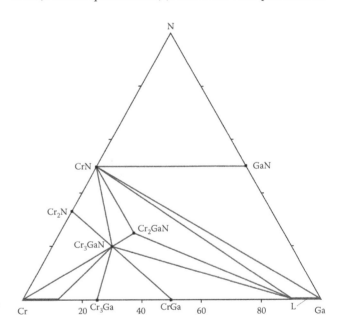

FIGURE 8.13   Isothermal section of the Ga–Cr–N ternary system at 800°C under 0.1 MPa. (From Mohney, S.E., et al., *Mater. Sci. Eng. B*, 49(2), 152, 1997. With permission.)

Gröbner et al. (1999). These discrepancies concern mainly the Cr-rich region of the ternary system and could be explained by the short time (only 9 h) of samples annealed by Farber and Barsoum (1999).

Two ternary compounds, **Cr_2GaN** and **Cr_3GaN**, are formed in the Ga–Cr–N system. Cr_2GaN crystallizes in the hexagonal structure with the lattice parameters $a = 287.5$ and $c = 1277$ pm and a calculated density of 6.82 g·cm^{-3} (Barsoum 2000) ($a = 288.1$ and $c = 1277$ pm [Beckmann et al. 1969]; $a = 287.5 \pm 0.07$ and $c = 1270.0 \pm 0.7$ pm [Farber and Barsoum 1999]; $a = 283.6$ and $c = 1210.6$ pm and bulk modulus $B = 214$ GPa and its pressure derivative $B' = 3.83$ [local density approximation] and $a = 289.5$ and $c = 1237.5$ pm and bulk modulus $B = 177$ GPa and its pressure derivative $B' = 3.77$ [generalized gradient approximation] [Bouhemadou 2009]). This compound was obtained by sintering CrN, Cr, and Ga in an evacuated quartz ampoule at 750°C for 5 days (Beckmann et al. 1969).

Cr$_3$GaN crystallizes in the cubic structure with the lattice parameter $a = 386.8$ (Farber and Barsoum 1999).

8.31 GALLIUM–MOLYBDENUM–NITROGEN

An isothermal section of the Ga–Mo–N ternary system is shown in Figure 8.14 (Venugopalan and Mohney 1998). No ternary phases were found at this temperature. Liquid Ga, Mo$_8$Ga$_{41}$, Mo$_3$Ga, and β-Mo$_2$N were observed to be in thermodynamic equilibrium with GaN. Equilibrium between γ-Mo$_2$N and GaN can also be inferred.

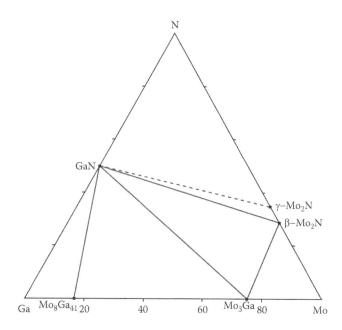

FIGURE 8.14 Isothermal section of the Ga–Mo–N ternary system at 800°C. (From Venugopalan, H.S., and Mohney, S.E., *Z. Metallkde*, 89(3), 184, 1998. With permission.)

The samples were annealed at 800°C for at least 1 week before they were quenched, ground into powder, pressed, encapsulated, and annealed for 4 more weeks. The annealed samples were quenched in water.

8.32 GALLIUM–TUNGSTEN–NITROGEN

Tungsten is expected to be in thermodynamic equilibrium with GaN at room temperature and at 600°C (Mohney and Lin 1996). No ternary compound is formed in the Ga–W–N system.

8.33 GALLIUM–CHLORINE–NITROGEN

GaN$_3$Cl$_2$ ternary compound is formed in the Ga–Cl–N system. It was obtained as a colorless air-sensitive solid that sublimes readily at 70°C–100°C in vacuum and melts at 210°C under atmospheric pressure (Kouvetakis et al. 1996, 1997). It was prepared by low-temperature decomposition of Me$_3$SiN$_3$·GaCl$_3$. This compound is a polymer in the solid state and has been used as a unimolecular precursor to deposit crystalline-oriented GaN heteroepitaxially on sapphire and Si substrates at 650°C–700°C by ultra-high-vacuum chemical vapor deposition (CVD).

8.34 GALLIUM–BROMINE–NITROGEN

GaN$_3$Br$_2$ ternary compound is formed in the Ga–Br–N system. To obtain it, GaN$_3$I$_2$ was introduced into a 50 mL flask, which was connected with a container with Br$_2$ (Dehnicke and Krüger 1978). The standing under the autogeneous vapor pressure Br$_2$ reacts slowly via the gas phase with GaI$_2$N$_3$ for 5 h. Then the rest of bromine was added to the reaction mixture, which was stirred for 12 h to achieve complete substance conversion. The residue was washed several times with small amounts of benzene and filtered, and after drying under vacuum, GaN$_3$Br$_2$ was obtained. Since the prime product still contained some iodine, it was cleaned as follows. Crude product (0.9 g) was dissolved in benzene (250 mL) and heated to 60°C. The mixture was filtered, concentrated under vacuum to 120 mL, and allowed to cool overnight. The precipitate was discarded and the filtrate was concentrated to 10 mL and finally left overnight in the refrigerator. After filtration and drying under vacuum, pure GaN$_3$Br$_2$ was prepared. The experiments require the exclusion of moisture.

8.35 GALLIUM–IODINE–NITROGEN

GaN₃I₂ ternary compound is formed in the Ga–I–N system. To obtain it, the suspension of GaI₃ (4.00 g) in benzene (40 mL) was slowly added to a solution of IN₃ (1.50 g) in benzene (100 mL) concentrated under vacuum to 10 mL and left overnight in a refrigerator (Dehnicke and Krüger 1978). The residue was washed several times with small amounts of benzene and filtered, and after drying under vacuum, the title compound was prepared.

8.36 GALLIUM–MANGANESE–NITROGEN

The calculated isothermal section of the Ga–Mn–N ternary system at 800°C is shown in Figure 8.15 (Sedmidubský et al. 2008). **Mn₃GaN** compound is formed in this system. It is in thermodynamic equilibrium with all phases occurring in the system. Ga₁₋ₓMnₓN solid solution, Mn₃GaN, and the ξ-phase of the Mn–N system are all in equilibrium with N₂ gas at these conditions. The equilibria with the liquid phase and the solid solutions in the Mn-rich part all correspond to significantly reduced nitrogen activity. The solubility of 2.4 at% Mn in the wurtzite Ga₁₋ₓMnₓN phase was predicted for the selected conditions.

The ternary phase diagram was constructed based on phase equilibria calculations and thermodynamic analysis of the relevant phases occurring in the studied system (Sedmidubský et al. 2008).

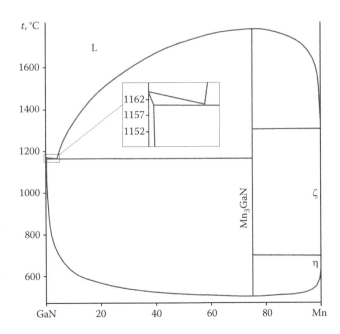

FIGURE 8.16 Calculated phase diagram of the GaN–Mn system. (From Sedmidubský, D., et al., *J. Alloys Compd.*, *452*(1), 105, 2008. With permission.)

GaN–Mn. The calculated phase diagram of this system is given in Figure 8.16 (Sedmidubský et al. 2008). The complete solubility observed below 502°C is most likely unrealistic.

GaN–MnN. The lattice constant increases when the Mn composition increases (Miyake et al. 2013).

8.37 GALLIUM–RHENIUM–NITROGEN

Rhenium is expected to be in thermodynamic equilibrium with GaN at room temperature and at 600°C (Mohney and Lin 1996). No ternary compound is formed in the Ga–Re–N system.

8.38 GALLIUM–IRON–NITROGEN

GaₓFe₄₋ₓN solid solutions are formed in the Ga–Fe–N ternary system (Houben et al. 2009; Burghaus et al. 2010, 2011). They crystallize in the cubic structure. There is a nonlinear behavior of the lattice parameter as a function of composition: $a = 379.74 \pm 0.01$, 379.74 ± 0.01, 379.58 ± 0.01, 379.41 ± 0.01, 379.34 ± 0.01, and 379.29 ± 0.01 pm and a calculated density of 7.62, 7.51, 7.47, 7.43, 7.38, and 7.33 g·cm⁻³ at $x = 0.90$, 0.75, 0.625, 0.50, 0.375, and 0.25, respectively. The decomposition of these solid solutions starts at about 540°C and is completed above 600°C.

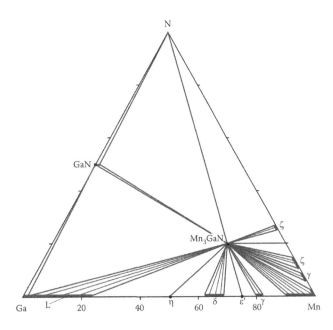

FIGURE 8.15 Calculated isothermal section of the Ga–Mn–N ternary system at 800°C. (From Sedmidubský, D., et al., *J. Alloys Compd.*, *452*(1), 105, 2008. With permission.)

Ga$_x$Fe$_{4-x}$N was synthesized by the two-step ammonolysis reaction starting from powdered Ga$_2$O$_3$ and Fe$_2$O$_3$ (Houben et al. 2009; Burghaus et al. 2010, 2011). The optimized synthesis uses a high-temperature step (1100°C, 1 min) and a subsequent nitridation reaction (530°C, 3 h). The ammonolysis gas was a NH$_3$/H$_2$ mixture with a 1:1 volume ratio.

8.39 GALLIUM–COBALT–NITROGEN

Gröbner et al. (1999) calculated the isothermal section of the Ga–Co–N ternary at 500°C under a pressure of 0.1 MPa. At these conditions, GaN is in thermodynamic equilibrium with CoGa$_3$ and CoGa, and CoGa is in equilibrium with nitrogen. The reaction of Co and GaN is very slow, and CoGa is the first formed phase. The samples were annealed at 500°C for different times up to 162 h.

Co$_6$Ga$_3$N ternary compound, which crystallizes in the cubic structure with the lattice parameter a = 368 pm, is formed in the Ga–Co–N system (Stadelmaier and Fraker 1962).

According to the data of Beznosikov (2003), **Co$_3$GaN** compound, which crystallizes in the cubic structure of perovskite type, could also form in this ternary system.

8.40 GALLIUM–NICKEL–NITROGEN

The calculated isothermal section of the Ga–Ni–N ternary system at 500°C is shown in Figure 8.17 (Gröbner et al. 1999). It was determined experimentally that Ni$_3$Ga is the only phase formed by the reaction of GaN and Ni. The three-phase equilibrium N$_2$ + GaN + NiGa at 0.1 MPa is shifted to the Ni-rich intermetallics. Finally, the equilibrium N$_2$ + GaN + (Ni) becomes stable at about 10 MPa, thus terminating any further reaction between GaN and Ni. The samples were annealed at 500°C for different times up to 162 h.

If the N$_2$ pressure in the annealing environments is reduced to 1 kPa, NiGa is no longer in equilibrium with GaN and a tie triangle is found between GaN, Ni$_2$Ga$_3$, and N$_2$ (Mohney and Lin 1996). A further decrease in the N$_2$ pressure to 10 Pa places a tie triangle between the gas phase, GaN, and liquid Ga with 5 at% N.

Ni$_2$GaN ternary compound is formed in the Ga–Ni–N system (Kamińska et al. 1997).

According to the data of Beznosikov (2003), **Ni$_3$GaN** compound, which crystallizes in the cubic structure of perovskite type, could also form in this ternary system.

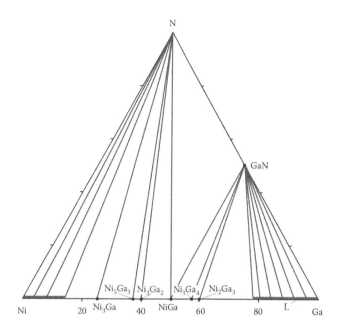

FIGURE 8.17   Calculated isothermal section of the Ga–Ni–N ternary system at 500°C under 0.1 MPa. (From Gröbner, J., et al., *J. Phase Equilibria*, *20*(6), 615, 1999. With permission.)

8.41 GALLIUM–RUTHENIUM–NITROGEN

At room temperature, Ru was predicted to be in thermodynamic equilibrium with GaN under 0.1 MPa of N$_2$, but at 600°C it is not in such equilibrium (Mohney and Lin 1996). Rather, tie-lines were predicted between ruthenium gallides and nitrogen. It is necessary to note that at 600°C, the tie-line between GaN and Ru is less stable than some of the competing tie-lines in this system by less than 10 kJ·g-atom^{-1}; hence, there is still considerable uncertainty in these predictions.

8.42 GALLIUM–RHODIUM–NITROGEN

At room temperature, Rh was predicted to be in thermodynamic equilibrium with GaN under 0.1 MPa of N$_2$, but there is considerable uncertainty in such a prediction since there are competing tie-lines that are less stable than the GaN–Rh tie-line (Mohney and Lin 1996). At 600°C, it is not in such equilibrium. Rather, tie-lines were predicted between rhodium gallides and nitrogen.

8.43 GALLIUM–PALLADIUM–NITROGEN

A calculated isothermal section of the Ga–Pd–N ternary system at 500°C is shown in Figure 8.18 (Gröbner et al. 1999). It was determined experimentally that Pd$_2$Ga is

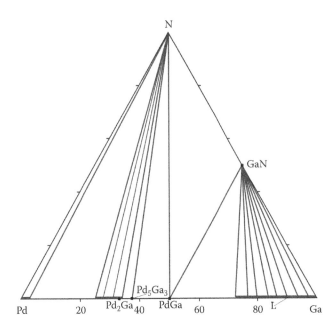

FIGURE 8.18 Calculated isothermal section of the Ga–Pd–N ternary system at 500°C under 0.1 MPa. (From Gröbner, J., et al., *J. Phase Equilibria*, *20*(6), 615, 1999. With permission.)

always formed as the first phase at the reaction of GaN and Ni. The samples were annealed at 500°C for different times up to 162 h.

8.44 GALLIUM–OSMIUM–NITROGEN

At room temperature, Os was predicted to be in thermodynamic equilibrium with GaN under 0.1 MPa of N_2, but at 600°C it is not in such equilibrium (Mohney and Lin 1996). Rather, tie-lines were predicted between osmium gallides and nitrogen. It is necessary to note that at 600°C, the tie-line between GaN and Os is less stable than some of the competing tie-lines in this system by less than 10 kJ·g-atom^{-1}; hence, there is still considerable uncertainty in these predictions.

8.45 GALLIUM–IRIDIUM–NITROGEN

At room temperature, Ir was predicted to be in thermodynamic equilibrium with GaN under 0.1 MPa of N_2, but there is considerable uncertainty in such prediction since there are competing tie-lines that are less stable than the GaN–Ir tie-line (Mohney and Lin 1996). At 600°C, it is not in such equilibrium. Rather, tie-lines were predicted between iridium gallides and nitrogen.

8.46 GALLIUM–PLATINUM–NITROGEN

Platinum is thermodynamically favored to react with GaN to form platinum gallides and release nitrogen, even under 0.1 MPa of N_2 at room temperature (Mohney and Lin 1996). Because of uncertainties in the thermodynamic data, it is not possible to predict precisely which of the platinum gallides is simultaneously in equilibrium with GaN and N_2 at 0.1 MPa, even though the lack of a tie-line between Pt and GaN at room temperature is strongly indicated. Higher temperatures increase the driving force for reaction between Pt and GaN.

CHAPTER 9

Systems Based on GaP

9.1 GALLIUM–SODIUM–PHOSPHORUS

Na_6GaP_3 ternary compound, which crystallizes in the triclinic structure with the lattice parameters $a = 797.1 \pm 0.4$, $b = 796.7 \pm 0.4$, and $c = 1691.9 \pm 0.8$ pm and $\alpha = 94.1 \pm 0.1°$, $\beta = 94.8 \pm 0.1°$, and $\gamma = 119.8 \pm 0.1°$, is formed in this system (Blase et al. 1993b). It was synthesized from the elements at 600°C in sealed Nb ampoules.

9.2 GALLIUM–POTASSIUM–PHOSPHORUS

K_2GaP_2 ternary compound, which crystallizes in the monoclinic structure with the lattice parameters $a = 1021.5 \pm 0.3$, $b = 1401.6 \pm 0.5$, and $c = 863.6 \pm 0.2$ pm and $\beta = 109.4 \pm 0.1°$, is formed in this system (Blase et al. 1991d). It was synthesized from the elements at 600°C in sealed Nb ampoules.

9.3 GALLIUM–RUBIDIUM–PHOSPHORUS

Rb_3GaP_2 ternary compound, which crystallizes in the orthorhombic structure with the lattice parameters $a = 1463.4 \pm 0.9$, $b = 2489.3 \pm 0.6$, and $c = 916.3 \pm 0.4$ pm, is formed in this system (Somer et al. 1990d). It was synthesized from the stoichiometric mixture of the elements at 730°C in sealed steel ampoules.

9.4 GALLIUM–CESIUM–PHOSPHORUS

$Cs_6Ga_2P_4$ ternary compound, which crystallizes in the monoclinic structure with the lattice parameters $a = 1117.3 \pm 0.3$, $b = 866.1 \pm 0.3$, and $c = 1893.9 \pm 0.3$ pm and $\beta = 99.64 \pm 0.02°$, is formed in this system (Somer et al. 1990h). It was synthesized from a stoichiometric mixture of the elements or from Cs_4P_6, Ga, and Cs in evacuated and sealed ampoules at 700°C.

9.5 GALLIUM–CALCIUM–PHOSPHORUS

Three ternary compounds are formed in the Ga–Ca–P system. $CaGa_2P_2$ crystallizes in the hexagonal structure with the lattice parameters $a = 382.39 \pm 0.07$ and $c = 1639.3 \pm 0.4$ pm and a calculated density of 3.853 g·cm^{-3} at 200 K (He et al. 2011). The title compound was found to decompose around 1050°C. To prepare its single crystals, Ca and P were taken in stoichiometric ratios, whereas Ga was used in a fivefold excess amount to serve as a self-flux. The reaction mixture was heated fast to 960°C, homogenized for 20 h, and then cooled at a rate of 6°C·h^{-1} to 500°C, at which point the molten Ga was decanted. The crystals of $CaGa_2P_2$ had a platelet shape with hexagonal facets. All manipulations were performed inside an Ar-filled glove box or under vacuum.

$Ca_3Ga_2P_4$ crystallizes in the monoclinic structure with the lattice parameters $a = 1262.3 \pm 0.3$, $b = 987.0 \pm 0.2$, $c = 644.35 \pm 0.15$ pm and $\beta = 91.050 \pm 0.003°$ and a calculated density of 3.174 g·cm^{-3} at 200 ± 2 K (He et al. 2012). To obtain it, the mixture of the elements with the stoichiometric ratio Ca/P of 3:4 and a 25-fold excess of Ga was loaded into an alumina crucible, which was then flame sealed in a fused silica tube under vacuum. The evacuated silica tube was heated at 960°C for 20 h, and then cooled to 500°C at a rate of 5°C·h^{-1}. The molten Ga was decanted at this temperature. Pb could be also used as a flux. The corresponding elements in the stoichiometric ratio and a large amount of Pb (20-fold excess) were used. Large, red single crystals were obtained from such an experiment. All synthetic and postsynthetic manipulations were performed inside an argon-filled glove box or under vacuum.

Ca$_{14}$GaP$_{11}$ crystallizes in the tetragonal structure with the lattice parameters $a = 1534.7 \pm 0.4$ and $c = 2076.2 \pm 0.8$ pm and a calculated density of 2.641 g·cm^{-3} (Vaughey and Corbett 1996). The initial crystals of the title compound were isolated from the composition "Ca$_3$Ga$_4$P" prepared by weighing stoichiometric amounts of Ca, P, and Ga in a Ta tube, of which one end had been previously welded. The Ta tube was then crimped and welded shut. This reaction vessel was then placed inside a silica jacket that was evacuated and sealed. The mixture was heated slowly to 1000°C, held for 3 days, and then slowly cooled to room temperature over a 1-week period. A single-phase sample of Ca$_3$Ga$_2$P$_4$ was subsequently synthesized by mixing stoichiometric amounts of the component elements and repeating the above procedure. All materials were handled in a N$_2$-filled glove box that had a moisture level below 0.1 ppm.

9.6 GALLIUM–STRONTIUM–PHOSPHORUS

Two ternary compounds were prepared in the Ga–Sr–P system. **Sr$_3$GaP$_3$** crystallizes in the orthorhombic structure with the lattice parameters $a = 1878.6 \pm 0.4$, $b = 638.67 \pm 0.12$, and $c = 1240.3 \pm 0.2$ pm at 200 K (Stoyko et al. 2015). For obtaining this compound, starting materials were Sr pieces, Ga ingots, and red P powder. Mixtures of the elements were loaded into alumina crucibles covered on the top with quartz wool and placed within fused silica tubes, which were evacuated and sealed. The tubes were heated to 960°C at a rate of 60°C·h^{-1}, held at that temperature for 20 h (samples with Al flux) or 40 h (samples with Pb flux), abd cooled to 750°C·h^{-1} at a rate of 5°C·h^{-1} (for samples with Al flux) or to 500°C·h^{-1} at a rate of 30°C·h^{-1} (for samples with Pb flux). The excess of metal flux was removed at 750°C·h^{-1} (in samples with Al flux) or at 500°C·h^{-1} (in samples with Pb flux) by using a centrifuge. All reagents and products were handled within an argon-filled glove box with controlled oxygen and a moisture level below 1 ppm.

Sr$_3$Ga$_2$P$_4$ crystallizes in the monoclinic structure with the lattice parameters $a = 1315.22 \pm 0.13$, $b = 1019.89 \pm 0.10$, and $c = 662.90 \pm 0.07$ pm and $\beta = 90.314 \pm 0.001°$ and a calculated density of 3.931 g·cm^{-3} at 200 $\pm$ 2 K (He et al. 2012). Large red single crystals were obtained the same way that Ca$_3$Ga$_2$P$_4$ was prepared using the mixture of the elements with the stoichiometric ratio Sr/P of 3:4.

9.7 GALLIUM–BARIUM–PHOSPHORUS

Some ternary compounds are formed in this ternary system. **BaGa$_2$P$_2$** crystallizes in the monoclinic structure with the lattice parameters $a = 733.63 \pm 0.13$, $b = 966.48 \pm 0.17$, and $c = 742.61 \pm 0.13$ pm and $\beta = 115.373 \pm 0.002°$ and a calculated density of 4.729 g·cm^{-3} at 200 $\pm$ 2 K (He et al. 2010). Two general synthetic methods were explored to prepare the title compound: on-stoichiometry reactions in sealed Nb tubes, and flux reactions in alumina crucibles using Ga as the reactive flux. Nb tubes were welded with an arc welder under a partial pressure of Ar. Before the heat treatment, both the Nb tubes and the crucibles were enclosed in fused silica tubes, which were subsequently flame sealed under vacuum. The molar ratio of Ba/Ga/P was 1:10:2 at the synthesis using the flux method.

Ba$_3$Ga$_3$P$_5$ crystallizes in the rhombohedral structure with the lattice parameters $a = 1461.3 \pm 0.3$ and $c = 2888.4 \pm 0.8$ pm (in a hexagonal setting), a calculated density of 4.343 g·cm^{-3} at 200 $\pm$ 2 K, and an energy gap of 1.4 eV (He et al. 2013). Crystals of the title compound were identified from a reaction of Ba, Ga, P, and Pb (molar ratio 2:1:2:20). The mixture was contained in an alumina crucible, where the excess amount of Pb was intended as a flux. The crucible was then jacketed in a fused silica tube, which was subsequently flame sealed under vacuum. Heat treatment included a ramp to 960°C over 14 h and a "soak" at that temperature for 40 h, followed by cooling to 500°C during a period of 16 h. Then, the crucible was taken out of the furnace, and the Pb flux was removed. The small, air-sensitive, dark red crystals were obtained. All manipulations of the starting materials and the reaction products were conducted inside an Ar-filled glove box or under vacuum.

Ba$_4$Ga$_5$P$_8$ crystallizes in the orthorhombic structure with the lattice parameters $a = 723.68 \pm 0.06$, $b = 1795.38 \pm 0.16$, and $c = 655.63 \pm 0.06$ pm and a calculated density of 4.467 g·cm^{-3} at 200 $\pm$ 2 K (He et al. 2011). This compound is thermally stable up to 850°C. It was obtained from a reaction mixture loaded with Ba, Ga, and P (molar ratio 1:10:2), in which the Ga excess was intended as a flux. This mixture was heated fast to 960°C, homogenized for 20 h, and then cooled to 750°C with a rate of 30°C·h^{-1}. At this point, the sample was removed and radiatively cooled to room temperature. Finally, the sealed tube was quickly reheated to 500°C (100°C·h^{-1}), before the Ga flux was decanted and the title compound was isolated. For obtaining the title compound, the use of Pb flux is an advantage. All manipulations were performed inside an Ar-filled glove box or under vacuum.

$Ba_6Ga_2P_6$ also crystallizes in the orthorhombic structure with the lattice parameters $a = 1947.5 \pm 0.4$, $b = 670.8 \pm 0.1$, and $c = 1301.7 \pm 0.3$ pm (Peters et al. 1996). It was obtained in the form of dark prisms from a stoichiometric mixture of Ba, BaP, and GaP in a sealed steel ampoule 1100°C.

$Ba_7Ga_4P_9$ also crystallizes in the orthorhombic structure with the lattice parameters $a = 987.7 \pm 0.2$, $b = 1654.8 \pm 0.4$, and $c = 663.95 \pm 0.15$ pm and a calculated density of 4.65 g·cm⁻³ (He et al. 2013, 2016). To obtain this compound, Ba, Ga, and P were used as starting materials. All reagents and products were handled within an Ar-filled glove box with controlled oxygen and a moisture level below 1 ppm. Single crystals were first isolated from reactions attempted to extend the $AETr_2Pn_2$ and $AE_3Tr_2Pn_4$ series (AE = Ca, Sr, Ba; Tr = Al, Ga, In; and Pn = P, As, Sb) through use of the Al or Pb fluxes. Mixtures of the elements were loaded into alumina crucibles covered on the top with quartz wool and placed within fused silica tubes, which were then evacuated and flame sealed. The tubes were heated to 960°C at a rate of 60°C·h⁻¹, held at that temperature for 20 h (samples with Al flux) or 40 h (samples with Pb flux), and cooled to 750°C at a rate of 5°C·h⁻¹ (for samples with Al flux) or to 500°C at a rate of 30°C·h⁻¹ (for samples with Pb flux). The molten metal fluxes were removed at 750°C (Al flux) or at 500°C (Pb flux) by centrifugation.

9.8 GALLIUM–ZINC–PHOSPHORUS

The liquidus surface of the Ga–Zn–P ternary system (Figure 9.1) was determined experimentally through differential thermal analysis (DTA) by Panish (1966e) and calculated by Jordan (1971a). It consists principally of two primary phase fields due to GaP and Zn_3P_2, respectively. The Zn_3P_2 phase field exists over part of the GaP–Zn quasi-binary section, and there is a ternary transition point at 407°C near the zinc corner of the diagram. By a combination of the ternary regular activity coefficients with the vapor pressures of the pure components and the liquidus isotherms, the component partial pressures along the liquidus isotherms were determined (Jordan 1971a).

The isothermal sections of this system were constructed at 900°C (Tuck and Jay 1976) and 1000°C (Panish 1966e), and the last of them is given in Figure 9.2.

GaP–Zn. The phase equilibria in this system are shown in Figure 9.3 (Panish 1966e). This system is a

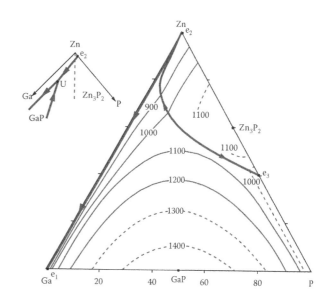

FIGURE 9.1 Liquidus surface of the Ga–Zn–P ternary system. (From Panish, M.B., *J. Electrochem. Soc.*, 113(3), 224, 1966. Reproduced with permission of The Electrochemical Society.)

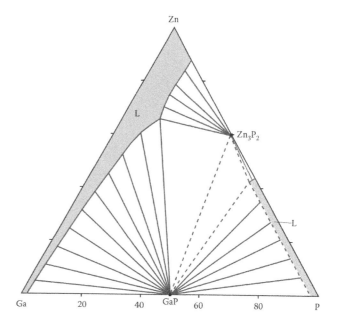

FIGURE 9.2 Isothermal section of the Ga–Zn–P ternary system at 1000°C. (From Panish, M.B., *J. Electrochem. Soc.*, 113(3), 224, 1966. Reproduced with permission of The Electrochemical Society.)

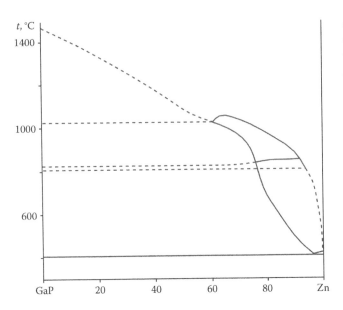

FIGURE 9.3 The phase equilibria in the GaP–Zn system. (From Panish, M.B., *J. Electrochem. Soc.*, *113*(3), 224, 1966. Reproduced with permission of The Electrochemical Society.)

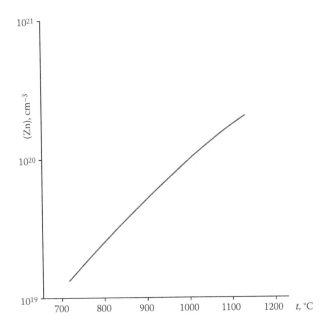

FIGURE 9.4 Experimental determined temperature dependence of Zn solubility in GaP. (From Chang, L.L., and Pearson, G.L., *J. Phys. Chem. Solids*, *25*(1), 23, 1964. With permission.)

non-quasi-binary section of the Ga–Zn–P ternary system. The solubility of Zn in GaP is retrograde with a maximum value of $4.3 \cdot 10^{20}$ cm^{-3} at 1350°C (Chang and Pearson 1964a, 1964b) ($4 \cdot 10^{20}$ cm^{-3} at 1200°C [Allison 1963]). The experimental determined temperature dependence of Zn solubility in GaP is shown in Figure 9.4 (Chang and Pearson 1964a, 1964b). Below 1100°C, this

dependence could be expressed by the following equation: c_{Zn} (cm^{-3}) $= 2.6 \cdot 10^{-23} \cdot \exp(-0.85/kT)$. The solid solubility isotherms for Zn in GaP have been also calculated by Jordan (1971b) as a function of Zn concentration along the Ga–Zn–P liquidus isotherms over a wide temperature range. The obtained solubility isotherms are in full agreement with the experimental data.

The equilibrium distribution coefficient of Zn in GaP is equal to 0.26 (Shchegoleva et al. 1971).

9.9 GALLIUM–CADMIUM–PHOSPHORUS

GaP–Cd. The phase diagram of this system is a eutectic type (Figure 9.5) (Kuznetsov et al. 1976a). The eutectic is degenerated, crystallizes at 320°C, and contains 0.0063 mol% GaP. This system was investigated through x-ray diffraction (XRD), DTA, and metallography. The liquidus was also calculated based on the model of quasi-regular solutions. The cadmium concentration in the liquid state decreases from 0.0432 to 0.0243 at% at the temperature increasing from 800°C to 1000°C.

9.10 GALLIUM–INDIUM–PHOSPHORUS

Liquidus temperatures in the In-rich region of the Ga–In–P ternary system have been determined in the temperature region 700°C–900°C through the method of direct observation of crystal dissolution by Astles (1974), and in the temperature region 800°C–1000°C through DTA, XRD, and metallography by Kuznetsov et al. (1976b) and

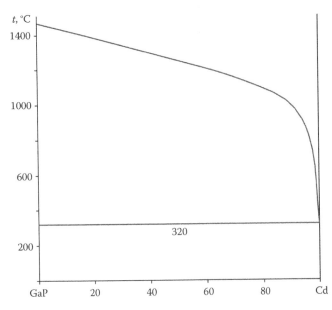

FIGURE 9.5 Phase diagram of the GaP–Cd system. (From Kuznetsov, G.M., et al., *Izv. AN SSSR Neorgan. Materialy*, *12*(7), 1168, 1976. With permission.)

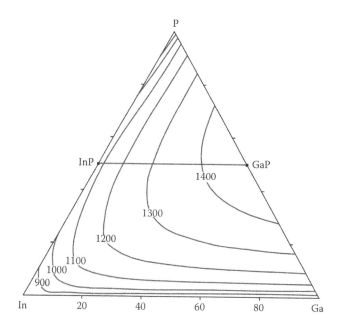

FIGURE 9.6   Calculated liquidus surface of the Ga–In–P ternary system. (From Stringfellow, G.B., *J. Electrochem. Soc.*, *117*(10), 1301, 1970. Reproduced with permission of The Electrochemical Society.)

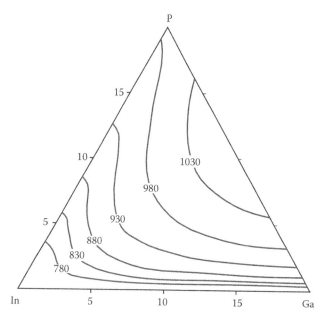

FIGURE 9.7   Calculated liquidus surface of the In-rich corner of the Ga–In–P ternary system. (From Khuber, D.V., in *Protsesy Rosta i Sinteza Poluprovodn. Kristallov i Plenok*, Part 2, Nauka Publish., Novosibirsk, 1975, 212–218. With permission.)

Shumilin et al. (1977) and calculated by Khuber (1975). The Ga–In–P phase diagram at temperatures below 800°C was experimentally obtained and calculated by Sugiura et al. (1980). Liquidus and solidus isotherms at 800°C were investigated using liquid phase epitaxy (LPE), and the solid phase composition was determined through XRD (Morrison and Bedair 1982). A full liquidus surface mainly in the GaP–InP–In–Ga region has been determined experimentally by Sugiura et al. (1980) and calculated on the basis of various solution models by Mabbitt (1970), Panish (1970b), Stringfellow (1970), Kajiyama (1971), Panish and Ilegems (1972), Sugiura et al. (1980), Ishida et al. (1989b), and Li et al. (2000); it is given in Figure 9.6. The calculated isothermal liquidus and isosolid concentration lines and P_2 and P_4 vapor pressure were found to be in agreement with published experimental data (Panish 1970b; Stringfellow 1970; Li et al. 2000). The calculated liquidus surface of the In-rich corner of the Ga–In–P ternary system is shown in Figure 9.7 (Khuber 1975). It was shown that quasi-chemical approximation is the best for theoretical description of this ternary system (Goliusov et al. 1980).

The isothermal sections of the Ga–In–P ternary system at 600°C, 1000°C, and 1100°C is shown in Figure 9.8 (Ishida et al. 1989a; Li et al. 2000).

GaP–In. This system is a non-quasi-binary section of the Ga–In–P ternary system (Figure 9.9) (Kuznetsov et al. 1976b). A $Ga_xIn_{1-x}P$ solid solution is formed at the interaction of GaP and In.

GaP–InP. A complete solid solubility in this system was suggested by Goryunova and Sokolova (1960a, 1960b). The solidus boundary in the GaP–InP system was determined by a technique that uses the electron microprobe to analyze the solid phase of compositions equilibrated in the solid–liquid phase field (Foster and Scardefield 1970). This boundary shows the substantial positive deviation from ideal behavior. A phase diagram of the GaP–InP system in the whole concentration range was constructed through DTA, XRD, and metallography by Shumilin et al. (1972), Bodnar et al. (1976), and Il'yin et al. (1976). This phase diagram was also calculated using various solution models by Blom (1971), Foster and Woods (1971), Osamura et al. (1972b), Panish and Ilegems (1972), Stringfellow (1972a), Cho (1976), Bublik and Leikin (1978), and Ishida et al. (1989a). A critically assessed phase diagram of the GaP–InP system is shown in Figure 9.10 (Li et al. 2000). The miscibility gap is asymmetric, leaning to the GaP side, and the critical temperature is about 660°C (482°C [Bodnar et al. 1976]; 624°C [Stringfellow 1982a]; 635°C [Stringfellow 1982b];

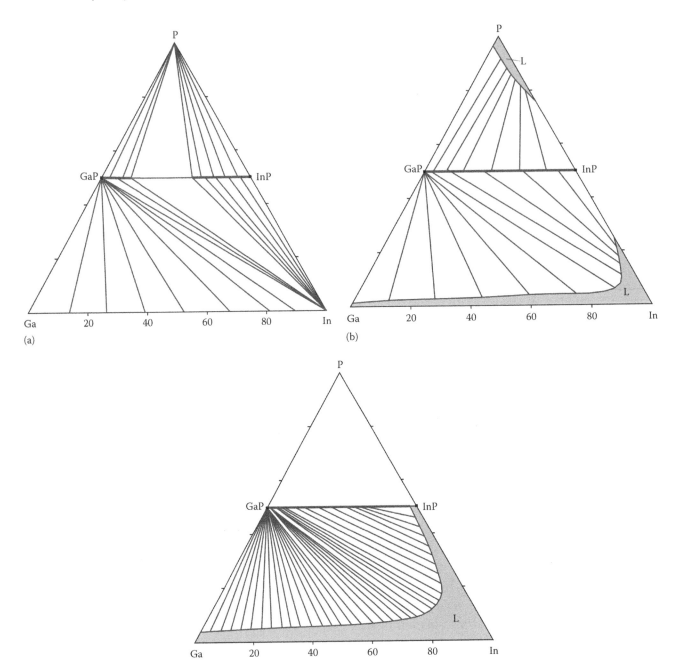

FIGURE 9.8 Isothermal sections of the Ga–In–P ternary system at (a) 600°C, (b) 1000°C, and (c) 1100°C. (From Ishida, K., et al., *J. Less Common Metals*, *155*(2), 193, 1989a; Li, Ch., et al., *J. Phase Equilibria*, *21*(4), 357, 2000. With permission.)

700°C [Stringfellow 1983]; 277 K and a critical composition of 60.3 mol% GaP [Wei et al. 1990]).

The *p-T-x* phase diagram of the GaP–InP system was determined by Ufimtsev et al. (1970). Integral excess Gibbs energy, enthalpy, and entropy at 27°C have been calculated, and a positive deviation was recorded (Kühn and Leonhardt 1972).

A first-principles calculation of a temperature composition epitaxial phase diagram of the GaP–InP system has been performed, and calculated values of the equilibrium molar volumes, the bulk moduli, and pressure derivatives for $GaInP_2$, $GaIn_3P_4$, and Ga_3InP_4 compositions were obtained (Wei et al. 1990).

Single crystals of the $Ga_xIn_{1-x}P$ solid solutions were grown by the chemical transport reaction using I_2 or InI as the transport agents (Bodnar et al. 1978, 1979). $Ga_{0.21}In_{0.79}P$ solid solution has been grown by gas source

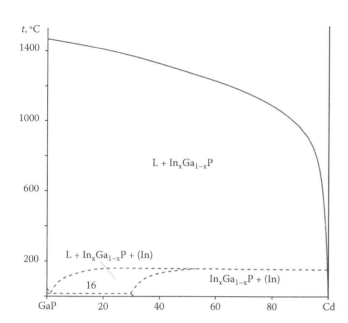

FIGURE 9.9 Phase relations in the GaP–In system. (From Kuznetsov, G.M., *Izv. AN SSSR Neorgan. Materialy*, *12*(8), 1352, 1976. With permission.)

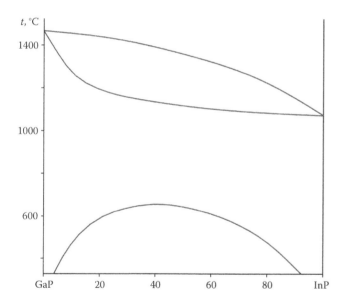

FIGURE 9.10 Phase diagram of the GaP–InP system. (From Li, Ch., et al., *J. Phase Equilibria*, *21*(4), 357, 2000. With permission.)

MBE (Asahia et al. 1996). Phase separation was not observed in this material by XRD.

InP–Ga. $Ga_xIn_{1-x}P$ solid solution containing 2.1 at% In and In with a small concentration of GaP (1.5 mol%) are formed at the interaction of InP and Ga (Batyrev et al. 1977).

9.11 GALLIUM–THALLIUM–PHOSPHORUS

GaP–TlP. The phase separation was observed in the $Ga_xTl_{1-x}P$ solid solutions grown by gas source MBE on GaAs substrate (Asahia et al. 1997). The thallium composition of phase separation has a lattice constant close to that of GaAs, ranging from about 0.47 to 0.53.

9.12 GALLIUM–EUROPIUM–PHOSPHORUS

Two ternary compounds, **$EuGa_2P_2$** and **$Eu_6Ga_2P_6$**, are formed in the Ga–Eu–P system. $EuGa_2P_2$ crystallizes in the monoclinic structure with the lattice parameters $a = 928.22 \pm 0.09$, $b = 389.67 \pm 0.04$, and $c = 1207.77 \pm 0.11$ pm and $\beta = 95.5220 \pm 0.0010°$ and a calculated density of 5.398 g·cm⁻³ at 100 ± 1 K (Goforth et al. 2009). It was synthesized using a 3:150:6 (Eu/Ga/P) molar ratio of the elements (11 g total mass). In this reaction, Ga is both a reactant and a flux. The elements were loaded into a 5 mL alumina crucible in the following order: Ga, Eu and P, and Ga. The crucible, held in place between two layers of tightly packed quartz wool, was subsequently sealed under circa 0.02 Pa of Ar in a fused silica ampoule. The reaction vessel was then heated in a box furnace in an upright position. The reaction contents were heated to 1100°C, kept at this temperature for 16 h, cooled to 600°C at 2°C·h⁻¹, and finally centrifuged to remove molten flux at 600°C. The ampoule was broken in air to reveal large, silver parallelepipeds of the title compound as the sole reaction product. All reagents were handled under an inert atmosphere.

$Eu_6Ga_2P_6$ crystallizes in the orthorhombic structure with the lattice parameters $a = 1854.3 \pm 0.3$, $b = 633.5 \pm 0.1$, and $c = 1227.6 \pm 0.2$ pm (Somer et al. 1996a). It was obtained from a stoichiometric mixture of GaP, EuP, and Eu in a sealed steel ampoule at 1100°C.

9.13 GALLIUM–SILICON–PHOSPHORUS

The liquidus surface (Figure 9.11) and thermodynamic activities for the Ga–Si–P ternary system have been calculated using the ternary regular solution model (Jordan and Weiner 1974). At 1097°C, the primary and secondary phase fields are tangential, while below that temperature, the three-phase equilibria of GaP, Si, and liquid take place. On the Ga-rich side, the ternary eutectic valley is shown from 1097°C to 800°C.

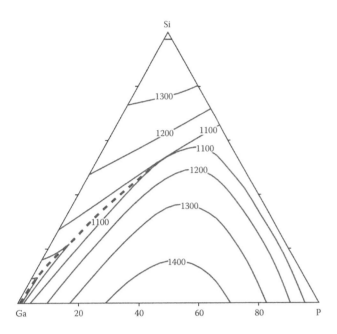

FIGURE 9.11 Calculated liquidus surface of the Ga–Si–P ternary system. (From Jordan, A.S., and Weiner, M.E., *J. Electrochem. Soc.*, *121*(12), 1634, 1974. Reproduced with permission of The Electrochemical Society.)

9.14 GALLIUM–GERMANIUM–PHOSPHORUS

The region of the solid solution based on Ge in the Ga–Ge–P ternary system at 600°C and 800°C is given in Figure 9.12 (Glazov and Malyutina 1963b). The samples were annealed at 600°C and 800°C and

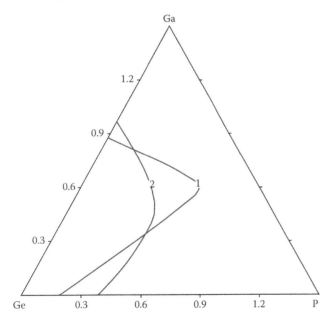

FIGURE 9.12 Solid solution based on Ge in the Ga–Ge–P ternary system at (1) 600°C and (2) 800°C. (From Glazov, V.M., and Malyutina, G.L., *Zhurn. Neorgan. Khimii*, 8(10), 2372, 1963. With permission.)

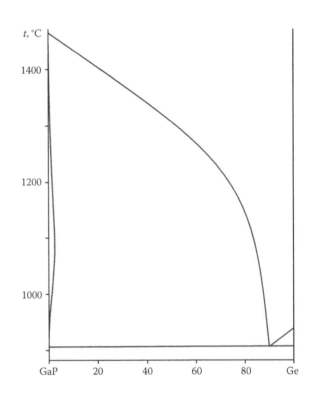

FIGURE 9.13 Phase diagram of the GaP-Ge system. (From Glazov, V.M., et al., *Izv. AN SSSR Neorgan. Materialy*, *16*(4), 595, 1980. With permission.)

investigated through metallography and measurement of microhardness.

GaP–Ge. The phase diagram of this system is a eutectic type (Figure 9.13) (Glazov et al. 1977b, 1980). The eutectic composition is 10 mol% GaP. The maximum solubility of GaP in Ge reaches 1.1 mol% at circa 800°C (Glazov and Malyutina 1963b).

Ultrafast cooling (10^7–10^8°C·s^{-1}) of the GaP–Ge melts leads to the diffusionless crystallization and formation of metastable solid solutions (Glazov et al. 1984).

9.15 GALLIUM–TIN–PHOSPHORUS

Theoretical analysis of the GaP–Ga–Sn subsystem of the Ga–Sn–P ternary system on the basis of the regular solution model has been presented by Leonhardt and Kühn (1974a). The solubilities of GaP in Sn in the temperature range of 600°C–1200°C were determined. The calculated liquidus surface of this subsystem is shown in Figure 9.14.

GaP–Sn. The phase diagram of this system is a eutectic type (Figure 9.15) (Leonhardt and Kühn 1974a; Kuznetsov et al. 1976a). The eutectic is degenerated, crystallizes at 231°C, and contains 8.7·10^{-5} mol% GaP. This system was investigated through XRD, DTA, and

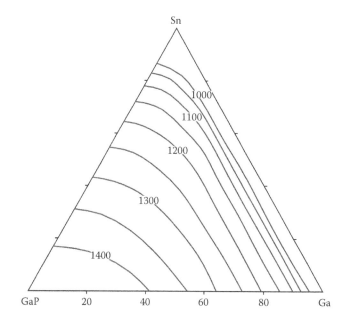

FIGURE 9.14 Calculated liquidus surface of the GaP–Ga–Sn subsystem. (From Leonhardt, A., and Kühn, G., *Krist. Tech.*, 9(1), 77, 1974. With permission.)

metallography. The liquidus was also calculated based on the model of quasi-regular solutions. Tin concentration in the liquid state decreases from 0.126 to 0.0336 at% at a temperature increasing from 800°C to 1000°C. The temperature dependence of the GaP limiting solubility in the Sn melt was determined by Vasil'yev et al. (1983) and is given in Figure 9.16.

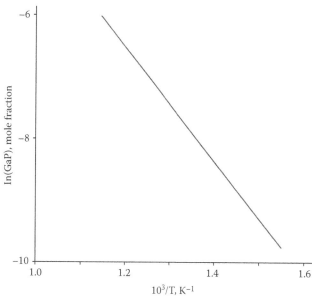

FIGURE 9.16 Temperature dependence of the GaP limiting solubility in the Sn melt. (From Vasil'yev, M.G., et al., *Izv. AN SSSR Neorgan. Materialy*, 19(11), 1933, 1983. With permission.)

9.16 GALLIUM–LEAD–PHOSPHORUS

GaP–Pb. The phase diagram of this system is a eutectic type (Figure 9.17) (Leonhardt and Kühn 1975; Kuznetsov et al. 1976a). The eutectic is degenerated, crystallizes at 327°C, and contains 0.0034 mol% GaP. This system was investigated through XRD, DTA, and metallography. The liquidus was also calculated based on the model of

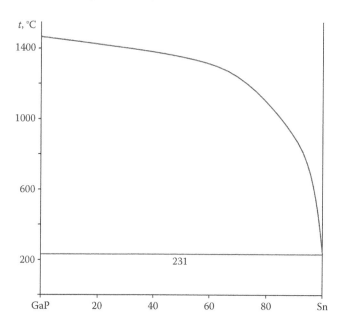

FIGURE 9.15 Phase diagram of the GaP–Sn system. (From Kuznetsov, G.M., et al., *Izv. AN SSSR Neorgan. Materialy*, 12(7), 1168, 1976. With permission.)

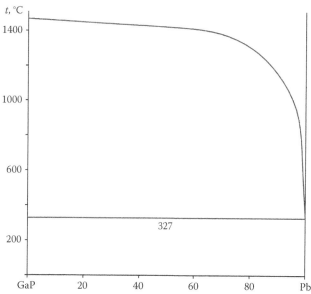

FIGURE 9.17 Phase diagram of the GaP–Pb system. (From Kuznetsov, G.M., et al., *Izv. AN SSSR Neorgan. Materialy*, 12(7), 1168, 1976. With permission.)

quasi-regular solutions. The solubility of GaP in the Pb melt at 1050°C–1275°C was determined by Leonhardt and Kühn (1974b).

9.17 GALLIUM–ARSENIC–PHOSPHORUS

A large part of the GaP–Ga–GaAs liquidus surface in the Ga-rich region has been determined using DTA and XRD (Osamura et al. 1972a). The liquidus surface has been also calculated using a quasi-chemical equilibrium model and the regular solution model. Both of these calculated phase diagrams were found to agree with the experimental data (Antypas 1970b; Osamura et al. 1972a; Panish and Ilegems 1972). Liquidus isotherms at 900°C, 1000°C, 1100°C, 1200°C, and 1300°C have been determined for this part of the Ga–As–P ternary system by Panish (1969). The corresponding solidus isotherms were determined at 900°C, 1100°C, 1200°C, and 1300°C, and crystallization paths for the solid and liquid were estimated. DTA experiments were also performed by Muszyński and Ryabcev (1975, 1976); however, original data points are not given, but only a mathematical description of the liquidus temperatures by a fourth-order polynomial of the compositions. The liquidus surface of the Ga–As–P ternary system has also been calculated by Ishida et al. (1989b).

The liquidus surface of the Ga–As–P ternary system given in Figure 9.18 displays the invariant equilibria and

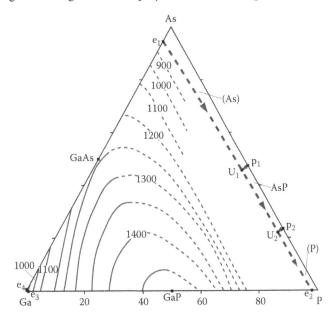

FIGURE 9.18 Calculated liquidus surface of the Ga–P–As system. (From Ishida, K., et al., *J. Cryst. Growth*, 98(1–2), 140, 1989. With permission.)

the huge primary phase field of GaP$_x$As$_{1-x}$ solid solutions (Ishida et al. 1989b). This field was taken in the GaP–Ga–GaAs region from the calculation of Osamura et al. (1972a), which is supported by all the experimental data up to 1300°C; a slight adaptation to the accepted binary liquidus line of GaAs has been made.

Isothermal sections of the Ga–As–P ternary system at 1200°C, 1300°C, and 1400°C are shown in Figure 9.19 (Kaufman et al. 1981; Kaufman and Ditchek 1991).

GaP–GaAs. Solid solutions are formed in this system at all concentrations (Folberth 1955; Pizzarello 1962; Straumanis et al. 1967). The solidus boundary in the GaP–GaAs system was determined by a technique that uses the electron microprobe to analyze the solid phase of compositions equilibrated in the solid–liquid phase field (Foster and Scardefield 1972b). The phase diagram of this system (Figure 9.20) was constructed experimentally through DTA and XRD (Osamura and Murakami 1969; Ovchinnikov et al. 1974) and calculated using various solution models (Antypas 1970b; Osamura et al. 1972b; Panish and Ilegems 1972; Stringfellow 1972a; Bublik et al. 1975; Cho 1976; Bublik and Leikin 1978; Kaufman et al. 1981; Kaufman and Ditchek 1991). The critical temperature is 688°C, and the critical composition is 67.6 mol% GaP (Wei et al. 1990). The concentration dependence of the lattice parameter of the GaP$_x$As$_{1-x}$ solid solution obeys Vegard's law (has a positive deviation from Vegard's low [Bublik et al. 1974]), and the concentration dependence of the energy gap of this solid solution has a small positive deviation from linearity (Pizzarello 1962; Straumanis et al. 1967; Foster et al. 1972b).

Integral excess Gibbs energy, enthalpy, and entropy at 27°C have been calculated, and a positive deviation was recorded (Kühn and Leonhardt 1972). The thermodynamic calculation on nonstoichiometric factor δ at the boundary of the homogeneity region in Al$_x$Ga$_{1-x}$As solid solutions is given in Figure 9.21 (Ivashchenko et al. 1990). These results may speak only about qualitative tendency. It is seen that increasing x does not essentially affect δ.

A first-principles calculation of a temperature composition epitaxial phase diagram of the GaP–GaAs system has been performed, and calculated values of the equilibrium molar volumes, the bulk moduli, and pressure derivatives for Ga$_2$AsP, Ga$_4$AsP$_3$, and Ga$_4$As$_3$P compositions were obtained (Wei et al. 1990).

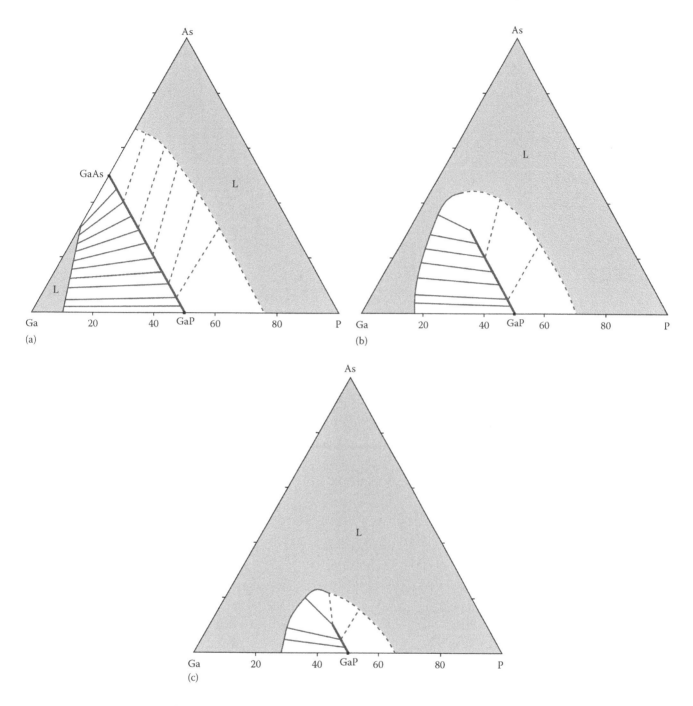

FIGURE 9.19 Calculated isothermal sections of the Ga–P–As ternary system at (a) 1200°C, (b) 1300°C, and (c) 1400°C. (From Kaufman, L., and Ditchek, B., et al., *J. Less Common Metals*, *168*(1), 115, 1991. With permission.)

Homogeneous solid solutions of gallium phosphide and gallium arsenide were synthesized using the chemical transport reactions using iodine as a transport agent (Pizzarello 1962).

9.18 GALLIUM–ANTIMONY–PHOSPHORUS

The solubility of GaP in liquid Sb at 800°C–1200°C (Figure 9.22) was determined by Kuznetsov and Sorokin

(1977). Using the quasi-regular liquid phase model, the interaction parameters have been calculated and the liquidus surface of the Ga–P–Sb ternary system has been constructed (Kuznetsov and Sorokin 1977, 1978). Liquidus equations have been derived from experimental data by Muszyński and Ryabcev (1975, 1976) using a simplex lattice method in order to interpolate the liquidus curves over the entire composition range. Liquidus

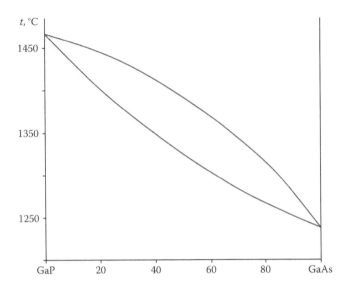

FIGURE 9.20 Phase diagram of the GaP–GaAs system. (From Foster, L.M., et al., *J. Electrochem. Soc.*, *119*(10), 1426, 1972. Reproduced with permission of The Electrochemical Society.)

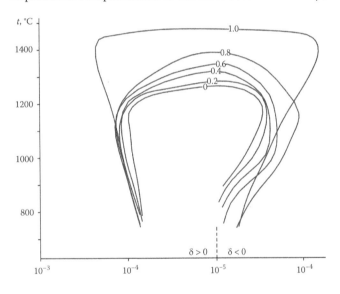

FIGURE 9.21 Calculated behavior of the nonstoichiometric factor δ at the boundary of the homogeneous region in the $GaP_{1-x}As_x$ ternary solid solution. (From Ivashchenko, A.I., et al., *Cryst. Res. Technol.*, *25*(6), 661, 1990. With permission.)

isotherms at temperatures between 1050°C and 1400°C are given in Figure 9.23.

Two isothermal sections at 650°C and 1000°C based on the thermodynamic calculations are given in Figure 9.24 (Ishida et al. 1989a). A slight difference between these sections is noted, in that the liquidus isotherms exhibit different curvatures on the phosphorus-rich side. No experimental data are available in that range to decide between these two different thermodynamic calculations.

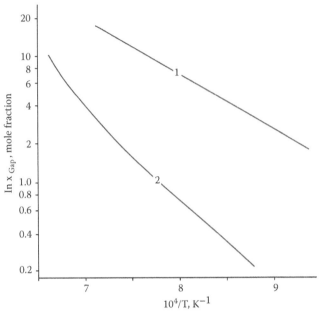

FIGURE 9.22 Temperature dependence of GaP solubility in (1) Sb and (2) Bi melts. (From Kuznetsov, V.V., and Sorokin, V.S., *Izv. AN SSSR Neorgan. Materialy*, *13*(5), 780, 1977. With permission.)

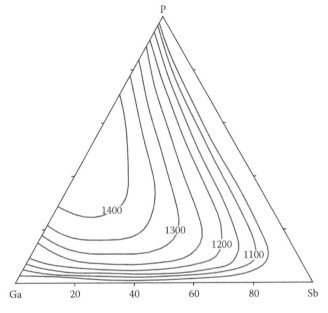

FIGURE 9.23 Calculated liquidus surface of the Ga–P–Sb ternary system. (From Kuznetsov, V.V., and Sorokin, V.S., *Izv. AN SSSR Neorgan. Materialy*, *13*(5), 780, 1977. With permission.)

GaP–GaSb. The phase diagram of this system is shown in Figure 9.25 (Ishida et al. 1989a, 1989b). The peritectic temperature is about 713°C. The mutual solubility between GaP and GaSb is very small. The miscibility gap obtained by means of electron probe microanalysis

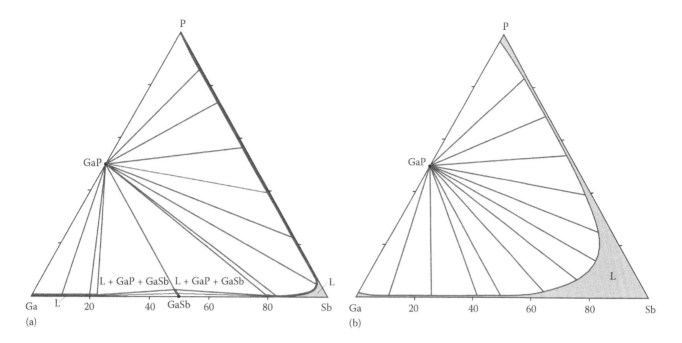

FIGURE 9.24   Isothermal sections of the Ga–P–As ternary system at (a) 650°C and (b) 1000°C. (From Ishida, K., et al., *J. Less Common Metals*, 155(2), 193, 1989. With permission.)

(EPMA) is asymmetric, leaning to the GaP side. The critical temperature for this miscibility gap is equal to 1692°C–1723°C (Stringfellow 1982b, 1983).

9.19 GALLIUM–BISMUTH–PHOSPHORUS

A theory of regular associated solutions has been used to provide a consistent representation of the ternary phase

equilibria in the Ga–Bi–P ternary system (Leonhardt and Kühn 1974a; Jordan 1976; Kuznetsov and Sorokin 1977). The liquidus isotherms for this system between 1000°C and 1400°C are presented in Figure 9.26 (Jordan 1976). On the P-rich side, the liquidus isotherms at and

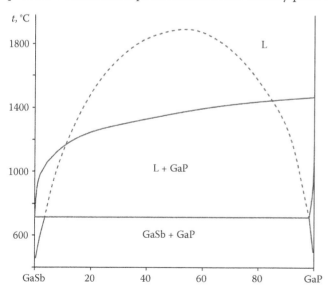

FIGURE 9.25   Phase diagram of the GaP–GaSb system: solid lines present the stable equilibria; dashed line is a metastable extrapolation of the GaP + GaSb miscibility gap. (From Ishida, K., et al., *J. Less Common Metals*, 155(2), 193, 1989. With permission.)

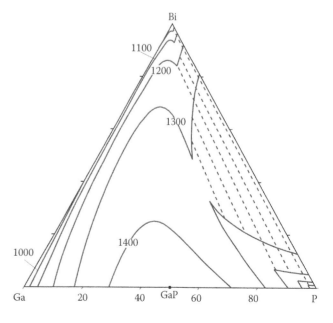

FIGURE 9.26   Calculated liquidus surface of the Ga–Bi–P ternary system. (From Jordan, A.S., *Met. Trans. B*, 7(2), 191, 1976. With permission.)

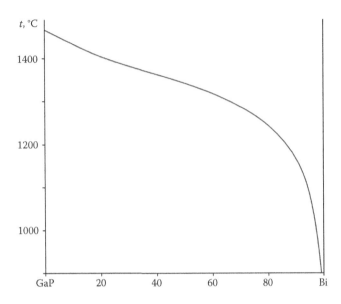

FIGURE 9.27 Phase diagram of the GaP–Bi system. (From Jordan, A.S., *Met. Trans. B*, *7*(2), 191, 1976. With permission.)

below 1300°C are intersected by a liquid–liquid miscibility gap originating from the Bi–P boundary binary system.

GaP–Bi. The phase diagram of this system is a eutectic type (Figure 9.27) (Leonhardt and Kühn 1974a; Akchurin et al. 1984). The eutectic is degenerated, crystallizes at 231°C, and contains $8.7 \cdot 10^{-5}$ mol% GaP. This system was investigated through XRD, DTA, and metallography. The liquidus was also calculated based on the model of quasi-regular solutions. The solubility of GaP in liquid antimony at 800°C–1200°C (Figure 9.22) was determined by Kuznetsov and Sorokin (1977). The solubility of GaP in the Bi melt at 1025°C–1275°C was also determined by Leonhardt and Kühn (1974b). The solidus curve in this system is retrograde (Jordan et al. 1976). The solubility of Bi in GaP increases with increasing temperature. Although the measured values at circa 1050°C are low (~10^{17} cm⁻³), it was predicted that doping levels as high as ~10^{19} cm⁻³ should be possible by a suitable high-temperature crystal growth technique. Bi-doped GaP crystals were grown by a thermal gradient transport technique.

At the Bi content in the liquid phase from 3 to 12 at%, the distribution coefficient of Bi in GaP is almost constant and equal to $(4.0 \pm 1.5) \cdot 10^{-5}$ (Saenko 1977). Thus, the Bi concentration in the grown crystals was within the limits $(1–5) \cdot 10^{16}$ cm⁻³.

9.20 GALLIUM–OXYGEN–PHOSPHORUS

The isothermal section of the Ga–O–P ternary system at room temperature is given in Figure 9.28 (Schwartz et al. 1980; Schwartz 1983). In the P-rich part of this system, there are inadequate data to establish the equilibrium tie-lines.

Three compounds, **GaPO$_4$**, **Ga(PO$_3$)$_3$**, and **Ga$_3$PO$_7$**, are formed in the Ga–O–P ternary system. GaPO$_4$ exists in various modifications and melts at 1670 ± 10°C (Shafer and Roy 1956) (at a temperature higher than 1600°C [Perloff 1956]). The stable GaPO$_4$ crystallizes in the trigonal structure with the lattice parameters $a = 490.08 \pm 0.11$, 490.61 ± 0.09, 491.23 ± 0.13, 491.97 ± 0.16, 492.82 ± 0.11, and 493.52 ± 0.18 and $c = 1104.9 \pm 0.1$, 1105.3 ± 0.2, 1105.9 ± 0.3, 1106.5 ± 0.4, 1107.0 ± 0.3, and 1107.9 ± 0.4 pm at 20°C, 96°C, 191°C, 285°C, 380°C, and 474°C, respectively (Nakae et al. 1995) ($a = 492$ and $c = 1110$ pm and a calculated density of 3.54 g·cm⁻³ [Mooney et al. 1954; Perloff 1956; Shafer and Roy 1956]; $a = 491.0 \pm 0.8$ and $c = 1107 \pm 2$ pm [Kosten and Arnold 1980]; $a = 490.1 \pm 0.1$, 493.4 ± 0.2, and 497.3 ± 0.2 and $c = 1104.8 \pm 0.1$, 1107.5 ± 0.4, and 1110.5 ± 0.6 pm at 20°C, 500°C, and 750°C, respectively [Baumgartner et al. 1984]; $a = 487.4 \pm 0.2$, 489.9 ± 0.1, and 489.7 ± 0.1 and $c = 1103.3 \pm 0.4$, 1103.4 ± 0.2, and 1102.1 ± 0.2 pm at –100°C, 20°C, and 100°C, respectively [Goiffon et al. 1986]; $a = 490.0 \pm$

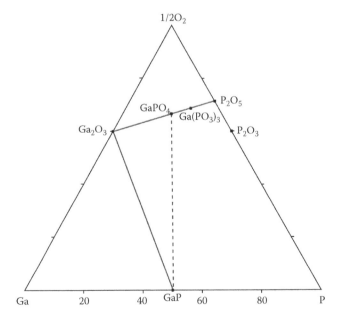

FIGURE 9.28 Isothermal section of the Ga–O–P ternary system at room temperature. (From Schwartz, G.P., *Thin Solid Films*, *103*(1–3), 3, 1983. With permission.)

0.1 and $c = 1104.8 \pm 0.3$ [Litvin et al. 1987]; $a = 490.43 \pm 0.07$ and $c = 1104.79 \pm 0.05$ pm at ambient conditions and $a = 464.21 \pm 0.08$ and $c = 1080.22 \pm 0.06$ pm at 6.80 ± 0.07 GPa [Sowa 1994]). A pressure dependence of the lattice parameters can be expressed as follows: a (pm) = $(490.2 \pm 0.3) - (4.7 \pm 0.2)p + (0.12 \pm 0.02)p^2$ and c (pm) $= 1104.6 - (4.9 \pm 0.3)p + (0.18 \pm 0.03)p^2$, where p is in gigapascals (Sowa 1994).

The inversion between the trigonal and tetragonal structure of $GaPO_4$ takes place at $933 \pm 4°C$ (Shafer and Roy 1956) (at around $1000°C$ [Hirano et al. 1986; Hirano and Kim 1989]). The lattice parameters of high-temperature modification follow: $a = 506$ and $c = 716$ pm (Mooney et al. 1954) ($a = 492$ and $c = 687$ pm and a calculated density of 3.27 g·cm^{-3} [Perloff 1956; Shafer and Roy 1956]). The tetragonal form of this compound shows a metastable inversion (Shafer and Roy 1956).

There is also a metastable modification of $GaPO_4$ that crystallizes in the orthorhombic structure (pseudotetragonal cell) with the lattice parameters $a = b = 695$ and $c = 687$ pm (Mooney et al. 1954) ($a = b = 696.7 \pm 0.3$ and $c = 686.6 \pm 0.3$ pm and a calculated density of 3.26 g·cm^{-3} [Mooney et al. 1956b]). Perloff (1956) indicated that at $1200°C$, this compound crystallizes in the cubic structure with the lattice parameter $a = 716$ pm.

According to the data of Dachille and Roy (1959), $GaPO_4$ transforms to a new form near 5.5 GPa at $450°C$. A pressure-induced amorphization takes place at about 9 GPa (Sowa 1994).

The quartz form of $GaPO_4$ with a trigonal structure was prepared by mixing Ga_2O_3 and 85% H_3PO_4 to moderately high temperatures ($200°C–300°C$) in a hydrothermal bomb for a week or more (Mooney et al. 1954; Perloff 1956; Shafer and Roy 1956; Kosten and Arnold 1980). $GaPO_4$ could also be obtained by evaporation of the solution of the title compound in 85% H_3PO_4 (Perloff 1956; Baumgartner et al. 1984). Powders of $GaPO_4$ were prepared by the solid-state reaction of a stoichiometric mixture of Ga_2O_3 and $NH_4H_2PO_4$ at $800°C$ or $1000°C$ for 24 h, and subsequently by hydrothermal treatment at $180°C$ or $300°C$ and 14.7 or 30 MPa for 24 h in 4 M H_3PO_4 or 8.25 M H_3PO_4 (Hirano et al. 1986; Hirano and Kim 1989).

$Ga(PO_3)_3$ crystallizes in the monoclinic structure with the lattice parameters $a = 829.3$, $b = 1519.6$, and $c = 618.8$ pm and $\beta = 105.89°$ (Chudinova et al. 1987). It was prepared from the mixture of Ga_2O_3 and H_3PO_4 with the presence of K^+ ions at the molar ratio P/K/Ga =

15:5:1 at a temperature higher than $350°C$ (Chudinova et al. 1977). The mixture of 85% H_3PO_4, K_2CO_3 and GaO_3 was heated in glassy carbon crucibles for 5–10 days.

Ga_3PO_7 crystallizes in the trigonal structure with the lattice parameters $a = 788.5 \pm 0.1$ and $c = 672.7 \pm 0.1$ pm and a calculated density of 4.843 g·cm^{-3} (Boudin and Lii 1998). Crystals of the title compound were obtained from a high-pressure hydrothermal synthesis. The starting materials, $NH_4H_2PO_4$ (76 mg) and Ga_2O_3 (124 mg) (molar ratio 1:1) and H_2O (50 mkL), were sealed in a 1 mL gold tube. The ampoule was placed in an autoclave in which pressure was applied by pumped water and compressed air. The autoclave was first heated at $500°C$ for 24 h at a pressure of 210 MPa, then cooled slowly to $270°C$ at a rate of $4°C·h^{-1}$, and finally quenched to room temperature by removing the autoclave from the furnace. The product was filtered, washed with H_2O and MeOH, and dried in air.

9.21 GALLIUM–SULFUR–PHOSPHORUS

$GaPS_4$ ternary compound with a homogeneity region of less than 10 mol%, which decomposes at $725°C$ (at $460°C$ [Nitsche and Wild 1970]), is formed in the Ga–S–P system (Galagovets et al. 1995). It crystallizes in the monoclinic structure with the lattice parameters $a = 860.3 \pm 0.4$, $b = 777.8 \pm 0.3$, and $c = 1185.8 \pm 0.5$ pm and $\beta = 135.46 \pm 0.04°$ and the calculated and experimental densities of 2.72 and 2.65 g·cm^{-3}, respectively (Buck and Carpentier 1973; Voroshilov et al. 2016) ($a = 861 \pm 3$, $b = 778 \pm 2$, and $c = 1185 \pm 5$ pm and $\beta = 135.4°$ [Buck and Nitsche 1971]). The crystal growth of the title compound was carried out by the chemical transport at $600°C–650°C$ using I_2 as a transport agent (Nitsche and Wild 1970; Buck and Nitsche 1971). $GaPS_4$ forms a colorless polyhedron.

$GaP–Ga_2S_3$. A high-solubility region based on GaP has been determined in this system through XRD, DTA, and measurement of microhardness (Negreskul 1964; Negreskul and Radautsan 1964; Radautsan and Negreskul 1964). The lattice parameter of this solid solution obeys Vegard's law.

9.22 GALLIUM–SELENIUM–PHOSPHORUS

$GaP–Ga_2Se_3$. According to the data of XRD, DTA, and measurement of microhardness, solid solutions are formed in this system at all concentrations (Radautsan et al. 1961a; Negreskul and Radautsan 1964).

9.23 GALLIUM–TELLURIUM–PHOSPHORUS

Part of the condensed ternary phase diagram for the Ga–Te–P system has been constructed on the basis of data obtained from DTA, XRD, and EPMA (Panish 1967b). No regions of extensive solid solubility of tellurium in GaP or GaP in Ga_2Te_3 were observed. The liquidus surface in the Ga-rich region of this ternary system is shown in Figure 9.29.

GaP–Ga_2Te_3. The earlier investigations indicated the formation of limited solid solubility based on both compounds (Goryunova and Radautsan 1964; Negreskul and Radautsan 1964; Pyshkin and Negreskul 1965), but a later (Panish 1967b) investigation did not confirm these data.

GaP–Te. This system is a quasi-binary section of the Ga–Te–P ternary system (Panish 1967b). DTA data were not obtained for mixtures on the GaP–Te system. XRD and EPMA for samples prepared along this cut yield data that indicate that a ternary compound may exist in the P-rich region, which decomposes to yield P and Ga_2Te_3 on cooling.

The equilibrium distribution coefficient of Te in GaP is equal to 0.041 (Shchegoleva et al. 1974).

9.24 GALLIUM–CHROMIUM–PHOSPHORUS

The maximum solubility of chromium in GaP at its melting temperature is equal to $1 \cdot 10^{18}$ cm^{-3} (Dement'ev et al. 1980). The calculated solidus for Cr solubility in GaP is shown in Figure 9.30. The distribution coefficient of Cr in GaP is 0.001.

9.25 GALLIUM–IODINE–PHOSPHORUS

A reaction of GaP with iodine begins at temperatures not exceeding 200°C (Sandulova and Ostrovskiy 1971). The reaction products are higher gallium iodides and phosphorus. A reaction of GaI_3 with GaP begins at 700°C. A GaP transport is carried out by the following reaction: $2GaP(s) + GaI_3(g) \longleftrightarrow 3GaI_2(g) + 0.5P_4(g)$. Free gallium interferes with the interaction of GaP and GaI_3. The equilibrium P pressure in the GaP–Ga–I system is not dependent on the iodine concentration. It is determined by the GaP dissociation pressure, which increases with increasing temperature.

9.26 GALLIUM–MANGANESE–PHOSPHORUS

The maximum solubility of manganese in GaP at its melting temperature is equal to $(1–3) \cdot 10^{18}$ cm^{-3} (Dement'ev et al. 1980). The calculated solidus for Mn solubility in

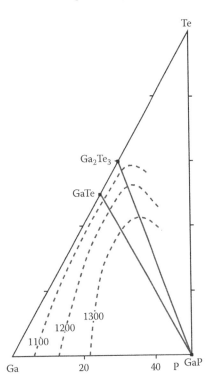

FIGURE 9.29 Liquidus surface in the Ga-rich region of the Ga–Te–P ternary system. (From Panish, M.B., *J. Electrochem. Soc.*, *114*(11), 1161, 1967. Reproduced with permission of The Electrochemical Society.)

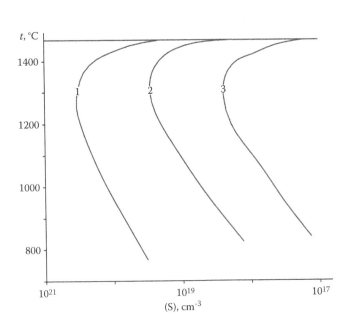

FIGURE 9.30 Calculated solubility for (1) Mn, (2) Cr, and (3) Fe in GaP. (From Dement'ev, Yu.S., et al., *Izv. AN SSSR Neorgan. Materialy*, *16*(7), 1164, 1980. With permission.)

GaP is shown in Figure 9.30. The distribution coefficient of Mn in GaP is 0.15.

9.27 GALLIUM–IRON–PHOSPHORUS

The maximum solubility of iron in GaP at its melting temperature is equal to $(5-8) \cdot 10^{17}$ cm^{-3} (Dement'ev et al. 1980). The calculated solidus for Fe solubility in GaP is shown in Figure 9.30. The distribution coefficient of Fe in GaP is 0.0001.

9.28 GALLIUM–COBALT–PHOSPHORUS

The isothermal section of the Ga–Co–P ternary system at 500°C (Figure 9.31) was constructed by Kodentsov et al. (2001). No ternary phases exist at this temperature. All binary Co-P and Co-Ga intermetallics are in equilibrium with GaP, which dissolves up to 4 at% Co (Swenson and Chang 1996a; Kodentsov et al. 2001).

The samples were sealed in quartz evacuated ampoules and first annealed at 500°C for 400 h (Kodentsov et al. 2001). Then, the sintered compacts were pulverized, again pressed into pellets, and annealed in vacuum at 500°C for another 400 h. For studying reactions between GaP and Co, the "sandwich" diffusion couples GaP/Co/GaP were prepared.

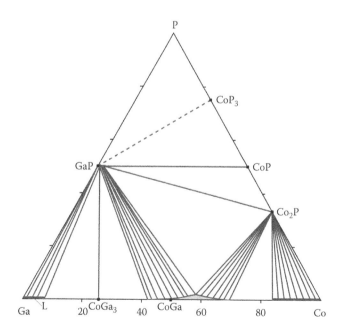

FIGURE 9.31 Isothermal section of the Ga–Co–P ternary system at 500°C. (From Kodentsov, A.A., et al., *J. Phase Equilibria*, 22(3), 227, 2001. With permission.)

9.29 GALLIUM–NICKEL–PHOSPHORUS

Phase equilibria in the Ga–Ni–P ternary system at 600°C were established through XRD (Swenson and Chang 1996a). A region of three-phase equilibrium was found to exist between GaP, Ni$_2$Ga$_3$, and Ni$_2$P. According to the phase rule, Ni$_2$P and Ni$_5$P$_4$ must also coexist with GaP, and NiGa is in equilibrium with Ni$_2$P. The ingots were annealed at 600°C for 14 days. Then, they were quenched in iced water and ground into powders, which were again pressed into pellets, encapsulated in evacuated quartz ampoules, and annealed for an additional 5 months

According to the data of Keimes et al. (1999), **GaNi$_{10}$P$_3$** ternary compound, which crystallizes in the trigonal structure with the lattice parameters $a = 762.9 \pm 0.1$ and $c = 944.7 \pm 0.2$ pm, is formed in the Ga–Ni–P ternary system. It was prepared by heating the stoichiometric mixtures of the elements in a corundum crucible at 1000°C–1100°C for 1 h in an Ar atmosphere.

9.30 GALLIUM–RHODIUM–PHOSPHORUS

Phase equilibria in the Ga–Rh–P ternary system at 700°C were established through XRD (Swenson and Chang 1996a). RhGa and RhP$_2$ were found to be in the thermodynamic equilibrium with GaP. All the rhodium gallides, as well as RhP$_3$, must be also in equilibrium with GaP. The samples were sealed in evacuated quartz ampoules and annealed at 1000°C for 1 week. The annealing temperature was then slowly lowered to 700°C over a course of a second week. The samples were subsequently quenched in ice water, ground into homogeneous powders, pressed into pellets, sealed in evacuated quartz ampoules, and annealed at 700°C for 2 additional months.

9.31 GALLIUM–PALLADIUM–PHOSPHORUS

Pd$_2$Ga$_{0.4}$P$_{0.6}$ and **Pd$_5$GaP** ternary compounds are formed in the Ga–Pd–P system. Pd$_2$Ga$_{0.4}$P$_{0.6}$ crystallizes in the hexagonal structure with the lattice parameters $a = 657.8 \pm 0.1$ and $c = 340.9 \pm 0.1$ pm (Ellner et al. 1985). The title compound was prepared from Pd, Ga, and P in evacuated quartz ampoules under an Ar atmosphere (40 kPa). The homogenization was carried out for heat treatment of alloys at 750°C for 1 (powder mixture) or 2 (mixture of globules) days.

Pd$_5$GaP crystallizes in the tetragonal structure with the lattice parameters $a = 387.1$ and $c = 673.5$ pm (El-Boragy and Schubert 1970).

9.32 GALLIUM–IRIDIUM–PHOSPHORUS

Phase equilibria in the Ga–Ir–P ternary system at 700°C were established through XRD (Swenson and Chang 1996a). IrP_2 is in thermodynamic equilibrium with GaP. According to the phase rule, Ir_2Ga_9, $IrGa_3$, and IrP_3 should also be in thermodynamic equilibrium with GaP. The samples were sealed in evacuated quartz ampoules and annealed at 1000°C for 1 week. The annealing temperature was then slowly lowered to 700°C over a course of a second week. The samples were subsequently quenched in ice water, ground into homogeneous powders, pressed into pellets, sealed in evacuated quartz ampoules, and annealed at 700°C for 2 additional months.

9.33 GALLIUM–PLATINUM–PHOSPHORUS

Phase equilibria in the Ga–Pt–P ternary system at 700°C were established through XRD (Swenson and Chang 1996a). Pt_2Ga_3 and PtP_2 were found to coexist with GaP at this temperature. The phases Pt_3Ga_7 and $PtGa_2$ must be in thermodynamic equilibrium with GaP as well, according to the phase rule. The samples were sealed in evacuated quartz ampoules and annealed at 700°C for 14 days. Then, they were quenched in iced water and ground into powders, and were again pressed into pellets, encapsulated in evacuated quartz ampoules, and annealed for an additional 2 months.

According to the data of El-Boragy and Schubert (1970), **Pt_5GaP** ternary compound, which crystallizes in the tetragonal structure with the lattice parameters $a = 389.7$ and $c = 665.1$ pm, is formed in this system.

CHAPTER 10

Systems Based on GaAs

10.1 GALLIUM–SODIUM–ARSENIC

$Na_2Ga_2As_3$ ternary compound, which crystallizes in the monoclinic structure with the lattice parameters $a = 1317.5 \pm 0.3$, $b = 670.5 \pm 0.2$, and $c = 1445.9 \pm 0.3$ pm and $\beta = 90.2 \pm 0.2°$, is formed in this system (Cordier et al. 1991k). It was synthesized from elements at 930°C.

10.2 GALLIUM–POTASSIUM–ARSENIC

Some ternary compounds are formed in this system. **K_2GaAs_2**, which crystallizes in the monoclinic structure with the lattice parameters $a = 1047.6 \pm 0.3$, $b = 1439.3 \pm 0.4$, and $c = 888.4 \pm 0.3$ pm and $\beta = 110.2 \pm 0.1°$, is formed in this system (Cordier et al. 1991b). It was synthesized from elements at 330°C.

$K_2Ga_2As_3$, which also crystallizes in the monoclinic structure with the lattice parameters $a = 1378.2 \pm 0.3$, $b = 637.9 \pm 0.2$, and $c = 1564.4 \pm 0.4$ pm and $\beta = 90.4 \pm 0.2°$, is formed in this system (Cordier et al. 1991e). It was synthesized from elements at 930°C.

$K_3Ga_3As_4$ crystallizes in the orthorhombic structure with the lattice parameters $a = 659.7 \pm 0.2$, $b = 1479.2 \pm 0.4$, and $c = 1058.9 \pm 0.3$ pm (Birdwhistell et al. 1990). This compound was prepared by direct combination of the elements at elevated temperatures. In a typical preparation, a dry quartz or Vycor tube was charged with K (11 mM), Ga (3.4 mM), and As (7.5 mM). The sample tube was evacuated at a pressure of approximately 0.7 Pa for at least 1.5 h and then sealed under vacuum. The sample tube was placed inside a larger-diameter tube, which was similarly evacuated and sealed. The sample was heated at 750°C in a furnace for 10 h and then cooled to ambient temperature over a 40 h period.

A metallic black crystalline product was obtained. Side reactions produce an orange powder that was readily isolated from the $K_3Ga_3As_4$ by mechanical separation. No attempt was made to identify the orange powder.

$K_{20}Ga_6As_{12.66}$ crystallizes in the hexagonal structure with the lattice parameters $a = 1678.0 \pm 0.5$ and $c = 524.5 \pm 0.2$ pm, and a calculated density of 4.793 g·cm^{-3} (Cordier and Ochmann 1990). To prepare this compound, Ga and As in stoichiometric ratio with a slight excess of K under Ar in a glove box in a dried corundum crucible were mixed. The mixture was slowly heated under protective gas at a rate of 100°C·h^{-1} to 300°C and maintained at this temperature for about 2 h. The mixture was then heated at the same rate to 1030°C, with excess potassium evaporates from the system, followed by cooling at a rate of 50°C·h^{-1} to room temperature. The obtained compound is very moisture sensitive.

10.3 GALLIUM–RUBIDIUM–ARSENIC

Rb_2GaAs_2 ternary compound, which crystallizes in the monoclinic structure with the lattice parameters $a = 1086.7 \pm 0.3$, $b = 1475.1 \pm 0.4$, and $c = 899.2 \pm 0.3$ pm and $\beta = 108.4 \pm 0.1°$, is formed in this system (Cordier et al. 1991h). It was synthesized from elements at 330°C.

10.4 GALLIUM–CESIUM–ARSENIC

Cs_3GaAs_2 ternary compound, which crystallizes in the monoclinic structure with the lattice parameters $a = 1137.1 \pm 0.5$, $b = 885.7 \pm 0.3$, and $c = 1946.0 \pm 0.7$ pm and $\beta = 99.225°$, is formed in this system (Somer et al. 1990c). It was synthesized from the stoichiometric mixture of the elements at 680°C in a sealed Nb ampoule.

10.5 GALLIUM–COPPER–ARSENIC

Estimates have been made of the thermodynamic properties of the Ga–Cu–As ternary liquids saturated with GaAs at 900°C from solubility measurements in the ternary system and available thermodynamic information on the Ga–As binary system (Furukawa and Thurmond 1965). The liquidus curve of this system at 900°C has been determined. The liquidus surface of this ternary system is shown in Figure 10.1 (Panish 1967a). The field of GaAs primary crystallization occupies the biggest part of this system.

The calculated isothermal section of the Ga–Cu–As ternary system at room temperature, which essentially remains unchanged to about 300°C, is given in Figure 10.2 (Schmid-Fetzer 1988). The dotted lines indicate possible alternative tie-lines. The predictive calculations were based on the following simplifications: ternary phases and solid solubilities were disregarded, and the Gibbs energy of formation of binary compounds was estimated by the enthalpy of formation and calculated from Miedema's model.

GaAs–Cu. This system is a non-quasi-binary section of the Ga–Cu–As ternary system (Kuznetsov et al. 1968). A considerable decrease in the solubility of copper in GaAs has been observed when diffusion into undoped GaAs is carried to saturation at 900°C under increasing arsenic vapor pressure (Furukawa

and Thurmond 1965). The solubility of copper in single crystals of GaAs has been determined from 700°C to 1160°C by means of ^{64}Cu (Fuller and Whelan 1958; Hall and Racette 1964). It was shown that this solubility can be expressed by the following relation: x_{Cu} (cm^{-3}) = $3.7 \cdot 10^{23} \cdot$ exp $(-1.3/kT)$. According to the data of Blanc and Weisberg (1964), the solubility of Cu in GaAs at 700°C is equal to $(1.8 \pm 0.5) \cdot 10^{17}$ cm^{-3} and increases to $(2.8 \pm 0.2) \cdot 10^{18}$ cm^{-3} at 1000°C. The solubility of Cu in solid GaAs is retrograde and is shown in Figure 10.3 (Kuznetsov and Kuznetsova 1966; Kuznetsov et al. 1968).

GaAs–Cu$_3$As. This system is also a non-quasi-binary section of the Ga–Cu–As ternary system (Luzhnaya et al. 1965). **Cu$_3$GaAs$_2$** ternary compound is not formed in this system.

10.6 GALLIUM–SILVER–ARSENIC

The liquidus surface of the Ga–Ag–As ternary system is shown in Figure 10.4 (Panish 1967a). The field of GaAs primary crystallization occupies the biggest part of this system.

Ag$_3$GaAs$_2$ ternary compound is not formed in the Ga–Ag–As system (Gubskaya et al. 1965).

GaAs–Ag. The phase diagram of this system is a eutectic type (Figure 10.5) (Kuznetsov and Kuznetsova 1968). The eutectic crystallizes at 740°C

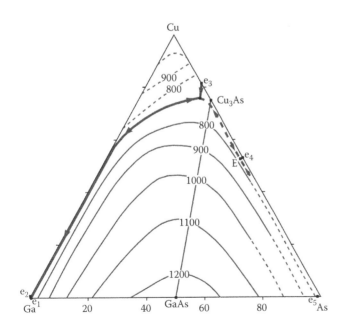

FIGURE 10.1 Liquidus surface of the Ga–Cu–As ternary system. (From Panish, M.B., *J. Electrochem. Soc.*, *114*(5), 516, 1967. Reproduced with permission of The Electrochemical Society.)

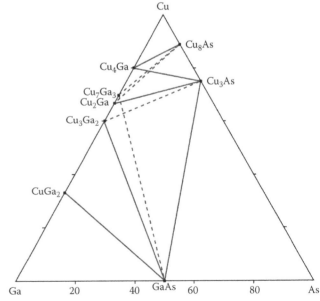

FIGURE 10.2 Calculated isothermal section of the Ga–Cu–As ternary system at room temperature. (From Schmid-Fetzer, R., *J. Electron. Mater.*, *17*(2), 193, 1988. With permission.)

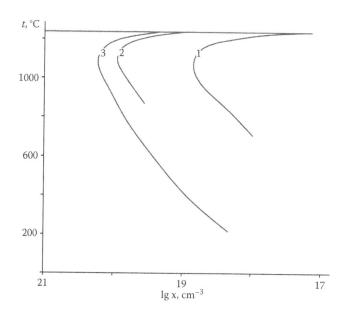

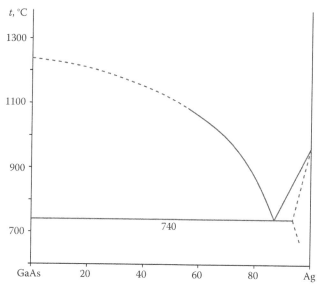

FIGURE 10.3 Solubility of (1) Cu, (2) Ge, and (3) Sn in solid GaAs. (From Kuznetsov, G.M., and Kuznetsova, S.K., *Izv. AN SSSR Neorgan. Mater.*, 2(4), 643, 1966. With permission.)

FIGURE 10.5 Phase diagram of the GaAs–Ag system. (From Kuznetsov, G.M., and Kuznetsova, S.K., *Izv. AN SSSR Neorgan. Mater.*, 4(8), 1243, 1968. With permission.)

x-ray diffraction (XRD), differential thermal analysis (DTA), and metallography. Panish (1967a) noted that the GaAs–Ag system is a non-quasi-binary section of the Ga–Ag–As ternary system.

10.7 GALLIUM–GOLD–ARSENIC

The liquidus surface of the Ga–Au–As ternary system is shown in Figure 10.6 (Panish 1967a). The field of GaAs primary crystallization occupies the biggest part of this system.

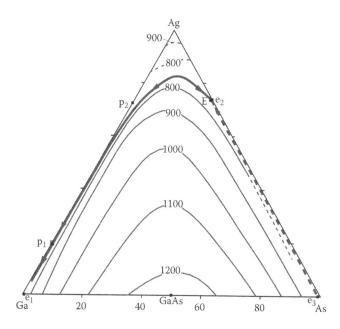

FIGURE 10.4 Liquidus surface of the Ga–Ag–As ternary system. (From Panish, M.B., *J. Electrochem. Soc.*, 114(5), 516, 1967. Reproduced with permission of The Electrochemical Society.)

and contains 13 mass% GaAs. The solubility of Ag in GaAs is negligible, and the solubility of GaAs in Ag is equal to 4.72, 6.3, and 6.5 mass% at 700°C, 720°C, and 740°C, respectively. The samples were annealed at 700°C for 50 h, and then again at 700° and 720°C for another 50 h. This system was investigated through

The calculated isothermal section of the Ga–Au–As ternary system at room temperature, which essentially remains unchanged to about 300°C, is given in Figure 10.7 (Tsai and Williams 1986; Schmid-Fetzer 1988). The dotted lines indicate possible alternative tie-lines. Quasi-binary tie-lines connected GaAs with four Au–Ga intermetallic compounds and with elemental Au. The predictive calculations were based on the following simplifications: ternary phases and solid solubilities were disregarded, and the Gibbs energy of formation of binary compounds was estimated by the enthalpy of formation and calculated from Miedema's model.

10.8 GALLIUM–MAGNESIUM–ARSENIC

GaAs–Mg. The calculated solidus of this system is retrograde (Kuznetsova 1975). The maximum solubility of Mg in GaAs takes place at 1000°C–1150°C and is equal to $5.00 \cdot 10^{20}$ cm^{-3}.

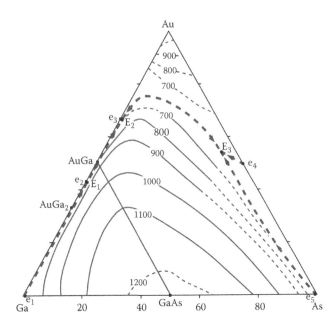

FIGURE 10.6 Liquidus surface of the Ga–Au–As ternary system. (From Panish, M.B., *J. Electrochem. Soc.*, *114*(5), 516, 1967. Reproduced with permission of The Electrochemical Society.)

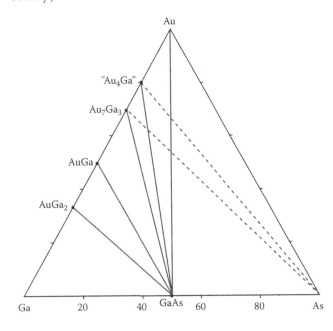

FIGURE 10.7 Calculated isothermal section of the Ga–Au–As ternary system at room temperature. (From Schmid-Fetzer, R., *J. Electron. Mater.*, *17*(2), 193, 1988. With permission.)

10.9 GALLIUM–CALCIUM–ARSENIC

Seven ternary compounds, **CaGa$_2$As$_2$, Ca$_2$Ga$_2$As$_4$, Ca$_3$GaAs$_3$, Ca$_4$Ga$_3$As$_5$, Ca$_5$Ga$_2$As$_6$, Ca$_6$Ga$_3$As$_7$,** and **Ca$_{14}$GaAs$_{11}$,** are formed in the Ga–Ca–As system. CaGa$_2$As$_2$ crystallizes in the trigonal structure with the

lattice parameters $a = 399.06 \pm 0.04$ and $c = 2482.0 \pm 0.4$ pm and a calculated density of 4.793 g·cm^{-3} at 200 K (He et al. 2010, 2011). The title compound was found to decompose around 1050°C. Two general synthetic methods were explored to prepare the title compound: on-stoichiometry reactions in sealed Nb tubes, and flux reactions in an alumina crucible using Ga as reactive flux. Ca and As were taken in stoichiometric ratios, whereas Ga was used in a fivefold excess amount to serve as a self-flux. The reaction mixture was heated fast to 960°C, homogenized for 20 h, and then cooled at a rate of 6°C·h^{-1} to 500°C, at which point the molten Ga was decanted. All manipulations were performed inside an Ar-filled glove box or under vacuum.

Black crystals of Ca$_2$Ga$_2$As$_4$ were obtained by the interaction of Ca and Ga arsenides at 1000°C for 24 h (Verdier et al. 1975a). This compound is easily hydrolyzed, forming amorphous products.

Ca$_3$GaAs$_3$ crystallizes in the orthorhombic structure with the lattice parameters $a = 1217.1 \pm 0.5$, $b = 419.7 \pm 0.2$, and $c = 1341.4 \pm 0.5$ pm and a calculated density of 4.02 g·cm^{-3} (Cordier et al. 1985b). It was synthesized from a mixture of Ca, Ga, and As (molar ratio 4:1:3) in an Ar atmosphere in a corundum crucible heated to 1000°C at a rate of 100°C·h^{-1} and then cooled at a rate of 50°C·h^{-1}.

Ca$_4$Ga$_3$As$_5$ crystallizes in the tetragonal structure with the lattice parameters $a = 396$ and $c = 3880$ pm (Verdier et al. 1975a, 1975b). It was prepared as black crystals at the heating of the binary arsenides and chemical elements up to 1000°C. This compound is easily hydrolyzed.

Ca$_5$Ga$_2$As$_6$ crystallizes in the orthorhombic structure with the lattice parameters $a = 1322.4 \pm 0.2$, $b = 1135.7 \pm 0.2$, and $c = 413.8 \pm 0.6$ pm and experimental and calculated densities of 4.11 and 4.218 g·cm^{-3}, respectively (Verdier et al. 1975a, 1975b, 1976) ($a = 1286.7$, $b = 1110.3$, and $c = 404.5$ pm [calculated values] and an energy gap of 0.37 eV [Yan et al. 2014]). Obtaining the crystals of the title compound requires heating at 1000°C for 15 h a mixture rich in As whose Ca/Ga molar ratio is equal to 3.

Black crystals of Ca$_6$Ga$_3$As$_7$ were obtained by the interaction of Ca and Ga arsenides at 1000°C for 24 h (Verdier et al. 1975b). This compound is easily hydrolyzed, forming amorphous products.

Ca$_{14}$GaAs$_{11}$ crystallizes in the tetragonal structure with the lattice parameters $a = 1564.2 \pm 0.2$ and $c = 2117.5 \pm 0.4$ pm at room temperature, and $a = 1562.0 \pm$

0.3 and $c = 2113.8 \pm 0.4$ pm at 130 K, a calculated density of 3.75 g·cm^{-3}, and an energy gap of 1.49 eV (Kauzlarich et al. 1991). The crystals of this compound could be synthesized by reacting stoichiometric combinations of the elements or by reacting stoichiometric combinations of Ca_3As_2, GaAs, and As or Ca_3As_2, Ga, and As sealed in a Nb tube with 0.02 MPa of Ar. The reaction was air quenched after 4 days at 1100°C. The air-sensitive product from the reaction primarily consisted of some well-formed hexagonal plates together with microcrystalline material, which could be identified as $Ca_{14}GaAs_{11}$. A very small amount of unidentified impurity was also present in the powder mixture.

GaAs–Ca. The calculated solidus of this system is retrograde (Kuznetsova 1975). The maximum solubility of Ca in GaAs takes place at 1000°C–1150°C and is equal to $6.83 \cdot 10^{18}$ cm^{-3}.

10.10 GALLIUM–STRONTIUM–ARSENIC

Four ternary compounds, **$SrGa_2As_2$**, **Sr_3GaAs_3**, **$Sr_3Ga_2As_4$**, and **$Sr_{14}GaAs_{11}$**, are formed in the Ga–Sr–As system. $SrGa_2As_2$ crystallizes in the monoclinic structure with the lattice parameters $a = 959.05 \pm 0.15$, $b = 404.74 \pm 0.06$, and $c = 1252.5 \pm 0.2$ pm and $\beta = 95.699 \pm 0.002°$ at 200 ± 2 K, a calculated density of 5.175 g·cm^{-3}, and an energy gap of ≈ 0.2 eV (He et al. 2011). The title compound appeared to melt incongruently at circa 1030°C. To obtain the title compound, Sr and As were taken in stoichiometric ratios, whereas Ga was used in a fivefold excess amount to serve as a self-flux. The reaction mixture was quickly heated to 960°C, homogenized for 20 h, and then cooled at a rate of 6°C·h^{-1} to 500°C, at which point the molten Ga was decanted. All manipulations were performed inside an Ar-filled glove box or under vacuum.

Sr_3GaAs_3 crystallizes in the orthorhombic structure with the lattice parameters $a = 1275.7 \pm 0.6$, $b = 1926.8 \pm 0.9$, and $c = 650.3 \pm 0.3$ pm and a calculated density of 4.63 g·cm^{-3} at 200 K (Stoyko et al. 2015). For obtaining this compound, starting materials were Sr pieces, Ga ingots, and As lumps. Mixtures of the elements were loaded into alumina crucibles covered on the top with quartz wool and placed within fused silica tubes, which were evacuated and sealed. The tubes were heated to 960°C at a rate of 60°C·h^{-1}, held at that temperature for 20 h, and cooled to 750°C·h^{-1} at a rate of 5°C·h^{-1}. The excess of metal flux was removed at 750°C·h^{-1} by using a centrifuge. All reagents and products were handled within an argon-filled glove box with controlled oxygen and a moisture level below 1 ppm.

$Sr_3Ga_2As_4$ crystallizes in the monoclinic structure with the lattice parameters $a = 1349.7 \pm 0.3$, $b = 1049.7 \pm 0.2$, and $c = 679.03 \pm 0.16$ pm and $\beta = 90.275 \pm 0.004°$ at 200 ± 2 K and a calculated density of 4.847 g·cm^{-3} (He et al. 2012). It was obtained as black crystals from the mixture of the elements with the stoichiometric ratio Sr/As of 3:4 and a 25-fold excess of Ga, which was used as a flux. The reaction was equilibrated at 960°C for 20 h, and then cooled to 750°C at a rate of 5°C·h^{-1}. The molten Ga was decanted at this temperature. Pb could be also used as a flux. The corresponding elements in the stoichiometric ratio and a large amount of Pb (20-fold excess) were used. All synthetic and postsynthetic manipulations were performed inside an argon-filled glove box or under vacuum.

$Sr_{14}GaAs_{11}$ crystallizes in the tetragonal structure with the lattice parameters $a = 1649.8 \pm 0.8$ and $c = 2213.2 \pm 1.2$ pm at 130 K, a calculated density of 4.676 g·cm^{-3}, and an energy gap of 1.44 eV (Kauzlarich and Kuromoto 1991). It was synthesized by reacting the elements, Sr, Ga, and As (molar ratio 14:1:11). The elements were sealed in a welded Nb tube, which was subsequently sealed under 0.02 MPa of purified Ar in a fused silica ampoule. The reaction was heated (1°C·min^{-1}) to 1200°C and remained at that temperature for 24 h before being rapidly cooled to room temperature. The air-sensitive product from the reaction consisted primarily of some well-formed highly reflective black needles, together with microcrystalline material that could be identified as $Sr_{14}GaAs_{11}$. The reactants and products were handled in N_2-filled dry boxes with typical water levels of less than 1 ppm.

10.11 GALLIUM–BARIUM–ARSENIC

Five ternary compounds, **$BaGa_2As_2$**, **$Ba_2Ga_5As_5$**, **Ba_3GaAs_3**, **$Ba_4Ga_5As_8$**, and **$Ba_7Ga_4As_9$**, are formed in the Ga–Ba–As system.

$BaGa_2As_2$ crystallizes in the monoclinic structure with the lattice parameters $a = 749.5 \pm 0.5$, $b = 990.1 \pm 0.6$, and $c = 764.3 \pm 0.5$ pm and $\beta = 115.381 \pm 0.008°$ at 200 ± 2 K and a calculated density of 5.530 g·cm^{-3} (He et al. 2010). Two general synthetic methods were explored to prepare the title compound: on-stoichiometry reactions in sealed Nb tubes, and flux reactions in an alumina crucible using Ga as reactive flux. Ba and As were taken in stoichiometric ratios, whereas Ga was used in a fivefold excess amount to serve as a self-flux. All manipulations were performed inside an Ar-filled glove box or under vacuum.

$Ba_2Ga_5As_5$ crystallizes in the orthorhombic structure with the lattice parameters $a = 1672.9 \pm 0.4$, $b = 410.42 \pm 0.09$, and $c = 1721.4 \pm 0.4$ pm, a calculated density of 5.607 $g \cdot cm^{-3}$ at 200 ± 2 K, and an energy gap of ≈ 0.4 eV (He et al. 2011). The title compound melts incongruently near 960°C. To obtain the title compound, Ba and As were taken in stoichiometric ratios, whereas Ga was used in a fivefold excess amount to serve as a self-flux. The reaction mixture was quickly heated to 960°C, homogenized for 20 h, and then cooled at a rate of 3°C·h⁻¹ to 750°C. At this point, the sample was removed and radiatively cooled to room temperature. Finally, the sealed tube was quickly reheated to 500°C (100°C·h⁻¹), before the Ga flux was decanted and the reaction products—lots of needle crystals of the title compound—were isolated. All manipulations were performed inside an Ar-filled glove box or under vacuum.

Ba_3GaAs_3 also crystallizes in the orthorhombic structure with the lattice parameters $a = 1335.89 \pm 0.10$, $b = 1997.88 \pm 0.15$, and $c = 680.08 \pm 0.05$ pm, a calculated density of 5.17 $g \cdot cm^{-3}$ at 200 ± 2 K, and an energy gap of 0.78 eV (Stoyko et al. 2015). For obtaining this compound, starting materials were Ba pieces, Ga ingots, and As lumps. Mixtures of the elements were loaded into alumina crucibles covered on the top with quartz wool and placed within fused silica tubes, which were evacuated and sealed. The tubes were heated to 960°C at a rate of 60°C·h⁻¹, held at that temperature for 20 h, and cooled to 750°C·h⁻¹ at a rate of 5°C·h⁻¹. The excess of metal flux was removed at 750°C·h⁻¹ by using a centrifuge. All reagents and products were handled within an argon-filled glove box with controlled oxygen and a moisture level below 1 ppm.

$Ba_4Ga_5As_8$ crystallizes in the monoclinic structure with the lattice parameters $a = 2019.6 \pm 0.2$, $b = 688.95 \pm 0.07$, and $c = 1334.61 \pm 0.14$ pm and $\beta = 110.010 \pm 0.002°$ at 200 ± 2 K and a calculated density of 5.700 $g \cdot cm^{-3}$ (He et al. 2011). It was obtained from a reaction mixture loaded with Ba, Ga, and As (molar ratio 1:10:2), in which the Ga excess was intended as a flux. This mixture was quickly heated to 960°C, homogenized for 20 h, and then cooled to 750°C at a rate of 30°C·h⁻¹. At this point, the sample was removed and radiatively cooled to room temperature. Finally, the sealed tube was quickly reheated to 500°C (100°C·h⁻¹), before the Ga flux was decanted and the title compound was isolated. For obtaining the title compound, the use of Pb flux is an advantage. All manipulations were performed inside an Ar-filled glove box or under vacuum.

$Ba_7Ga_4As_9$ also crystallizes in the orthorhombic structure with the lattice parameters $a = 1018.4 \pm 0.2$, $b = 1698.7 \pm 0.4$, and $c = 680.12 \pm 0.14$ pm and a calculated density of 5.40 $g \cdot cm^{-3}$ (He et al. 2010, 2016). To obtain this compound, Ba, Ga, and As were used as starting materials. All reagents and products were handled within an Ar-filled glove box with controlled oxygen and a moisture level below 1 ppm. Single crystals were first isolated from reactions attempted to extend the $AETr_2Pn_2$ and $AE_3Tr_2Pn_4$ series (AE = Ca, Sr, Ba; Tr = Al, Ga, In; and Pn = P, As, Sb) through use of the Al or Pb fluxes. Mixtures of the elements were loaded into alumina crucibles covered on the top with quartz wool and placed within fused silica tubes, which were then evacuated and flame sealed. The tubes were heated to 960°C at a rate of 60°C·h⁻¹, held at that temperature for 20 h (samples with Al flux) or 40 h (samples with Pb flux), and cooled to 750°C at a rate of 5°C·h⁻¹ (for samples with Al flux) or to 500°C at a rate of 30°C·h⁻¹ (for samples with Pb flux). The molten metal fluxes were removed at 750°C (Al flux) or at 500°C (Pb flux) by centrifugation.

10.12 GALLIUM–ZINC–ARSENIC

The liquidus surface of the Ga–Zn–As ternary system was assessed by Ghasemi and Johansson (2015) and is shown in Figure 10.8. Earlier, this liquidus surface was constructed experimentally through DTA and XRD (Köster and Ulrich 1958a; Panish 1966a, 1966d; Pelevin et al. 1972b) and calculated using various solution models (Jordan 1971a; Kharif et al. 1984). There are six regions of primary crystallization in this system: (As), (Ga), (Zn), GaAs, Zn_3As_2, and $ZnAs_2$. This ternary system consists only of three congruently melting binary compounds, and there are only eutectic reactions in the constituting binaries. There are three ternary eutectic reactions (E_1 at 746°C, E_2 at 722°C, and E_3 at 25°C) and one transition point (U at 416°C) in this system. By a combination of the ternary regular activity coefficients with the vapor pressures of the pure components and the liquidus isotherms, the component partial pressures

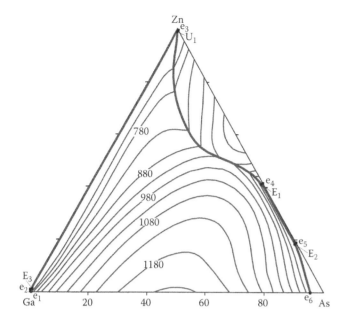

FIGURE 10.8 Liquidus surface of the Ga–Zn–As ternary system. (From Ghasemi, M., and Johansson, J., *J. Alloys Compd.*, *638*, 95, 2015. With permission.)

along the liquidus isotherms were determined (Jordan 1971a). The partial pressure of As and Zn along the 900°C, 1000°C, and 1050°C isotherms was measured using an optical adsorption method (Shih et al. 1968). The activity coefficients of the components have been determined. It was found that these coefficients are in agreement with the Gibbs–Duhem equations.

The homogeneity region in the Ga–Zn–As ternary system based on GaAs is elongated along the GaAs–$ZnAs_2$ and GaAs–Zn_3As_2 quasi-binary sections (Pelevin and Ufimtseva 1974).

GaAs–Zn. Assessed phase relations in this system are given in Figure 10.9 (Ghasemi and Johansson 2015). In agreement with Panish (1966c), Pelevin et al. (1972c), and Kharif et al. (1984), along the GaAs–Zn cut three primary fields appear upon crystallization: GaAs (0–71.0 at% Zn), Zn_3As_2 (71.0–100 at% Zn), and Zn (ca. 100 at% Zn). The calculations show that there is only one thermal event at the temperature range 610°C–645°C, similar to that predicted by Panish (1966c). However, Pelevin et al. (1972c) have registered two thermal effects at this temperature range. Moreover, according to Panish (1966c) and Pelevin et al. (1972c), Zn_3As_2 should precipitate along the GaAs–Zn cut at a slightly lower Zn composition and a wider temperature range, while the calculations contradict that (Ghasemi and Johansson 2015).

According to the data of Köster and Ulrich (1958a), the phase diagram of the GaAs–Zn system is a eutectic type. The eutectic crystallizes at 414°C and contains 1.5 at% Zn.

The experimentally determined temperature dependence of Zn solubility in GaAs is shown in Figure 10.10 (Chang and Pearson 1964b). The solid solubility isotherms for Zn in GaAs have also been calculated by Jordan (1971b) as a function of Zn concentration along

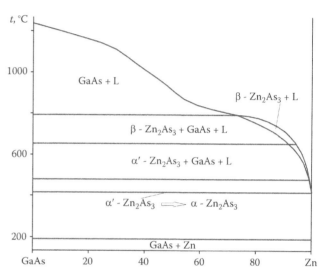

FIGURE 10.9 Phase relations in the GaAs–Zn system. (From Ghasemi, M., and Johansson, J., *J. Alloys Compd.*, *638*, 95, 2015. With permission.)

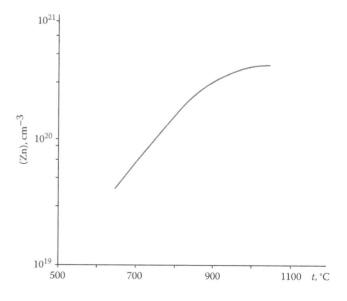

FIGURE 10.10 Experimentally determined temperature dependence of Zn solubility in GaAs. (From Chang, L.L., and Pearson, G.L., *J. Phys. Chem. Solids*, *25*(1), 23, 1964. With permission.)

the Ga–Zn–As liquidus isotherms over a wide temperature range. The obtained solubility isotherms are in full agreement with the experimental data. The solubility of Zn in GaAs is 0.7 at% ($3.5 \cdot 10^{20}$ cm^{-3}) at 900°C and is 1.0 at% ($4.5 \cdot 10^{20}$ cm^{-3}) at 1000°C (Pelevin and Ufimtseva 1974; Tuck 1976). According to the data of McCaldin (1963), the solubility of Zn in GaAs is $4 \cdot 10^{18}$ cm^{-3} at 1000°C.

The distribution coefficient of Zn in GaAs is equal to 0.415–0.416 (Mil'vidskiy and Pelevin 1965; Mil'vidskiy et al. 1966; Pelevin and Mil'vidskiy 1968). When the critical Zn concentration ($[4–5] \cdot 10^{19}$ cm^{-3}) was reached, there was a sharp reduction of the distribution coefficient to the value of circa 0.1 (Mil'vidskiy et al. 1966).

GaAs–ZnAs$_2$. The phase diagram of this system is a eutectic type (Figure 10.11) (Köster and Ulrich 1958a; Ghasemi and Johansson 2015). The eutectic crystallizes at 754°C and contains approximately 1 mol% GaAs.

GaAs–Zn$_3$As$_2$. The phase diagram of this system is a eutectic type (Figure 10.12) (Köster and Ulrich 1958a; Panish 1966a; Ghasemi and Johansson 2015). The eutectic crystallizes at 932°C and contains 16 mol% GaAs. The solubility of Zn$_3$As$_2$ has a maximum at eutectic temperature and is equal (for the Zn recalculation) to $5.5 \cdot 10^{20}$ cm^{-3} (Pelevin and Ufimtseva 1974).

10.13 GALLIUM–CADMIUM–ARSENIC

The liquidus surface of the Ga–Cd–As ternary system is shown in Figure 10.13 (Marenkin et al. 1987a; Marenkin 1993). Six fields of primary crystallization of the Cd, Ga, Cd$_3$As$_2$, CdAs$_2$, GaAs, and As phases exist in this system. The field of GaAs primary crystallization is the biggest among them. GaAs defines the triangulations

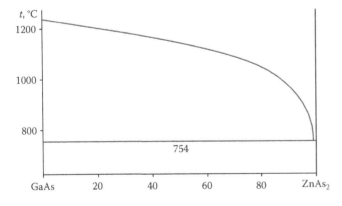

FIGURE 10.11 Phase diagram of the GaAs–ZnAs$_2$ system. (From Ghasemi, M., and Johansson, J., *J. Alloys Compd.*, 638, 95, 2015. With permission.)

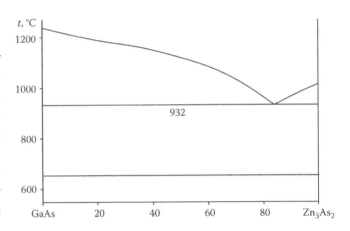

FIGURE 10.12 Phase diagram of the GaAs–Zn$_3$As$_2$ system. (From Ghasemi, M., and Johansson, J., *J. Alloys Compd.*, 638, 95, 2015. With permission.)

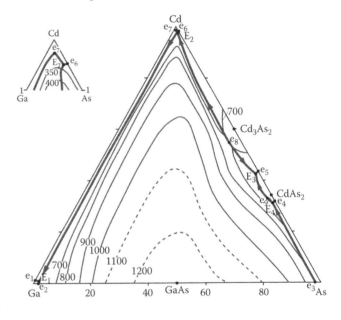

FIGURE 10.13 Liquidus surface of the Ga–Cd–As ternary system. (From Marenkin, S.F., et al., *Izv. AN SSSR Neorgan. Mater.*, 23(8), 1241, 1987. With permission.)

of this ternary system. Four ternary eutectics are in the Ga–Cd–As ternary system: E_1 (29.3°C), L $\longleftrightarrow$ GaAs + Cd + Ga; E_2 (317°C), L $\longleftrightarrow$ GaAs + Cd + Cd$_3$As$_2$; E_3 (604°C), L $\longleftrightarrow$ GaAs + CdAs$_2$ + Cd$_3$As$_2$; and E_4 (612°C), L $\longleftrightarrow$ GaAs + CdAs$_2$ + As. No ternary phase was determined in this system.

GaAs–Cd. The phase diagram of this system is a eutectic type (Figure 10.14) (Marenkin et al. 1986a, 1987a; Marenkin 1993). The eutectic crystallizes at 318°C and contains 0.2 mol% GaAs. The mutual solubility of GaAs and Cd is negligible. From the diffusion profiles, the solubility of Cd in GaAs has been obtained

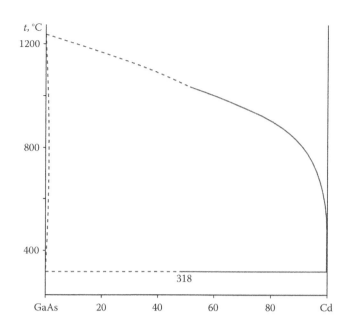

FIGURE 10.14 Phase diagram of the GaAs–Cd system. (From Marenkin, S.F., et al., *Izv. AN SSSR Neorgan. Mater.*, 23(8), 1241, 1987. With permission.)

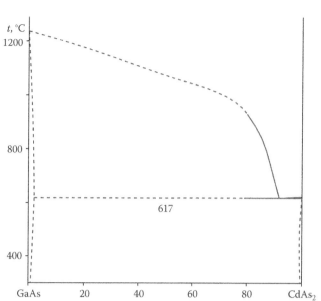

FIGURE 10.16 Phase diagram of the GaAs–CdAs₂ system. (From Marenkin, S.F., et al., *Izv. AN SSSR Neorgan. Mater.*, 23(8), 1241, 1987. With permission.)

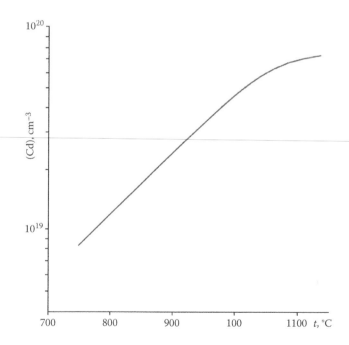

FIGURE 10.15 Solubility of Cd in GaAs. (From Fujimoto, M., *Rev. Phys. Chem. Jpn.*, 40(1), 34, 1970. With permission.)

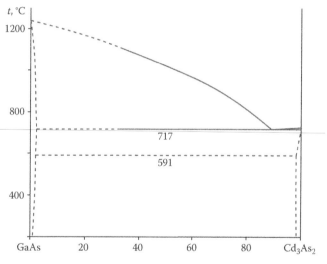

FIGURE 10.17 Phase diagram of the GaAs–Cd₃As₂ system. (From Marenkin, S.F., et al., *Izv. AN SSSR Neorgan. Mater.*, 23(8), 1241, 1987. With permission.)

at a temperature between 800°C and 1100°C (Fujimoto 1970). The experimentally measured total solubility of Cd in GaAs is given in Figure 10.15.

GaAs–CdAs₂. The phase diagram of this system is a eutectic type (Figure 10.16) (Marenkin et al. 1986c, 1987a; Marenkin 1993). The eutectic crystallizes at 617°C and contains 8.2 mol% GaAs. The samples were annealed at 550°C for 200 h.

GaAs–Cd₃As₂. The phase diagram of this system is a eutectic type (Figure 10.17) (Marenkin et al. 1986b, 1987a; Marenkin 1993). The eutectic crystallizes at 717°C and contains 10.7 mol% GaAs. The mutual solubility of GaAs and Cd is negligible (~1 mol%).

The Ga–Cd–As ternary system was investigated trough DTA, XRD, metallography, and measurement of microhardness (Marenkin et al. 1986a, 1986b, 1986c, 1987a; Marenkin 1993).

10.14 GALLIUM–INDIUM–ARSENIC

The Ga–In–As ternary system has been the subject of intense study for the last 60 years owing to interest in the optical and semiconducting properties of the GaAs and InAs compounds. The part of the liquidus surface of this system in the In-rich corner was constructed by Kuznetsov et al. (1969) and Wu and Pearson (1972) through DTA, XRD, metallography, and the mass loss of the seed. This part of the liquidus surface was also calculated by Stringfellow and Greene (1969) using a quasi-chemical equilibria model. Liquidus and solidus data determined *in situ* by the method of direct visual observation and electron probe microanalysis (EPMA) were presented for the 800°C, 850°C, and 900°C isotherms in the In-rich corner by Pollak et al. (1975). The liquidus isotherms at 600°C, 650°C, and 700°C in this part of the Ga–In–As system were also experimentally determined by an improved seed dissolution technique using InP seeds (Nakajima et al. 1979; Nakajima and Okazaki 1985).

The existence of miscibility gaps in the binary, ternary, and quaternary systems based on III–V semiconductor compounds is a general feature. A calculated liquid–solid equilibrium phase diagram in the miscibility gap region is shown in Figure 10.18 (Onabe 1983).

Solid lines and thin broken lines represent solidus and liquidus, respectively. The bold broken line designates a trace of binodal points. An open circle corresponds to a critical point for solid phase stability. As the Ga and As atomic fraction in the liquid decreases, isoconcentration solidus lines become dense, while the liquidus temperature rapidly decreases. The solidus line for $Ga_xIn_{1-x}As$ at $x = 0.5$ terminates at 482°C. This temperature is the calculated critical temperature for solid phase stability.

A full liquidus surface of the Ga–In–As ternary system was constructed experimentally through DTA, XRD, metallography, and measurement of microhardness (Kovaleva et al. 1969). It was also calculated using various solution models (Antypas 1970a; Panish 1970a; Panish and Ilegems 1972; Ishida et al. 1989b; Litvak and Charykov 1991). Liquidus equations have been derived from experimental data by Muszyński and Ryabcev (1975, 1976) using a simplex lattice method in order to interpolate the liquidus curves over the entire composition range. Figure 10.19 shows liquidus curves calculated by Shen et al. (1995), based on an assessment of the phase diagram and thermodynamic properties of the system.

The solid recovered from the filtrates at the high-temperature filtration technique of GaAs and In indicated the formation of $Ga_xIn_{1-x}As$ solid solution from 22 mol% InAs for the filtrations at 500°C to 7 mol% InAs for the filtrations at 1200°C (Rubinstein 1966).

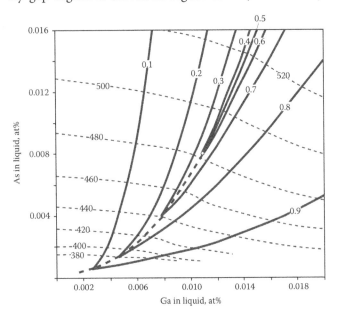

FIGURE 10.18 Calculated liquid–solid equilibrium phase diagram for Ga–In–As system in In-rich region, showing the existence of a solid miscibility gap. (From Onabe, K., *Jpn. J. Appl. Phys.*, 22(1), Pt. 1, 201, 1983. With permission.)

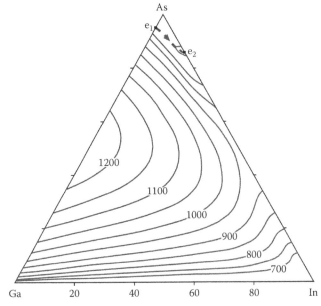

FIGURE 10.19 Calculated liquidus surface of the Ga–In–As ternary system. (From Shen, J.-Y., *CALPHAD*, 19(2), 215, 1995. With permission.)

Vaypolin et al. (2001) obtained a Ga$_4$InAs$_3$ super-structure at the XRD investigations of the epitaxial layers of the composition Ga$_{0.82}$In$_{0.18}$As.

GaAs–InAs. A complete solid solubility in this system was suggested by Goryunova (1955), Goryunova and Fedorova (1955), Woolley and Smith (1958a), and van Hook and Lenker (1963). The equilibrium condition of the solid solution throughout the whole range of composition was established only after a considerable time of annealing (at 900°C for 300 h) (Woolley and Smith 1957). Homogeneous alloys were also produced across the entire system by using the method of zone leveling (Abrahams et al. 1959). The phase diagram of the GaAs–InAs system was constructed experimentally through DTA, XRD, and metallography (Hockings et al. 1966; Bodnar 1975) and calculated using various solution models (Stringfellow and Greene 1969; Antypas 1970a; Steininger 1970; Foster and Woods 1971; Osamura et al. 1972b; Panish and Ilegems 1972; Stringfellow 1972a; Cho 1976; Bublik and Leikin 1978; Ishida et al. 1989b). Figure 10.20 shows the assessed GaAs–InAs phase diagram (Shen et al. 1995). A miscibility gap is also depicted. The critical temperature is 544°C (456°C–462°C [Stringfellow 1982a, 1982b, 1983]; 357°C and a critical composition of 77 mol% GaAs [Wei et al. 1990]). Within the limits of experimental error, a linear relationships in the variation of lattice parameters as a function of composition was obtained, showing that Vegard's law is satisfied (Woolley and Smith 1957, 1958a; Katayama et al. 1989) (van Hook and Lenker [1963] indicated a slight positive deviation from Vegard's law). The band gap varies continuously with composition for the entire system (Abrahams et al. 1959; Woolley et al. 1961; Wu and Pearson 1972). A concave upward dependence is observed upon going from InAs to GaAs.

A p-T-x phase diagram of the GaAs–InAs quasi-binary system was constructed by Rakov and Ufimtsev (1969) and Mirtskhulava and Sakvarelidze (1977).

The activity of Ga in the solid solutions of Ga$_{1-x}$In$_x$As coexisting with the As-poor ternary liquid phase was derived as a function of the composition and temperature from electromotive force (EMF) measurements of galvanic cells with the zirconia base solid electrolyte at 700°C–900°C (Katayama et al. 1989). The activity of Ga is small, changes linearly with the concentration up to 70 mol% GaAs, and increases very sharply in the GaAs-rich region. Integral excess Gibbs energy, enthalpy, and entropy at 27°C have been calculated, and a positive deviation was recorded (Kühn and Leonhardt 1972).

A first-principles calculation of a temperature composition epitaxial phase diagram of the GaAs–InAs system has been performed, and calculated values of the equilibrium molar volumes, the bulk moduli, and pressure derivatives for GaInAs$_2$, GaIn$_3$As$_4$, and Ga$_3$InAs$_4$ compositions were obtained (Wei et al. 1990).

A Calvet microcalorimeter has been used with liquid tin solution techniques to measure the mixing enthalpy of the GaAs–InAs system (Rugg et al. 1995). It was shown that the enthalpy of formation of Ga$_x$In$_{1-x}$As solid solutions is small and positive. Conditions for congruent vaporization of these solid solutions under vacuum were calculated using thermodynamic data of basic binary and ternary systems (Shen and Chatillon 1990). Partial congruent vaporization was defined that is useful for molecular beam epitaxy (MBE) to avoid the droplet appearance at the compound surface. The maximal temperature for partial congruent vaporization is calculated as a function of the composition varying from pure InAs to pure GaAs. The curve has a maximum at 719°C for the Ga$_{0.99}$In$_{0.01}$As composition.

The As total pressure in the GaAs–InAs system was determined by Peuschel et al. (1981), and it was shown that the good agreement was obtained between calculated and experimental data.

Phase equilibria in Ga$_x$In$_{1-x}$As thin films grown on GaAs and InP substrates were calculated, taking into consideration the elastic contribution caused by the lattice mismatch between the film and substrate (Ohtani et al.

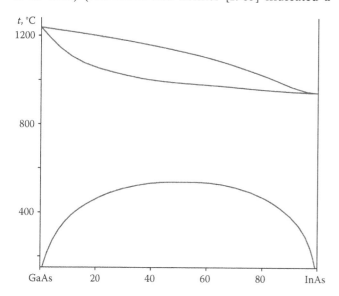

FIGURE 10.20   Phase diagram of the GaAs–InAs system. (From Shen, J.-Y., *CALPHAD*, *19*(2), 215, 1995. With permission.)

2001). The basic concept used in describing the elastic free energy is that strain due to a lattice mismatch accumulates in a thin film, and interfacial misfit dislocations form when the strain energy exceeds a certain energy barrier.

It has been shown that the $Ga_xIn_{1-x}As$ solid solutions with small x could be prepared in the form of large crystals by means of a gradient freezing method (Tombs et al. 1963). Single crystals of these solid solutions have been prepared using chemical transport reactions with iodine as the transport agent (Khlystovskaya and Chuveleva 1973).

10.15 GALLIUM–THALLIUM–ARSENIC

It was found that at low temperatures (ca. 200°C), an air-stable $Ga_{1-x}Tl_xAs$ film with uniform composition could be achieved with a thallium concentration of 5 at% (Antonell et al. 2000). Increasing the Tl flux produced layers containing as high as 15 at% Tl. $Ga_{1-x}Tl_xAs$ solid solutions could also be grown by MBE at temperatures lower than 350°C (Kajikawa et al. 2002). At 250°C, the incorporation of Tl into GaAs was limited to $x = 0.012$, while the limit was extended beyond 2.8 at% Tl at the growth temperature of 200°C.

10.16 GALLIUM–SAMARIUM–ARSENIC

The solubility of Sm in GaAs is retrograde and is given in Figure 10.21 (Romanenko and Kheyfets 1973).

10.17 GALLIUM–EUROPIUM–ARSENIC

Two ternary compounds, **$EuGa_2As_2$** and **$Eu_3Ga_2As_4$**, are formed in the Ga–Eu–As system. $EuGa_2As_2$ crystallizes in the monoclinic structure with the lattice parameters $a = 949.53 \pm 0.07$, $b = 402.94 \pm 0.03$, and $c = 1242.37 \pm 0.09$ pm and $\beta = 95.3040 \pm 0.0010°$ at 100 ± 1 K and a calculated density of 6.192 g·cm⁻³ (Goforth et al. 2009). It was synthesized using a 3:51:6 (Eu/Ga/As) molar ratio of the elements (9 g total mass). In this reaction, Ga is both a reactant and a flux. The elements were loaded into a 5 mL alumina crucible in the following order: Ga, Eu and As, and Ga. The crucible, held in place between two layers of tightly packed quartz wool, was subsequently sealed under circa 0.02 Pa of Ar in a fused silica ampoule. The reaction vessel was then heated in a box furnace in an upright position. The sealed ampoule was heated at a rate of 60°C·h⁻¹ to 1000°C, and kept at this temperature for 6 h before it was cooled at a rate of 200°C·h⁻¹ to 850°C. The reaction vessel was maintained at 850°C for 6 h before it was slow cooled at a rate of 1°C·h⁻¹ to 500°C, at which point it was

removed from the furnace, inverted, and centrifuged to remove the flux. The vessel was subsequently broken in air to reveal large silver parallelepipeds of the title compound. All reagents were handled under an inert atmosphere.

$Eu_3Ga_2As_4$ also crystallizes in the monoclinic structure with the lattice parameters $a = 1338.6 \pm 0.4$, $b = 1042.6 \pm 0.2$, and $c = 673.2 \pm 0.2$ pm and $\beta = 90.742 \pm 0.004°$ at 200 ± 2 K and a calculated density of 6.327 g·cm⁻³ (He et al. 2012). It was obtained as black crystals from the mixture of the elements with the stoichiometric ratio Eu/As of 3:4 and a 25-fold excess of Ga, which was used as a flux. The reaction was equilibrated at 960°C for 20 h, and then cooled to 750°C at a rate of 5°C·h⁻¹. The molten Ga was decanted at this temperature. Pb could also be used as a flux. The corresponding elements in the stoichiometric ratio and a large amount of Pb (20-fold excess) were used. All synthetic and postsynthetic manipulations were performed inside an argon-filled glove box or under vacuum.

10.18 GALLIUM–GADOLINIUM–ARSENIC

The solubility of Gd in GaAs is retrograde and is given in Figure 10.21 (Romanenko and Kheyfets 1973).

10.19 GALLIUM–DYSPROSIUM–ARSENIC

The solubility of Dy in GaAs is retrograde and is given in Figure 10.21 (Romanenko and Kheyfets 1973).

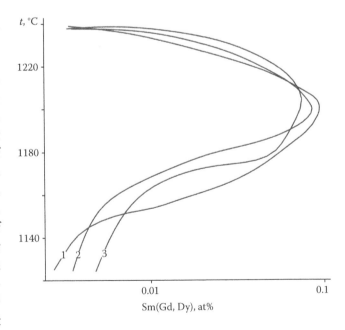

FIGURE 10.21 Solubility of (1) Gd, (2) Sm, and (3) Dy in GaAs. (From Romanenko, V.N., and Kheyfets, V.S., *Izv. AN SSSR Neorgan. Mater.*, 9(2), 190, 1973. With permission.)

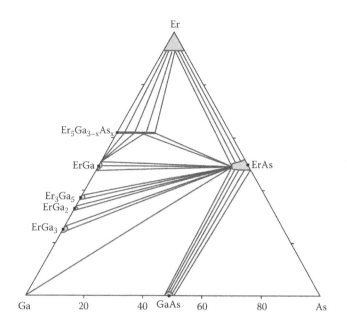

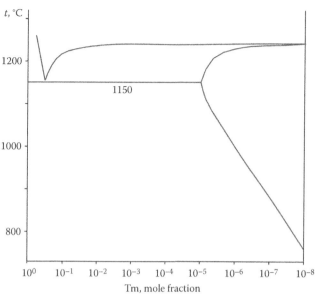

FIGURE 10.22 Isothermal section of the Ga–Er–As ternary system at 800°C. (From Députier, S., et al., *J. Alloys Compd.*, *202*(1–2), 95, 1993. With permission.)

FIGURE 10.23 Phase diagram of the GaAs–Tm system. (From Casey, H.C., Jr., and Pearson G.L., *J. Appl. Phys.*, *35*(11), 3401, 1964. With permission.)

10.20 GALLIUM–ERBIUM–ARSENIC

The solid-state equilibria in the Ga–Er–As ternary system were determined at 800°C with the use of XRD, EPMA, and scanning electron microscopy (SEM) (Figure 10.22) (Députier et al. 1993). No ternary phases were found, and very limited solid solubilities were measured in the constituent binary Er–Ga and Er–As compounds, with the exception of Er_5Ga_3, which showed a broad homogeneity range with an As-rich limit corresponding to the formula Er_5Ga_2As. GaAs, ErAs, and Ga form a three-phase region that dominates the GaAs side of the phase diagram. ErAs dissolves 3 at% Ga and is a "key" compound around which the Er/GaAs interaction progresses (Députier et al. 1994). Samples were first annealed at 800°C for 48 h and then, after cooling, ground to powder, cold pressed into pellets, sealed again, and annealed at 800°C for 20 days to ensure homogeneity (Députier et al. 1993). Finally, samples were quenched in ice water.

10.21 GALLIUM–THULIUM–ARSENIC

GaAs–Tm. The phase diagram of this system is a eutectic type (Figure 10.23) (Casey and Pearson 1964). The eutectic crystallizes at 1150°C. The solubility region of Tm in GaAs was determined using radioactive [170]Tm. A maximum solid solution solubility of $4·10^{17}$ cm^{-3} Tm in GaAs was found at the eutectic temperature.

The liquidus was calculated through an ideal solution model.

10.22 GALLIUM–CARBON–ARSENIC

The temperature dependence of carbon solubility in GaAs is retrograde (Figure 10.24), with a maximum ($1.9·10^{19}$ cm^{-3}) at 1100°C (Borisova et al. 1978).

10.23 GALLIUM–SILICON–ARSENIC

The liquidus surface of the Ga–Si–As ternary system has been studied by DTA (Panish 1966b). A ternary eutectic L $\longleftrightarrow$ GaAs + SiAs + (Si) at 1031°C was suggested. However, the eutectic liquid composition is outside of the triangle GaAs–SiAs–(Si), and if this eutectic exists, both GaAs–Si and GaAs–SiAs cuts should be quasibinary (Han and Schmid-Fetzer 1993a), which is not the case for the latter (Panish 1966b). Both inconsistencies are removed if a transition-type reaction U_1, L + (Si) $\longleftrightarrow$ GaAs + SiAs at 1031°C, is assumed instead of a ternary eutectic (Han and Schmid-Fetzer 1993a). Together with other reported reactions, U_2 (908°C), L + SiAs = GaAs + SiAs$_2$; E_1 (770°C), L $\longleftrightarrow$ GaAs + (As) + SiAs$_2$, and E_2 (29.75°C), L $\longleftrightarrow$ GaAs + (Ga) + Si, which is almost degenerated into the melting point of Ga, there are a total of four invariant four-phase reactions. The liquidus surface in the subsystem GaAs–Ga–Si has also been studied by Panish and Sumski (1970) and Pelevin (1975)

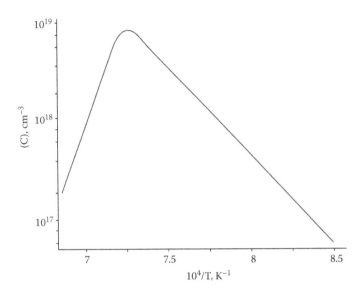

FIGURE 10.24 Temperature dependence of the carbon solubility in GaAs. (From Borisova, L.A., et al., *Izv. AN SSSR Neorgan. Mater.*, *14*(10), 1790, 1978. With permission.)

with similar results. Using the data of Panish (1966b) and Panish and Sumski (1970), a calculation of liquidus isotherms and thermodynamic activities was presented by Jordan and Weiner (1974), based on the assumption of a ternary regular solution model. The polythermal projection of the liquidus surface of the Ga–Si–As ternary system (Figure 10.25) combines the data of Panish (1966b), Panish and Sumski (1970), Pelevin (1975), and Han and Schmid-Fetzer (1993a). XRD data confirm

the tie-lines GaAs–Si, GaAs–SiAs, and GaAs–SiAs$_2$ at 800°C (Figure 10.26) (Han and Schmid-Fetzer 1993a).

GaAs–Si. The phase diagram of this system is a eutectic type (Figure 10.27) (Panish 1966b). The eutectic crystallizes at 1125°C and contains 32 at% Si (at 1135°C, contains 47.5 at% Si [Pelevin et al. 1973]; Pelevin et al. [1971] indicated the existence of a degenerated eutectic). The solubility of Si in GaAs reaches 4 at% at the eutectic temperature (Borisova and Valisheva 1974) (according

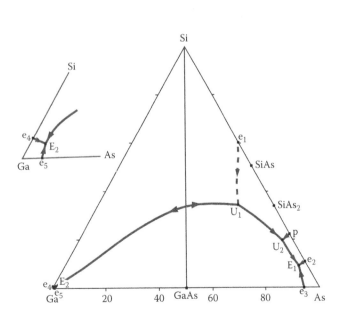

FIGURE 10.25 Polythermal projection of the liquidus surface of the Ga–Si–As ternary system. (From Han, Q., and Schmid-Fetzer, R., *J. Mater. Sci.*, *4*(2), 113, 1993. With permission.)

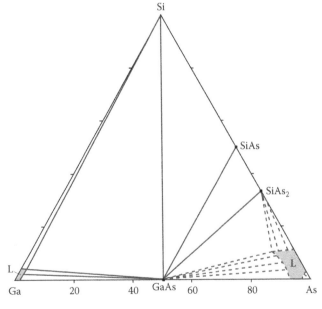

FIGURE 10.26 Isothermal projection of the Ga–Si–As ternary system at 800°C. (From Han, Q., and Schmid-Fetzer, R., *J. Mater. Sci.*, *4*(2), 113, 1993. With permission.)

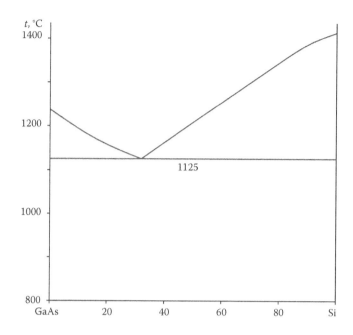

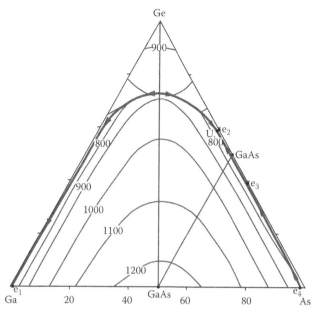

FIGURE 10.27 Phase diagram of the GaAs–Si system. (From Panish, M.B., *J. Electrochem. Soc.*, *113*(11), 1226, 1966. Reproduced with permission of The Electrochemical Society.)

FIGURE 10.28 Liquidus surface of the Ga–Ge–As ternary system. (From Panish, M.B., *J. Less Common Metals*, *10*(6), 416, 1966. With permission.)

to the data of Kolm et al. [1957], this solubility is small, being between 0.50 and 0.60 at%). The calculated solidus of this system is retrograde (Kuznetsova 1975). The maximum solubility of Si in GaAs takes place at 1000°C–1150°C and is equal to $7.5 \cdot 10^{20}$ cm^{-3}.

The distribution coefficient of Si in GaAs is equal to 0.136 ± 0.012 (Pelevin et al. 1971), and the equilibrium distribution coefficient is 0.04 (Borisova and Valisheva 1974).

10.24 GALLIUM–GERMANIUM–ARSENIC

The liquidus surface of the Ga–Ge–As ternary system is shown in Figure 10.28 (Panish 1966c; Ansara and Dutartre 1984). It consists principally of the primary field of GaAs crystallization. The liquidus surface in the subsystem GaAs–Ga–Ge has also been studied by Pelevin (1975). The triangulation of this system is defined by the GaAs–Ge, GaAs–GeAs, and GaAs–GeAs$_2$ sections (Pelevin et al. 1972a). The solid solution region based on Ge in this system determined through DTA and metallography is given in Figure 10.29 (Glazov and Malyutina 1963a). The samples for solid solubility determination were annealed at 600°C and 800°C for 500 h.

GaAs–Ge. The phase diagram of this system is a eutectic type (Figure 10.30) (Panish 1966c, 1966f; Ansara and Dutartre 1984). The eutectic crystallizes at 863°C ± 3°C and contains 16 mol% GaAs (at 888°C ± 3°C and 18.7 mol% GaAs [Glazov and Malyutina 1963a]; at 865°C ±

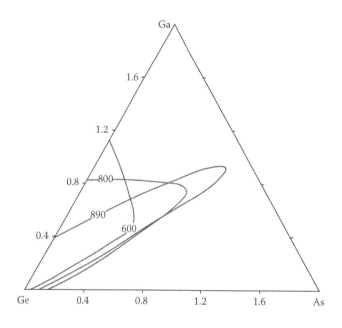

FIGURE 10.29 Solid solutions based on Ge in the Ga–Ge–As ternary system. (From Glazov, V.M., and Malyutina, G.L., *Zhurn. Neorgan. Khimii*, *8*(8), 1921, 1963. With permission.)

2°C and 15 mol% GaAs [Takeda et al. 1965]; at 860°C ± 2°C and 25 mol% GaAs [Lieth and Heyligers 1966]; at 860°C ± 2°C and 14.3 mol% GaAs [Pelevin et al. 1973]).

The solubility of Ge in solid GaAs is retrograde and is shown in Figure 10.3 (Kuznetsov and Kuznetsova 1966; Lavrishchev and Khludkov 1971). According to

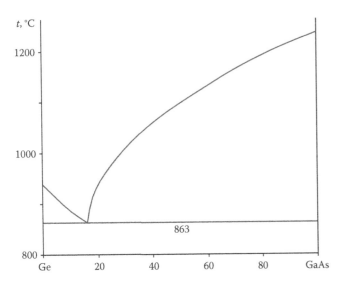

FIGURE 10.30 Phase diagram of the GaAs–Ge system. (From Panish, M.B., *J. Electrochem. Soc.*, *113*(6), 626, 1966. Reproduced with permission of The Electrochemical Society.)

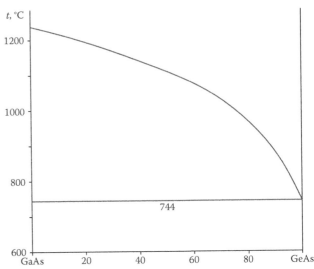

FIGURE 10.31 Phase diagram of the GaAs–GeAs system. (Panish, M.B., *J. Less Common Metals*, *10*(6), 416, 1966. With permission.)

the data of Jenny and Braunstein (1958), the solid solubility of Ge in GaAs, when crystallized from a melt, is less than 2 at%. The presence of the maximum soluble amount of Ge in GaAs reduces the band gap by about 0.1 eV. The maximum solubility of GaAs in Ge reaches 1.9 mol% (Glazov and Malyutina 1963a). Takeda et al. (1965) noted that the solubility of Ge in GaAs is not higher than 0.5 at%, and the solubility of GaAs in Ge is less than 0.25 mol%. According to the data of Kolm et al. (1957), the solubility of Ge in GaAs is small, being between 0.50 and 0.60 at%.

The distribution coefficient of Sn in GaAs is equal to $(1.46 \pm 0.25) \cdot 10^{-2}$ (Pelevin et al. 1971).

As a result of ultrafast crystallization ($10^{7}°C \cdot s^{-1}$) of the GaAs–Ge melts, metastable solid solutions were fixed in the entire range of concentration (Glazov et al. 1977a).

GaAs–GeAs. The phase diagram of this system constructed experimentally through DTA and XRD is a eutectic type (Figure 10.31) (Panish 1966c). The eutectic crystallizes at 744°C and is degenerated from the GeAs side.

10.25 GALLIUM–TIN–ARSENIC

The liquidus surface of the Ga–Sn–As ternary system was constructed experimentally by Panish 1966c, 1973) and calculated by Morozov et al. (1987) and is given in Figure 10.32. It consists principally of the primary field of GaAs crystallization. The liquidus surface in the subsystem GaAs–Ga–Si has also been studied by Pelevin

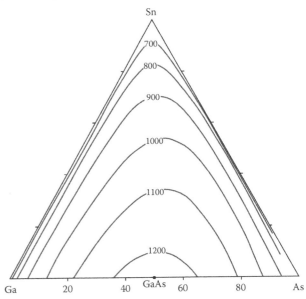

FIGURE 10.32 Liquidus surface of the Ga–Sn–As ternary system. (From Panish, M.B., *J. Appl. Phys.*, *44*(6), 2659, 1973. With permission.)

(1975). The triangulation of this system is defined by the GaAs–Sn, GaAs–Sn$_3$As$_2$, and GaAs–SnAs sections (Pelevin et al. 1972a).

The doping of GaAs with Sn leads to a narrowing of the low-temperature boundaries of the GaAs homogeneity region from both an excess of Ga and an excess of As (Figure 10.33) (Morozov et al. 1987). The composition of the GaAs solid phase, corresponding to a maximum melting point, with increasing Sn content remains constant with respect to Ga and As concentration.

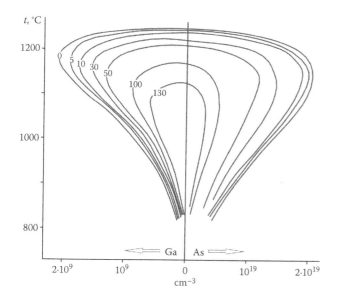

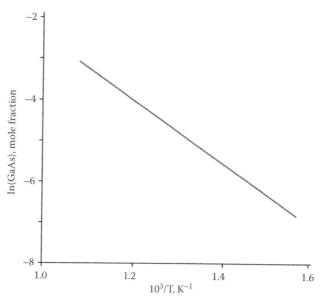

FIGURE 10.33 Influence of doping by Sn on the GaAs homogeneity region: the numbers on the curves indicate the Sn concentration in the solid phase ($\times 10^{18}$ cm^{-3}). (From Morozov, A.N., et al., *Izv. AN SSSR Neorgan. Mater.*, *23*(9), 1429, 1987. With permission.)

FIGURE 10.35 Temperature dependence of the GaAs limiting solubility in the Sn melt. (From Vasil'yev, M.G., et al., *Izv. AN SSSR Neorgan. Mater.*, *19*(11), 1933, 1983. With permission.)

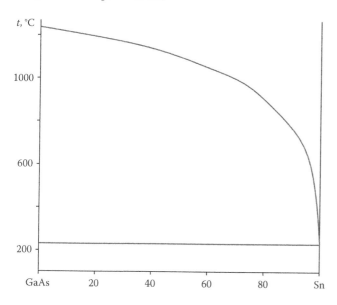

FIGURE 10.34 Phase diagram of the GaAs–Sn system. (From Pelevin, O.V., et al., *Izv. AN SSSR Neorgan. Mater.*, *8*(3), 446, 1972. With permission.)

GaAs–Sn. The phase diagram of this system is a eutectic type (Figure 10.34) (Vasil'yev and Vyatkin 1965; Kuznetsov and Kuznetsova 1967; Pelevin et al. 1972c). The eutectic is degenerated from the Sn side and crystallizes at 227.5°C–231.2°C.

The solubility of Ge in solid GaAs is retrograde and is shown in Figure 10.3 (Kuznetsov and Kuznetsova

1966, 1967; Pelevin et al. 1972c). The temperature dependence of the GaAs limiting solubility in the Sn melt was determined by Vasil'yev et al. (1983) and is given in Figure 10.35. According to the data of Kolm et al. (1957), the solubility of Sn in GaAs is small, being between 0.50 and 0.60 at%. The solubility of GaAs in liquid Sn as a function of temperature (500°C–1000°C) was obtained using a high-temperature filtration technique by Rubinstein (1966).

The distribution coefficient of Sn in GaAs is equal to $(5.1 \pm 0.9) \cdot 10^{-3}$ (Pelevin et al. 1971).

GaAs–Sn$_3$As$_2$. The phase diagram of this system constructed experimentally through DTA and XRD is a eutectic type (Figure 10.36) (Panish 1966c).

10.26 GALLIUM–LEAD–ARSENIC

GaAs–Pb. The phase diagram of this system is a eutectic type (Figure 10.37) (Kuznetsov and Kuznetsova 1968; Leonhardt and Kühn 1975). The eutectic is degenerated from the Pb side. The solubility of GaAs in liquid Pb as a function of temperature (700°C–1000°C) was obtained using a high-temperature filtration technique by Rubinstein (1966). According to the data of Kolm et al. (1957), the solubility of Pb in GaAs is small, being between 0.50 and 0.60 at%. The solubility of GaAs in the Pb melt at 700°C–1050°C was determined by Leonhardt and Kühn (1974b). The system was investigated through

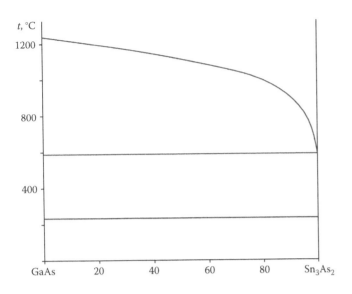

FIGURE 10.36 Phase diagram of the GaAs–Sn₃As₂ system. (From Panish, M.B., *J. Less Common Metals*, *10*(6), 416, 1966. With permission.)

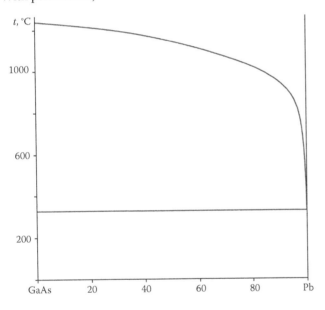

FIGURE 10.37 Phase diagram of the GaAs–Pb system. (From Kuznetsov, G.M., and Kuznetsova, S.K., *Izv. AN SSSR Neorgan. Mater.*, *4*(8), 1243, 1968. With permission.)

XRD, DTA, and metallography (Kuznetsov and Kuznetsova 1968).

According to the calculations, the miscibility region in this system is metastable (Litvak 1992).

10.27 GALLIUM–TITANIUM–ARSENIC

The calculated isothermal section of the Ga–Ti–As ternary system at room temperature, which essentially remains unchanged to about 300°C, is given in Figure 10.38 (Schmid-Fetzer 1988). The dotted lines

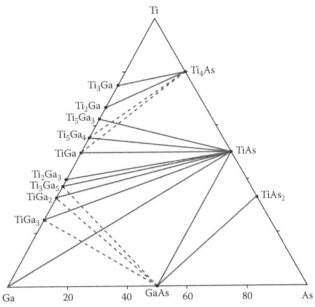

FIGURE 10.38 Calculated isothermal section of the Ga–Ti–As ternary system at room temperature. (From Schmid-Fetzer, R., *J. Electron. Mater.*, *17*(2), 193, 1988. With permission.)

indicate possible alternative tie-lines. The predictive calculations were based on the following simplifications: ternary phases and solid solubilities were disregarded, and the Gibbs energy of formation of binary compounds was estimated by the enthalpy of formation and calculated from Miedema's model.

10.28 GALLIUM–ZIRCONIUM–ARSENIC

The calculated isothermal section of the Ga–Zr–As ternary system at room temperature, which essentially remains unchanged to about 300°C, is given in Figure 10.39 (Schmid-Fetzer 1988). The dotted line indicates a possible alternative tie-line. The predictive calculations were based on the following simplifications: ternary phases and solid solubilities were disregarded, and the Gibbs energy of formation of binary compounds was estimated by the enthalpy of formation and calculated from Miedema's model.

10.29 GALLIUM–HAFNIUM–ARSENIC

The calculated isothermal section of the Ga–Hf–As ternary system at room temperature, which essentially remains unchanged to about 300°C, is given in Figure 10.40 (Schmid-Fetzer 1988). The dotted line indicates a possible alternative tie-line. The predictive calculations were based on the following simplifications: ternary phases and solid solubilities were disregarded, and the Gibbs energy of formation of binary compounds

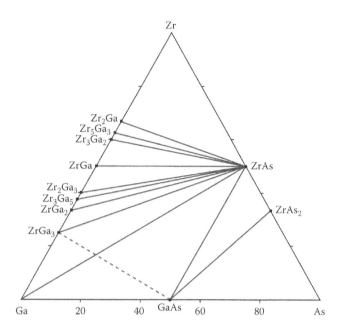

FIGURE 10.39 Calculated isothermal section of the Ga–Zr–As ternary system at room temperature. (From Schmid-Fetzer, R., *J. Electron. Mater.*, 17(2), 193, 1988. With permission.)

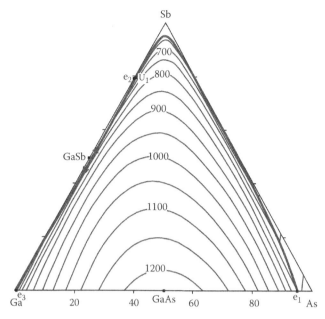

FIGURE 10.41 Calculated liquidus surface of the Ga–Sb–As ternary system. (From Ishida, K., et al., *J. Less Common Metals*, 142, 135, 1988. With permission.)

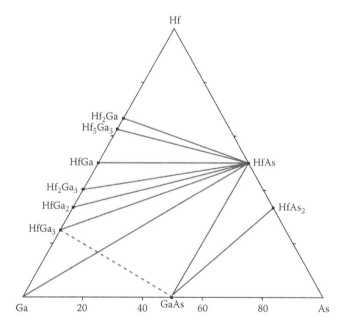

FIGURE 10.40 Calculated isothermal section of the Ga–Hf–As ternary system at room temperature. (From Schmid-Fetzer, R., *J. Electron. Mater.*, 17(2), 193, 1988. With permission.)

was estimated by the enthalpy of formation and calculated from Miedema's model.

10.30 GALLIUM–ANTIMONY–ARSENIC

The Ga–Sb–As ternary system exhibits a eutectic valley extending from the Sb-rich side of the binary Ga–Sb

phase diagram to the As-rich region of the As–Ga phase diagram. Due to the high vapor pressure of As, the phase diagram determinations have been limited to As-dilute compositions. The liquidus surface of this system was calculated using various solution models and is shown in Figure 10.41 (Gratton and Woolley 1980; Lendvay 1984; Ishida et al. 1988a, 1989b; Litvak and Charykov 1991; Li et al. 1998a). The field of GaAs primary crystallization occupies the biggest part of the ternary system. The calculated isothermal sections of this system at 600°C and 900°C is given in Figure 10.42 (Ishida et al. 1988a).

Liquidus and solidus isotherms have been determined experimentally at 720°C–775°C in the Ga-rich region by Sugiyama and Oe (1977). The liquidus isotherms were determined using the saturation technique at 725°C, 750°C, and 775°C (Figure 10.43a). For solidus isotherm determination, $GaAs_{1-x}Sb_x$ epitaxial layers were performed on GaAs (100) substrates by the step-cooling or supercooling technique. The solid compositions are shown in Figure 10.43b as a function of Sb molar fraction in solutions.

The phase diagram of the Ga-rich corner of the Ga–As–Sb system in the range 500°C–700°C was determined by experiments using liquids of constant Sb concentration $x^L_{Sb} = 0.0615$ (Mani et al. 1986). The experimental data concerning liquidus temperatures

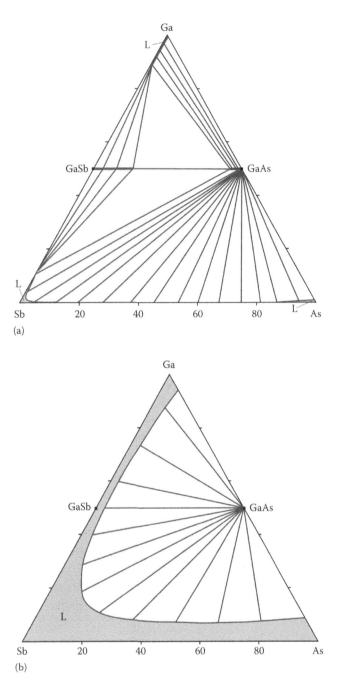

FIGURE 10.42 Calculated isothermal sections of the Ga–Sb–As ternary system at (a) 600°C and (b) 900°C. (From Ishida, K., et al., *J. Less Common Metals, 142*, 135, 1988. With permission.)

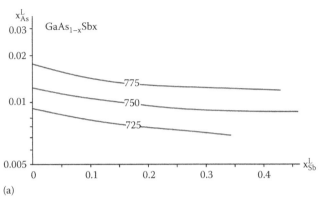

(a)

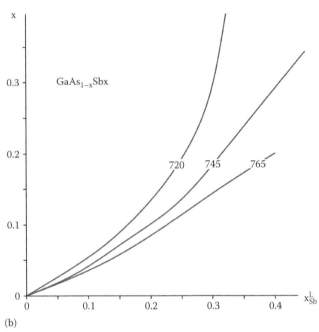

(b)

FIGURE 10.43 Liquidus (a) and solidus (b) isotherms in the Ga corner of the Ga–Sb–As ternary system. (From Sugiyama, K., and Oe, K., *Jpn. J. Appl. Phys., 16*(1), 197, 1977. With permission.)

and the composition of the GaSb-rich solid phase could be fitted by calculated curves on the basis of the simple solution model.

GaAs–GaSb. Goryunova (1955) and Goryunova and Fedorova (1955) suggested that the mutual solid solubility in this system is not higher than 3 mol%. Later, it was determined that the solid solutions formed in the GaAs–GaSb system in the entire range of concentration

(Woolley and Smith 1958a). Inoue and Osamura (1971) have reported a eutectic-type behavior at a composition of 3.0 mol% GaAs and a temperature of 708°C, and the mutual solubility was found to be 0.14 ± 0.06 mol% GaAs and 1.24 ± 0.09 mol% GaSb at the eutectic temperature. Other investigators of this system have indicated an existence of a peritectic reaction with a miscibility gap in the solid state (Gratton and Woolley 1973, 1980; Sirota and Matyas 1973; Gratton et al. 1979; Lendvay 1984; Mani et al. 1986; Ishida et al. 1988a). According to the data of Gratton and Woolley (1973), a maximum solid miscibility gap for the GaAs$_x$Sb$_{1-x}$ solid solutions within the region $0.38 < x < 0.61$ ($0.38 < x < 0.68$ [Gratton et al. 1979; Gratton and Woolley

1980]; $0.45 < x < 0.69$ [Bublik and Leikin 1978]) exists at the peritectic temperature (745°C ± 5°C). It was also shown that as the temperature is lowered, the miscibility gap widens rapidly, being from 30 to 95 mol% GaAs at 700°C (Gratton et al. 1979; Gratton and Woolley 1980). The full phase diagram of the GaAs–GaSb has been calculated by Panish and Ilegems (1972), Stringfellow (1972), Cho (1976), Bublik and Leikin (1978), and Ishida et al. (1988a, 1989b), and assessed by Li et al. (1998a). It is shown in Figure 10.44. The miscibility gap was found to be very asymmetric even at temperatures a little below the peritectic horizontal (the bimodal is quite symmetrical about $x = 0.5$ [Pessetto and Stringfellow 1983]). The critical temperature is 1106°C, and the critical composition is 64 mol% GaAs (El Haj Hassan et al. 2010) (729.5°C [Nahory et al. 1977]; 583°C [Stringfellow 1982a, 1982b]; 759°C [Mani et al. 1986]; 807°C and 59.5 mol% GaAs [Wei et al. 1990]). The GaAs–GaSb phase diagram appears to indicate strong positive deviations from ideality in the solid solution (Foster and Woods 1972). The large deviation from ideality in the solid state is attributed to lattice strain resulting from the large lattice mismatch between GaAs and GaSb.

In the range $0 < x < 0.38$ ($x < 0.5$ [Straumanis and Kim 1965]), the lattice parameter follows Vegard's law (Gratton and Woolley 1973). The room temperature band gap is given by the following equation: E_g (eV) = $1.43 - 1.9x + 1.2 x^2$ (Nahory et al. 1977).

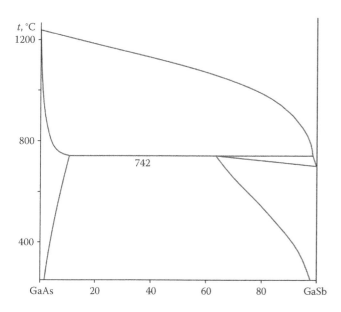

FIGURE 10.44 Phase diagram of the GaAs–GaSb system. (From Ishida, K., et al., *J. Less Common Metals*, 142, 135, 1988. With permission.)

Integral excess Gibbs energy, enthalpy, and entropy of the $GaAs_xSb_{1-x}$ solid solution at 27°C have been calculated, and a positive deviation was recorded (Kühn and Leonhardt 1972). Phase equilibria in $GaAs_xSb_{1-x}$ thin films grown on GaAs and InP substrates were calculated, taking into consideration the elastic contribution caused by the lattice mismatch between the film and substrate (Ohtani et al. 2001). The basic concept used in describing the elastic free energy is that strain due to a lattice mismatch accumulates in a thin film, and interfacial misfit dislocations form when the strain energy exceeds a certain energy barrier. The Sb distribution coefficient in this solid solution has been determined by Pessetto and Stringfellow (1983). According to the mass spectrometric measurements, in the vapor phase over the $GaAs_xSb_{1-x}$ solid solutions at 661°C–729°C there are mainly Sb_4 molecules (Sirota et al. 1978). Arsenic was detected only from the GaAs side. Chemical analysis of the condensate after the evaporation of the test substance from the Knudsen cell showed that the Sb content in the condensate is several orders of magnitude higher than that of As.

A first-principles calculation of a temperature composition epitaxial phase diagram of the GaAs–GaSb system has been performed, and calculated values of the equilibrium molar volumes, the bulk moduli, and pressure derivatives for Ga_2SbAs, Ga_4SbAs_3, and Ga_4Sb_3As compositions were obtained (Wei et al. 1990).

10.31 GALLIUM–BISMUTH–ARSENIC

Theoretical analysis of the GaAs–Ga–Bi subsystem of the Ga–Bi–As ternary system on the basis of the regular solution model has been presented by Leonhardt and Kühn (1974a), Akchurin et al. (1984), Yevgen'ev and Ganina (1984), and Yakusheva and Chikichev (1987). The solubilities of GaAs in Bi in the temperature range of 600°C–1200°C were determined. The calculated liquidus surface of this subsystem is shown in Figure 10.45.

The vapor pressure in the Ga–Bi–As system was investigated in the temperature range 730°C–885°C (Kao Le Din' et al. 1984). A positive deviation of the vapor pressure over the melt of this system has been found from Raoult's law.

The incorporation of Bi was investigated in the MBE growth of $GaBi_xAs_{1-x}$ films on a GaAs substrate (Lewis et al. 2012; Bennarndt et al. 2016). It was shown that the Bi content increases rapidly as the As_2/Ga flux ratio is lowered to 0.5 and then saturates for lower flux ratios.

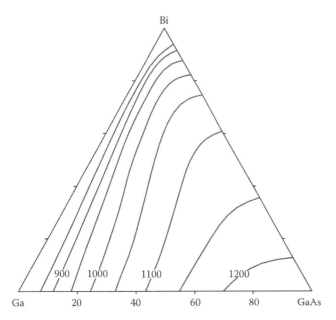

FIGURE 10.45 Calculated liquidus surface of the GaAs–Ga–Bi subsystem. (From Leonhardt, A., and Kühn, G., *Krist. Tech.*, 9(1), 77, 1974. With permission.)

Growth under Ga- and Bi-rich conditions shows that the Bi content increases strongly with decreasing temperature. Highly crystalline films containing up to 20–22 at% Bi were grown at 200°C–400°C. GaBi$_x$As$_{1-x}$ solid solutions have also been grown by metal-organic vapor phase epitaxy (MOVPE) (Oe and Okamoto 1998). A low growth temperature, such as 365°C, is required to obtain the solid solutions with *x* up to 0.02.

A rapid decrease of the band gap with Bi concentration was found (Francoeur et al. 2003; Tixier et al. 2003). The linearized band gap reduction is 83–84 meV for 1 at% Bi.

GaAs–Bi. The phase diagram of this system is a eutectic type (Figure 10.46) (Leonhardt and Kühn 1974a; Akchurin et al. 1984; Yevgen'ev and Ganina 1984). The eutectic is degenerated and crystallizes at 270°C. The maximum solubility of Bi in GaAs reaches up to $8 \cdot 10^{18}$ cm^{-3} at 1050°C–1150°C (Akchurin et al. 1986). The solubility of GaAs in Bi (in atomic fractions of arsenic in the liquid phase c_{As} vs. temperature in the range of 600°C–900°C) is a straight line described by the following equation: $c_{As} = 1.2 \cdot 10^{-5} \exp(9.45 \cdot 10^{-3} t)$, where *t* is the temperature in °C (Yakusheva and Chikichev 1987). The distribution coefficient changes from $3 \cdot 10^{-5}$ at 700°C to $5 \cdot 10^{-4}$ at 800°C. The solubility of GaAs in the Bi melt at 700°C–1050°C was determined by Rubinstein (1966) and Leonhardt and Kühn (1974b).

This system was investigated through XRD, DTA, and metallography (Leonhardt and Kühn 1974a; Akchurin et al. 1984; Yevgen'ev and Ganina 1984).

10.32 GALLIUM–VANADIUM–ARSENIC

The calculated isothermal section of the Ga–V–As ternary system at room temperature, which essentially remains unchanged to about 300°C, is given in Figure 10.47 (Schmid-Fetzer 1988). The dotted lines

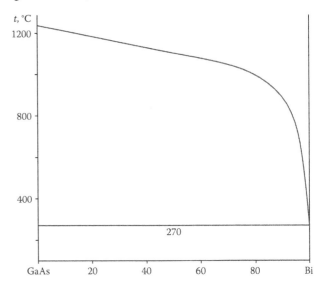

FIGURE 10.46 Phase diagram of the GaAs–Bi system. (From Leonhardt, A., and Kühn, G., *Krist. Tech.*, 9(1), 77, 1974. With permission.)

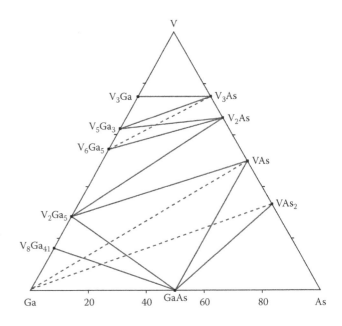

FIGURE 10.47 Calculated isothermal section of the Ga–V–As ternary system at room temperature. (From Schmid-Fetzer, R., *J. Electron. Mater.*, 17(2), 193, 1988. With permission.)

indicate a possible alternative tie-lines. The predictive calculations were based on the following simplifications: ternary phases and solid solubilities were disregarded, and the Gibbs energy of formation of binary compounds was estimated by the enthalpy of formation and calculated from Miedema's model.

GaAs–VAs. The phase diagram of this system is a eutectic type (Reiss and Renner 1966). The eutectic contains 8.4 mol% VAs.

10.33 GALLIUM–NIOBIUM–ARSENIC

The isothermal section of the Ga–Nb–As ternary system at 600°C was determined with the use of XRD, EPMA, and SEM by Schulz et al. (1989) and is given in Figure 10.48. No ternary phases were found, and limited solid solubilities were measured in the constituent binary Nb–Ga and Nb–As compounds. The phases GaAs, NbGa₃, and NbAs coexist with each other to form a three-phase equilibrium that dominates the GaAs side of the phase diagram. Nb₃Ga dissolves 3–4 at% As. A great deal of uncertainty exists in the Nb corner of the phase diagram. Positive identification of the Nb–As phases in this region was difficult. It is believed that this isothermal section would change very little between room temperature and 1000°C. Samples first were annealed at 600°C for 30 days, and then pulverized, pressed again, and sealed in quartz ampoules. The ampoules were annealed

at 950°C for 30 days, followed by a final annealing at 600°C for 20 days. The samples were removed from the furnace and quenched in ice water.

The isothermal section of this ternary system at room temperature was also calculated by Schmid-Fetzer (1988) with practically the same results.

10.34 GALLIUM–TANTALUM–ARSENIC

The isothermal section of the Ga–Ta–As ternary system at 850°C is given in Figure 10.49 (Han and Schmid-Fetzer 1993b). The results were confirmed by the investigation of the Ta/GaAs diffusion couple annealed at 850°C for 3 or 5 days. The isothermal section of this ternary system at room temperature was also calculated by Schmid-Fetzer (1988) with practically the same results. These results were confirmed by Lahav and Eizenberg (1984), who determined that Ta interacts with GaAs at 650°C, forming TaAs.

10.35 GALLIUM–OXYGEN–ARSENIC

The isothermal section of the Ga–O–As ternary system at room temperature is shown in Figure 10.50 (Schwartz et al. 1980; Schwartz 1983; Thurmond et al. 1980). **GaAsO₄** ternary compound is formed in this system. It crystallizes in the trigonal structure with the lattice parameters $a = 498.6 \pm 0.3$, 499.3 ± 0.1, and 499.4 ± 0.1 and $c = 1137.5 \pm 0.7$, 1136.6 ± 0.2, and

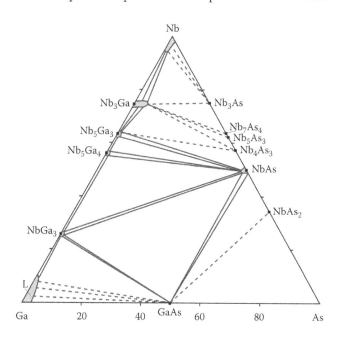

FIGURE 10.48 Isothermal section of the Ga–Nb–As ternary system at 600°C. (From Schulz, K.J., and Zheng, X.-Y., *Bull. Alloy Phase Diagr.*, *10*(4), 314, 1989. With permission.)

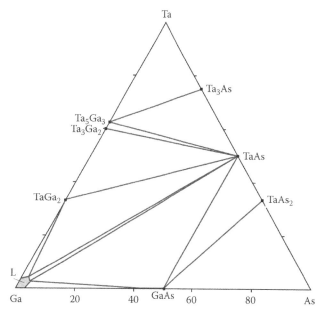

FIGURE 10.49 Isothermal section of the Ga–Ta–As ternary system at 850°C. (From Han, Q., and Schmid-Fetzer, R., *Mater. Sci. Eng. B*, *17*(1–3), 147, 1993. With permission.)

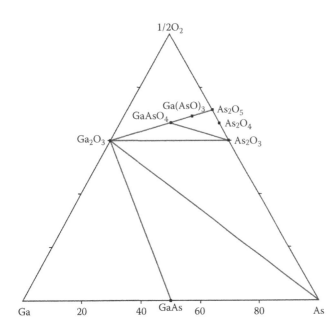

FIGURE 10.50 Isothermal section of the Ga–O–As ternary system at room temperature. (From Schwartz, G.P., *Thin Solid Films*, *103*(1–3), 3, 1983. With permission.)

1136.1 ± 0.2 pm at –100°C, 20°C, and 100°C, respectively (Goiffon et al. 1986) (a = 499.6 ± 0.5 and c = 1139 ± 2 pm [Kosten and Arnold 1980]), or in a hexagonal structure with the lattice parameters a = 500.0 ± 0.2 and c = 1140.0 ± 0.5 pm (Matar et al. 1990) (a = 500.0 and c = 1136 pm and a calculated density of 4.20 g·cm⁻³ [Shafer and Roy 1956]).

According to the data of Dachille and Roy (1959), there are no new crystalline forms of $GaAsO_4$ at heating at 600°C up to 6 MPa, but Matar et al. (1990) obtained new high-pressure varieties of this compound at the pressure range from 3 to 9 GPa. The first modification obtained at 3–7 GPa crystallizes in the monoclinic structure with the lattice parameters a = 783.0, b = 846.4, and c =511.4 pm and β = 110.5°; the second modification obtained at 4–7 GPa crystallizes in the orthorhombic structure with the lattice parameters a = 553.2, b = 828.4, and c =602.5 pm; and the third modification obtained at 9 GPa and circa 900°C crystallizes in the tetragonal structure with the lattice parameters a = 436.7 ± 0.3 and c = 564.0 ± 0.5 pm.

The quartz form of $GaAsO_4$ with a trigonal structure was prepared by mixing Ga_2O_3 and 85% H_3AsO_4 at 260°C and 5 MPa (Kosten and Arnold 1980), or by mixing Ga_2O_3 and H_3AsO_4 into a slurry in a Pt crucible and heating to 600°C (Shafer and Roy 1956), or under 0.1 GPa at 800°C for 4 h starting either from a homogenized mixture of Ga_2O_3 and As_2O_3 under a pressure of dry O_2, or from a homogenized mixture of Ga_2O_3 and As_2O_5 under a pressure of a mixture of dry O_2/N_2 (volume ratio 1:9) (Matar et al. 1990).

The solubility of oxygen in GaAs was determined experimentally by Borisova et al. (1977a, 1977b) and is given in Figure 10.51. The temperature dependence of oxygen solubility is retrograde with a maximum value at circa 1100°C.

10.36 GALLIUM–SULFUR–ARSENIC

GaAs–Ga₂S₃. This system is a non-quasi-binary section of the Ga–S–As ternary system (Kozhina et al. 1962). The samples were annealed at 900°C for

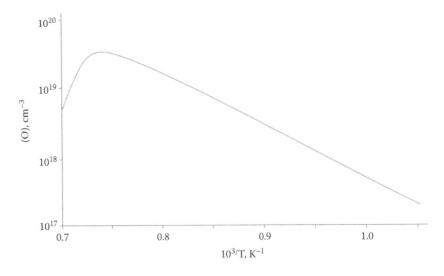

FIGURE 10.51 Temperature dependence of oxygen solubility in GaAs. (From Borisova, L.A., et al., *Izv. AN SSSR Neorgan. Mater.*, *13*(5), 908, 1977. With permission.)

880–1440 h. The system was investigated through DTA, XRD, metallography, and measurement of microhardness.

GaS–As₂S₃. $Ga_3As_4S_9$ ternary compound, which melts incongruently at 750°C, is formed in this system (Il'yasov and Rustamov 1982; Rustamov et al. 1982).

10.37 GALLIUM–SELENIUM–ARSENIC

The liquidus surface of the Ga–Se–As ternary system is shown in Figure 10.52 (Lakeenkov et al. 1977). The field of GaAs primary crystallization occupies the biggest part of this system.

GaAs–Se. This system is a non-quasi-binary section of the Ga–Se–As ternary system (Rubinstein 1966; Lakeenkov et al. 1977). In heavily doped solid solutions of Se in GaAs, the formation of Ga_2Se_3 was observed (Rubinstein 1966; Kuznetsov et al. 1973). The solubility of Se in GaAs in the temperature range 600°C–1100°C could be expressed by the following equation: c (cm^{-3}) $= 9.5 \cdot 10^{23} \cdot \exp(-1.23 \pm 0.02 \text{ eV})/kT$, where T is the temperature in kelvin (Lidov et al. 1978). The maximum solubility of Se takes place along the GaAs–Ga₂Se₃ section and is equal to 4.5 at% at 1170°C (Dashevskiy et al. 1974a).

GaAs–GaSe. The phase diagram of this system is a eutectic type (Figure 10.53) (Lakeenkov et al. 1977). The eutectic crystallizes at 920°C.

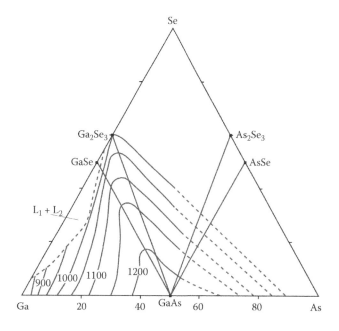

FIGURE 10.52 Liquidus surface of the Ga–Se–As ternary system. (From Lakeenkov, V.M., et al., *Izv. AN SSSR Neorgan. Mater.*, 13(8), 1373, 1977. With permission.)

GaAs–Ga₂Se. This system is a non-quasi-binary section of the Ga–Se–As ternary system as Ga₂Se melts incongruently (Lakeenkov et al. 1977).

GaAs–Ga₂Se₃. A continuous series of solid solutions are formed in this system (Goryunova 1955; Goryunova and Grigor'eva 1956; Woolley and Smith

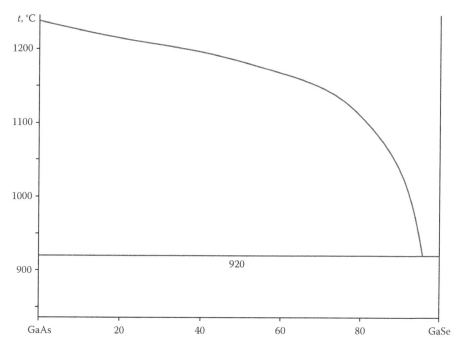

FIGURE 10.53 Phase diagram of the GaAs–GaSe system. (From Lakeenkov, V.M., et al., *Izv. AN SSSR Neorgan. Mater.*, 13(8), 1373, 1977. With permission.)

1958b; Lakeenkov et al. 1977). The alloys were annealed at 900°C for 1–2 months (Woolley and Smith 1958b). There is a considerable deviation from Vegard's law at the Ga_2Se_3 end of the diagram (Goryunova and Grigor'eva 1956; Woolley and Smith 1958b).

This ternary system was investigated through DTA, XRD, and metallography (Goryunova and Grigor'eva 1956; Lakeenkov et al. 1977).

GaSe–As_2Se_3. $Ga_3As_4Se_9$ ternary compound, which melts incongruently at 600°C, is formed in this system (Rustamov et al. 1979; Il'yasov and Rustamov 1982).

10.38 GALLIUM–TELLURIUM–ARSENIC

The liquidus surface of the Ga–Te–As ternary system constructed through DTA, XRD, and EPMA is shown in Figure 10.54 (Panish 1967c). There is some evidence for a ternary compound in the As-rich region, the composition of which is not known. The triangulation of the Ga–Te–As system is defined by the GaAs–GaTe, GaAs–Ga_2Te_3, and GaAs–As_2Te_3 sections (Gimel'farb et al. 1972; Pelevin et al. 1972a). The maximum solubility of Te takes place along the GaAs–Ga_2Te_3 section and is equal to 3.2 at% at 1170°C (Dashevskiy et al. 1974a). The glasses were obtained in this system at the heating of the mixtures of elements at 1000°C ± 10°C for 12 h with vigorous stirring, followed by quenching in water (Minaev 1983).

GaAs–As_2Te_3. The phase diagram of this system is a eutectic type (Figure 10.55) (Babaeva and Rustamov 1972; Rustamov et al. 1974b). The eutectic crystallizes at 350°C and contains 20 mol% GaAs. The solubility of GaAs in As_2Te_3 at room temperature is 1 mol%. This system was investigated through DTA, XRD, metallography, and measurement of microhardness. According to the data of Panish (1967c), the GaAs–As_2Te_3 system is not a true quasi-binary and has a considerable region in which the liquid is in equilibrium with a ternary solid solution with a structure based on Ga_2Te_3.

GaAs–Ga_2Te_3. The results approximating the single-phase condition were obtained with all specimens containing more than 35 mol% Ga_2Te_3 (the system is seen as 3GaAs–Ga_2Te_3) (Woolley and Smith 1958b). No indication of any ordering was observed in these alloys.

GaAs–Te. This system is non-quasi-binary section of the Ga–Te–As ternary system (Gimel'farb et al. 1972). There is evidence to believe that some Ga_2Te_3 is formed at the interaction of GaAs and Te (Rubinstein 1966; Ufimtsev and Gimel'farb 1973). The distribution coefficient of Te in GaAs is equal to 0.46 (Mil'vidskiy and Pelevin 1965).

GaTe–As_2Te_3. $GaAs_2Te_4$ ternary compound, which melts incongruently at 385°C, is formed in this system (Il'yasov and Rustamov 1982).

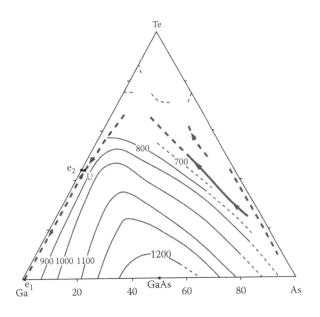

FIGURE 10.54 Liquidus surface of the Ga–Te–As ternary system. (From Panish, M.B., *J. Electrochem. Soc., 114*(1), 91, 1967. Reproduced with permission of The Electrochemical Society.)

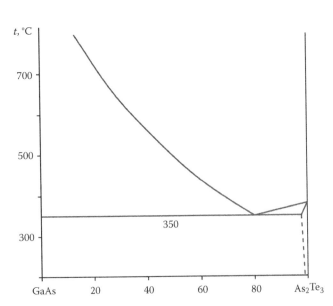

FIGURE 10.55 Phase diagram of the GaAs–As_2Te_3 system. (From Rustamov, P.G., et al., in *Khalkogenidy*, vyp. 3, Nauk. Dumka Publish., Kiev, 1974, 106. With permission.)

10.39 GALLIUM–CHROMIUM–ARSENIC

The solubility of GaAs in the Ga–Cr melts was experimentally determined by André and Le Duc (1969) and is given in Figure 10.56. A set of curves represent the crystallization temperature as a function of Cr content in Ga solution at a constant percentage of As.

Based on the XRD and DTA results, isothermal sections of the Ga–Cr–As ternary system from room temperature to 1300°C were constructed (Deal and Stevenson 1984; Deal et al. 1985). Several of these (at room temperature, at 670°C, and at 1000°C) are shown in Figure 10.57. No ternary phases were determined in this ternary system. The samples were annealed at 600°C–1050°C for 10 h, followed by quenching in water. Isothermal sections of this ternary system were also calculated at room temperature by Schmid-Fetzer (1988) and at 1130°C, 1200°C, and 1210°C by Kaufman and Ditchek (1991).

GaAs–Cr. The calculated solidus of this system is retrograde (Kuznetsova 1975). The maximum solubility of Cr in GaAs takes place at 1000°C–1150°C and is equal to $1.85 \cdot 10^{18}$ cm^{-3} ([0.8–5.0]$\cdot 10^{17}$ cm^{-3} [Sokolov et al. 1982]; $2 \cdot 10^{17}$ cm^{-3} [Deal and Stevenson 1984]). According to the data of Deal and Stevenson (1984), the solubility of Cr in GaAs at 800°C is $3 \cdot 10^{16}$ cm^{-3}. The temperature dependence of Cr solubility in GaAs is shown in Figure 10.58.

Whatever the doping technique, at concentrations of Cr more than $4 \cdot 10^{16}$ cm^{-3} the solid solution was observed

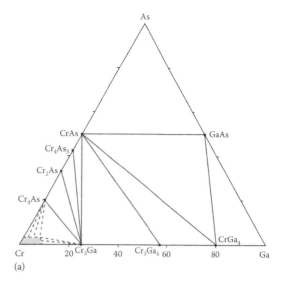

(a)

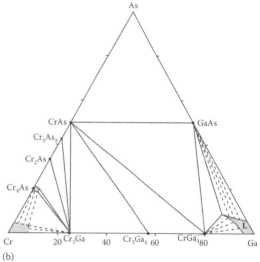

(b)

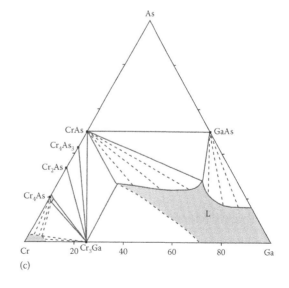

(c)

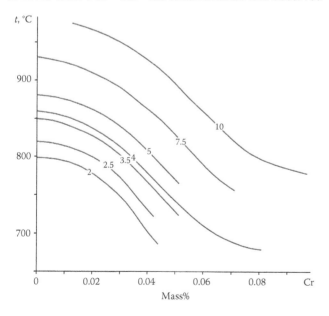

FIGURE 10.56 Solubility of GaAs in the solutions of Ga in Cr: the numbers on the curves indicate the As concentration in the solid phase (mass%). (From André, E., and Le Duc, J.M., *Mater. Res. Bull.*, 4(3), 149, 1969. With permission.)

FIGURE 10.57 Isothermal sections of the Ga–Cr–As ternary system at (a) 25°C, (b) 670°C, and (c) 1000°C. (From Deal, M.D., et al., *J. Phys. Chem. Solids*, 46(7), 859, 1985. With permission.)

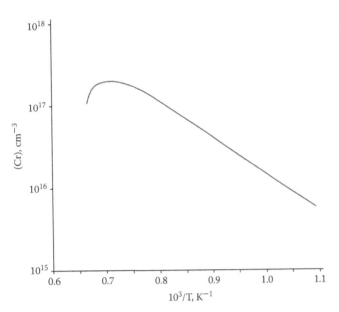

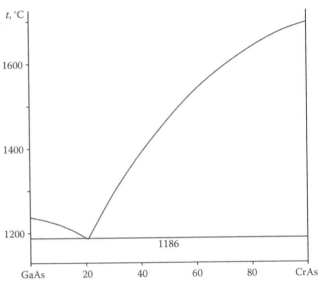

FIGURE 10.58 Temperature dependence of Cr solubility in GaAs. (From Deal, M.D., and Stevenson, D.A., *J. Electrochem. Soc.*, **131**(10), 2343, 1984. Reproduced with permission of The Electrochemical Society.)

FIGURE 10.59 Phase diagram of the GaAs–CrAs system. (From Kaufman, L., and Ditchek, B., *J. Less Common Metals*, **168**(1), 115, 1991. With permission.)

to contain precipitates that grew, at least partially, due to the migration of substitutional atoms and their transformation into the precipitates (Bochkareva et al. 1982). The distribution coefficient of Cr in GaAs at 700°C, 800°C, 900°C, and 1000°C is $9 \cdot 10^{-7}$, $4.5 \cdot 10^{-6}$, $1.2 \cdot 10^{-5}$, and $2.5 \cdot 10^{-5}$, respectively (Deal and Stevenson 1984). According to the data of Sokolov et al. (1982), the distribution coefficient of Cr in GaAs is equal to $(1.2 \pm 0.5) \cdot 10^{-3}$.

GaAs–CrAs. The phase diagram of this system is a eutectic type (Figure 10.59) (Kaufman and Ditchek 1991). According to the calculations, the eutectic crystallizes at 1186°C and contains 20.76 mol% CrAs. According to the experiments, the eutectic contains 35.4 mol% (Reiss and Renner 1966). The single-phase $Ga_{1-x}Cr_xAs$ films were synthesized at 250°C–400°C with x up to 0.07 (Lu et al. 2008). For the films with $x > 0.02$, the aggregation of Cr atoms is strongly enhanced as both growth temperature and x increase.

10.40 GALLIUM–MOLYBDENUM–ARSENIC

The liquidus surface is shown in Figure 10.60 using solid lines for the condensed equilibria of liquids and solids stable below 100 kPa pressure, which are extrapolated as dashed lines into the higher-pressure region (Han and Schmid-Fetzer 1992). These lines correspond to the experimental condition in the sealed capsule. They

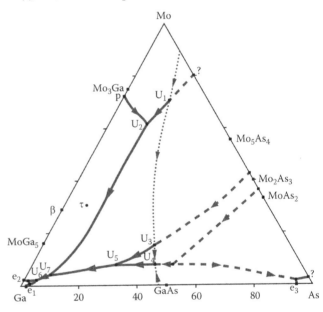

FIGURE 10.60 Liquidus surface of the Ga–Mo–As ternary system with the condensed equilibria stable at 100 kPa and their extensions to $p > 100$ kPa, intersected by the L + S + G (100 kPa) equilibria. (From Han, Q., and Schmid-Fetzer, R., *J. Phase Equilibria*, **13**(6), 588, 1992. With permission.)

are intersected by the (L + S + G) equilibria shown as a dotted line at 100 kPa pressure, extending from the well-known (L + G ⟷ GaAs) equilibrium at 1237°C to the [L ⟷ (Mo) + G] equilibrium crudely estimated at about 2200°C. Viewing the entire ternary system at 100 kPa pressure, this dotted line presents the stability limit of the liquid phase, and all the dashed extensions are cut off. The $Ga_{61}Mo_{31}As_8$ (τ) ternary phase was detected, and also the following five nonvariant equilibria: L + $Mo_2As_3 \longleftrightarrow Mo_5As_4$ + GaAs (1167°C), L + $Mo_5As_4 \longleftrightarrow Mo_3Ga$ + GaAs (912°C), L + Mo_3Ga + GaAs + τ (883°C), L + $Mo_3Ga \longleftrightarrow$ "$MoGa_5$" + τ (825°C), and L + $\tau \longleftrightarrow MoGa_5$ + GaAs (820°C). The decomposition temperature of τ was found to be 883°C. The consistency of all data was confirmed by the reaction scheme (Han and Schmid-Fetzer 1992).

Seven isothermal sections of the Ga–Mo–As ternary system in the range 600°C–1300°C have been constructed (Han and Schmid-Fetzer 1992). At 25°C, GaAs is in thermodynamic equilibrium with $MoGa_5$, β-phase (this phase is in the Ga–Mo system at 70 at% Ga), Mo_3Ga, Mo_5As_4, Mo_2As_3, and $MoAs_2$, and Mo_3Ga coexists with Mo_5As_4 (Schmid-Fetzer 1988; Han and Schmid-Fetzer 1992). Isothermal sections of this system at 600°C, 800°C, 1000°C, and 1300°C are shown in Figure 10.61.

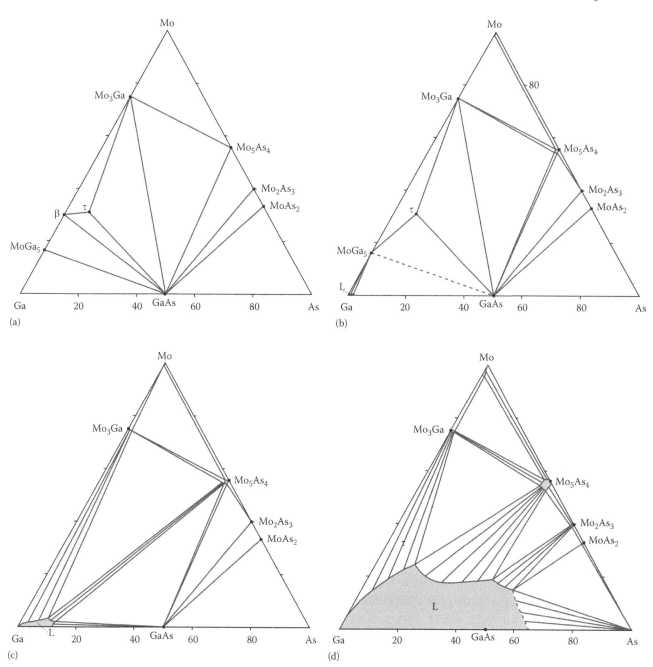

FIGURE 10.61 Isothermal sections of the Ga–Mo–As ternary system at (a) 600°C, (b) 800°C, (c) 1000°C, and (d) 1300°C. (From Han, Q., and Schmid-Fetzer, R., *J. Phase Equilibria*, *13*(6), 588, 1992. With permission.)

The solubility limits estimated at 1300°C are 7 at% As in Mo_3Ga, 6 at% As in Mo_5Ga_4, 2 at% As in Mo_2As_3, and 2 at% As in $MoAs_2$.

The ternary Ga–Mo–As phase equilibria have been analyzed using DTA, XRD, metallography, and SEM with EDX (Han and Schmid-Fetzer 1992).

$Ga_2Mo_5As_4$ ternary compound, which is a semiconductor and crystallizes in the tetragonal structure with the lattice parameters $a = 964.1 \pm 0.2$ and $c = 328.2 \pm 0.3$ pm, was prepared by reaction of the elements in the required stoichiometry in evacuated sealed silica ampoules at 1000°C–1050°C for about 15 days, with one grinding in between (Nanjundaswamy and Gopalakrishnan 1988). Dronkowski et al. (1991) noted that it is necessary to reject the existence of this compound. Together with the findings that the lattice constants of Mo_5As_4 obtained in the absence or presence of Ga are identical within the standard deviation, and that no characteristic x-ray emission of Ga has been found, it was concluded that Mo_5As_4 does not incorporate a significant amount of Ga.

GaAs–MoAs. The phase diagram of this system is a eutectic type (Reiss and Renner 1966). The eutectic contains 9.5 mol% MoAs.

10.41 GALLIUM–TUNGSTEN–ARSENIC

An isothermal section of the Ga–W–As ternary system at room temperature was calculated by Schmid-Fetzer (1988), and at 800°C was determined experimentally by Han and Schmid-Fetzer (1993a). At both temperatures, GaAs is in thermodynamic equilibrium with W, W_2As_3, and WAs_2 (Pelevin et al. 1972a; Schmid-Fetzer 1988; Han and Schmid-Fetzer 1993a).

10.42 GALLIUM–MANGANESE–ARSENIC

$Mn_2Ga_{0.5}As_{0.5}$ metastable compound is formed in the Ga–Mn–As ternary system (Ellner and El-Boragy 1986). It crystallizes in the hexagonal structure with the lattice parameters $a = 406.2 \pm 0.3$ and $c = 504.8 \pm 0.6$ pm for $Mn_{67}Ga_{19}As_{14}$ composition, $a = 406.5 \pm 0.1$ and $c = 505.0 \pm 0.3$ pm for $Mn_{66}Ga_{20}As_{14}$ composition, and $a = 406.3 \pm 0.3$ and $c = 504.5 \pm 0.6$ pm for $Mn_{65}Ga_{21}As_{14}$ composition. Mn, Ga, and As were used for its preparation. The stoichiometric mixture of the elements was sealed in an initially evacuated quartz ampoule that was then filled with Ar. Then it was melted gradually and homogenized.

GaAs–MnAs. Alloys' phase stability of zinc blende $Ga_{1-x}Mn_xAs$ was investigated by using the first-principles full-potential linearized augmented plane-wave method and the cluster expansion method (Hatano et al. 2007; Nakamura et al. 2007). All approximations demonstrate that the GaAs–MnAs system has a tendency to segregate into GaAs and MnAs, and so to inherently favor clustering. The calculated lattice constant was found to increase when the substitutional Mn composition increases. GaAs-based diluted magnetic semiconductor $Ga_{1-x}Mn_xAs$ with x up to 0.07 was prepared by MBE on a GaAs substrate at temperatures ranging from 160°C to 320°C (Shen et al. 1997). The lattice constant showed a linear increase with the increase of Mn composition. According to the data of Sadowski et al. (2000), $Ga_{1-x}Mn_xAs$ with Mn content varying from 0.25 to 3.5 at% has been grown by the same method. The maximum concentration of Mn in such films given at 250°C, at which no MnAs precipitations have been detected, is about 4 at%.

GaAs–Mn. The calculated solidus of this system is retrograde (Kuznetsova 1975). The maximum solubility of Mn in GaAs takes place at 1000°C–1150°C and is equal to $8.4 \cdot 10^{19}$ cm^{-3} ($[0.5–1.0] \cdot 10^{19}$ cm^{-3} [Sokolov et al. 1982]). According to the data of Sokolov et al. (1982), the distribution coefficient of Mn in GaAs is equal to $(1.67 \pm 0.20) \cdot 10^{-2}$.

10.43 GALLIUM–RHENIUM–ARSENIC

An isothermal section of the Ga–Re–As ternary system at room temperature was calculated by Schmid-Fetzer (1988), and at 800°C was determined experimentally by Han and Schmid-Fetzer (1993a). At both temperatures, GaAs is in thermodynamic equilibrium with Re and Re_3As_7. The additional tie-line GaAs–ReGa$_5$ develops below 764°C.

10.44 GALLIUM–IRON–ARSENIC

Solid-state phase equilibria in the Ga–Fe–As ternary system have been established at 600°C using XRD, and SEM with EDX (Figure 10.62) (Députier et al. 1997, 1998). The existence of a ternary phase $Fe_3Ga_{2-x}As_x$ ($0.20 \leq x \leq 1.125$) has been evidenced (Harris et al. 1987, 1989; Greaves et al. 1990; Députier et al. 1997, 1998). The original feature of the isothermal section is the occurrence of a tie-line between $Fe_3Ga_{2-x}As_x$ and GaAs (Députier et al. 1997, 1998; Raghavan 2004a). Therefore, upon the doping of GaAs single crystals with Fe at compositions

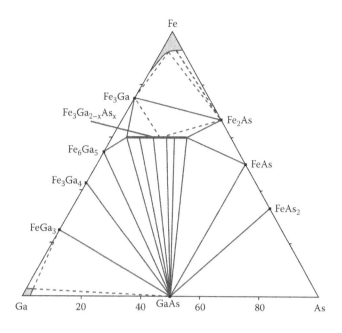

FIGURE 10.62 Isothermal section of the Ga–Fe–As ternary system at 600°C. (From Députier, S., et al., *J. Alloys Compd.*, *262–263*, 416, 1997. With permission.)

greater than its solid solubility limit, the $Fe_3Ga_{2-x}As_x$ phase is formed (Harris et al. 1987). For $0.85 \leq x \leq 1.125$ this phase crystallizes in a disordered hexagonal structure with the lattice parameters $a = 400.9 \pm 0.2$ and $c = 504.6 \pm 0.1$ pm ($a = 400.6 \pm 0.1$ and $c = 504.0 \pm 0.1$ pm and an experimental density of 7.78 g·cm^{-3} [Ellner and El-Boragy 1986]) for Fe_3GaAs (Députier et al. 1997; Raghavan 2004a). The ordered phase exists at $0.20 \leq x \leq 0.85$, and the lattice parameters are $a = 813.3 \pm 0.2$ and $c = 501.0 \pm 0.1$ pm for $Fe_3Ga_{1.6}As_{0.4}$ ($a = 812.6$ and $c = 498.8$ pm for $Fe_{60}Ga_{34}As_6$ [Moze et al. 1994]). Any existence of a two-phase field in the narrow range around $x = 0.85$ has not be detected (Députier et al. 1998). From the triangulations of the Ga–Fe–As ternary system proposed by Pelevin et al. (1972a), only one tie-line (GaAs–FeAs) coincides with the investigations of Députier et al. (1997) because GaAs is not in thermodynamic equilibrium with Fe and Fe_2As.

The samples were given a final annealing at 600°C for 10 days and quenched in an ice–water mixture (Députier et al. 1997, 1998).

GaAs–Fe. The calculated solidus of this system is retrograde (Boltaks et al. 1975; Kuznetsova 1975). The maximum solubility of Fe in GaAs takes place at 1000°C–1150°C and is equal to $3.38 \cdot 10^{18}$ cm^{-3} (Kuznetsova 1975) ($1.3 \cdot 10^{18}$ cm^{-3} at 1100°C [Boltaks

et al. 1975]; $[0.8–1.0] \cdot 10^{18}$ cm^{-3} [Sokolov et al. 1982]). According to the data of Sokolov et al. (1982), the distribution coefficient of Fe in GaAs is equal to $(1.1 \pm 0.1) \cdot 10^{-3}$ ($1.5 \cdot 10^{-3}$ [Pelevin et al. 1969]).

The decomposition of a solid solution of Fe in GaAs at 900°C has been studied by magnetic force microscopy and transmission electron microscopy (TEM) (Prudaev et al. 2012). The results demonstrate that annealing of Fe-doped GaAs for 3 h leads to the formation of disc-shaped ferromagnetic inclusions 50–500 nm in diameter and 1.5–50 nm in thickness. The average inclusion concentration is 10^{11} cm^{-3}. The estimated decomposition time of the solid solution at 900°C is within 1 h.

10.45 GALLIUM–COBALT–ARSENIC

Phase equilibria in the Ga–Co–As ternary system at 600°C and 800°C have been investigated by XRD, optical microscopy, SEM, and EPMA (Figures 10.63 and 10.64) (Shiau et al. 1988, 1989; Lindeberg and Andersson 1991). No ternary phases were found at 600°C. The so-called ternary phase Co_2GaAs is a supersaturated solid solution of α-CoAs, and it may be represented by the formula $CoGa_xAs_{1-x}$, with x varying from 0.77 to 1.0. The solubility of As in CoGa and α-Co is small at 600°C, about 1–2 at% (Shiau et al. 1988, 1989). The solubility of Ga in CoAs is significant, about 19 mol% CoGa in CoAs. There is a slight solubility of Co in GaAs.

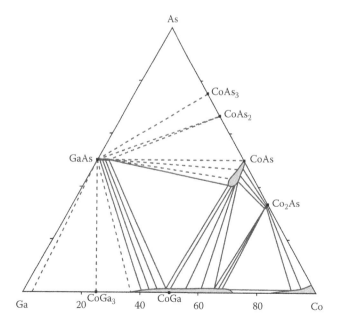

FIGURE 10.63 Isothermal section of the Ga–Co–As ternary system at 600°C. (From Shiau, F.-Y., et al., *Z. Metallkde*, *80*(8), 544, 1989. With permission.)

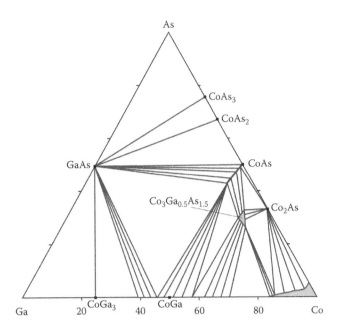

FIGURE 10.64 Isothermal section of the Ga–Co–As ternary system at 800°C. (From Lindeberg, I., and Andersson, Y., *J. Less Common Metals*, *175*(1), 155, 1991. With permission.)

At 800°C, GaAs forms equilibria with $CoGa_3$, CoGa, $CoAs_3$, $CoAs_2$, and CoAs (Lindeberg and Andersson 1991). The presence of a ternary phase with a homogeneity range around the composition $Co_3Ga_{0.5}As_{1.5}$ has been established. Below 800°C, the title compound decomposes into CaAs, Co_2As, and CoGa. It crystallizes in the hexagonal structure with the lattice parameters a = 381.77 ± 0.04 and c = 509.51 ± 0.07 pm.

Samples were first annealed at 600°C for 10 days, and then pulverized, pressed, and again sealed in silica ampoules (Shiau et al. 1989). The ampoules were then annealed at 600°C for 30 days. They were subsequently removed from the furnace and quenched in ice water. Lindeberg and Andersson (1991) annealed the samples at 800°C for at least 1 month and then quenched them in water.

According to the data of Ellner and El-Boragy (1986), $Co_2Ga_{0.5}As_{0.5}$ compound is formed in the Ga–Co–As ternary system. It crystallizes in the hexagonal structure with the lattice parameters a = 396.8 ± 0.1 and c = 502.1 ± 0.1 pm for $Co_{65}Ga_{20}As_{15}$ composition. Co, Ga, and As were used for its preparation. The stoichiometric mixture of the elements was sealed in an initially evacuated quartz ampoule that was then filled with Ar. Then it was melted gradually and homogenized.

The maximum solubility of Co in GaAs is equal to $(2–6)\cdot10^{17}$ cm^{-3}, and the distribution coefficient of Co in GaAs is $(3.6 \pm 0.5)\cdot10^{-3}$ (Sokolov et al. 1982).

10.46 GALLIUM–NICKEL–ARSENIC

The isothermal section of the Ga–Ni–As ternary system at 600°C is given in Figure 10.65 (Ingerly et al. 1996). NiGa was found to possess a smaller range (1 at%) of As solubility than was reported by Zheng et al. (1989a) (10 at%). GaAs dissolves up to 2 at% Ni. In contrast to the earlier studies, there is at most one ternary phase, **Ni_2GaAs** located near the $Ni_{13}Ga_9$ binary, at this temperature. This compound decomposes at 600°C, forming NiGa and NiAs (Ogawa 1980; Lahav and Eizenberg 1984; Lahav et al. 1985, 1986). All the other reported ternary phases were shown to be specific compositions of the NiAs solid solution, which has extensive solubility in the ternary system. Ternary eutectic L ⟷ GaAs + NiGa + NiAs crystallizes at 810°C ± 5°C.

The isothermal section at room temperature was also determined experimentally by Guérin and Guivarc'h (1989) and calculated by Schmid-Fetzer (1988), and at 600°C established using XRD, EPMA, and SEM by Zheng et al. (1989a).

From the triangulations of the Ga–Ni–As ternary system proposed by Pelevin et al. (1972a), only one

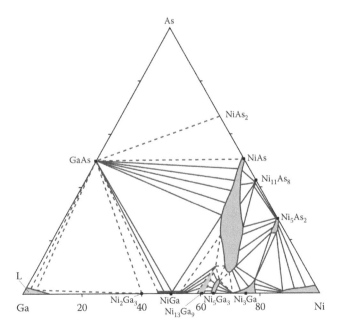

FIGURE 10.65 Isothermal section of the Ga–Ni–As ternary system at 600°C. (From Ingerly, D.B., et al., *J. Appl. Phys.*, *80*(1), 543, 1996. With permission.)

tie-line (GaAs–NiAs) coincides with the investigations of Ingerly et al. (1996) because GaAs is not in thermodynamic equilibrium with Ni and Ni_5As_2.

Ni_2GaAs crystallizes in the hexagonal structure with the lattice parameters $a = 392.5$ and $c = 480.7$ pm (Lahav et al. 1985) ($a = 384$ and $c = 496$ pm [Ogawa 1980]; $a = 383$ and $c = 504$ pm [Lahav et al. 1986]; $a = 390 \pm 1$ and $c = 501 \pm 2$ pm [annealing at 220°C], $a = 388 \pm 1$ and $c = 507 \pm 2$ pm [annealing at 315°C], and $a = 379 \pm 1$ and $c = 501 \pm 2$ pm [annealing at 410°C] [these authors indicated the composition of the compound as **Ni₃GaAs**] [Sands et al. 1986b]; $a = 387 \pm 1$ and $c = 503 \pm 1$ pm [these authors also indicated the composition of the compound as Ni₃GaAs] [Zheng et al. 1989a]).

According to the data of Guérin and Guivarc'h (1989), **Ni₃GaAs** ternary compound, which crystallizes in the hexagonal structure with the lattice parameters $a = 668.9 \pm 0.2$ and $c = 1518.4 \pm 0.3$ pm, is formed in this system.

To prepare the samples of the Ga–Ni–As ternary system, powder mixtures of GaAs, Ni, Ga, and As were uniaxially pressed into pellets (Ingerly et al. 1996). All pellets were sealed in quartz evacuated ampoules. The pellets were first annealed at 600°C for 10 days, and then pulverized, pressed, and resealed in quartz ampoules. The pellets were then reannealed at 600°C for 30 days. After the second annealing, the samples were quenched in ice water.

GaAs–Ni. This system is a non-quasi-binary section of the Ga–Ni–As ternary system (Testova et al. 1986). The calculated solidus of this system is retrograde (Kuznetsova 1975). The maximum solubility of Ni in GaAs takes place at 1000°C–1150°C and is equal to $1.10 \cdot 10^{17}$ cm^{-3} ([1.5–2.0]$\cdot 10^{17}$ cm^{-3} [Sokolov et al. 1982]). The distribution coefficient of Ni in GaAs is $(2.3 \pm 0.7) \cdot 10^{-4}$ (Sokolov et al. 1982).

GaAs–NiGa. The phase relations in this system are shown in Figure 10.66 (Ingerly et al. 1996).

GaAs–NiAs. The phase diagram of this system is a eutectic type (Figure 10.67) (Ingerly et al. 1996). The eutectic crystallizes at 910°C ± 5°C and contains 17.5 mol% GaAs.

10.47 GALLIUM–RHODIUM–ARSENIC

Solid-state equilibrium in the Ga–Rh–As ternary system at 800°C (Figure 10.68) was determined with the use of XRD, EPMA, and SEM (Schulz et al. 1991). No ternary phases were found, and very limited solid solubility was measured in GaAs and the constituent binary

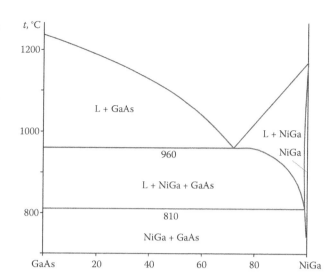

FIGURE 10.66 Phase relations in the GaAs–NiGa system. (From Ingerly, D.B., et al., *J. Appl. Phys.*, *80*(1), 543, 1996. With permission.)

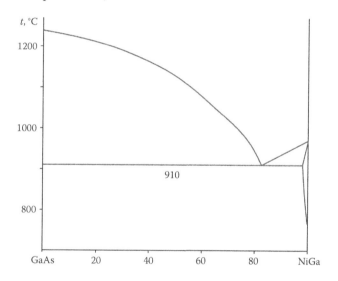

FIGURE 10.67 Phase diagram of the GaAs–NiAs system. (From Ingerly, D.B., et al., *J. Appl. Phys.*, *80*(1), 543, 1996. With permission.)

Rh–Ga and Rh–As compounds. Measured solid solubility is probably less than 1 at% As in RhGa and Ga in RhAs. GaAs, RhGa, and $RhAs_2$ form a three-phase region that dominates the GaAs side of the phase diagram. The samples were annealed at 800°C for 30 days and then quenched in ice water.

The isothermal section of this ternary system at 1000°C is given in Figure 10.69 (Guérin et al. 1990; Deputier et al. 1991). **Ga$_x$RhAs$_{1-x}$**, **Ga$_x$Rh$_2$As$_{1-x}$** and **Ga$_2$Rh$_5$(Ga$_x$As$_{1-x}$)** solid solutions exist at this temperature. Ga_2Rh_5As crystallizes in the orthorhombic structure with the lattice parameters $a = 546.3 \pm 0.4$, $b =$

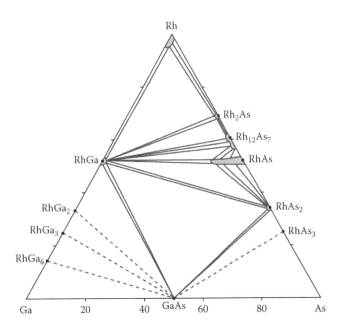

FIGURE 10.68 Isothermal section of the Ga–Rh–As ternary system at 800°C. (From Schulz, K.J., et al., *J. Phase Equilibria*, *12*(1), 10, 1991. With permission.)

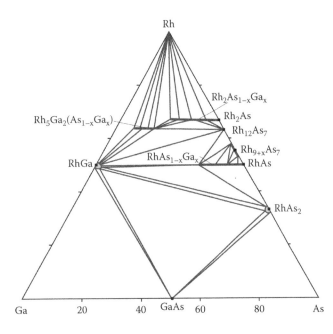

FIGURE 10.69 Isothermal section of the Ga–Rh–As ternary system at 1000°C. (From Guérin, R., et al., *Rev. Phys. Appl.*, *25*(5), 411, 1990. With permission.)

1021.5 ± 0.8, and c = 401.9 ± 0.3 pm and a calculated density of 8.24 g·cm⁻³ (Deputier et al. 1991).

The samples were annealed at 100°C for 72 h with the next quenching in cold water (Guérin et al. 1990).

The isothermal section of the Ga–Rh–As ternary system at room temperature was calculated by Schmid-Fetzer (1988).

10.48 GALLIUM–PALLADIUM–ARSENIC

The isothermal section of the Ga–Pd–As ternary system at 800°C is given in Figure 10.70 (El-Boragy and Schubert 1981; J.-C. Lin et al. 1988). There is a ternary phase τ represented by the formula $\mathbf{Pd_5(Ga_{1-x}As_x)_3}$ (0.16 ≤ x ≤ 0.47). Two other binary phases, Pd_2As and Pd_2Ga, exhibit extensive solubility. Pd_2As (hexagonal structure) dissolves a considerable amount of Ga to replace As. This phase may be represented by the formula $\mathbf{Pd_2Ga_xAs_{1-x}}$. The other solution phase has the same formula, $\mathbf{Pd_2Ga_{1-x}As_x}$, but it is based on the Pd_2Ga phase with the orthorhombic structure.

The isothermal section of the Ga–Rh–As ternary system at room temperature was calculated by Schmid-Fetzer (1988).

Some compounds were obtained in this ternary system. According to the data of El-Boragy and Schubert (1970), $\mathbf{Pd_5GaAs}$ ternary compound crystallizes in the tetragonal structure with the lattice parameters a = 394.9 and c = 687.6 pm.

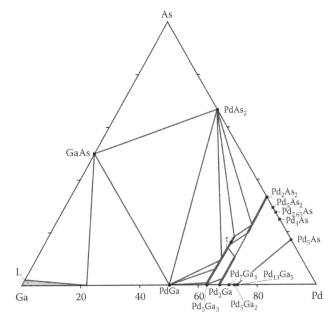

FIGURE 10.70 Isothermal section of the Ga–Pd–As ternary system at 600°C. (From Lin, J.-C., et al., *J. Mater. Res.*, *3*(1), 148, 1988. With permission.)

The results of a transmission electron microscopy study of the Pd–GaAs reaction are presented in Sands et al. (1985, 1986a) and Zhao and Wu (1991). It was shown that the first two reaction products are ternary phases **Pd₅(GaAs)₂** and **Pd₄GaAs**, which crystallize in the hexagonal structure with the lattice parameters $a = 673 \pm 2$ and 920 ± 10 and $c = 338 \pm 1$ and 370 ± 5 pm, respectively (Sands et al. 1985, 1986a).

Pd₁₂Ga₄.₅As₂.₅ and **Pd₂₅Ga₄.₅As₄.₅** crystallize also in the hexagonal structure with the lattice parameters $a = 942.5 \pm 0.1$ and $c = 370.8 \pm 0.1$ pm and a calculated density of 10.35 g·cm⁻³ for the first compound (Deputier et al. 1991) and $a = 736.8 \pm 0.1$ and $c = 1060.3 \pm 0.1$ pm for the second (Ellner and El-Boragy 1990).

The reactions of Pd on atomically clean or air-exposed (100) and (110) GaAs surfaces at temperatures between 20° to 500°C in different ambiences were investigated by TEM (Kuan et al. 1985). Interfacial reactions quite different from previous XRD results were observed, and two ternary phases were identified. At lower temperatures (≤250°C), the formation of a ternary phase **PdGa~0.3~As~0.2~**, which has a hexagonal structure with $a = 672$ and $c = 340$ pm, was determined. At temperatures between 350°C and 500°C, only one phase, PdGa, was observed to form in a high-vacuum environment, whereas in a gas-forming ambience, either a mixture of PdAs₂ and another ternary phase **PdGa~0.6~As~0.4~** (at 350°C) or a mixture of PdAs₂ and PdGa (at 500°C) was observed. The ternary phase PdGa~0.6~As~0.4~ is also hexagonal in structure, with $a = 947$ nm and $c = 374$ pm.

The solid-state reaction of thin Pd films with GaAs substrates has been investigated using Auger sputter profiling, XRD, He-ion backscattering, and sheet resistivity measurements (Olowolafe et al. 1979; Zeng and Chung 1982). The As and Ga compounds formed as a result of contact reactions were identified to be PdAs₂ and PdGa at 250°C; PdAs₂, PdGa, and Pd₂Ga at 350°C; and PdGa at 500°C.

10.49 GALLIUM–OSMIUM–ARSENIC

The isothermal section of the Ga–Os–As ternary system at room temperature, which essentially remains unchanged to about 300°C, was calculated by Schmid-Fetzer (1988). GaAs is in thermodynamic equilibrium with OsGa and OsAs₂, and OsGa coexists with OsAs₂. The predictive calculations were based on the following simplifications: ternary phases and solid solubilities

were disregarded, and the Gibbs energy of formation of binary compounds was estimated by the enthalpy of formation and calculated from Miedema's model.

10.50 GALLIUM–IRIDIUM–ARSENIC

The Ga–Ir–As solid-state equilibrium phase diagram was determined at 600°C (Figure 10.71) with the use of XRD, EPMA, and SEM (Schulz et al. 1990). No ternary phases were found, and very limited solid solubility was measured in GaAs and the constituent binary Ir–Ga and Ir–As compounds. GaAs, Ir₃Ga₅, and IrAs₂ were found to form a three-phase region that dominates the GaAs side of the phase diagram. Measured solid solubilities are 5 at% As in Ga₅Ir₃ and probably less than 1 at% As in GaIn and 1 at% Ga in IrAs₂.

The powders of GaAs and Ir were intimately mixed and pressed into pellets, which were sealed in evacuated fused quartz ampoules (Schulz et al. 1990). They were annealed at 800°C for 7 days, and then pulverized, pressed, and again sealed in quartz ampoules. These ampoules were annealed at 600°C for 30 days, and then quenched in ice water.

The isothermal section of the Ga–Ir–As ternary system at room temperature was calculated by Schmid-Fetzer (1988).

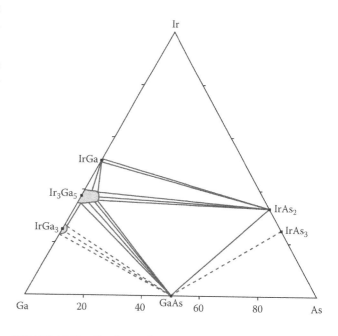

FIGURE 10.71 Isothermal section of the Ga–Ir–As ternary system at 600°C. (From Schulz, K.J., et al., *Bull. Alloy Phase Diagr.*, *11*(3), 211, 1990. With permission.)

10.51 GALLIUM–PLATINUM–ARSENIC

The isothermal section of the Ga–Pt–As ternary system at 600°C was determined with the use of XRD, EPMA, and SEM and is shown in Figure 10.72 (Zheng et al. 1989b). No ternary compounds were found, and limited solid solubility was measured in the constituent binary Pt–Ga and Pt–As compounds. GaAs, PtGa, and PtAs form a region of three-phase equilibrium that dominates the GaAs side of the isothermal section. The maximum solubility of As is 4 at% in Pt_3Ga, 1 at% in PtGa, and less than 1 at% in the other Pt–Ga phases. No measurable solubility of Ga was found in $PtAs_2$ or Pt in GaAs.

The samples were first annealed at 600°C for 30 days, and then pulverized, pressed, and again sealed in quartz ampoules (Zheng et al. 1989b). The ampoules were then annealed at 850°C for 20 days, followed by final annealing at 600°C for 20 days. The samples were removed from the furnace and quenched in ice water.

The isothermal section of the Ga–Pt–As ternary system at room temperature was calculated by Schmid-Fetzer (1988).

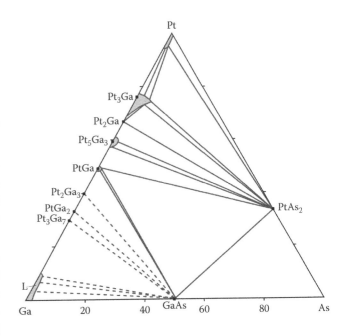

FIGURE 10.72 Isothermal section of the Ga–Pt–As ternary system at 600°C. (From Zheng, X.-Y., et al., *J. Less Common Metals*, *146*(1–2), 233, 1989. With permission.)

CHAPTER 11

Systems Based on GaSb

11.1 GALLIUM–SODIUM–ANTIMONY

$Na_2Ga_2Sb_3$ ternary compound, which crystallizes in the orthorhombic structure with the lattice parameters $a = 2689.0 \pm 0.8$, $b = 428.3 \pm 0.2$, and $c = 723.6 \pm 0.3$ pm and a calculated density of 4.94 g·cm⁻³, is formed in this system (Cordier et al. 1986a). It was prepared by the melting of a stoichiometric mixture of elements in a corundum crucible at 1000°C.

11.2 GALLIUM–POTASSIUM–ANTIMONY

Some ternary compounds are formed in this system. **$KGaSb_2$** (**$K_2Ga_2Sb_4$**) crystallizes in the orthorhombic structure with the lattice parameters $a = 765.0 \pm 0.3$, $b = 1804.8 \pm 0.7$, and $c = 2964 \pm 1$ pm (Cordier and Ochmann 1991l). It was prepared by direct combination of the elements at elevated temperatures (Wu et al. 1995; Cordier and Ochmann 1991l). The mixture was heated in an evacuated quartz ampoule at 750°C for 10 h and then cooled to ambient temperature over a 40 h period. A metallic gray crystalline product was obtained. The title compound is a semiconductor with an energy gap of 2 eV.

$KGaSb_4$ also crystallizes in the orthorhombic structure with the lattice parameters $a = 1034.8 \pm 0.3$, $b = 420.3 \pm 0.2$, and $c = 1782.3 \pm 0.5$ pm (Cordier and Ochmann 1991n). It was prepared by the melting of a stoichiometric mixture of elements at 650°C.

K_2GaSb_2 also crystallizes in the orthorhombic structure with the lattice parameters $a = 1506.2 \pm 0.3$, $b = 1037.3 \pm 0.2$, and $c = 914.5 \pm 0.2$ pm and a calculated density of 3.64 g·cm⁻³ (Cordier et al. 1986b). It was synthesized from the elements taken in a stoichiometric ratio. The mixture was heated in an Ar atmosphere in an alumina crucible at a rate of 100°C·h⁻¹ up to 300°C and held at this temperature for 12 h. Thereafter, the mixture

was heated to 1000°C with the same rate and cooled at a rate of 20°C·h⁻¹. The title compound is characterized by silvery metallic luster and is air and moisture sensitive.

$K_2Ga_2Sb_3$ crystallizes in the monoclinic structure with the lattice parameters $a = 1474.3 \pm 0.3$, $b = 718.5 \pm 0.2$, and $c = 1658.4 \pm 0.4$ pm and $\beta = 90.5 \pm 0.2°$ (Cordier and Ochmann 1991c). It was prepared by the melting of a stoichiometric mixture of elements at 1030°C.

$K_{20}Ga_6Sb_{12.66}$ crystallizes in the hexagonal structure with the lattice parameters $a = 1780.0 \pm 0.5$ and $c = 543.8 \pm 0.2$ pm and a calculated density of 3.05 g·cm⁻³ (Cordier and Ochmann 1990). To prepare this compound, Ga and Sb in stoichiometric ratio with a slight excess of K under Ar in a glove box in a dried corundum crucible were mixed. The mixture was slowly heated under protective gas at a rate of 100°C·h⁻¹ to 300°C and maintained at this temperature for about 2 h. The mixture was then heated at the same rate to 1030°C, with excess potassium evaporating from the system, followed by the cooling at a rate of 50°C·h⁻¹ to room temperature. The obtained compound is very moisture sensitive.

11.3 GALLIUM–RUBIDIUM–ANTIMONY

Rb_2GaSb_2 crystallizes in the orthorhombic structure with the lattice parameters $a = 1540.8 \pm 0.4$, $b = 1077.6 \pm 0.3$, and $c = 930.2 \pm 0.3$ pm (Cordier and Ochmann 1991g). It was synthesized the same way as K_2GaSb_2 was prepared.

11.4 GALLIUM–CESIUM–ANTIMONY

Two compounds, **Cs_2GaSb_2** and **Cs_6GaSb_3**, are formed in this system. Cs_2GaSb_2 crystallizes in the orthorhombic structure with the lattice parameters $a = 1806.0 \pm 0.8$, $b = 1116.7 \pm 0.5$, and $c = 835.8 \pm 0.4$ pm (Cordier and Ochmann 1991a). It was synthesized from the elements taken in stoichiometric ratio at 300°C.

Cs_6GaSb_3 crystallizes in the monoclinic structure with the lattice parameters $a = 1085.8 \pm 0.4$, $b = 649.0 \pm 0.2$, and $c = 1272.9 \pm 0.4$ pm and $\beta = 101.1 \pm 0.1°$ (Blase et al. 1992a). Cs_6GaSb_3 was synthesized from the elements at 600°C in sealed Nb ampoules.

11.5 GALLIUM–COPPER–ANTIMONY

The chemical potential diagram for the Ga–Cu–Sb system is given in Figure 11.1 (Šesták et al. 1995). A $\log [a(Ga)/a(Cu)]$ versus $\log p(Sb_2)$ plot was adopted because the Sb partial pressure is one of the major controlling factors in experiments, and also because the logarithmic activity ratio of Ga to Cu is appropriate to treat both elements in an equivalent way. In this plot, the stability polygon of GaSb has a certain range of the Sb potential and can also be represented by the activity ratios a_{Cu}/a_{Ga} versus a_{Sb}/a_{Ga}. This corresponds to different states of the doped copper GaSb. Although these diagrams are constructed by using only the

thermodynamic data of compounds, the stability polygon of the respective compounds yields information on impurities or dopants within the infinitely dilute solution approximation. This means that each point of the polygon of GaSb implicitly indicates its respective dopant level of copper. The distribution coefficient of copper in GaSb is 0.0021 ± 0.0006.

Cu_3GaSb_2 ternary compound is not formed in this system (Luzhnaya et al. 1965).

11.6 GALLIUM–SILVER–ANTIMONY

Ag_3GaSb_2 ternary compound is not formed in the Ga–Ag–Sb system (Gubskaya et al. 1965).

GaSb–Ag. The phase diagram of this system is a eutectic type (Figure 11.2) (Glazov and Pavlova 1977). The eutectic composition and temperature are 19 mol% Ag and 475°C, respectively. This system was investigated using differential thermal analysis (DTA), metallography, and measurement of microhardness.

11.7 GALLIUM–GOLD–ANTIMONY

The calculated isothermal section of the Ga–Au–Sb ternary system at room temperature is given in Figure 11.3

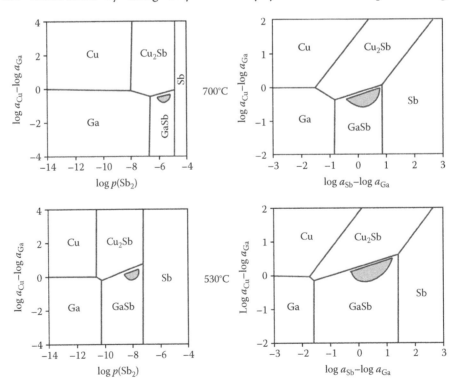

FIGURE 11.1 Chemical potential diagrams for the Ga–Cu–Sb ternary system at 530°C and 700°C where the stability area of Cu, Ga, GaSb, Cu_2Sb, and Sb is bounded, while the possible existence of the intermetallics $Cu:Ga_{ss}$, Cu_3Sb, and Cu-doped GaSb is marked by vertical, horizontal, and intercrossed lines, respectively. (From Šesták, J., et al., *J. Therm. Anal.*, 43(2), 389, 1995. With permission.)

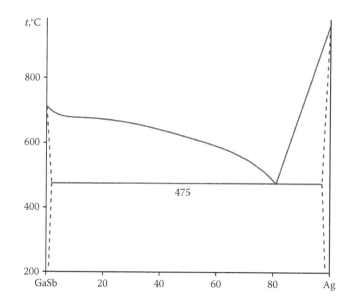

FIGURE 11.2 Phase diagram of the GaSb–Ag system. (From Glazov, V.M., and Pavlova L.M., *Izv. AN SSSR Neorgan. Mater.*, *13*(1), 15, 1977. With permission.)

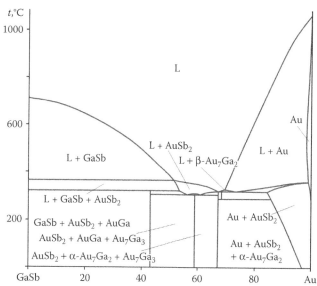

FIGURE 11.4 Phase relations in the GaSb–Au system. (From Manasijević, D., et al., *J. Phys. Chem. Solids*, *74*(2), 285, 2013. With permission.)

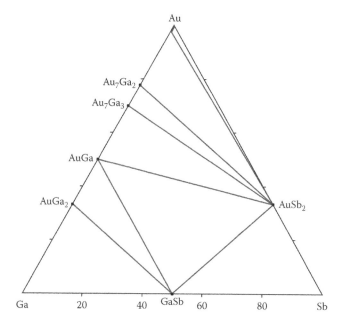

FIGURE 11.3 Isothermal section of the Ga–Au–Sb ternary system at 25°C. (From Manasijević, D., et al., *J. Phys. Chem. Solids*, *74*(2), 285, 2013. With permission.)

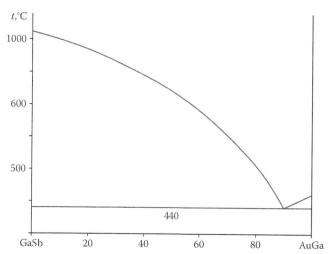

FIGURE 11.5 Phase diagram of the GaSb–AuGa system. (From Manasijević, D., et al., *J. Phys. Chem. Solids*, *74*(2), 285, 2013. With permission.)

(Tsai and Williams 1986; Manasijević et al. 2013). Six vertical sections were constructed. GaSb is in thermodynamic equilibrium with AuGa, AuGa$_2$, and AuSb$_2$, and AuSb$_2$ coexists with AuGa. The calculated phase diagram and differential scanning calorimetry (DSC) results are in good agreement.

GaSb–Au. Au is not in thermodynamic equilibrium with GaSb (Figure 11.4) (Manasijević et al. 2013). The

calculated vertical section includes four primary crystallization areas: GaSb, AuSb$_2$, β-Au$_7$Ga$_2$, and (Au). The DSC signal, related to the predicted transition reaction L + GaSb ⟷ AuSb$_2$ + AuGa at 322.5°C, was detected on the heating curves.

GaSb–AuGa. The phase diagram of this system is a eutectic type (Figure 11.5) (Manasijević et al. 2013). The experimental and calculated eutectic temperatures are equal to 454.3°C and 440.6°C, respectively.

GaSb–AuGa$_2$. The phase diagram of this system is a eutectic type (Figure 11.6) (Manasijević et al. 2013). The

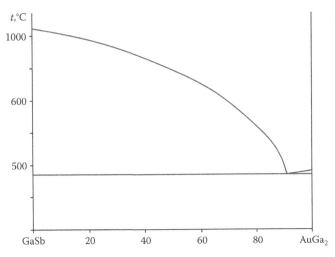

FIGURE 11.6 Phase diagram of the GaSb–AuGa$_2$ system. (From Manasijević, D., et al., *J. Phys. Chem. Solids*, 74(2), 285, 2013. With permission.)

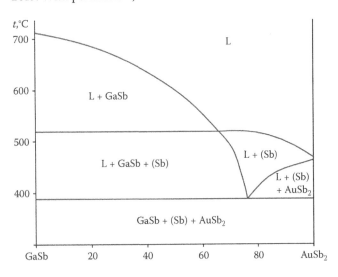

FIGURE 11.7 Phase relations in the GaSb–AuSb$_2$ system. (From Manasijević, D., et al., *J. Phys. Chem. Solids*, 74(2), 285, 2013. With permission.)

experimental and calculated eutectic temperatures are equal to 485°C and 481.5°C, respectively.

GaSb–AuSb$_2$. The phase relations in this system are shown in Figure 11.7 (Manasijević et al. 2013).

11.8 GALLIUM–CALCIUM–ANTIMONY

Two compounds, **Ca$_5$Ga$_2$Sb$_6$** and **Ca$_{11}$GaSb$_9$**, are formed in this ternary system. Ca$_5$Ga$_2$Sb$_6$ crystallizes in the orthorhombic structure with the lattice parameters $a = 1402.1 \pm 0.5$, $b = 1210.6 \pm 0.4$, and $c = 445.2 \pm 0.2$ pm ($a = 1369.7$, $b = 1183.3$, and $c = 434.6$ pm [Yan et al. 2014]) and a calculated density

of 4.70 g·cm^{-3} (4.52 g·cm^{-3} [Zevalkink et al. 2012]) (Cordier et al. 1985c). Ca$_5$Ga$_2$Sb$_6$ is a semiconductor with an energy gap of 0.43 eV (Zevalkink et al. 2012) (0.088 eV [Yan et al. 2014]). It was obtained from stoichiometric mixtures of the elements (Cordier et al. 1985c). These mixtures were heated up to 1000°C at a rate of 100°C·h^{-1} and then cooled at a rate of 50°C·h^{-1}. The obtained compound decomposes under moist air. It could be also prepared if stoichiometric amounts of GaSb, Ca, and Sb were loaded into stainless steel vials with stainless steel balls in an Ar-filled glove box (Zevalkink et al. 2012). The content was dried by ball milling for 1 h. The resulting fine powder was hot pressed into high-density graphite dies in Ar using 110 MPa of pressure. To ensure complete reaction, the powders were first held at the minimum load at 600°C for 2 h. The samples were then consolidated for 3 h at 700°C under 110 MPa of pressure, followed by a 2 h stress-free cooldown.

The electronic structure of Ca$_5$Ga$_2$Sb$_6$ was investigated by using first-principles calculations (Yan et al. 2014).

Ca$_{11}$GaSb$_9$ also crystallizes in the orthorhombic structure with the lattice parameters $a = 1183.9 \pm 0.2$, $b = 1253.6 \pm 0.3$, and $c = 1671.6 \pm 0.1$ pm at room temperature and $a = 1180.5 \pm 0.3$, $b = 1246.3 \pm 0.3$, and $c = 1665.1 \pm 0.2$ pm and a calculated density of 4.355 g·cm^{-3} at 130 K (Young and Kauzlarich 1995). This compound was prepared by adding stoichiometric amounts of the elements in a Nb tube that was sealed on one end. This tube was then sealed in a quartz ampoule under vacuum. Heating the reactants at a rate of 60°C·h^{-1} to 850°C for 2 weeks and subsequently cooling at the same rate to room temperature provided the highest yield of the title compound, which is air sensitive. Single crystals of the Ca$_{11}$GaSb$_9$ were obtained in one reaction at 1000°C.

11.9 GALLIUM–STRONTIUM–ANTIMONY

Three compounds, **Sr$_3$GaSb$_3$**, **Sr$_7$Ga$_2$Sb$_6$**, and **Sr$_7$Ga$_8$Sb$_8$**, are formed in this ternary system. Sr$_3$GaSb$_3$ crystallizes in the monoclinic structure with the lattice parameters $a = 1176.2 \pm 0.4$, $b = 1450.9 \pm 0.5$, and $c = 1174.9 \pm 0.4$ pm and $\beta = 110.0 \pm 0.1°$ and a calculated density of 4.92 g·cm^{-3} (Cordier et al. 1987). The preparation of this compound was achieved by direct synthesis from elements. The stoichiometric mixture was additionally sealed in quartz ampoules under an Ar atmosphere and heated to 1000°C at a rate of 100°C·h⁻¹, maintained at this temperature for 1 h, and cooled at a rate of 50°C·h⁻¹.

$Sr_7Ga_2Sb_6$ crystallizes in the cubic structure with the lattice parameter $a = 991.47 \pm 0.09$ pm and a calculated density of 5.049 g·cm⁻³ at 120 ± 2 K (Xia et al. 2008). To obtain it, the flux-synthesis method using alumina crucibles sealed in evacuated fused silica jackets was employed. The elements were loaded with the Sr/Ga/Sb molar ratio of 1.1:10:1, and the optimized temperature profile was the following: (1) quick heating to 950°C (300°C·h⁻¹), (2) homogenization at 950°C for 27 h, and (3) cooling to 400°C (10°C·h⁻¹). At this point, the sealed ampoule was removed from the furnace and the molten Ga was decanted.

$Sr_7Ga_8Sb_8$ is metastable and crystallizes in the hexagonal structure with the lattice parameters $a = 454.09 \pm 0.08$ and $c = 1748.6 \pm 0.4$ pm and a calculated density of 5.70 g·cm⁻³ at 120 ± 2 K (Bobev et al. 2010). To prepare it, the starting materials were loaded in alumina crucibles, which were then enclosed in fused silica jackets and flame sealed under vacuum. The nominal molar ratio Sr/Sb was in the range of 0.9:1 to 1.1:1. Ga was typically used in large excess (five times or greater than the stoichiometric amount) and served as flux. Handling of the reactants was done in an Ar-filled glove box or under vacuum.

11.10 GALLIUM–BARIUM–ANTIMONY

Some compounds are formed in this ternary system. $BaGa_2Sb_2$ melts congruently at 767°C and crystallizes in the orthorhombic structure with the lattice parameters $a = 2545.4 \pm 0.5$, $b = 444.21 \pm 0.09$, and $c = 1027.3 \pm 0.6$ pm, a calculated density of 5.950 g·cm⁻³, and an energy gap of circa 0.35 eV (Kim and Kanatzidis 2001). To prepare it, the mixture of Ba (1 mM), Ga (2 mM), and Sb (2 mM) was handled in a N_2-filled glove box, placed in a graphite tube, and sealed in an evacuated silica tube. The sealed mixture was heated slowly up to 950°C, kept at that temperature for 1 day, and subsequently cooled to room temperature for 1 day. Rod-shaped crystals of the title compound were obtained. When the air-exposed sample is heated at circa 500°C, it slowly decomposes to Ga and unknown phases.

Ba_3GaSb_3 crystallizes in the orthorhombic structure with the lattice parameters $a = 1411.7 \pm 0.5$, $b = 2116.7 \pm 0.7$, and $c = 712.8 \pm 0.3$ pm and a calculated density of 5.28 g·cm⁻³ (Cordier et al. 1985b). It was

synthesized from a mixture of Ba, Ga, and Sb (molar ratio 4:1:3) in an Ar atmosphere in a corundum crucible heated to 1000°C at a rate of 100°C·h⁻¹ and then cooled at a rate of 50°C·h⁻¹.

$Ba_3Ga_4Sb_5$ melts congruently at 714°C and also crystallizes in the orthorhombic structure with the lattice parameters $a = 1324.8 \pm 0.3$, $b = 450.85 \pm 0.09$, and $c = 2437.4 \pm 0.5$ pm and a calculated density of 5.930 g·cm⁻³ (Park et al. 2003b). It was obtained from the reaction of a mixture of Ba, Ga, and Sb (molar ratio 5:11:9). The excess Ga acted as flux. The reaction mixture was placed in a graphite tube and sealed in an evacuated silica tube. The sealed mixture was heated up slowly to 1000°C for 3 days, and kept at 500°C for 3 days, and subsequently cooled to room temperature over 1 day. The reaction led to the formation of a few rod-shaped black crystals, along with gray featureless pieces of unknown compound. $Ba_3Ga_4Sb_5$ is stable at room temperature in air.

$Ba_7Ga_2Sb_6$ crystallizes in the cubic structure with the lattice parameters $a = 1031.90 \pm 0.09$ pm and a calculated density of 5.536 g·cm⁻³ at 120 ± 2 K (Xia et al. 2008). It was prepared the same way as $Sr_7Ga_2Sb_6$ was prepared, using Ba instead of Sr.

$Ba_7Ga_4Sb_9$ crystallizes in the orthorhombic structure with the lattice parameters $a = 1802.4 \pm 0.7$, $b = 1086.1 \pm 0.4$, and $c = 710.0 \pm 0.3$ pm and experimental and calculated densities of 5.54 and 5.58 g·cm⁻³, respectively (Cordier et al. 1986c). It was prepared by the melting of a stoichiometric mixture of elements in a corundum crucible at 1000°C.

$Ba_7Ga_8Sb_8$ is metastable and crystallizes in the hexagonal structure with the lattice parameters $a = 470.45 \pm 0.12$ and $c = 1812.3 \pm 0.9$ pm and a calculated density of 5.96 g·cm⁻³ at 120 ± 2 K (Bobev et al. 2010). To prepare it, the starting materials were loaded in alumina crucibles, which were then enclosed in fused silica jackets and flame sealed under vacuum. The nominal molar ratio Ba/Sb was in the range of 0.9:1 to 1.1:1. Ga was typically used in large excess (five times or greater than the stoichiometric amount) and served as flux. Handling of the reactants was done in an Ar-filled glove box or under vacuum.

11.11 GALLIUM–ZINC–ANTIMONY

Mirgalovskaya and Komova (1975) have determined liquidus projection of the Ga–Zn–Sb ternary system with seven small primary crystallization fields [(Zn), (Ga), α-Zn$_3$Sb$_2$, β-Zn$_3$Sb$_2$, Zn$_4$Sb$_3$, ZnSb, and (Sb)] and one large crystallization field of GaSb. Five invariant reactions were determined, two of them were E-type and three of them were U-type reactions.

The most reliable liquidus surface of the Ga–Zn–Sb ternary system (Figure 11.8) and the isothermal section of this system at 400°C (Figure 11.9) were calculated

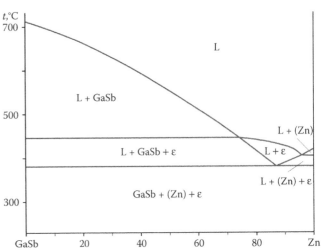

FIGURE 11.10 Phase relations in the GaSb–Zn system. (From Derviševič, I., et al., *J. Mater. Sci.*, 45(10), 2725, 2010. With permission.)

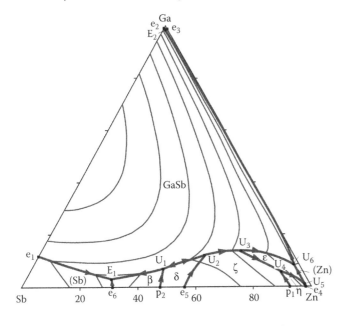

FIGURE 11.8 Liquidus surface of the Ga–Zn–Sb ternary system. (From Derviševič, I., et al., *J. Mater. Sci.*, 45(10), 2725, 2010. With permission.)

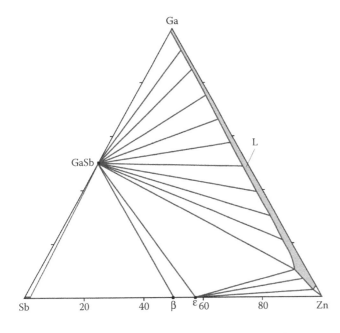

FIGURE 11.9 Isothermal section of the Ga–Zn–Sb ternary system at 400°C. (From Derviševič, I., et al., *J. Mater. Sci.*, 45(10), 2725, 2010. With permission.)

applying the CALPHAD method (Derviševič et al. 2010). Phase transition temperatures were also determined by DTA and DSC, and scanning electron microscopy (SEM) with energy-dispersive x-ray spectroscopy (EDX) was used for microstructure investigations. In the Ga–Zn–Sb ternary system, there are eight invariant reactions, two eutectics and six transition points: E_2 (25°C), L $\longleftrightarrow$ GaSb + (Zn) + (Ga); U_6 (376°C), L + ε $\longleftrightarrow$ GaSb + (Zn); U_5 (408°C), L + η $\longleftrightarrow$ ε + (Zn); U_4 (440°C), L + ζ $\longleftrightarrow$ ε + η; U_3 (466°C), GaSb + ζ $\longleftrightarrow$ L + ε; E_1 (499°C), L $\longleftrightarrow$ GaSb + ZnSb + (Sb); U_2 (512°C), L + δ $\longleftrightarrow$ GaSb + ζ; and U_1 (530°C), L + δ $\longleftrightarrow$ GaSb + ZnSb (δ, ε, ζ, η—the phases in the Ga–Zn binary system). At 350°C, CdSb dissolves 0.75 at% Zn, 0.11 mol% Zn$_3$Sb$_2$, 0.05 mol% Zn$_4$Sb$_3$, and 0.21 mol% ZnSb (Mirgalovskaya and Komova 1975).

GaSb–Zn. This system is a non-quasi-binary section of the Ga–Zn–Sb ternary system (Figure 11.10) (Mirgalovskaya and Komova 1975; Derviševič et al. 2010).

11.12 GALLIUM–CADMIUM–ANTIMONY

The liquidus projection of the Ga–Cd–Sb ternary system is characterized by seven fields of primary crystallization of GaSb (the biggest field), Sb, CdSb, Cd$_4$Sb$_3$, Cd$_3$Sb$_2$, Cd, and Ga and the miscibility region near the Cd–Ga binary system (Mirgalovskaya and Alekseeva 1965). There are two eutectics and two transition points in this system.

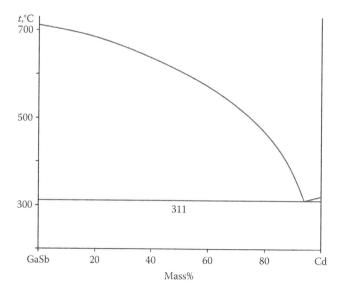

FIGURE 11.11 Phase diagram of the GaSb–Cd system. (From Mirgalovskaya, M.S., and Alekseeva, E.M., *Izv. AN SSSR Neorgan. Mater.*, 1(2), 193, 1965. With permission.)

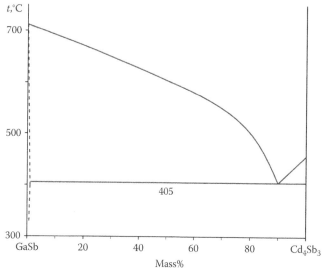

FIGURE 11.12 Phase diagram of the GaSb–Cd$_4$Sb$_3$ system. (From Mirgalovskaya, M.S., and Alekseeva, E.M., *Izv. AN SSSR Neorgan. Mater.*, 1(2), 193, 1965. With permission.)

CdSb–Cd. The phase diagram of this system is a eutectic type (Figure 11.11) (Mirgalovskaya and Alekseeva 1965). The eutectic crystallizes at 311°C and contains 6 mol% GaSb. The solubility of Cd in GaSb is 0.42 at%. According to the data of Fistul' et al. (1982), the solubility of Cd in GaSb is retrograde with a maximum value (2.3 at%) within the interval from 525°C to 575°C.

GaSb–CdSb. This system is a non-quasi-binary section of the Ga–Cd–Sb ternary system (Mirgalovskaya and Alekseeva 1965). The solubility of CdSb in GaSb reaches up to 1.8 mol% at 450°C and 2.8 mol% at 600°C (Fistul' et al. 1982).

GaSb–Cd$_3$Sb$_2$. This system is a non-quasi-binary section of the Ga–Cd–Sb ternary system (Mirgalovskaya and Alekseeva 1965).

CdSb–Cd$_4$Sb$_3$. The phase diagram of this system is a eutectic type (Figure 11.12) (Mirgalovskaya and Alekseeva 1965). The eutectic crystallizes at 405°C and contains 10 mol% GaSb. The solubility of Cd$_4$Sb$_3$ in GaSb is 0.016 mol%. It is necessary to note that Cd$_4$Sb$_3$ is not formed in the Cd–Sb binary system.

The Ga–Cd–Sb ternary system was investigated through DTA, metallography, and measurement of microhardness (Mirgalovskaya and Alekseeva 1965). The samples were annealed at 350°C for 100 h.

11.13 GALLIUM–INDIUM–ANTIMONY

The liquidus surface of the Ga–In–Sb ternary system was constructed for the first time by Köster and Thoma (1955). Liquidus isotherms of the whole system (Blom and Plaskett 1971) or of the part with less than 50 at% Sb (Antypas 1972; Joullié et al. 1974) were determined by saturation of the liquid with GaSb. A pure GaSb sample was dipped a few hours into the liquid. From its mass loss, the composition change of the liquid was calculated. The liquidus surface of this system up to 50 at% Sb, solidus isotherms, and isoconcentration lines have also been experimentally determined by means of the techniques of annealing samples in the two-phase liquid-solid field and quenching (Gratton and Woolley 1978). The solidus isotherm at 400°C and liquidus isotherms at 400°C, 500°C, and 600°C were determined experimentally by Miki et al. (1978). Liquidus data were also obtained by direct observation of the melting points of the ternary solids formed by heating the melt mixtures to about 700°C to completely melt all species, homogenizing the melt by keeping the temperature fixed at 700°C for about 1 h, and then cooling the furnace until a ternary crust began to form (Abrokwah and Gershenzon 1981).

The liquidus surfaces in the (Ga + In)-rich and Sb-rich regions were calculated using the models of regular associated solutions by Szapiro (1980) and Yunusov et al. (1989), respectively. The full liquidus surface of the Ga–In–Sb ternary system was also calculated using various solution models (Antypas 1970a; Panish 1970a; Panish and Ilegems 1972; Ansara et al. 1976; Liao et al. 1982; Lendvay 1984; Ishida et al. 1989b; Litvak and

Charykov 1991; Sharma and Mukerjee 1992). The liquidus surface, calculated from the data set of Jianrong and Watson (1994), is shown in Figure 11.13. It is dominated by the area of primary crystallization of $Ga_xIn_{1-x}Sb$ solid solution. In the Sb corner, a small area of primary crystallization of (Sb) exists. Primary crystallization of (In) or (Ga) is practically restricted to the Ga–In binary system.

Calculated solidus isoconcentration lines in the (Ga + In)-rich region of the Ga–In–Sb ternary system

are given in Figure 11.14 (Szapiro 1980). Solidus isotherms of this ternary system have been calculated at 300°C, 350°C, and 410°C by Joullié et al. (1974) and are shown in Figure 11.15.

The isothermal sections of this ternary system were constructed at 550°C by Ansara et al. (1976); at 400°C, 500°C, 550°C, 600°C, and 675°C by Kaufman et al. (1981); at 400°C, 500°C, 600°C, and 650°C by Sharma and Mukerjee (1992); and at 530°C by Jianrong and Watson (1994). The isothermal section at 500°C from Sharma and Mukerjee (1992) is shown in Figure 11.16. A three-phase equilibrium region has been found in the Sb-rich side at temperatures from 495°C to 587°C (Jianrong and Watson 1994).

The activity of Ga in liquid Ga–In–Sb alloys was measured by an electromotive force (EMF) method with zirconia as the solid electrolyte between 780°C and 880°C over the entire composition range, and the iso-activity curves were obtained for this ternary system (Figure 11.17) (Katayama et al. 1993b). With a high-temperature microcalorimeter, the molar excess enthalpy of mixing for liquid Ga–In–Sb alloys has been measured at 722°C (Ansara et al. 1976). Using these results, the whole set of thermodynamic functions in this ternary system has been calculated. The mixing enthalpies at 637°C were determined using quantitative differential thermal analysis (Figure 11.18) (Vecher et al. 1974). The maximum mixing enthalpy is equal to 3430 ± 105 J·g-at.$^{-1}$ for the alloy containing 39 at% In, 50 at% As, and 11 at% Ga.

Calculation of the thermodynamic characteristics for this ternary system was also performed by Gomidželović

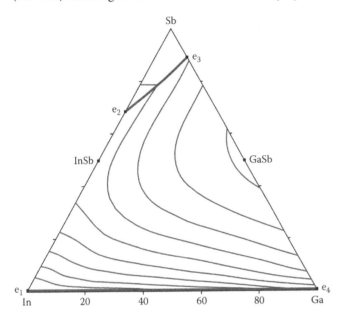

FIGURE 11.13 Liquidus surface of the Ga–In–Sb ternary system. (From Jianrong, Y., and Watson, A., *CALPHAD*, 18(2), 165, 1994. With permission.)

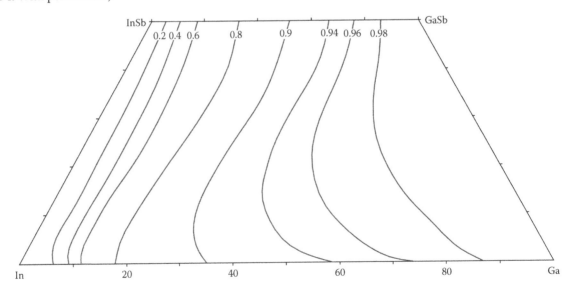

FIGURE 11.14 Calculated solidus isoconcentration lines in the (Ga + In)-rich region of the Ga–In–Sb ternary system. (From Szapiro, S., *J. Phys. Chem. Solids*, *41*(3), 279, 1980. With permission.)

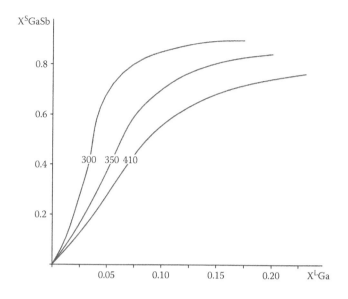

FIGURE 11.15 Solidus isotherms of the Ga–In–Sb ternary system at 300°C, 350°C, and 410°C. (From Joullié, A., et al., *J. Cryst. Growth*, *24–25*, 276, 1974. With permission.)

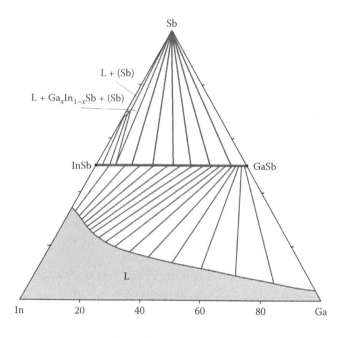

FIGURE 11.16 Isothermal section of the Ga–In–Sb ternary system at 500°C. (From Sharma, R.C., and Mukerjee, I., *J. Phase Equilibria*, *13*(1), 5, 1992. With permission.)

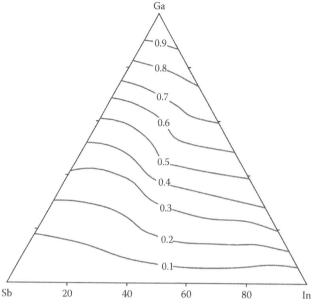

FIGURE 11.17 Isoactivity curves of Ga in liquid Ga–In–Sb alloys at 800°C. (From Katayama, I., et al., *J. Non-Cryst. Solids*, *156–158*, 393, 1993. With permission.)

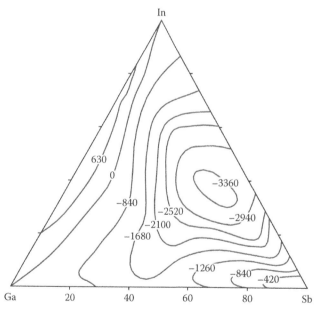

FIGURE 11.18 Mixing enthalpies (J•g-at.$^{-1}$) in the Ga–In–Sb ternary system at 637°C. (From Vecher, A.A., et al., *Zhurn. Fiz. Khimii*, *48*(4), 1007, 1974. With permission.)

et al. (2010) using the Redlich–Kister–Muggianu model. It was shown that the addition of Ga in the In–Sb binary system leads to the reduction of the negative deviation of the In and Sb activities, thus concluding that Ga reduces the miscibility of both metals in the ternary system.

GaSb–InSb. The first two studies of the quasi-binary section GaSb–InSb (Goryunova 1955; Goryunova and Fedorova 1955; Köster and Thoma 1955) had different results. Goryunova (1955) and Goryunova and Fedorova (1955) concluded complete solid solubility between GaSb and InSb, whereas Köster and Thoma (1955) interpreted their results of metallography and DTA to suggest that the compounds GaSb and InSb do not exhibit mutual solid solubilities and form a quasi-binary system with a degenerate eutectic at pure InSb. In a later paper (Köster and Ulrich 1958b), however, this interpretation was corrected and confirmed the complete solid solubility between these two compounds. Due to the low diffusivities in the $Ga_xIn_{1-x}Sb$ phase, the segregation is so heavy that first crystallizes nearly pure GaSb and finally nearly pure InSb, which do not react significantly within moderate annealing times. After annealing for 1000 h at 500°C, the samples showed nearly sharp x-ray lines, giving lattice parameters between those of GaSb and InSb, but they were still not yet completely homogenized. Other authors (Woolley et al. 1956; Gorshkov and Goryunova 1958; Woolley and Smith 1958a; Woolley and Lees 1959; Trumbore et al. 1962; Sirota et al. 1968; Blom and Plaskett 1971; Ufimtsev et al. 1971; Bucher et al. 1986) confirmed complete solid miscibility between GaSb and InSb after a long time annealing. Kolm et al. (1957) assumed a partial mutual solubility of about 10 mol% and interpreted alloys between these limits as two-phase, but they did not report details of the heat treatment of the samples. Woolley et al. (1956) studied the progress of homogenization during annealing at temperatures starting at 500°C and increasing slowly in small steps in order to always remain below the solidus. More than 1000 h of total annealing time was needed to get fairly homogeneous samples.

The phase diagram of the GaSb–InSb system was also calculated using various solution models for the liquid and solid phases (Steininger 1970; Foster and Woods 1971; Osamura et al. 1972b; Panish and Ilegems 1972; Stringfellow 1972a; Ansara et al. 1976; Cho 1976; Szapiro 1980; Kaufman et al. 1981; Liao et al. 1982; Lendvay 1984; Chang et al. 1985; Ishida et al. 1989b; Sharma and Mukerjee 1992). Figure 11.19 shows the assessed GaSb–InSb phase diagram (Jianrong and Watson 1994). A miscibility gap is also depicted. The critical temperature is 193°C (160°C [Chang et al. 1985]). The presence of the miscibility gap has not been confirmed experimentally owing to sluggish diffusion rates at the relatively low temperature. Within the limits of experimental error, a linear relationship in the variation of lattice parameters as a function of composition was obtained, showing that Vegard's law is satisfied (Woolley et al. 1956; Gorshkov and Goryunova 1958; Woolley and Smith 1958a).

The mixing enthalpy of liquid GaSb–InSb alloys has been determined with the aid of a high-temperature calorimeter (Gerdes and Predel 1979b). The maximum

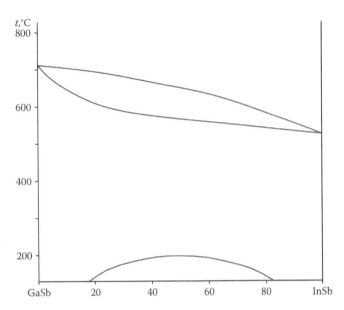

FIGURE 11.19 Phase diagram of the GaSb–InSb system. (From Jianrong, Y., and Watson, A., *CALPHAD*, 18(2), 165, 1994. With permission.)

deviation of the mixing enthalpy value from those linearly interpolated between the mixing enthalpies of the GaSb and InSb melts is $\delta(\Delta H)^{max}_{exp} = +162$ J·M^{-1}. A Calvet microcalorimeter has been used with liquid tin solution techniques to measure the mixing enthalpy of the GaSb–InSb system (Rugg et al. 1995). It was confirmed that the enthalpy of formation of $Ga_xIn_{1-x}Sb$ solid solutions is small and positive. The Ga activity in $Ga_xIn_{1-x}Sb$ liquid mixtures was obtained from EMF measurements across a solid electrolyte electrochemical cell (Chang et al. 1985). The measured values of the a_{Ga} showed moderate negative deviations from ideal behavior and are directly proportional to the Ga molar fraction.

By extremely rapid cooling of the liquid alloys, solid solutions in the GaSb–InSb were obtained (Bucher et al. 1986). In order to attain the high cooling rates (10^7–10^8 °C·s^{-1}), a shock wave tube was used.

InSb–Ga. Upon the interacting of InSb with Ga, a new structure consisting of a layer of GaSb in a volume of InSb and pure indium on its surface has been formed (Ivin 1977). The calculated effective distribution coefficient of Ga in InSb is 2.1 ± 0.5 (Kevorkov et al. 1988).

11.14 GALLIUM–LANTHANUM–ANTIMONY

Three compounds, **LaGaSb$_2$**, **La$_{12}$Ga$_4$Sb$_{23}$**, and **La$_{13}$Ga$_8$Sb$_{21}$**, are formed in the Ga–La–Sb ternary system. LaGaSb$_2$ crystallizes in the orthorhombic structure with the lattice parameters $a = 438.2 \pm 0.3$,

$b = 2277.5 \pm 1.3$, and $c = 447.4 \pm 0.3$ pm (Mills and Mar 2001). The starting materials for obtaining it were powders of La and Sb, and Ga granules. The reactions were carried out in evacuated fused silica tubes. Stoichiometric reactions at the composition $LaGaSb_2$ led to the formation of products that were significantly contaminated with the $Pr_{12}Ga_4Sb_{23}$-type phase. Use of an excess of Ga minimizes the formation of $Pr_{12}Ga_4Sb_{23}$-type impurity. A mixture of La, Ga, and Sb (molar ratio 1:2:2) was heated at 900°C for 3 days, cooled to 500°C over 4 days, and then cooled to 20°C over 18 h.

$La_{12}Ga_4Sb_{23}$ also crystallizes in the orthorhombic structure with the lattice parameters $a = 434.4 \pm 0.2$, $b = 1975.0 \pm 0.7$, and $c = 2686.0 \pm 1.1$ pm (Mills and Mar 2000). It was obtained from a reaction of La, Ga, and Sb (molar ratio 12:4:23) at 1000°C for 3 days, followed by cooling to 20°C over 18 h.

$La_{13}Ga_8Sb_{21}$ crystallizes in the hexagonal structure with the lattice parameters $a = 1765.7 \pm 0.2$ and $c = 433.78 \pm 0.07$ pm and a calculated density of 6.903 g·cm⁻³ (Mills and Mar 2000). Single, gray, needlelike crystals of this compound were isolated from a reaction of La, Ga, and Sb (molar ratio 1:2:2). The sample was heated at 900°C for 3 days, cooled slowly to 500°C over 4 days, and then cooled to 20°C over 12 h.

11.15 GALLIUM–CERIUM–ANTIMONY

Two compounds, **CeGaSb₂** and **Ce₁₂Ga₄Sb₂₃**, are formed in the Ga–Ce–Sb ternary system. $CeGaSb_2$ crystallizes in the tetragonal structure with the lattice parameters $a = 437.08 \pm 0.16$ and $c = 4507 \pm 2$ pm (Mills and Mar 2001). The starting materials for obtaining it were powders of Ce and Sb, and Ga granules. The reactions were carried out in evacuated fused silica tubes. Stoichiometric reactions at the composition $CeGaSb_2$ led to the formation of products that were significantly contaminated with the $Pr_{12}Ga_4Sb_{23}$-type phase. Use of an excess of Ga minimizes the formation of $Pr_{12}Ga_4Sb_{23}$-type impurity. A mixture of Ce, Ga, and Sb (molar ratio 1:2:2) was heated at 900°C for 3 days, cooled to 500°C over 4 days, and then cooled to 20°C over 18 h.

$Ce_{12}Ga_4Sb_{23}$ crystallizes in the orthorhombic structure with the lattice parameters $a = 430.8 \pm 0.2$, $b = 1950.9 \pm 0.9$, and $c = 2666.7 \pm 1.2$ pm (Mills and Mar 2000). It was obtained from a reaction of Ce, Ga,

and Sb (molar ratio 12:4:23) at 1000°C for 3 days, followed by cooling to 20°C over 18 h.

11.16 GALLIUM–PRASEODYMIUM– ANTIMONY

Two compounds, **PrGaSb₂** and **Pr₁₂Ga₄Sb₂₃**, are formed in the Ga–Pr–Sb ternary system. $PrGaSb_2$ crystallizes in the tetragonal structure with the lattice parameters $a = 435.99 \pm 0.16$ and $c = 4481 \pm 2$ pm (Mills and Mar 2001). The starting materials for obtaining it were powders of Pr and Sb, and Ga granules. The reactions were carried out in evacuated fused silica tubes. Stoichiometric reactions at the composition $PrGaSb_2$ led to the formation of products that were significantly contaminated with the $Pr_{12}Ga_4Sb_{23}$. Use of an excess of Ga minimizes the formation of such impurity. A mixture of Pr, Ga, and Sb (molar ratio 1:2:2) was heated at 900°C for 3 days, cooled to 500°C over 4 days, and then cooled to 20°C over 18 h.

$Pr_{12}Ga_4Sb_{23}$ crystallizes in the orthorhombic structure with the lattice parameters $a = 426.12 \pm 0.03$, $b = 1940.70 \pm 0.13$, and $c = 2639.72 \pm 0.18$ pm and a calculated density of 7.236 g·cm⁻³ (Mills and Mar 2000). It was obtained from a reaction of Pr, Ga, and Sb (molar ratio 13:8:21) at 1000°C for 3 days, followed by cooling to 20°C over 18 h.

11.17 GALLIUM–NEODYMIUM–ANTIMONY

Two compounds, **NdGaSb₂** and **Nd₁₂Ga₄Sb₂₃**, are formed in the Ga–Nd–Sb ternary system. $NdGaSb_2$ crystallizes in the tetragonal structure with the lattice parameters $a = 434.86 \pm 0.03$ and $c = 4457.9 \pm 0.8$ pm and a calculated density of 7.209 g·cm⁻³ (Mills and Mar 2001). The starting materials for obtaining it were powders of Nd and Sb, and Ga granules. The reactions were carried out in evacuated fused silica tubes. Stoichiometric reactions at the composition $NdGaSb_2$ led to the formation of products that were significantly contaminated with the $Pr_{12}Ga_4Sb_{23}$-type phase. Use of an excess of Ga minimizes the formation of $Pr_{12}Ga_4Sb_{23}$-type impurity. Gray, rectangular plates of the title compound were isolated from a reaction of Nd, Ga, and Sb (molar ratio 1:2:2) heated at 800°C for 3 days and then cooled to 20°C over 18 h.

$Nd_{12}Ga_4Sb_{23}$ crystallizes in the orthorhombic structure with the lattice parameters $a = 426.8 \pm 0.1$, $b = 1930.8 \pm 0.6$, and $c = 2642.5 \pm 0.8$ pm (Mills and Mar 2000). It was obtained from a reaction of Nd, Ga,

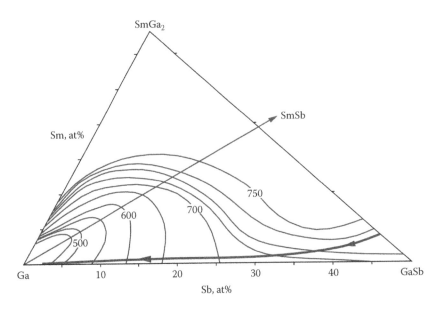

FIGURE 11.20 Liquidus surface in the region of GaSb–Ga–SmGa$_2$ of the Ga–Sm–Sb ternary system. (From Arbenina, V.V., et al., *Izv. AN SSSR Neorgan. Mater.*, *21*(4), 541, 1985. With permission.)

and Sb (molar ratio 12:4:23) at 1000°C for 3 days, followed by cooling to 20°C over 18 h.

11.18 GALLIUM–SAMARIUM–ANTIMONY

The liquidus surface in the region of GaSb–Ga–SmGa$_2$ of the Ga–Sm–Sb ternary system is given in Figure 11.20 (Arbenina et al. 1985). The field of SmGa$_2$ primary crystallization occupies the biggest part of this region. The liquidus isotherms in the region of the Ga–Sm–Sb ternary system near GaSb are shown in Figure 11.21, and the solubility isotherms of Sm in GaSb are given in Figure 11.22.

Two compounds, **SmGaSb$_2$** and **Sm$_{12}$Ga$_4$Sb$_{23}$**, are formed in the Ga–Sm–Sb ternary system. SmGaSb$_2$ melts at 1050°C ± 20°C (Kuliev et al. 1990) and crystallizes in the orthorhombic structure with the lattice parameters $a = 430.87 \pm 0.05$, $b = 2209.3 \pm 0.4$, and $c = 433.19 \pm 0.04$ pm and a calculated density of 7.467 g·cm^{-3} (Mills and Mar 2001) (in the cubic structure with the lattice parameter $a = 608.96$ [Kuliev et al. 1990]). The starting materials for obtaining it were powders of Sm and Sb, and Ga granules. The reactions were carried out in evacuated fused silica tubes. Single, black, platelike crystals of the title compound were obtained from a reaction of Sm, Ga, and Sb in the ratio 1:1:2 heated at 900°C for 3 days, cooled to 500°C over 4 days, and then cooled to 20°C over 18 h.

Sm$_{12}$Ga$_4$Sb$_{23}$ also crystallizes in the orthorhombic structure with the lattice parameters $a = 421.3 \pm 0.3$,

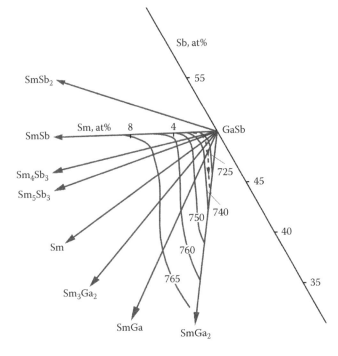

FIGURE 11.21 Liquidus isotherms in the region of the Ga–Sm–Sb ternary system near GaSb: the dotted line indicates a peritectic line. (From Arbenina, V.V., et al., *Izv. AN SSSR Neorgan. Mater.*, *21*(4), 541, 1985. With permission.)

$b = 1912.0 \pm 1.0$, and $c = 2609.9 \pm 1.5$ pm (Mills and Mar 2000). It was obtained from a reaction of Sm, Ga, and Sb (molar ratio 12:4:23) at 1000°C for 3 days, followed by cooling to 20°C over 18 h.

GaSb–Sm. This system is a non-quasi-binary section of the Ga–Sm–Sb ternary system (Arbenina et al. 1985).

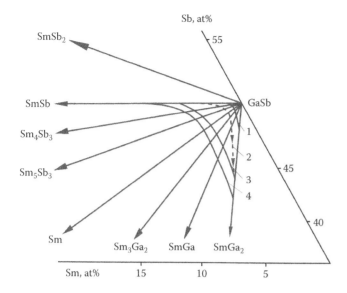

FIGURE 11.22 Solubility isotherms of Sm in GaSb at (1) room temperature, (3) 502°C, and (4) peritectic temperature; the dotted line (2) indicates a peritectic line. (From Arbenina, V.V., et al., *Izv. AN SSSR Neorgan. Mater.*, 21(4), 541, 1985. With permission.)

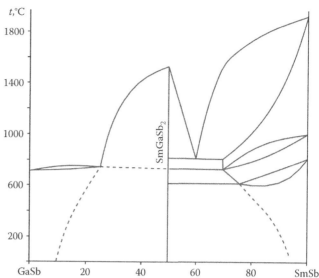

FIGURE 11.23 Phase diagram of the GaSb–SmSb system. (From Kuliev, A.N., et al., *Izv. AN SSSR Neorgan. Mater.*, 26(3), 500, 1990. With permission.)

GaSb–SmGa₂. The phase diagram of this system is a peritectic type (Arbenina et al. 1985). The peritectic crystallizes at 738°C ± 3°C. The solubility of Sm in GaSb along this section reaches up to 5 at%.

GaSb–SmSb. This system is a peritectic-type system (Arbenina et al. 1985). The peritectic crystallizes at 736°C ± 3°C and contains 3–4 at% Sb. The solubility of Sm in GaSb along this section reaches up to 7 at%.

The phase diagram of the GaSb–SmSb system was also constructed by Kuliev et al. (1990) and is shown in Figure 11.23. SmGaSb₂ ternary compound is formed in this system. The solubility of SmSb in GaAs reaches up to 10 mol% (Guliev 1973; Kuliev et al. 1990). This diagram causes some doubts, because SmSb has no phase transformations. According to the authors, this phase diagram needs additional studies to clarify the region of 60–100 mol% SmSb.

GaSb–SmSb₂. The solubility of Sm in GaSb at room temperature along this section is 0.5–0.7 at% (Arbenina et al. 1985).

This ternary system was investigated through DTA, x-ray diffraction (XRD), metallography, measurement of microhardness, and local x-ray analysis (Guliev 1973; Arbenina et al. 1985; Kuliev et al. 1990). The ingots were annealed at 500°C for 500 h (Arbenina et al. 1985).

11.19 GALLIUM–EUROPIUM–ANTIMONY

Three compounds, **Eu₇Ga₂Sb₆**, **Eu₇Ga₆Sb₈**, and **Eu₇Ga₈Sb₈**, are formed in this ternary system. Eu₇Ga₂Sb₆ crystallizes in the cubic structure with the lattice parameter $a = 978.66 \pm 0.08$ pm and a calculated density of 6.852 g·cm⁻³ at 120 ± 2 K (Xia et al. 2008). To obtain it, the flux-synthesis method using alumina crucibles sealed in evacuated fused silica jackets was employed. The elements were loaded with the molar ratio Eu/Ga/Sb = 1.1:10:1, and the optimized temperature profile was the following: (1) quick heating to 950°C (300°C·h⁻¹), (2) homogenization at 950°C for 27 h, and (3) cooling to 400°C (10°C·h⁻¹). At this point, the sealed ampoule was removed from the furnace and the molten Ga was decanted.

Eu₇Ga₆Sb₈ melts at 670°C and crystallizes in the orthorhombic structure with the lattice parameters $a = 1564.70 \pm 0.17$, $b = 1728.76 \pm 0.19$, and $c = 1792.00 \pm 0.19$ pm and a calculated density of 6.731 g·cm⁻³ at 173 ± 2 K (Park et al. 2004). It is stable at room temperature in air; however, when an air-exposed sample was heated at about 670°C, it slowly decomposed to Ga and unknown phases. The title compound was obtained from the reaction of a mixture of Eu, Ga, and Sb (molar ratio 3:2:4). The reaction mixture was placed in a graphite tube and sealed in an evacuated silica tube. The mixture was then heated slowly up to 900°C for 2 days, kept at that temperature for 1 day, and subsequently cooled to room temperature over 1 day. The reaction gave a few rod-shaped black crystals, along with gray featureless pieces of unknown compound.

$Eu_7Ga_8Sb_8$ is metastable and crystallizes in the hexagonal structure with the lattice parameters $a = 449.42 \pm 0.08$ and $c = 1727.4 \pm 0.6$ pm and a calculated density of 7.13 g·cm^{-3} at 120 ± 2 K (Bobev et al. 2010). To prepare it, the starting materials were loaded in alumina crucibles, which were then enclosed in fused silica jackets and flame sealed under vacuum. The nominal molar ratio Eu/Sb was in the range 0.9:1–1.1:1. Ga was typically used in large excess (five times or greater than the stoichiometric amount) and served as flux. Handling of the reactants was done in an Ar-filled glove box or under vacuum.

11.20 GALLIUM–THULIUM–ANTIMONY

GaSb–TmSb. The phase diagram of this system is shown in Figure 11.24 (Aliev et al. 2007). **TmGaSb$_2$** ternary compound, which melts incongruently at 907°C, is formed in this system. The solubility of TmSb in GaSb reaches up 3.5 mol%, and TmSb dissolves circa 1.5 mol% GaSb at room temperature. The system was investigated through DTA, XRD, metallography, and measurement of microhardness and density. The samples were annealed at 530°C for 250 h.

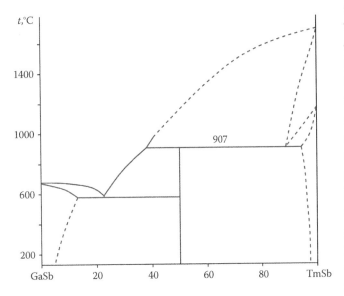

FIGURE 11.24 Phase diagram of the GaSb–TmSb system. (From Aliev, I.I., et al., *Azerb. Khim. Zhurn.*, (2), 111, 2007. With permission.)

11.21 GALLIUM–YTTERBIUM–ANTIMONY

Two ternary compounds, **Yb$_5$Ga$_2$Sb$_6$** and **Yb$_{11}$GaSb$_9$**, are formed in the Ga–Yb–Sb system. Yb$_5$Al$_2$Sb$_6$ crystallizes in the orthorhombic structure with the lattice parameters $a = 728.64 \pm 0.07$, $b = 2290.2 \pm 0.2$, and $c = 440.20 \pm 0.05$ pm and a calculated density 7.68 g·cm^{-3} (Aydemir et al. 2015) ($a = 728.4 \pm 0.2$, $b = 2292.5 \pm 0.6$, and $c = 440.3 \pm 0.2$ pm [Fornasini and Manfrinetti 2009]; $a = 727.69 \pm 0.2$, $b = 2291.02 \pm 0.05$, and $c = 439.840 \pm 0.010$ pm, a calculated density of 7.859 g·cm^{-3}, and an energy gap of 0.36 eV [Subbarao et al. 2013]). The title compound could be obtained if stoichiometric amounts of Yb, Ga, and Sb were sealed in Ta crucibles and melted in an induction furnace, raising the temperature up to about 1310°C while shaking (Fornasini and Manfrinetti 2009). The alloys were then annealed at 900°C for 7 days. Yb$_5$Ga$_2$Sb$_6$ is very brittle and air sensitive. It was also synthesized by ball milling, followed directly by hot pressing (Aydemir et al. 2015). Stoichiometric amounts of small-cut Yb ingot, Sb shot, and GaSb were loaded under Ar into stainless steel vials with stainless steel balls. The contents were ball milled for 1 h. The resulting fine powder was hot pressed under Ar using a maximum pressure and temperature of 45 MPa and 550°C, respectively, for 1 h. The samples were cooled down to room temperature slowly under no load.

Yb$_5$Al$_2$Sb$_6$ could be also prepared by the metal flux method and high-frequency induction heating method (Subbarao et al. 2013). According to the first method, well-shaped single crystals of the title compound were obtained by combining Yb (0.3 g), Ga (2 g), and Sb (0.4 g) in an alumina crucible. The crucible was placed in a fused silica tube, which was flame sealed under vacuum. The reactants were then heated to 1000°C over 10 h, maintained at that temperature for 5 h, then cooled down to 850°C in 2 h, and kept at this temperature for 72 h. Finally, the sample was allowed to cool slowly to 30°C over 48 h. The reaction product was isolated from the excess Ga flux by heating at 400°C and subsequently centrifuging through a coarse frit. Extra gallium was removed by immersion and sonication in a 2–4 M solution of iodine in dimethylformamide (DMF) over 12–24 h at room temperature. The product was rinsed with hot water and DMF and dried with acetone and ether. The gray, rodlike crystals of Yb$_5$Ga$_2$Sb$_6$ were obtained. This compound is stable in air, and no decomposition was observed even after several months.

According to the second method, Yb, Ga, and Sb (molar ratio 5:2:6) were mixed and sealed in a Ta ampoule under an Ar atmosphere in an arc melting apparatus (Subbarao et al. 2013). The Ta ampoule was subsequently placed in a water-cooled sample chamber of an induction furnace, first rapidly heated to circa 930°C–1080°C and kept at that temperature for 30 min. Finally, the reaction was rapidly cooled to room temperature by switching off the power supply.

$Yb_{11}GaSb_9$ also crystallizes in the orthorhombic structure with the lattice parameters $a = 1173.20 \pm 0.13$, $b = 1231.99 \pm 0.13$, and $c = 1664.01 \pm 0.18$ pm (Yi et al. 2010) ($a = 1172.57 \pm 0.12$, $b = 1232.04 \pm 0.13$, and $c = 1663.3 \pm 0.2$ pm and a calculated density of 8.483 g·cm⁻³ at 90 ± 3 K [Bobev et al. 2005]). To obtain the title compound, the starting materials were loaded in an alumina crucible with a stoichiometric ratio of Yb/Ga/Sb at 11:76:9 (Yi et al. 2010). The crucible was put in a fused silica ampoule with quartz wool on the top and at the bottom, which was then sealed under a vacuum. The reaction was heated in a furnace under the following temperature profile: heating to 500°C from room temperature in 2 h and dwelling at 500°C for 6 h, and then ramping to 1000°C within 2 h and dwelling for 6 h at this temperature, followed by cooling to 600°C at a rate of 2°C·h⁻¹. At this point, the ampoule was quickly removed from the furnace, inverted, and centrifuged to remove the extra flux. It could also be synthesized by direct fusion of the corresponding elements, and large single crystals were prepared from high-temperature flux synthesis (Bobev et al. 2005). All materials were handled in a nitrogen-filled glove box or with other inert atmosphere techniques.

$Yb_{11}AlSb_9$ is thermally stable up to 1061°C but shows surface oxidation after a few hours' exposure in air (Yi et al. 2010). It is a small-band-gap semiconductor (Bobev et al. 2005; Yi et al. 2010).

GaSb–Yb. The phase diagram of the GaSb–Yb system is a eutectic type (Figure 11.25) (Yevgen'ev et al. 1990). The eutectic crystallizes at 675°C and contains 10 at% Yb. The solubility of Yb in GaSb is retrograde and at 680°C is equal to 0.12–0.3 at%. The system was investigated through DTA, XRD, metallography, and measurement of microhardness. The samples were annealed at 680°C for 500 h with the next quenching in water.

11.22 GALLIUM–GERMANIUM–ANTIMONY

The isotherms of the Ga and Sb solubility in Ge are shown in Figure 11.26 (Akopyan et al. 1988). The samples were

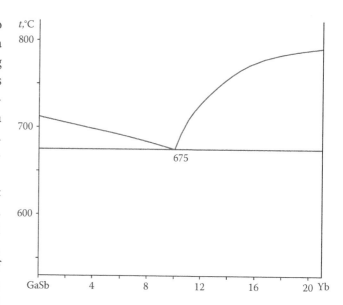

FIGURE 11.25 Part of the phase diagram of the GaSb–Yb system. (From Yevgen'ev, S.B., et al., *Izv. AN SSSR Neorgan. Mater.*, 26(6), 1323, 1990. With permission.)

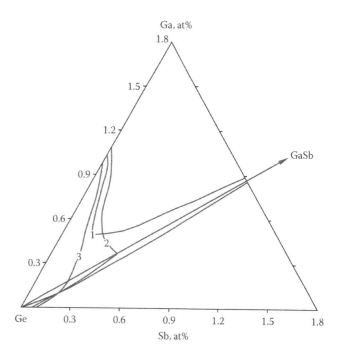

FIGURE 11.26 Isotherms of Ga and Sb solubility in Ge at (1) 600°C, (2) 725°C, and (3) 800°C. (From Akopyan, R.A., et al., *Zhurn. Neorgan. Khimii*, 33(1), 190, 1988. With permission.)

annealed at 725°C for 800 h and then at 600°C and 800°C for 1200 and 900 h, respectively. The system was investigated through metallography and measurement of microhardness.

The EMF of galvanic cells with zirconia solid electrolytes was measured from 785°C to 963°C on ternary

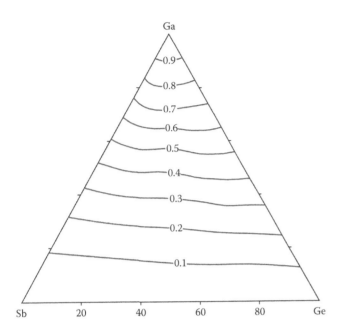

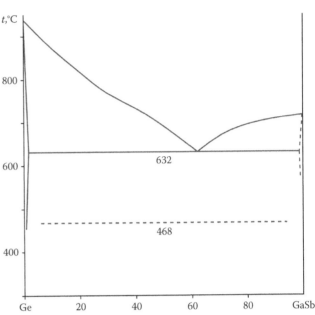

FIGURE 11.27 Isoactivity curves of gallium in liquid Ga–Ge–Sb alloys at 800°C. (From Katayama, I., et al., *Ber. Bunsenges. Phys. Chem.*, *102*(9), 1235, 1998. With permission.)

FIGURE 11.28 Phase diagram of the GaSb–Ge system. (From Akopyan, R.A., et al., *Izv. AN SSSR Neorgan. Mater.*, *13*(9), 1570, 1977. With permission.)

Ga–Ge–Sb alloys by use of Ga–Ga_2O_3 as a reference electrode (Katayama et al. 1998a). From the EMF values, the Ga activity in the liquid alloys was determined. Isoactivity curves were derived by combining the activity data of Ga–Sb, Ga–Ge, and the ternary alloys. The shape of the curves at 800°C is simple in the Ga-poor region, but shows a very small wavy form with increasing Ga concentration (Figure 11.27). Isovalue curves in the excess Gibbs energies of mixing in the ternary alloys are derived from Darken's method.

Activities, activity coefficients, and partial molar properties (excess Gibbs energy and Gibbs energy of mixing for Ga) in the Ga–$GeSb_{0.855}$ section in the Ga–Ge–Sb ternary system at 1000°C were obtained experimentally and determined using thermodynamic predicting methods (Kostov et al. 2000, 2001). The calculated values for the Ga activity show a negative deviation from Raoult's law in the whole concentration area. With x_{Ga} increasing, the activity of Ga uniformly increases, which means that gallium is mixed well with other constituents in the investigated section. The negativity of the partial molar excess Gibbs energy and partial molar Gibbs energy of mixing for Ga decreases with increasing molar content. A negative deviation from Raoult's law was also noticed for Sb, while Ge showed a positive deviation from Raoult's law. This indicates better miscibility between Ga and Sb than Ge; that is, Sb is

more dissolved into Ga than Ge into Ga. The experimental results obtained by the use of DTA, XRD, and SEM with EDX for investigation of the Ga–$GeSb_{0.855}$ section showed a structure consisting of primary crystals of Ge and Sb and eutectic (Ga + Sb) for content from $x_{Ga} = 0$ to $x_{Ga} = 0.452$; for $x_{Ga} > 0.452$, the structure consists of primary crystals of Ga and eutectic (Ga + Sb).

GaSb–Ge. The phase diagram of this system is a eutectic type (Figure 11.28) (Glazov et al. 1959; Glazov 1961; Akopyan et al. 1977). The eutectic crystallizes 632°C and contains 62 mol% GaSb. The solubility of GaSb in Ge is equal to 0.80, 1.20, 2.30, 2.50, 2.00, and 1.00 mass% at 400°C, 500°C, 600°C, 632°C, 670°C, and 725°C, respectively (Glazov et al. 1959). The samples were annealed at 600°C, 700°C, and 725°C for 300 h and investigated through metallography and measurement of microhardness.

As a result of ultrafast crystallization (10^7°C·s^{-1}) of the AlSb–Ge melts, metastable continuous solutions were obtained (Akopyan et al. 1977; Glazov et al. 1977a; Bucher et al. 1986).

Epitaxial metastable $(GaSb)_{1-x}Ge_x$ alloys with composition across the quasi-binary phase diagram have been grown on GaAs substrates (Cadien et al. 1980). The lattice constants of the polycrystalline films obey Vegard's law. These films were found to be stable to decomposition during long anneals. Annealing studies carried out on $(GaSb)_{0.38}Ge_{0.62}$ showed that such films transform from the as-deposited alloy to GaSb-rich and Ge-rich phases in which the solute concentration continuously decreases until an equilibrium two-phase alloy is obtained.

11.23 GALLIUM–TIN–ANTIMONY

The liquidus surface of the Ga–Sn–Sb ternary system is shown in Figure 11.29 (Arbenina and Beylin 1986). The GaSb field of primary crystallization is the biggest field in this system.

The EMF of galvanic cells with zirconia solid electrolytes was measured to determine the activity of gallium in liquid alloys in the 780°C–880°C temperature range (Katayama et al. 1998b). A mixture of Ga and Ga_2O_3 was used as a reference electrode. The activity curves of Ga show negative-to-positive deviations from the ideal behavior, with increasing Sn concentrations in the lower concentration range of Ga, whereas in the $x_{Ga} \geq 0.6$ concentration range, the activity of Ga in the ternary alloys shows positive deviations from the ideal behavior. Isoactivity curves at 800°C (Figure 11.30) were derived by combining the activity data for Ga–Sb, Ga–Sn, and ternary alloys. The shape of the curves is simple in the Ga-poor region, but shows a wavy form with increasing Ga concentration. Isovalue curves in the excess free

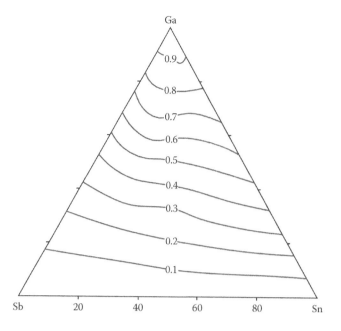

FIGURE 11.30 Isoactivity curves of gallium in liquid Ga–Sn–Sb alloys at 800°C. (From Katayama, I., et al., *Thermochim. Acta*, *314*(1–2), 175, 1998. With permission.)

energies of mixing in the ternary alloys are derived by Darken's method.

GaSb–Sn. The phase diagram of this system is a eutectic type (Figure 11.31) (Gerdes and Predel 1981a; Arbenina and Beylin 1986; Živković et al. 2003). The eutectic is degenerated from the Sn side and contains

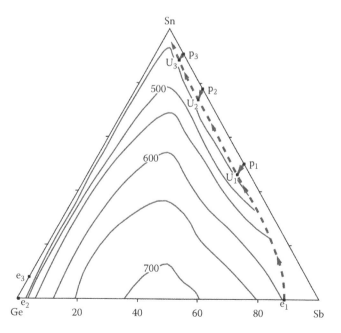

FIGURE 11.29 Liquidus surface of the Ga–Sn–Sb ternary system. (From Arbenina, V.V., and Beylin, Yu.A., *Izv. AN SSSR Neorgan. Mater.*, *22*(9), 1425, 1986. With permission.)

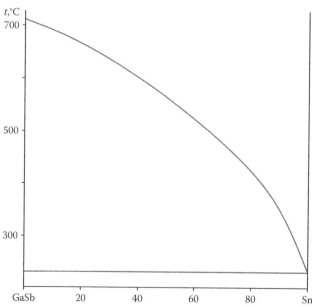

FIGURE 11.31 Phase diagram of the GaSb–Sn system. (From Živković, D., et al., *J. Therm. Anal. Calorim.*, *71*(2), 567, 2003. With permission.)

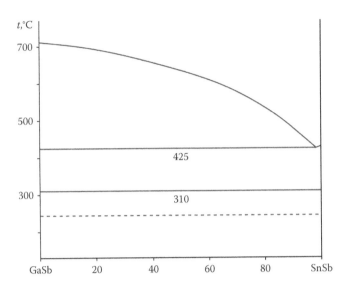

FIGURE 11.32 Phase diagram of the GaSb–SnSb system. (From Arbenina, V.V., and Beylin, Yu.A., *Izv. AN SSSR Neorgan. Mater.*, 22(9), 1425, 1986. With permission.)

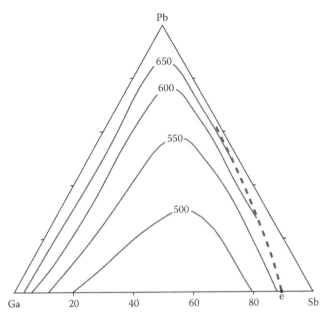

FIGURE 11.33 Liquidus surface of the Ga–Pb–Sb ternary system. (From Baranov, A.N., et al., *Izv. AN SSSR Neorgan. Mater.*, 25(6), 922, 1989. With permission.)

0.0058 mol% GaSb. The solubility of Sn in GaSb is circa $5 \cdot 10^{17}$ cm^{-3} (Popkov et al. 1980).

Using a high-temperature calorimeter, the change $\delta(\Delta H)$ in enthalpy upon mixing the molten GaSb with Sn in the entire concentration range was determined (Gerdes and Predel 1981c). It was shown that the concentration dependence of the $\delta(\Delta H)$ values is negative at low Sn contents and positive at high Sn concentrations.

By extremely rapid cooling of the liquid alloys, continuous solid solutions in the GaSb–Sn were obtained (Bucher et al. 1986). In order to attain the high cooling rates (10^7–10^8 °C·s^{-1}), a shock wave tube has been used.

GaSb–SnSb. The phase diagram of this system is shown in Figure 11.32 (Arbenina and Beylin 1986). The region near SnSb (0–10 mol% GaSb) is not studied in detail. At 310°C, a phase transformation in SnSb takes place (according to the literature data, this compound has no phase transformation). It is possible that this system is a non-quasi-binary section of the Ga–Sn–Sb ternary system.

11.24 GALLIUM–LEAD–ANTIMONY

The liquidus surface of the Ga–Pb–Sb ternary system is given in Figure 11.33 (Baranov et al. 1989), and the liquidus isotherms in the Pb corner of this system are shown in Figure 11.34 (Hernández-Zarazúa et al. 2004). The liquidus points were determined as the temperatures where the last solid disappeared while slowly heating the solutions.

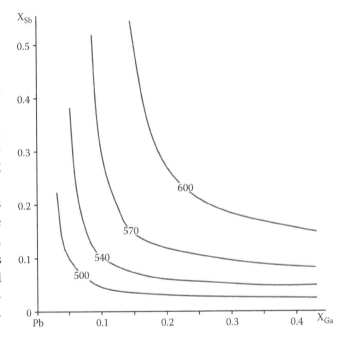

FIGURE 11.34 Liquidus isotherms in the Pb corner of the Ga–Pb–Sb ternary system. (From Hernández-Zarazúa, R., et al., *Thin Solid Films*, 461(2), 233, 2004. With permission.)

Integral enthalpy of mixing and integral excess Gibbs energy dependences on composition were obtained by Manasijević et al. (2003). These results were used for further calculations of partial molar quantities for every component of the Ga–Pb–Sb system. Calculated activity

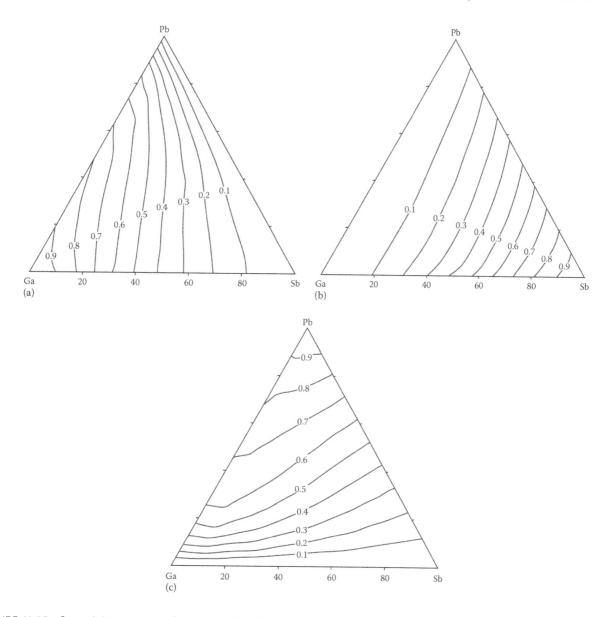

FIGURE 11.35 Isoactivity curves in the ternary Ga–Pb–Sb system at 800°C for (a) Ga, (b) Sb, and (c) Pb. (From Manasijević, et al., *CALPHAD*, 27(4), 361, 2003. With permission.)

values at 800°C for all components are given in the form of isoactivity diagrams (Figure 11.35).

GaSb–Pb. The phase diagram of this system is a eutectic type (Figure 11.36) (Manasijević et al. 2004). The eutectic crystallizes at 312°C and contains 2 mol% GaSb (according to the data of Baranov et al. [1989], the eutectic is degenerated from the Pb side). The solubility of GaSb in liquid Pb as a function of temperature is shown in Figure 11.37 (Hernández-Zarazúa et al. 2004).

Along this quasi-binary section, the change in enthalpy δ(ΔH) upon mixing liquid Pb with the molten GaSb was determined using a high-temperature calorimeter (Gerdes and Predel 1981b). Positive δ(ΔH) values

were found, the concentration dependence of which can be represented in terms of an association model and also largely on the basis of the regular solution model.

According to the calculations, the miscibility region in this system is metastable (Litvak 1992).

11.25 GALLIUM–TITANIUM–ANTIMONY

GaTi$_5$Sb$_2$ ternary compound, which crystallizes in the tetragonal structure with the lattice parameters $a = 1064.2 \pm 0.2$ and $c = 532.7 \pm 0.1$ pm (Kozlov and Pavlyuk 2003) ($a = 1047.19$ and $c = 519.82$ pm and enthalpy of formation of −53.1 kJ·M^{-1} [calculated values] [Colinet and Tedenac 2015]), is formed in the Ga–Ti–Sb

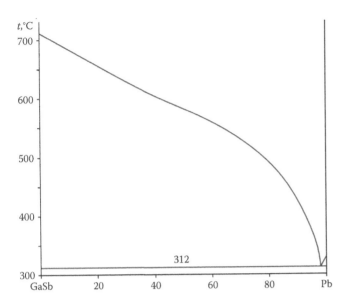

FIGURE 11.36 Phase diagram of the GaSb–Pb system. (From Manasijević, D., et al., *Thermochim. Acta*, *419*(1–2), 295, 2004. With permission.)

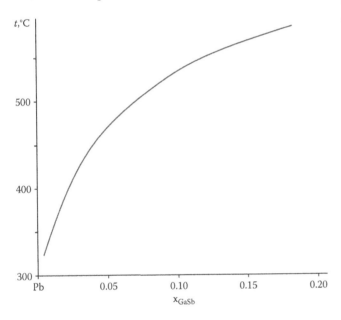

FIGURE 11.37 Solubility of GaSb in liquid Pb as a function of temperature. (From Hernández-Zarazúa, R., et al., *Thin Solid Films*, *461*(2), 233, 2004. With permission.)

system. The title compound was prepared by arc melting of a mixture of Ga, Ti, and Sb in stoichiometric ratio in an Ar atmosphere, annealed at 400°C for 720 h, and quenched in cold water (Kozlov and Pavlyuk 2003).

11.26 GALLIUM–HAFNIUM–ANTIMONY

The isothermal section at 600°C of the Ga–Hf–Sb ternary system was constructed in the whole concentration range,

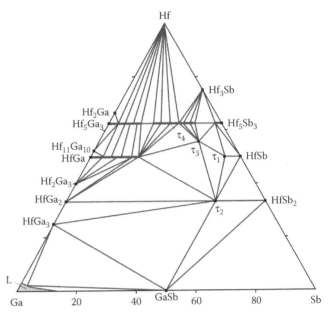

FIGURE 11.38 Isothermal section of the Ga–Hf–Sb ternary system at 600°C. (From Tokaychuk, I., et al., *Chem. Met. Alloys*, *6*(1–2), 75, 2013. With permission.)

using XRD, and is shown in Figure 11.38 (Tokaychuk et al. 2013). Limited solid solutions based on the binary compounds HfGa (17 at% Sb), Hf_5Ga_3 (11 at% Sb), and Hf_5Sb_3 (3 at% Ga) were observed. Four ternary compounds, $\mathbf{HfGa_{0.1}Sb_{0.9}}$ (τ_1), $\mathbf{Hf_2GaSb_3}$ (τ_2), $\mathbf{Hf_5GaSb_3}$ (τ_3), and $\mathbf{Hf_5Ga_{1.84-0.72}Sb_{1.16-2.28}}$ (τ_4), form at 600°C in this system. $HfGa_{0.1}Sb_{0.9}$ crystallizes in the cubic structure with the lattice parameter $a = 557.52 \pm 0.03$ pm and a calculated density of 11.313 g·cm⁻³.

Hf_2GaSb_3 crystallizes in the tetragonal structure with the lattice parameters $a = 389.841 \pm 0.008$ and $c = 862.650 \pm 0.019$ pm (Tokaychuk et al. 2012, 2013). It was synthesized from the elements by arc melting in a water-cooled Cu crucible with a W electrode under a purified Ar atmosphere, using Ti as a getter. To ensure homogeneity, the sample was remelted twice. The ingot was wrapped in Ta foil, annealed at 600°C in a quartz ampoule under vacuum for 1 month, and subsequently quenched in cold water.

Hf_5GaSb_3 crystallizes in the hexagonal structure with the lattice parameters $a = 847.47 \pm 0.05$ and $c = 571.90 \pm 0.05$ pm and a calculated density of 12.398 g·cm⁻³ (Tokaychuk et al. 2013).

$Hf_5Ga_{1.84-0.72}Sb_{1.16-2.28}$ crystallizes in the tetragonal structure with the lattice parameters $a = 1084.972 \pm 0.015$ and $c = 550.154 \pm 0.008$ pm and a calculated density of

12.0978 g·cm⁻³ for composition $Hf_5Ga_{1.51}Sb_{1.49}$ (Voznyak et al. 2012; Tokaychuk et al. 2013).

The samples were annealed at 600°C for 720 h (Tokaychuk et al. 2013). Finally, the ampoules were quenched into cold water.

11.27 GALLIUM–BISMUTH–ANTIMONY

The liquidus surface of the Ga–Bi–Sb ternary system is shown in Figure 11.39 (Akchurin et al. 1984; Sorokina and Morgun 1985). By analyzing four investigated vertical sections, it can be concluded that the liquidus surface is dominated by the area of GaSb primary crystallization (Minić et al. 2011). On the Bi–Sb side, a small area of primary crystallization of Bi_xSb_{1-x} solid solution exists. A calculated degenerate invariant reaction at 29.6°C exists in this ternary system.

The isothermal section of the Ga–Bi–Sb ternary system at 500°C is given in Figure 11.40 (Minić et al. 2011). The samples were annealed at 500°C for 2 weeks and investigated through DTA, and SEM with EDX.

The integral excess Gibbs energy of the Ga–Bi–Sb ternary system was calculated and compared with the experimental results (Živković et al. 1999). The calculated data show agreement between results obtained by different models, as well as with experimental literature data. The activity of Ga in liquid Ga–Bi–Sb alloys was measured by the EMF method, with zirconia as

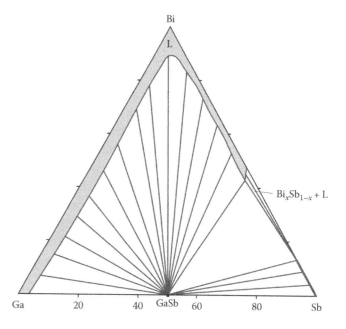

FIGURE 11.40   Isothermal section of the Ga–Bi–Sb ternary system at 500°C. (From Minić, D., et al., *Mater. Sci. Technol.*, 27(5), 884, 2011. With permission.)

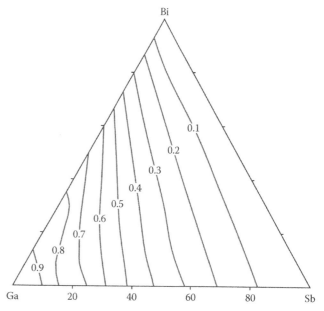

FIGURE 11.41   Isoactivity curves of gallium in liquid Ga–Bi–Sb alloys at 800°C. (From Katayama, I., et al., *Mater. Trans.*, 34(9), 792, 1993. With permission.)

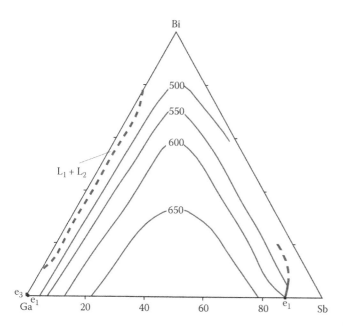

FIGURE 11.39   Liquidus surface of the Ga–Bi–Sb ternary system. (From Sorokina, O.V., and Morgun, A.I., *Zhurn. Neorgan. Khimii*, 30(12), 3174, 1985. With permission.)

the solid electrolyte at 800°C in the whole composition range (Figure 11.41) (Katayama et al. 1993a). The shape of the isoactivity curves is simple and shows a very small deviation from the straight lines that connect the starting points in the Ga–Sb system and the ending points in the Ga–Bi system.

The phase relations in the $Ga_{0.1}Sb_{0.9}$–Bi section have been calculated by Živković et al. (2010). The results obtained showed good agreement in comparison with the experimental data. Activities, activity coefficients, partial molar quantities for Bi, and integral molar quantities for the $Ga_{0.1}Sb_{0.9}$–Bi section were obtained at 800°C using experimental calorimetric investigations (Živković et al. 2000). A vertical section $Ga_{0.1}Sb_{0.9}$–Bi was constructed using DTA and SEM. A positive deviation from Raoult's law was determined in the composition range up to 60 at% Bi, while for a further increase of Bi content, a negative deviation from the ideal behavior was noticed. Negative values for the integral molar excess Gibbs energies were obtained in the whole concentration range of the $Ga_{0.1}Sb_{0.9}$–Bi section.

GaSb–Bi. The phase diagram of this system is a eutectic type (Figure 11.42) (Akchurin et al. 1984; Sorokina and Morgun 1985; Živković et al. 2001; Minić et al. 2011). The eutectic crystallizes at 270.8°C (Minić et al. 2011) (at 266°C, contains 2 mol% GaSb [Živković et al. 2001]; according to the data of Akchurin et al. [1984] and Sorokina and Morgun [1985], the eutectic is degenerated from the Bi side). The solubility of Bi in GaSb is not higher than 0.1 at% (Akchurin et al. 1984) (ca. 0.8 at% at the eutectic temperature [Zinov'ev et al. 1993]). The solubility of GaSb in liquid Bi was determined within the temperature interval from 350°C to 660°C and is given in Figure 11.43 (Golubev et al. 1972).

Thermodynamic properties of the alloys of the GaSb–Bi system were experimentally and analytically determined and compared (Živković et al. 2001). A positive

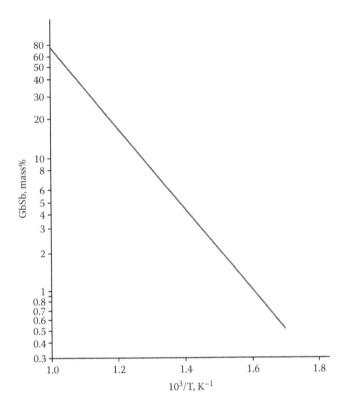

FIGURE 11.43 Solubility of GaSb in liquid Bi as a function of temperature. (From Golubev, L.V., et al., *Elektronnaya Tekhnika. Ser. 6 Materialy*, (5), 156, 1972. With permission.)

deviation from Raoult's law can be noticed for Bi activities at a given temperature, which causes the positive values for the partial molar excess Gibbs energy. The enthalpy change due to the mixing of molten GaSb with liquid Bi has been determined by using a high-temperature calorimeter (Gerdes and Predel 1979a). Melting of a quasi-binary system yields a maximum value of the enthalpy change of $+2370$ J·mol^{-1} at 747°C. The calculated result is consistent within an order of magnitude with experimental data.

Molecular beam epitaxy (MBE) has been used to grow $GaBi_xSb_{1-x}$ solid solutions with x up to 0.05 (Rajpalke et al. 2013). The Bi content increases as the growth temperature decreases and reaches a plateau corresponding to $x = 0.05$ for growth temperatures below $\approx 275°C$ (Rajpalke et al. 2014). The Bi incorporation into GaSb at the epitaxial growth is varied in the range 0–9.6 at%, with the growth rate varying from 0.31 to 1.33 mkm·h^{-1} at two growth temperatures: 250°C and 275°C. The Bi content is inversely proportional to the growth rate, but with higher Bi contents achieved at 250°C than at 275°C. The lattice parameter varies linearly following Vegard's law (Germogenov et al. 1989; Rajpalke et al. 2013, 2014).

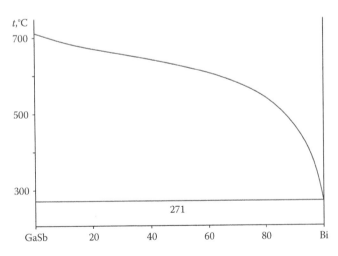

FIGURE 11.42 Phase diagram of the GaSb–Bi system. (From Minić, D., et al., *Mater. Sci. Technol.*, 27(5), 884, 2011. With permission.)

The band gap reduction corresponds to 35–36 meV for 1 at% Bi (Rajpalke et al. 2013, 2014).

11.28 GALLIUM–VANADIUM–ANTIMONY

The isothermal section at 500°C of the Ga–V–Sb ternary system was constructed in the whole concentration range, using XRD, and is shown in Figure 11.44 (Antonyshyn et al. 2008). Two compounds, **V₂GaSb₂** (τ_1) and **V₆GaSb** (τ_2), are formed in this system. The structure of the first compound is unknown, and V₆GaSb crystallizes in the cubic structure with the lattice parameter $a = 489.62 \pm 0.07$ pm and a calculated density of 7.127 ± 0.003 g·cm⁻³ for the composition V₆Ga₀.₈₇Sb₁.₁₃ ($a = 489.2 \pm 0.2$ pm [Rasmussen and Hazell 1978]). This compound was synthesized from GaSb and V (Rasmussen and Hazell 1978). The samples were annealed at 500°C for 1000 h and quenching in cold water (Antonyshyn et al. 2008).

GaSb–V₂Ga₅. The phase diagram of this system is a eutectic type (Müller and Wilhelm 1967c). The eutectic crystallizes at 707°C and contains 4.4 mass% V₂Ga₅.

GaSb–V₃GaSb₅. The phase diagram of this system is a eutectic type (Müller and Wilhelm 1967c). The eutectic crystallizes at 710°C and contains 4.9 mass% V₃GaSb₅.

11.29 GALLIUM–OXYGEN–ANTIMONY

The isothermal section of the Ga–O–Sb ternary system at room temperature is shown in Figure 11.45 (Schwartz

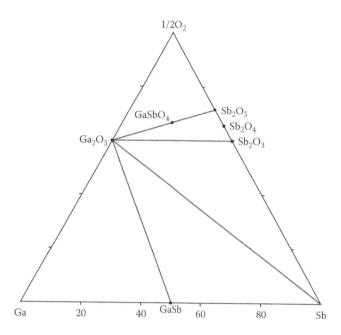

FIGURE 11.45 Isothermal section of the Ga–O–Sb ternary system at room temperature. (From Schwartz, G.P., *Thin Solid Films*, *103*(1–3), 3, 1983. With permission.)

et al. 1980; Schwartz 1983). **GaSbO₄** ternary compound is formed in this system. It crystallizes in the tetragonal structure with the lattice parameters $a = 460$ and $c = 303.4$ pm and an experimental density of 6.54 g·cm⁻³ (Varfolomeev et al. 1975a) ($a = 460$ and $c = 303$ pm and a calculated density of 6.57 g·cm⁻³ [Shafer and Roy 1956]). To prepare the title compound, Ga₂O₃ and Sb₂O₃ were mixed together under EtOH and then transferred to a Pt crucible and heated up to 800°C (Shafer and Roy 1956). The samples of GaSbO₄ were also prepared by coprecipitation of hydrated gallium and antimony oxides with aqueous ammonia solution from chloride-nitrate solutions containing the above-mentioned elements (Varfolomeev et al. 1975a). The resulting precipitates were dried and calcined with an intermediate trituration at 800°C and then at 1000°C for 100–200 h.

According to the data of Dachille and Roy (1959), there are no new crystalline forms of GaSbO₄ at heating at 600°C up to 6 MPa.

11.30 GALLIUM–SULFUR–ANTIMONY

The chemical potential diagrams of the Ga–S–Sb ternary system have been estimated by Šesták et al. (1994). The equilibrium composition of coexisting phases in this system has been evaluated (Figure 11.46). It is seen that the stability polygon of GaSb has a certain range of the sulfur potential that corresponds to different

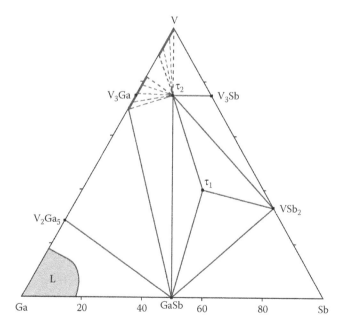

FIGURE 11.44 Isothermal section of the Ga–V–Sb ternary system at 500°C. (From Antonyshyn, I., et al., *Visnyk L'viv Un-Tu. Ser. Khim.*, (53), Pt. 1, 50, 2008. With permission.)

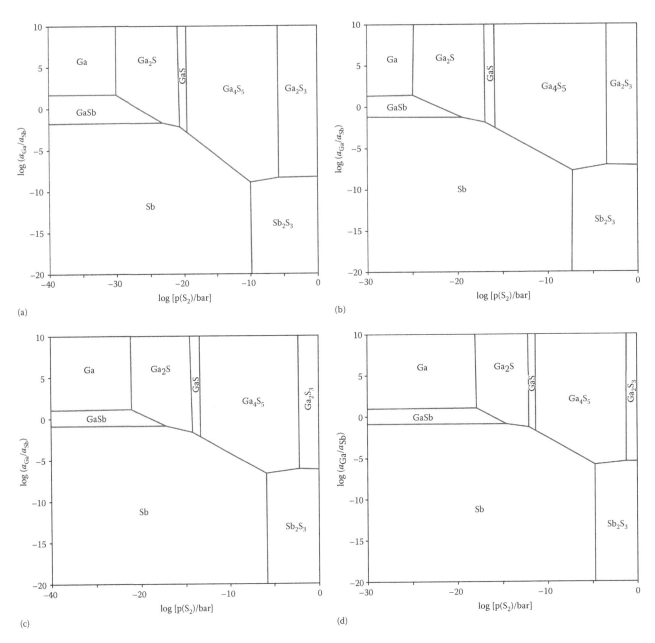

FIGURE 11.46 Chemical potential diagrams in the Ga–S–Sb system for (a) 512°C, (b) 612°C, (c) 712°C, and (d) 812°C. (From Šesták J., et al., *Thermochim. Acta*, *245*, 189, 1994. With permission.)

states of the S-doped GaSb. Although these diagrams are constructed by using only the thermodynamic data of compounds, stability regions of the respective compounds bear certain information about the third component within the approximation of an infinitely dilute solution. This means that each point of the polygon of GaSb implicitly indicates its dopant level of sulfur. When the thermodynamic effect of doping is to be explicitly considered, the GaSb polygon is slightly modified so that the three-phase GaSb–Ga₂S-Sb junction moves to the higher-sulfur-potential region. In other words, the high-sulfur-potential side corner of the GaSb polygon will be extended in the direction of Ga₂S. In Figure 11.46, the stability polygons of GaSb and Ga₂S are separated by several orders of the sulfur potential.

The isothermal sections of the Ga–S–Sb ternary system at 712°C and 812°C obtained on the basis of chemical potential diagrams are given in Figure 11.47 (Šesták et al. 1994). Since the composition of the liquid phase changes with temperature, the appearance of the phase diagram changes dramatically, while the same features

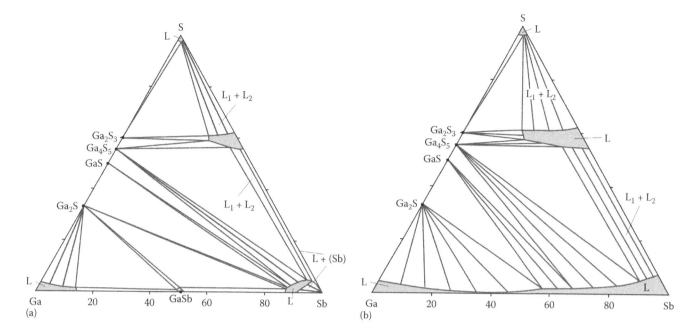

FIGURE 11.47 Isothermal sections of the Ga–S–Sb ternary system at 712°C and 812°C. (From Šesták J., et al., *Thermochim. Acta*, 245, 189, 1994. With permission.)

appear without large differences from those obtained under the stoichiometric assumptions.

GaSb–GaS. The phase diagram of this system is a eutectic type with a monotectic transformation (Figure 11.48) (Azhdarova et al. 1974). The eutectic crystallizes at 675°C and contains 10 mol% GaS. GaSb dissolves 0.7 mol% GaS and the solubility of GaSb in GaAs is 1 mol%. The system was investigated through DTA and metallography. The samples were annealed at 675°C for 300 h.

GaSb–Ga₂S. This system is a non-quasi-binary section of the Ga–S–Sb ternary system as Ga_2S melts incongruently (Azhdarova et al. 1974). A miscibility region exists in this section. GaSb dissolves 1.5 mol% Ga_2S. The system was investigated through DTA and metallography. The samples were annealed at 675°C for 300 h.

GaSb–Ga₂S₃. The phase diagram of this system is a peritectic type (Guseinova and Rustamov 1969). The peritectic crystallizes at 575°C. This system needs additional studies.

GaSb–S. This system is a non-quasi-binary section of the Ga–S–Sb ternary system (Kompanichenko et al. 2003). At the interaction of GaSb and S beside Ga_2S_3 and Sb_2S_3, the compound Ga_2S_2 was identified.

GaSb–Sb₂S₃. This system is a non-quasi-binary section of the Ga–S–Sb ternary system, which intersects the field of primary crystallization of GaSb, GaS, Ga_4S_5,

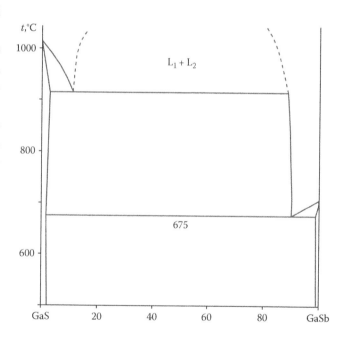

FIGURE 11.48 Phase diagram of the GaSb–GaS system. (From Azhdarova, D.S., et al., *Uch. Zap. Azerb. Un-T. Ser. Khim. N.*, (2), 47, 1974. With permission.)

Ga_2S_3, and Sb_2S (Rustamov et al. 1974a). The solubility of Sb_2S_3 in GaSb is 0.1 mol%. An isothermal line at 515°C is defined by the nonvariant eutectic equilibrium in the GaSb–GaS–Sb subsystem. The crystallization of the melts in the GaS–Ga_2S_3–Sb subsystem is complex. One part of the melts crystallizes at the temperature of a transition point (630°C), and another part of the melts crystallizes at the temperature of the ternary eutectic (610°C). Full crystallization takes place at the temperature of the ternary eutectic at 475°C.

11.31 GALLIUM–SELENIUM–ANTIMONY

The liquidus surface of the Ga–Se–Sb ternary system includes eight fields of primary crystallization and the miscibility region, which occupies the biggest part of this system (Rustamov et al. 1970). Five ternary eutectics exist in the system. They crystallize at 570°C, 590°C, 490°C, 200°C, and 29°C. The miscibility region in this system at 850°C with the liquidus projection was constructed by Fedorov et al. (1972). All these data need further investigations, as Ga_2Se compound does not form in the Ga–Se bordering system.

GaSb–GaSe. The phase diagram of this system is a eutectic type (Figure 11.49) (Fedorov et al. 1970). The eutectic crystallizes at 700°C and contains 9 mol% GaSe (at 710°C, contains 5 mol% GaSe [Rustamov et al. 1969, 1970]). According to the data of Rustamov et al. (1969, 1970), a miscibility region exists in this system in the region of 8–77 mol% GaSb, but Fedorov et al. (1970) did not determine the miscibility gap in this system. This system was investigated through DTA, XRD, metallography, and measurement of microhardness (Rustamov et al. 1969, 1970; Fedorov et al. 1970). The samples were annealed at 620°C for 300 h (Fedorov et al. 1970).

GaSb–Ga_2Se. This system is a non-quasi-binary section of the Ga–Se–Sb ternary system (Rustamov et al. 1971). A miscibility region exists in this system in the region of 8–80 mol% GaSb. All the melts end crystallization at 520°C. The solubility of Ga_2Se in GaSb is 2 mol%.

GaSb–Ga_2Se_3. This system is a non-quasi-binary section of the Ga–Se–Sb ternary system (Guseinova et al. 1968; Guseinova and Rustamov 1969; Rustamov et al. 1970). The solubility of Ga_2Se_3 in 3GaSb at 500°C is 10 mol% (this system is considered 3GaSb–Ga_2Se_3) (Woolley and Blazey 1964).

GaSb–Se. This system is a non-quasi-binary section of the Ga–Se–Sb ternary system (Guseinova and Rustamov 1970; Rustamov et al. 1970). The solubility of S in GaSb is 2 at%.

11.32 GALLIUM–TELLURIUM–ANTIMONY

The liquidus surface of the Ga–Te–Sb ternary system was constructed by Rustamov and Geydarova (1977). This surface includes nine fields of primary crystallization and a miscibility region. It is necessary to note that this surface needed further investigations, as the data on the Ga–Te binary system are not consistent with the results given by Rustamov and Geydarova (1977).

The region of the solid solutions based on GaSb in the Ga–Te–Sb ternary system at 650°C is given in Figure 11.50 (Mirgalovskaya et al. 1977). According to the data of Dashevskiy et al. (1980), the maximum solubility of Te at 700°C is 6.43 at% and 3.15 at% at 650°C. The region of solid solutions in this ternary system is elongated along the GaSb–Ga_2Te_3 system.

The activity of Ga in liquid Ga–Te–Sb alloys was measured by the EMF method with zirconia solid electrolyte at 730°C–880°C (Katayama et al. 1995). Combining the binary data of Ga–Sb and Ga–Te systems with the ternary data, isoactivity curves at 850°C were obtained in the whole ternary system (Figure 11.51).

GaSb–GaTe. The phase diagram of this system is a eutectic type (Figure 11.52) (Kurata and Hirai 1966). The eutectic crystallizes at 695°C and contains 17 mol% GaTe.

GaSb–Ga_2Te_3. This system is a non-quasi-binary section of the Ga–Te–Sb ternary system (Rustamov and Geydarova 1977). The mutual solubility of GaSb and Ga_2Te_3 is not higher than 2 mol%. According to the data of Woolley and Blazey (1964), the solubility of Ga_2Te_3 in 3GaSb at 500°C is 36 mol% (this system is considered 3GaSb–Ga_2Te_3).

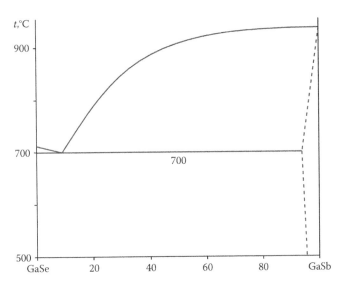

FIGURE 11.49 Phase diagram of the GaSb–GaSe system. (From Fedorov, P.I., et al., *Izv. AN SSSR Neorgan. Mater.*, 6(10), 1877, 1970. With permission.)

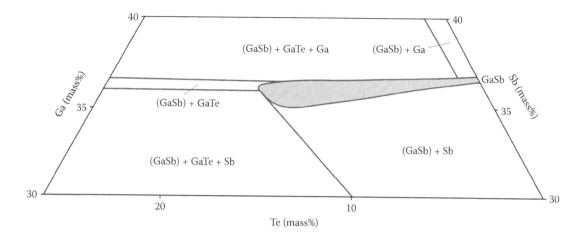

FIGURE 11.50   Region of the solid solutions based on GaSb in the Ga–Te–Sb ternary system at 650°C. (From Mirgalovskaya, M.S., et al., *Izv. AN SSSR Neorgan. Mater.*, *13*(9), 1574, 1977. With permission.)

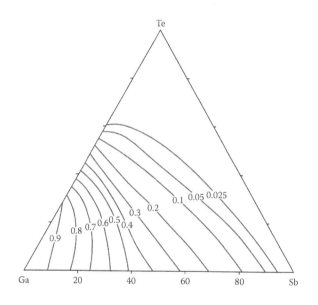

FIGURE 11.51   Isoactivity curves of gallium in liquid Ga–Te–Sb alloys at 850°C. (From Katayama, I., et al., *Mater. Trans.*, *36*(1), 41, 1995. With permission.)

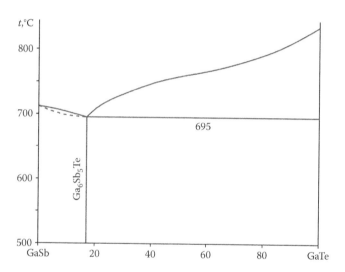

FIGURE 11.52   Phase diagram of the GaSb–GaTe system. (From Kurata, K., and Hirai, T., *Solid State Electron.*, *9*(6), 633, 1966. With permission.)

GaSb–Sb₂Te₃. The phase diagram of this system is a eutectic type with peritectic transformation (Figure 11.53) (Rustamov and Geydarova 1977). **Ga₄Sb₆Te₃** ternary compound, which melts incongruently at 600°C, is formed in this system. The eutectic crystallizes at 525°C.

11.33 GALLIUM–CHROMIUM–ANTIMONY

GaSb–CrSb. The phase diagram of this system is a eutectic type. The eutectic crystallizes at 690°C ± 2°C and contains 13.4 mass% CrSb (Müller and Wilhelm 1965a) (at 417°C, contains 4 mass% CrSb [Aliev et al. 1981]).

CrGa₂Sb₂ ternary compound, which crystallizes in the orthorhombic structure with the lattice parameters $a = 1178.1 \pm 0.5$, $b = 596.8 \pm 0.2$, and $c = 590.2 \pm 0.2$ pm and a calculated density of 6.926 g·cm⁻³, is formed in this system (Sakakibara et al. 2010). Dark gray crystals of this compound have been synthesized by a direct reaction from Cr and GaSb at 6.3 GPa and 500°C–700°C for 30 min using a belt-type, high-pressure apparatus. The title compound is metastable at ambient pressure conditions and decomposes into Cr and GaSb by heating above 370°C under an inert atmosphere.

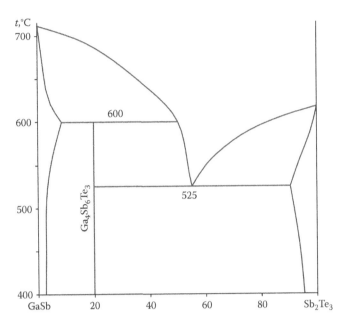

FIGURE 11.53 Phase diagram of the GaSb–Sb₂Te₃ system. (From Rustamov, P.G., and Geydarova, E.A., *Azerb. Khim. Zhurn.*, (3), 108, 1977. With permission.)

11.34 GALLIUM–TUNGSTEN–ANTIMONY

An isothermal section of the Ga–W–Sb ternary system at room temperature was calculated by Liu and Mohney (2003a). It was determined that GaSb is in thermodynamic equilibrium with W.

11.35 GALLIUM–MANGANESE–ANTIMONY

GaSb–Mn. The effect of high pressure (6 GPa) on the formation of new phases in a polycrystalline mixture of GaSb and Mn (molar ratio 1:1) upon heating was studied by Popova et al. (2006). It was established that a sphalerite-type solid solution with a small amount of Mn forms at temperatures below 250°C–330°C. At higher temperatures, new crystalline phases were synthesized: a phase with a cubic structure with the lattice parameter $a = 294.6 \pm 0.1$ pm and an experimental density of 6.5 ± 0.1 g·cm⁻³ (at 350°C–400°C) and a phase with a tetragonal structure with the lattice parameters $a = 642.6 \pm 0.4$ and $c = 534.9 \pm 0.4$ pm and experimental and calculated densities of 7.40 ± 0.15 and 7.40 g·cm⁻³, respectively (at 420°C–600°C). These phases are metastable under normal conditions.

MnGa₂Sb₂ ternary compound, which crystallizes in the orthorhombic structure with the lattice parameters $a = 1180 \pm 7$, $b = 596.8 \pm 0.2$, and $c = 585.8 \pm 0.2$ pm and a calculated density of 6.988 g·cm⁻³, is formed in this system (Sakakibara et al. 2009). Dark gray crystals

of this compound have been synthesized by a direct reaction from Mn and GaSb at 4–6 GPa and 400°C–700°C for 30 min using a belt-type high-pressure apparatus. A nearly single phase was obtained by repeating the reaction at 6 GPa and 500°C two times. The title compound is metastable at ambient pressure conditions and decomposes into Mn and GaSb by heating above 260°C under an inert atmosphere.

GaSb–MnSb. The phase diagram of this system is a eutectic type (Figure 11.54) (Aliev et al. 1974; Safaraliev et al. 1991; Marenkin et al. 2013). MnSb melts incongruently. The eutectic crystallizes at 632°C and contains 41 mol% MnSb (Marenkin et al. 2013) (at 647°C, contains 60 mol% MnSb [Guliev 1973; Aliev et al. 1974; Safaraliev et al. 1991]). According to the data of Aliev et al. (1974) and Safaraliev et al. (1991), the solubility of MnSb in GaSb reaches up to 30 mol% and MnSb dissolves 7 mol% GaSb. This system was investigated through DTA, XRD, metallography, and measurement of microhardness (Aliev et al. 1974; Safaraliev et al. 1991; Marenkin et al. 2013). The ingots were annealed at 550°C for 200 h.

Polycrystalline ingots of Ga₁₋ₓMnₓSb ($0 < x < 0.15$) were grown by the vertical Bridgman method (Basu and Adhikari 1994). The lattice parameter of the samples was found to decrease linearly with increasing Mn concentration and could be expressed by the following analytical expression: a (pm) $= 610.4 - 0.528x$. The band gap of these solid solutions decreases with increasing x, and the variation followed a nonlinear-to-linear trend (Basu and Adhikari 1995). The band gap decreases rapidly with x at lower Mn concentrations and then gradually with a

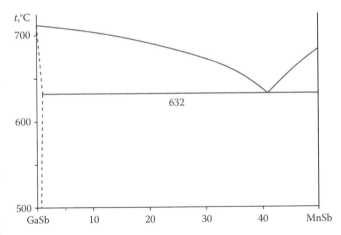

FIGURE 11.54 Part of the phase diagram of the GaSb–MnSb system. (From Marenkin, S.F., et al., *Russ. J. Inorg. Chem.*, 58(11), 1324, 2013. With permission.)

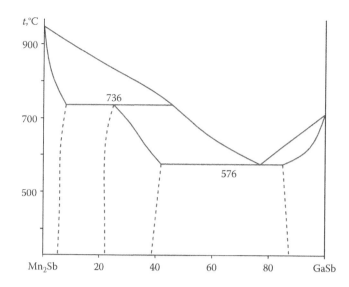

FIGURE 11.55 Phase diagram of the GaSb–Mn₂Sb system. (From Safaraliev, G.I., and Vagabova, L.K., *Izv. AN SSSR Neorgan. Mater.*, 24(4), 550, 1988. With permission.)

smaller slope. The nonlinearity at lower Mn concentration was explained on the basis of magnetic exchange interactions.

GaSb–Mn₂Sb. The phase diagram of this system is a eutectic type with a peritectic transformation (Figure 11.55) (Safaraliev and Vagabova 1988). **Mn₆GaSb₄** ternary compound, which melts incongruently at 736°C and crystallizes in the tetragonal structure, is formed in this system. The eutectic crystallizes at 576°C and contains 25 mol% Mn₂Sb. The solubility of GaSb in Mn₂Sb is not higher than 5 mol%, and GaSb dissolves up to 8 mol% Mn₂Sb.

11.36 GALLIUM–RHENIUM–ANTIMONY

An isothermal section of the Ga–Re–Sb ternary system at room temperature was calculated by Liu and Mohney (2003a). It was determined that GaSb is in thermodynamic equilibrium with Re and ReGa₅.

11.37 GALLIUM–IRON–ANTIMONY

The isothermal section of the Ga–Fe–Sb ternary system at 600°C is shown in Figure 11.56 (Députier et al. 2002; Raghavan 2004b). The ordered form of *bcc* Fe (α′′′) and the Sb-rich liquid, omitted by Députier et al. (2002), were schematically indicated by Raghavan (2004b). The main feature of this section is the large solubility of Ga in the Fe₃Sb₂ (ε) phase. The Ga-rich end has the composition range FeₜGa₀.₈Sb₁.₂ (2.15 ≤ t ≤ 2.80), which corresponds to an iron range of 52–58 at%. This range does

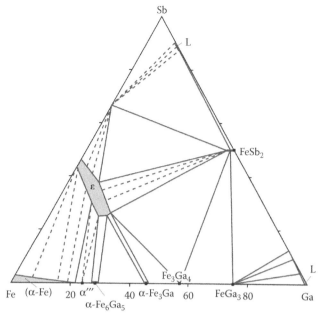

FIGURE 11.56 Isothermal section of the Ga–Fe–Sb ternary system at 600°C. (From Raghavan, V., *J. Phase Equilibria Diffus.*, 25(1), 85, 2004. With permission.)

not cover the composition found at 800°C by Moze et al. (1994). In addition to **FeₜGa₂₋ₓSbₓ**, GaSb forms tie-lines with several other binary phases. The solubility range of GaSb corresponds to about 2 at% Fe. The samples were annealed at 600°C for a long time and then quenched in an ice + water mixture (Députier et al. 2002). The phase equilibria were studied by XRD and SEM with EDX.

FeₜGa₂₋ₓSbₓ crystallizes in the hexagonal structure with the lattice parameters a = 410.3–412.7 and c = 503.0–514.7 pm for t = 3 and 0.1 ≤ x ≤ 1.5 (the parameters display a decrease with increasing Ga content) (Smith et al. 1992) (a = 409.4 and c = 508.4 pm for Fe₃Ga₁.₁₅Sb₀.₈₅ composition [Moze et al. 1994]). The Fe₃Ga₂₋ₓSbₓ alloys were prepared by melting together the appropriate proportions of Fe, Ge, and Sb (Smith et al. 1992; Moze et al. 1994). The melts were solidified by furnace cooling. The alloys were annealed at 800°C for 14 days.

Fe₄GaSb₁₂ metastable compound, which crystallizes in the cubic structure with the lattice parameter a = 916.5 ± 0.1 pm, could be obtained in the Ga–Fe–Sb ternary system (Sellinschegg et al. 1998). This compound forms at 181°C and decomposes at 362°C. Modulated elemental reactants were used to form amorphous ternary reaction intermediates of the desired compositions. The amorphous reaction intermediate crystallizes exothermically below 200°C, forming kinetically stable "filled" skutterudite. This metastable compound can only be prepared by controlling the reaction intermediates, avoiding the formation of more thermodynamically stable binary compounds.

GaSb–Fe. The kinetics of Fe dissolution in GaSb melt was investigated by Denisov and Dubovikov (1987) using the rotating disc method. The parameters of dissolution and diffusion in this system were determined. The solubility of Fe in the GaSb melt was also determined.

GaSb–Fe₃Ga₄. The phase diagram of this system is a eutectic type. The eutectic crystallizes at 695°C ± 2°C (at 422°C [Aliev et al. 1981]) and contains 3 mass% Fe (Müller and Wilhelm 1965a). Fe₃Ga₄ melts incongruently.

11.38 GALLIUM–COBALT–ANTIMONY

The liquidus projection of the Ga–Co–Sb ternary system was determined based on the experimental results and the phase diagrams of its three constituent binary systems (Figure 11.57) (Chen et al. 2015). The liquidus projection has 10 primary solidification phase regions: Co, CoSb, CoSb₂, CoSb₃, Sb, GaSb, Ga, CoGa₃, CoGa, and a ternary compound **Co₃Sb₂Ga₄** (τ). The ternary solubilities in most of the binary compounds are not significant. Nine invariant reactions are in this ternary system: E_1 (temperature was not determined), L ⟷ Co

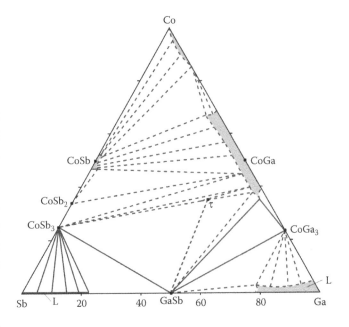

FIGURE 11.58 Isothermal section of the Ga–Co–Sb ternary system at 650°C. (From Chen, S.-W., et al., *J. Alloys Compd.*, 637, 98, 2015. With permission.)

+ CoSb + CoGa; E_2 (588°C), L ⟷ CoSb₃ + Sb + GaSb; E_3 (29.77°C), L ⟷ Ga + CoGa₃ + GaSb; U_1 (755°C), L + CoSb ⟷ CoSb₂ + τ; U_2 (818°C), L + CoGa ⟷ CoSb + τ; U_3 (747°C), L + CoSb₂ ⟷ CoSb₃ + τ; U_4 (686°C), L + τ ⟷ CoSb₃ + GaSb; U_5 (temperature was not determined), L + CoGa ⟷ τ + GaSb; and U_6 (687°C), L + CoGa ⟷ CoGa₃ + GaSb.

The isothermal section of the Ga–Co–Sb ternary system at 650°C is shown in Figure 11.58 (Chen et al. 2015). The isothermal section of this system at the same temperature was also constructed by Qiu et al. (2013), but these authors did not determine the existence of the Co₃Sb₂Ga₄ ternary compound.

GaSb–Co. The kinetics of Co dissolution in the GaSb melt was investigated by Denisov and Dubovikov (1987) using the rotating disc method. The parameters of dissolution and diffusion in this system were determined. The solubility of Co in the GaSb melt was also determined.

GaSb–Co₃Ga₄. The phase diagram of this system is a eutectic type. The eutectic crystallizes at 697°C ± 2°C (at 422°C [Aliev et al. 1981]) and contains 3 mass% Co (Müller and Wilhelm 1965a). Co₃Ga₄ melts incongruently.

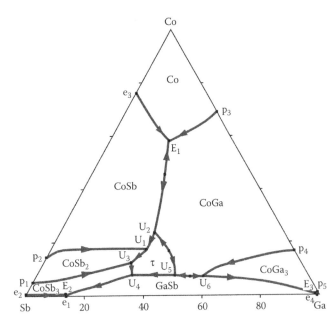

FIGURE 11.57 Projection of the liquidus surface of the Ga–Co–Sb ternary system. (From Chen, S.-W., et al., *J. Alloys Compd.*, 637, 98, 2015. With permission.)

11.39 GALLIUM–NICKEL–ANTIMONY

The liquidus surface of the Ga–Ni–Sb ternary system was constructed through DTA and XRD and is given

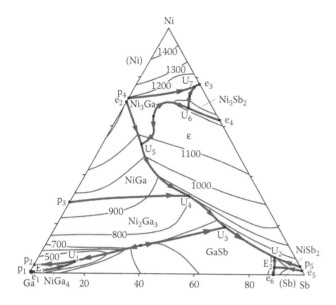

FIGURE 11.59 Liquidus surface of the Ga–Ni–Sb ternary system. (From Micke, K., et al., *Z. Metallkde, 92*(1) 14, 2001. With permission.)

in Figure 11.59 (Micke et al. 2001). Four vertical sections, including GaSb–Ni, through the ternary system were also constructed, and the reaction scheme was presented. The central part of the liquidus surface is dominated by the fields of primary crystallization of NiGa and NiSb, whereas at low Ni contents GaSb and Ni_2Ga_3 are dominant. The range of Ni_3Ga primary crystallization expands drastically into the ternary system. This is in line with considerable solubility of Sb in Ni_3Ga. Nine invariant reactions are in this ternary system: E_1 (<30°C), L $\longleftrightarrow$ (Ga) + $NiGa_4$ + GaSb; E_2 (579°C), L $\longleftrightarrow$ GaSb + $NiSb_2$ + (Sb); U_1 (<363°C), L + Ni_2Ga_3 $\longleftrightarrow$ $NiGa_4$ + GaSb; U_2 (593°C), L + NiSb $\longleftrightarrow$ GaSb + $NiSb_2$; U_3 (634°C), L + Ni_2Ga_3 $\longleftrightarrow$ GaSb + NiSb; U_4 (804°C), L + NiGa $\longleftrightarrow$ Ni_2Ga_3 + NiSb; U_5 (1060°C), L + Ni_3Ga $\longleftrightarrow$ NiGa + NiSb; U_6 (1092°C), L + Ni_3Ga $\longleftrightarrow$ Ni_5Sb_2 + NiSb; and U_7 (1115°C), L + Ni_8Ga $\longleftrightarrow$ (Ni) + Ni_5Sb_2.

The isothermal sections of this ternary system were constructed at 500°C, 600°C, and 900°C (Le Clanche et al. 1994; Micke et al. 1998; Markovski et al. 2000; Richter and Ipser 2002; Sutopo 2011a). At 900°C (Figure 11.60a), in the Ni-rich part of the system two three-phase equilibria, (Ni) + Ni_3Ga + Ni_5Sb_2 and Ni_3Ga + Ni_5Sb_2 + ε, were found (Markovski et al. 2000). On the Ni-deficient side of the ε-phase, an extensive two-phase field between NiGa and ε was found. The existence of the three-phase fields L + NiGa + Ni_2Ga_3 and L + NiGa + ε follows from the reaction sequence in the ternary system.

At 600°C (Figure 11.60b), as at 900°C, two three-phase equilibria, (Ni) + Ni_3Ga + Ni_5Sb_2 and Ni_3Ga + Ni_5Sb_2 + ε, were found in the Ni-rich part of the ternary phase diagram (Markovski et al. 2000). The ε corner point of the three-phase triangle Ni_3Ga + Ni_5Sb_2 + ε shifted toward the Sb-rich side in comparison with the section at 900°C. A continuous solid solution between Ni_3Ga_2 and NiSb cannot exist anymore because Ni_3Ga_2 decomposes above this temperature. Two three-phase fields, ε + NiSb + Ni_5Sb_2 and ε + NiSb + GaNi, form near the ε-phase.

At 500°C (Figure 11.60c), a negligible solubility of Ni in GaSb was detected (Markovski et al. 2000). Compared with the section at 600°C, several new three-phase triangles have appeared, mainly due to the changes in the constituent binary subsystems at temperatures lower than 600°C. In the Ni-rich part of the isothermal section at 500°C, the equilibria (Ni) + Ni_7Sb_3 + NiGa and (Ni) + Ni_7Sb_3–Ni_3Sb are formed as a consequence of the appearance of the phase Ni_7Sb_3 at 570°C. The homogeneity range of the ε-phase narrows with decreasing temperature. As can be seen from the phase diagram, the ε-phase retreats further from the NiSb phase and at 500°C exists only in a restricted composition range. At the expense of the ε-phase, the three-phase fields NiGa + NiSb + $Ni_{13}Ga_9$ and $Ni_{13}Ga_9$ + NiSb + ε, as well as Ni_3Ga + NiSb + ε and Ni_3Ga + NiSb–Ni_7Sb_3, are formed. Finally, the existence of the Ni_3Ga_4 phase in the Ga–Ni subsystem at 500°C leads to two new three-phase equilibria, Ni_3Ga_4 + Ni_3Ga_2 + NiSb and Ni_3Ga_4 + NiGa + NiSb. In the Sb-rich region, the three-phase fields GaSb + NiSb + Sb and GaSb + $NiSb_2$ + NiSb were found.

The samples were annealed at 500°C for 3 months, at 600°C for 4–10 weeks, and at 900°C for 2–5 weeks and afterwards quenched in cold water (Markovski et al. 2000). The isothermal sections were constructed through XRD, DTA, and electron probe microanalysis (EPMA).

Ni_3GaSb (ε-phase) ternary compound is formed in this system (Jan and Chang 1991; Le Clanche et al. 1994; Micke et al. 1998; Richter and Ipser 2002; Suriwong et al. 2011). According to the data of Sutopo (2011a), Ni_3GaSb is not a true ternary compound, but it represents a solid solution based on NiSb. This phase melts at 1066°C and crystallizes in the hexagonal structure with the lattice parameters $a = 401.35$ and $c = 511.36$ pm and experimental and calculated densities of 8.18 and

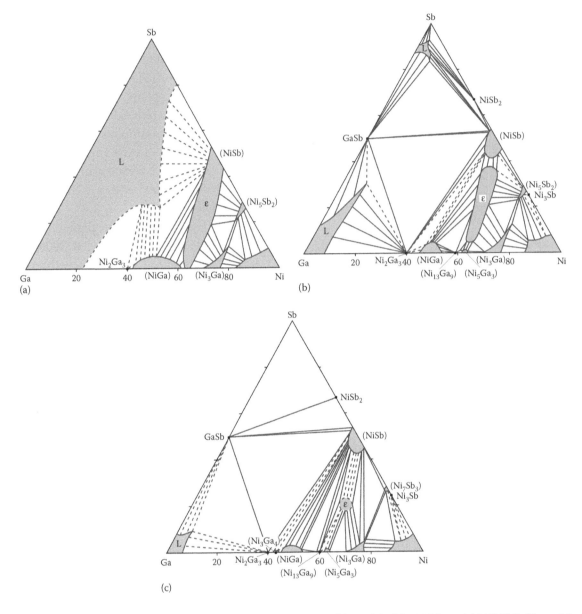

FIGURE 11.60   Isothermal sections of the Ga–Ni–Sb ternary system at (a) 900°C, (b) 600°C, and (c) 500°C. (From Markovski, S.L., et al., *J. Alloys Compd.*, *302*(1–2), 128, 2000. With permission.)

8.56 g·cm⁻³, respectively (Suriwong et al. 2011) (*a* = 400 and *c* = 509 pm [Jan and Chang 1991]; *a* = 402.8 ± 0.1 and *c* = 514.1 ± 0.2 pm [Le Clanche et al. 1994]).

GaSb–Ni. From XRD and measurement of microhardness, it was observed that the lattice constant of the solid solution $Ga_{1-x}Ni_xSb$ decreases linearly with Vegard's law up to 5 at% Ni (Kamilla et al. 2006).

The kinetics of Ni dissolution in the GaSb melt was investigated by Denisov and Dubovikov (1987) using the rotating disc method. The parameters of dissolution and diffusion in this system were determined. The solubility of Ni in the GaSb melt was also determined.

11.40 GALLIUM–PALLADIUM–ANTIMONY

The liquidus surface of the Ga–PdGa–PdSb–Sb subsystem of the Ga–Pd–Sb ternary system was derived from the results of DTA and EPMA and is shown in Figure 11.61 (Richter and Ipser 1998b). The field of PdGa primary crystallization, which is the most extended, is surrounded by the primary crystallization fields of PdSb, $PdSb_2$, Sb, GaSb, and finally Pd_3Ga_7, which only cover a very small area in the vicinity of the binary Ga–Pd phase diagram. The investigated part of the Ga–Pd–Sb system may be considered to consist of the three simple subsystems GaSb–Ga–PdGa, GaSb–PdGa–Sb, and PdGa–Sb–PdSb.

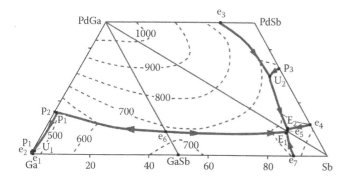

FIGURE 11.61 Liquidus surface of the Ga–Pd–Sb ternary system. (From Richter, K.W., and Ipser, H., *Ber. Bunsenges. Phys. Chem.*, *102*(9), 1245, 1998. With permission.)

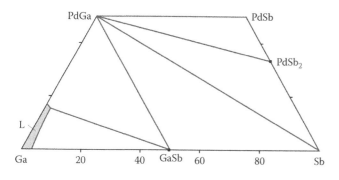

FIGURE 11.62 Isothermal section of the Ga–Pd–Sb ternary system at 500°C. (From Richter, K.W., and Ipser, H., *Ber. Bunsenges. Phys. Chem.*, *102*(9), 1245, 1998. With permission.)

The isothermal section of this ternary system at 500°C is given in Figure 11.62 (Richter and Ipser 1998b, 2002). No ternary phases have been found in the composition range between 0 and 50 at% Pd. PdGa is the only binary phase in equilibrium with GaSb. The solid solubility of the third element within the various binary phases was found to be negligible for all phases except PdSb and PdGa. For the former, the solubility of Ga in PdSb was found to be approximately 1 at% at 500°C.

The highest solubility—up to 8 at% Sb—was found for Pd_2Ga (Matselko et al. 2016). According to the data of Ellner and El-Boragy (1990), $Pd_{74}Ga_{13}Sb_{13}$ ($Pd_{25}Ga_{4.5}Sb_{4.5}$ [Matselko et al. 2016]), which is stable up to 750°C and crystallizes in the hexagonal structure with the lattice parameters $a = 750.0 \pm 0.1$ and $c = 1073.2 \pm 0.1$ pm, is formed in the Ga–Pd–Sb ternary system. The liquid field existing in the Ga corner of the phase diagram is limited by the bordering binary system Pd–Ga (~18 at% Pd) and Ga–Sb (~4 at% Sb) (Matselko et al. 2016). Three phases were revealed in the Pd-rich part of the phase diagram: $Pd_2Ga_{1-x}Sb_x$ ($x = 0.5–0.7$) with

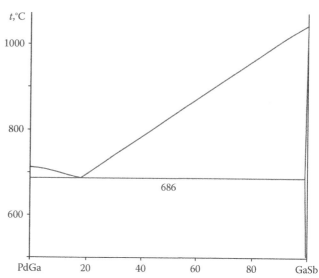

FIGURE 11.63 Phase diagram of the GaSb–PdGa system. (From Richter, K.W., and Ipser, H., *Ber. Bunsenges. Phys. Chem.*, *102*(9), 1245, 1998. With permission.)

the hexagonal structure and two phases with tetragonal primitive unit cells and approximate composition close to $Pd_3Ga_{0.5}Sb_{0.5}$ (Matselko et al. 2016). $α$-$Pd_3Ga_{0.5}Sb_{0.5}$ exists until 600°C, whereas $β$-$Pd_3Ga_{0.5}Sb_{0.5}$ is stable until 700°C.

GaSb–Pd. This system is a non-quasi-binary section of the Ga–Pd–Sb ternary system (Richter and Ipser 1998b, 2002). A liquid phase is formed at 572°C corresponding to the ternary eutectic reaction $L \longleftrightarrow GaSb + PdGa + Sb$.

GaSb–PdGa. The phase diagram of this system is a eutectic type (Figure 11.63) (Richter and Ipser 1998b). The eutectic crystallizes at 686°C and contains 82 mol% GaSb.

11.41 GALLIUM–OSMIUM–ANTIMONY

The calculated isothermal section of the Ga–Os–Sb ternary system at room temperature is given in Figure 11.64 (Liu and Mohney 2003a). The dotted lines indicate possible alternative tie-lines.

11.42 GALLIUM–PLATINUM–ANTIMONY

The calculated liquidus surface of the Ga–Pt–Sb ternary system is shown in Figure 11.65 (Guo et al. 2016). Compared with the liquidus surface determined by Richter and Ipser (1998a), the main difference is that a transition reaction $L + Pt_2Ga_3 \longleftrightarrow PtGa_2 + PtSb_2$ at 894°C is changed to a eutectic reaction $L \longleftrightarrow PtGa_2 + PtSb_2 + Pt_2Ga_3$, which is caused by the highest temperature

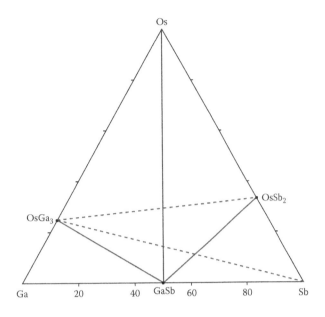

FIGURE 11.64 Isothermal section of the Ga–Os–Sb ternary system at room temperature. (From Liu, W.E., and Mohney S.E., *J. Electron. Mater.*, *32*(10), 1090, 2003. With permission.)

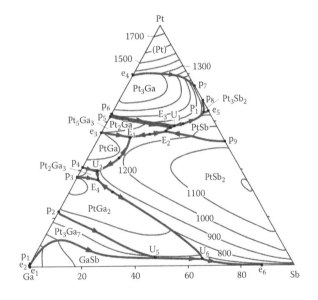

FIGURE 11.65 Calculated liquidus surface of the Ga–Pt–Sb ternary system. (From Guo, C., et al., *CALPHAD*, 52, 169, 2016. With permission.)

point in the monovariant line L ⟷ Pt₂Ga + PtSb₂. PtSb₂, PtGa₂, and Pt₃Ga₇ show extended fields of primary crystallization, whereas the primary crystallization

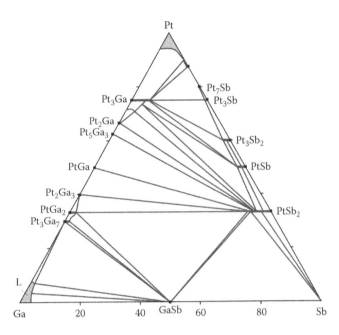

FIGURE 11.66 Calculated isothermal sections of the Ga–Pt–Sb ternary system at 500°C. (From Guo, C., et al., *CALPHAD*, *52*, 169, 2016. With permission.)

fields of PtGa and Pt₂Ga₃ are rather small. Furthermore, primary crystallization of GaSb and Sb occurs only in a very narrow composition range near the limiting binary Ga–Sb. There are 14 invariant reactions in this ternary system: 5 ternary eutectics, 2 ternary peritectics, and 7 transition reactions.

The isothermal sections of the Ga–Pt–Sb ternary system at room temperature and at 500°C were determined by Markovski et al. (1998), Richter and Ipser (1998a, 2002), and Guo et al. (2016). One of them, at 500°C, is given in Figure 11.66 (Guo et al. 2016). The three-phase region GaSb + PtGa₂ + PtSb₂ dominates in the system. The small solubilities of Sb (<1 at%) in PtGa₂ and Pt₅Ga₃ were detected (Markovski et al. 1998). PtSb₂ is in equilibrium with Pt₂Ga₃, PtGa, Pt₅Ga₃, and Pt₃Ga. No solubility of Ga in PtSb₂ was found. Pt₃Ga is in equilibrium with all Pt–Sb compounds. A clear solubility of Sb was detected in Pt₃Ga (3–4 at%). According to the calculation (Guo et al. 2016), the solid solubility of Ga in PtSb₂, PtSb, and Pt₃Sb is about 2–4 at% larger than the experimental values (Markovski et al. 1998).

CHAPTER 12

Systems Based on InN

12.1 INDIUM–CALCIUM–NITROGEN

Some compounds are formed in the In–Ca–N ternary system. **Ca$_2$InN** crystallizes in the orthorhombic structure with the lattice parameters $a = 353.2 \pm 0.3$, $b = 2020.4 \pm 1.6$, and $c = 496.9 \pm 0.4$ pm at 160 K and a calculated density of 3.915 g·cm^{-3} (Bailey and DiSalvo 2003). Dark red metallic-looking needles of the title compound were synthesized in sealed Nb tubes from the elements in molten Na with Na$_3$N as the nitrogen source. Inside an Ar-filled glove box, Na$_3$N, Ca, In, and Na were placed into a Nb tube, the amounts of which were chosen such that the atomic ratios of Na/In/Ca/N$_2$ were 6:1.5:1.4:1. The tube was sealed under 0.1 MPa of Ar and then itself sealed inside a fused silica sheath. These starting materials were heated to 800°C in 15 h, remained at temperature for 24 h, and slow cooled to 600°C at 0.2°C·h^{-1} and then to 200°C at 0.4°C·h^{-1}. At this point, the furnace was shut off and allowed to cool naturally. Unreacted Na was removed by sublimation from a mixture of NaIn, In, and dark red needle crystals by heating the niobium tube to 325°C under a pressure of 0.1 Pa for 8 h. This compound was not detected during the experiments by Kirchner et al. (2005).

Ca$_3$InN crystallizes in the cubic structure with the lattice parameter $a = 498.59 \pm 0.05$ pm (Kirchner et al. 2004).

Ca$_4$In$_2$N crystallizes in the tetragonal structure with the lattice parameters $a = 491.14 \pm 0.04$ and $c = 2907.7 \pm 0.3$ pm (Kirchner et al. 2005) ($a = 491.07 \pm 0.02$ and $c = 2911.4 \pm 0.2$ pm [Kirchner et al. 2004]; $a = 491.5 \pm 0.1$ and $c = 2910.2 \pm 0.5$ pm and a calculated density of 3.82 g·cm^{-3} [Cordier and

Rönninger 1987]). The synthesis of single-phase samples of this compound was carried out within two steps (Kirchner et al. 2005). First, a melt bead with a Ca/In molar ratio of 2:1 was obtained in an arc furnace from the elements and reacted under ambient back-pressure of nitrogen at 550°C for 1 week (Ta crucibles). For homogenization, the product was reacted under the same conditions. In a second step, the grayish powder was heated under a back-pressure of Ar at 800°C for 1 week. All manipulations were carried out in Ar-filled glove boxes.

Ca$_7$In$_{1.02-1.08}$N$_4$ crystallizes in the orthorhombic structure with the lattice parameters $a = 1168.1 \pm 0.2$, $b = 1212.9 \pm 0.2$, and $c = 364.3 \pm 0.1$ pm and a calculated density of 2.93 g·cm^{-3} (Höhn et al. 2004, 2006). Single-phase powder samples of this compound were synthesized by the reaction of appropriate mixtures of Ca$_3$N$_2$ and In in a Ta crucible at 850°C for 48 h under an Ar atmosphere. Repeated reactions with intermediate regrinding and reforming of pellets were employed. Small amounts of needlelike single crystals were prepared by using a small excess of Ca in the reaction mixture. The title compound readily hydrolyzes when exposed to air.

Ca$_{19}$In$_8$N$_7$ crystallizes in the cubic structure with the lattice parameter $a = 1471.65 \pm 0.03$ pm (Kirchner et al. 2004, 2005). Single-phase powders of the title compound were obtained by reaction of melt beads of Ca and In in the appropriate molar ratio (fused in an arc furnace) under a back-pressure of N$_2$ at 850°C for 1 week, with subsequent grinding and rereacting. This compound was first reported with the composition Ca$_{18.5}$In$_8$N$_7$.

12.2 INDIUM–STRONTIUM–NITROGEN

Two compounds, **Sr$_4$In$_2$N** and **Sr$_{19}$In$_8$N$_7$**, are formed in the In–Sr–N ternary system. Sr$_4$In$_2$N crystallizes in the tetragonal structure with the lattice parameters $a = 524.0 \pm 0.2$ and $c = 3067.0 \pm 0.7$ pm and a calculated density of 4.69 g·cm^{-3} (Cordier and Rönninger 1987). It was prepared from the elements under a N$_2$ atmosphere at 1150°C.

Sr$_{19}$In$_8$N$_7$ crystallizes in the cubic structure with the lattice parameter $a = 1561.0 \pm 0.1$ pm (Kirchner et al. 2004, 2005). Single-phase powders of the title compound were obtained by reaction of melt beads of Sr and In in the appropriate molar ratio (fused in an arc furnace) under a back-pressure of N$_2$ at 700°C for 1 week, with subsequent grinding and rereacting.

12.3 INDIUM–BARIUM–NITROGEN

Four compounds are formed in the In–Ba–N ternary system. **Ba$_6$In$_{4.78}$N$_{2.72}$** crystallizes in the cubic structure with the lattice parameter $a = 1521.6 \pm 0.1$ pm (Bailey et al. 2005). For obtaining the title compound, Ba, In, and Na (2:1:6 molar ratio) were weighed and loaded in a BN crucible. This crucible was placed into a stainless steel container and sealed under Ar, and the whole assembly was connected to N$_2$ feed line. After heating the sample to 750°C, N$_2$ was introduced into the container and the pressure was maintained at 7 MPa. The sample was heated at this temperature for 1 h and then cooled to 550°C at a rate of 2°C·h^{-1} under 7 MPa of N$_2$. Below 550°C, the sample was cooled to room temperature in the furnace by shutting off the power. The products in the crucible were washed in liquid NH$_3$. Unreacted sodium was removed by sublimation from the products by heating to 350°C under a pressure of approximately 0.1 Pa for 8 h.

Ba$_6$In$_5$N crystallizes in the tetragonal structure with the lattice parameters $a = 823.4 \pm 0.3$ and $c = 4412 \pm 2$ pm (Schlechte et al. 2004). Single crystals with a metallic luster of this compound were obtained from the reaction of melt beads of the general composition "Ba$_3$In" with a N$_2$ atmosphere at 700°C next to further ternary phases. Near-single-phase material was prepared under the same conditions, starting from melt beads with the bulk composition "Ba$_6$In$_5$" after two reannealing cycles at 700°C.

Ba$_{19}$In$_8$N$_7$ crystallizes in the cubic structure with the lattice parameter $a = 1673.6 \pm 0.1$ pm (Kirchner et al. 2004).

Ba$_{19}$In$_9$N$_9$ crystallizes in the monoclinic structure with the lattice parameters $a = 5733.4 \pm 0.6$, $b = 791.01 \pm 0.08$, and $c = 1019.91 \pm 0.10$ pm and $\beta = 97.237 \pm 0.002°$ (Yamane et al. 2004). Single crystals of the title compound were synthesized by the Na flux method under N$_2$ pressure. In an Ar-filled glove box, Ba, In, and Na (molar ratio 2:1:6) were loaded into a BN crucible. The crucible was placed into a stainless steel container and sealed with Ar gas. After heating to 750°C in a furnace, N$_2$ was introduced into the container and the pressure was maintained at 7 MPa. The sample was heated at this temperature for 1 h and then cooled to 550°C at a rate of 2°C·h^{-1} under 7 MPa of N$_2$. Below 550°C, the sample was cooled to room temperature in the furnace by shutting off the furnace power. The products in the crucible were then washed in liquid NH$_3$ to dissolve the Na flux. Single platelet crystals of Ba$_{19}$In$_9$N$_9$ with a black metallic luster were obtained, together with crystals of Ba$_6$In$_{4.77}$N$_{2.7}$.

12.4 INDIUM–THALLIUM–NITROGEN

InN–TlN. The lattice parameter of zinc blende Tl$_x$In$_{1-x}$N alloys, calculated within a supercell approach, reveals the behavior of Vegard's law (Winiarski et al. 2014). The modified Becke–Johnson potential was used for the band structure calculations. The calculations predict a linear decrease of band gap with increasing Tl content, as well as semimetallic character above $x \approx 0.25$.

12.5 INDIUM–SCANDIUM–NITROGEN

Sc$_3$InN ternary compound, which crystallizes in the cubic structure with the lattice parameter $a = 444.82 \pm 0.02$ pm, is formed in the In–Sc–N system (Kirchner et al. 2003). It is thermodynamically stable and readily reacts with N$_2$ at elevated temperatures, forming ScN and In. For obtaining the title compound, the starting materials ScN and Sc$_2$In were ground to fine powders, mixed in the appropriate molar ratio, and pressed to a pellet. The pellet was sealed in a Ta ampoule, which itself was sealed in an evacuated quartz tube, and annealed at 1400°C for 1 week. To obtain single-phase material, the resulting multiphase sample (ScN, Sc$_2$N, and Sc$_3$InN) was again ground, pelletized, and annealed at 1400°C. All manipulations were carried out under inert conditions in an Ar-filled glove box.

InN–ScN. The full-potential linearized augmented plane-wave method within the density functional theory has been applied to investigate the structural properties of $In_xSc_{1-x}N$ solid solutions in NaCl and wurtzite structures (Pérez et al. 2007, 2009). It was found that the lattice parameter a increases with the increment of the In composition in both structures, while the bulk modulus diminishes with the increase of x. The lattice constant and bulk modulus present a small bowing in both structures. It was also determined that the NaCl structure is the ground state phase for an In-composition range of $0 \leq x < 0.5$, while the wurtzite structure is most stable in the range of $0.5 \leq x \leq 1$ (Pérez et al. 2007). It was predicted that the band gap of the $In_xSc_{1-x}N$ alloys is wider when the In concentration decreases (Pérez et al. 2007). $In_xSc_{1-x}N$ alloys are direct semiconductors, while ScN is an indirect semiconductor in the wurtzite phase. A small deviation of the energy gap from the linear composition dependence was observed. The band gap was predicted to range from infrared to violet for $In_xSc_{1-x}N$ alloys in the wurtzite-like structure.

12.6 INDIUM–LANTHANUM–NITROGEN

La_3InN ternary compound, which crystallizes in the cubic structure with the lattice parameter $a = 512.75 \pm 0.02$ pm ($a = 510.2$ pm and a calculated density of 6.821 g·cm^{-3} [Zhao et al. 1995]), is formed in the In–La–N system (Kirchner et al. 2003). It is thermodynamically stable and readily reacts with N_2 at elevated temperatures, forming LaN and In. For preparation of the title compound, the starting materials LaN and La_2In were fused together in an arc furnace. The resulting bead was wrapped in Mo foil and sealed in an evacuated quartz ampoule. This assembly was annealed at 950°C for at least 1 week. All manipulations were carried out under inert conditions in an Ar-filled glove box.

12.7 INDIUM–CERIUM–NITROGEN

Ce_3InN ternary compound, which crystallizes in the cubic structure with the lattice parameter $a = 504.724 \pm 0.009$ pm, is formed in the In–Ce–N system (Kirchner et al. 2003) ($a = 504.32 \pm 0.01$ pm for $Ce_3InN_{0.9}$ [Auffermann et al. 2001]). It is thermodynamically stable and readily reacts with N_2 at elevated temperatures, forming CeN and In. For preparation of the title compound, the starting materials CeN and Ce_2In were fused together in an arc furnace. The resulting bead was wrapped in Mo foil and sealed in an evacuated quartz ampoule. This assembly was annealed at 950°C for at

least 1 week. All manipulations were carried out under inert conditions in an Ar-filled glove box.

12.8 INDIUM–PRASEODYMIUM–NITROGEN

Pr_3InN ternary compound, which crystallizes in the cubic structure with the lattice parameter $a = 500.95 \pm 0.04$ pm, is formed in the In–Pr–N system (Kirchner et al. 2003). It is thermodynamically stable and readily reacts with N_2 at elevated temperatures, forming PrN and In. For preparation of the title compound, the starting materials PrN and Pr_2In were fused together in an arc furnace. The resulting bead was wrapped in Mo foil and sealed in an evacuated quartz ampoule. This assembly was annealed at 950°C for at least 1 week. All manipulations were carried out under inert conditions in an Ar-filled glove box.

12.9 INDIUM–NEODYMIUM–NITROGEN

Nd_3InN ternary compound, which crystallizes in the cubic structure with the lattice parameter $a = 497.96 \pm 0.02$ pm ($a = 494.9$ pm [Haschke et al. 1967]), is formed in the In–Nd–N system (Kirchner et al. 2003). It is thermodynamically stable and readily reacts with N_2 at elevated temperatures, forming NdN and In. For preparation of the title compound, the starting materials NdN and Nd_2In were fused together in an arc furnace. The resulting bead was wrapped in Mo foil and sealed in an evacuated quartz ampoule. This assembly was annealed at 950°C for at least 1 week. All manipulations were carried out under inert conditions in an Ar-filled glove box.

12.10 INDIUM–SAMARIUM–NITROGEN

Sm_3InN ternary compound, which crystallizes in the cubic structure with the lattice parameter $a = 490.77 \pm 0.01$ pm, is formed in the In–Sm–N system (Kirchner et al. 2003). It is thermodynamically stable and readily reacts with N_2 at elevated temperatures, forming SmN and In. For obtaining the title compound, the starting materials SmN and Sm_2In were ground to fine powders, mixed in the appropriate molar ratio, and pressed to a pellet. The pellet was sealed in a Ta ampoule, which itself was sealed in an evacuated quartz tube, and annealed at 950°C for 1 week. To obtain single-phase material, the resulting multiphase sample (SmN, Sm_2N, and Sm_3InN) was again ground, pelletized, and annealed at 950°C. All manipulations were carried out under inert conditions in an Ar-filled glove box.

12.11 INDIUM–GADOLINIUM–NITROGEN

Gd$_3$InN ternary compound, which crystallizes in the cubic structure with the lattice parameter a = 486.27 ± 0.03 pm, is formed in the In–Gd–N system (Kirchner et al. 2003). It is thermodynamically stable and readily reacts with N$_2$ at elevated temperatures, forming GdN and In. For obtaining the title compound, the starting materials GdN and Gd$_2$In were ground to fine powders, mixed in the appropriate molar ratio, and pressed to a pellet. The pellet was sealed in a Ta ampoule, which itself was sealed in an evacuated quartz tube, and annealed at 950°C for 1 week. To obtain single-phase material, the resulting multiphase sample (GdN, Gd$_2$N, and Gd$_3$InN) was again ground, pelletized, and annealed at 950°C. All manipulations were carried out under inert conditions in an Ar-filled glove box.

12.12 INDIUM–TERBIUM–NITROGEN

Tb$_3$InN ternary compound, which crystallizes in the cubic structure with the lattice parameter a = 481.26 ± 0.03 pm, is formed in the In–Tb–N system (Kirchner et al. 2003). It is thermodynamically stable and readily reacts with N$_2$ at elevated temperatures, forming TbN and In. For obtaining the title compound, the starting materials TbN and Tb$_2$In were ground to fine powders, mixed in the appropriate molar ratio, and pressed to a pellet. The pellet was sealed in a Ta ampoule, which itself was sealed in an evacuated quartz tube, and annealed at 950°C for 1 week. To obtain single-phase material, the resulting multiphase sample (TbN, Tb$_2$N, and Tb$_3$InN) was again ground, pelletized, and annealed at 950°C. All manipulations were carried out under inert conditions in an Ar-filled glove box.

12.13 INDIUM–DYSPROSIUM–NITROGEN

Dy$_3$InN ternary compound, which crystallizes in the cubic structure with the lattice parameter a = 478.59 ± 0.06 pm, is formed in the In–Dy–N system (Kirchner et al. 2003). It is thermodynamically stable and readily reacts with N$_2$ at elevated temperatures, forming DyN and In. For obtaining the title compound, the starting materials DyN and Dy$_2$In were ground to fine powders, mixed in the appropriate molar ratio, and pressed to a pellet. The pellet was sealed in a Ta ampoule, which itself was sealed in an evacuated quartz tube, and annealed at 950°C for 1 week. To obtain single-phase material, the resulting multiphase sample (DyN, Dy$_2$N, and Dy$_3$InN)

was again ground, pelletized, and annealed at 950°C. All manipulations were carried out under inert conditions in an Ar-filled glove box.

12.14 INDIUM–HOLMIUM–NITROGEN

Ho$_3$InN ternary compound, which crystallizes in the cubic structure with the lattice parameter a = 476.11 ± 0.03 pm, is formed in the In–Ho–N system (Kirchner et al. 2003). It is thermodynamically stable and readily reacts with N$_2$ at elevated temperatures, forming HoN and In. For obtaining the title compound, the starting materials HoN and Ho$_2$In were ground to fine powders, mixed in the appropriate molar ratio, and pressed to a pellet. The pellet was sealed in a Ta ampoule, which itself was sealed in an evacuated quartz tube, and annealed at 950°C for 1 week. To obtain single-phase material, the resulting multiphase sample (HoN, Ho$_2$N, and Ho$_3$InN) was again ground, pelletized, and annealed at 950°C. All manipulations were carried out under inert conditions in an Ar-filled glove box.

12.15 INDIUM–ERBIUM–NITROGEN

Er$_3$InN ternary compound, which crystallizes in the cubic structure with the lattice parameter a = 473.66 ± 0.03 pm, is formed in the In–Er–N system (Kirchner et al. 2003). It is thermodynamically stable and readily reacts with N$_2$ at elevated temperatures, forming ErN and In. For obtaining the title compound, the starting materials ErN and Er$_2$In were ground to fine powders, mixed in the appropriate molar ratio, and pressed to a pellet. The pellet was sealed in a Ta ampoule, which itself was sealed in an evacuated quartz tube, and annealed at 950°C for 1 week. To obtain single-phase material, the resulting multiphase sample (ErN, Er$_2$N, and Er$_3$InN) was again ground, pelletized, and annealed at 950°C. All manipulations were carried out under inert conditions in an Ar-filled glove box.

12.16 INDIUM–THULIUM–NITROGEN

Tm$_3$InN ternary compound, which crystallizes in the cubic structure with the lattice parameter a = 470.80 ± 0.02 pm, is formed in the In–Tm–N system (Kirchner et al. 2003). It is thermodynamically stable and readily reacts with N$_2$ at elevated temperatures, forming TmN and In. For obtaining the title compound, the starting materials TmN and Tm$_2$In were ground to fine powders, mixed in the appropriate molar ratio, and pressed to a pellet. The pellet was sealed in a Ta ampoule, which itself was sealed in an evacuated quartz tube, and annealed at 950°C for 1 week. To obtain single-phase material, the resulting multiphase sample (TmN, Tm_2N, and Tm_3InN) was again ground, pelletized, and annealed at 950°C. All manipulations were carried out under inert conditions in an Ar-filled glove box.

12.17 INDIUM–LUTETIUM–NITROGEN

Lu_3InN ternary compound, which crystallizes in the cubic structure with the lattice parameter $a = 467.16 \pm 0.03$ pm, is formed in the In–Lu–N system (Kirchner et al. 2003). It is thermodynamically stable and readily reacts with N_2 at elevated temperatures, forming LuN and In. For preparation of the title compound, the starting materials LuN and Lu_2In were fused together in an arc furnace. The resulting bead was wrapped in Mo foil and sealed in an evacuated quartz ampoule. This assembly was annealed at 950°C for at least 1 week. All manipulations were carried out under inert conditions in an Ar-filled glove box.

12.18 INDIUM–CARBON–NITROGEN

$In(CN)_3$ and $In_{2.24}(NCN)_3$ ternary compounds are formed in the In–C–N system. $In(CN)_3$ ternary compound, which sublimes at 340°C–350°C and exists in two polymorphic modifications, is formed in this system (Goggin et al. 1966; Williams et al. 1998).

The *first modification* crystallizes in the cubic structure with the lattice parameter $a = 562.7 \pm 0.1$ pm (Williams et al. 1998). For obtaining it, a solution of Me_3SiCN (17.2 mM) in hexane (20 mL) at 0°C was added dropwise to a suspension of $InCl_3$ (5.2 mM) in diethyl ether (30 mL). The mixture was stirred at room temperature for 12 h, during which time it gradually darkened to a light brown color. The precipitate was isolated by filtration and was pumped at 40°C under vacuum for several hours. The colorless product was heated in hexane for 12 h, and subsequently, it was extracted several times with warm hexane to remove the soluble impurities. Heating at 160°C–200°C under vacuum for 12 h removed the remaining volatile components to yield a solid cubic $In(CN)_3$.

The *second modification* crystallizes in the monoclinic structure with the lattice parameters $a = 680 \pm 5$, $b = 1190 \pm 5$, and $c = 697 \pm 5$ pm and $\beta = 116.8 \pm 0.2°$ (Goggin et al. 1966). In moist air, this modification is extremely deliquescent and hydrolyzed. It could be prepared by the following four methods:

1. InOI (0.5 g) was placed in a porcelain boat and heated at 300°C in a stream of cyanogen gas for 3 h. ICN sublimed from the solid, which changed from the red color of InOI at 300°C to a colorless product. The temperature of the reaction tube was increased to 350°C for a further 3 h, and a white sublimate began to form at the cooler end. The reaction mixture was allowed to cool in a stream of N_2, and the tube was then stoppered. The white crystalline sublimate was scraped from the surface of the glass in a dry box.

2. $Hg(CN)_2$ (0.5 g) was dissolved in liquid NH_3 (20 mL), and In shavings (1.0 g) were added. The tube was closed with a Bunsen valve and maintained at 70°C for 12 h. Fresh In shavings were added, and the reaction was allowed to proceed for a further 2 days. The liquid was decanted and the ammonia allowed to evaporate at room temperature. The solid residue was heated at 100°C under a vacuum for 2 h, after which time the solid was formed. It was transferred to a combustion tube and heated at 350°C in a stream of N_2. Some white crystalline sublimate of monoclinic modification was formed.

3. The reaction of In with HCN at 350°C for 5 h leads to the formation of a small quantity of this modification.

4. The reaction of $In(OH)_3$ and HCN also gives the monoclinic modification of the title compound. Air was bubbled through liquid HCN and acted as a carrier gas. If contact between the air–HCN mixture was allowed to take place for 16 h at 100°C and the temperature was then increased to 350°C in the reaction zone, two white sublimates could be obtained in the cooler parts of the combustion tube. One of these sublimates, which consisted of colorless platelike crystals, was the title modification of $In(CN)_3$.

$In_{2.24}(NCN)_3$ crystallizes in the rhombohedral structure with the lattice parameters $a = 606.09 \pm 0.04$ and $c = 2884 \pm 2$ pm (in hexagonal setting) and a calculated density of 4.099 g.cm^{-3} (Dronskowski 1995). Single crystals of this compound can be synthesized from a reaction between InBr and NaCN at 400°C, followed by chemical transport at 400°C–500°C. These crystals seem to exhibit a moderate sensitivity with respect to air or humidity. Therefore, all subsequent operations were performed under protective gas (argon).

12.19 INDIUM–SILICON–NITROGEN

Phase equilibria in the In–Si–N ternary system at 120°C were investigated by means of x-ray diffraction (XRD) (Weitzer et al. 1990). No ternary compound was observed. InN coexists at this temperature with Si_3N_4 and Si_3N_4 and is in thermodynamic equilibrium with In.

12.20 INDIUM–TITANIUM–NITROGEN

Ti_3InN ternary compound, which has two polymorphic modifications, is formed in the In–Ti–N system. The *first modification* crystallizes in the cubic structure with the lattice parameter a = 419.0 pm and a calculated density of 6.15 g·cm^{-3} (Jeitschko et al. 1964b). It was produced by sintering of TiN, Ti, and In in closed quartz tube at 750°C for 600 h.

The *second modification* crystallizes in the hexagonal structure with the lattice parameters a = 307.4 and c = 1397.5 pm and a calculated density of 6.52 g·cm^{-3} (Jeitschko et al. 1964a) (a = 308 and c = 1408 pm [Ivanovskii et al. 2000]). It was obtained by the interaction of InN with Ti in the quartz ampoules at 500°C for 850 h (Jeitschko et al. 1964a).

According to the calculations, **Ti_2InN** could also exist in this system (Benayad et al. 2011). It will be crystallized in the hexagonal structure with the calculated lattice parameters a = 303.3 and c = 1372.7 pm (a = 307 and c = 1397 pm and a calculated density of 6.54 g·cm^{-3} [Barsoum 2000]) and calculated bulk modulus B = 121.78 GPa. The structural, elastic, and electronic properties of this compound have been calculated using the full-potential linear muffin tin orbital method.

12.21 INDIUM–ZIRCONIUM–NITROGEN

Zr_2InN ternary compound, which crystallizes in the hexagonal structure with the lattice parameters a = 327 and c = 1483 pm and a calculated density of 7.53 g·cm^{-3} (a = 327.7 and c = 1484 pm and a calculated density of 7.49 g·cm^{-3} for Zr_3InN compound [Jeitschko et al. 1964a]), is formed in the In–Zr–N system (Barsoum 2000). It was obtained by the interaction of InN with Zr in the quartz ampoules at 500°C for 850 h (Jeitschko et al. 1964a).

12.22 INDIUM–HAFNIUM–NITROGEN

Hf_2InN ternary compound is formed in the In–Hf–N system (Jeitschko et al. 1964d).

12.23 INDIUM–PHOSPHOR–NITROGEN

The calculated temperature dependence of nitrogen solubility in InP is shown in Figure 12.1 (Ho and Stringfellow 1996a). The calculations indicate that the solubility of nitrogen increases significantly with increasing temperature.

The N incorporation behavior in InP grown by gas source molecular beam epitaxy (MBE) using a nitrogen radical beam source has been investigated (Bi and Tu 1996, 1997). The nitrogen composition was found to be different from the N_2 flow rate fraction in the gas phase. With an increasing N_2 flow rate fraction, the nitrogen composition increases up to certain point and then saturates. The nitrogen incorporation can be increased by lowering the growth temperature. As the substrate temperature was increased from 310°C to 420°C, the nitrogen concentration decreased from 0.93 to 0.44 at%. Optical studies on these samples show that adding a small amount of nitrogen results in a large band gap bowing.

12.24 INDIUM–ARSENIC–NITROGEN

The calculated temperature dependence of nitrogen solubility in InAs is shown in Figure 12.1 (Ho and Stringfellow 1996a). The calculations indicate that the solubility of nitrogen increases significantly with increasing temperature.

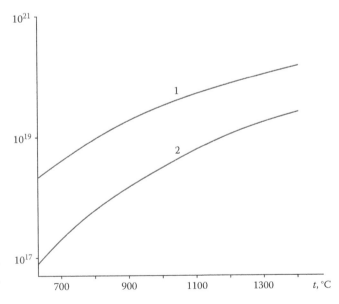

FIGURE 12.1 Calculated temperature dependence of nitrogen solubility in (1) InP and (2) InAs. (From Ho, I.H., and Stringfellow, G.B., *Mater. Res. Soc. Symp. Proc.*, *449*, 871, 1996. With permission.)

InN–InAs. InN_xAs_{1-x} solid solution films have been successfully grown using active nitrogen supplied by a radio-frequency-activated plasma source (Foxon et al. 1995). Significant amounts of nitrogen (≈ 20 at%) were incorporated in such films. It was shown than the band gap decreases or increases rapidly with increasing or decreasing compositions of nitrogen (Benaissa et al. 2009). As a consequence, the optical band gap bowing was found to be strong and composition dependent. The obtained solid solutions may change from semiconducting to metal (passing to a negative bowing). The calculation was based on the density functional theory in the local density approximation.

12.25 INDIUM–OXYGEN–NITROGEN

In_3O_3N ternary compound could exist in the In–O–N system. It could crystallize in the cubic structure with the lattice parameter $a = 903.7$ pm (local density approximation) and $a = 928.5$ pm (generalized gradient approximation) and an energy gap of 0.40 eV (Okeke and Lowthe 2008).

12.26 INDIUM–CHROMIUM–NITROGEN

According to the data of Beznosikov (2003), Cr_3InN compound, which crystallizes in the cubic structure of perovskite type, could form in the In-Mn-N ternary system.

12.27 INDIUM–MANGANESE–NITROGEN

According to the data of Beznosikov (2003), Mn_3InN compound, which crystallizes in the cubic structure of perovskite type, could form in the In–Mn–N ternary system.

12.28 INDIUM–IRON–NITROGEN

Fe_3InN ternary compound exists in the In-Fe-N ternary system. To produce the title compound, mechanical alloying can be used (Kuhnen et al. 2000). It decomposes under high vacuum very quickly.

12.29 INDIUM–COBALT–NITROGEN

Two ternary compounds, $Co_{71}In_{18}N_{11}$ and $Co_{6.1}In_{20.2}N_{18.3}$, which crystallize in the cubic structure with the lattice parameter $a = 385$ pm and $a = 386$ pm, respectively, is formed in the In–Co–N system (Stadelmaier and Fraker 1962).

According to the data of Beznosikov (2003), Co_3InN compound, which crystallizes in the cubic structure of perovskite type, could form in the Ga–Co–N ternary system.

12.30 INDIUM–NICKEL–NITROGEN

$Ni_{65.7}In_{21.9}N_{12.4}$ ternary compound, which crystallizes in the cubic structure with the lattice parameter $a = 384.4$ pm, is formed in the In–Ni–N system (Stadelmaier and Fraker 1962).

According to the data of Beznosikov (2003), Ni_3InN compound, which crystallizes in the cubic structure of perovskite type, could form in the In–Ni–N ternary system.

CHAPTER 13

Systems Based on InP

13.1 INDIUM–SODIUM–PHOSPHORUS

Na₃InP₂ ternary compound, which crystallizes in the monoclinic structure with the lattice parameters $a = 940.1 \pm 0.5$, $b = 737.1 \pm 0.4$, and $c = 1535.8 \pm 0.4$ pm and $\beta = 92.4 \pm 0.1°$, is formed in the In–Na–P system (Blase et al. 1991h). It was prepared from the elements at 600°C in sealed Nb ampoules.

13.2 INDIUM–POTASSIUM–PHOSPHORUS

Two ternary compounds, **K₃InP₂** and **K₆InP₃**, are formed in the In–K–P system. K₃InP₂ crystallizes in the orthorhombic structure with the lattice parameters $a = 1448.9 \pm 0.6$, $b = 765.8 \pm 0.4$, and $c = 681.6 \pm 0.5$ pm and a calculated density of 2.58 g·cm⁻³ (Blase et al. 1991g; Ohse et al. 1993). It was prepared from the elements at 600°C–730°C in sealed Nb ampoules.

K₆InP₃ ternary compound crystallizes in the triclinic structure with the lattice parameters $a = 888.0 \pm 0.5$, $b = 893.5 \pm 0.4$, and $c = 995.2 \pm 0.6$ pm and $\alpha = 82.0 \pm 0.1°$, $\beta = 73.1 \pm 0.1°$, and $\gamma = 60.5 \pm 0.1°$ (Blase et al. 1991f). It was prepared from the elements at 600°C in sealed Nb ampoules.

13.3 INDIUM–RUBIDIUM–PHOSPHORUS

Rb₁₂In₄P₈ ternary compound, which crystallizes in the triclinic structure with the lattice parameters $a = 939.7 \pm 0.1$, $b = 1250.0 \pm 0.1$, and $c = 1592.7 \pm 0.3$ pm and $\alpha = 97.16 \pm 0.01°$, $\beta = 107.00 \pm 0.01°$, and $\gamma = 106.72 \pm 0.01°$, is formed in the In–Rb–P system (Somer et al. 1998b). It was prepared from a mixture of Rb, InP, and P (molar ratio 4.5:1:1) in sealed Nb ampoules

at 700°C as black plates. The excess of Rb was removed in high vacuum at 200°C.

13.4 INDIUM–CESIUM–PHOSPHORUS

Cs₁₂In₄P₈ ternary compound, which crystallizes in the triclinic structure with the lattice parameters $a = 966.2 \pm 0.6$, $b = 1288.4 \pm 0.5$, and $c = 1584.0 \pm 0.8$ pm and $\alpha = 81.1 \pm 0.1°$, $\beta = 81.6 \pm 0.1°$, and $\gamma = 70.7 \pm 0.1°$, is formed in the In–Cs–P system (Blase et al. 1991c). It was prepared from the elements at 600°C in sealed Nb ampoules.

13.5 INDIUM–COPPER–PHOSPHORUS

An isothermal section of the In–Cu–P ternary system at 500°C is given in Figure 13.1 (Mohney 1998). The dashed lines represent the most probable tie-lines where the phase equilibria were not conclusively demonstrated. Phase equilibria were experimentally investigated by mixing Cu and InP to provide samples of desired compositions. Samples were annealed at 500°C for 2 months.

According to the data of Lange et al. (2008a, 2008b), **Cu₅InP₁₆** ternary compound, which crystallizes in the monoclinic structure with the lattice parameters $a = 1112.4 \pm 0.3$, $b = 966.3 \pm 0.3$, and $c = 753.3 \pm 0.2$ pm and $\beta = 109.96 \pm 0.01°$ and a calculated density of 4.07 g cm⁻³, is formed in this system. It was synthesized as dark red, nontransparent crystals by reacting Cu, In, and red P (molar ratio 5:1:16) in evacuated silica ampoules. CuI (10 mg) was added to a total of 500 mg of starting materials. After an initial heating step to 280°C for 8 h, the mixture was kept at 550°C for 14 days.

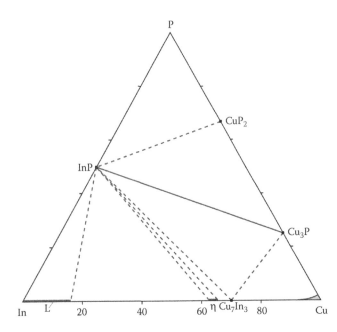

FIGURE 13.1 Isothermal section of the In–Cu–P system at 500°C. (From Mohney, S.E., *J. Electron. Mater.*, 27(1), 26, 1998. With permission.)

13.6 INDIUM–GOLD–PHOSPHORUS

The thermal reaction between Au and InP causes the decomposition of InP at the interface between the contact and the substrate (Piotrowska et al. 1981). Initial reactions cause a progressive dissolution of In in the Au film and formation of the $Ag_{0.9}In_{0.1}$ solid solution. During further annealing, two crystalline phases are formed: Au_2P_3 and Au_3In.

The metallurgical structure of the Au/InP contacts annealed at 320°C–360°C was investigated by the combined use of x-ray diffraction (XRD), transmission electron microscopy (TEM), scanning electron microscopy (SEM), and 2 MeV He⁺ backscattering spectrometry.

13.7 INDIUM–CALCIUM–PHOSPHORUS

Two ternary compounds, **$CaIn_2P_2$** and **Ca_3InP_3**, are formed in the In–Ca–P system. $CaIn_2P_2$ crystallizes in the hexagonal structure with the lattice parameters $a = 402.20 \pm 0.17$ and $c = 1740.8 \pm 0.8$ pm at 90 K (Rauscher et al. 2009) ($a = 403.82$ and $c = 1734.71$ pm, an energy gap of 0.39 eV, a bulk modulus of 58.44 GPa, and a calculation density of 4.50 g·cm⁻³ [Guechi et al. 2013]).

Crystals of the title compound were grown by reaction of the pure elements in an alumina crucible with a Ca/In/P molar ratio 3:110:6 (Rauscher et al. 2009). Ca and In were cut into small pieces and layered in a

5 mL crucible with In on the top and bottom, and sealed under vacuum in quartz tubes. All materials were handled under an inert atmosphere or under vacuum. The reaction contents were heated in a furnace at a rate of 60°C·h⁻¹ to 1100°C, maintained at that temperature for 16 h, cooled at a rate of 2°C·h⁻¹ to 873°C, maintained at that temperature for 24 h, and centrifuged to remove molten In flux. The crystals were removed in air. The reaction yielded a few highly reflective platelike crystals on the sides and bottom of the crucible. These crystals were found to be air stable.

Structural, elastic, electronic, and optical properties of $CaIn_2P_2$ were studied by means of first-principles calculations based on the density functional theory (Guechi et al. 2013).

Ca_3InP_3 crystallizes in the orthorhombic structure with the lattice parameters $a = 1201.9 \pm 0.5$, $b = 413.8 \pm 0.2$, and $c = 1346.0 \pm 0.5$ pm and a calculated density of 3.25 g·cm⁻³ (Cordier et al. 1985b). It was synthesized from a mixture of Ca, In, and P (molar ratio 4:1:3) in an Ar atmosphere in a corundum crucible heated to 1000°C at a rate of 100°C·h⁻¹ and then cooled at a rate of 50°C·h⁻¹.

13.8 INDIUM–STRONTIUM–PHOSPHORUS

Three compounds, **$SrIn_2P_2$**, **Sr_3InP_3**, and **$Sr_3In_2P_4$**, are formed in the In–Sr–P ternary system.

$SrIn_2P_2$ ternary compound crystallizes in the hexagonal structure with the lattice parameters $a = 409.45 \pm 0.16$ and $c = 1781.2 \pm 0.7$ pm at 90 K (Rauscher et al. 2009) ($a = 410.99$ and $c = 1777.09$ pm, an energy gap of 0.28 eV, a bulk modulus of 55.80 GPa, and a calculation density of 4.84 g·cm⁻³ [Guechi et al. 2013]).

Crystals of the title compound were grown by reaction of the pure elements in an alumina crucible with a Sr/In/P molar ratio 3:110:6 (Rauscher et al. 2009). Sr and In were cut into small pieces and layered in a 5 mL crucible with In on the top and bottom, and sealed under vacuum in quartz tubes. All materials were handled under an inert atmosphere or under vacuum. The reaction contents were heated in a furnace at a rate of 60°C·h⁻¹ to 1100°C, maintained at that temperature for 16 h, cooled at a rate of 2°C·h⁻¹ to 873°C, maintained at that temperature for 24 h, and centrifuged to remove molten In flux. The crystals were removed in air. The reaction yielded a few highly reflective platelike crystals on the sides and bottom of the crucible. These crystals were found to be air stable.

Structural, elastic, electronic, and optical properties of $SrIn_2P_2$ were studied by means of first-principles calculations based on the density functional theory (Guechi et al. 2013).

Sr_3InP_3 crystallizes in the orthorhombic structure with the lattice parameters $a = 1280.9 \pm 0.5$, $b = 433.0 \pm 0.2$, and $c = 1386.6 \pm 0.5$ pm and a calculated density of 4.06 g·cm^{-3} (Cordier et al. 1987). The preparation of this compound was achieved by direct synthesis from elements. The stoichiometric mixture was additionally sealed in quartz ampoules under Ar atmosphere and heated to 1000°C at a rate of 15°C·h^{-1}, maintained at this temperature for 1 h, and cooled at a rate of 50°C·h^{-1}.

$Sr_3In_2P_4$ crystallizes in the orthorhombic structure with the lattice parameters $a = 1632.3 \pm 0.6$, $b = 682.8 \pm 0.3$, and $c = 428.9 \pm 0.2$ pm and a calculated density of 4.28 g·cm^{-3} (Cordier et al. 1986d). It was prepared from the elements in an Ar atmosphere at 980°C.

13.9 INDIUM–BARIUM–PHOSPHORUS

Four compounds, $BaIn_2P_2$, $Ba_2In_2P_5$, $Ba_3In_2P_4$, and $Ba_{14}InP_{11}$, are formed in the In–Ba–P ternary system.

$BaIn_2P_2$ crystallizes in the monoclinic structure with the lattice parameters $a = 996.52 \pm 0.08$, $b = 417.89 \pm 0.03$, and $c = 1298.34 \pm 0.10$ pm and $\beta = 95.326 \pm 0.002°$ at 90 K (Rauscher et al. 2009). Crystals of the title compound were grown by reaction of the pure elements in an alumina crucible with an Ba/In/P molar ratio 3:110:6 (Rauscher et al. 2009). Ba and In were cut into small pieces and layered in a 5 mL crucible with In on the top and bottom, and sealed under vacuum in quartz tubes. All materials were handled under an inert atmosphere or under vacuum. The reaction contents were heated in a furnace at a rate of 60°C·h^{-1} to 1100°C, maintained at that temperature for 16 h, cooled at a rate of 2°C·h^{-1} to 873°C, maintained at that temperature for 24 h, and centrifuged to remove molten In flux. The crystals were removed in air. The reaction yielded black needle-shaped crystals. These crystals were found to be air stable.

$Ba_2In_2P_5$ melts incongruently at 935°C and crystallizes in the orthorhombic structure with the lattice parameters $a = 1725.7 \pm 0.3$, $b = 415.99 \pm 0.06$, and $c = 1749.0 \pm 0.6$ pm, a calculated density of 5.31 g·cm^{-3}, and an energy gap of 0.6 eV (Mathieu et al. 2008). This compound was originally made in a reaction of the elements Ba, In, Ge, and P (molar ratio 0.5:15:1:1), with In acting as a flux. All elements were combined in an alumina crucible, with half the In flux loaded on the bottom and half on top of the rest of the reactants. The crucible was placed in a fused silica tube; another alumina crucible was filled with ceramic fiber and inverted on top of the reaction crucible in the silica tube to act as a filter during centrifugation. The fused silica tube was sealed under a vacuum of 1.3 Pa, and then heated at 60°C·h^{-1} to 500°C and held at this temperature for 1 h to facilitate mixing; heating continued at 60°C·h^{-1} to 1000°C, where the reaction was kept for 48 h. It was then cooled at a rate of 2°C·h^{-1} to 600°C and centrifuged to remove excess molten flux from the mass of needle-shaped crystalline product. Because Ge was not incorporated into the product, the synthesis was done again without this element. The title compound formed as very thin silvery needles with metallic luster and is stable under N_2 until about 900°C.

$Ba_3In_2P_4$ crystallizes in the monoclinic structure with the lattice parameters $a = 1397.4 \pm 0.1$, $b = 1086.9 \pm 0.1$, and $c = 706.4 \pm 0.1$ pm and $\beta = 90.27 \pm 0.01°$ ($\beta = 89.73°$ [Carrillo-Cabrera et al. 1996]) (Somer et al. 1998a). It was obtained as a by-product during attempts to synthesize $Ba_{14}InP_{11}$ by varying the ratio of BaP, In, and P, as well as the reaction conditions. Pure title compound could be prepared from a mixture of BaP and InP (molar ratio 3:2) in a sealed Nb ampoule at 1050°C.

$Ba_{14}InP_{11}$ crystallizes in the tetragonal structure with the lattice parameters $a = 1709.7 \pm 0.2$ and $c = 2294.0 \pm 0.5$ pm, an energy gap of 1.6 eV, and a calculated density of 4.711 g·cm^{-3} (Carrillo-Cabrera et al. 1996). It was prepared from elements in a sealed stainless steel ampoule at 1070°C for 2–5 days. In the starting nominal composition, an excess of In and P was necessary in order to get a single-phase final product. The title compound is black, brittle, and very sensitive to air and moisture.

13.10 INDIUM–ZINC–PHOSPHORUS

InP–Zn. Transmission electron microscopy and evolved gas analysis (EGA) were used to study an interaction between Zn metallization and InP substrate (Nakahara et al. 1984). In addition to the rapid Zn lattice diffusion into InP crystal, the formation of Zn_3P_2 near the interface region has been detected. Because of its poor thermal stability in the studied temperature range, the formation of this phase was found not to be readily reproducible. Due to the intrinsic irreproducible nature, the exact condition for this phase formation has not been established. EGA has indicated, however, that this phase, if formed, can be decomposed at a temperature over 450°C.

The effective distribution coefficient of Zn in the crystals of InP is equal to approximately 1 (Nashelskiy et al. 1982).

13.11 INDIUM–CADMIUM–PHOSPHORUS

The liquidus surface of the In–Cd–P ternary system is characterized by nine fields of primary crystallization of InP, Cd_3P_2, Cd_6P_9, Cd_7P_{10}, CdP_2, CdP_4, Cd, P, and In (Figure 13.2) (Marenkin et al. 1988a; Marenkin 1993). InP occupies the biggest field of primary crystallization. Seven nonvariant points exist in this system, four of which are the eutectics (E_1 at 125.3°C, E_2 at 315.8°C, E_3 at 696°C, and E_4 at 597°C) and three the transition points (U_1 at 723°C, U_2 at 731°C, and U_3 763°C). There are also two Van Rein settle points (e_7 at 316°C and e_8 at 768°C) and one synthectic point (S at 700°C) in this system.

InP–Cd. The phase diagram of this system is a eutectic type (Figure 13.3) (Marenkin et al. 1987b, 1988a; Marenkin 1993). The eutectic contains 0.35 mol% InP and crystallizes at 316°C.

The Cd solubility in InP was investigated using radioactive isotopes. Within the temperature interval from

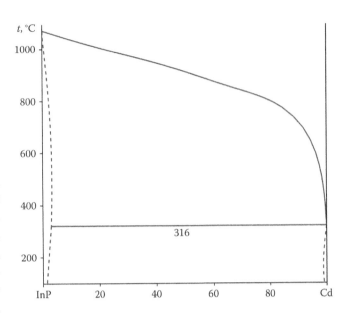

FIGURE 13.3 Phase diagram of the InP–Cd system. (From Marenkin, S.F., et al., *Izv. AN SSSR Neorgan. Mater.*, 24(7), 1075, 1988. With permission.)

650°C to 900°C, the solubility changes from $8.8 \cdot 10^{18}$ cm^{-3} to $6.8 \cdot 10^{19}$ cm^{-3} (Figure 13.4) (Arseni et al. 1967). The temperature dependence of Cd solubility in InP is retrograde with a maximum near 800°C.

The liquidus temperatures along the InP–Cd tie-line have been measured for dilute solutions corresponding to a linear relationship $lnx_{InP} = -(\Delta H^s/RT) + C$, where x_{InP} is the molar fraction of InP. Figure 13.5 shows a plot

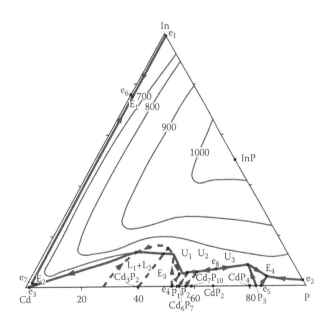

FIGURE 13.2 Liquidus surface of the In–Cd–P ternary system. (From Marenkin, S.F., et al., *Izv. AN SSSR Neorgan. Mater.*, 24(7), 1075, 1988. With permission.)

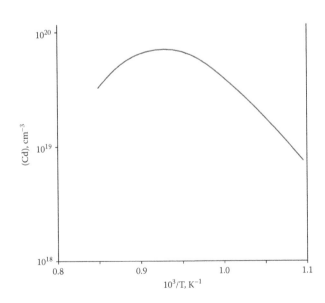

FIGURE 13.4 Temperature dependence of the Cd solubility in InP. (From Arseni, K.A., et al., *Izv. AN SSSR Neorgan. Mater.*, 3(9), 1679, 1967. With permission.)

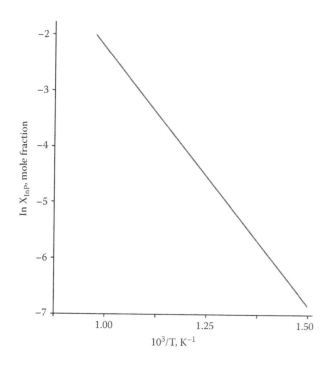

FIGURE 13.5   Solubility of InP in Cd as a function of temperature. (From Buehler, E., and Bachmann, K.J., *J. Cryst. Growth*, 35(1), 60, 1976. With permission.)

of x_{InP} versus T^{-1} for solutions of InP in Cd (Buehler and Bachmann 1976).

The effective distribution coefficient of Cd in the crystals of InP is equal to 0.15 (Nashelskiy et al. 1982).

InP–CdP₂. The phase diagram of this system is a eutectic type (Figure 13.6) (Marenkin et al. 1988a;

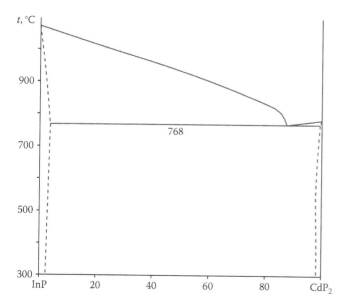

FIGURE 13.6   Phase diagram of the InP–CdP₂ system. (From Marenkin, S.F., et al., *Izv. AN SSSR Neorgan. Mater.*, 24(7), 1075, 1988. With permission.)

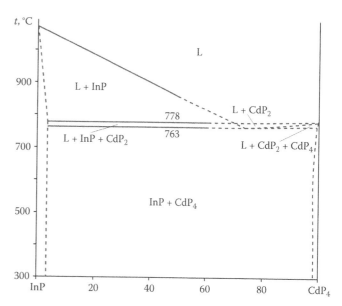

FIGURE 13.7   Phase relations in the InP–CdP₄ system. (From Marenkin, S.F., et al., *Izv. AN SSSR Neorgan. Mater.*, 24(7), 1075, 1988. With permission.)

Marenkin 1993). The eutectic contains 12.5 mol% InP and crystallizes at 768°C.

InP–CdP₄. The phase relations in this system are shown in Figure 13.7 (Marenkin et al. 1988a; Marenkin 1993). InP is the primary phase in the alloys containing 0–80 mol% CdP₄. For these alloys, InP and CdP₂ cocrystallization takes place at 778°C. Cocrystallization of CdP₂ and CdP₄ for the alloys with a composition from 80 to 100 mol% CdP₄ takes place within the interval from 780°C to 763°C. The crystallization of all alloys ends at 763°C.

InP–Cd₃P₂. The phase relations in this system are shown in Figure 13.8 (Marenkin et al. 1988a; Marenkin 1993). The crystallization of all alloys of this section ends at 700°C according to the synthectic reaction L₁ + L₂ ⟷ InP + Cd₃P₂. The synthectic contains 73 mol% Cd₃P₂ and crystallizes at 741°C within the interval from 27 to 100 mol% InP. A miscibility gap with the critical temperature of 810°C exists within the interval from 71 to 100 mol% Cd₃P₂.

The In–Cd–P ternary system was investigated using differential thermal analysis (DTA), XRD, metallography, chemical analysis, and measurement of microhardness (Marenkin et al. 1987b, 1988a; Marenkin 1993).

13.12 INDIUM–THALLIUM–PHOSPHORUS

InP–TlP. In$_x$Tl$_{1-x}$P solid solutions (0 < x < 1) were grown by gas source molecular beam epitaxy (MBE)

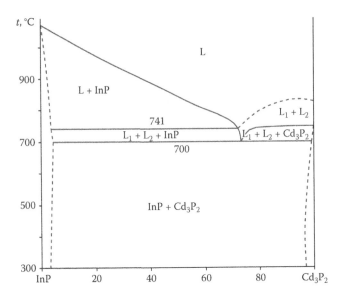

FIGURE 13.8 Phase relations in the InP–Cd$_3$P$_2$ system. (From Marenkin, S.F., et al., *Izv. AN SSSR Neorgan. Mater.*, 24(7), 1075, 1988. With permission.)

on InP substrate (Asahia et al. 1997). No phase separation was observed. A number of key optical and structural properties of In$_{1-x}$Tl$_x$P solid solutions were studied within the local density functional theory (van Schilfgaarde et al. 1994). In$_{1-x}$Tl$_x$P at $x = 0.67$ was estimated to have an energy gap of 0.1 eV. This solid solution possesses a direct gap for all x, with a significant bowing.

13.13 INDIUM–SCANDIUM–PHOSPHORUS

InP–ScP. The structural, electronic, and thermodynamic properties of cubic Sc$_{1-x}$In$_x$P semiconductor alloys have been studied using first-principles total energy calculations (López-Pórez et al. 2013). It was found that the lattice parameters increase with indium concentration, showing a negative deviation from Vegard's law. The bulk modulus decreases with composition x, showing a small deviation from the linear concentration dependence approach. The obtained results predict that the band gap shows an x-dependent nonlinear behavior.

The calculated phase diagrams indicate a significant phase miscibility gap with a critical temperature of 1834.5°C (local density approximation) and 1342.6°C (Perdew–Burke–Ernzerhof functional). These results indicate that the Sc$_{1-x}$In$_x$P alloy is unstable over a wide range of intermediate compositions at normal growth temperatures and stable at high temperature.

13.14 INDIUM–EUROPIUM–PHOSPHORUS

Three ternary compounds, **EuIn$_2$P$_2$**, **Eu$_3$InP$_3$**, and **Eu$_3$In$_2$P$_4$**, are formed in the In–Eu–P system.

EuIn$_2$P$_2$ crystallizes in the hexagonal structure with the lattice parameters $a = 408.29 \pm 0.06$ and $c = 1759.5 \pm 0.4$ pm at 90 K and a calculated density of 5.799 g·cm^{-3} (Jiang and Kauzlarich 2006). It has a semiconductor–metal transition, along with a magnetic transition, at 24 K.

The black and air-stable crystals of the title compound were grown by reaction of pure elements in a cylindrical crucible (Eu/In/P molar ratio of 3:110:6), where In acts as both a reactant and a flux (Jiang and Kauzlarich 2006). All reactants were handled in a glove box with a N$_2$ atmosphere. Eu and P were packed into an alumina crucible between layers of granular In. Another cylindrical crucible stuffed with quartz wool was inverted and capped on the first crucible. These two crucibles were sealed in a quartz jacket under an Ar pressure of 20 kPa. The reaction contents were heated to 1100°C, allowed to stand for 16 h, cooled to 600°C at 2°C·h^{-1}, allowed to stand for 2 h, and centrifuged at 600°C.

Eu$_3$InP$_3$ crystallizes in the orthorhombic structure with the lattice parameters $a = 1265.17 \pm 0.15$, $b = 426.83 \pm 0.05$, and $c = 1356.43 \pm 0.14$ pm and a calculated density of 6.018 g·cm^{-3} (Jiang et al. 2005b). It is a magnetic, semiconducting Zintl compound with a small band gap.

Needle-shaped single crystals of the title compound were formed from a reaction of elemental Eu, P, and In (molar ratio 14:11:551), where In was both a reactant and the flux (Jiang et al. 2005b). All reactants and products were handled under a N$_2$ atmosphere. Eu and P were packed into an alumina crucible between layers of granular In. A second crucible containing quartz wool was inverted over the reaction crucible, and the entire reaction vessel was sealed in a fused silica jacket that was backfilled with Ar at 20 kPa. The reaction sequence involved heating the vessel to 500°C, holding for 1 h at that temperature, heating to 1100°C, holding for 6 h at 1100°C, slow cooling at 3°C·h^{-1} to 850°C, holding for 19 h at this temperature, and then centrifuging at 850°C. The obtained crystals are extremely reactive, decomposing into a yellow powder upon exposure to air.

Eu$_3$In$_2$P$_4$ crystallizes also in an orthorhombic structure with the lattice parameters $a = 1609.7 \pm 0.3$, $b = 669.92 \pm 0.13$, and $c = 427.12 \pm 0.09$ pm at 90 ± 2 K, a calculated density of 5.836 g·cm⁻³, and an energy gap of 0.452 ± 0.004 eV at room temperature (Jiang et al. 2005a).

To obtain this compound, Eu, red P, and In were mixed in a mole ratio 3:4:120 under an N_2 atmosphere. The mixture was placed in a 2 mL cylindrical alumina crucible with the Eu (136.8 mg) and P (37.2 mg) between two layers of In (4.1335 g). Another crucible filled with quartz wool was inverted and covered the reaction crucible, and the entire system was sealed in a quartz ampoule under an Ar pressure of 20 kPa. The sealed reaction container was heated accordingly: ramp to 500°C over a period of 1 h and dwell for 1 h, and ramp to 1100°C over a period of 1 h, dwell for 6 h, cool at 3°C·h⁻¹ to 850°C, and dwell for 15 h. The reaction vessel was removed from the furnace at 850°C, inverted, and centrifuged. When exposed to air, the black crystals decompose into a yellow powder.

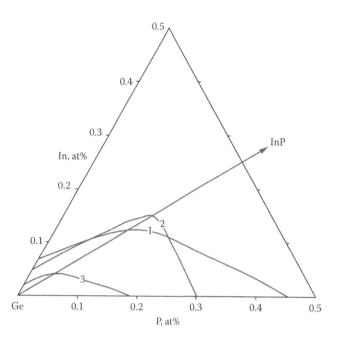

FIGURE 13.9 Solid solution based on Ge in the In–Ge–P ternary system at (1) 600°C, (2) 700°C, and (3) 800°C. (From Glazov, V.M., and Malyutina, G.L., *Zhurn. Neorgan. Khimii,* 8(10), 2372, 1963. With permission.)

13.15 INDIUM–SILICON–PHOSPHORUS

InP–Si. The phase diagram of this system is a eutectic type) (Sokolov and Goncharov 1971). The eutectic contains 25 mol% InP and crystallizes at 945°C. The mutual solubility of InP and Si is negligible. Within the interval of 25–90 mol% InP, there is a thermal effect, the nature of which was not established. This system was investigated using DTA and metallography.

13.16 INDIUM–GERMANIUM–PHOSPHORUS

The region of the solid solution based on Ge in the In–Ge–P ternary system at 600°C, 700°C, and 800°C is given in Figure 13.9 (Glazov and Malyutina 1963b). The samples were annealed at 600°C and 800°C and investigated through metallography and measurement of microhardness.

InP–Ge. The phase diagram of this system is a eutectic type (Figure 13.10) (Glazov et al. 1977b, 1980). The eutectic crystallizes at 804°C and contains 20 mol% GaP. The maximum solubility of InP in Ge reaches 0.18 mol% at the eutectic temperature.

The solubility limit of Ge in InP is shown in Figure 13.11 (Pavlova et al. 1982; Glazov et al. 1983). The solidus line is retrograde, and the maximum solubility is circa 0.18 at% at 900°C. At the eutectic temperature, the solubility of Ge in InP is equal to approximately 0.1 at% and decreases upon lowering the temperature.

The ingots were annealed at 700°C, 800°C, 850°C, 900°C, and 1000°C for 750, 700, 650, 600, and 500 h,

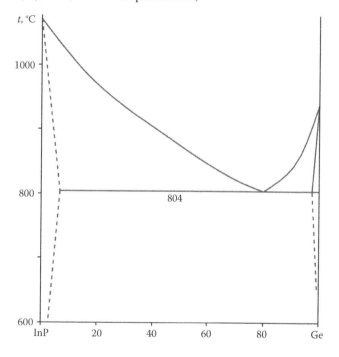

FIGURE 13.10 Phase diagram of the InP–Ge system. (From Glazov, V.M., et al., *Izv. AN SSSR Neorgan. Mater.,* 13(2), 209, 1977. With permission.)

respectively (Pavlova et al. 1982; Glazov et al. 1983). This system was investigated using DTA and metallography.

The distribution coefficient of Ge in crystals of InP depends on its content in the melt and changes from 0.26

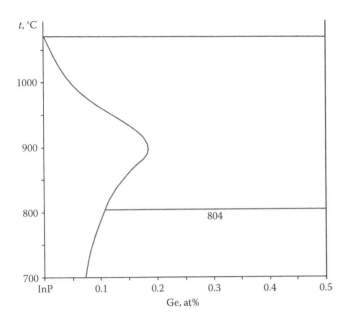

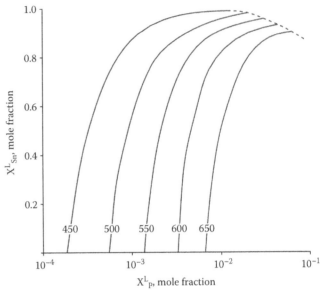

FIGURE 13.11 Solid solution based on Ge in the InP–Ge system. (From Pavlova, L.M., et al., *Izv. AN SSSR Neorgan. Mater.*, 18(9), 1444, 1982. With permission.)

to 0.17 at 0.05–1.5 mass% Ge in the melt (Zakharenkov et al. 1985).

Ultrafast cooling (10^7–10^8°C·s^{-1}) of the GaP–Ge melts leads to diffusionless crystallization and the formation of metastable solid solutions (Glazov et al. 1984).

13.17 INDIUM–TIN–PHOSPHORUS

To estimate the equilibrium compositions of liquid and solid phases, a computational approach was used, followed by verification of its adequacy for some experimental points. Isotherms of the liquidus surface of the In–Sn–P ternary system for the Sn-rich melts are given in Figure 13.12 (Vasil'yev et al. 1985).

InP–Sn. The phase diagram of this system is a eutectic type, with the eutectic degenerated from the Sn side (Figure 13.13) (Buehler et al. 1974; Leonhardt and Kühn 1975). The liquidus temperature increases monotonically with an increasing molar fraction of InP. A eutectic exists very close to the melting point of Sn. X-ray fluorescence analysis of InP epitaxial layers grown from solution containing 3 mol% InP in Sn reveals a considerable solid solubility of Sn in InP (ca. 2 mol%). The liquidus temperatures along the InP–Sn tie-line have been measured for dilute solutions corresponding to a linear relationship $ln x_{InP} = -(\Delta H^s/RT) + C$, where x_{InP} is the molar fraction of InP. Figure 13.14 shows a plot of x_{InP} versus T^{-1} for solutions of InP in Sn (Buehler and Bachmann 1976). The temperature dependence of the InP limiting

FIGURE 13.12 Isotherms of the liquidus surface of the In–Sn–P ternary system for the Sn-rich melts. (From Vasil'yev M.G., et al., in *Legir. Poluprovodn. Materialy*, Nauka Publish., Moscow, 1985, 61. With permission.)

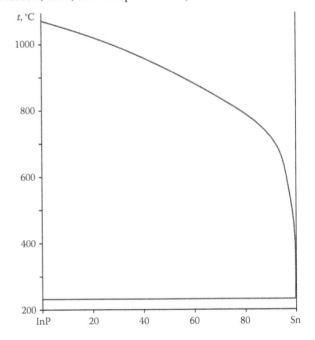

FIGURE 13.13 Phase diagram of the InP–Sn system. (From Buehler, E., et al., *J. Cryst. Growth*, 24–25, 260, 1974. With permission.)

solubility in the Sn melt was determined by Vasil'yev et al. (1983) with the same results. The solubility of InP in the Sn melt at 600°C–900°C was also determined by Leonhardt and Kühn (1974b).

The distribution coefficient of Sn in crystals of InP depends on its content in the melt and changes

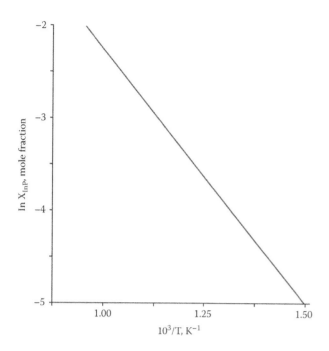

FIGURE 13.14 Solubility of InP in Sn as a function of temperature. (From Buehler, E., and Bachmann, K.J., *J. Cryst. Growth*, *35*(1), 60, 1976. With permission.)

from 0.1 to 0.02 at 0.005–1.0 mass% Sn in the melt (Zakharenkov et al. 1985). The effective distribution coefficient of Sn in the crystals of InP is equal to 0.14 (Nashelskiy et al. 1982).

13.18 INDIUM–LEAD–PHOSPHORUS

InP–Pb. The phase diagram of this system is a eutectic type with the eutectic degenerated from Pb side (Figure 13.15) (Leonhardt and Kühn 1975).

The liquidus temperatures along the InP–Pb tie-line have been measured for dilute solutions corresponding to a linear relationship $lnx_{InP} = -(76.5 \text{ kJ·mol}^{-1}/RT) + 5.980$, where x_{InP} is the molar fraction of InP (Buehler and Bachmann 1976).

The solubility of InP in the Pb melt at 750°C–975°C was also determined by Leonhardt and Kühn (1974b).

13.19 INDIUM–ARSENIC–PHOSPHORUS

The liquidus surface of the In–P–As ternary system was studied experimentally by DTA in the composition range 30–80 at% In (Figure 13.16) (Semenova et al. 1988). The calculation based on a regular solution was also given by these authors, as well as calculated isobars. Thermodynamic calculations of the liquidus surface in the InP–In–InAs subsystem have been performed by Panish and Ilegems (1972) and Baranov et al. (1990b), and

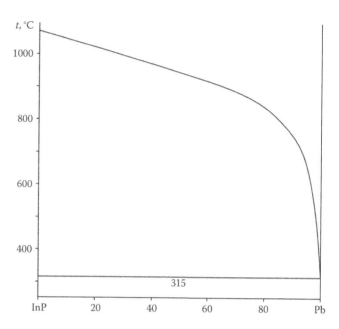

FIGURE 13.15 Phase diagram of the InP–Pb system. (From Leonhardt, A., and Kühn, G., *J. Less Common Metals*, *39*(2), 247, 1975. With permission.)

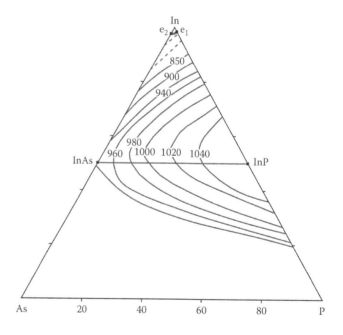

FIGURE 13.16 Liquidus surface of the In–P–As ternary system. (From Semenova, G.V., et al., *Zhurn. Neorgan. Khimii*, *33*(4), 1000, 1988. With permission.)

the liquidus surface in the entire range of compositions has been calculated using regular-type solution models (Kaufman et al. 1981; Ishida et al. 1989b). However, a comparison with the experimental data is either entirely missing (Ishida et al. 1989b) or only given for the quasi-binary system (Kaufman et al. 1981). Liquidus equations have been derived from experimental data by Muszyński and Ryabcev (1975, 1976) using a simplex lattice method in order to interpolate the liquidus curves over the entire composition range. The pressure and vapor composition in equilibrium with the liquid were measured by static manometric and two-temperature methods, and isobaric lines and calculated solid-state isoconcentration lines were reported by Semenova et al. (1990, 1992).

Isothermal sections of the In–As–P ternary system at 950°C, 975°C, 1000°C, and 1025°C are shown in Figure 13.17 (Kaufman et al. 1981).

InP–InAs. In early studies, a complete series of solid solutions was found in this system (Folberth 1955; Weiss 1956). It was shown that the alloys were homogeneous after melting and slow cooling without further heat treatment (Köster and Ulrich 1958b). However,

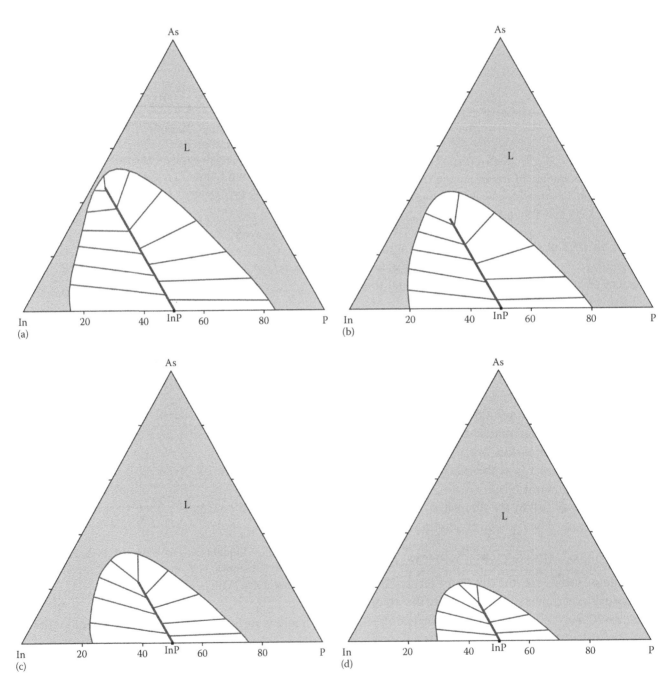

FIGURE 13.17 Calculated isothermal sections of the In–P–As ternary system at (a) 950°C, (b) 975°C and 1000°C, and (c) 1025°C. (From Kaufman, L., and Ditchek, B., et al., *J. Less Common Metals*, *168*(1), 115, 1991. With permission.)

considerable segregation was detected in alloys with 10–50 mol% InP, prepared from the pure elements in a sealed silica tube with two temperature zones, one for the liquid in a silica boat and the other for the volatile constituents (P, As) (Bogorodskiy et al. 1961). The segregation may be reduced to some extent by zone leveling. XRD showed broadening of diffraction lines and a nonlinear and scattering dependence of the lattice parameter of $InAs_{1-x}P_x$ crystals from composition, which is probably still due to nonhomogeneity.

The phase equilibria have been studied experimentally in the quasi-binary InP–InAs system by Ugay et al. (1968, 1970a, 1970b), Thompson and Wagner (1971) and Bodnar and Matyas (1977). Ugay et al. (1970a, 1970b) also reported pressure data of the L + InP_xAs_{1-x} equilibrium, obtained from a second temperature zone in the furnace where the volatile components were kept in an elongation of the sealed system.

Thompson and Wagner (1971) performed slow bulk crystallization and measured the composition in the first part of the crystal, but away from the seed, and took this and the initial liquid composition as the tie-line compositions. The data from both the liquid-encapsulated Czochralski growth and the sealed capsule crystallization agree with each other, and the two-phase field is much narrower than that of Köster and Ulrich (1958b), obviously due to segregation in the latter case. The data of Thompson and Wagner (1971) are also in good agreement with the DTA data of Bodnar and Matyas (1977), where the L + InP_xAs_{1-x} field is shifted slightly to the InP-rich side.

Vertical sections of InP–As and InAs–P were given by Ugay et al. (1971) from thermal analysis and metallography, and Ugay et al. (1970b) gave the liquidus profile along the vertical sections $In–As_{0.25}P_{0.75}$, $In–As_{0.5}P_{0.5}$, $In–As_{0.6}P_{0.4}$, and $In–As_{0.75}P_{0.25}$ from thermal analysis. All the above sections contradict the InP–InAs quasi-binary section and the assessed liquidus surface.

The phase diagram of this quasi-binary system was also calculated by many investigators (Steininger 1970; Panish and Ilegems 1972; Stringfellow 1972a; Brebrick and Panlener 1974; Kaufman et al. 1981; Ishida et al. 1989b). The model of regular solutions could be used for the theoretical description of this system (Semenova et al. 2008).

The phase diagram of the InP–InAs system is given in Figure 13.18 from Thompson and Wagner (1971). It is in good agreement with the data of the experimental and

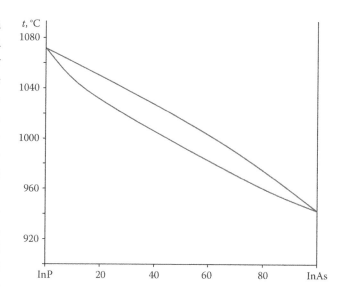

FIGURE 13.18 Phase diagram of the InP–InAs system. (From Thompson, A.G., and Wagner, J.W., *J. Phys. Chem. Solids*, 32(11), 2613, 1971. With permission.)

calculated data of other authors. The homogeneity region of the InP_xAs_{1-x} solid solutions is shown in Figure 13.19 (Semenova et al. 1993). It was obtained by the calculation using the method of quasi-chemical reactions. This region is asymmetric and depends on the solid solution compositions. The solubility of excess components in the solid solution is retrograde. The deviation from stoichiometry for the $InP_{0.33}As_{0.67}$ solid solution at 850°C determined using the method of "frozen" equilibria is $1.2 \cdot 10^{19}$ cm⁻³ (Semenova et al. 1999).

Comparison of the results of the temperature dependence of the dissociation pressure of the alloys from the quasi-binary InP–InAs system with the composition of the saturated vapor allowed constructing the *p-T-x* diagram for this system (Ugay et al. 1985, 1986, 1991). The vapor phase above the InP–InAs melts is enriched by phosphorus as a more volatile component.

The concentration dependence of the lattice parameter for the InP_xAs_{1-x} solid solutions obeys Vegard's law (Antyukhov 1986). The single crystals of these solid solutions were obtained by chemical transport reactions (Yegorov et al. 1970).

13.20 INDIUM–ANTIMONY–PHOSPHORUS

Little experimental information is reported on the phase equilibria in the In–P–Sb ternary system. The liquidus surface of this system was calculated by Ishida et al. (1989b) and Semenova et al. (2005) and is given in Figure 13.20. The field of the InP primary crystallization

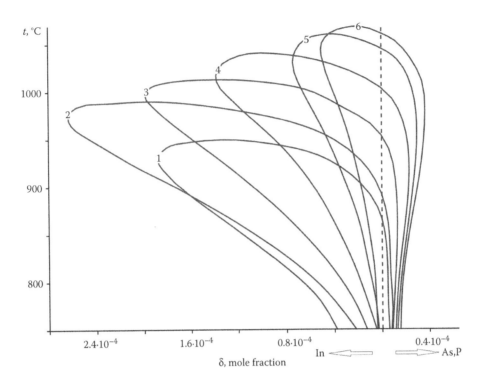

FIGURE 13.19 Homogeneity region of the InP_xAs_{1-x} solid solutions for x equal to (1) 0, (2) 0.3, (3) 0.5, (4) 0.7, (5) 0.9, and (6) 1. (From Semenova, G.V., et al., *Zhurn. Neorgan. Khimii*, *38*(10), 1717, 1993. With permission.)

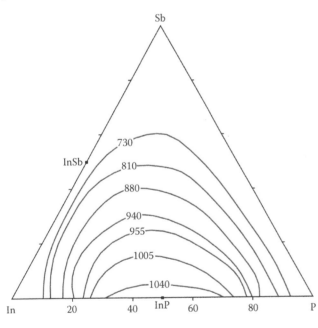

FIGURE 13.20 Liquidus surface of the In–P–Sb ternary system. (From Semenova, G.V., et al., *Zhurn. Neorgan. Khimii*, *50*(4), 710, 2005. With permission.)

occupies the biggest part of the concentration triangle. The vapor phase above the melt consists essentially of phosphorus, whose pressure is several orders of magnitude higher than the pressure of indium and antimony (Semenova et al. 2005); the projections of isobaric

sections of the liquidus surface in the region of InP primary crystallization were also given. The liquidus surface of this system in the In-rich corner was also determined by Baranov et al. (1990b) using thermodynamic calculations. The calculated dependences of phosphorus equilibrium solubility, x^P_L, versus the antimony concentration in the liquid phase, x^{Sb}_L, at 490°C–650°C are shown in Figure 13.21 (Golubev et al. 1990).

Two isothermal sections at temperatures of 500°C and 750°C are presented in Figure 13.22 as proposed by Ishida et al. (1989a). These sections result from consistent thermodynamic calculations.

InP–InSb. The liquidus and solidus lines for the InP–InSb quasi-binary section were completely described by Nakajima et al. (1973), Lendvay (1984), Ishida et al. (1989a, 1989b), and Semenova et al. (2005). For these determinations, they used DTA, metallography, and electron probe microanalysis (EPMA) experiments and thermodynamic calculation that appear to be in good agreement. The calculated phase diagram is presented in Figure 13.23 (Nakajima et al. 1973). It could not be determined whether the invariant reaction is eutectic or peritectic because the invariant temperature is so close to the melting temperature of InSb (Ishida et al. 1989a). According to the data of Nakajima et al. (1973) and Semenova et al. (2005), this nonvariant reaction is

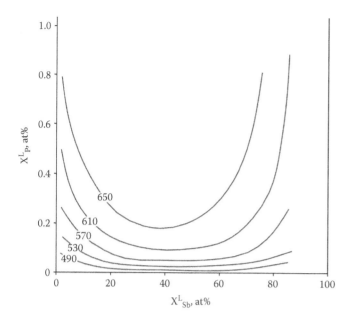

FIGURE 13.21 Calculated dependences of phosphorus equilibrium solubility, x^P_L, versus the antimony concentration in the liquid phase, x^{Sb}_L. (From Golubev, L.V., et al., *Izv. AN SSSR Neorgan. Mater.*, 26(6), 1321, 1990. With permission.)

eutectic, with eutectic temperatures of 525°C ± 3.7°C and 500°C ± 3°C, respectively. In the thermodynamic calculations, the mutual solubility limits of InSb and InP at the eutectic temperature were shown to be 0.02 mol% InP and 1.85 mol% InSb (Nakajima et al. 1973). According to the data of Goryunova and Sokolova (1960a), solid solutions did not form in this system. A miscibility gap is abnormal in the InP–InSb system, leaning to the side of the component (InSb) with a lower melting temperature (Nakajima et al. 1973). The critical temperature is 1033°C–1046°C (Stringfellow 1982b, 1983) (1927°C [Stringfellow 1972a]).

13.21 INDIUM–BISMUTH–PHOSPHORUS

The liquidus surface of the InP–In–Bi subsystem of the In–Bi–P ternary system is given in Figure 13.24 (Islamov et al. 1984).

InP–InBi. The phase diagram of this system is a eutectic type with the eutectic degenerated from the InBi side (Figure 13.25) (Akchurin et al. 1984; Islamov et al. 1984; Yevgen'ev 1987). The eutectic crystallizes at 88°C (Akchurin et al. 1984). The solubility from the InP side is not higher than 0.1 at% Bi. Using Knudsen's method, the temperature and concentration dependences of the vapor pressure in this system have been studied (Yevgen'ev 1987). Based on the experimental results and the results of the calculation of the interaction

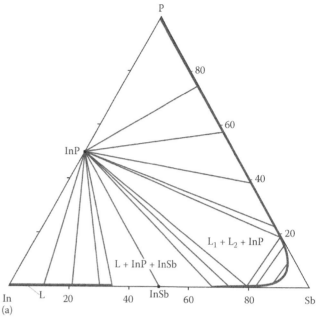

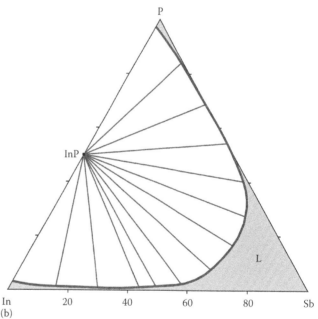

FIGURE 13.22 Isothermal sections of the In–P–Sb ternary system at (a) 500°C and (b) 750°C. (From Ishida, K., et al., *J. Less Common Metals*, 155(2), 193, 1989. With permission.)

parameter and the activity coefficient, it has been shown that this system is disposed to delamination. Significant positive deviations from Raoult's law exist in the InP–InBi system.

$InBi_xP_{1-x}$ single crystal films were obtained by gas source MBE (Gu et al. 2014; K. Wang et al. 2014). A maximum Bi composition of 2.4 at% was determined. The band gap reduction caused by the Bi incorporation was estimated to be about 56 meV per 1 at% Bi.

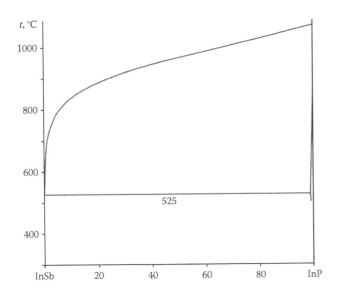

FIGURE 13.23 Phase diagram of the InP–InSb system. (From Nakajima, K., et al., *J. Jpn. Inst. Met.*, *37*(12), 1276, 1973. With permission.)

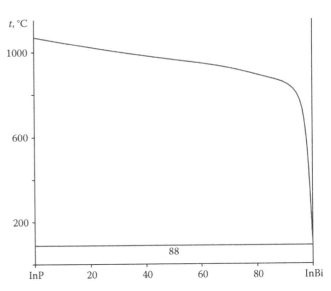

FIGURE 13.25 Phase diagram of the InP–InBi system. (From Islamov, S.A., et al., *Zhurn. Neorgan. Khimii*, *29*(9), 2369, 1984. With permission.)

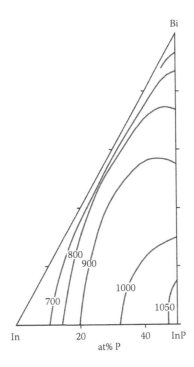

FIGURE 13.24 Liquidus surface of the InP–In–Bi subsystem of the In–Bi–P ternary system. (From Islamov, S.A., et al., *Zhurn. Neorgan. Khimii*, *29*(9), 2369, 1984. With permission.)

The effects of Bi flux, PH_3 pressure, and InP growth rate on surface, structural, and transport properties of $InBi_xP_{1-x}$ thin films grown by gas source MBE have been studied by Pan et al. (2016). The improvement in crystalline and surface quality with Bi content up to $x = 0.024$ was demonstrated. Electron mobility reveals

no significant change for $x < 0.024$ but decreases for higher Bi concentrations accompanied by some structural degradation. Changing the PH_3 pressure was found to have little influence on the Bi concentration, but a strong influence on the crystalline quality and transport property, especially for the sample grown with a proper PH_3 pressure. Bismuth incorporation was found to be inversely proportional to the InP growth rate, and up to $x = 0.037$ bismuth incorporation has been demonstrated at a low growth rate of 0.50 mm·h⁻¹.

InP–In₂Bi. The phase diagram of this system is a eutectic type with the eutectic degenerated from the In_2Bi side (Figure 13.26) (Akchurin et al. 1984; Islamov et al. 1984; Yevgen'ev 1987). The eutectic crystallizes at 70°C (Akchurin et al. 1984). The solubility from the InP side is not higher than 0.1 at% Bi. Using Knudsen's method, the temperature and concentration dependences of the vapor pressure in this system have been studied (Yevgen'ev 1987). Based on the experimental results and the results of the calculation of the interaction parameter and the activity coefficient, it has been shown that this system is disposed to delamination. Significant positive deviations from Raoult's law exist in the InP–In₂Bi system.

InP–Bi. The phase diagram of this system is a eutectic type, with the eutectic degenerated from the Bi side (Figure 13.27) (Leonhardt and Kühn 1975; Akchurin et al. 1984; Islamov et al. 1984; Yevgen'ev 1987). The solubility from the InP side is not higher than 0.1 at%

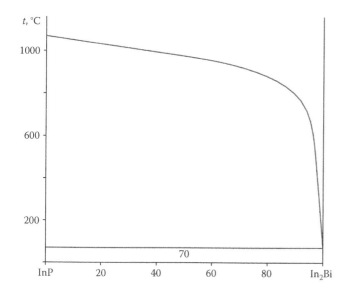

FIGURE 13.26 Phase diagram of the InP–In$_2$Bi system. (From Islamov, S.A., et al., *Zhurn. Neorgan. Khimii*, *29*(9), 2369, 1984. With permission.)

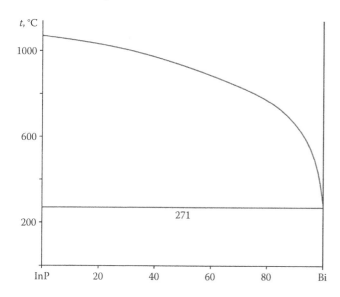

FIGURE 13.27 Phase diagram of the InP–Bi system. (From Islamov, S.A., et al., *Zhurn. Neorgan. Khimii*, *29*(9), 2369, 1984. With permission.)

Bi. The solubility of InP in the Bi melt at 600°C–975°C was determined by Leonhardt and Kühn (1974b). Using Knudsen's method, the temperature and concentration dependences of the vapor pressure in this system have been studied (Yevgen'ev 1987). Based on the experimental results and the results of the calculation of the interaction parameter and the activity coefficient, it has been shown that this system is disposed to delamination. Significant positive deviations from Raoult's law exist in the InP–Bi system.

13.22 INDIUM–OXYGEN–PHOSPHORUS

The isothermal section of the In–O–P ternary system at 20°C is presented in Figure 13.28 (Schwartz et al. 1982b; Schwartz 1983). Two ternary compounds, **InPO$_4$** and **In(PO$_3$)$_3$**, are formed in this system. InPO$_4$ crystallizes in the orthorhombic structure with the lattice parameters $a = 530.8 \pm 0.5$, $b = 785.1 \pm 0.8$, and $c = 676.7 \pm 0.7$ pm and a calculated density of 4.91 g·cm^{-3} (Mooney et al. 1956a) ($a = 531$, $b = 785$, and $c = 677$ pm [Mooney 1954]). This compound was prepared by heating a mixture of H$_3$PO$_4$ and In$_2$O$_3$ to 350°C–400°C in a hydrothermal bomb for a week or more (Mooney et al. 1954; Mooney 1956a). In general, the products were well-crystallized powders, although small single crystals were grown in some cases. It was also obtained by precipitation of an aqueous solution of In(III) salts with sodium phosphate (Ensslin et al. 1947). The white powder thus obtained has an experimental density of 3.91 g·cm^{-3}.

In(PO$_3$)$_3$ crystallizes in the monoclinic structure with the lattice parameters $a = 1087.6 \pm 0.2$, $b = 1958.1 \pm 0.2$, and $c = 965.8 \pm 0.2$ pm and $\beta = 97.77 \pm 0.01°$ and calculated and experimental densities of 3.44 and 3.41 g·cm^{-3}, respectively (Bentama et al. 1988) ($a = 1355 \pm 2$, $b = 653.8 \pm 0.1$, and $c = 1089.4 \pm 0.5$ pm and $\beta = 135.03 \pm 0.09°$ [Palkina et al. 1993]). The powder of the title compound was obtained by long heating of the InPO$_4$ solution in the excess H$_3$PO$_4$ (Ensslin et al.

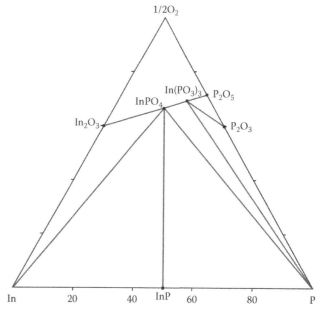

FIGURE 13.28 Isothermal section of the In–O–P ternary system at 20°C. (From Schwartz, G.P., *Thin Solid Films*, *103*(1–3), 3, 1983. With permission.)

1947; Bentama et al. 1988). It was also synthesized upon slow heating (5 h) in a Pt crucible of a mixture of In_2O_3, $NaH_2PO_4\cdot2H_2O$, $(NH_4)_2HPO_4$, and H_3PO_4 with a P/Na/NH_4/In molar ratio of 75:30:30:1 to 360°C, followed by holding at this temperature for 5 days (Palkina et al. 1993).

According to the data of Peltier et al. (1996), once more compound, **$In_3P_2O_8$**, exists in the In–O–P ternary system. It crystallizes in the cubic structure with the lattice parameter $a = 1115.2 \pm 0.1$ and a calculated density of 5.12 g·cm⁻³. Powdered $In_3P_2O_8$ can be synthesized from a mixture of $InPO_4$ and In_2O_3 (molar ratio 4:1) after thermal reduction under a H_2 atmosphere at 400°C. Its single crystals have been obtained after heating a mixture of powdered InN and PON pellets (molar ratio In/P = 1) in a silica ampoule, sealed under vacuum at 700°C. Upon opening of the ampoule after 30 days, colorless and transparent crystals in the form of needles had grown on the surfaces of the pellets.

13.23 INDIUM–SULFUR–PHOSPHORUS

The isothermal sections of the In–S–P ternary system at room temperature and at 500°C are shown in Figures 13.29 and 13.30 (Gendry et al. 1986; Gebesh et al. 1991; Potoriy and Milyan 2016). Three compounds, **$InPS_4$** (τ_1), **In_3PS_3** (τ_2), and **$In_4(P_2S_6)_3$** (τ_3), are formed in this system at 500°C. $InPS_4$ crystallizes in the tetragonal structure with the lattice parameters $a = 562.3 \pm 0.1$

and $c = 905.8 \pm 0.2$ pm and calculated and experimental densities of 3.177 and 3.16 ± 0.01 g·cm⁻³ (3.18 g·cm⁻³ [Potoriy and Milyan 2016]), respectively (Diehl and Carpentier 1978) ($a = 560 \pm 1$ and $c = 902 \pm 1$ pm [Carpentier et al. 1970; Voroshilov et al. 2016]; $a = 559$ and $c = 900$ pm [Nitsche and Wild 1970]). The title compound decomposes at 647°C (at 450°C–700°C [Nitsche and Wild 1970; Bubenzer et al. 1975; Gendry et al. 1986]) without melting (Galagovets et al. 1995). It was synthesized upon reacting the pure elements in evacuated quartz ampoules at 600°C (Bubenzer et al. 1975). Single crystals of this compound as colorless needles were obtained by chemical vapor transport experiments using iodine as a transporting agent (Carpentier et al. 1970; Nitsche and Wild 1970; Diehl and Carpentier 1978).

In_3PS_3 crystallizes in the cubic structure with the lattice parameter $a = 1078.9$ pm (Robbins and Lambrecht 1975; Voroshilov et al. 2016) ($a = 1035 \pm 1$ pm and the experimental density is 4.99 g·cm⁻³ [Gebesh et al. 1991; Potoriy and Milyan 2016]). This compound was prepared from a mixture of InP, In, and S (Robbins and Lambrecht 1975). The pressed pellets of the mixture were sealed in evacuated quartz tubes and heated at a rate of 10°C–15°C·h⁻¹ to 800°C. The samples were held at this temperature for 48 h and cooled to room temperature.

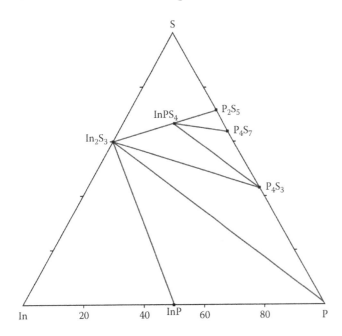

FIGURE 13.29 Isothermal section of the In–S–P ternary system at room temperature. (From Gendry, M., et al., *Thin Solid Films*, 139(1), 53, 1986. With permission.)

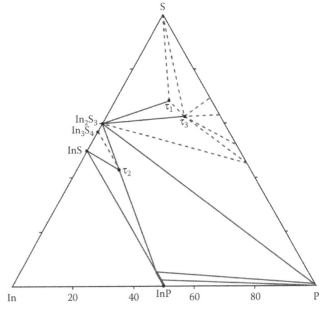

FIGURE 13.30 Isothermal section of the In–S–P ternary system at 500°C. (From Gebesh, V.Yu., et al., *Ukr. Khim. Zhurn.*, 57(8), 803, 1991. With permission.)

$In_4(P_2S_6)_3$ is formed according to the syntectic reaction at 787°C and exists in two polymorphic modifications (Gebesh et al. 1991; Potoriy and Milyan 2016). Low-temperature modification crystallizes in the monoclinic structure with the lattice parameters $a = 683.3 \pm 0.5$, $b = 1826 \pm 9$, and $c = 1054 \pm 2$ and $\beta = 107.60 \pm 0.02°$ ($a = 684.2 \pm 0.2$, $b = 1052.8 \pm 0.2$, and $c = 1826.6 \pm 0.5$ and $\beta = 107.67 \pm 0.02°$ and calculated and experimental densities of 3.237 and 3.24 ± 0.4 g·cm⁻³, respectively [Diehl and Carpentier 1978]; $a = 1823$, $b = 1050$, and $c = 683$ and $\beta = 107.5°$ and calculated and experimental densities of 3.26 and 3.24 g·cm⁻³, respectively [Soled and Wold 1976; Voroshilov et al. 2016]). High-temperature modification crystallizes in the rhombohedral structure with lattice parameters $a = 610 \pm 4$ and $c = 1954 \pm 4$ pm (in a hexagonal setting) (Gebesh et al. 1991; Voroshilov et al. 2016). Single crystals of this compound as yellow needles were obtained by chemical vapor transport experiments using chlorine or iodine as a transporting agent (Soled and Wold 1976; Diehl and Carpentier 1978). $In_4(P_2S_6)_3$ is also formed at the decomposition of $InPS_4$ (Gendry et al. 1986).

According to the data of Radautsan and Madan (1970), **InP_2S_4** ternary compound is also formed in the In–S–P ternary system. It melts at 815°C and crystallizes in the monoclinic structure with the lattice parameters $a = 680$, $b = 1050$, and $c = 1835$ pm and $\beta = 108°$ and an energy gap of 2.42 eV. Gebesh et al. (1991) did not confirm the existence of this compound. Galagovets et al. (1995) indicated the formation of one more compound, **$In_4(P_2S_7)_3$**, which forms according to the syntectic reaction $L_1 + L_2 \longleftrightarrow In_4(P_2S_6)_3$ on the In_2S_3–P_2S_5 section of the In–S–P ternary system.

InP–In_2S_3. According to the data of Goryunova and Sokolova (1960a) and Radautsan and Madan (1962), solid solutions did not form in this system, but Gebesh et al. (1991) and Potoriy and Milyan (2016) indicated that InP dissolves up to 10 mol% In_2S_3.

InP–S. The effective distribution coefficient of S in crystals of InP is equal to 0.46–0.48 (Nashelskiy et al. 1982).

13.24 INDIUM–SELENIUM–PHOSPHORUS

The isothermal section of the In–Se–P ternary system at 455°C is shown in Figure 13.31 (Voroshilov et al. 1991; Potoriy and Milyan 2016). **$In_4(P_2Se_6)_3$** (τ) ternary compound is formed according to the syntectic reaction at

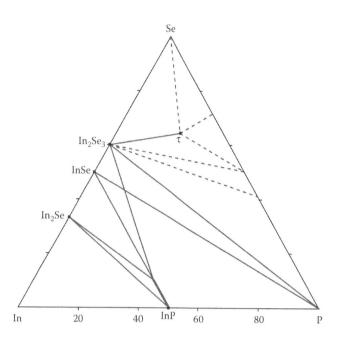

FIGURE 13.31 Isothermal section of the In–Se–P ternary system at 455°C. (From Voroshilov, Yu.V., et al., *Izv. AN SSSR Neorgan. Mater.*, *27*(12), 2495, 1991. With permission.)

607°C ± 5°C and crystallizes in the hexagonal structure with the lattice parameters $a = 636.2 \pm 0.3$ and $c = 1992.9 \pm 0.6$ pm and calculated and experimental densities of 4.90 and 4.85–4.89 g·cm⁻³, respectively (Voroshilov et al. 1991; Potoriy and Milyan 2016) (in the tetragonal structure, with the lattice parameters $a = 637 \pm 3$ and $c = 1986 \pm 8$ pm, an energy gap of 1.9 ± 0.1 eV, and calculated and experimental densities of 4.91 and 4.88 ± 0.05 g·cm⁻³, respectively [Katty et al. 1977]; in the orthorhombic structure, with the lattice parameters $a = 1919.7 \pm 0.9$, $b = 1109.0 \pm 0.9$, and $c = 2006.7 \pm 0.7$ pm and calculated and experimental densities of 4.818 and 4.80 ± 0.04 g·cm⁻³, respectively [Diehl and Carpentier 1978; Voroshilov et al. 2016]). Single crystals of this compound as thin red plates were obtained by chemical vapor transport reactions using iodine as a transporting agent (Katty et al. 1977; Diehl and Carpentier 1978).

According to the data of Radautsan and Madan (1970), **InP_2Se_4** ternary compound, which melts at 635°C and crystallizes in the monoclinic structure, is also formed in the In–Se–P system. It decomposes upon heating in air up to 300°C and sublimes at temperatures higher than 450°C. The title compound is the *n*-type semiconductor with an energy gap of 1.29 eV.

InP–$InSe$. Limited solubility exists on the basis of both compounds (Radautsan and Madan 1970).

InP–In₂Se₃. The solubility of In_2Se_3 in InP reaches 50 mol% (Goryunova and Sokolova 1960a; Hahn and Thiele 1960; Radautsan et al. 1960; Radautsan 1962; Radautsan and Madan 1962; Voroshilov et al. 1991; Potoriy and Milyan 2016), and In_2Se_3 dissolves up to 10 mol% InP (Potoriy and Milyan 2016). A boundary homogeneity alloy corresponds to the composition **In₃PSe₃**, interpreted by Robbins and Lambrecht (1975) as an individual ternary compound (cubic structure, $a = 578.7$ pm). In_2Se_3 dissolves 10 mol% InP (Voroshilov et al. 1991).

InP–Se. The solubility of Se in InP reaches 7 at% (Radautsan and Madan 1970).

13.25 INDIUM–TELLURIUM–PHOSPHORUS

InP–In₂Te₃. According to the data of Goryunova and Sokolova (1960a) and Radautsan and Madan (1962), solid solutions are apparently formed only in a narrow range of concentrations near the starting compounds.

InP–Te. There is an evidence to believe that some In_2Te_3 is formed at the interaction of InP and Te (Ufimtsev and Gimel'farb 1973). The effective distribution coefficient of Te in crystals of InP is equal to 0.38–0.43 (Nashelskiy et al. 1982).

13.26 INDIUM–CHROMIUM–PHOSPHORUS

In the In–Cr–P ternary system, InP is in thermodynamic equilibrium with CrP and $Cr_{12}P_7$, and $Cr_{12}P_7$ forms a tie-line with In at 600°C (Andersson-Söderberg 1992; Andersson 1993). Solid phase equilibria were determined at ambient pressure using XRD. The samples were homogenized at 1000°C for 3 days and then slowly cooled to 600°C. The times of heat treatment varied from 1 week to 3 months. All samples were finally quenched in ice water.

13.27 INDIUM–MOLYBDENUM–PHOSPHORUS

In the In–Mo–P ternary system, InP is in thermodynamic equilibrium with MoP_2 and MoP, and MoP forms a tie-line with In at 600°C (Andersson-Söderberg 1992; Andersson 1993). Solid phase equilibria were determined at ambient pressure using XRD. The samples were homogenized at 1000°C for 3 days and then slowly cooled to 600°C. The times of heat treatment varied from 1 week to 3 months. All samples were finally quenched in ice water.

13.28 INDIUM–TUNGSTEN–PHOSPHORUS

In the In–W–P ternary system, InP is in thermodynamic equilibrium with WP_2 and WP, and WP forms a tie-line with In at 600°C (Andersson-Söderberg 1992; Andersson 1993). Solid phase equilibria were determined at ambient pressure using XRD. The samples were homogenized at 1000°C for 3 days and then slowly cooled to 600°C. The times of heat treatment varied from 1 week to 3 months. All samples were finally quenched in ice water.

13.29 INDIUM–FLUORINE–PHOSPHORUS

InPF₆ ternary compound is formed in the In–F–P system. It crystallizes in the cubic structure with the lattice parameter $a = 796.6 \pm 1.8$ pm at 200 K (Goreshnik and Mazej 2008) ($a = 807 \pm 2$ pm [Mazej 2005]). To obtain this compound, a reaction mixture of In (2.96 mM) and approximately 1.00 g of PF_5 in 4 mL of anhydrous HF was loaded into translucent fluorocarbon polymer reaction vessels in a glove box (Mazej 2005). Anhydrous HF and, when necessary, PF_5 were condensed onto the solid reactants at 77 K. The reaction vessels were brought to ambient temperature. The reaction mixtures were vigorously agitated, and after some time, the volatiles were pumped off. After 7 days, pieces of In metal were still present. Cubic-shaped crystals and a colorless powdered solid were observed beside unreacted In metal. After isolation, the colorless powdered solid turned gray. Single crystals of the title compound were obtained from In and excess PF_5 in anhydrous HF as a solvent (Goreshnik and Mazej 2008).

13.30 INDIUM–MANGANESE–PHOSPHORUS

InP–Mn. The solubility and distribution coefficient of Mn in InP at 27°C is equal to $3 \cdot 10^{19}$ cm^{-3} and 0.11 ± 0.02, respectively (Zakharenkov et al. 1983). The effective distribution coefficient of Mn in crystals of InP is equal to 0.02 (Nashelskiy et al. 1982).

13.31 INDIUM–IRON–PHOSPHORUS

The isothermal section of the In–Fe–P ternary system at room temperature is shown in Figure 13.32 (Andersson-Söderberg 1990; Raghavan 1998). Indium forms tie-lines with Fe_3P, Fe_2P, and FeP, and InP is in thermodynamic equilibrium with FeP, FeP_2, and FeP_4. No ternary compounds were found. Samples quenched from 900°C indicated that the isothermal section at 900°C is similar to that at room temperature, except that FeP_4 is not present. The samples were heated in evacuated silica tubes at

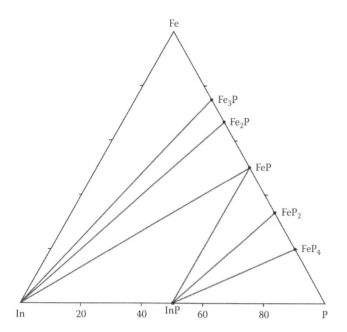

FIGURE 13.32   Isothermal section of the In–Fe–P ternary system at room temperature. (From Raghavan, V., *J. Phase Equilibria Diffus.*, 19(3), 284, 1998. With permission.)

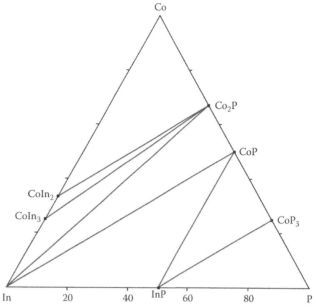

FIGURE 13.33   Isothermal section of the In–Co–P ternary system at 450°C. (From Swenson, D., and Chang, Y.A., *Mater. Sci. Eng. B*, 8(3), 225, 1991. With permission.)

900°C for several weeks. After annealing, the samples were either quenched in water or slowly cooled to room temperature. The phase equilibria were studied by XRD and DTA.

InP–Fe. The solubility of Fe in InP was investigated using radioactive Fe isotope (Shishiyanu et al. 1977). It was shown that solubility increases from $4 \cdot 10^{17}$ cm^{-3} at 815°C to $2 \cdot 10^{18}$ cm^{-3} at 940°C and could be expressed by the following equation: $N_{Fe} = 3 \cdot 10^{21}$ exp[(−0.8 ± 0.2)/kT]. The effective distribution coefficient of Fe in crystals of InP is equal to 0.002 (Nashelskiy et al. 1982).

13.32 INDIUM–COBALT–PHOSPHORUS

Phase equilibria were established in the In–Co–P ternary system at 450°C using XRD and EPMA (Figure 13.33) (Swenson and Chang 1991). No ternary phases were detected in the system, and the ternary solubility of the binary phases was found to be negligible. InP was demonstrated to be in thermodynamic equilibrium with liquid indium metal, CoP$_3$, and CoP. The experimentally determined isothermal section was identical to a ternary isotherm calculated by making certain simplifying assumptions and utilizing thermodynamic data found in the literature for the constituent binary phases. The samples were annealed in evacuated quartz ampoules at 450°C for 45 days, with finally quenching in ice water.

InP–Co. The solubility of Co in InP was investigated using radioactive Co isotope (Shishiyanu et al. 1977). It was shown that solubility increases from $3.5 \cdot 10^{15}$ cm^{-3} at 610°C to $1.2 \cdot 10^{17}$ cm^{-3} at 950°C and could be expressed by the following equation: $N_{Co} = 1.3 \cdot 10^{21}$ exp[(−1.0 ± 0.2)/kT].

13.33 INDIUM–NICKEL–PHOSPHORUS

The solid-state equilibria in the In–Ni–P ternary system at 800°C have been determined at ambient pressure using XRD and DTA (Figure 13.34) (Andersson-Söderberg 1991). One ternary phase, **Ni$_{21}$In$_2$P$_6$** (τ_1), was found at this temperature (Andersson-Söderberg and Andersson 1990; Andersson-Söderberg 1991, 1992). It crystallizes in the cubic structure with the lattice parameter $a = 1111.20 \pm 0.04$ pm. No phase transformation was observed for this compound up to 1000°C. InP forms equilibria with Ni$_2$P, Ni$_5$P$_4$, NiP$_2$, and NiP$_3$, and indium is in thermodynamic equilibrium with Ni$_{12}$P$_2$ and Ni$_2$P. The samples were homogenized at 1000°C for 3 days, slowly cooled to 800°C, and annealed for about 3 weeks (Andersson-Söderberg 1991). Some of the samples needed further heat treatment to reach equilibrium. The samples were finally quenched in water. Ni$_{21}$In$_2$P$_6$ was prepared by heating Ni, InP, and red P in an evacuated silica tube. The temperature was raised from room

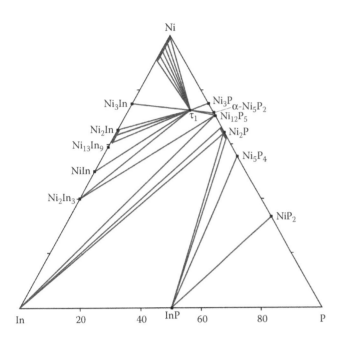

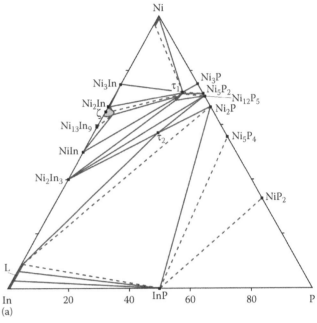

FIGURE 13.34 Isothermal section of the In–Ni–P ternary system at 800°C. (From Andersson-Söderberg, M., *J. Less Common Metals*, *171*(2), 179, 1991. With permission.)

temperature to 1000°C, and after a few days, it was slowly cooled to 800°C and kept there for a month.

The isothermal sections of the In–Ni–P ternary system at 470°C and 600°C are given in Figure 13.35 (Mohney and Chang 1992). Two ternary phases were found. A phase with composition $Ni_{57}In_{22}P_{21}$ (τ_2) is present at both 470°C and 600°C. It melts at 736°C. Ni_2InP (τ_3) was found to be in equilibrium with InP at 470°C. It melts at 526°C and crystallizes in the monoclinic structure with the lattice parameters $a = 681$, $b = 529$, and $c = 1280$ pm and $\beta = 95°$ (Sands et al. 1987). The phases (In), Ni_2P, Ni_5P_4, and NiP_2 are in equilibrium with InP at both temperatures (Mohney and Chang 1992). The samples were annealed at 600°C or 470°C for at least 1 week before they were quenched, ground into powder, pressed, encapsulated, and returned to the furnace for 1–3 months. The Ni-rich and lower-temperature samples were allowed more time to reach equilibrium. The annealed samples were then quenched in water.

Three ternary phases have been observed to be the primary reaction products of the Ni/InP reaction (Sands et al. 1987). The first phase, amorphous Ni_xInP ($x \approx 2.7$), forms at the interface by a solid-state amorphization process at low temperatures (<200°C). Amorphous Ni_xInP crystallizes at circa 300°C to form a hexagonal Ni_xInP phase with the lattice parameters

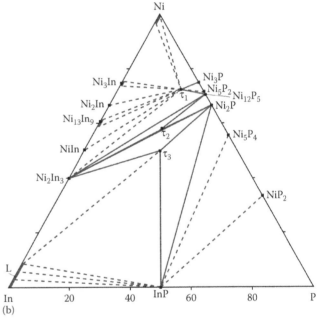

FIGURE 13.35 Isothermal sections of the In–Ni–P ternary system at (a) 470°C and (b) 600°C. (From Mohney, S.E., and Chang, Y.A., *J. Mater. Res.*, *7*(4), 955, 1992. With permission.)

$a = 412$ and $c = 483$ pm. A third ternary phase, Ni_2InP, nucleates at higher temperatures and is the final product. This phase is stable up to 500°C in samples capped with SiO_2. Andersson-Söderberg (1991) did not confirm the existence of any of these three ternary phases. However, when dealing with interface reactions, it cannot be assumed that the system is in equilibrium. The existence of these phases can most likely be explained by interfacial and/or diffusion constraints.

13.34 INDIUM–RUTHENIUM–PHOSPHORUS

The phase equilibria in the In–Ru–P ternary system at 600°C were experimentally investigated by mixing Ru and InP to provide samples of the desired overall compositions (Mohney 1998). The three-phase equilibria between InP, RuP_2, and In were observed.

13.35 INDIUM–RHODIUM–PHOSPHORUS

The phase equilibria in the In–Rh–P ternary system at 600°C were experimentally investigated by mixing Rh and InP to provide samples of the desired overall compositions (Mohney 1998). The three-phase equilibria between InP, RhP_2, and In were observed.

13.36 INDIUM–PALLADIUM–PHOSPHORUS

The isothermal section of the In–Pd–P ternary system at 600°C is shown in Figure 13.36 (Mohney and Chang 1993). Four ternary phases were identified. **$Pd_5In_2P_2$** (τ_1) appears to be related to the ternary phase identified as **Pd_2InP** (Ivey et al. 1991; Andersson-Söderberg 1992). It exists in two polymorphic modifications, and both of them crystallize in the tetragonal structure with the lattice parameters $a = 592.1$ and $c = 815.8$ pm and $a = 1156$ and $c = 857.8$ pm for α-$Pd_5In_2P_2$ and β-$Pd_5In_2P_2$, respectively. **$Pd_{3.83}InP$** (τ_2), **Pd_5InP** (τ_3),

and **$Pd_{70.2}In_{7.6}P_{22.1}$** ($\tau_4$) were also obtained in this system. Four binary phases were confirmed to be in equilibrium with InP: PdP_2, Pd_3In_7, Pd_2In_3, and PdIn at the stoichiometric composition or when the phase is In-rich. The pressed samples were sealed in evacuated quartz ampoules and annealed at 600°C for at least 1 week before they were quenched, ground into powder, pressed, encapsulated, and returned to the furnace for approximately 1 more month. After annealing, all samples were quenched in H_2O.

According to the data of El-Boragy and Schubert (1970), Pd_5InP crystallizes in the tetragonal structure with the lattice parameters $a = 392.8$ and $c = 691.7$ pm.

InP–Pd. The interaction of 40 nm of Pd layers on InP(100) substrates has been examined by Caron-Popowich et al. (1988). The samples were isochronally annealed in an Ar + 5 vol% H_2 ambience at temperatures from 175°C to 650°C. TEM, XRD, and Auger electron spectroscopy (AES) were used to study the phases that formed. The reaction began upon deposition. With subsequent annealing at 175°C, an amorphous ternary phase of approximate composition $Pd_{4.8}InP_{0.7}$ was formed. For samples annealed at 215°C and 250°C, the tetragonal ternary phase Pd_5InP ($a = 392.8$ and $c = 691.7$ pm) was found. After high-temperature annealing (up to 650°C), the cubic PdIn was observed.

The reaction between a thin layer of Pd (≈100 nm) and an InP substrate has also been studied by Ivey et al. (1991) at annealing temperatures between 250°C and 450°C for up to 30 s. Palladium reacts readily with InP, initially forming an amorphous ternary phase, which transforms to crystalline Pd_2InP upon annealing (cubic structure, $a = 830$ pm).

13.37 INDIUM–OSMIUM–PHOSPHORUS

The phase equilibria in the In–Os–P ternary system at 700°C were experimentally investigated by mixing Os and InP to provide samples of the desired overall compositions (Mohney 1998). The three-phase equilibria between InP, OsP_2, and In were observed.

13.38 INDIUM–IRIDIUM–PHOSPHORUS

The phase equilibria in the In–Ir–P ternary system at 700°C were experimentally investigated by mixing Ir and InP to provide samples of the desired overall compositions (Mohney 1998). The three-phase equilibria between InP, IrP_2, and In were observed.

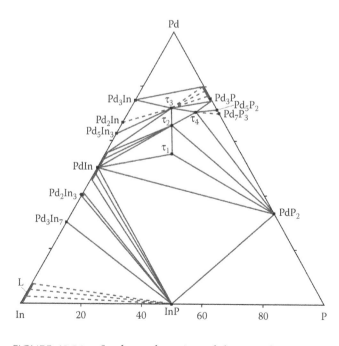

FIGURE 13.36 Isothermal section of the In–Pd–P ternary system at 600°C. (From Mohney, S.E., and Chang, Y.A., *Mater. Sci. Eng. B*, *18*(1), 94, 1993. With permission.)

13.39 INDIUM–PLATINUM–PHOSPHORUS

The isothermal sections of the In–Pt–P ternary system at 600°C and 725°C, constructed with the use of XRD and EPMA, are presented in Figure 13.37 (Lin et al. 1993). Three ternary phases, **Pt₅InP** (τ_1), **Pt₅₀.₇In₄₀.₇P₈.₆** or approximately **Pt₆In₅P** (τ_2), and **Pt₈In₂P** (τ_3), were found. The major difference between the 725°C and 600°C isothermal sections is the phases in equilibrium with InP. At 600°C, Pt₃In₇ and PtP₂ are in equilibrium with InP. However, a tie-line transition

from InP–Pt₃In₇ to In–PtP₂ was found between 600°C and 725°C. The samples were annealed at 725°C for at least 1 month.

Pt₅InP and Pt₈In₂P crystallize in the tetragonal structure with the lattice parameters $a = 393$ and $c = 695$ pm and $a = 395$ and $c = 1102$ pm, respectively (Lin et al. 1993; El-Boragy and Schubert 1970). Pt₅₀.₇In₄₀.₇P₈.₆ or Pt₆In₅P has a negligible range of homogeneity and crystallizes in the hexagonal structure with the lattice parameters $a = 683.6$ and $c = 552.0$ pm.

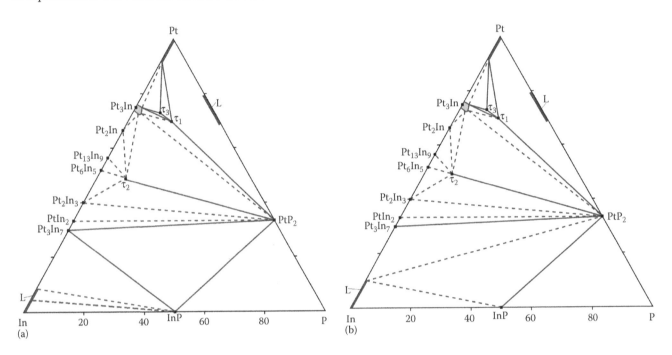

FIGURE 13.37 Isothermal sections of the In–Pt–P ternary system at (a) 600°C and (b) 725°C. (From Lin, C.F., et al., *J. Appl. Phys.*, 74(7), 4398, 1993. With permission.)

CHAPTER 14

Systems Based on InAs

14.1 INDIUM–SODIUM–ARSENIC

Na$_3$InAs$_2$ ternary compound, which crystallizes in the monoclinic structure with the lattice parameters $a = 967.7 \pm 0.4$, $b = 754.7 \pm 0.3$, and $c = 1573.1 \pm 0.6$ pm and $\beta = 92.6 \pm 0.1°$, is formed in the In–Na–As system (Cordier and Ochmann 1991q). It was synthesized from elements at 630°C.

14.2 INDIUM–POTASSIUM–ARSENIC

Three compounds, **K$_2$In$_2$As$_3$ (K$_4$In$_4$As$_6$)**, **K$_3$In$_2$As$_3$**, and **K$_6$InAs$_3$**, are formed in the In–K–As ternary system.

K$_2$In$_2$As$_3$ (K$_4$In$_4$As$_6$) crystallizes in the monoclinic structure with the lattice parameters $a = 1432.3 \pm 0.3$, $b = 710.6 \pm 0.2$, and $c = 1579.5 \pm 0.3$ pm and $\beta = 90.30 \pm 0.02°$ at 100 ± 5 K and a calculated density of 4.44 g·cm^{-3} (Birdwhistell et al. 1991) ($a = 1434.4 \pm 0.3$, $b = 711.2 \pm 0.2$, and $c = 1585.0 \pm 0.4$ pm and $\beta = 90.3 \pm 0.2°$ at room temperature [Cordier and Ochmann 1991o]). It was prepared by adding K (25.6 mM), In (8.5 mM), and As (17.0 mM) to a quartz tube that was then sealed under vacuum (Birdwhistell et al. 1991). The sample was heated at 750°C in a furnace for 10 h, and then cooled to ambient temperature over a 40 h period. A metallic gray crystalline title compound was obtained.

K$_3$In$_2$As$_3$ crystallizes in the orthorhombic structure with the lattice parameters $a = 1967.4 \pm 0.9$, $b = 678.4 \pm 0.3$, and $c = 1487.1 \pm 0.6$ pm (Cordier and Ochmann 1991f). It was synthesized from elements at 930°C.

K$_6$InAs$_3$ crystallizes in the triclinic structure with the lattice parameters $a = 913.0 \pm 0.2$, $b = 911.2 \pm 0.2$, and $c = 1960.9 \pm 0.5$ pm and $\alpha = 94.2 \pm 0.1°$, $\beta = 94.5 \pm 0.1°$, and $\gamma = 119.9 \pm 0.1$ (Blase et al. 1993a). It was synthesized from elements at 600°C in sealed Nb ampoules.

14.3 INDIUM–CESIUM–ARSENIC

Two compounds, **Cs$_5$In$_3$As$_4$** and **Cs$_6$InAs$_3$**, are formed in the In–Cs–As ternary system. Cs$_5$In$_3$As$_4$ crystallizes in the monoclinic structure with the lattice parameters $a = 1703.7 \pm 0.6$, $b = 1225.3 \pm 0.1$, and $c = 1810.5 \pm 0.4$ pm and $\beta = 117.24 \pm 0.02°$ and a calculated density of 5.175 g·cm^{-3} (Gascoin and Sevov 2001). To obtain the title compound, the mixture of Cs, In, and As was loaded in a Nb container that was then sealed by arc welding under Ar. The containers were enclosed in a fused silica ampoule that was flame sealed under vacuum. This assembly was heated at 500°C for 4 days and was then cooled to room temperature at a rate of 5°C·h^{-1}. All manipulations were performed inside a nitrogen-filled glove box with a moisture level below 1 ppm.

Cs$_6$InAs$_3$ also crystallizes in the monoclinic structure with the lattice parameters $a = 1046.9 \pm 0.3$, $b = 635.6 \pm 0.1$, and $c = 1220.8 \pm 0.2$ pm and $\beta = 101.3 \pm 0.1°$ (Blase et al. 1991a, 1991b, 1991e). It was obtained as shiny metallic crystals from elements at 600°C–680°C in sealed Nb ampoules.

14.4 INDIUM–COPPER–ARSENIC

The isothermal sections of the In–Cu–As ternary system at room temperature were approximately calculated by Klingbeil and Schmid-Fetzer (1989), and at 600°C determined experimentally (Figure 14.1) (Klingbeil and Schmid-Fetzer 1994). At this temperature, InAs is in

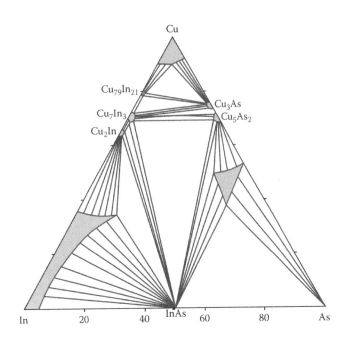

FIGURE 14.1   Isothermal section of the In–Cu–As ternary system at 600°C. (From Klingbeil, J., *CALPHAD*, *18*(4), 429, 1994. With permission.)

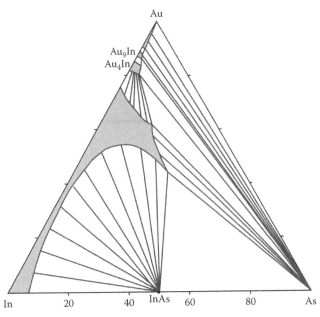

FIGURE 14.2   Isothermal section of the In–Au–As ternary system at 600°C. (From Klingbeil, J., *CALPHAD*, *18*(4), 429, 1994. With permission.)

thermodynamic equilibrium with liquid based on In, (Cu_2In), (Cu_7In_3), (Cu_5As_2), and ternary liquid near the Cu–As binary system. "**Cu_3InAs_2**" was found not to be a single phase (Luzhnaya et al. 1965). The InAs–Cu section of this ternary system was studied by Koppel et al. (1967b).

14.5 INDIUM–SILVER–ARSENIC

The isothermal sections of the In–Ag–As ternary system at room temperature were approximately calculated by Klingbeil and Schmid-Fetzer (1989), and at 500°C determined experimentally (Klingbeil and Schmid-Fetzer 1994). The experimental results confirm the calculated phase equilibria. At 500°C, InAs is in thermodynamic equilibrium with liquid based on In, (Ag_2In), (Ag), and Ag_9As. "**Ag_3InAs_2**" was found not to be a single phase (Gubskaya et al. 1965). The InAs–Ag section of this ternary system was studied by Koppel et al. (1967b), and the InAs–"Ag_3As" vertical section was constructed by Gubskaya et al. (1965).

14.6 INDIUM–GOLD–ARSENIC

The isothermal sections of the In–Au–As ternary system at room temperature were approximately calculated by Tsai and Williams (1986) and Klingbeil and Schmid-Fetzer (1989), and at 600°C determined experimentally (Figure 14.2) (Klingbeil and Schmid-Fetzer 1994).

At room temperature, five Au–In phases, $AuIn_2$, $AuIn$, Au_7In_3, Au_3In, and $Au_{10}In_3$, form quasi-binary cuts with InAs. The sections $Au_{10}In_3$–As, Au_4In–As, and Au_9In–As are also quasi-binary cuts of this ternary system. At 600°C, InAs is in thermodynamic equilibrium only with liquid based on In. "**Au_3InAs_2**" was found to not be a single phase (Luzhnaya et al. 1965).

14.7 INDIUM–CALCIUM–ARSENIC

$Ca_3In_2As_4$ ternary compound which crystallizes in the orthorhombic structure with the lattice parameters $a = 1621.2 \pm 0.6$, $b = 659.5 \pm 0.3$, and $c = 430.6 \pm 0.2$ pm and a calculated density of 4.69 g·cm^{-3} is formed in the In–Ca–As system (Cordier et al. 1986d). It was prepared from the elements in an Ar atmosphere at 980°C.

14.8 INDIUM–BARIUM–ARSENIC

Two compounds, **$Ba_2In_2As_5$** and **Ba_3InAs_3**, are formed in the In–Ba–As ternary system. $Ba_2In_2As_5$ appears to melt at 889°C and crystallizes in the orthorhombic structure with the lattice parameters $a = 1746.1 \pm 0.2$, $b = 429.05 \pm 0.04$, and $c = 1796.1 \pm 0.2$ pm, a calculated density of 6.04 g·cm^{-3}, and an energy gap of 0.4 eV (Mathieu et al. 2008). This compound was originally made in a reaction of the elements Ba, In, Ge, and As (molar ratio 0.5:15:1:1) with In acting as a flux. All elements were combined in an alumina crucible with half the In flux loaded on the bottom and half on top of the rest of the reactants. The crucible was placed in a fused silica tube; another alumina crucible was filled with ceramic fiber and inverted on top of the reaction crucible in the silica tube to act as a filter during centrifugation. The fused silica tube was sealed under a vacuum of 1.3 Pa, and then heated to 1000°C in 10 h, held at this temperature for 48 h, cooled to 850°C in 150 h, held for 24 h, cooled to 800°C in 50 h, held for 24 h, cooled to 700°C in 50 h, held for 24 h, and finally cooled to 600°C. At this temperature, the silica tubes were removed from the furnace, inverted, and quickly placed in a centrifuge to remove excess molten flux from the mass of the needle-shaped crystalline product. Because Ge was not incorporated into the product, the synthesis was done again without this element, heating at 600°C·h^{-1} to 1000°C, holding for 48 h, cooling at 2°C·h^{-1} to 600°C, and then centrifuging. The title compound was formed as very thin silvery needles with metallic luster and is stable under N_2 until about 900°C.

Ba_3InAs_3 also crystallizes in the orthorhombic structure with the lattice parameters $a = 1384.0 \pm 0.3$, $b = 460.1 \pm 0.1$, and $c = 1445.0 \pm 0.3$ pm (Somer et al. 1996b). It was prepared from a mixture of BaAs, In, and As (molar ratio: 3:1:1.5) in a sealed Nb ampoule at 1052°C.

14.9 INDIUM–STRONTIUM–ARSENIC

$SrIn_2As_2$ ternary compound, which crystallizes in the hexagonal structure with the lattice parameters $a = 421.45 \pm 0.07$ and $c = 1806.7 \pm 0.6$ pm, is formed in the In–Sr–As system (Mathieu et al. 2008).

14.10 INDIUM–ZINC–ARSENIC

The liquidus surface of the In–Zn–As ternary system is presented in Figure 14.3 (Koppel et al. 1967b). Four ternary eutectics exist in this system. The fields of α- and β-Zn_3As_2 primary crystallization occupy the biggest part of the liquidus surface. No ternary compounds were found.

InAs–Zn. This system is a non-quasi-binary section of the In–Zn–As ternary system (Koppel et al. 1967b). The solubility of Zn in InAs at 700°C, 800°C, 850°C, and 900°C is $1.2 \cdot 10^{20}$, $1.6 \cdot 10^{20}$, $2.2 \cdot 10^{20}$, and $2.9 \cdot 10^{20}$ cm^{-3}, respectively (Glazov et al. 1979). The alloys were annealed at 700°C for 1200 h, at 800°C for 1000 h, at 850°C for 800 h, and at 900°C for 500 h and studied through metallography and measurement of microhardness.

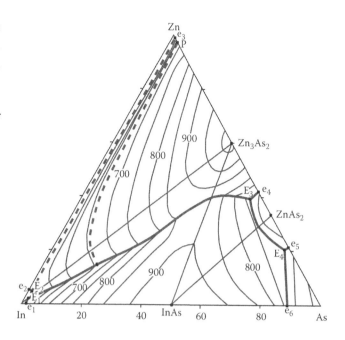

FIGURE 14.3 Liquidus surface of the In–Zn–As ternary system. (From Koppel, Kh.D., et al., *Izv. AN SSSR Neorgan. Mater.*, 3(2), 300, 1967. With permission.)

InAs–ZnAs$_2$. The phase diagram of this system is a eutectic type (Figure 14.4) (Koppel et al. 1967b; Marenkin et al. 1988b). The eutectic crystallizes at 735°C and contains 19.5 mol% InAs (Marenkin et al. 1988b) (at 725°C, contains 17.5 mol% InAs [Koppel et al. 1967b]).

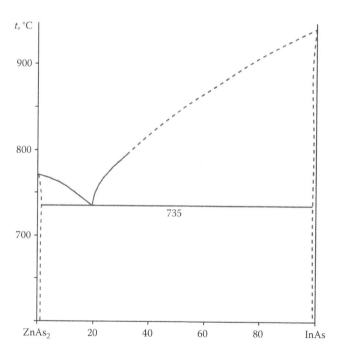

FIGURE 14.4 Phase diagram of the InAs–ZnAs$_2$ system. (From Marenkin, S.F., et al., *Zhurn. Neorgan. Khimii*, 33(4), 988, 1988. With permission.)

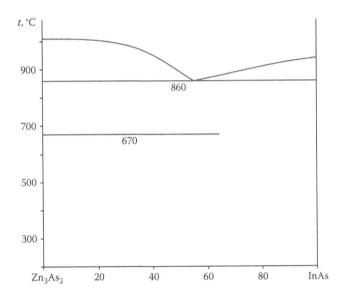

FIGURE 14.5 Phase diagram of the InAs–Zn$_3$As$_2$ system. (From Koppel, Kh.D., et al., *Izv. AN SSSR Neorgan. Mater.*, 3(2), 300, 1967. With permission.)

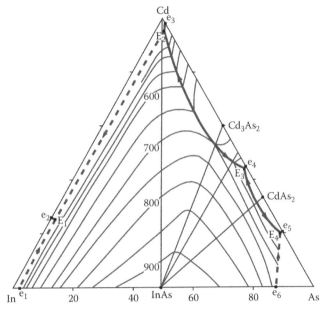

FIGURE 14.6 Liquidus surface of the In–Cd–As ternary system. (From Koppel, Kh.D., et al., *Zhurn. Neorgan. Khimii*, 10(10), 2315, 1965. With permission.)

InAs–Zn$_3$As$_2$. The phase diagram of this system is a eutectic type (Figure 14.5) (Koppel et al. 1967b). The eutectic crystallizes at 860°C and contains 55 mol% InAs. The phase transformation of Zn$_3$As$_2$ takes place at 670°C.

14.11 INDIUM–CADMIUM–ARSENIC

The liquidus surface of the In–Cd–As ternary system is presented in Figure 14.6 (Koppel et al. 1965; Lushnaja et al. 1966). Four ternary eutectics exist in this system. The field of InAs primary crystallization occupies the biggest part of the liquidus surface. There are also the small fields of primary crystallization of In, Cd, Cd$_3$As$_2$, CdAs$_2$, and As. No ternary compounds were found.

InAs–Cd. The phase diagram of this system is a eutectic type (Figure 14.7) (Koppel et al. 1965). The eutectic crystallizes at 318°C and contains 4 mol% InAs.

InAs–CdAs$_2$. The phase diagram of this system is a eutectic type (Figure 14.8) (Koppel et al. 1965; Marenkin et al. 1988c). The eutectic crystallizes at 611°C and contains 9 mol% InAs.

InAs–Cd$_3$As$_2$. The phase diagram of this system is a eutectic type (Figure 14.9) (Koppel et al. 1965). The eutectic crystallizes at 702°C and contains 29 mol% InAs.

14.12 INDIUM–THALLIUM–ARSENIC

InAs–Tl. The solubility of Tl in InAs at the temperature region from 500°C to 800°C was determined by

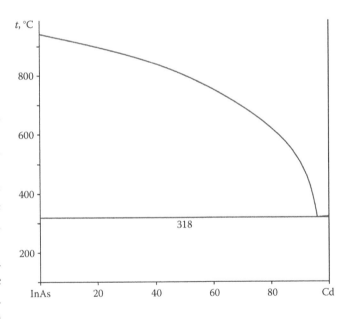

FIGURE 14.7 Phase diagram of the InAs–Cd system. (From Koppel, Kh.D., et al., *Zhurn. Neorgan. Khimii*, 10(10), 2315, 1965. With permission.)

Mavlonov and Makhmatkulov (1983). It was shown that this solubility reaches the maximum value ($5 \cdot 10^{20}$ cm^{-3}) at ca. 750°C.

InAs–TlAs. A number of key optical and structural properties of In$_{1-x}$Tl$_x$As solid solutions were studied within local density functional theory (van Schilfgaarde

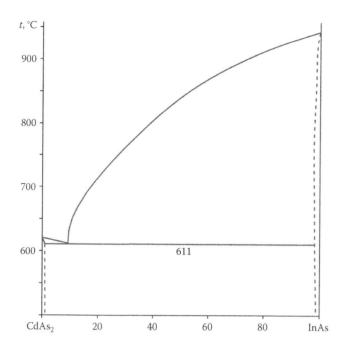

FIGURE 14.8   Phase diagram of the InAs–CdAs$_2$ system. (From Marenkin, S.F., et al., *Izv. AN SSSR Neorgan. Mater.*, *24*(9), 1440, 1988. With permission.)

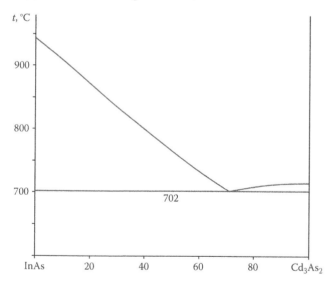

FIGURE 14.9   Phase diagram of the InAs–Cd$_3$As$_2$ system. (From Koppel, Kh.D., et al., *Zhurn. Neorgan. Khimii*, *10*(10), 2315, 1965. With permission.)

et al. 1994). In$_{1-x}$Tl$_x$As at $x = 0.15$ was estimated to have an energy gap of 0.1 eV. This solid solution possesses a direct energy gap for all x, with a significant bowing.

14.13 INDIUM–EUROPIUM–ARSENIC

EuIn$_2$As$_2$ ternary compound, which crystallizes in the hexagonal structure with the lattice parameters $a = 420.55 \pm 0.03$ and $c = 1788.7 \pm 0.2$ pm, is formed in the In–Eu–As system (Mathieu et al. 2008) ($a = 420.67 \pm 0.03$ and $c = 1788.9 \pm 0.2$ pm and a calculated density of 6.438 g·cm^{-3} at 90 ± 2 K [Goforth et al. 2008]). Large single crystals of the title compound were synthesized from the elements in a reactive indium flux using a 3:36:9 (Eu/In/As) molar ratio (10 g total mass). The mixture was heated at the rate of 60°C·h^{-1} to 1100°C and maintained at this temperature for 12 h before it was slowly cooled to 700°C at a rate of 2°C·h^{-1}. The ampoule was removed from the furnace at 700°C, inverted, and centrifuged to remove the molten indium flux. The reaction vessel was broken in air to yield large, black hexagonal plate crystals of EuIn$_2$As$_2$, in addition to lesser quantities of large, silver block crystals of **Eu$_3$In$_2$As$_4$** and large, black block crystals of **Eu$_3$InAs$_3$**.

14.14 INDIUM–SILICON–ARSENIC

InAs–Si. The phase diagram of this system is a eutectic type (Klimova et al. 1969). The eutectic crystallizes at circa 900°C and contains 20 at% Si.

14.15 INDIUM–GERMANIUM–ARSENIC

The solid solution region based on Ge in this system determined through differential thermal analysis (DTA) and metallography is given in Figure 14.10 (Glazov and Malyutina 1963a). The samples for solid solubility determination were annealed at 500°C, 600°C, 700°C, and 800°C for 1000, 860, 650, and 600 h, respectively.

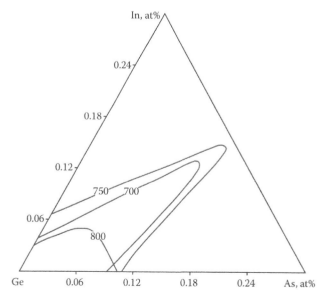

FIGURE 14.10   Solid solutions based on Ge in the Ge–In–As ternary system. (From Glazov, V.M., and Malyutina, G.L., *Zhurn. Neorgan. Khimii*, *8*(8), 1921, 1963. With permission.)

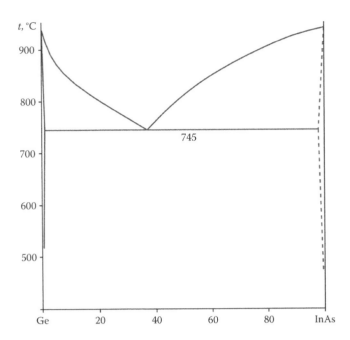

FIGURE 14.11 Phase diagram of the InAs–Ge system. (From Glazov, V.M., and Malyutina, G.L., *Zhurn. Neorgan. Khimii*, 8(8), 1921, 1963. With permission.)

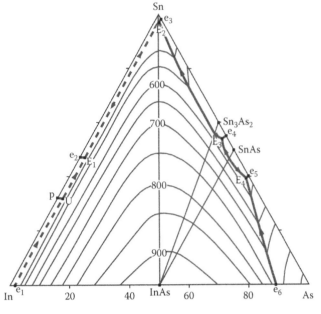

FIGURE 14.12 Liquidus surface of the In–Sn–As ternary system. (From Koppel, Kh.D., et al., *Izv. AN SSSR Neorgan. Mater.*, 3(2), 300, 1967. With permission.)

InAs–Ge. The phase diagram of this system is a eutectic type (Figure 14.11) (Glazov and Malyutina 1963a). The eutectic crystallizes at 745°C ± 3°C and contains 37 mol% InAs.

As a result of ultrafast crystallization (10^{7}°C·s^{-1}) of the InAs–Ge melts, a metastable phase with unknown structure was fixed (Glazov et al. 1977a).

14.16 INDIUM–TIN–ARSENIC

The liquidus surface of the In–Sn–As ternary system is presented in Figure 14.12 (Koppel et al. 1967b). The field of InAs primary crystallization occupies the biggest part of the liquidus surface. There are also the fields of primary crystallization of In, the β-phase of the In–Sn system, Sn, Sn$_3$As$_2$, SnAs, and As. Four ternary eutectics exist in this system. No ternary compounds were determined.

The liquidus isotherms for the In–Sn–As system were calculated in the temperature range 440°C–740°C, the tin content in the solvent being varied from 0 to 0.90 molar fractions (Hitova et al. 1991). The theoretical curves were compared with the experimental data. The results showed that the liquid phase in the In–Sn–As system can be considered a quasi-regular solution only for the 560°C–620°C temperature range. The calculated liquidus isotherms for the In–Sn–As system near the In–Sn binary system are shown in Figure 14.13

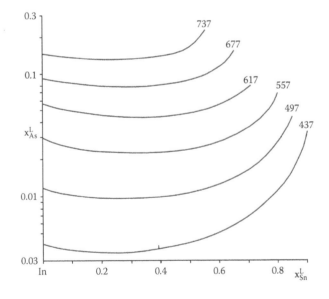

FIGURE 14.13 Calculated liquidus isotherms for the In–Sn–As system near the In–Sn binary system. (From Hitova, L., et al., *Thin Solid Films*, 198(1–2), 387, 1991. With permission.)

(Hitova et al. 1991). The liquidus isotherms are plotted as a dependence of the arsenic solubility, x_{As}^{L}, on the composition of the two-component (In + Sn) solvent for each temperature under consideration. The liquidus curves are asymmetric with respect to the tin content in the solution and are similar for all investigated temperatures. With rising temperature, the distance between the curves tends to decrease.

The dissolution of the InAs in In–Sn melts at 440°C–740°C was experimentally studied by Yakimova and Hitova (1992). It was shown that the increase of InAs solubility by a factor of almost 2 and the decrease in the liquidus slope by a factor of about 3 when Sn is added in quantity from 10 to 50 mol% at 440°C were obtained. The enthalpies of InAs dissolution in the In–Sn melts were obtained. It was shown that the determined values decrease as the tin concentration increases.

InAs–Sn. The phase diagram of this system is a eutectic type (Figure 14.14) (Leonhardt and Kühn 1975; Koppel et al. 1967b). The eutectic crystallizes at 220°C–230°C and is degenerated from the Sn side. The calculated solubility of Sn in InAs is shown in Figure 14.15 (Morozov et al. 1985). The temperature dependence of the InAs limiting solubility in the Sn melt was determined by Vasil'ev et al. (1983) and is given in Figure 14.16. The solubility of InAs in the Sn melt at 450°C–750°C was also determined by Leonhardt and Kühn (1974b).

InAs–SnAs. The phase diagram of this system is a eutectic type (Figure 14.17) (Koppel et al. 1967b). The eutectic crystallizes at 580°C and contains ~7 mol% InAs. Calculated homogeneity regions of $In_{1-x}Sn_xAs$ solid solution at tin concentrations up to $1.5 \cdot 10^{20}$ cm^{-3} are presented in Figure 14.18 (Morozov et al. 1985).

InAs–Sn$_3$As$_2$. The phase diagram of this system is a eutectic type (Figure 14.19) (Koppel et al. 1967b). The

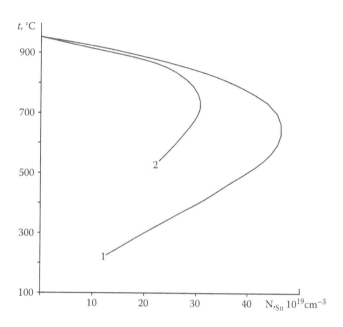

FIGURE 14.15　Calculated solubility of Sn in InAs along the (1) InAs–Sn and (2) InAs–Sn$_3$As$_2$ sections. (From Morozov, A.N., et al., *Kristallografiya*, 30(3), 548, 1985. With permission.)

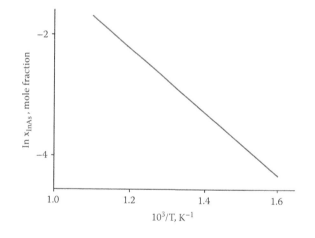

FIGURE 14.16　Temperature dependence of the InAs limiting solubility in the Sn melt. (From Vasil'yev, M.G., et al., *Izv. AN SSSR Neorgan. Mater.*, 19(11), 1933, 1983. With permission.)

eutectic crystallizes at 568°C and contains ~10 mol% InAs.

14.17 INDIUM–LEAD–ARSENIC

The liquidus surface of the In–Pb–As ternary system is presented in Figure 14.20 (Koppel et al. 1967b). The field of InAs primary crystallization occupies the biggest part of the liquidus surface. There are also the fields of primary crystallization of In, the α-phase of the In–Pb system, Pb, and As. No ternary compounds were determined. The calculated liquidus isotherms for

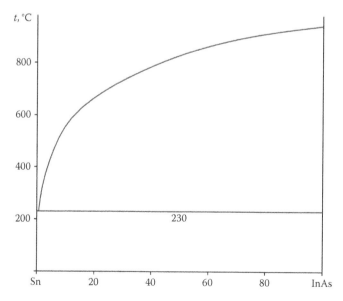

FIGURE 14.14　Phase diagram of the InAs–Sn system. (From Koppel, Kh.D., et al., *Izv. AN SSSR Neorgan. Mater.*, 3(2), 300, 1967. With permission.)

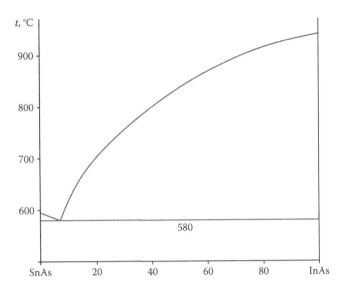

FIGURE 14.17   Phase diagram of the InAs–SnAs system. (From Koppel, Kh.D., et al., *Izv. AN SSSR Neorgan. Mater.*, 3(2), 300, 1967. With permission.)

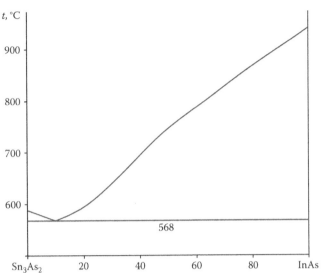

FIGURE 14.19   Phase diagram of the InAs–Sn₃As₂ system. (From Koppel, Kh.D., et al., *Izv. AN SSSR Neorgan. Mater.*, 3(2), 300, 1967. With permission.)

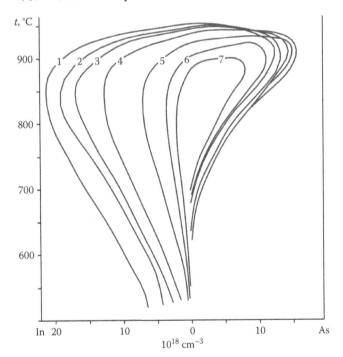

FIGURE 14.18   Calculated homogeneity region of $In_{1-x}Sn_xAs$ solid solution at tin concentrations of (1) 0, (2) $5 \cdot 10^{18}$, (3) 10^{19}, (4) $2 \cdot 10^{19}$, (5) $5 \cdot 10^{19}$, (6) 10^{20}, and (7) $1.5 \cdot 10^{20}$ cm^{-3}. (From Morozov, A.N., et al., *Kristallografiya*, 30(3), 548, 1985. With permission.)

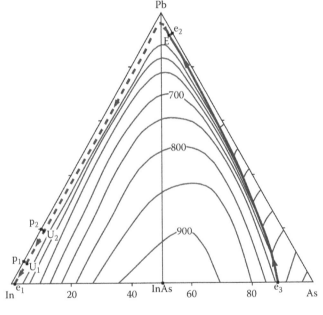

FIGURE 14.20   Liquidus surface of the In–Pb–As ternary system. (From Koppel, Kh.D., et al., *Izv. AN SSSR Neorgan. Mater.*, 3(2), 300, 1967. With permission.)

the In–Pb–As system near the In–Pb binary system are shown in Figure 14.21 (Baranov et al. 1992).

InAs–Pb. The phase diagram of this system is a eutectic type (Figure 14.22) (Koppel et al. 1967b; Leonhardt and Kühn 1975; Baranov et al. 1992). The eutectic crystallizes

at 318°C–320°C and is degenerated from the Pb side. The solubility of InAs in the Pb melt at 525°C–825°C was determined by Leonhardt and Kühn (1974b).

14.18 INDIUM–TITANIUM–ARSENIC

The isothermal section of the In–Ti–As ternary system at room temperature was approximately calculated by Klingbeil and Schmid-Fetzer (1989) (Figure 14.23). The

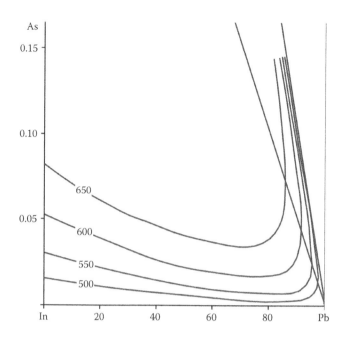

FIGURE 14.21 Calculated liquidus isotherms for the In–Pb–As system near the In–Pb binary system. (From Baranov, A.N., et al., *Zhurn. Neorgan. Khimii*, 37(2), 448, 1992. With permission.)

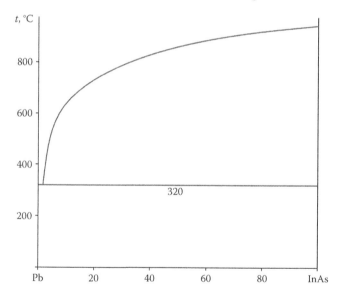

FIGURE 14.22 Phase diagram of the InAs–Pb system. (From Koppel, Kh.D., et al., *Izv. AN SSSR Neorgan. Mater.*, 3(2), 300, 1967. With permission.)

calculations are based on the following approximations: ternary phases and solid solubility are disregarded, and the Gibbs energy of formation of binary compounds is mostly estimated by the enthalpy of formation calculated from Miedema's model. The dotted lines indicate possible alternative tie-lines. According to x-ray diffraction (XRD), at 600°C InAs is in thermodynamic equilibrium with TiAs and TiAs₂ and TiAs coexists with

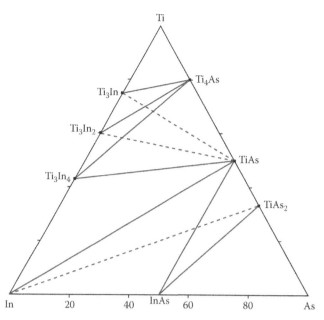

FIGURE 14.23 Calculated isothermal section of the In–Ti–As ternary system at room temperature. (From Klingbeil, J., and Schmid-Fetzer, R., *CALPHAD*, 13(4), 367, 1989. With permission.)

liquid based on In, Ti_3In_4, and Ti_3In_2. There are also Ti_3In_2–Ti_5As_3, Ti_3In_2–Ti_3As, and (Ti_3In)–Ti_3As tie-lines (Klingbeil and Schmid-Fetzer 1994). The samples were slowly heated to 600°C and annealed at this temperature for at least 30 days.

14.19 INDIUM–ZIRCONIUM–ARSENIC

The isothermal section of the In–Zr–As ternary system at room temperature was approximately calculated by Klingbeil and Schmid-Fetzer (1989) (Figure 14.24). The calculations are based on the following approximations: ternary phases and solid solubility are disregarded, and the Gibbs energy of formation of binary compounds is mostly estimated by the enthalpy of formation calculated from Miedema's model. The dotted line indicates a possible alternative tie-line.

14.20 INDIUM–HAFNIUM–ARSENIC

The isothermal section of the In–Hf–As ternary system at room temperature was approximately calculated by Klingbeil and Schmid-Fetzer (1989) (Figure 14.25). The calculations are based on the following approximations: ternary phases and solid solubility are disregarded, and the Gibbs energy of formation of binary compounds is mostly estimated by the enthalpy of formation calculated from Miedema's model. The dotted line indicates a possible alternative tie-line.

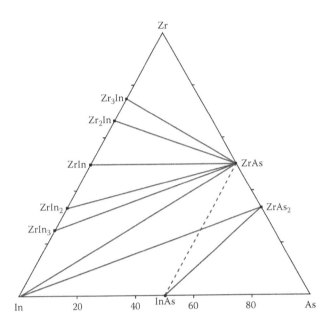

FIGURE 14.24 Calculated isothermal section of the In–Zr–As ternary system at room temperature. (From Klingbeil, J., and Schmid-Fetzer, R., *CALPHAD*, *13*(4), 367, 1989. With permission.)

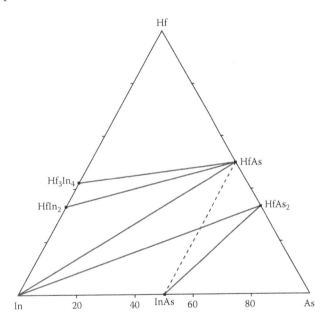

FIGURE 14.25 Calculated isothermal section of the In–Hf–As ternary system at room temperature. (From Klingbeil, J., and Schmid-Fetzer, R., *CALPHAD*, *13*(4), 367, 1989. With permission.)

14.21 INDIUM–ANTIMONY–ARSENIC

Several determinations of the liquidus surface of the In–As–Sb ternary system have been reported in the literature. Extensive measurements of the constitution of this system were performed by Shih and Peretti (1954, 1956)

using thermal analysis and metallographic techniques. In these studies, the In–As–Sb (Shih and Peretti 1954) section, as well as three sections in the In–As–In–InSb, four sections in the InSb–In–As–Sb, and two sections in the In–As–As–Sb subsystems (Shih and Peretti 1956), were examined. Dissolution experiments were performed by (Stringfellow and Greene 1971) for solutions that dilute in As at temperatures typically encountered in liquid phase epitaxy (LPE). Liquidus data were also obtained by Abrokwah and Gershenzon (1981) at 500°C and 520°C by direct visual observation of the dissolution of a ternary crust. In addition to the quasi-binary liquidus measurements mentioned previously by Gromova et al. (1983), these authors also investigated six compositions dilute in As. It is well established in the literature that a significant amount of supercooling (as great as 50°C) can exist, even when using low cooling rates or a solid substrate, as in the LPE experiments. In view of this difficulty, the As-dilute liquidus data are in surprisingly good agreement. The full liquidus surface of the In–As–Sb ternary system was calculated using various solution models (Stringfellow and Greene 1969; Panish and Ilegems 1972; Lendvay 1984; Ishida et al. 1989b; Li et al. 1998b; Litvak and Charykov 1991) and is presented in Figure 14.26. The liquidus surface of the InAs–In–InSb subsystem, as well as the isobars of this liquidus surface, was determined by Sorokina et al. (1984). The

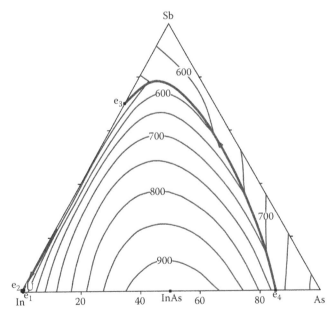

FIGURE 14.26 Calculated liquidus surface of the In–As–Sb ternary system. (From Li, J.-B., et al., *J. Phase Equilibria*, *19*(5), 473, 1998. With permission.)

liquidus and solidus isotherms of the In–As–Sb ternary system in the In corner were investigated experimentally by Baranov et al. (1990a) within the temperature range of 460°C–600°C and calculated in the temperature range of 300°C–400°C by Popov et al. (1998). The calculations have been made by applying the rare simple solution model. The results of calculations are presented in Figures 14.27 and 14.28. Figure 14.27 shows the variation of the Sb atomic fraction, x_{Sb}, in the liquid versus the As atomic fraction, x_{As}, for several values of the temperature (°C). Figure 14.28 shows the corresponding solidus composition x of the $InAs_{1-x}Sb_x$ as a function of the Sb atomic fraction in the liquid. The simple phase diagram model agrees with the obtained experimental data quite well and can be a good guide for low-temperature LPE growth of $InAs_{1-x}Sb_x$ solid solutions over the 300°C–400°C temperature range.

The isothermal sections of the In–As–Sb ternary system at 575°C, 750°C, 850°C, and 900°C were constructed by Kaufman et al. (1981), at 450°C and 700°C by Ishida et al. (1989a), and at 450°C by Li et al. (1998b). The isothermal section at 450°C from Li et al. (1998b) is given in Figure 14.29. Three-phase equilibrium regions (I and II) occur below the critical temperature (492.2°C). The difference between Ishida et al. (1989a) and Li et al. (1998b) is in the presence at 450°C of a narrow liquid region adjacent to the As–Sb binary in Ishida

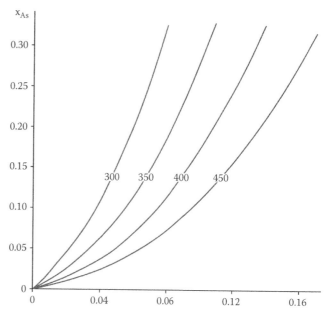

FIGURE 14.28   Solid compositions of $InAs_{1-x}Sb_x$ layers versus the Sb atom fraction in the liquid phase. (From Popov, A.S., et al., *J. Cryst. Growth*, 186(3), 338, 1998. With permission.)

et al. (1989a), whereas the calculations (Li et al. 1998b) revealed no liquid at this temperature.

InAs–InSb. A number of investigators have examined the quasi-binary InAs–InSb section. Thermal analysis, XRD, and metallographic studies by Shih and Peretti (1953, 1954, 1956) indicate a eutectic-type phase

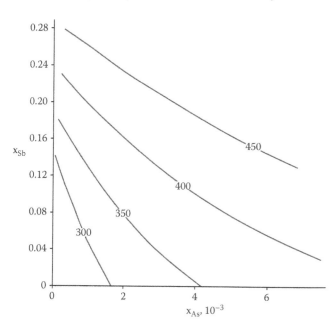

FIGURE 14.27   Liquidus data of the In–As–Sb phase diagram in the range 300°C–450°C. (From Popov, A.S., et al., *J. Cryst. Growth*, 186(3), 338, 1998. With permission.)

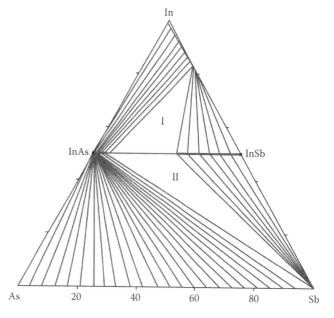

FIGURE 14.29   Calculated isothermal section of the In–As–Sb ternary system at 450°C. (From Li, J.-B., et al., *J. Phase Equilibria*, 19(5), 473, 1998. With permission.)

diagram that is nearly degenerate with respect to InSb. The extrapolated InAs-rich solidus gives an estimated 2 at.% InSb maximum solubility at the eutectic temperature. According to the data of Goryunova (1955) and Goryunova and Fedorova (1955), the mutual solubility of InAs and InSb is not higher than 3 mol%. In contrast, the solidus determination by XRD (Woolley and Smith 1958a) indicates an isomorphous-type phase diagram with a broad liquidus–solidus gap. This finding was later confirmed by near-equilibrium liquid phase epitaxial growth studies at constant temperature (Stringfellow and Greene 1971). With the exception of the solidus data of Shih and Peretti (1953, 1954, 1956), the measured values of the more recent solidus data are in good agreement. Several works (Shih and Peretti 1953; Woolley and Smith 1958a; Stringfellow and Greene 1971; Gromova et al. 1983) have reported measurements of the liquidus temperatures in the alloys of the quasi-binary section. Although Gromova et al. (1983) extrapolated their DTA measurements to a zero cooling rate to account for supercooling of the melt, their four liquidus temperatures for Sb-rich melts are somewhat below the values reported by other investigators. Further evidence for the formation of the continuous solid solution is found in Goryunova and Takhtareva (1961).

Part of the reason for the discrepancy in the solidus is related to the difficulty in achieving equilibrium at the relatively low temperatures of solid solution existence. Although no solid solution thermodynamic data are available, semiempirical and assessment procedures have been used to estimate the interaction energy in the solid solution. Using these procedures, solid solution immiscibility was predicted above room temperature with a calculated critical temperature ranging from well below to just above the solidus curve (Stringfellow and Greene 1969; Stringfellow 1972a, 1982b; Kaufman et al. 1981). The existence of solid solution immiscibility has recently been experimentally confirmed by electron probe microanalysis (EPMA) (Ishida et al. 1989a; Li et al. 1998b). Their data indicate an asymmetric miscibility gap with the critical temperature located just below the solidus curve at an InAs-rich composition.

The phase diagram of the InAs–InSb quasi-binary system was calculated by many investigators (Steininger 1970; Osamura et al. 1972b; Panish and Ilegems 1972; Stringfellow 1972a; Kaufman et al. 1981; Ishida et al.

1989a, 1989b; Li et al. 1998b). The model of regular solutions could be used for the theoretical description of this system (Semenova et al. 2008). The phase diagram of this system is given in Figure 14.30 from Li et al. (1998b). According to the data of Stringfellow (1982b), the critical temperature for the miscibility gap is 299°C (approximately 230°C [Semenova et al. 2008]).

The concentration dependence of the lattice parameter for the $InAs_xSb_{1-x}$ solid solutions shows a slight deviation from Vegard's law (Woolley and Warner 1964). Slight negative deviations were observed for InAs compositions greater than 70 mol%, while somewhat larger positive deviations were reported for InSb-rich compositions.

A Calvet microcalorimeter has been used with liquid tin solution techniques to measure the mixing enthalpy of the InAs–InSb system (Rugg et al. 1995). It was shown that the enthalpy of formation of $InAs_xSb_{1-x}$ solid solutions is relatively small and probably positive.

Bulk $InAs_xSb_{1-x}$ crystals have been grown by using the vertical Bridgman method under the solidification condition (Haris et al. 2013). Structural analysis reveals that the as-grown $InAs_xSb_{1-x}$ crystals have the incorporation of arsenic inside the InSb matrix. Chemically homogeneous solid solutions in this system have been obtained by an ultrafast cooling rate (10^6–10^8°C·s^{-1}) (Glazov and Poyarkov 1999, 2000).

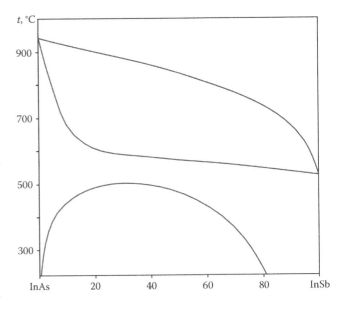

FIGURE 14.30 Calculated phase diagram of the InAs–InSb system. (From Li, J.-B., et al., *J. Phase Equilibria*, *19*(5), 473, 1998. With permission.)

14.22 INDIUM–BISMUTH–ARSENIC

The liquidus surface of the InAs–In–Bi subsystem of the In–Bi–P ternary system is given in Figure 14.31 (Islamov et al. 1984).

InAs–InBi. The phase diagram of this system is a eutectic type with the eutectic degenerated from the InBi side (Figure 14.32) (Akchurin et al. 1984; Islamov et al. 1984; Yevgen'ev et al. 1988; Ma et al. 1990, 1991). The eutectic crystallizes at 110°C (Akchurin et al. 1984; Ma et al. 1990, 1991). The solubility from the InAs side is not higher than 0.1 at% Bi. Using Knudsen's method, the temperature and concentration dependences of the vapor pressure in this system have been studied (Yevgen'ev et al. 1988). Based on the experimental results and the results of the calculation of the interaction parameter and the activity coefficient, it has been shown that this system is disposed to delamination. Significant positive deviations from Raoult's law exist in the InAs–InBi system.

Thermodynamic calculations of the phase diagram of this system were carried out using the delta-lattice-parameter model (Ma et al. 1990, 1991). The obtained results agree well with the experimental data of Akchurin et al. (1984), Islamov et al. (1984), and Yevgen'ev et al. (1988). The calculations predict that the solid solubility limit of Bi in InAs is less than 0.025 at%. A tremendously

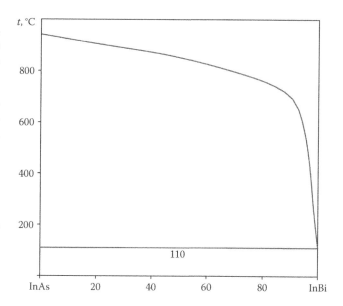

FIGURE 14.32 Phase diagram of the InAs–InBi system. (From Islamov, S.A., et al., *Zhurn. Neorgan. Khimii*, 29(9), 2369, 1984. With permission.)

large miscibility gap exists in this system. The critical temperature was predicted to be 2569°C at 41.8 mol% InBi.

High-quality crystals of $InBi_xAs_{1-x}$ solid solutions have been grown using the metal-organic vapor phase epitaxy (MOVPE) technique over a temperature range from 600°C to 275°C (Ma et al. 1989, 1991, 1992). It was demonstrated that lowering the growth temperature is the most effective approach for increasing the maximum Bi content in the solid solutions. $InBi_xAs_{1-x}$ with a Bi concentration as high as 6.1 at% has been successfully grown at 275°C.

InAs–In₂Bi. The phase diagram of this system is a eutectic type, with the eutectic degenerated from the In₂Bi side (Figure 14.33) (Akchurin et al. 1984; Islamov et al. 1984; Yevgen'ev et al. 1988). The eutectic crystallizes at 70°C (Akchurin et al. 1984). The solubility from the InAs side is not higher than 0.1 at% Bi. Using Knudsen's method, the temperature and concentration dependences of the vapor pressure in this system have been studied (Yevgen'ev et al. 1988). Based on the experimental results and the results of the calculation of the interaction parameter and the activity coefficient, it has been shown that this system is disposed to delamination. Significant positive deviations from Raoult's law exist in the InAs–In₂Bi system.

InAs–Bi. The phase diagram of this system is a eutectic type, with the eutectic degenerated from the Bi side

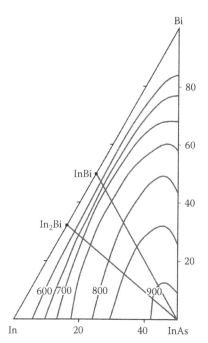

FIGURE 14.31 Liquidus surface of the InAs–In–Bi subsystem of the In–Bi–As ternary system. (From Islamov, S.A., et al., *Zhurn. Neorgan. Khimii*, 29(9), 2369, 1984. With permission.)

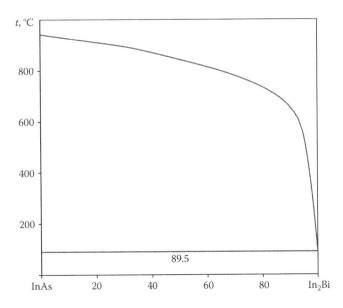

FIGURE 14.33   Phase diagram of the InAs–In$_2$Bi system. (From Islamov, S.A., et al., *Zhurn. Neorgan. Khimii*, *29*(9), 2369, 1984. With permission.)

(Figure 14.34) (Leonhardt and Kühn 1975; Akchurin et al. 1984; Islamov et al. 1984; Yevgen'ev et al. 1988). The eutectic crystallizes at 245°C (Akchurin et al. 1984). The solubility from the InAs side is not higher than 0.1 at% Bi. The solubility of InAs in the Bi melt at 475°C–900°C was determined by Leonhardt and Kühn (1974b). Using Knudsen's method, the temperature and concentration dependences of the vapor pressure in this system have been studied (Yevgen'ev et al. 1988). Based on the experimental results and the results of the calculation of the interaction parameter and the activity coefficient, it has been shown that this system is disposed to delamination. Significant positive deviations from Raoult's law exist in the InAs–Bi system.

14.23 INDIUM–VANADIUM–ARSENIC

The isothermal section of the In–V–As ternary system at room temperature was approximately calculated by Klingbeil and Schmid-Fetzer (1989) (Figure 14.35). The calculations are based on the following approximations: ternary phases and solid solubility are disregarded, and the Gibbs energy of formation of binary compounds is mostly estimated by the enthalpy of formation calculated from Miedema's model. The dotted lines indicate possible alternative tie-lines. According to XRD, the isothermal section at 600°C is the same as at room temperature: InAs is in thermodynamic equilibrium with VAs and VAs$_2$, and the liquid based on In coexists with V$_3$As, V$_2$As, V$_5$As$_3$, V$_3$As$_2$, V$_4$As$_3$, and VAs (Klingbeil and Schmid-Fetzer 1994). The samples were slowly heated to 600°C and annealed at this temperature for at least 30 days.

14.24 INDIUM–NIOBIUM–ARSENIC

The isothermal section of the In–Nb–As ternary system at room temperature was approximately calculated by

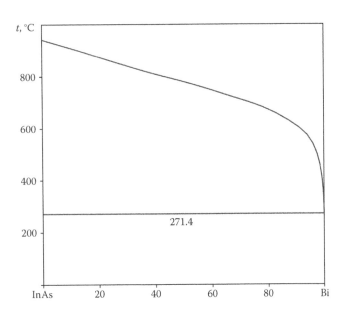

FIGURE 14.34   Phase diagram of the InAs–Bi system. (From Islamov, S.A., et al., *Zhurn. Neorgan. Khimii*, *29*(9), 2369, 1984. With permission.)

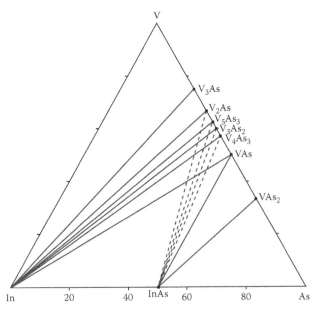

FIGURE 14.35   Calculated isothermal section of the In–V–As ternary system at room temperature. (From Klingbeil, J., and Schmid-Fetzer, R., *CALPHAD*, *13*(4), 367, 1989. With permission.)

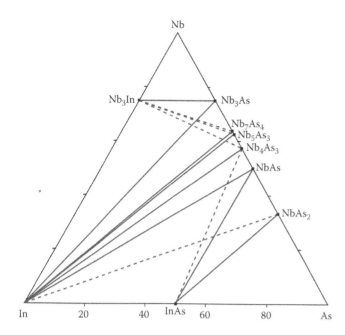

FIGURE 14.36 Calculated isothermal section of the In–Nb–As ternary system at room temperature. (From Klingbeil, J., and Schmid-Fetzer, R., *CALPHAD*, *13*(4), 367, 1989. With permission.)

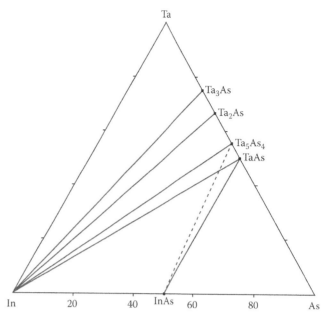

FIGURE 14.37 Calculated isothermal section of the In–Ta–As ternary system at room temperature. (From Klingbeil, J., and Schmid-Fetzer, R., *CALPHAD*, *13*(4), 367, 1989. With permission.)

Klingbeil and Schmid-Fetzer (1989) (Figure 14.36). The calculations are based on the following approximations: ternary phases and solid solubility are disregarded, and the Gibbs energy of formation of binary compounds is mostly estimated by the enthalpy of formation calculated from Miedema's model. The dotted lines indicate possible alternative tie-lines.

14.25 INDIUM–TANTALUM–ARSENIC

The isothermal section of the In–Ta–As ternary system at room temperature was approximately calculated by Klingbeil and Schmid-Fetzer (1989) (Figure 14.37). The calculations are based on the following approximations: ternary phases and solid solubility are disregarded, and the Gibbs energy of formation of binary compounds is mostly estimated by the enthalpy of formation calculated from Miedema's model. The dotted line indicates a possible alternative tie-line.

14.26 INDIUM–OXYGEN–ARSENIC

The isothermal section of the In–O–P ternary system at room temperature is given in Figure 14.38 (Schwartz et al. 1982b; Schwartz 1983). The major area of uncertainty is contained in the region of the diagram bounded by **InAsO$_4$–As$_2$O$_3$–As$_2$O$_5$**. In particular, a complete characterization of **In(AsO$_3$)$_3$** remains to be

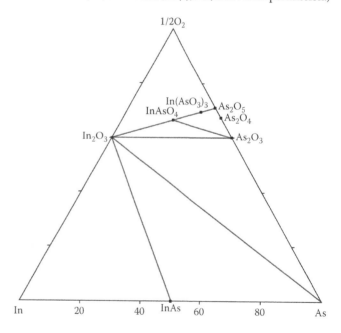

FIGURE 14.38 Isothermal section of the In–O–As ternary system at room temperature. (From Schwartz, G.P., *Thin Solid Films*, *103*(1–3), 3, 1983. With permission.)

accomplished, along with the establishment of the tie-lines to this material and As$_2$O$_4$.

14.27 INDIUM–SULFUR–ARSENIC

The calculated isotherms of the liquidus surface of the In–S–As ternary system are presented in Figure 14.39

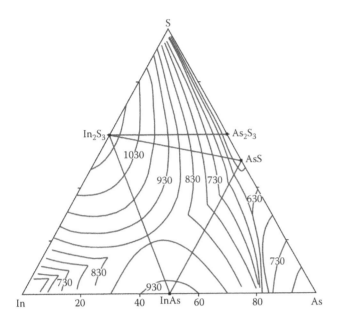

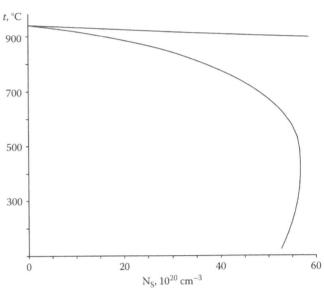

FIGURE 14.39   Calculated isotherms of the liquidus surface of the In–S–As ternary system. (From Talis, L.D., et al., *Izv. AN SSSR Neorgan. Mater.*, *25*(10), 1609, 1989. With permission.)

FIGURE 14.40   Solubility of sulfur in InAs along the InAs–In$_2$S$_3$ quasi-binary section. (From Talis, L.D., et al., *Izv. AN SSSR Neorgan. Mater.*, *25*(10), 1609, 1989. With permission.)

(Talis et al. 1989). The field of In$_2$S$_3$ primary crystallization occupies the biggest part of this surface. There are also the fields of primary crystallization of InAs and As and the degenerated fields of primary crystallization of AsS, As$_2$S$_3$, In, and S. The triangulation of this ternary system is determined by the InAs–In$_2$S$_3$, InAs–AsS, In$_2$S$_3$–AsS, and In$_2$S$_3$–As$_2$S$_3$ quasi-binary system.

According to the data of Safarov et al. (1992), **InAs$_2$S$_4$** and **In$_3$As$_2$S$_6$** ternary compounds, which melt incongruently at 357°C and 267°C, respectively, are formed in the In–S–As system.

Under normal pressure, In$_2$S$_3$ has a trigonal structure between 754°C and its melting point (γ-In$_2$S$_3$). This modification cannot be quenched to room temperature. Substitution of about 5 at% indium by arsenic stabilizes the structure at room temperature with the lattice parameters $a = 380.6 \pm 0.1$ and $c = 904.4 \pm 0.3$ pm for **In$_{1.9}$As$_{0.1}$S$_3$** composition ($a = 380$ and $c = 904$ pm and an experimental density of 4.67 g·cm^{-3} for In$_{1.93}$As$_{0.07}$S$_3$ composition [Diehl et al. 1970]), an experimental density of 4.75 $\pm$ 0.08 g·cm^{-3}, and an energy gap of 2.01 eV (Diehl and Nitsche 1973; Diehl et al. 1976). Dark red hexagonal plates of In$_{1.9}$As$_{0.1}$S$_3$ have been grown in a closed system by iodine transport in the presence of As$_2$S$_3$ vapor (Diehl and Nitsche 1973; Diehl and Nitsche 1975). They decompose at 320°C.

InAs–In$_2$S$_3$. The limited solid solution is formed in this system based on InAs up to 25 mol% In$_2$S$_3$ (Radautsan 1962; Radautsan and Madan 1962; Zhitar' 1965). The calculations indicated that solubility of sulfur along this section is retrograde with a maximum of circa 5.6·10^{21} cm^{-3} at approximately 420°C (Figure 14.40) (Talis et al. 1989). According to the data of Robbins and Lambrecht (1975), **In$_3$AsS$_3$** ternary compound, which crystallizes in the cubic structure with the lattice parameter $a = 1079.6$ pm, is formed in this system.

InAs–S. This system is a non-quasi-binary section of the In–S–As ternary system (Kompanichenko et al. 2003). At the interaction of InAs and sulfur, In$_2$S$_3$ and As$_2$S$_3$ were identified. The homogeneity region of InAs doped with sulfur is given in Figure 14.41 (Talis et al. 1989). Doping with sulfur up to circa 5·10^{18} cm^{-3} has no significant influence on the width of the homogeneity region; however, the temperatures of maximum solubility of excess In and As vary, respectively, from ~810°C and ~870°C for the undoped material to ~850°C and ~910°C for the doped material. Upon doping with S up to ~(1–2)·10^{19} cm^{-3}, the boundary of the homogeneity region shifted toward an excess of As; however, the maximum temperature of the homogeneity region is hardly shifted. With a further increase of the doping level, the configuration of the homogeneity region continues to change. There is a sharp narrowing of the homogeneity region, which weakens the

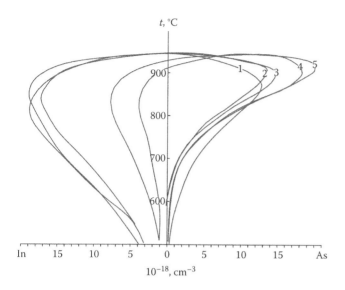

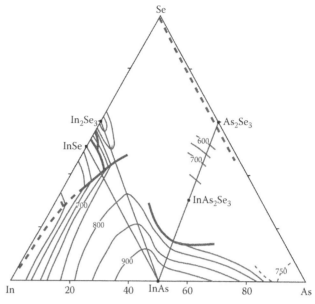

FIGURE 14.41 Calculated homogeneity region of InS_xAs_{1-x} solid solution at sulfur concentrations of (1) 0, (2) $5 \cdot 10^{18}$, (3) 10^{19}, (4) $5 \cdot 10^{19}$, and (5) 10^{20} cm^{-3}. (From Talis, L.D., et al., *Izv. AN SSSR Neorgan. Mater.*, 25(10), 1609, 1989. With permission.)

FIGURE 14.42 Part of the liquidus surface of the In–Se–As ternary system. (From Luzhnaya, N.P., et al., *Zhurn. Neorgan. Khimii*, 9(5), 1174, 1964. With permission.)

retrograde nature of the solubility of excess In. On the side of excess As, the homogeneity region expands monotonically with increasing doping level. The maximum melting temperature of InAs doped with S up to $(1–2) \cdot 10^{20}$ cm^{-3} is almost the same as the melting temperature of undoped InAs. However, the point of the maximum melting temperature shifts in the composition toward excess As. The retrograde nature of the solidus from the side of excess As increases with increasing S concentration.

According to the data of Glazov et al. (1979), the solubility of S in InAs at 700°C, 800°C, 850°C, and 900°C is $1.6 \cdot 10^{20}$, $2.5 \cdot 10^{20}$, $5.0 \cdot 10^{20}$, and $7.2 \cdot 10^{20}$ cm^{-3}, respectively. The alloys were annealed at 700°C for 1200 h, at 800°C for 1000 h, at 850°C for 800 h, and at 900°C for 500 h and studied through metallography and measurement of microhardness. The solubility of S in InAs immediately after the synthesis without annealing is 2 at% (Kotrubenko 1965).

According to the data of Demchenko et al. (1975), **InAsS₃** ternary compound, which melts incongruently at 740°C ± 5°C, is formed in this system.

14.28 INDIUM–SELENIUM–ARSENIC

The part of the liquidus surface of the In–Se–As ternary system is given in Figure 14.42 (Luzhnaya et al. 1964). The field of InAs primary crystallization occupies the biggest part of the liquidus surface. The projection of the

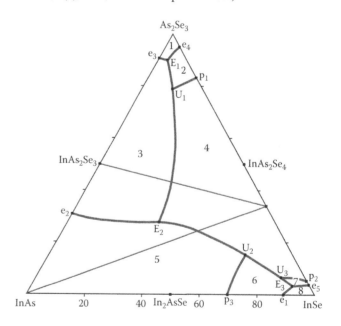

FIGURE 14.43 Projection of the liquidus surface of the InAs–As₂Se₃–InSe subsystem: 1–8, the fields of primary crystallization of As₂Se₃, InAs₂Se₄, InAs₃Se₃, In₂Se₃, InAs, In₂AsSe, In₅Se₆, and InSe, respectively. (From Safarov, M.G., et al., *Zhurn. Neorgan. Khimii*, 35(2), 556, 1990. With permission.)

liquidus surface of the InAs–As₂Se₃–InSe subsystem is shown in Figure 14.43 (Safarov et al. 1990). There eight fields of primary crystallizations of As₂Se₃, **InAs₂Se₄**, **InAs₃Se₃**, InAs, **InAs₂Se**, In₅Se₆, and InSe on the liquidus surface.

According to the data of Katty et al. (1978), **In$_{1.9}$As$_{0.1}$Se$_3$** solid solution, which crystallizes in the hexagonal structure with the lattice parameters $a = 397$ and $c = 1890$ pm, calculated and experimental densities of 5.95 and 5.90 ± 0.04 g·cm^{-3}, respectively, and an energy gap of 1.4 eV, is formed in this system. Its crystals were grown by chemical vapor transport with Cl$_2$ as the transport agent.

InAs–As$_2$Se$_3$. The phase diagram of this system is shown in Figure 14.44 (Luzhnaya et al. 1964). InAs$_3$Se$_3$ ternary compound, which melts congruently at 800°C, is formed in the InAs–As$_2$Se$_3$ system. Two invariant eutectic equilibria exist in this system. One of the eutectics crystallizes at 750°C and contains 25 mol% As$_2$Se$_3$, and another crystallizes at 290°C and contains 97 mol% As$_2$Se$_3$. The solubility of As$_2$Se$_3$ in InAs is not higher than 10 mol%. This system was investigated through DTA, XRD, and metallography. The samples were annealed at 180°C for 5000 h and at 300°C for 1500 h.

InAs–InSe. The phase diagram of this system is shown in Figure 14.45 (Safarov et al. 1988). **In$_4$As$_3$Se** ternary compound, which melts incongruently at 595°C and crystallizes in the cubic structure with the lattice parameter $a = 586$ pm and an experimental density of 5.82 g·cm^{-3}, is formed in the InAs–InSe system. The eutectic crystallizes at 540°C and contains 13 mol% InAs. The solubility of InAs in InSe and InSe in InAs is 5 and 6 mol%, respectively (Hahn and Thiele 1960; Safarov et al. 1988). This system was investigated through DTA,

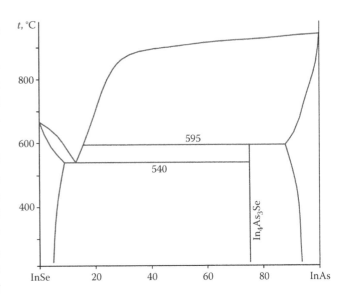

FIGURE 14.45   Phase diagram of the InAs–InSe system. (From Safarov, M.G., et al., *Zhurn. Neorgan. Khimii*, 33(2), 524, 1988. With permission.)

XRD, metallography, and measurement of microhardness and density. The samples were annealed at 430°C for 200 h. Safarov et al. (1990) noted that instead of the In$_4$As$_3$Se ternary compound, the In$_2$AsSe ternary phase, which melts incongruently at 800°C, is formed in the InAs–InSe system. This compound crystallizes in the cubic structure with the lattice parameter $a = 591.4$ pm (Hahn 1957).

InAs–Se. The solubility of Se in InAs in the temperature region from 500°C to 800°C was determined by Mavlonov and Makhtumkulov (1983). It was shown that this solubility reaches the maximum value ($4 \cdot 10^{20}$ cm^{-3}) at ca. 750°C. According to the data of Glazov et al. (1979), the solubility of Se in InAs at 700°C, 800°C, 850°C, and 900°C is $1.4 \cdot 10^{20}$, $2.0 \cdot 10^{20}$, $3.9 \cdot 10^{20}$, and $4.7 \cdot 10^{20}$ cm^{-3}, respectively. The alloys were annealed at 700°C for 1200 h, at 800°C for 1000 h, at 850°C for 800 h, and at 900°C for 500 h and studied through metallography and measurement of microhardness. The solubility of Se in InAs immediately after the synthesis without annealing is 11 at% (Kotrubenko 1965).

InAs–In$_2$Se$_3$. The phase diagram of this system is presented in Figure 14.46 (Radautsan 1959). The solubility of In$_2$Se$_3$ in InAs reaches 60 mol% (Goryunova and Radautsan 1958b; Goryunova et al. 1959a; Hahn and Thiele 1960; Woolley and Keating 1961a, 1961b; Radautsan 1959, 1962; Radautsan and Madan 1962). The samples were annealed at 600°C for 145–200 h (Goryunova and Radautsan 1958b) (at 600°C for 1100 h

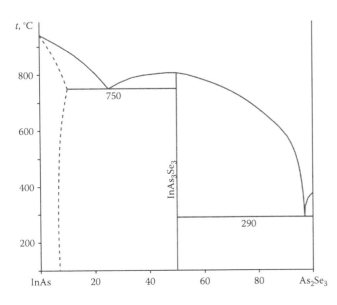

FIGURE 14.44   Phase diagram of the InAs–As$_2$Se$_3$ system. (From Luzhnaya, N.P., et al., *Zhurn. Neorgan. Khimii*, 9(5), 1174, 1964. With permission.)

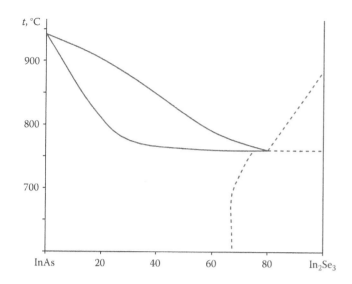

FIGURE 14.46 Phase diagram of the InAs–In$_2$Se$_3$ system. (From Radautsan S.I., *Zhurn. Neorgan. Khimii*, 4(5), 1121, 1959. With permission.)

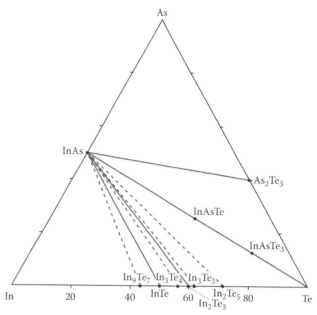

FIGURE 14.47 Isothermal section of the In–Te–As ternary system at room temperature. (From Peretti, E.A.J., et al., *J. Less Common Metals*, 35(2), 293, 1974. With permission.)

[Radautsan 1959, 1962]; at 720°C for 350 h [Woolley and Keating 1961a, 1961b]) and investigated through DTA, XRD, and metallography.

According to the data of Hahn (1957) and Robbins and Lambrecht (1975), **In$_3$AsSe$_3$** ternary compound, which crystallizes in the cubic structure with the lattice parameter $a = 582.6$ or 579.4 pm, is formed in the InAs–In$_2$Se$_3$ system.

InSe–As$_2$Se$_3$. InAs$_2$Se$_4$ ternary compound, which melts incongruently at 605°C, is formed in this system (Safarov and Gamidov 1990; Safarov et al. 1990).

14.29 INDIUM–TELLURIUM–ARSENIC

The isothermal section of the In–Te–As ternary system at room temperature is shown in Figure 14.47 (Peretti et al. 1974). It shows the valid tie-lines in this system as deduced from XRD. InAs is in thermodynamic equilibrium with all binary compounds and with Te. The glasses were obtained in this system upon heating of the mixtures of elements at 1000°C ± 10°C for 12 h with vigorous stirring, followed by quenching in water (Minaev 1983).

InAs–As$_2$Te$_3$. The solubility of As$_2$Te$_3$ in 3InAs reaches up to 2 mol% (Kotrubenko 1966).

InAs–InTe. The phase diagram of this system is given in Figure 14.48 (Peretti et al. 1974). The eutectic crystallizes at 679.5°C and contains 3.5 mass% InAs. The existence of **In$_8$As$_5$Te$_3$** compound could not be confirmed. Thermal arrests at 425°C and 455°C upon

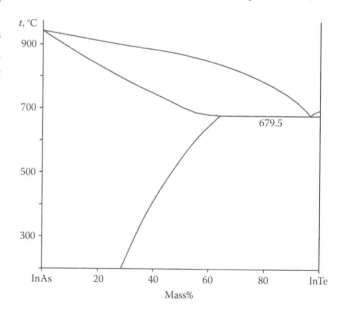

FIGURE 14.48 Phase diagram of the InAs–InTe system. (From Peretti, E.A.J., et al., *J. Less Common Metals*, 35(2), 293, 1974. With permission.)

cooling alloys from the melt were also detected. These arrests were ascribed to nonequilibrium effects due to the large solid solubility field in the InAs–InTe system, which was also indicated by Hahn and Thiele (1960). The alloys were thermally equilibrated at 600°C for 800 h (Peretti et al. 1974).

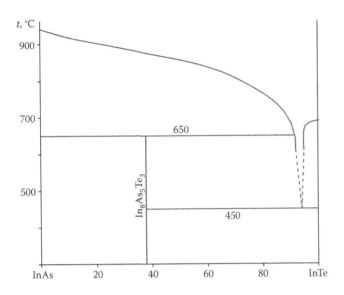

FIGURE 14.49 Phase diagram of the InAs–InTe system. (From Kurata, K., and Hirai, T., *Solid State Electron.*, 9(6), 633, 1966. With permission.)

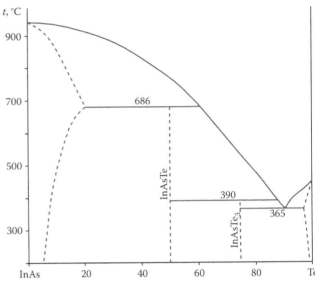

FIGURE 14.50 Phase diagram of the InAs–Te system. (From Bobrov, V.I., et al., *Izv. AN SSSR Neorgan. Mater.*, 3(7), 1273, 1967. With permission.)

According to the data of Kurata and Hirai (1966), the phase diagram of the InAs–InTe system is a eutectic type with peritectic transformation (Figure 14.49). This system has a eutectic line at 450°C, and a peritectic exists as $In_8As_5Te_3$ with the peritectic temperature of 650°C. The eutectic composition was not determined.

Earlier investigations indicated the formation of solid solutions based on InAs up to 50 mol% InTe (Lyalikov et al. 1964) and the **In_2AsTe** ternary compounds, which crystallize in the cubic structure with the lattice parameter $a = 608.7$ pm (Hahn 1957).

InAs–In_2Te_3. This system is a quasi-binary section of the In–Te–As ternary system (Peretti et al. 1974), but according to the data of Woolley and Smith (1958b), the phase diagram of this system cannot be treated as quasibinary. Solid solutions occur throughout the whole range of composition (Goryunova and Burdiyan 1958; Woolley and Smith 1958b; Radautsan 1962; Radautsan and Madan 1962). According to the data of XRD, mixed crystal phases of the zinc blende type were observed in this system (Hahn and Thiele 1960).

InAs–Te. The phase diagram of this system is of eutectic type (Figure 14.50) (Bobrov et al. 1967). The eutectic crystallizes at 365°C and contains 8 mol% InAs. Two compounds, **InAsTe** and **InAsTe₃**, are formed in this system. InAsTe melts incongruently at 680°C (Bobrov et al. 1967) and crystallizes in the cubic structure with the lattice parameter $a = 613$ pm, calculated and experimental densities of 6.1 and 5.56 g·cm⁻³, respectively

(Kotrubenko et al. 1967), and an energy gap of 0.49–0.58 eV (Yemel'yanenko et al. 1968). The title compound could be prepared by quenching of the melt or by a zone recrystallization (Kotrubenko et al. 1967).

InAsTe₃ melts incongruently at 390°C (Bobrov et al. 1967).

The solubility of tellurium in InAs at 680°C reaches up to 20 at% and is equal to 5 at% at room temperature (Bobrov et al. 1967). According to the data of Kotrubenko et al. (1964), the region of the solid solutions based on InAs contains up to 10 at% Te. The solubility of Te in InAs immediately after the synthesis without annealing is 9–10 at% (Kotrubenko 1965, 1966). Additional annealing at 600°C for 700 h increases the solubility up to 15 at% Te.

There is evidence to believe that some In_2Te_3 is formed at the interaction of InAs and Te (Ufimtsev and Gimel'farb 1973).

InTe–As_2Te_3. **$InAs_2Te_4$** ternary compound, which melts incongruently at 370°C and crystallizes in the hexagonal structure with the lattice parameters $a = 672$ and $c = 914$ pm and calculated and experimental densities of 6.52 and 6.38 g·cm⁻³, respectively, is formed in this system (Safarov 1988).

14.30 INDIUM–CHROMIUM–ARSENIC

The isothermal section of the In–Cr–As ternary system at room temperature was approximately calculated by Klingbeil and Schmid-Fetzer (1989) (Figure 14.51). The

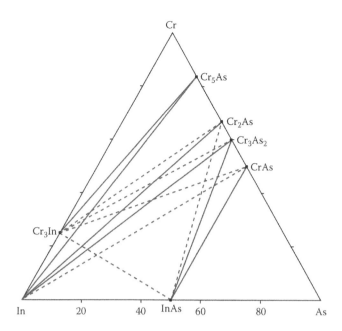

FIGURE 14.51　Calculated isothermal section of the In–Cr–As ternary system at room temperature. (From Klingbeil, J., and Schmid-Fetzer, R., *CALPHAD*, 13(4), 367, 1989. With permission.)

calculations are based on the following approximations: ternary phases and solid solubility are disregarded, and the Gibbs energy of formation of binary compounds is mostly estimated by the enthalpy of formation calculated from Miedema's model. The dotted lines indicate possible alternative tie-lines.

InAs–CrAs. The phase diagram of this system is a eutectic type. The eutectic crystallizes at 937°C ± 2°C and contains 1.7 mass% CrAs (Müller and Wilhelm 1965a).

14.31　INDIUM–MOLYBDENUM–ARSENIC

The isothermal section of the In–Mo–As ternary system at room temperature was approximately calculated by Klingbeil and Schmid-Fetzer (1989) (Figure 14.52). The calculations are based on the following approximations: ternary phases and solid solubility are disregarded, and the Gibbs energy of formation of binary compounds is mostly estimated by the enthalpy of formation calculated from Miedema's model. The dotted line indicates a possible alternative tie-line. According to XRD, at 600°C InAs is in thermodynamic equilibrium with Mo_5As_4, Mo_2As_3, $MoAs_2$, and $MoAs_5$, the existence of which is still questionable, and Mo_5As_4 coexists with liquid In (Klingbeil and Schmid-Fetzer 1994). The samples were slowly heated to 600°C and annealed at this temperature for at least 30 days.

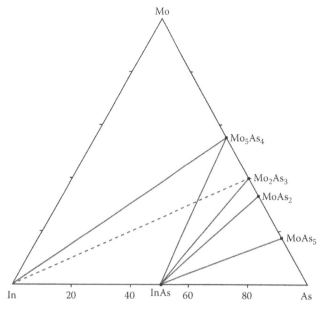

FIGURE 14.52　Calculated isothermal section of the In–Mo–As ternary system at room temperature. (From Klingbeil, J., and Schmid-Fetzer, R., *CALPHAD*, 13(4), 367, 1989. With permission.)

14.32　INDIUM–TUNGSTEN–ARSENIC

The isothermal sections of the In–W–As ternary system at room temperature were approximately calculated by Klingbeil and Schmid-Fetzer (1989), and at 600°C determined experimentally (Klingbeil and Schmid-Fetzer 1994). The calculations are based on the following approximations: ternary phases and solid solubility are disregarded, and the Gibbs energy of formation of binary compounds is mostly estimated by the enthalpy of formation calculated from Miedema's model. The experimental results confirm the calculated phase equilibria. At both temperatures, InAs is in thermodynamic equilibrium with W, WAs_2, and possibly W_2As_3.

14.33　INDIUM–FLUORINE–ARSENIC

$InAsF_6$ ternary compound is formed in the In–F–As system. It crystallizes in the rhombohedral structure with the lattice parameters $a = 750.29 \pm 0.13$ and $c = 774.69 \pm 0.19$ pm (in hexagonal settings) at 200 K (Goreshnik and Mazej 2008) ($a = 758 \pm 2$ and $c = 790 \pm 1$ pm [Mazej 2005]). To obtain this compound, reaction mixture of $InBF_4$ (2.32 mM) and AsF_5 (2.44 mM) in 6 mL of anhydrous HF was loaded into translucent fluorocarbon polymer reaction vessels in a glove box (Mazej 2005). Anhydrous HF and, when necessary, AsF_5 were condensed onto the solid reactants at 77 K. The reaction vessels were brought to ambient temperature. The reaction mixtures were vigorously agitated, and after some time, the volatiles were pumped off. A white precipitate formed immediately, and the reaction was left to proceed for 10 min. To decrease the loss of $InAsF_6$ because of its solubility in anhydrous HF, the reaction vessel was cooled to 243 K and, while cold, the solution was decanted to remove any possible traces of $InBF_4$. Single crystals of the title compound were obtained by recrystallization from anhydrous HF (Goreshnik and Mazej 2008).

14.34 INDIUM–CHLORINE–ARSENIC

InAs–InCl. Only thermal effects of InAs and InCl melting and InCl boiling were determined at the heating of the mixture containing 10 mass% InAs (Koppel et al. 1967a).

14.35 INDIUM–IODINE–ARSENIC

InAs–InI. Only thermal effects of InAs and InI melting and InI boiling were determined at the heating of the mixture containing 10 mass% InAs (Koppel et al. 1967a).

InAs–I₂. Iodine began to interact with InAs at 80°C (Yegorov and Kochnev 1969). The molecules of I_2, InI_3, and arsenic predominate in the gas phase up to 410°C. InI is formed within the temperature interval from 410°C to 710°C. With the temperature increasing, the content of InI in the gas phase also increases. The InI, As_4, and As_2 molecules predominate in the gas phase at temperatures higher than 710°C.

14.36 INDIUM–MANGANESE–ARSENIC

The isothermal section of the In–Mn–As ternary system at room temperature was approximately calculated by Klingbeil and Schmid-Fetzer (1989) (Figure 14.53). The calculations are based on the following approximations: ternary phases and solid solubility are disregarded, and the Gibbs energy of formation of binary compounds is mostly estimated by the enthalpy of formation calculated from Miedema's model. The dotted lines indicate possible alternative tie-lines.

InAs–MnAs. $In_{1-x}Mn_xAs$ ($x \leq 0.18$) solid solutions have been produced by molecular beam epitaxy (MBE) at temperatures ranging from 200°C to 300°C (Munekata et al. 1989, 1991). The lattice constant decreases with increasing Mn composition.

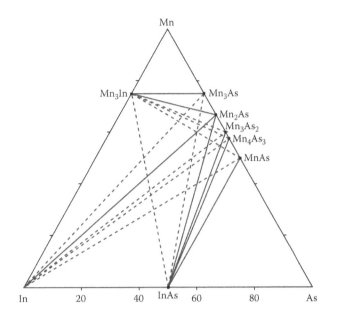

FIGURE 14.53   Calculated isothermal section of the In–Mn–As ternary system at room temperature. (From Klingbeil, J., and Schmid-Fetzer, R., *CALPHAD*, 13(4), 367, 1989. With permission.)

14.37 INDIUM–RHENIUM–ARSENIC

The isothermal section of the In–Re–As ternary system at room temperature was approximately calculated by Klingbeil and Schmid-Fetzer (1989). At this temperature, InAs is in thermodynamic equilibrium with Re and Re_3As_7. The calculations are based on the following approximations: ternary phases and solid solubility are disregarded, and the Gibbs energy of formation of binary compounds is mostly estimated by the enthalpy of formation calculated from Miedema's model.

14.38 INDIUM–IRON–ARSENIC

The isothermal section of the In–Fe–As ternary system at room temperature was approximately calculated by Klingbeil and Schmid-Fetzer (1989) (Figure 14.54). The calculations are based on the following approximations: ternary phases and solid solubility are disregarded, and the Gibbs energy of formation of binary compounds is mostly estimated by the enthalpy of formation calculated from Miedema's model. The dotted line indicates a possible alternative tie-line.

InAs–FeAs. The phase diagram of this system is a eutectic type. The eutectic crystallizes at 930°C ± 2°C and contains 10.5 mass% FeAs (Müller and Wilhelm 1965a; Khlystovskaya et al. 1975).

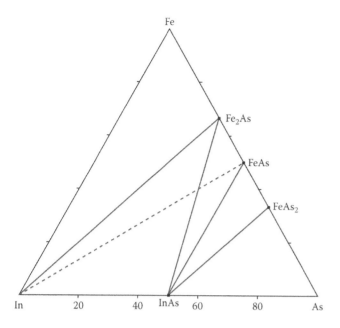

FIGURE 14.54   Calculated isothermal section of the In–Fe–As ternary system at room temperature. (From Klingbeil, J., and Schmid-Fetzer, R., *CALPHAD*, 13(4), 367, 1989. With permission.)

14.39 INDIUM–COBALT–ARSENIC

The isothermal section of the In–Co–As ternary system at room temperature was approximately calculated by Klingbeil and Schmid-Fetzer (1989) (Figure 14.55). The calculations are based on the following approximations: ternary phases and solid solubility are disregarded, and

the Gibbs energy of formation of binary compounds is mostly estimated by the enthalpy of formation calculated from Miedema's model. The dotted lines indicate possible alternative tie-lines.

14.40 INDIUM–NICKEL–ARSENIC

Klingbeil and Schmid-Fetzer (1994) have reported the phase equilibria in the In–Ni–As ternary system at 600°C, based on XRD analysis of five samples. These results were then compared with a 25°C isotherm, which was calculated from experimental thermodynamic data by Klingbeil and Schmid-Fetzer (1989). According to Klingbeil and Schmid-Fetzer (1994), the intermediate phases Ni_2In_3, NiIn, $NiAs_2$, and NiAs are in thermodynamic equilibrium with InAs, and the three-phase region InAs + NiIn + NiAs bridges the constituent binary systems of the In–Ni–As system. Although Klingbeil and Schmid-Fetzer (1994) fabricated a sample corresponding to the composition Ni_3InAs, they did not find it to be single phase; rather, they observed the presence of three phases: Ni_2In, Ni_3In, and NiAs.

The phase equilibria in the In–Ni–As ternary system at 600°C were established using XRD and EPMA by Swenson and Chang (1996b). An In-rich liquid phase, $NiAs_2$, and NiAs were demonstrated to be in thermodynamic equilibrium with InAs (Figure 14.56). The most prominent of the system is a solid solution phase

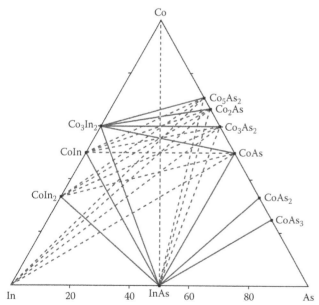

FIGURE 14.55   Calculated isothermal section of the In–Co–As ternary system at room temperature. (From Klingbeil, J., and Schmid-Fetzer, R., *CALPHAD*, 13(4), 367, 1989. With permission.)

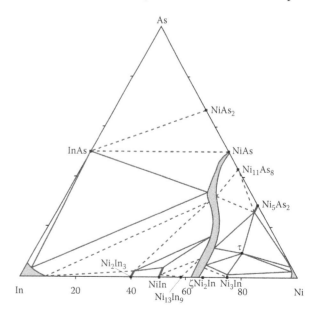

FIGURE 14.56   Isothermal section of the In–Ni–As ternary system at 600°C. (From Swenson, D., and Chang, Y.A., et al., *Mater. Sci. Eng. B*, 39(3), 232, 1996.)

denoted the ζ-phase, which probably traverses the isothermal section from NiAs to the phase ζ in the In–Ni binary system. A ternary phase with the average composition $Ni_{0.75}In_{0.16}As_{0.09}$ (τ) was found to exist in this system as well. It crystallizes in the cubic structure with the lattice parameter $a = 584.37 \pm 0.08$ pm. The compound Ni_3InAs may be seen to represent a specific composition of the ζ-phase. The ternary solubilities of the remaining constituent binary phases were found to be small, except for NiIn, which dissolves approximately 4 at% As.

Initially, the samples were annealed at 600°C for 2 weeks and subsequently quenched in iced water (Swenson and Chang 1996b). After that, they were ground into powders, pressed into pellets, sealed in evacuated quartz ampoules, and annealed at 600°C for at least 3 months and quenched in iced water.

14.41 INDIUM–RUTHENIUM–ARSENIC

The isothermal section of the In–Ru–As ternary system at room temperature was approximately calculated by Klingbeil and Schmid-Fetzer (1989) (Figure 14.57). The calculations are based on the following approximations: ternary phases and solid solubility are disregarded, and the Gibbs energy of formation of binary compounds is mostly estimated by the enthalpy of formation calculated from Miedema's model. The dotted lines indicate possible alternative tie-lines.

14.42 INDIUM–RHODIUM–ARSENIC

Klingbeil and Schmid-Fetzer (1989) have calculated a room temperature phase diagram isotherm of the In–Rh–As system, using experimental thermodynamic data for InAs and estimated thermodynamic data for the binary Rh–In and Rh–As phases. Their calculations predicted that $RhIn_3$, $RhAs_3$, $RhAs_2$, and RhAs would be in thermodynamic equilibrium with InAs. However, as was discussed by Klingbeil and Schmid-Fetzer (1989), such calculated diagrams, while often qualitatively correct, must be confirmed experimentally. For example, if an uncertainty of ±8 kJ g-atom[-1] was assumed for the estimated thermodynamic data, the authors found nine alternative tie-lines to be possible in the In–Rh–As system. Furthermore, the occurrence of ternary phases, which may drastically alter the topologies of such calculated diagrams, cannot be predicted *a priori*.

Phase equilibria in the In–Rh–As ternary system at 600°C were established, using XRD, EPMA, scanning electron microscopy (SEM), and DTA, by Swenson et al. (1994b) and are given in Figure 14.58. InAs was demonstrated to be in thermodynamic equilibrium with $RhIn_3$, $\mathbf{Rh_3In_5As_2}$, $RhAs_2$, and $RhAs_3$. $Rh_3In_5As_2$ ternary compound forms peritectically at 829°C ± 5°C according to the reaction RhIn + L ⟷ $Rh_3In_5As_2$, has a narrow range of homogeneity, and crystallizes in the cubic structure with the lattice parameter $a = 916.53 \pm$

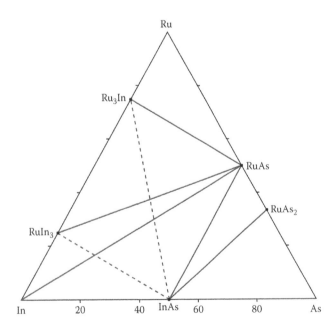

FIGURE 14.57 Calculated isothermal section of the In–Ru–As ternary system at room temperature. (From Klingbeil, J., and Schmid-Fetzer, R., *CALPHAD*, 13(4), 367, 1989. With permission.)

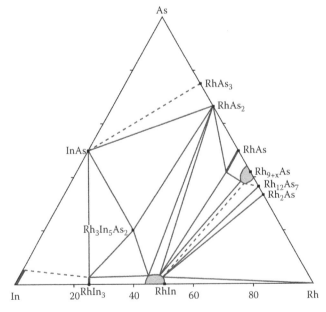

FIGURE 14.58 Isothermal section of the In–Rh–As ternary system at 600°C. (From Swenson, D., et al., *J. Alloys Compd.*, 216(1), 67, 1994. With permission.)

0.06 pm. RhAs was found to dissolve 8 at% In. $Rh_{9+x}As_7$, $RhIn_3$, and RhIn also showed some ternary solubility. RhIn exhibited an appreciable range of binary homogeneity, extending from at least 44–49 at% Rh.

InAs–RhIn. The phase equilibria in this system are presented in Figure 14.59 (Swenson et al. 1994b).

14.43 INDIUM–PALLADIUM–ARSENIC

Klingbeil and Schmid-Fetzer (1989) have calculated a room temperature phase diagram isotherm of the In–Pd–As system, using experimental thermodynamic data for InAs and estimated thermodynamic data for the binary Pd–In and Pd–As phases. Their calculations predicted that $PdIn_3$, Pd_2In_3, and $PdAs_2$ would be in thermodynamic equilibrium with InAs. However, as was discussed by Klingbeil and Schmid-Fetzer (1989), such calculated diagrams, while often qualitatively correct, must be confirmed experimentally. For example, if an uncertainty of ±8 kJ g-atom^{-1} was assumed for the estimated thermodynamic data, the authors found six alternative tie-lines to be possible in the In–Pd–As ternary system. Furthermore, the occurrence of ternary phases, which may drastically alter the topologies of such calculated diagrams, cannot be predicted *a priori*.

Phase equilibria in the In–Rh–As ternary system at 600°C were established by El-Boragy and Schubert (1970) and are given in Figure 14.60. Two ternary compounds,

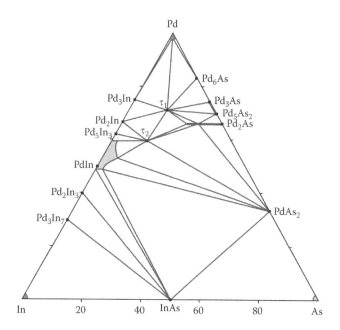

FIGURE 14.60 Isothermal section of the In–Pd–As ternary system at 600°C. (From El-Boragy, M., and Schubert, K., *Z. Metallkde*, 61(8), 579, 1970. With permission.)

Pd_5InAs (τ_1) and **Pd_6In_3As** (τ_2), are formed in this system. Pd_5InAs (τ_1) crystallizes in the tetragonal structure with the lattice parameters $a = 396.6$ and $c = 693.1$ pm, and Pd_6In_3As (τ_2) crystallizes in the rhombohedral structure with the lattice parameters $a = 884.2$ and $c = 2160.1$ pm (in a hexagonal setting).

14.44 INDIUM–OSMIUM–ARSENIC

The isothermal section of the In–Os–As ternary system at room temperature was approximately calculated by Klingbeil and Schmid-Fetzer (1989). At this temperature, InAs is in thermodynamic equilibrium with $OsAs_2$ and probably Os. The calculations are based on the following approximations: ternary phases and solid solubility are disregarded, and the Gibbs energy of formation of binary compounds is mostly estimated by the enthalpy of formation calculated from Miedema's model.

14.45 INDIUM–IRIDIUM–ARSENIC

The isothermal section of the In–Ir–As ternary system at room temperature was approximately calculated by Klingbeil and Schmid-Fetzer (1989). The calculations are based on the following approximations: ternary phases and solid solubility are disregarded, and the Gibbs energy of formation of binary compounds is mostly estimated by the enthalpy of formation calculated from Miedema's model. According to XRD and

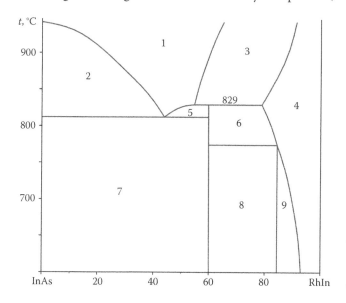

FIGURE 14.59 Phase equilibria in the InAs–RhIn system: 1, L; 2, L + InAs; 3, L + RhIn; 4, RhIn; 5, L + $Rh_3In_5As_2$; 6, $Rh_3In_5As_2$ + RhIn; 7, InAs + $Rh_3In_5As_2$; 8, $Rh_3In_5As_2$ + $RhAs_2$ + RhIn; and 9, $RhAs_2$ + RhIn. (From Swenson, D., et al., *J. Alloys Compd.*, 216(1), 67, 1994. With permission.)

EPMA, the isothermal section at 600°C is the same as at room temperature (Swenson et al. 1994a). InAs was shown to be in thermodynamic equilibrium with $IrAs_3$ and $IrAs_2$, and $InAs_2$ forms tie-lines with In, $IrIn_3$, and $IrIn_2$. No ternary compounds were found to exist in this ternary system

14.46 INDIUM–PLATINUM–ARSENIC

The isothermal sections of the In–Pt–As ternary system at room temperature were approximately calculated by Klingbeil and Schmid-Fetzer (1989), and at 600°C determined experimentally by Klingbeil and Schmid-Fetzer (1994) and Swenson and Chang (1994). The isothermal section at 600°C constructed through XRD and EPMA by Swenson and Chang (1994) is given in Figure 14.61. InAs was found to be in thermodynamic equilibrium with Pt_3In_7, $PtIn_2$, and $PtAs_2$. Two ternary compounds, **Pt_5InAs** (τ_1) and **Pt_8In_2As** (τ_2), were found to exist in this system. These phases appear to have very narrow ranges of homogeneity. Both compounds crystallize in the tetragonal structure with the lattice parameters $a = 399.7$ and $c = 705.5$ pm for Pt_5InAs (El-Boragy and Schubert 1970) and $a = 398.7$ and $c = 1107.9$ pm for Pt_8In_2As (Swenson and Chang 1994). The ternary solubilities in the constituent binary phases were found to be negligible, with the possible exception of Pt_3In, which may dissolve 2–3 at% As at compositions near 75 at% Pt (Swenson and Chang 1994). The samples were annealed at 600°C for 14 days. The ampoules were then quenched in ice water, and the pellets were ground into powder

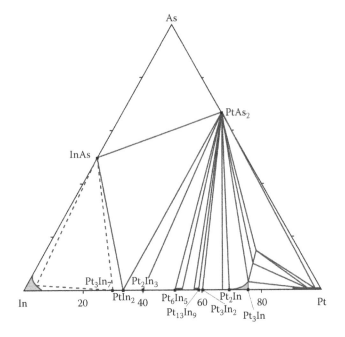

FIGURE 14.61   Isothermal section of the In–Pt–As ternary system at 600°C. (From Swenson, D., and Chang, Y.A., *Mater. Sci. Eng. B*, 22(2–3), 267, 1994. With permission.)

and again pressed into pellets, encapsulated in evacuated quartz ampoules, and annealed for a minimum of 3 additional months. Subsequently, the ampoules were quenched in ice water.

According to the data of Klingbeil and Schmid-Fetzer (1994), at 600°C InAs is also in thermodynamic equilibria with Pt_2In_3 and Pt_6In_5 and the ternary compounds were not found in the In–Pt–As ternary system.

CHAPTER 15

Systems Based on InSb

15.1 INDIUM–LITHIUM–ANTIMONY

The isothermal section of the In–Li–Sb ternary system at 400°C is shown in Figure 15.1 (Sitte and Weppner 1985). Two ternary compounds, **Li₃InSb₂** and nominally "**Li₆InSb₃**," are formed in this system. Both compounds are stable in equilibrium with elemental indium and antimony. Li₃InSb₂ has a small homogeneity region from 49 to 51 at% Li. Li₆InSb₃ shows a large variation of the composition from 59 to 71 at% Li in the direction of the InSb–Li₃Sb quasi-binary system. The lithium activities for the Li₃InSb₂ and Li₆InSb₃ compounds are limited to ranges from $6.6 \cdot 10^{-8}$ to $3.6 \cdot 10^{-7}$ and $9.3 \cdot 10^{-8}$ to $1.1 \cdot 10^{-5}$, respectively, at 400°C. The standard Gibbs energies of formation of Li₃InSb₂ and Li₆InSb₃ are –296.2 and –568.8 kJ·mol⁻¹, respectively, at 400°C and ideal stoichiometry. The activity ranges of Li, In, and Sb are given for the stability of all phases of the ternary system.

15.2 INDIUM–SODIUM–ANTIMONY

Two ternary compounds, **Na₂In₂Sb₃** and **Na₃InSb₂**, are formed in the In–Na–Sb system.

Na₂In₂Sb₃ crystallizes in the monoclinic structure with the lattice parameters $a = 1459.5 \pm 0.4$, $b = 753.2 \pm 0.3$, and $c = 1563.5 \pm 0.4$ pm and $\beta = 90.0 \pm 0.2°$ (Cordier and Ochmann 1991i). This compound was obtained from the elements at 1130°C.

Na₃InSb₂ also crystallizes in the monoclinic structure with the lattice parameters $a = 1028.5 \pm 0.5$, $b = 796.3 \pm 0.4$, and $c = 1665.2 \pm 0.7$ pm and $\beta = 92.6 \pm 0.1°$ (Cordier and Ochmann 1991p). This compound was obtained from the elements at 630°C.

15.3 INDIUM–POTASSIUM–ANTIMONY

K₂In₂Sb₃ (K₄In₄Sb₆) ternary compound, which crystallizes in the monoclinic structure with the lattice parameters $a = 1528.8 \pm 0.5$, $b = 754.4 \pm 0.1$, and $c = 1679.0 \pm 0.3$ pm and $\beta = 90.54 \pm 0.02°$ (Birdwhistell et al. 1991) ($a = 1525.4 \pm 0.3$, $b = 753.4 \pm 0.2$, and $c = 1679.8 \pm 0.4$ pm and $\beta = 90.5 \pm 0.1°$ [Cordier and Ochmann 1991d]), is formed in the In–K–Sb system. It was prepared by adding K (25.6 mM), In (8.5 mM), and Sb (17.0 mM) to a quartz tube that was then sealed under vacuum (Birdwhistell et al. 1991). The sample was heated at 750°C in a furnace for 10 h, and then cooled to ambient temperature over a 40 h period. A metallic gray crystalline title compound was obtained.

15.4 INDIUM–RUBIDIUM–ANTIMONY

Rb₂In₂Sb₃ ternary compound, which crystallizes in the monoclinic structure with the lattice parameters $a = 1555.5 \pm 0.2$, $b = 756.92 \pm 0.06$, and $c = 1736.2 \pm 0.7$ pm and $\beta = 90.598 \pm 0.014°$, is formed in the In–Rb–Sb system (Gourdon et al. 1996). This compound was prepared by direct combination of the elements. A dry quartz tube was charged with Rb (3 mM), In (2 mM), and Sb (3 mM). The tube was evacuated for 1.5 h at 0.1 Pa and sealed under vacuum. The sample was heated at 600°C for 10 h and then cooled to ambient temperature over a 96 h period.

15.5 INDIUM–CESIUM–ANTIMONY

Cs₂In₂Sb₃ ternary compound, which crystallizes in the monoclinic structure with the lattice parameters $a = 1582.0 \pm 0.1$, $b = 756.8 \pm 0.2$, and $c = 1799.8 \pm 0.2$ pm and $\beta = 90.7°$, is formed in the In–Cs–Sb system (Blase

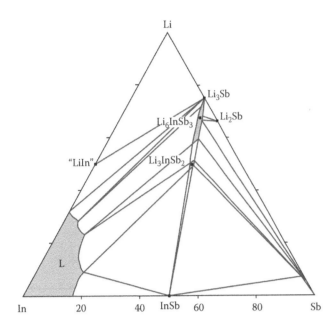

FIGURE 15.1 Isothermal section of the In–Li–Sb ternary system at 400°C. (From Sitte, W., and Weppner, W., *Appl. Phys. A*, *38*(1), 31, 1985. With permission.)

et al. 1995). This compound was obtained from the elements at 830°C.

15.6 INDIUM–COPPER–ANTIMONY

The liquidus surface (Figure 15.2), invariant equilibria, and four vertical sections of the In–Cu–Sb ternary system were calculated by the CALPHAD method using binary thermodynamic parameters (Manasijević et al.

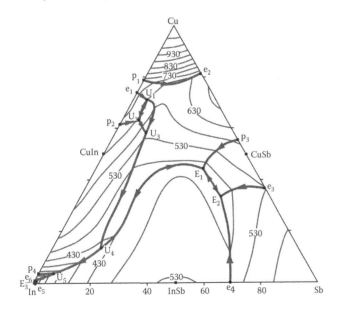

FIGURE 15.2 Calculated liquidus surface of the In–Cu–Sb ternary system. (From Manasijević, D., et al., *CALPHAD*, *33*(1), 221, 2009. With permission.)

2009). Experimental data were obtained using differential thermal analysis (DTA) and differential scanning calorimetry (DSC). The alloys were annealed at 300°C for 1 week, cooled to 100°C with additional annealing for 1 day, and after that slowly cooled down to room temperature. Eight invariant reactions, 17 monovariant lines, and 10 primary crystallization areas are observed in this ternary system. Predicted ternary eutectic reactions at 152°C (E_3) and 447°C (E_2), as well as a ternary liquid transition reaction at 293°C (U_5), are in agreement with the obtained experimental results.

$Cu_{5.33}In_{2.29}Sb_{1.71}$, $Cu_{6.743}In_{2.725}Sb$, $Cu_{6.861}In_{2.691}Sb$, and $Cu_2In_{0.75}Sb_{0.25}$ ternary compounds are formed in the In–Cu–Sb system. $Cu_{5.33}In_{2.29}Sb_{1.71}$ crystallizes in the orthorhombic structure with the lattice parameters $a = 1018.13 \pm 0.04$, $b = 845.62 \pm 0.04$, and $c = 737.74 \pm 0.02$ pm and a calculated density of 8.3019 g·cm^{-3} (Müller et al. 2012). For obtaining it, Cu, In, and Sb were reacted in a molar ratio 6:3:2 in an Ar-purged, evacuated silica ampoule at 900°C $\pm$ 5°C in a muffle furnace for 3 days. After furnace cooling, the samples were placed in an alumina block and annealed at 300°C $\pm$ 5°C for 30 days.

$Cu_{6.743}In_{2.725}Sb$ obtained at 500°C also crystallizes in the orthorhombic structure with the lattice parameters $a = 517.70 \pm 0.02$, $b = 740.93 \pm 0.04$, and $c = 856.59 \pm 0.04$ pm and a calculated density of 8.723 g·cm^{-3} (Müller and Lidin 2015). It was prepared by the interaction of Cu, In, and Sb (molar ratio 7:2.5:0.5) in an Ar-purged, evacuated silica ampoule at 900°C $\pm$ 5°C for 3 days. After furnace cooling, the samples were annealed at 300°C–500°C for 220 days before quenching.

$Cu_{6.861}In_{2.691}Sb$ obtained at 400°C also crystallizes in the orthorhombic structure with the lattice parameters $a = 518.07 \pm 0.01$, $b = 741.51 \pm 0.01$, and $c = 857.29 \pm 0.02$ pm and a calculated density of 8.740 g·cm^{-3} (Müller and Lidin 2015). It was prepared by the interaction of Cu, In, and Sb (molar ratio 6.6:2.3:0.8) in an Ar-purged, evacuated silica ampoule at 900°C $\pm$ 5°C for 3 days. After furnace cooling, the samples were annealed at 400°C–500°C for 220 days before quenching. This phase decomposes at temperatures below 400°C. By quenching samples, its crystal structure may be maintained at room temperature.

"Cu_3InSb_2" was found not to be a single phase (Luzhnaya et al. 1965).

InSb–Cu. This system is a non-quasi-binary section of the In–Cu–Sb ternary system (Kuznetsov and Bobrov 1971).

15.7 INDIUM–SILVER–ANTIMONY

The liquidus surface of the In–Ag–Sb ternary system, omitting the β-phase from the Ag–In binary system, was calculated by Liu et al. (2004) based of experimental work of Kuznetsov and Bobrov (1971) along the InSb–Ag vertical section and on their own DSC work along the AgIn–Sb vertical section. In Figure 15.3, the β-phase is added and the liquidus surface has been slightly modified in order to meet the accepted binary phase boundaries. There are two eutectic and three transition reactions in this system.

The isothermal sections of the In–Ag–Sb ternary system were constructed at 200°C by Buchtová et al. (2005) and at 400°C by Borisov et al. (2007) on the basis of the experimental results and thermodynamic calculations. The sections at 350°C and 400°C given in Figure 15.4 are based on those calculated by Liu et al. (2004). The three-phase fields are drawn by dashed lines, since tie-lines are not sufficient to describe boundaries. It is seen that ζ-phase continually exists from Ag–In to the Ag–Sb side. The surface tension and viscosity of the liquid phase were also estimated (Liu et al. 2004).

Itabashi et al. (2001) measured the activity of indium in the In–Ag–Sb system using a galvanic cell with a zirconia solid electrolyte in temperature ranges between 697°C and 1007°C. Isoactivity curves at 827 and 927°C

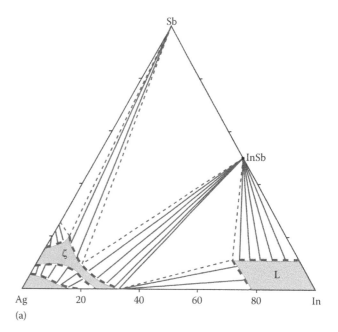

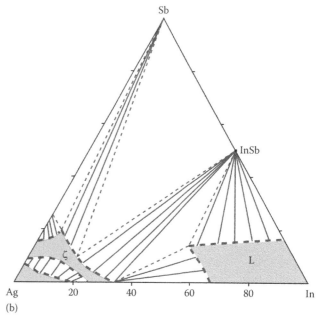

FIGURE 15.4 Isothermal sections of the In–Ag–Sb ternary system at (a) 350°C and (b) 400°C. (From Liu, X.J., et al., *Mater. Trans.*, 45(3), 637, 2004. With permission.)

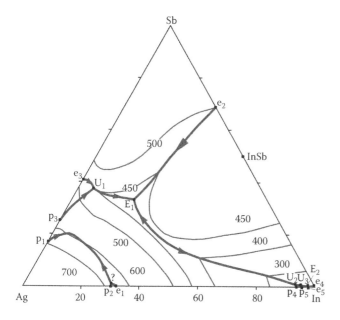

FIGURE 15.3 Calculated liquidus surface of the In–Ag–Sb ternary system. (From Liu, X.J., et al., *Mater. Trans.*, 45(3), 637, 2004. With permission.)

are given in Figure 15.5. They are derived from combining the activity data of the In–Sb and Ag–In binaries. The shape of the curves exhibits a slight bend to the Ag–Sb binary side.

Gather et al. (1987) determined the enthalpy of mixing of the In–Ag–Sb ternary liquid in a heat flow calorimeter. The measured ternary mixing isoenthalpy curves are presented in Figure 15.6.

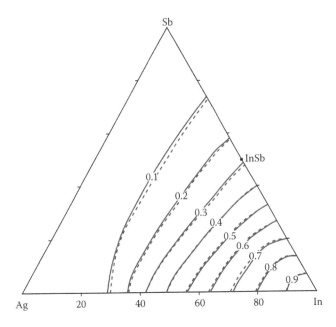

FIGURE 15.5   Isoactivity curves of In in the In–Ag–Sb ternary system at 827°C (dotted lines) and 927°C (solid lines). (From Itabashi, S., et al., *J. Jpn. Inst. Metals*, 65(9), 888, 2001. With permission.)

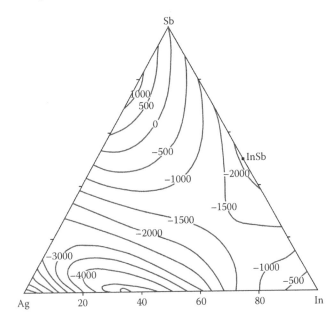

FIGURE 15.6   Ternary mixing isoenthalpy curves in the In–Ag–Sb system at 980°C. (From Gather, B., et al., *Z. Metallkde*, 78(4), 280, 1987. With permission.)

InSb–Ag. This system is a non-quasi-binary system of the In–Ag–Sb ternary system (Kuznetsov and Bobrov 1971; Liu et al. 2004; Buchtová et al. 2005). In the early study of Gubskaya et al. (1965), a sample of composition Ag₃InSb was fused directly from the elements, studied

by DTA and metallography, and found not to be a single-phase material.

15.8 INDIUM–GOLD–ANTIMONY

The calculated liquidus surface of the In–Au–Sb ternary system is shown in Figure 15.7 (Gierlotka 2013). A new description reproduces experimental data more accurately than the interpolation from binary systems, as proposed Liu et al. (2003).

The isothermal section of this ternary system at room temperature was approximately calculated by Tsai and Williams (1986) and is given in Figure 15.8. Eight quasi-binary tie-lines were identified: InSb–AuIn₂, InSb–AuIn, AuIn–Sb, Au₇In₃–AuSb₂, Au₃In–AuSb₂, Au₁₀In₃–AuSb₂, ζ–AuSb₂, and α₁–AuSb₂. The Au₇In₃–Sb cut was tentatively assumed to also be quasi-binary. The isothermal sections were also constructed at 200°C–300°C by Kubiak and Schubert (1980), at 230°C by Liu and Mohney (2003b), and at 250°C by Gierlotka (2013). The last section is presented in Figure 15.9. According to the data of Kubiak and Schubert (1980), the **AuIn₀.₉Sb₀.₁** and **AuIn₀.₇Sb₀.₇** phases were determined in the In–Au–Sb ternary system. AuIn₀.₉Sb₀.₁ crystallizes in the orthorhombic structure with the lattice parameters $a = 430.0 \pm 0.2$, $b = 1058.2 \pm 0.4$, and $c = 358.0 \pm 0.3$ pm, and AuIn₀.₇Sb₀.₇ crystallizes in the hexagonal structure with the lattice parameters $a = 430.14 \pm 0.04$ and $c = 553.47 \pm$

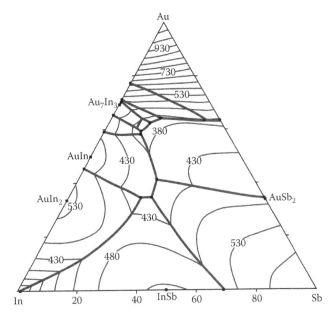

FIGURE 15.7   Calculated liquidus surface of the In–Au–Sb ternary system. (From Gierlotka, W., *J. Alloys Compd.*, 579, 533, 2013. With permission.)

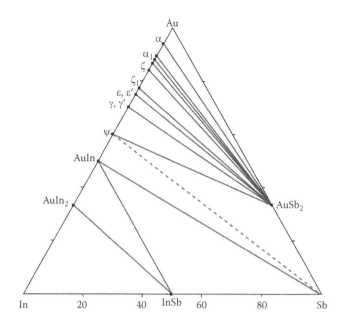

FIGURE 15.8 Isothermal section of the In–Au–Sb ternary system at room temperature. (From Tsai, C.T., and Williams, R.S., *J. Mater. Res.*, 1(2), 352, 1986. With permission.)

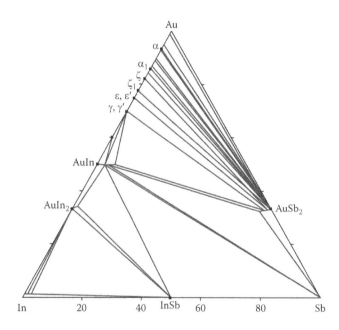

FIGURE 15.9 Isothermal section of the In–Au–Sb ternary system at 250°C. (From Gierlotka, W., *J. Alloys Compd.*, 579, 533, 2013. With permission.)

0.16 pm. "**Au₃InSb₂**" was found not to be a single phase (Luzhnaya et al. 1965).

By using a high-temperature Calvet microcalorimeter, molar enthalpies of mixing of the liquid In–Au–Sb ternary alloys have been determined at 700°C and

are presented in Figure 15.10 (Hassam et al. 2012). The calculated results are in good agreement with obtained experimental data. The mixing enthalpy calculation for this ternary system was also done by Gomidzelović et al. (2007a) using two different approaches, and integral molar enthalpies of mixing were obtained in the temperature interval 30°C–730°C. Strong negative values for integral mixing enthalpies were obtained in the whole concentration range, which points to the considerable attraction existing between constitutive components and their good mutual solubility. The thermodynamic properties of the In–Au–Sb ternary system were also calculated by Gomidzelovic et al. (2006) using the general solution model, and Redlich–Kister ternary interaction parameters were obtained in the temperature interval from 600°C to 1400°C.

AuIn–Sb. Earlier investigation indicated that the phase diagram of this system is a eutectic type (Nikitina and Lobanova 1974). The eutectic crystallizes at 421°C and contains 28 at% Sb. The solubility of AuIn in Sb at the eutectic temperature reaches 10 and 5 mol% at room temperature, and the solubility of Sb in AuIn at the eutectic temperature is slightly higher than 10 at% and circa 7 at% at room temperature. According to the data of Gomidzelović et al. (2007b, 2008), at the interaction of AuIn and antimony, AuIn + AuIn₂ + AuSb₂, InSb

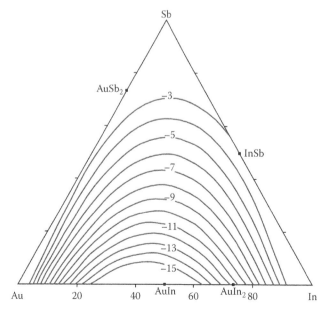

FIGURE 15.10 Calculated isoenthalpy curves of liquid In–Au–Sb alloys at 700°C (values in kJ·mol⁻¹). Standard states: undercooled liquid Au, liquid In, and liquid Sb. (From Hassam, S., et al., *J. Alloys Compd.*, 520, 65, 2012. With permission.)

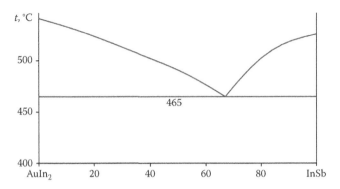

FIGURE 15.11 Calculated phase diagram of the InSb–AuIn$_2$ system. (From Liu, H.S., et al., *J. Electron. Mater.*, 32(2), 81, 2003. With permission.)

+ AuIn$_2$ + AuSb$_2$, and InSb + AuSb$_2$ + (Sb) three-phase regions are formed. Therefore, this system is a nonquasi-binary section of the In–Au–Sb ternary system.

InSb–AuIn. This system is a non-quasi-binary section of the In–Au–Sb ternary system (Babitsyna and Luzhnaya 1972; Liu et al. 2003; Gierlotka 2013). It crosses the fields of InSb, AuIn$_2$, and AuIn primary crystallization. InSb dissolves 0.5 at% Au.

InSb–AuIn$_2$. The phase diagram of this system was determined experimentally by Nikitina et al. (1967) and calculated by Liu et al. (2003) and is presented in Figure 15.11. The eutectic crystallizes at 465°C ± 5°C and contains 33 mol% AuIn$_2$ (Nikitina and Lobanova 1974). The solubility of AuIn$_2$ in InSb reaches 10 mol%, and AuIn$_2$ dissolves 5 mol% InSb. The samples were annealed at 400°C for 1330 h, and the system was investigated through DTA, x-ray diffraction (XRD), and metallography (Nikitina et al. 1967). According to the calculations of Liu et al. (2003), the mutual solubility of InSb and AuIn$_2$ is negligible.

15.9 INDIUM–CALCIUM–ANTIMONY

Two ternary compounds, **Ca$_5$In$_2$Sb$_6$** and **Ca$_{11}$InSb$_9$**, are formed in the In–Ca–Sb system. Ca$_5$In$_2$Sb$_6$ crystallizes in the orthorhombic structure with the lattice parameters $a = 1425.6 \pm 0.5$, $b = 1213.3 \pm 0.4$, and $c = 457.2 \pm 0.2$ pm and a calculated density of 4.87 g·cm^{-3} (4.90 g·cm^{-3} [Zevalkink et al. 2012]) (Cordier et al. 1985c) ($a = 1397.8$, $b = 1181.7$, and $c = 447.3$ pm [calculated values] and an energy gap of 0.32 eV [Yan et al. 2014]). Ca$_5$In$_2$Sb$_6$ is a semiconductor with an energy gap of 0.64 eV (Zevalkink et al. 2012). It was obtained from stoichiometric mixtures of the elements (Cordier et al. 1985c). These mixtures were heated up to 1000°C at a rate of 100°C·h^{-1} and

then cooled at a rate of 50°C·h^{-1}. The obtained compound decomposes under moist air. It could also be prepared if stoichiometric amounts of InSb, Ca, and Sb were loaded into stainless steel vials with stainless steel balls in an Ar-filled glove box (Zevalkink et al. 2012). The content was dried by ball milling for 1 h. The resulting fine powder was hot pressed in high-density graphite dies in Ar using 110 MPa of pressure. To ensure complete reaction, the powders were first held at the minimum load at 450°C for 2 h. The samples were then consolidated for 3 h at 700°C under 110 MPa of pressure, followed by a 2 h stress-free cooldown.

Ca$_{11}$InSb$_9$ also crystallizes in the orthorhombic structure with the lattice parameters $a = 1189.9 \pm 0.2$, $b = 1259.6 \pm 0.2$, and $c = 1672.2 \pm 0.3$ pm (Young and Kauzlarich 1995) ($a = 1189.4 \pm 0.5$, $b = 1259.4 \pm 0.5$, and $c = 1673.0 \pm 0.7$ pm and a calculated density of 4.38 g·cm^{-3} [Cordier et al. 1985a]). This compound was prepared by adding stoichiometric amounts of the elements in a Nb tube that was sealed on one end. This tube was then sealed in a quartz ampoule under vacuum. Heating the reactants at a rate of 60°C·h^{-1} to 850°C for 2 weeks and subsequently cooling at the same rate to room temperature provided the highest yield of the title compound, which is air sensitive (Young and Kauzlarich 1995). It was also synthesized from a mixture of Ga, In, and Sb in an Ar atmosphere in a corundum crucible at 1000°C (Cordier et al. 1985a).

15.10 INDIUM–STRONTIUM–ANTIMONY

Two ternary compounds, **Sr$_5$In$_2$Sb$_6$** and **Sr$_{11}$InSb$_9$**, are formed in the In–Sr–Sb system. Sr$_5$In$_2$Sb$_6$ crystallizes in the orthorhombic structure with the lattice parameters $a = 1474.9 \pm 0.5$, $b = 1269.6 \pm 0.4$, and $c = 466.0 \pm 0.2$ pm and a calculated density of 5.32 g·cm^{-3} (Cordier et al. 1985c) ($a = 1452.5$, $b = 1245.7$, and $c = 458.1$ pm [calculated values] and an energy gap of 0.29 eV [Yan et al. 2014]). It was obtained from stoichiometric mixtures of the elements. These mixtures were heated up to 1000°C at a rate of 100°C·h^{-1} and then cooled at a rate of 50°C·h^{-1}. The obtained compound decomposes under moist air.

Sr$_{11}$InSb$_9$ also crystallizes in the orthorhombic structure with the lattice parameters $a = 1238.85 \pm 0.13$, $b = 1310.03 \pm 0.14$, and $c = 1749.66 \pm 0.18$ pm at 120 ± 2 K and a calculated density of 5.086 g·cm^{-3} (Hullmann et al. 2007). To obtain the title compound, the reaction was carried out by loading Sr, In, and Sb (molar ratio

11:75:9) in an alumina crucible. The large excess of In was intended as a metal flux. The crucible with the reaction mixture was then flame sealed under vacuum in a silica ampoule, which was then placed in a furnace and heated to 1000°C at a rate of 300°C·h⁻¹. The reaction proceeded at this temperature for 24 h before being cooled to 600°C at a rate of 10°C·h⁻¹. At 600°C, the ampoule was removed and the In flux was decanted. The main product of the reaction consisted of black crystals with irregular shapes, which were determined to be $Sr_{11}InSb_9$. This compound decomposes in air. Handling of the raw materials and the reaction products was done inside an Ar-filled glove box.

15.11 INDIUM–BARIUM–ANTIMONY

Four ternary compounds, $Ba_4In_8Sb_{16}$, $Ba_5In_2Sb_6$, $Ba_7In_2Sb_6$, and $Ba_7In_8Sb_8$, are formed in the In–Ba–Sb system. $Ba_4In_8Sb_{16}$ crystallizes in the orthorhombic structure with the lattice parameters $a = 1016.6 \pm 0.3$, $b = 452.39 \pm 0.14$, and $c = 1949.5 \pm 0.6$ pm, a calculated density of 6.327 g·cm⁻³, and an energy gap of 0.10–0.12 eV (Kim et al. 1999). To prepare this compound, the reaction mixture of Ba, In, and Sb (molar ratio 1:2:4) was placed in a graphite tube and sealed in an evacuated quartz tube. The sealed mixture was heated slowly up to 700°C, kept at that temperature for 1 day, and subsequently cooled to room temperature over 1 day. The reaction led to the formation of a few bar-shaped black crystals. The title compound is stable at room temperature in air. However, when heated at about 500°C, it slowly decomposes to InSb and a Ba-rich amorphous phase. $Ba_4In_8Sb_{16}$ does not melt congruently.

$Ba_5In_2Sb_6$ also crystallizes in the orthorhombic structure with the lattice parameters $a = 1530.7 \pm 0.6$, $b = 1335.8 \pm 0.5$, and $c = 478.6 \pm 0.2$ pm and a calculated density of 5.59 g·cm⁻³ (Cordier and Stelter 1988) ($a = 1513.0$, $b = 1320.6$, and $c = 470.3$ pm [calculated values] and an energy gap of 0.27 eV [Yan et al. 2014]).

$Ba_7In_2Sb_6$ crystallizes in the cubic structure with the lattice parameter $a = 1047.6 \pm 0.5$ pm at 120 ± 2 K (Xia et al. 2008).

$Ba_7In_8Sb_8$ is metastable and crystallizes in the hexagonal structure with the lattice parameters $a = 482.74 \pm 0.03$ and $c = 1842.1 \pm 0.2$ pm and a calculated density of 6.37 g·cm⁻³ at 120 ± 2 K (Bobev et al. 2010). To prepare it, the starting materials were loaded in alumina crucibles, which were then enclosed in fused silica jackets and

flame sealed under vacuum. The nominal molar ratio In/Sb was in the range of 0.9:1 to 1.1:1. Indium was typically used in large excess (five times or greater than the stoichiometric amount) and served as flux. Handling of the reactants was done in an Ar-filled glove box or under vacuum.

15.12 INDIUM–ZINC–ANTIMONY

The fields of the primary crystallizations of the phases in the In–Zn–Sb ternary system are shown in Figure 15.12 (Skudnova et al. 1964). There are seven fields of primary crystallizations in this system. No ternary compound was found. The region of the solid solution based on InSb in the In–Zn–Sb system at 480°C is given in Figure 15.13 (Lapkina et al. 1987).

InSb–Zn. This system is a non-quasi-binary section of the In–Zn–Sb ternary system (Skudnova et al. 1964). The solubility of Zn in InSb is retrograde (Figure 15.14) and has a maximum value (1.51 at% Zn) at 480°C (Lapkina et al. 1987). According to the data of metallography, InSb dissolves 0.72 at% Zn at 280°C (Skudnova et al. 1964).

InSb–ZnSb. This system is a non-quasi-binary section of the In–Zn–Sb ternary system (Skudnova et al. 1964). InSb dissolves 0.78 mol% ZnSb at 440°C (1.3 at% Zn at 480°C [Lapkina et al. 1987]).

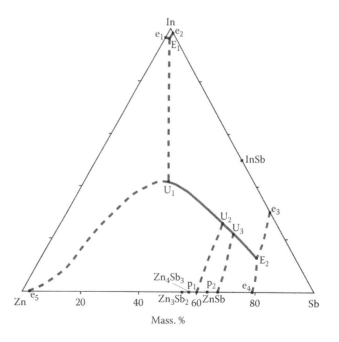

FIGURE 15.12 Fields of the primary crystallizations of the phases in the In–Zn–Sb ternary system. (From Skudnova, E.V., et al., *Zhurn. Neorgan. Khimii*, 9(2), 367, 1964. With permission.)

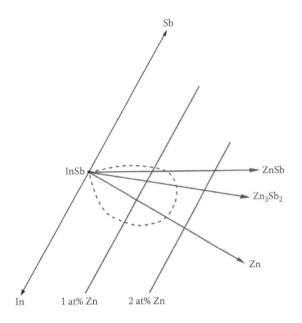

FIGURE 15.13   The region of the solid solution based on InSb in the In–Zn–Sb ternary system at 480°C. (From Lapkina, I.A., et al., *Izv. AN SSSR Neorgan. Mater.*, 23(4), 533, 1987. With permission.)

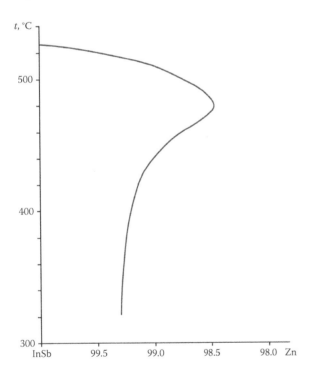

FIGURE 15.14   Temperature dependence of Zn solubility in InSb. (From Lapkina, I.A., et al., *Izv. AN SSSR Neorgan. Mater.*, 23(4), 533, 1987. With permission.)

InSb–Zn₃Sb₂. The phase diagram of this system is a eutectic type (Figure 15.15) (Skudnova et al. 1964). The eutectic crystallizes at 465°C and contains circa 50 mass% InSb. The region of the solid solution based on

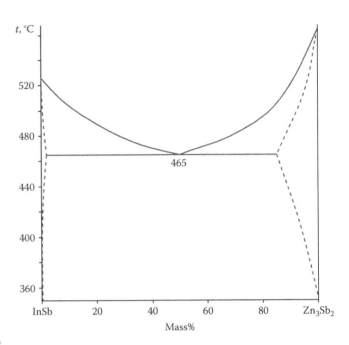

FIGURE 15.15   Phase diagram of the InSb–Zn₃Sb₂ system. (From Lapkina, I.A., et al., *Izv. AN SSSR Neorgan. Mater.*, 23(4), 533, 1987. With permission.)

Zn₃Sb₂ reaches 13.8 mass% InSb at the eutectic temperature, and InSb dissolves at this temperature ~0.6 mass% Zn₃Sb₂ (InSb dissolves 1.29 at% Zn at 480°C along this section [Lapkina et al. 1987]).

InSb–Zn₄Sb₃. This system is a non-quasi-binary section of the In–Zn–Sb ternary system (Skudnova et al. 1964).

15.13 INDIUM–CADMIUM–ANTIMONY

The liquidus surface of the In–Cd–Sb ternary system includes five fields of the primary crystallization of InSb, Sb, CdSb, Cd, and In and is given in Figure 15.16 (Belotskiy et al. 1978; Storonkin et al. 1986). The field of InSb primary crystallization occupies the biggest part of this surface. Ternary eutectics E_1, E_2, and E_3 crystallize at 410°C, 265°C, and 121°C, respectively. Storonkin et al. (1986) presented two variants of the liquidus surface: one of them for the primary crystallization from the melt of CdSb and the other for such crystallization of Cd₄Sb₃. The obtained results do not differ significantly from the results of Belotskiy et al. (1978). The region of the solid solution based on InSb in the In–Cd–Sb system at 350°C, 400°C, and 480°C is given in Figure 15.17 (Lapkina et al. 1989).

The thermodynamic functions of the mixing of the melt alloys of the In–Cd–Sb ternary system were

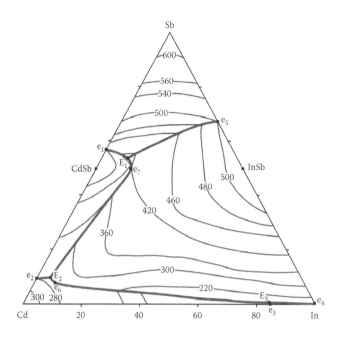

FIGURE 15.16 Liquidus surface of the In–Cd–Sb ternary system. (From Belotskiy, D.P., et al., *Izv. AN SSSR Neorgan. Mater.*, *14*(5), 838, 1978. With permission.)

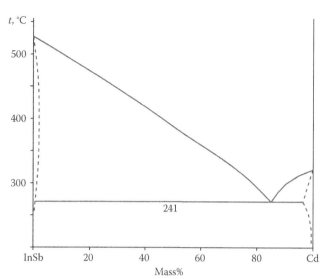

FIGURE 15.18 Phase diagram of the InSb–Cd system. (From Kuznetsov, G.M., and Bobrov, A.P., *Izv. AN SSSR Neorgan. Mater.*, *7*(5), 766, 1971. With permission.)

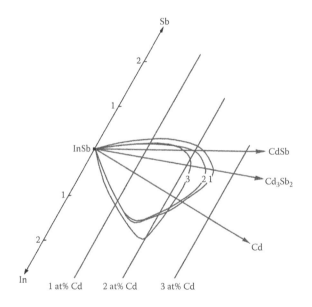

FIGURE 15.17 Solid solution regions based on InSb in the In–Cd–Sb ternary system at (1) 350°C, (2) 400°C, and (3) 480°C. (From Lapkina, I.A., et al., *Izv. AN SSSR Neorgan. Mater.*, *25*(3), 360, 1989. With permission.)

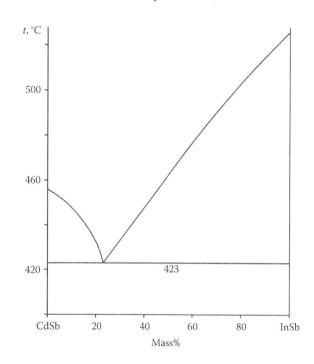

FIGURE 15.19 Phase diagram of the InSb–CdSb system. (From Kiriy, V.G., et al., *Zhurn. Neorgan. Khimii*, *41*(10), 1720, 1996. With permission.)

determined using the electromotive force (EMF) method (Mamedov et al. 1987).

InSb–Cd. The phase diagram of this system is a eutectic type (Figure 15.18) (Kuznetsov and Bobrov 1971). The eutectic crystallizes at 271°C and contains 85 mass% Cd. The solubility of Cd in InSb is retrograde (according to

the data of Lapkina et al. [1989], it is not retrograde) and is equal to 1.85 at% at the eutectic temperature (1.8 at% at 350°C [Lapkina et al. 1989]).

InSb–CdSb. The phase diagram of this system is a eutectic type (Figure 15.19) (Kiriy et al. 1996). The eutectic crystallizes at 423°C and contains 23 mass% InSb. The alloys of this system tend to a metastable crystallization

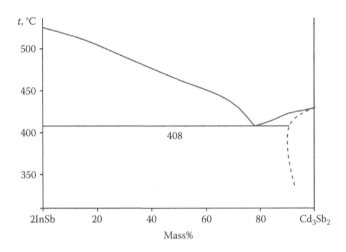

FIGURE 15.20 Phase diagram of the InSb–Cd₃Sb₂ system. (From Belotskiy, D.P., and Kotsumakha, M.P., *Dop. AN URSR*, (2), 222, 1965. With permission.)

with the formation of primary crystals of the Cd₄Sb₃ metastable phase and Cd₄Sb₃ + Sb and Cd₄Sb₃ + Sb + InSb metastable binary and ternary eutectics (Abrikosov et al. 1966; Kiriy et al. 1996). The solubility of Cd in InSb along this section reaches 2.1 at% at 350°C (Lapkina et al. 1989).

InSb–Cd₃Sb₂. The phase diagram of this system is a eutectic type (Figure 15.20) (Belotskiy and Kotsumakha 1965). The eutectic crystallizes at 408°C and contains 22 mol% 2InSb. Cd₃Sb₂ dissolves at the eutectic temperature 10 mol% 2InSb. It is possible that a part of Cd₃Sb₂ transforms to CdSb at 260°C–270°C. According to the data of Lapkina et al. (1989), the solubility of Cd in InSb along this section reaches 2.4 at% at 350°C.

15.14 INDIUM–THALLIUM–ANTIMONY

The liquidus surface of the In–Tl–Sb ternary system is presented in Figure 15.21 (Sorokina et al. 1985; Yevgen'ev et al. 1985). No ternary compound was found. The field of InSb primary crystallization occupies the biggest part of this surface. The liquidus isotherms are elongated along the InSb–Tl section.

InSb–Tl. The phase diagram of this system is a eutectic type (Figure 15.22) (Sorokina et al. 1983; Yevgen'ev et al. 1985). The eutectic crystallizes at 247°C and contains 88 at% Tl. The solid solutions based on Tl are characterized by a phase transformation at 147°C. Solubility of InSb in β-Tl reaches 8 mol% at the eutectic temperature. The solubility of Tl in InSb is retrograde (Figure 15.23), and the maximum solubility is 0.22 at% at 500°C

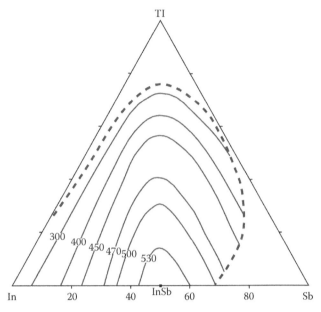

FIGURE 15.21 Liquidus surface of the In–Tl–Sb ternary system. (From Sorokina, O.V., et al., *Zhurn. Neorgan. Khimii*, *30*(10), 2642, 1985. With permission.)

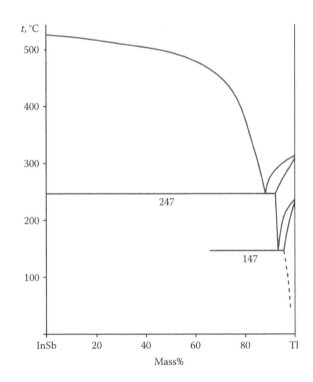

FIGURE 15.22 Phase diagram of the InSb–Tl system. (From Yevgen'ev, S.B., et al., *Izv. AN SSSR Neorgan. Mater.*, *21*(12), 2000, 1985. With permission.)

(Sorokina et al. 1983, 1985; Yevgen'ev et al. 1984, 1985). The distribution coefficient of Tl in InSb is 0.028.

InSb–TlSb. The calculated phase diagram of this system is given in Figure 15.24 (van Schilfgaarde et al.

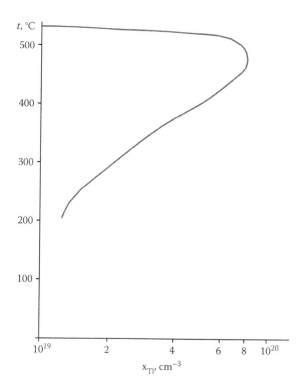

FIGURE 15.23 Solidus of the InSb–Tl system. (From Yevgen'ev, S.B., et al., *Izv. AN SSSR Neorgan. Mater.*, *21*(12), 2000, 1985. With permission.)

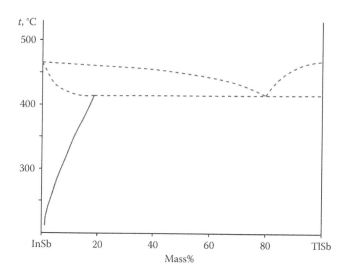

FIGURE 15.24 Calculated phase diagram of the InSb–TlSb system. (From van Schilfgaarde, M., et al., *Appl. Phys. Lett.*, *62*(16), 1857, 1993. With permission.)

1993). As the melting temperature of TlSb is unknown, it was speculated that the melting points of InSb and TlSb may be comparable. The properties of $In_{1-x}Tl_xSb$ solid solution were calculated within the framework of the density functional theory, and it was shown that the alloy at $x = 0.09$ could have an energy gap of 0.1 eV.

15.15 INDIUM–LANTHANUM–ANTIMONY

$LaIn_{1-x}Sb_2$ ternary compound is formed in the In–La–Sb system (Ferguson et al. 1999). It crystallizes in the monoclinic structure with the lattice parameters $a = 452.1 \pm 0.3$, $b = 433.1 \pm 0.3$, and $c = 1191.3 \pm 0.7$ pm and $\beta = 99.6680 \pm 0.0011°$ and a calculated density of 6.850 g·cm^{-3} for $x = 0.8$. To prepare this compound, La, In, and Sb were loaded into fused silica tubes (molar ratio 1:8:2, total mass 250 mg). The tubes were then evacuated, sealed, and heated at 570°C for 1 day and at 950°C for 2 days, cooled to 500°C over 1 day, and finally cooled to room temperature over 5 h.

15.16 INDIUM–CERIUM–ANTIMONY

$CeIn_{1-x}Sb_2$ ternary compound is formed in the In–Ce–Sb system (Ferguson et al. 1999). It crystallizes in the monoclinic structure with the lattice parameters $a = 447.8 \pm 0.2$, $b = 432.3 \pm 0.2$, and $c = 1179.6 \pm 0.5$ pm and $\beta = 99.36 \pm 0.02°$ for $x = 0.8$. To obtain the title compound, Ce, In, and Sb were loaded into fused silica tubes (molar ratio 1:8:2, total mass 250 mg). The tubes were then evacuated, sealed, and heated at 570°C for 1 day and at 950°C for 2 days, cooled to 500°C over 1 day, and finally cooled to room temperature over 5 h.

15.17 INDIUM–PRASEODYMIUM–ANTIMONY

Two compounds, **$PrIn_{1-x}Sb_2$** and **Pr_6InSb_{15}**, are formed in the In–Pr–Sb system. $PrIn_{1-x}Sb_2$ crystallizes in the monoclinic structure with the lattice parameters $a = 446.5 \pm 0.2$, $b = 430.3 \pm 0.2$, and $c = 1173.3 \pm 0.5$ pm and $\beta = 99.45 \pm 0.02°$ for $x = 0.8$ (Ferguson et al. 1999). For obtaining it, Pr, In, and Sb were loaded into fused silica tubes (molar ratio 1:8:2, total mass 250 mg). The tubes were then evacuated, sealed, and heated at 570°C for 1 day and at 950°C for 2 days, cooled to 500°C over 1 day, and finally cooled to room temperature over 5 h.

Pr_6InSb_{15} crystallizes in the orthorhombic structure with the lattice parameters $a = 423.73 \pm 0.03$, $b = 1503.7 \pm 0.1$, and $c = 1933.7 \pm 0.1$ pm and a calculated density of 7.477 g·cm^{-3} (Tkachuk et al. 2008). To prepare this compound, the mixtures of the elements were placed within the evacuated fused silica tubes and heated in a furnace. The tubes were heated at 570°C for 24 h and at 950°C for 48 h, cooled to 500°C over 24 h, and then cooled to room temperature over 5 h. Needle-shaped crystals of the title compound were obtained.

15.18 INDIUM–NEODYMIUM–ANTIMONY

NdIn$_{1-x}$Sb$_2$ ternary compound is formed in the In–Nd–Sb system (Ferguson et al. 1999). It crystallizes in the monoclinic structure with the lattice parameters a = 444.5 ± 0.4, b =429.7 ± 0.4, and c = 1167.7 ± 0.9 pm and β = 99.22 ± 0.05° for x = 0.8. To obtain the title compound, Nd, In, and Sb were loaded into fused silica tubes (molar ratio 1:8:2, total mass 250 mg). The tubes were then evacuated, sealed, and heated at 570°C for 1 day and at 950°C for 2 days, cooled to 500°C over 1 day, and finally cooled to room temperature over 5 h.

15.19 INDIUM–EUROPIUM–ANTIMONY

Three compounds, **Eu$_5$In$_2$Sb$_6$**, **Eu$_{11}$InSb$_9$**, and **Eu$_{14}$InSb$_{11}$**, are formed in the In–Eu–Sb system. Eu$_5$In$_2$Sb$_6$ is a narrow band-gap semiconductor and crystallizes in the orthorhombic structure with the lattice parameters a = 1251.0 ± 0.3, b =1458.4 ± 0.3, and c = 462.43 ± 0.09 pm and a calculated density of 6.770 g·cm^{-3} (Park et al. 2002). It was obtained from a direct element combination reaction (molar ratio 5:2:6) in a sealed graphite tube at 870°C.

Eu$_{11}$InSb$_9$ also crystallizes in the orthorhombic structure with the lattice parameters a = 1222.4 ± 0.2, b =1287.4 ± 0.2, and c = 1731.5 ± 0.3 pm and a calculated density of 7.025 g·cm^{-3} at 120 ± 2 K (Xia et al. 2007). To prepare this compound, the reaction was carried out using a mixture of Eu and Sb (molar ratio 11:9) and a 100-fold excess of In metal. The elements were loaded in an alumina crucible, which was subsequently enclosed in an evacuated fused silica ampoule. The reaction was heated from 100°C to 1000°C at a rate of 300°C·h^{-1}, allowed to dwell at this temperature for 20 h, and subsequently slowly cooled down to 600°C at a rate of 3°C·h^{-1}. At this temperature, the ampoule was taken out from the furnace and the excess of molten In was removed by centrifugation. All manipulations were performed inside an argon-filled glove box with controlled oxygen and moisture levels below 1 ppm or under vacuum.

Eu$_{14}$InSb$_{11}$ crystallizes in the tetragonal structure with the lattice parameters a = 1728.0 ± 0.5 and c = 2312.9 ± 0.8 pm and a calculated density of 6.621 g·cm^{-3} at 143 K and a = 1728.9 ± 0.8 and c = 2277 ± 2 pm at room temperature (Chan et al. 1997). It was prepared by weighing stoichiometric amounts of the elements in a dry box and loading into a clean Ta tube with a sealed end and subsequently sealing in an argon-filled arc welder. The sealed Ta tube was further sealed in a fused silica tube under purified argon. A high yield of polycrystalline title compound was obtained by heating the mixtures to temperatures at 1000°C for periods between 24 h and 5 days and cooling the reaction to room temperature at rates of 5–60°C·h^{-1}. Needle-shaped single crystals, along with dendritic crystallites, were produced when the reaction temperature was 1250°C for 3 days.

15.20 INDIUM–YTTERBIUM–ANTIMONY

Two ternary compounds, **Yb$_5$In$_2$Sb$_6$** and **Yb$_{11}$InSb$_9$**, are formed in the In–Yb–Sb system. Yb$_5$In$_2$Sb$_6$ is stable in air at room temperature, melts congruently at 767°C, and crystallizes in the orthorhombic structure with the lattice parameters a = 739.92 ± 0.05, b = 2300.1 ± 0.6, and c = 451.39 ± 0.04 pm, a calculated density of 7.891 g·cm^{-3}, and an energy gap of 0.2–0.4 eV (Kim et al. 2000) (a = 739.50 ± 0.07, b = 2298.5 ± 0.1, and c = 451.19 ± 0.02 pm and a calculated density of 7.89 g·cm^{-3} [Aydemir et al. 2015]). To prepare the title compound, a reaction mixture composed of Yb (5 mM), In (2 mM), and Sb (6 mM) was placed in a graphite tube and then sealed in an evacuated silica tube (Kim et al. 2000). The mixture was heated slowly (over 4 days) up to 700°C, kept at that temperature for 1 day, and subsequently cooled to room temperature over 1 day. A few bar-shaped black crystals were obtained. It was also synthesized by ball milling, followed directly by hot pressing (Aydemir et al. 2015). Stoichiometric amounts of small-cut Yb ingot, Sb shot, and InSb were loaded under Ar into stainless steel vials with stainless steel balls. The contents were ball milled for 1 h. The resulting fine powder was hot pressed under Ar using a maximum pressure and temperature of 45 MPa and 550°C, respectively, for 1 h. The samples were cooled down to room temperature slowly under no load.

Yb$_{11}$InSb$_9$ also crystallizes in the orthorhombic structure with the lattice parameters a = 1178.91 ± 0.24, b = 1242.28 ± 0.25, and c = 1667.35 ± 0.33 pm (Yi et al. 2010) (a = 1178.86 ± 0.11, b = 1241.51 ± 0.12, and c = 1667.43 ± 0.15 pm and a calculated density of 8.475 g·cm^{-3} at 120 ± 2 K [Xia et al. 2007]). To obtain the title compound, the starting materials were loaded in an alumina crucible with a stoichiometric ratio of Yb/In/Sb at 11:(76–100):9 (Xia et al. 2007; Yi et al. 2010). The crucible was put in a fused silica ampoule with quartz wool on the top and at the bottom, which was then sealed under a vacuum. The reaction was heated in a furnace under the following temperature profile: heating to 500°C from room temperature in 2 h and dwelling at 500°C for 6 h, and then ramping to 1000°C within 2 h and dwelling 6 h at this temperature, followed by cooling to 600°C at a rate of 2°C·h⁻¹. At this point, the ampoule was quickly removed from the furnace, inverted, and centrifuged to remove the extra flux. Xia et al. (2007) did not use the quartz wool, and the temperature treatment was as follows: heating from 100°C to 1000°C at a rate of 300°C·h⁻¹, dwelling at this temperature for 20 h, and subsequently slowly cooling down to 600°C at a rate of 3°C·h⁻¹. All materials were handled in a nitrogen-filled glove box or with other inert atmosphere techniques (Xia et al. 2007; Yi et al. 2010).

$Yb_{11}InSb_9$ is thermally stable up to 1094°C but shows surface oxidation after a few hours' exposure in air (Yi et al. 2010). It appears to be a degenerate semiconductor or semimetal.

15.21 INDIUM–SILICON–ANTIMONY

The solubility of Si in InSb is less than 1 at% (Kolm et al. 1957).

15.22 INDIUM–GERMANIUM–ANTIMONY

The liquidus surface of the In–Ge–Sb ternary system is given in Figure 15.25 (Al'fer et al. 1983). The fields of InSb, Ge, Sb, and In primary crystallizations exist in this system (the field of In primary crystallization is degenerated). Two ternary eutectics are on the liquidus surface:

one of them crystallizes at 514.5°C, and the other is degenerated from the In corner.

InSb–Ge. This system is a non-quasi-binary section of the In–Ge–Sb ternary system (Woolley and Lees 1959; Zemskov et al. 1962). The temperature region of the existing three-phase field is small (not higher than 4°C) (Woolley and Lees 1959). The solubility of InSb in Ge is equal to 0.35, 0.50, 0.65, 0.70, 0.40, and 0.10 mass% at 300°C, 400°C, 500°C, 506°C, 700°C, and 900°C, respectively (Glazov et al. 1959). The samples were annealed at each temperature for 300 h and investigated through metallography and measurement of microhardness. According to the data of Kolm et al. (1957), the solubility of Ge in InSb is less than 1 at%. Based on the measurement of viscosity, it was concluded on increasing the degree of dissociation of InSb dissolved in the molten germanium at higher temperatures and at the dilution of the solution (Glazov et al. 1966).

As a result of ultrafast crystallization (10^7°C·s⁻¹) of the InSb–Ge melts, a metastable phase with cubic structure ($a = 853$ pm) was fixed (Glazov et al. 1977a).

15.23 INDIUM–TIN–ANTIMONY

The liquidus surface of the In–Sn–Sb ternary system was constructed by Löhberg (1968); Ishihara et al. (1999); Ohnuma et al. (2000); Legendre et al. (2001); and Dinsdale et al. (2008). The field of InSb primary crystallization occupies the biggest part of this system. There are also the fields of primary crystallization of (Sb), SbSn, Sb_2Sn_3, and (Sn), and degenerated fields of (In) and β- and γ-phases of the In–Sn binary system. According to the data of Rayson et al. (1959), the solubility of Sb in the γ-phase is small and the stable hard phase X exists in this ternary system. It is probably based on InSb, and may be expected to affect the homogeneity ranges of other phases of the system.

The isothermal section of the In–Sn–Sb ternary system at room temperature is shown in Figure 15.26 (Vasil'yev and Legendre 2007). No ternary compounds have been identified. In this system, the InSb + (β-SnSb) two-phase region is of special interest. (β-SnSb) represents a series of homologous phases separated by narrow two-phase regions. Each of these phases may form an $In_xSn_ySb_z$ solid solution with In, whose molar fraction is within 0.05. Under nonequilibrium conditions, the (β-SnSb) phase fields merge into the β-$In_xSn_ySb_z$ region. The thermodynamic properties of alloys and phase

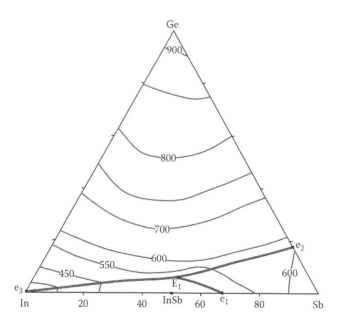

FIGURE 15.25 Liquidus surface of the In–Ge–Sb ternary system. (From Al'fer, S.A., et al., *Zhurn. Fiz. Khimii*, *57*(5), 1292, 1983. With permission.)

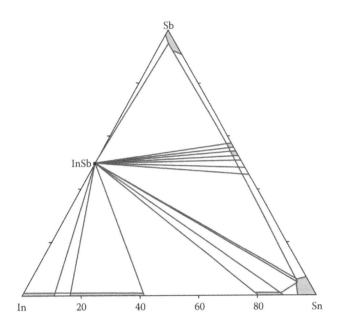

FIGURE 15.26 Isothermal section of the In–Sn–Sb ternary system at room temperature. (From Vasil'yev, V.P., and Legendre, B., *Inorg. Mater.*, 43(8), 803, 2007. With permission.)

equilibria in the In–Sn–Sb system were studied by EMF measurements at temperatures from 330°C to 560°C.

Phase equilibria in this system have also been studied experimentally and calculated by the CALPHAD method by Manasijević et al. (2008) and Dinsdale et al. (2008), and isothermal sections at 100°C and 300°C were constructed (Figure 15.27). It is seen that SbSn dissolves about 4 at% In. There is no solubility of Sn in indium antimonide. Also, no solubility of indium in the solid solution of Sn in Sb was found. The samples were annealed at 300°C for 200 and 850 h (depending on composition) and studied through DTA and scanning electron microscopy (SEM) with energy-dispersive x-ray spectroscopy (EDX). The isothermal sections of the In–Sn–Sb ternary systems at 119°C, 140°C, 215°C, 240°C, 242°C, 330°C, and 407°C were constructed, and the reaction scheme was presented by Legendre et al. (2001). The isothermal sections of this system at 100°C and 200°C were also determined by Ohnuma et al. (2000).

The EMF measurements were carried out on the liquid In–Sn–Sb alloys, and excess thermodynamic quantities were determined by Emi and Shimoji (1967). It was shown that in this system, excess molar entropies and enthalpies of In showed abrupt changes near the composition where the atomic ratio of In to Sb is equal to about

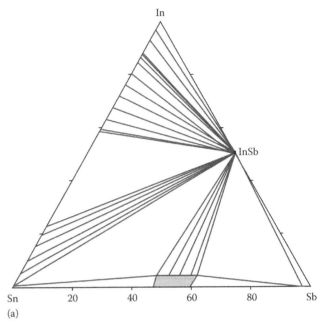

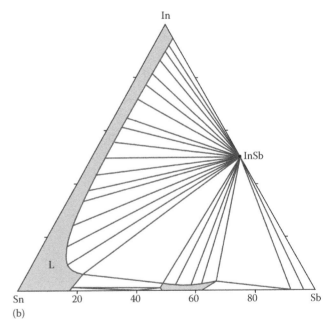

FIGURE 15.27 Isothermal sections of the In–Sn–Sb ternary system at (a) 100°C and (b) 300°C. (From Manasijević, D., et al., *J. Alloys Compd.*, 450(1–2), 193, 2008. With permission.)

unity. The isoactivity curves of indium in the liquid In–Sn–Sb alloys at 630°C are presented in Figure 15.28.

The enthalpies of mixing in the liquid state of the In–Sn–Sb ternary system were determined in a heat flow calorimeter at 663°C and are given in Figure 15.29 (Gather et al. 1985). Enthalpies of formation of the In–Sn–Sb melts at 632°C were also determined by quantitative thermographic analysis (Vecher et al. 1977).

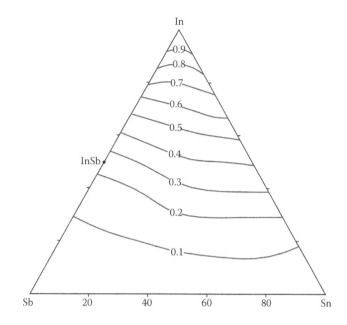

FIGURE 15.28   Isoactivity curves of In in the liquid In–Sn–Sb alloys at 630°C. (From Emi, T., and Shimoji, M., *Ber. Bunsenges. Phys. Chem.*, 71(4), 367, 1967. With permission.)

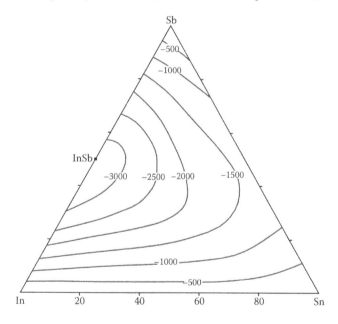

FIGURE 15.29   Enthalpies of mixing in the liquid state of the In–Sn–Sb ternary system. (From Gather, B., et al., *Z. Metallkde*, 76(8), 523, 1985. With permission.)

Using a molten salt method, EMF measurements have been obtained for In–Sn–Sb alloys (Vassiliev et al. 2001). From the three descriptions of the Gibbs function for the binary liquid phases, a ternary calculation was done on the basis of Muggianu's model. Fair agreement was observed between the experimental and calculated values of In activity for all the compositions studied.

This confirms the absence of ternary interactions in the liquid state. Calculation of the thermodynamic characteristics for this ternary system was also performed by Gomidželović et al. (2010) using the Redlich–Kister–Muggianu model. It was shown that the addition of Sn in the In–Sb binary system leads to the reduction of the negative deviation of the In and Sb activities.

InSb–Sn. The phase relations in this system were studied by many scientists. Baruch and Desse (1955) indicated that the phase diagram of this system is a eutectic type, with the eutectic coordinates 230°C ± 2°C and 34 at% Sn. Transformation in the solid state at 212°C was determined, but its nature is unknown. According to the data of Woolley and Lees (1959) and Zitter (1958), this system shows a peritectic form with a peritectic line at 235°C. The liquidus of this system was also determined by Papadakis et al. (1965) and Gerdes and Predel (1981a). As was noted by Woolley and Lees (1959), this section cannot be treated as quasi-binary. These data were confirmed by many investigations (Wipf and Coles 1961; Löhberg 1968; Ishihara et al. 1999; Dinsdale et al. 2008; Manasijević et al. 2008), where a three-phase region was determined in the InSb–Sn system.

The solubility of Sn in InSb is less than 0.5–1 at% (Kolm et al. 1957; Zitter 1958), and the solubility of InSb in Sn reaches 2 mol% (Baruch and Desse 1955) (4 mol% [Zitter 1958]).

Using a high-temperature calorimeter, the change δ(ΔH) in enthalpy upon mixing the molten InSb with Sn in the entire concentration range was determined (Gerdes and Predel 1981c). The temperature dependence of the δ(ΔH) values was also measured. It was shown that the concentration dependence of the δ(ΔH) values above 580°C is negative at low Sn contents and positive at high Sn concentrations. The concentration dependence of the δ(ΔH) values has been described using a simple association model.

By extremely rapid cooling of the liquid alloys, continuous solid solutions in the InSb–Sn were obtained (Bucher et al. 1986). In order to attain the high cooling rates (10^7–10^8°C·s^{-1}), a shock wave tube has been used.

The high-pressure forms of InSb and Sn are soluble in all proportions (Banus et al. 1965a, 1965b). Such solid solutions were obtained as metastable phases at atmospheric pressure if they were cooled to 77 K before the high pressure was released. They are retained indefinitely if kept at 77 K, but were transformed into stable two-phase mixtures at sufficiently high temperatures.

Earlier superconductivity data (Nye et al. 1964) suggested the existence of a miscibility gap between 5 and 65 at% Sn because the samples in that region had not been annealed sufficiently to become single phase.

InSb–SnSb. This section of the In–Sn–Sb ternary system is not quasi-binary, since SnSb is formed according to a peritectic reaction (Abrikosov et al. 1966; Löhberg 1968; Ishihara et al. 1999; Vasil'yev and Legendre 2007; Dinsdale et al. 2008). On this section, up to 75 mol% SnSb, primary crystallization of InSb is observed. Antimony primary crystallizes on the SnSb side.

15.24 INDIUM–LEAD–ANTIMONY

The liquidus surface of the In–Pb–Sb ternary system was constructed experimentally by Geis and Peretti (1962a, 1963) and calculated using binary thermodynamic data by Minić et al. (2008); it is shown in Figure 15.30. InSb is the primary phase of precipitation over most of the system. A ternary eutectic melting at 249°C is located at 0.97 mass% In, 87 mass% Pb, and 12.03 mass% Sb. No evidence of ternary compounds was encountered, and only small terminal solid solubilities were found.

Predicted isothermal sections of this ternary system at 300°C and 350°C are given in Figure 15.31 (Minić et al. 2008). The isothermal section at 350°C includes one three-phase region, seven two-phase regions, and three single-phase regions. At 300°C, the phase diagram

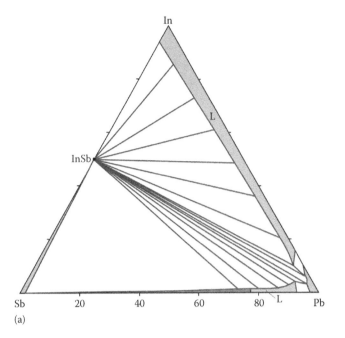

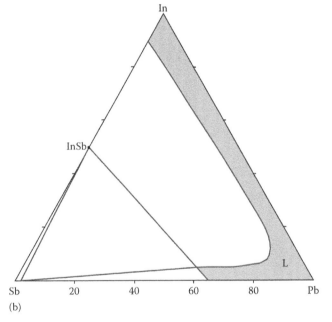

FIGURE 15.31 Isothermal sections of the In–Pb–Sb ternary system at (a) 300°C and (b) 350°C. (From Minić, D., et al., *J. Serb. Chem. Soc.*, *73*(3), 377, 2008. With permission.)

includes three three-phase region, seven two-phase regions, and three single-phase regions.

According to the results of the thermodynamic and structural investigations, weak interactions of Pb with In and Sb could be noticed through positive deviation from Raoult's law (Minić et al. 2003). Structural investigations confirmed that the alloys in this ternary system are two-phased; one phase is based on Pb, and the other is based on In–Sb.

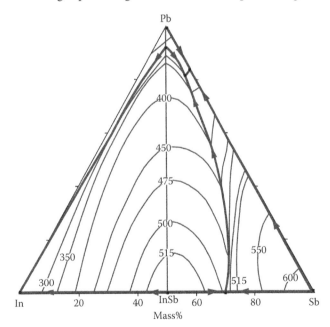

FIGURE 15.30 Liquidus surface of the In–Pb–Sb ternary system. (From Geis, D.R., and Peretti, E.A., *J. Chem. Eng. Data*, *8*(3), 470, 1963. With permission.)

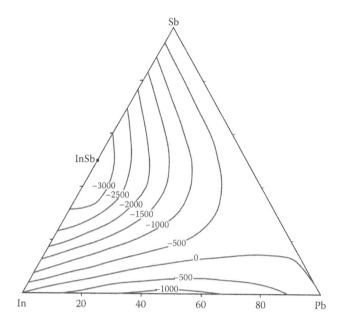

FIGURE 15.32 Ternary mixing isoenthalpy curves in the In–Pb–Sb system at 980°C. (From Gather, B., et al., *Z. Metallkde,* 78(4), 280, 1987. With permission.)

Gather et al. (1987) determined enthalpy of mixing of the In–Pb–Sb ternary liquid in a heat flow calorimeter. The measured ternary mixing isoenthalpy curves are presented in Figure 15.32.

InSb–Pb. The phase diagram of this system is shown in Figure 15.33 (Geis and Peretti 1962b; Predel and Gerdes 1978; Peuschel et al. 1981; Minić et al. 2008). The eutectic crystallizes at 297°C and contains 10 mol%

InSb (Minić et al. 2008) (at 298.5°C, contains 6.5 mass% InSb [Geis and Peretti 1962b]). The solubility of Pb in InSb is less than 1 at% (Kolm et al. 1957).

Along this quasi-binary section, the change in enthalpy $\delta(\Delta H)$ upon mixing liquid Pb with the molten InSb was determined using a high-temperature calorimeter (Predel and Gerdes 1978; Gerdes and Predel 1981b). Positive $\delta(\Delta H)$ values were found, the concentration dependence of which can be represented in terms of an association model and also largely on the basis of the regular solution model.

15.25 INDIUM–TITANIUM–ANTIMONY

The calculated isothermal section of the In–Ti–Sb ternary system at room temperature is shown in Figure 15.34 (Liu and Mohney 2003b). Binary phases were treated as stoichiometric compounds, and ternary phases and ternary solubilities were disregarded. The dotted lines indicate possible alternative tie-lines.

InTi₅Sb₂. $InTi_5Sb_2$ ternary compound, which crystallizes in the tetragonal structure with the lattice parameters $a = 1073.8 \pm 0.1$ and $c = 550.1 \pm 0.2$ pm (Kozlov and Pavlyuk 2003) ($a = 1057.9$ and $c = 538.51$ pm and enthalpy of formation of -39.6 kJ·M⁻¹ [calculated values] [Colinet and Tedenac 2015]), is formed in the In–Ti–Sb system. The title compound was prepared by arc melting of a mixture of In, Ti, and Sb in stoichiometric ratio in an Ar

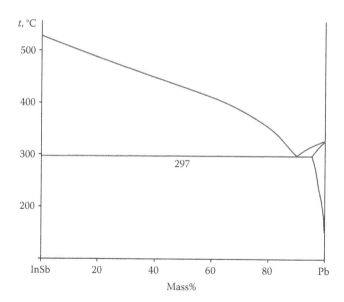

FIGURE 15.33 Phase diagram of the InSb–Pb system. (From Minić, D., et al., *J. Serb. Chem. Soc.,* 73(3), 377, 2008. With permission.)

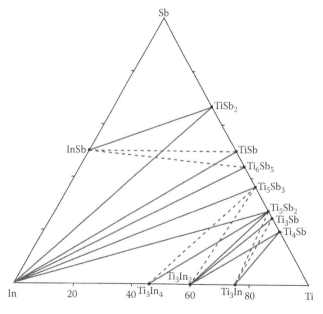

FIGURE 15.34 Calculated isothermal section of the In–Ti–Sb ternary system at room temperature. (From Liu, W.E., and Mohney, S.E., *Mater. Sci. Eng. B,* 103(2), 189, 2003. With permission.)

atmosphere, annealed at 400°C for 720 h, and quenched in cold water (Kozlov and Pavlyuk 2003).

15.26 INDIUM–ZIRCONIUM–ANTIMONY

The calculated isothermal section of the In–Zr–Sb ternary system at room temperature is shown in Figure 15.35 (Liu and Mohney 2003b). Binary phases were treated as stoichiometric compounds, and ternary phases and ternary solubilities were disregarded. The dotted lines indicate possible alternative tie-lines.

15.27 INDIUM–HAFNIUM–ANTIMONY

The calculated isothermal section of the In–Hf–Sb ternary system at room temperature is shown in Figure 15.36 (Liu and Mohney 2003b). Binary phases were treated as stoichiometric compounds, and ternary phases and ternary solubilities were disregarded. The dotted line indicates a possible alternative tie-line.

According to the data of Boller and Parthé (1963), **Hf₅(In₀.₅Sb₀.₅)₃** ternary compound, which crystallizes in the hexagonal structure with the lattice parameters $a = 846$ and $c = 579$ pm, is formed in this system. To obtain the title compound, the well-mixed metal powders were pressed into pellets and induction melted in BN crucibles under an inert atmosphere of Ar. The alloys formed were homogenized in sealed-off evacuated silica tubes at 1000°C for 24 h.

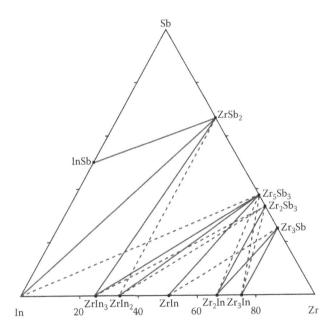

FIGURE 15.35 Calculated isothermal section of the In–Zr–Sb ternary system at room temperature. (From Liu, W.E., and Mohney, S.E., *Mater. Sci. Eng. B*, *103*(2), 189, 2003. With permission.)

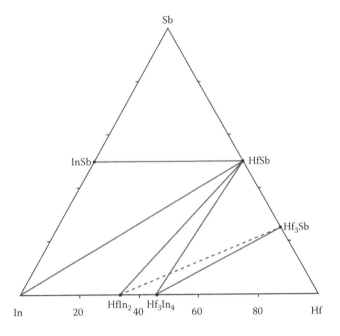

FIGURE 15.36 Calculated isothermal section of the In–Hf–Sb ternary system at room temperature. (From Liu, W.E., and Mohney, S.E., *Mater. Sci. Eng. B*, *103*(2), 189, 2003. With permission.)

15.28 INDIUM–BISMUTH–ANTIMONY

A thermodynamic description of the In–Bi–Sb ternary system was developed using the CALPHAD method (Cui et al. 2002). Phase equilibria information, such as vertical sections, isothermal sections, and liquidus projection, was calculated and compared with experimental data. It was shown that calculated (Cui et al. 2002) and experimental (Peretti 1960, 1961; Goryacheva et al. 1985) data are in excellent agreement in most cases. The calculated liquidus surface is presented in Figure 15.37. The liquidus temperatures increase significantly with increasing Sb concentration. It can be seen that, totally, six ternary invariant reactions exist in this ternary system: E_1 (109°C), L $\longleftrightarrow$ Sb$_x$Bi$_{1-x}$ + InBi$_x$Sb$_{1-x}$ + InBi; U_1 (93°C), L + (In) $\longleftrightarrow$ InBi$_x$Sb$_{1-x}$ + In$_{1-x}$Bi$_x$; U_2 (89°C), L + InBi $\longleftrightarrow$ InBi$_x$Sb$_{1-x}$ + In$_5$Bi$_3$; E_2 (88°C), L $\longleftrightarrow$ InBi$_x$Sb$_{1-x}$ + In$_2$Bi + In$_5$Bi$_3$; E_3 (73°C), L $\longleftrightarrow$ InBi$_x$Sb$_{1-x}$ + In$_{1-x}$Bi$_x$ + In$_2$Bi; and D_1 (49°C), In$_{1-x}$Bi$_x$ $\longleftrightarrow$ In$_2$Bi + InBi$_x$Sb$_{1-x}$ + (In). The solid phase equilibrium D_1 is totally degenerated and exactly equals the binary eutectoid In$_{1-x}$Bi$_x$ $\longleftrightarrow$ In$_2$Bi + (In), with the InBi$_x$Sb$_{1-x}$ phase being in equilibrium but not participating in the reaction.

The calculated isothermal sections of the In–Bi–Sb ternary system at 75°C, 120°C, and 300°C are presented in Figures 15.38 and 15.39 (Cui et al. 2002; Minić et al. 2006).

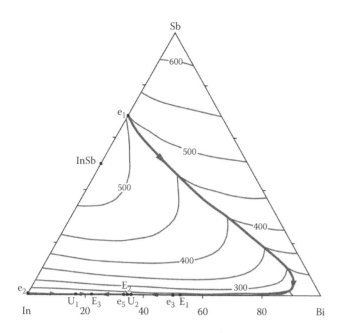

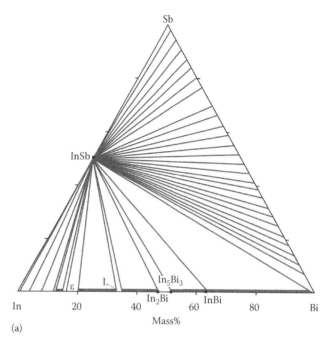

(a)

FIGURE 15.37 Liquidus surface of the In–Bi–Sb ternary system. (From Cui, Y., et al., *Mater. Trans*, 43(8), 1879, 2002. With permission.)

The activity of In has been determined by EMF measurements using a zirconia electrolyte at 613°C–906°C (Kameda et al. 1997). The isoactivity curves of In at 727°C and 827°C were obtained in the whole composition ranges and are given in Figure 15.40.

InSb–InBi. The phase diagram of this system is a eutectic type (Figure 15.41) (Peretti 1958; Karavaev 1974; Ufimtsev et al. 1979; Akchurin et al. 1984; Ma et al. 1990; Cui et al. 2002; Vakiv et al. 2011). The eutectic is degenerated from the InBi side. The solubility of InBi in InSb at 114°C is 1.9 mol% (Latypov et al. 1977) (the maximum solubility of InBi in InSb is 2.4 mol% [Joukoff and Jean-Louis 1972]). According to the data of Raukhman et al. (1977), the solubility of InBi in InSb is retrograde. Thermodynamic calculations of the phase diagram of this system were carried out using the delta-lattice-parameter model (Ma et al. 1990). The obtained results agree well with the experimental data. A tremendously large miscibility gap exists in this system. The critical temperature was predicted to be 496°C at 45.5 mol% InBi.

Epitaxial metastable InBi$_x$Sb$_{1-x}$ alloys with InBi concentrations up to 12 mol% have been grown on GaAs substrates (Zilko and Greene 1978; Cadien et al. 1980). The lattice constants of the polycrystalline films obey Vegard's law. These films were found to be stable to decomposition during long anneals, which is within 100°C of the equilibrium liquidus temperature.

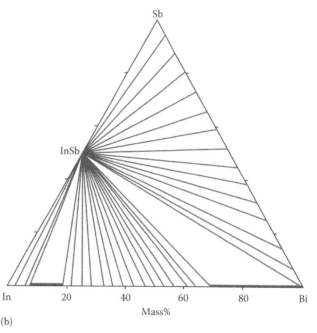

(b)

FIGURE 15.38 Calculated isothermal sections of the In–Bi–Sb ternary system at (a) 75°C and (b) 120°C. (From Cui, Y., et al., *Mater. Trans.*, 43(8), 1879, 2002. With permission.)

InSb–In$_2$Bi. The phase diagram of this system is a eutectic type (Figure 15.42) (Ufimtsev et al. 1979; Akchurin et al. 1984; Cui et al. 2002). The eutectic is degenerated from the In$_2$Bi side. The solubility of In$_2$Bi in InSb at 96°C is 2.8 mol% (Latypov et al. 1977).

InSb–Bi. This system is a non-quasi-binary section of the In–Bi–Sb ternary system (Akchurin et al. 1984). The

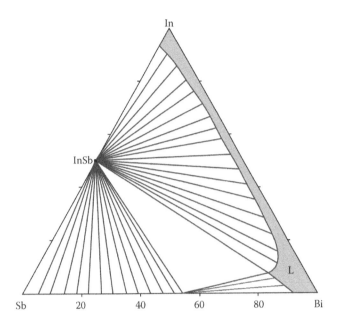

FIGURE 15.39   Calculated isothermal section of the In–Bi–Sb ternary system at 300°C. (From Minić, D., et al., *J. Serb. Chem. Soc.*, *71*(7), 843, 2006. With permission.)

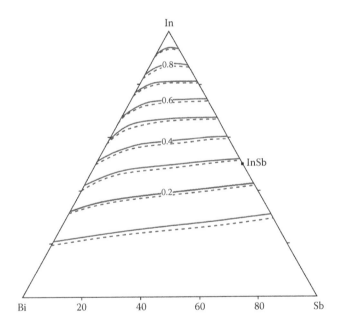

FIGURE 15.40   Isoactivity curves at 727°C (solid lines) and 827°C (dotted lines) of indium in the liquid In–Bi–Sb alloys. (From Kameda, K., et al., *J. Jpn. Inst. Met.*, *61*(5), 444, 1997. With permission.)

solubility of Bi in InSb at 273°C is 0.9 mol% (Latypov et al. 1977).

The enthalpy change due to the mixing of molten InSb with liquid Bi has been determined by using a high-temperature calorimeter (Predel and Gerdes 1978; Gerdes and Predel 1979a). Melting of this system yields a maximum

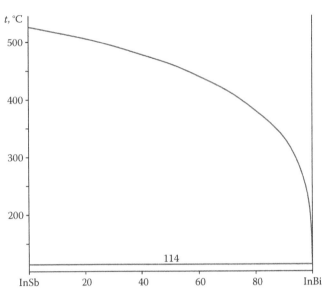

FIGURE 15.41   Phase diagram of the InSb–InBi system. (From Cui, Y., et al., *Mater. Trans.*, *43*(8), 1879, 2002. With permission.)

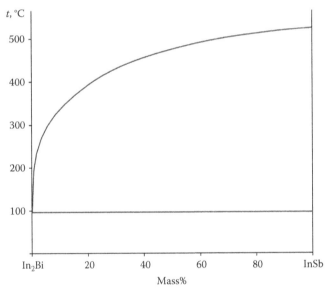

FIGURE 15.42   Phase diagram of the InSb–In₂Bi system. (From Ufimtsev, V.B., et al., *Izv. AN SSSR Neorgan. Mater.*, *15*(10), 1740, 1979. With permission.)

value of the enthalpy change of +536 J·mol⁻¹ at 612°C. The thermodynamic properties of the liquid InSb–Bi alloys are distinctly defined by the formation of associates (Predel and Gerdes 1978). The calculated result is consistent within an order of magnitude with experimental data.

15.29 INDIUM–VANADIUM–ANTIMONY

The calculated isothermal section of the In–V–Sb ternary system at room temperature is shown in Figure 15.43

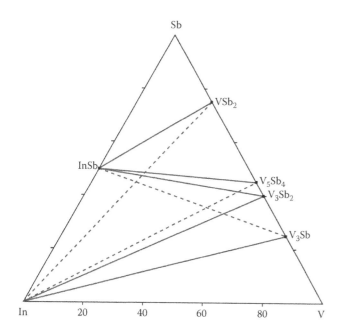

FIGURE 15.43   Calculated isothermal section of the In–V–Sb ternary system at room temperature. (From Liu, W.E., and Mohney, S.E., *Mater. Sci. Eng. B*, *103*(2), 189, 2003. With permission.)

(Liu and Mohney 2003b). Binary phases were treated as stoichiometric compounds, and ternary phases and ternary solubilities were disregarded. The dotted lines indicate possible alternative tie-lines.

15.30 INDIUM–NIOBIUM–ANTIMONY

The calculated isothermal section of the In–Nb–Sb ternary system at room temperature is shown in Figure 15.44 (Liu and Mohney 2003b). Binary phases were treated as stoichiometric compounds, and ternary phases and ternary solubilities were disregarded. The dotted line indicates a possible alternative tie-line.

15.31 INDIUM–TANTALUM–ANTIMONY

The calculated isothermal section of the In–Ta–Sb ternary system at room temperature is shown in Figure 15.44 (Liu and Mohney 2003b). Binary phases were treated as stoichiometric compounds, and ternary phases and ternary solubilities were disregarded. The dotted line indicates a possible alternative tie-line.

15.32 INDIUM–OXYGEN–ANTIMONY

The isothermal section of the Ga–O–Sb ternary system at 25°C–400°C for 0.1 MPa oxygen pressure is shown in Figure 15.45 (Smirnova et al. 1981; Schwartz 1983). **InSbO$_4$** (**In$_{2-x}$Sb$_x$O$_{3+x}$** at $x \approx 1$) and **In$_2$Sb$_4$O$_9$** ternary compounds are formed in this system. InSbO$_4$

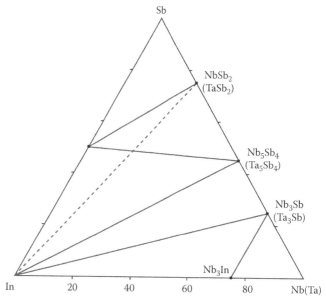

FIGURE 15.44   Calculated isothermal sections of the In–Nb(Ta)–Sb ternary systems at room temperature. (From Liu, W.E., and Mohney, S.E., *Mater. Sci. Eng. B*, *103*(2), 189, 2003. With permission.)

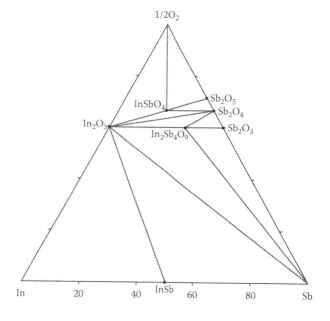

FIGURE 15.45   Isothermal section of the Ga–O–Sb ternary system at 25°C–400°C for 0.1 MPa oxygen pressure. (From Smirnova, T.P., et al., *Thin Solid Films*, *76*(1), 11, 1981. With permission.)

crystallizes in the monoclinic structure with the lattice parameters $a = 470 \pm 1$, $b = 478 \pm 1$, and $c = 322 \pm 1$ pm and $\beta = 90°24 \pm 5$ and an experimental density of 6.89 g·cm^{-3} (Varfolomeev et al. 1975a, 1975b) (in the tetragonal structure, with the lattice parameters $a = 474$ and $c = 321.5$ pm and calculated and experimental densities of 6.89 and 6.60 g·cm^{-3}, respectively [Bayer 1963]). The samples of the title compound were prepared by coprecipitation of hydrated indium and antimony oxides with aqueous ammonia solution from chloride-nitrate solutions containing the above-mentioned elements. The resulting precipitates were dried and calcined with an intermediate trituration at 800°C, and then at 1000°C for 100–200 h. InSbO$_4$ could also be obtained by a solid-state reaction between In$_2$O$_3$ and Sb$_2$O$_3$ (Bayer 1963). The compacted, stoichiometric mixtures were heated slowly to about 850°C in order to oxidize Sb$_2$O$_3$ to Sb$_2$O$_5$, and finally heated for 20 h at 1000°C and quenched to room temperature.

In$_2$Sb$_4$O$_9$ was prepared by the interaction of In$_2$O$_3$ and Sb$_2$O$_3$ (molar ratio 1:2) in an inert atmosphere at circa 650°C as a yellow, finely crystalline substance (Varfolomeev et al. 1975b). The experimental density of this compound is 4.80 ± 0.05 g·cm^{-3}. It decomposes at temperatures higher than 700°C, forming In$_2$O$_3$ and Sb$_2$O$_3$.

15.33 INDIUM–SULFUR–ANTIMONY

InSbS$_3$ and **InSb$_3$S$_6$** ternary compounds are formed in the In–S–Sb ternary system. InSbS$_3$ melts incongruently at 630°C ± 5°C (Magerramov et al. 1975; Guliev et al. 1977b) (at 475°C ± 5°C [Kompanichenko et al. 1973]) and crystallizes in the orthorhombic structure with the lattice parameters $a = 390$, $b = 960$, and $c = 1400$ pm and an energy gap of 1.88–1.97 eV (Moldovyan 1990). Single crystals of this compound were grown by the Bridgman method and by chemical transport reactions with I$_2$ or Br$_2$ as the transport agent (Guliev et al. 1977b; Moldovyan 1990; Guliev 1997).

InSb$_3$S$_6$ melts incongruently at 530°C ± 5°C and decomposes at 475°C ± 5°C, forming InSbS$_3$ (Kompanichenko et al. 1973).

Under normal pressure, In$_2$S$_3$ has a trigonal structure between 754°C and its melting point (γ-In$_2$S$_3$). This modification cannot be quenched to room temperature. Substitution of about 5 at% of indium by antimony stabilizes the structure at room temperature with the lattice parameters $a = 383.1 ± 0.1$ and $c = 904.9 ± 0.6$ pm for **In$_{1.8}$Sb$_{0.2}$S$_3$** composition, an experimental density of 4.80 ± 0.08 g·cm^{-3}, and an energy gap of 1.55 eV (Diehl and Nitsche 1973; Diehl et al. 1976). Black hexagonal plates of In$_{1.8}$Sb$_{0.2}$S$_3$ have been grown in a closed system by iodine transport in the presence of Sb$_2$S$_3$ vapor (Diehl and Nitsche 1973; Diehl and Nitsche 1975). They decompose at 540°C.

InSb–In$_2$S$_3$. Limited solid solutions based on InSb were determined in this system (Radautsan and Madan 1962).

InSb–Sb$_2$S$_3$. According to the data of metallography, InSb dissolves 0.1 mass% Sb$_2$S$_3$ (Skudnova and Mirgalovskaya 1965). Upon the growth of InSb single crystals in the [$\bar{1}\bar{1}\bar{1}$] and [111] directions, the effective distribution coefficients of sulfur are 1.20 ± 0.18 and 0.40 ± 0.06, respectively.

15.34 INDIUM–SELENIUM–ANTIMONY

The liquidus surface of the In–Se–Sb ternary system was constructed by Ragimova et al. (1985). Eight fields of the primary crystallizations and three miscibility regions exist in this system. In$_2$Se$_3$ occupies the biggest part of the liquidus surface. It is necessary to note that the liquidus isotherms contradict the binary systems, except one, at 330°C. Therefore, this system needs further investigations.

InSbSe$_3$ and **In$_{1.8}$Sb$_{0.2}$Se$_3$** ternary compounds are formed in this system. InSbSe$_3$ could be obtained upon melting of the stoichiometric quantities of InCl$_3$ and Sb$_2$Se$_3$ at 640°C–680°C (Rustamov et al. 1974b). It melts incongruently at 765°C–775°C (at 550°C [Rustamov et al. 1974b]; at 750°C [Guliev 1997]) and has a polymorphic transformation at 525°C ± 5°C (at 550°C [Guliev 1997]) (Guliev et al. 1977a, 1977b). α-InSbSe$_3$ crystallizes in the orthorhombic structure with the lattice parameters $a = 945$, $b = 1390$, and $c = 392$ pm and an experimental density of 6.16 g·cm^{-3} (5.4 g·cm^{-3} [Rustamov et al. 1974b]) (Guliev et al. 1977a, 1977b) ($a = 943 ± 1$, $b = 1402 ± 5$, and $c = 396 ± 1$ pm, calculated and experimental densities of 6.11 and 6.07 ± 0.02 g·cm^{-3}, respectively, and an energy gap of 1.1 eV [Spiesser et al. 1978]). According to the data of Rustamov et al. (1974b), the energy gap of this compound is 1.66 eV. β-InSbSe$_3$ crystallizes in the monoclinic structure with the lattice parameters $a = 686$, $b = 393$, and $c = 938$ pm and $β = 94°7'$ (Guliev et al. 1977a, 1977b). Single crystals of this compound were grown by the Bridgman method and by chemical transport reactions with I$_2$ or Br$_2$ as a transport agent (Guliev et al. 1977b; Spiesser et al. 1978; Guliev 1997).

In$_{1.8}$Sb$_{0.2}$Se$_3$ crystallizes in the hexagonal structure with the lattice parameters $a = 397 ± 1$ and $c = 1887 ± 1$ pm, calculated and experimental densities of 6.02 and 5.98 ± 0.03 g·cm^{-3}, respectively, and an energy gap of 0.92 eV (Spiesser et al. 1978). Single crystals of this compound were grown by the Bridgman method.

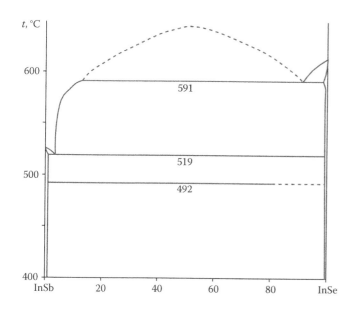

FIGURE 15.46 Phase diagram of the InSb–InSe system. (From O'Kane, D.F., and Stemple, N.R., *J. Electrochem. Soc.*, *113*(3), 289, 1966. Reproduced with permission of The Electrochemical Society).

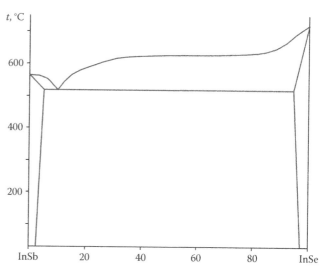

FIGURE 15.47 Phase diagram of the InSb–InSe system. (From Ragimova, V.M., et al., *Zhurn. Neorgan. Khimii*, *29*(11), 2888, 1984. With permission.)

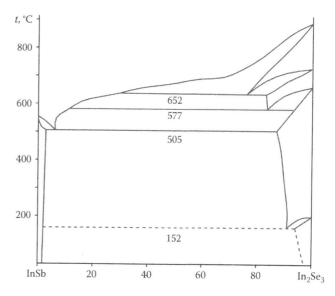

FIGURE 15.48 Phase diagram of the InSb–In₂Se₃ system. (From Ragimova, V.M., et al., *Zhurn. Neorgan. Khimii*, *29*(11), 2888, 1984. With permission.)

InSb–InSe. The phase diagram of this system is shown in Figure 15.46 (O'Kane and Stemple 1966). A monotectic exists at 591°C and 91 mol% InSe, and the composition of the InSb-rich melt at the monotectic temperature is 13 mol% InSe. A eutectic occurs at 519°C and 3 mol% InSe, and two other phase transitions at 492°C and 460°C were present in the system. The solubility of InSe in InSb is less than 1 mol%, and less than 1 mol% of InSb is soluble in InSe. DTA and XRD have been used to characterize the InSb–InSe quasi-binary phase diagram. No ternary compounds were obtained in this system (Radautsan et al. 1961b; Radautsan and Madan 1962).

The phase diagram of this system according to the data of Ragimova et al. (1984, 1985) is a eutectic type without monotectic reaction (Figure 15.47). The eutectic crystallizes at 500°C and contains 10 mol% InSe. It is necessary to note that InSb melts at 526°C, but these authors gave the melting temperature as 564°C (taken from the original figure). The solubilities of InSe in InSb and InSb in InSe at the eutectic and room temperatures are 4 and 2 mol% and 5 and 3 mol%, respectively.

The In–Se–Sb ternary system was investigated through DTA, XRD, metallography, and measurement of microhardness (Ragimova et al. 1984, 1985). The ingots were annealed at 330°C for 250 h.

InSb–In₂Se₃. The phase diagram of this system is given in Figure 15.48 (Ragimova et al. 1984, 1985). The eutectic crystallizes at 505°C and contains 5 mol% In₂Se₃. The phase transformations of In₂Se₃ at 152°C, 577°C, and 652°C were fixed on the phase diagram. The limited solid solution is formed in this system based on InSb and In₂Se₃ (Woolley and Keating 1961a, 1961b; Radautsan and Madan 1962).

InSb–Sb₂Se₃. This system is a non-quasi-binary section of the In–Se–Sb ternary system (Ragimova et al. 1984, 1985). Solid solutions based on InSb, In₂Se₃, and

Sb$_2$Se$_3$ crystallize primarily from the liquid phase. A miscibility region in the liquid state within the concentration range from 35 to 55 mol% Sb$_2$Se$_3$ exists in this section.

InSb–Se. This system is also a non-quasi-binary section of the In–Se–Sb ternary system (Ragimova et al. 1983, 1985). Solid solutions based on InSb, In$_2$Se$_3$, and Se crystallize primarily from the liquid phase.

15.35 INDIUM–TELLURIUM–ANTIMONY

The liquidus surface of the In–Te–Sb ternary system was constructed by Rosenberg and Strauss (1961); Deneke and Rabenau (1964); and Bobrov et al. (1970). It includes 13 fields of primary crystallizations of In, InSb, Sb, the γ- and δ-phases of the Sb–Te system, Sb$_2$Te$_3$, Te, In$_2$Te$_3$, In$_2$Te$_5$, In$_3$Te$_5$, InTe, In$_9$Te$_7$, and In$_3$SbTe$_2$. The field of In$_3$Te$_4$ primary crystallization is degenerated. A miscibility gap near the In corner exists in this system. The liquidus surface of the InSb–InTe–Sb subsystem was constructed by Bobrov et al. (1971), and that of the InSb–In–InTe subsystem was determined by Dashevskiy et al. (1974b).

The liquidus surface of the InTe–InSb–Sb subsystem of this ternary system is given in Figure 15.49 (Stegman and Peretti 1966b). The liquidus and solidus surfaces of this subsystem have been determined by DTA, XRD and

metallography. InTe is the primary phase of precipitation over most of the system.

The region of the solid solutions based on InSb at 400°C and 500°C was determined by Dashevskiy et al. (1974b, 1975). At 500°C, this region is elongated along the InSb–In$_2$Te$_3$ section and the solubility of Te reaches 7 at% (Figure 15.50).

InSbTe$_3$ (In$_7$SbTe$_6$ according to the data of Kurata and Hirai [1966]) ternary compound is formed in this system (Rosenberg and Strauss 1961; Rustamov et al. 1974b). This compound could be obtained upon melting of the stoichiometric quantities of InCl$_3$ and Sb$_2$Te$_3$ at 640°C–680°C (Rustamov et al. 1974b). It melts incongruently at 600°C, and its experimental density is 4.3 g·cm^{-3}.

Itabashi et al. (2001) measured the activity of indium in the In–Ag–Sb system using a galvanic cell with a zirconia solid electrolyte in a temperature range between 597°C and 987°C. Isoactivity curves at 727°C and 827°C are given in Figure 15.51. They are derived from combining the activity data of the In–Sb and In–Te binaries. The isoactivity curves gather from the whole composition range of the In–Sb binary alloy to the intermediate composition of the In–Te system.

InSb–InTe. There are some discrepancies in the experimental results concerning this quasi-binary system. According to the data of Kurata and Hirai (1966), the phase diagram is a eutectic type with peritectic transformation (Figure 15.52). The eutectic crystallizes at 510°C. The ternary compound In$_7$SbTe$_6$, which

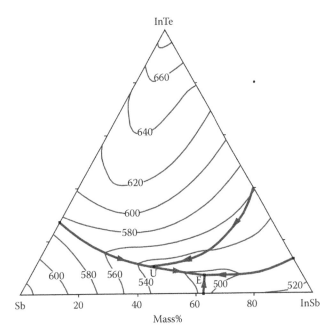

FIGURE 15.49 Liquidus surface of the InTe–InSb–Sb subsystem of the In–Te–Sb ternary system. (From Stegman, R.L., and Peretti, E.A., *J. Chem. Eng. Data*, 11(4), 496, 1966. With permission.)

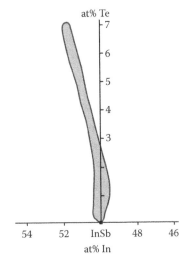

FIGURE 15.50 Region of the solid solutions based on InSb in the In–Te–Sb ternary system at 500°C. (From Dashevskiy, M.Ya., et al., *Izv. AN SSSR Neorgan. Mater.*, 11(9), 1564, 1975. With permission.)

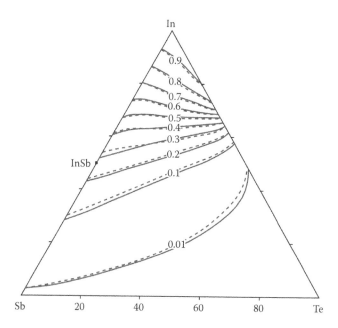

FIGURE 15.51　Isoactivity curves of In in the In–Te–Sb ternary system at 727°C (dotted lines) and 827°C (solid lines). (From Itabashi, S., et al., *J. Jpn. Inst. Metals*, 65(9), 888, 2001. With permission.)

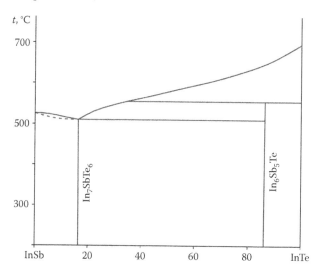

FIGURE 15.52　Phase diagram of the InSb–InTe system. (From Kurata, K., and Hirai, T., *Solid State Electron.*, 9(6), 633, 1966. With permission.)

melts incongruently at 555°C, is formed in this system. Another intermediate phase, **In₆Sb₅Te**, is apparently the limiting composition of the solid solution based on InSb.

The phase diagram of the InSb–InTe system constructed by Deneke and Rabenau (1964) is presented in Figure 15.53. It is a eutectic type with the eutectic crystallizing at 512°C and containing 85 mol% InSb (506.4°C and 85.9 mol% InSb [Stegman and Peretti 1966a]; 505°C and 80 mol% InSb [Bobrov et al. 1970]).

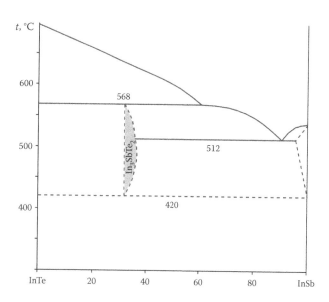

FIGURE 15.53　Phase diagram of the InSb–InTe system. (From Deneke, K., and Rabenau, A., *Z. Anorg. Allg. Chem.*, 333(4–6), 201, 1964. With permission.)

The solid solubility of InSb in InTe was found to be immeasurably small (Stegman and Peretti 1966a), and the solubility of InTe in InSb at 490°C is 7 mol% (Bobrov et al. 1970) (InSb dissolves up to 15 mol% InTe [Lyalikov et al. 1964; Molodyan 1964; Molodyan and Radautsan 1964a, 1964b]; 9.6 mol% InTe at 500°C and 2.9 mol% InTe at 488°C [Stegman and Peretti 1966a]). **In₃SbTe₂** ternary compound is formed in this system (Deneke and Rabenau 1964; Woolley 1966; Bobrov et al. 1970). It melts at 568°C (553.5°C [Stegman and Peretti 1966a]) and decomposes at about 420°C into InSb and InTe (Deneke and Rabenau 1964; (Stegman and Peretti 1966a). By quenching, it can be obtained metastable at room temperature. The title compound crystallizes in the cubic structure with the lattice parameter $a = 612.6$ pm. Apparently, the ternary compound **In₄SbTe₃**, which crystallizes in the cubic structure with the lattice parameter $a = 612.8 \pm 0.3$ pm and an experimental density of 6.16 g·cm⁻³ (Goryunova et al. 1959b; Kiosse et al. 1960; Molodyan et al. 1961; Radautsan 1962; Radautsan and Madan 1962; Lyalikov et al. 1964; Molodyan 1964; Molodyan and Radautsan 1964a, 1964b; Bobrov et al. 1970), corresponds to the In₃SbTe₂ ternary compound. According to the data of Stegman and Peretti (1966a), the compositional range of this compound was found to be very narrow.

InSb–In₂Te₃. This system is a non-quasi-binary section of the In–Te–Sb ternary system (Woolley 1966). The solubility of In₂Te₃ in 3InSb reaches 14–15 mol% (Kiosse

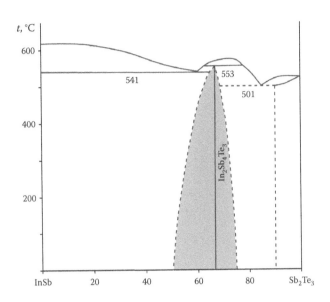

FIGURE 15.54   Phase diagram of the InSb–Sb_2Te_3 system. (From Vassilev, V., et al., *J. Phase Equilibria Diffus.*, 37(5), 524, 2016. With permission.)

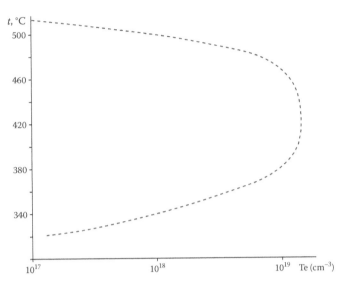

FIGURE 15.55   Temperature dependence of Te solubility in InSb. (From Rekalova, G.I., et al., *Izv. AN SSSR Neorgan. Mater.*, 7(8), 1314, 1971. With permission.)

et al. 1960; Radautsan and Molodyan 1960; Woolley et al. 1960; Radautsan and Madan 1962; Lyalikov et al. 1964; Woolley 1966).

InSb–Sb_2Te_3. The phase diagram of this system (Figure 15.54) was determined based on XRD, DTA, and microhardness and density measurements (Vassilev et al. 2016). Two invariant eutectic equilibria exist in the InSb–Sb_2Te_3 system, with the eutectic point coordinates at compositions of 60 and circa 85 mol% InSb and eutectic temperatures of 541 and ≈501°C, respectively. A wide region of solid solutions exists in this system, and the miscibility gap exists in the region from 62 to 78 mol% InSb with the monotectic temperature of 553 ± 5°C. **2InSb·Sb_2Te_3 ($In_2Sb_4Te_3$)** is formed as a result of a syntectic reaction. It melts incongruently at 553°C and crystallizes in the orthorhombic structure with the lattice parameters $a = 439.37$, $b = 420.35$, and $c = 354.33$ pm and $\beta = 93.354°$ and an experimental density of 5.71 g·cm^{-3}. This compound has a variable composition of $(2 + \delta)$InSb·Sb_2Te_3 ($-1 < \delta < 1$), with a homogeneity region of 50–75 mol% InSb.

InSb–Te. This system is a non-quasi-binary section of the In–Te–Sb ternary system (Woolley 1966; Bobrov et al. 1970). This section crosses six fields of primary crystallization. The temperature dependence of Te solubility in InSb is presented in Figure 15.55 (Rekalova et al. 1971). The maximum solubility takes place in the temperature region from 350°C to 480°C and is $(1–2)·10^{19}$

cm^{-3}. According to the data of Woolley (1966), InSb dissolves 7 at% Te (less than 2 at% Te [Bobrov et al. 1970]).

The distribution coefficient of Te in InSb remains constant (0.8) until the Te concentration in the solid solution is less than $2·10^{18}$ cm^{-3} (Belaya and Zemskov 1973). Upon further increasing the Te concentration, the distribution coefficient is reduced to 0.2 at the Te concentration of $8·10^{19}$ cm^{-3}. The calculated effective distribution coefficient of Te in InSb is 0.52–0.79 (Kevorkov et al. 1988). There is evidence to believe that some In_2Te_3 is formed at the interaction of InSb and Te (Ufimtsev and Gimel'farb 1973).

15.36 INDIUM–CHROMIUM–ANTIMONY

The calculated isothermal section of the In–Cr–Sb ternary system at room temperature is shown in Figure 15.56 (Liu and Mohney 2003b). Binary phases were treated as stoichiometric compounds, and ternary phases and ternary solubilities were disregarded. The dotted lines indicate possible alternative tie-lines. It is seen that the InSb–Cr tie-line possesses a notable uncertainty.

InSb–CrSb. The phase diagram of this system is a eutectic type (Müller and Wilhelm 1965b). The eutectic crystallizes at 516°C ± 2°C and contains 0.18 mass% Cr.

15.37 INDIUM–MOLYBDENUM–ANTIMONY

The calculated isothermal section of the In–Mo–Sb ternary system was constructed by Liu and Mohney

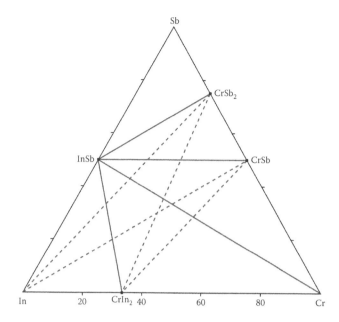

FIGURE 15.56   Calculated isothermal section of the In–Cr–Sb ternary system at room temperature. (From Liu, W.E., and Mohney, S.E., *Mater. Sci. Eng. B*, *103*(2), 189, 2003. With permission.)

(2003b). Binary phases were treated as stoichiometric compounds, and ternary phases and ternary solubilities were disregarded. InSb is in thermodynamic equilibrium with Mo_3Sb_7 and possibly Mo (the InSb–Mo tie-line possesses a notable uncertainty).

15.38 INDIUM–TUNGSTEN–ANTIMONY

The calculated isothermal section of the In–W–Sb ternary system was constructed by Liu and Mohney (2003b). Binary phases were treated as stoichiometric compounds, and ternary phases and ternary solubilities were disregarded. InSb is in thermodynamic equilibrium with W.

15.39 INDIUM–FLUORINE–ANTIMONY

InSbF$_6$ ternary compound is formed in the In–F–Sb system (Mazej 2005). To obtain this compound, a reaction mixture of InBF$_4$ (2.23 mM) and SbF$_5$ (2.23 mM) in 5 mL of anhydrous HF was loaded into translucent fluorocarbon polymer reaction vessels in a glove box (Mazej 2005). Anhydrous HF and, when necessary, SbF$_5$ were condensed onto the solid reactants at 77 K. The reaction vessels were brought to ambient temperature. The reaction mixtures were vigorously agitated. A white precipitate formed immediately. After 1 day, the volatiles were pumped off. The isolated title compound was of very low crystallinity.

15.40 INDIUM–IODINE–ANTIMONY

InSb–I$_2$. As a result of the thermodynamic analysis of the reactions occurring in the InSb–I$_2$ system, it was shown that the primary reaction responsible for supplying the InSb transport is the reaction $2InSb + InI_3 \longleftrightarrow 3InI + \frac{1}{4}Sb_4$ (Sandulova et al. 1972).

15.41 INDIUM–MANGANESE–ANTIMONY

The calculated isothermal section of the In–Mn–Sb ternary system at room temperature is shown in Figure 15.57 (Liu and Mohney 2003b). Binary phases were treated as stoichiometric compounds, and ternary phases and ternary solubilities were disregarded. The dotted lines indicate possible alternative tie-lines. The $Mn_3In + MnSb + InSb$ three-phase equilibrium dominates the phase diagram. However, the prediction of the dominant tie-line configuration in this ternary system is not very conclusive due to the existence of InSb–Mn, MnSb–In, and Mn_2Sb–In alternative tie-lines.

In$_x$MnSb$_{1-x}$ ($0 \leq x \leq 0.25$) solid solutions were prepared by sintering the pellets of stoichiometric mixtures repeatedly at temperatures in the range from 600°C to 850°C for 48 h, followed by quenching to room temperature (Bai and Rao 1982).

InSb–Mn. The solubility of Mn in InSb is 0.25 at% (Sanygin et al. 2010). Starting at this Mn concentration, the In$_{1-x}$Mn$_x$Sb solid solutions are supersaturated.

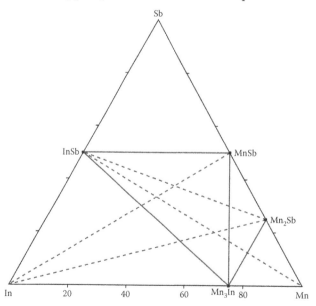

FIGURE 15.57   Calculated isothermal section of the In–Mn–Sb ternary system at room temperature. (From Liu, W.E., and Mohney, S.E., *Mater. Sci. Eng. B*, *103*(2), 189, 2003. With permission.)

According to the data of Pashkova et al. (2006), the Mn concentration in the synthesized $In_{1-x}Mn_xSb$ solid solutions is less than 0.01 mass%.

InSb–MnSb. The phase diagram of this system is a eutectic type (Figure 15.58) (Novotortsev et al. 2011). The eutectic crystallizes at 513°C and contains 6.5 mol% MnSb (at 510°C ± 2°C, contains 2.02 mass% Mn [Müller and Wilhelm 1965b]; at 512°C, contains 6.5 mass% MnSb [Aliev et al. 1981]).

15.42 INDIUM–RHENIUM–ANTIMONY

The calculated isothermal section of the In–Re–Sb ternary system was constructed by Liu and Mohney (2003b). Binary phases were treated as stoichiometric compounds, and ternary phases and ternary solubilities were disregarded. InSb is in thermodynamic equilibrium with Re.

15.43 INDIUM–IRON–ANTIMONY

The calculated isothermal section of the In–Fe–Sb ternary system at room temperature is shown in Figure 15.59 (Liu and Mohney 2003b). Binary phases were treated as stoichiometric compounds, and ternary phases and ternary solubilities were disregarded. The InSb–Fe tie-line was predicted with a margin of only

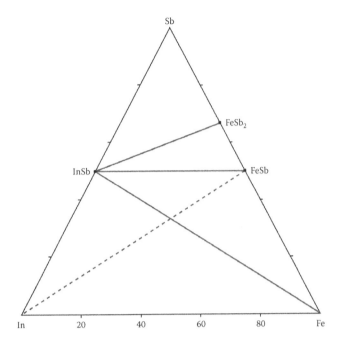

FIGURE 15.59 Calculated isothermal section of the In–Fe–Sb ternary system at room temperature. (From Liu, W.E., and Mohney, S.E., *Mater. Sci. Eng. B*, *103*(2), 189, 2003. With permission.)

3.6 kJ·M⁻¹; therefore, the dotted line indicates a possible alternative tie-line.

Fe_4InSb_{12} metastable compound, which crystallizes in the cubic structure with the lattice parameter $a = 920.1 \pm 0.1$ pm, could be obtained in the In–Fe–Sb ternary system (Sellinschegg et al. 1998). This compound forms at 146°C and decomposes at 463°C. Modulated elemental reactants were used to form amorphous ternary reaction intermediates of the desired compositions. The amorphous reaction intermediate crystallizes exothermically below 200°C, forming kinetically stable "filled" skutterudite. This metastable compound can only be prepared by controlling the reaction intermediates, avoiding the formation of more thermodynamically stable binary compounds.

InSb–Fe. The kinetics of Fe dissolution in InSb melt was investigated by Denisov and Dubovikov (1987) using the disc rotation method. The parameters of dissolution and diffusion in this system were determined. The solubility of Fe in the InSb melt was also determined.

InSb–FeSb. The phase diagram of this system is a eutectic type (Müller and Wilhelm 1965b). The eutectic crystallizes at 520°C ± 2°C and contains 0.21 mass% Fe (at 510°C, contains 0.65 mass% FeSb [Aliev et al. 1981]).

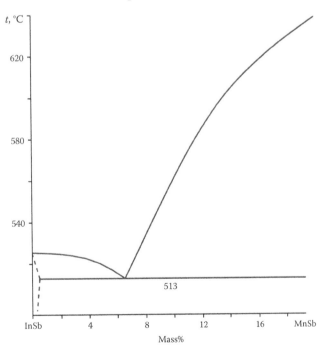

FIGURE 15.58 Part of the phase diagram of the InSb–MnSb system (Novotortsev, V.M., et al., *Zhurn. Neorgan. Khimii*, *56*(12), 2038, 2011.)

15.44 INDIUM–COBALT–ANTIMONY

The projection of the liquidus surface of the In–Co–Sb ternary system near the In–Sb binary is presented in Figure 15.60 (Grytsiv et al. 2013). The primary crystallization field of CoSb covers an extended area on the liquidus surface. Four ternary invariant reactions exist in this ternary system: E_1 (492°C), L $\longleftrightarrow$ (Sb) + CoSb$_3$ + InSb; U_1 (484°C), L + CoSb$_3$ $\longleftrightarrow$ CoSb$_2$ + InSb; U_2 (484°C), L + CoSb$_2$ $\longleftrightarrow$ CoSb + InSb; and E_2 (~154°C), L $\longleftrightarrow$ (In) + CoSb + InSb. The maximal solubility of the third component in the binary compounds CoSb, CoSb$_2$, and InSb does not exceed 0.1 at%. The maximal solubility of indium in CoSb$_3$ corresponds to 1.4 at%. The solubility limit shows a small dependence in the temperature range from 475°C to 700°C. Solid solutions of **In$_x$Co$_4$Sb$_{12}$** based on CoSb$_3$ crystallize in the cubic structure with the lattice parameter $a = 903.86 \pm 0.02$, 904.41 ± 0.02, and 905.22 ± 0.02 pm for $x = 0.13$ and $a = 904.43$, 905.07, and 905.95 pm for $x = 0.21$ at 100, 200, and 295 K, respectively ($a = 905.294 \pm 0.006$ pm for $x = 0.2$ at room temperature [He et al. 2006]).

As-cast alloys were annealed in evacuated quartz ampoules for 4–32 days at temperatures in the range from 375°C to 800°C, followed by quenching in cold water (Grytsiv et al. 2013). The samples containing CoSb were powdered to a grain size below 50 mkm, filled into graphite cartridges, and then hot pressed under Ar at 475°C and 42 MPa for 2 h, followed by annealing at desired temperatures. In$_x$Co$_4$Sb$_{12}$ ($x = 0.17$, 0.3, and 0.5) samples were synthesized from CoSb, InSb, and Sb ball milled in tungsten carbide vessels. Afterwards, the powders were hot pressed at 475°C ($x = 0.3$ and 0.5) and 500°C ($x = 0.15$ and 0.5) and 56 MPa for 2 h. Polycrystalline samples of these solid solutions were also prepared by solid-state reaction (He et al. 2006). High-purity powders of Co, Sb, and In were mixed thoroughly in a stoichiometric ratio and loaded into an alumina crucible. The powders were calcined at 610°C for 12 h and then at 675°C for 36 h under a gas mixture of 5 vol% H$_2$ and 95 vol% Ar. The obtained powders were reground and pressed into discs. These discs were sintered at 675°C for 4 h under the same gas mixture. The samples with $x \leq 0.22$ were single-phase materials, while InSb starts to appear for $x = 0.25$.

In$_x$Co$_4$Sb$_{12}$ ($x = 0.16$, 0.25, 0.4, 0.8, and 1.2) samples were also prepared by a vacuum induction melting process (Mallik et al. 2009). Structural characterization by XRD revealed that the In$_{0.16}$Co$_4$Sb$_{12}$ sample is polycrystalline and of the single δ-phase skutterudite structure. The XRD results of In$_x$Co$_4$Sb$_{12}$ ($x = 0.25$, 0.4, 0.8, and 1.2) show a polycrystalline structure with secondary phases (γ-CoSb$_2$ and InSb). SEM results of annealed In$_{0.25}$Co$_4$Sb$_{12}$ and In$_{0.4}$Co$_4$Sb$_{12}$ show the presence of clearly separated phase domains of Co–Sb and InSb. The Co–Sb domains mainly consist of CoSb$_3$ and finely distributed CoSb$_2$. Both phases contain In dissolved close to the filling fraction limits ($x = 0.22$).

The calculated isothermal section of the In–Co–Sb ternary system at room temperature is shown in Figure 15.61 (Liu and Mohney 2003b). Binary phases were treated as stoichiometric compounds, and ternary phases and ternary solubilities were disregarded. The existence of the InSb–CoSb, InSb–CoSb$_2$, and InSb–CoSb$_3$ tie-lines was proved experimentally by Grytsiv et al. (2013). The dotted line indicates a possible alternative tie-line.

InSb–Co. The kinetics of Co dissolution in the InSb melt was investigated by Denisov and Dubovikov (1987) using the disc rotation method. The parameters of dissolution and diffusion in this system were determined. The solubility of Co in the InSb melt was also determined.

15.45 INDIUM–NICKEL–ANTIMONY

The calculated projection of the liquidus surface of the In–Ni–Sb ternary system is presented in Figure 15.62 (Cao et al. 2013). Thirteen invariant four-phase reactions and one invariant three-phase reaction exist in

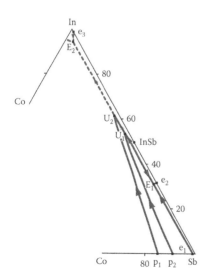

FIGURE 15.60 Projection of the liquidus surface of the In–Co–Sb ternary system near In–Sb binary. (From Grytsiv, A., et al., *J. Electron. Mater.*, 42(10), 2940, 2013. With permission.)

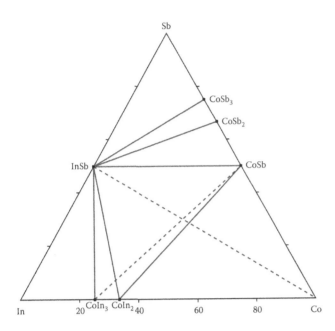

FIGURE 15.61 Calculated isothermal section of the In–Co–Sb ternary system at room temperature. (From Liu, W.E., and Mohney, S.E., *Mater. Sci. Eng. B*, 103(2), 189, 2003. With permission.)

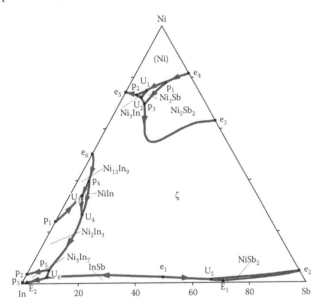

FIGURE 15.62 Calculated projection of the liquidus surface of the In–Ni–Sb ternary system. (From Cao, Z., et al., *J. Electron. Mater.*, 42(8), 2615, 2013. With permission.)

this system: P_1 (1064°C), L + Ni$_5$Sb$_2$ + (Ni) ⟷ Ni$_3$Sb; P_2 (1044°C), L + Ni$_5$Sb$_2$ + Ni$_3$Sb ⟷ ζ; U_1 (1010°C), L + Ni$_3$Sb ⟷ Ni + Ni$_3$In; U_2 (1007°C), L + Ni$_3$Sb ⟷ ζ + Ni$_3$In; P_3 (968°C), L + Ni$_3$In + (Ni) ⟷ ζ; P_4 (896°C), L + ζ + δ ⟷ NiIn; U_3 (875°C), L + δ ⟷ NiIn + Ni$_2$In$_3$; U_4 (804°C), L + NiIn ⟷ Ni$_2$In$_3$ + ζ; U_5 (517°C), L + ζ ⟷ InSb + NiSb$_2$; e_1 (500°C), InSb + ζ; E_1 (500°C), L ⟷

InSb + NiSb$_2$ + Sb; P_5 (896°C), L + ζ + Ni$_2$In$_3$ ⟷ Ni$_3$In$_7$; U_6 (336°C), L + ζ ⟷ InSb + Ni$_3$In$_7$; and E_2 (155°C), L ⟷ InSb + Ni$_3$In$_7$ + In, where δ is the high-temperature phase in the In–Ni system and ζ is a continuous solid solution based on NiSb and Ni$_2$In. As can be seen, the liquidus projection is dominated by the region of the primary crystallization for the ζ-phase because of its high stability. Two three-phase valleys leave the saddle at the quasi-binary eutectic point e_1: one of them passes the transition reaction U_5 and finally ends in the ternary eutectic E_1; the other one first reaches the transition reaction U_6 and finally ends in the ternary eutectic E_2. Invariant reactions E_1, E_2, U_5, and U_6 are very close to the In–Sb binary subsystem. Reasonable agreement is obtained with the experimental results for the investigated composition range (Richter 1998; Richter et al. 1998b). Due to the lack of experimental information on the Ni-rich region at high temperature, further experimental data are needed to verify the liquidus projection in the Ni-rich region.

The isothermal sections of the In–Ni–Sb ternary system were constructed at room temperature (Liu and Mohney 2003b), at 300°C and 440°C (Richter et al. 1998b; Richter and Ipser 2002), at 450°C (Sutopo 2011b), and at 750°C, 825°C, and 900°C (Richter 1998). Burkhardt and Schubert (1959) constructed an isothermal section of the Ni-rich part of this ternary system at 500°C and reported a significant extension of Ni$_3$In and Ni$_3$Sb into the ternary system. The isothermal sections at 300°C, 440°C, 750°C, and 900°C are presented in Figures 15.63 and 15.64. At 300°C, the solubility limits of Sb in the binary In–Ni compounds were found to be 1.5 at% for Ni$_2$In$_3$ and 3 at% for Ni$_3$In$_7$. In contrast, NiSb$_2$, InSb, and Sb did not show any significant solubility for an additional element (Richter et al. 1998b). Starting from the Sb-rich part of the isothermal section, the following three-phase fields could be detected at 440°C: InSb + NiSb$_2$ + Sb, InSb + NiSb$_2$ + ζ, L + InSb + ζ, L + Ni$_3$In$_7$ + ζ, and Ni$_2$In$_3$ + Ni$_3$In$_7$ + ζ. NiIn was found to be stable up to 9 at% Sb at 750°C (Richter 1998). The isothermal section at 900°C shows the absence of ε-NiIn and Ni$_2$In$_3$. The solubility of Sb in the nonstoichiometric binary δ-NiIn was found to be rather low (<1 at%).

No ternary compound was found in the In–Ni–Sb ternary system (Richter and Ipser 2002). **Ni$_3$InSb** (hexagonal structure, a = 411 and c = 519 pm [Jan and Chang 1991]; a = 411.11 and c = 518.82 pm and calculated and experimental densities of 8.97 and 8.60 g·cm^{-3},

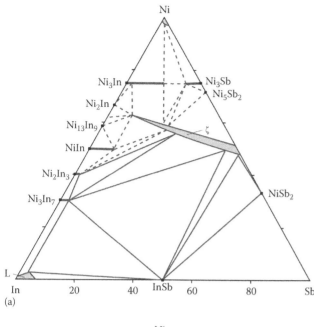

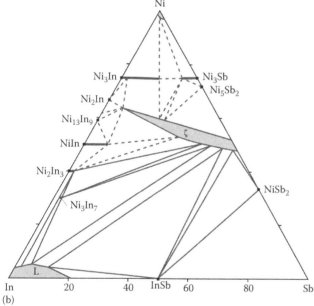

FIGURE 15.63 Isothermal sections of the In–Ni–Sb ternary system at (a) 300°C and (b) 440°C. (From Richter, K.W., et al., *Mater. Sci. Eng. B*, 55(1-2), 44, 1998. With permission.)

and partial molar enthalpies of Sb were derived for the ternary ζ-phase, which crystallizes in a NiAs-type structure. The composition dependence of the activities is given for 900°C, and isoactivity curves are constructed for the same temperature. The variation of the ζ activity along the ζ/((ζ + L) phase boundary was estimated.

Indium activities within the InSb–NiSb–Sb subsystem in the temperature range 370°C–590°C have been

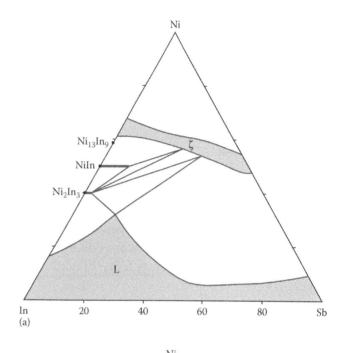

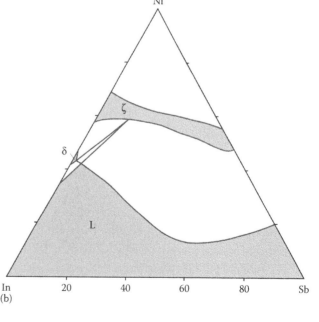

FIGURE 15.64 Isothermal sections of the In–Ni–Sb ternary system at (a) 750°C and (b) 900°C. (From Richter, K.W., *J. Phase Equilibria*, 19(5), 455, 1998. With permission.)

respectively [Suriwong et al. 2011]) is not a true ternary compound but rather a specific composition of extensive solid solution in the Ni₂In–NiSb system (Cao et al. 2013).

Antimony vapor pressures were determined in this system along an isopleth with x_{In}/x_{Ni} = 1:9 using an isopiestic method (Richter et al. 1998a). The measurements were made between 707°C and 1077°C, covering a composition range 43–48 at% Sb, with a few samples at higher antimony content. Thermodynamic activities

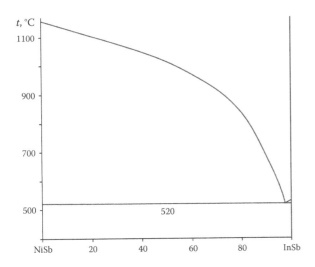

FIGURE 15.65 Phase diagram of the InSb–NiSb system. (From Müller, and Wilhelm, M., *Mater. Res. Bull.*, 2(7), 531, 1967. With permission.)

measured by the EMF method (Vassiliev et al. 2003). The values of the partial molar thermodynamic functions for the liquid InSb alloy; for the three solid heterogeneous regions InSb–NiSb$_2$–Sb, InSb–NiSb$_1\pm\delta$–NiSb$_2$, and InSb– NiSb$_1\pm\delta$; and for the six ternary liquid–solid alloys have been calculated.

InSb–Ni. The kinetics of Ni dissolution in the InSb melt was investigated by Denisov and Dubovikov (1987) using the disc rotation method. The parameters of dissolution and diffusion in this system were determined. The solubility of Ni in the InSb melt was also determined.

InSb–NiSb. The phase diagram of this system is a eutectic type (Figure 15.65) (Müller and Wilhelm 1965a, 1965b, 1967a). The eutectic crystallizes at 520°C and contains 2.3 mol% NiSb.

15.46 INDIUM–RUTHENIUM–ANTIMONY

The calculated isothermal section of the In–Ru–Sb ternary system at room temperature is shown in Figure 15.66 (Liu and Mohney 2003b). Binary phases were treated as stoichiometric compounds, and ternary phases and ternary solubilities were disregarded. The InSb–Ru tie-line was predicted with a margin of only 0.7 kJ·M^{-1}. The dotted lines indicate possible alternative tie-lines.

15.47 INDIUM–RHODIUM–ANTIMONY

The calculated isothermal section of the In–Rh–Sb ternary system at room temperature is shown in Figure 15.67 (Liu and Mohney 2003b). Binary phases were treated as stoichiometric compounds, and ternary

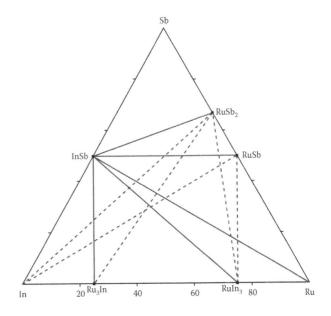

FIGURE 15.66 Calculated isothermal section of the In–Ru–Sb ternary system at room temperature. (From Liu, W.E., and Mohney, S.E., *Mater. Sci. Eng. B*, 103(2), 189, 2003. With permission.)

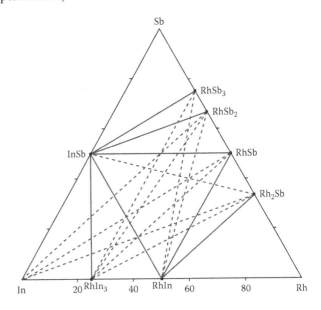

FIGURE 15.67 Calculated isothermal section of the In–Rh–Sb ternary system at room temperature. (From Liu, W.E., and Mohney, S.E., *Mater. Sci. Eng. B*, 103(2), 189, 2003. With permission.)

phases and ternary solubilities were disregarded. The dotted lines indicate possible alternative tie-lines.

According to the data of Eilertsen et al. (2010, 2011), **In$_x$Rh$_4$Sb$_{12}$** ternary phase is formed in this system. The In solubility limits approached 0.15. All samples are semiconductors. **In$_{0.1}$Rh$_4$Sb$_{12}$** crystallizes in the cubic structure with the lattice parameter $a = 923.75 \pm 0.1$

pm. Polycrystalline samples of this phase ($0 \leq x \leq 0.2$) were prepared by standard solid-state reaction. The mixture of In, Rh, and Sb powders was loaded into alumina crucibles and reacted in a tube furnace at 610°C for 10 h and at 675°C for 36 h under a constant flow of Sb vapor and 95 vol%/5 vol% N_2/H_2 gas. The furnace cooled samples were grounded and loaded into a graphite die and sintered in a uniaxial hot press at 600°C for 60 min with a pressure of 200 MPa under a dynamic vacuum.

15.48 INDIUM–PALLADIUM–ANTIMONY

The liquidus surface of the In–Pd–Sb ternary system is given in Figure 15.68 (Luef et al. 2003). Nine invariant four-phase reactions exist in this system—two ternary eutectics, six transition points, and one ternary peritectic: U_6 (748°C), L + $Pd_{58.1}In_{24.8}Sb_{17.1}$ (τ_2) $\longleftrightarrow$ PdIn + PdSb; U_5 (580°C), L + PdSb $\longleftrightarrow$ PdIn + $PdSb_2$; P (537°C), L + PdIn + $PdSb_2$ $\longleftrightarrow$ $Pd_{33.3}In_{42.1}Sb_{24.6}$ (τ_1); U_4 (518°C), L + PdIn $\longleftrightarrow$ τ_1 + Pd_2In_3; U_3 (494°C), L + τ_1 $\longleftrightarrow$ Pd_2In_3 + InSb; U_2 (490°C), L + Pd_2In_3 $\longleftrightarrow$ InSb + (Pd_3In_7); U_1 (480°C), L + τ_1 $\longleftrightarrow$ InSb + $PdSb_2$; E_2 (470°C), L $\longleftrightarrow$ InSb + $PdSb_2$ + (Sb); and E_2 (<154°C), L $\longleftrightarrow$ InSb + (Pd_3In_7) + (In). In addition to these, one maximum could be found in the L + InSb + τ_1 three-phase equilibrium. The field of the PdIn primary crystallization is the most extended

in the investigation composition range. The monovariant equilibrium line, which connects U_1 with U_3, apparently runs through a maximum. The liquidus valley U_6-U_5-P separates the region of PdIn primary crystallization from those of PdSb and $PdSb_2$. On the In-rich side of this line, the liquidus surface ascends quite steeply, whereas the Sb-rich side is relatively flat. All the other three-phase equilibria run in the vicinity of the In–Sb bordering system.

Two ternary compounds, $Pd_{33.3}In_{42.1}Sb_{24.6}$ (τ_1) and $Pd_{58.1}In_{24.8}Sb_{17.1}$ (τ_2), are formed in the In–Pd–Sb system. τ_1 crystallizes in the tetragonal structure with the lattice parameters a = 643.29 and c = 2452.6 pm (Luef et al. 2003) (a = 644.2 and c = 2439.2 pm for $Pd_3In_4Sb_2$ composition [El-Boragy and Schubert 1971]; a = 643.50 ± 0.01 and c = 2436.38 ± 0.03 pm and a calculated density of 8.995 g·cm⁻³ for $PdIn_{1.26}Sb_{0.74}$ composition [Flandorfer et al. 2002]). This compound melts incongruently (Flandorfer et al. 2002).

τ_2 crystallizes in the monoclinic structure with the lattice parameters a = 1520.8, b = 880.29, and c = 1361.6 pm and β = 123.83° (Luef et al. 2003) (a = 1518.9 ± 0.2, b = 879.9 ± 0.1, and c = 1360.2 ± 0.2 pm and β = 123.83 ± 0.01° and a calculated density of 10.744 g·cm⁻³ for $Pd_{13}In_{5.25}Sb_{3.75}$ composition [Flandorfer et al. 2002]).

Samples of τ_1 and τ_2 were prepared from In rods, Pd, and Sb lumps (Flandorfer et al. 2002). The calculated amounts of the pure elements were weighed to a total mass of about 2 g, sealed in evacuated quartz ampoules, and heated slowly to 1100°C for 6 h. This procedure was repeated once after powdering the sample. Samples were annealed at 850°C (14 days) and at 500°C (21 days) for τ_1 and τ_2, respectively, and quenched in cold water.

The isothermal sections of the In–Pd–Sb ternary system were constructed at 300°C (Richter and Ipser 2002; Liu and Mohney 2003b), at 500°C (El-Boragy and Schubert 1971), and at 300°C, 450°C, and 700°C (Luef et al. 2003). Isothermal sections at 300°C, 450°C, and 700°C are presented in Figures 15.69 and 15.70. Ternary phase τ_1, (Pd_3In_7), Pd_2In_3, and $PdSb_2$ were found to be in thermodynamic equilibrium with InSb at 300°C (Figure 15.69) (Richter and Ipser 2002). At 450°C, Pd_5In_3 and Pd_3In_7 dissolve approximately 3 at% Sb (Figure 15.70a) (Luef et al. 2003). InPd does not dissolve much Sb (<1 at%). PdSb shows a significant solubility for In; up to 12 at% Sb can be replaced by indium.

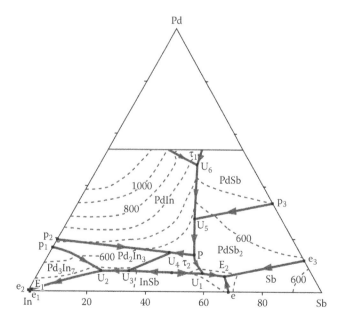

FIGURE 15.68 Liquidus surface of the In–Pd–Sb ternary system. (From Luef, C., et al., *J. Electron. Mater.*, 32(2), 43, 2003. With permission.)

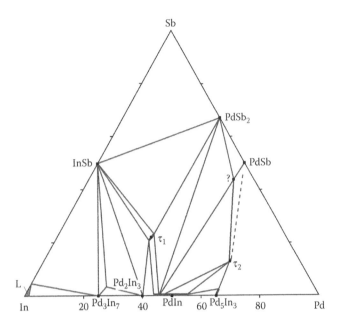

FIGURE 15.69 Isothermal section of the In–Pd–Sb ternary system at 300°C. (From Richter, K.W., and Ipser, H., *Acta Metallurg. Sin.*, *15*(2), 143, 2002. With permission.)

The ternary compound τ_1 has a solubility range of $PdIn_{1.22-1.31}Sb_{0.78-0.69}$. The second ternary compound (τ_2) is located at $Pd_{58}In_{25}Sb_{17}$ with a significant, but much smaller, homogeneity range.

On the isothermal section at 700°C (Figure 15.70b), the three-phase field PdIn + PdSb + τ_2 and two

two-phase fields PdIn + Pd_5In_3 and PdSb + τ_2 could be observed (Luef et al. 2003). τ_1 does not exist anymore at this temperature, so one must conclude that it decomposes between 450°C and 700°C.

15.49 INDIUM–OSMIUM–ANTIMONY

The calculated isothermal section of the In–Os–Sb ternary system was constructed by Liu and Mohney (2003b). Binary phases were treated as stoichiometric compounds, and ternary phases and ternary solubilities were disregarded. InSb is in thermodynamic equilibrium with $OsSb_2$ and Os.

15.50 INDIUM–IRIDIUM–ANTIMONY

The calculated isothermal section of the In–Ir–Sb ternary system at room temperature is shown in Figure 15.71 (Liu and Mohney 2003b). Binary phases were treated as stoichiometric compounds, and ternary phases and ternary solubilities were disregarded. The dotted lines indicate possible alternative tie-lines.

15.51 INDIUM–PLATINUM–ANTIMONY

The calculated projection of the liquidus surface of the In–Pt–Sb ternary system is given in Figure 15.72 (Patrone et al. 2006; Guo et al. 2014). Fifteen invariant four-phase reactions exist in this system—six ternary

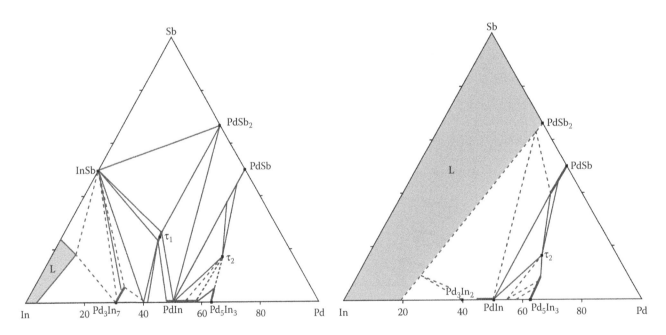

FIGURE 15.70 Isothermal sections of the In–Pd–Sb ternary system at (a) 450°C and (b) 700°C. (From Luef, C., et al., *J. Electron. Mater.*, *32*(2), 43, 2003. With permission.)

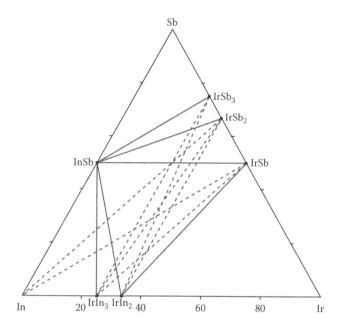

FIGURE 15.71   Calculated isothermal section of the In–Ir–Sb ternary system at room temperature. (From Liu, W.E., and Mohney, S.E., *Mater. Sci. Eng. B*, *103*(2), 189, 2003. With permission.)

eutectics and nine transition points: U_1 (984°C), L + PtIn ⟷ Pt_6In_5 + Pt_2In_3; U_2 (965°C), L + Pt_3In ⟷ PtSb + β-Pt_3In_2; E_1 (964°C), L ⟷ β-Pt_3In_2 + $Pt_{13}In_9$ + PtSb; U_3 (960°C), L + $Pt_{13}In_9$ ⟷ Pt_6In_5 + PtSb; U_4 (948°C), L + PtSb ⟷ Pt_2In_3 + $PtSb_2$; E_2 (939°C), L ⟷ $PtIn_2$ + Pt_2In_3 + PtSb; E_3 (935°C), L ⟷ Pt_6In_5 + Pt_2In_3 + PtSb; U_5 (675°C), L + Pt_5Sb ⟷ (Pt) + Pt_3Sb; U_6 (674°C), L + (Pt) ⟷ Pt_3In + Pt_3Sb; E_4 (638°C), L ⟷ Pt_3Sb + Pt_3In + $PtSb_2$; U_7 (533°C), L + PtSb ⟷ Pt_3In + Pt_3Sb_2; U_8 (533°C), L + Pt_2In ⟷ Pt_3In_7 + Pt_3Sb_2; U_9 (498°C), L + $PtSb_2$ ⟷ Pt_3In_7 + InSb; E_5 (494°C), L ⟷ (Sb) + InSb + $PtSb_2$; and E_6 (152°C), L ⟷ (In) + InSb + Pt_3In_7.

The isothermal sections of the In–Pt–Sb ternary system were constructed at room temperature (Liu and Mohney 2003b) and at 400°C and 700°C (Ipser and Richter 2003; Patrone et al. 2006; Guo et al. 2014). The calculated isothermal sections at 400°C and 700°C are presented in Figure 15.73. No ternary compound was found, but most binary phases exhibit noticeable solid solubilities of the third component.

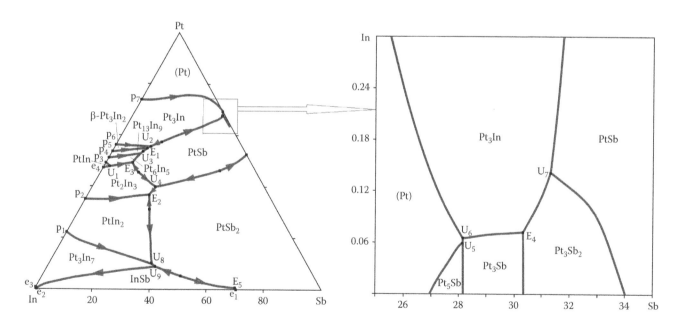

FIGURE 15.72   Calculated liquidus surface of the In–Pt–Sb ternary system. (From Guo, C., et al., *Int. J. Mater. Res.*, *105*(6), 525, 2014. With permission.)

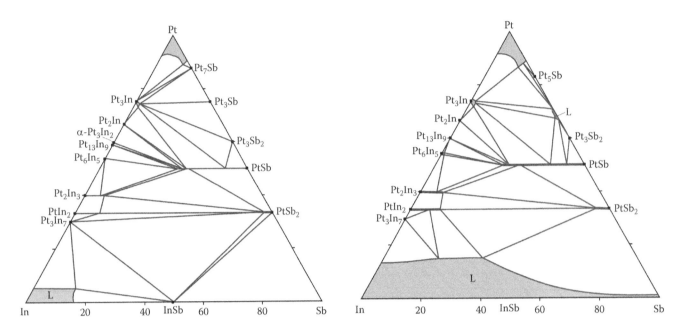

FIGURE 15.73   Calculated isothermal sections of the In–Pt–Sb ternary system at (a) 400°C and (b) 700°C. (From Guo, C., et al., *Int. J. Mater. Res.*, *105*(6), 525, 2014. With permission.)

References

Abrahams M.S., Braunstein R., Rosi F.D. "Thermal, electrical and optical properties of (In,Ga)As alloys." *J. Phys. Chem. Solids*, *10*(2–3), 204–210 (1959).

Abrikosov N.Kh., Glazov V.M., Li Chzhen'-yuan'. "Investigation of separate and joint solubility of aluminum and phosphorus in germanium" [in Russian]. *Zhurn. Neorgan. Khimii*, *7*(4), 831–835 (1962).

Abrikosov N.Kh., Skudnova E.V., Poretskaya L.V., Pavlova N.G. "Investigation of the In–Sb–Cd–Sn quaternary system for the determination of the phase equilibria in the vInSb–(CdSnSb$_2$) section" [in Russian]. *Izv. AN SSSR Neorgan. Mater.*, *2*(8), 1416–1428 (1966).

Abrokwah J.K., Gershenzon M. "Liquid phase epitaxial growth and characterization of InAs$_x$Sb$_{1-x}$ and In$_{1-y}$Ga$_y$Sb on (111)B InSb substrates." *J. Electron. Mater.*, *10*(2), 379–421 (1981).

Akchurin R.Kh., Evgen'ev S.B., Zinov'ev V.G., Islamov S.A., Lapkina I.A., Sorokina O.V., Ufimtsev V.B. "Heterogeneous equilibria in the A^{III}B^V–Bi systems" [in Russian]. *Elektonnaya Tekhnika. Ser. 6. Materialy*, *10*(195), 62–65 (1984).

Akchurin R.Kh., Kao Le Din', Nishanov D.N., Fistul' V.I. "Heterogeneous equilibria in the Bi–GaAs quasi-binary system" [in Russian]. *Izv. AN SSSR Neorgan. Mater.*, *22*(1), 9–12 (1986).

Akopyan R.A., Abdullaev A.A., Mamedova S.Kh., Makhmurova N.T. "Physical-chemical analysis of the ternary solid solutions based on germanium in the Ge–Al–Sb system" [in Russian]. *Zhurn. Neorgan. Khimii*, *32*(5), 1219–1222 (1987).

Akopyan R.A., Evdokimov R.A., Pavlova L.M. "Metastable continuous solid solutions in the Ge–GaSb system" [in Russian]. *Izv. AN SSSR Neorgan. Mater.*, *13*(9), 1570–1573 (1977).

Akopyan R.A., Mamedova C.Kh., Makhmurova N.T., Simonyan S.L. "Ge–Ga–Sb system" [in Russian]. *Zhurn. Neorgan. Khimii*, *33*(1), 190–193 (1988).

Aleshko-Ozhevskaya L.A., Bagaratyan N.V., Emelyanov A.M., L'vov A.Yu., Makarov A.V., Pronchatov A.N. "Investigation of the vapor phase of the boron nitride–silicon melt system" [in Russian]. *Vysokochist. Veshchestva*, (4), 145–152 (1993).

Al'fer S.A., Mechkovskiy L.A., Vecher A.A. "Investigation of the liquidus surface of the germanium–indium–antimony through differential thermal analysis using a simplex-lattice simulation of the experiment" [in Russian]. *Zhurn. Fiz. Khimii*, *57*(5), 1292–1293 (1983).

Aliev I.I., Kuliev A.N., Mamedov F.M., Aliev F.G. "Physicochemical investigation of the GaSb–TmSb system" [in Russian]. *Azerb. Khim. Zhurn.*, (2), 111–112 (2007).

Aliev M.I., Safaraliev G.I., Guliev A.N., Dadashev I.Sh., Mardakhaev B.N. "Phase diagram amd physical properties of the GaSb–MnSb system" [in Russian]. *Izv. AN SSSR Neorgan. Mater.*, *10*(10), 1778–1782 (1974).

Aliev M.I., Suleymanov Z.I., Dzhafarov Z.A., Dzhabbarov P.M., Agashev A.E. "Investigation of the eutectic alloys based on indium and gallium antimonide" [in Russian]. *Izv. AN SSSR Neorgan. Mater.*, *17*(11), 1969–1971 (1981).

Alling B., Ruban A.V., Karimi A., Peil O.E., Simak S.I., Hultman L., Abrikosov I.A. "Mixing and decomposition thermodynamics of c-Ti$_{1-x}$Al$_x$N from first-principles calculations." *Phys. Rev. B*, *75*(4), 045123_1–045123_13 (2007).

Allison H.W. "Solubility and diffusion of zinc in gallium phosphide." *J. Appl. Phys.*, *34*(1), 231–233 (1963).

Ameri M., Bentouaf A., Doui-Aici M., Khenata K., Boufadi F., Touia A. "Structural and electronic properties calculations of Al$_x$In$_{1-x}$P alloy." *Mater. Sci. Appl.*, *2*(7), 729–738 (2011).

Anderbouhr S., Gilles S., Blanquet E., Bernard C., Madar R. "Thermodynamic modeling of the Ti–Al–N system and application to the simulation of CVD processes of the (Ti,Al)N metastable phase." *Chem. Vap. Deposition*, *5*(3), 109–115 (1999).

Andersson M. "Ternary phase equilibria in the (Cr,Mo,W)–In–P systems at 600°C." *J. Alloys Compd.*, *198*(1–2), L15–L18 (1993).

Andersson-Söderberg M. "Solid phase equilibria in the Fe–In–P system." *J. Less Common Metals*, *159*(1–2), L13–L16 (1990).

Andersson-Söderberg M. "Phase relationships in the Ni–In–P system." *J. Less Common Metals*, *171*(2), 179–186 (1991).

Andersson-Söderberg M. "Phase analysis and crystal chemistry of some transition metal–indium–phosphorus systems." *Acta Univ. Ups. Compr. Summ. Upsala Diss. Fac. Sci.*, (376), 1–28 (1992).

Andersson-Söderberg M., Andersson Y. "The crystal structure of $Ni_{21}In_2P_6$." *J. Solid State Chem.*, 85(2), 315–317 (1990).

André E., Le Duc J.M. "Arséniure de gallium dopé au chrome obtenu par épitaxie en phase liquid." *Mater. Res. Bull.*, 4(3), 149–152 (1969).

Andrievskiy R.A., Anisimova I.A. "Phase diagram calculation for TiN based pseudobinary nitride system" [in Russian]. *Izv. AN SSSR Neorgan. Mater.*, 27(7), 1450–1453 (1991).

Ansara I., Dutartre D. "Thermodynamic study of the Al–Ga–As–Ge system." *CALPHAD*, 8(4), 323–342 (1984).

Ansara I. "The Al–As–Ga system (aluminum–arsenic–gallium)." *J. Phase Equilibria*, 13(6), 624–628 (1992).

Ansara I., Gambino M., Bros J.P. "Étude thermodynamique du système ternaire gallium–indium–antimoine." *J. Cryst. Growth*, 32(1), 101–110 (1976).

Antonell M.J., Gila B., Powers K., Abernathy C.R. "Growth and characterization of GaTlAs." *J. Vac. Sci. Technol.*, A18(5), 2448–2451 (2000).

Antonyshyn I., Oryshchyn S., Zhak O., Babizhetskyi V. "The V–Ga–Sb system at 500°C" [in Ukrainian]. *Visnyk L'viv Un-Tu Ser. Khim.*, (49), Pt. 1, 50–57 (2008).

Antypas G.A. "Liquid-phase epitaxy of $In_xGa_{1-x}As$." *J. Electrochem. Soc.*, 117(11), 1393–1397 (1970a).

Antypas G.A. "The Ga–GaP–GaAs ternary phase diagram." *J. Electrochem. Soc.*, 117(5), 700–703 (1970b).

Antypas G.A. "Liquidus and solidus data at 500°C for the In–Ga–Sb system." *J. Cryst. Growth*, 16(2), 181–182 (1972).

Antyukhov A.M. "Vegard's rule in the case of the solid solutions in the InP–InAs system" [in Russian]. *Izv. AN SSSR Neorgan. Mater.*, 22(3), 494–496 (1986).

Arbenina V.V., Beylin Yu.A. "Interaction of GaSb with Sn and SnSb" [in Russian]. *Izv. AN SSSR Neorgan. Mater.*, 22(9), 1425–1431 (1986).

Arbenina V.V., Cher Pak, Skakovskiy S.I., Tarasov A.V. "Investigation of equilibria in the GaSb–AlSb–Sn system" [in Russian]. *Neorgan. Mater.*, 28(3), 513–517 (1992).

Arbenina V.V., Grafova M.I., Fistul' V.I. "Interaction of GaSb with samarium and its compounds" [in Russian]. *Izv. AN SSSR Neorgan. Mater.*, 21(4), 541–546 (1985).

Arseni K.A., Boltaks B.I., Gordin V.L., Ugay Ya.A. "Diffusion and solubility of indium in indium phosphide" [in Russian]. *Izv. AN SSSR Neorgan. Mater.*, 3(9), 1679–1681 (1967).

Asahia H., Fushida M., Yamamoto K., Iwata K., Koh H., Asami K., Gonda S., Oe K. "New semiconductors TlInGaP and their gas source MBE growth." *J. Cryst. Growth*, 175/176, Pt. 2, 1195–1199 (1997).

Asahia H., Yamamoto K., Iwata K., Gonda S.-I., Oe K. "New III-V compound semiconductors TlInGaP for 0.9 μm to over 10 μm wavelength range laser diodes and their first successful growth." *Jpn. J. Appl. Phys.*, 35(7B), Pt. 2, L876–L879 (1996).

Asaka T., Banno H., Funahashi S., Hirosaki N., Fukuda K. "Electron density distribution and crystal structure of 27R-AlON, $Al_9O_3N_7$." *J. Solid State Chem.*, 204, 21–26 (2013a).

Asaka T., Kudo T., Banno H., Funahashi S., Hirosaki N., Fukuda K. "Electron density distribution and crystal structure of 21R-AlON, $Al_7O_3N_5$." *Powder Diffr.*, 28(3), 171–177 (2013b).

Astles M.G. "Liquidus isotherms for Ga + In + P mixtures." *J. Chem. Thermodyn.*, 6(2), 105–110 (1974).

Auffermann G., Kniep R., Prots Yu., Niewa R., Tovar M. "Metal nitrides." *Hahn-Meitner-Inst. Berlin*, (584), 102 (2001).

Aulombard R.L., Joullie A. "Melt growth and some electrical properties of GaSb–AlSb system." *Mater. Res. Bull.*, 14(3), 349–359 (1979).

Aydemir U., Kokal I., Prots Yu., Förster T., Sichelschmidt J., Schappacher F.M., Pöttgen R., Ormeci A., Somer M. "A novel europium (III) nitridoborate $Eu_3[B_3N_6]$: Synthesis, crystal structure, magnetic properties, and Raman spectra." *J. Solid State Chem.*, 239, 75–83 (2016).

Aydemir U., Zevalkink A., Ormeci A., Wang H., Ohno S., Bux S., Snyder G.J. "Thermoelectric properties of the Zintl phases $Yb_5M_2Sb_6$ (M = Al, Ga, In)." *Dalton Trans.*, 44(15), 6767–6774 (2015).

Azevedo S. "Energetic stability of B–C–N monolayer." *Phys. Lett. A*, 351(1–2), 109–112 (2006).

Azhdarova D.S., Dzhalil-zade T.A., Rustamov P.G. "Interaction of gallium antimonide with gallium silfides" [in Russian]. *Uch. Zap. Azerb. Un-T Ser. Khim. N.*, (2), 47–51 (1974).

Babaeva B.K., Rustamov P.G. "Investigation of the interaction in the $GaAs–As_2Te_3$ and $Ga_2Te_3–As_2Te_3$ systems" [in Russian]. *Izv. AN SSSR Neorgan. Mater.*, 8(4), 761–762 (1972).

Babitsyna A.A., Luzhnaya N.P. "Investigation of indium arsenide interaction with AuIn" [in Russian]. *Zhurn. Neorgan. Khimii*, 17(6), 1741–1743 (1972).

Bachheimer J.P., Berge B., Dolino G., Saint-Gregoire P., Zeyen C.M.E. "Calorimetric and neutron scattering studies of the incommensurate phase of berlinite ($AlPO_4$)." *Solid State Commun.*, 51(1), 55–58 (1984).

Badzian A. "Cubic boron nitride–diamond mixed crystals." *Mater. Res. Bull.*, 16(11), 1385–1393 (1981).

Bai V.S., Rao K.V.S.R. "Ferromagnetic resonance investigations in $MnSb_{1-x}In_x$ films." *Phys. Stat. Solidi (A)*, 73(2), K303–K305 (1982).

Bailey M.S., DiSalvo F.J. "The synthesis and structure of Ca_2InN, a novel ternary indium nitride." *J. Alloys Compd.*, 353, 146–152 (2003).

Bailey M.S., Shen D.Y., McGuire M.A., Fredrickson D.C., Toby B.H., DiSalvo F.J., Yamane H., Sasaki S., Shimada M. "The indium subnitrides $Ae_6In_4(In_xLi_y)N_{3-z}$ (Ae = Sr and Ba)." *Inorg. Chem.*, 44(19), 6680–6690 (2005).

Balakumar T., Medraj M. "Thermodynamic modeling of the Mg–Al–Sb system." *CALPHAD*, 29(1), 24–36 (2005).

Banno B., Funahashi S., Asaka T., Hirosaki N., Fukuda K. "Disordered crystal structure of 20H-AlON, $Al_{10}O_3N_8$." *J. Solid State Chem.*, 230, 149–154 (2015).

Banus M.D., Farrell L.B., Strauss A.J. "Resistivities and transformation rates of high-pressure InSb-β Sn alloys." *J. Appl. Phys.*, 36(7), 2186–2188 (1965a).

Banus M.D., Vernon S.N., Gatos H.C. "Superconducting characteristics of the high-pressure InSb-β tin system." *J. Appl. Phys.*, 36(3), 864 (1965b).

Baranov A.N., Gorelenok A.A., Litvak A.M., Sherstnev V.V., Yakovlev Yu.P. "Investigation of the phase equilibria in the In–As–Pb system" [in Russian]. *Zhurn. Neorgan. Khimii*, 37(2), 448–453 (1992).

Baranov A.N., Litvak A.M., Cherneva T.V., Sherstnev V.V., Yastrebov S.G. "Analysis of the phase equilibria in the In–As–Sb system using the model of quasi-regular associated solutions" [in Russian]. *Izv. AN SSSR Neorgan. Mater.*, 26(10), 2021–2025 (1990a).

Baranov A.N., Litvak A.M., Moiseev K.D., Charykov N.A., Sherstnev V.V. "Phase equilibria melt—Solid state in the In–Ga–As–Sb and In–As–P–Sb quaternary systems" [in Russian]. *Zhurn. Fiz. Khimii*, 64(6), 1651–1654 (1990b).

Baranov A.N., Litvak A.M., Sherstnev V.V. "Phase diagram of the Ga–Sb–Pb system" [in Russian]. *Izv. AN SSSR Neorgan. Mater.*, 25(6), 922–925 (1989).

Baranov B.V., Goryunova N.A. "Preparation of homogeneous solid solutions in the AlSb–InSb system" [in Russian]. *Fiz. Tv. Tela*, 2(2) 262–266 (1960).

Baranov B.V., Prochukhan V.D., Goryunova N.A. "Thermal analysis of some solid solutions" [in Russian]. *Izv. AN LatvSSR Ser. Khim.*, (3), 301–308 (1965).

Baranov B.V., Prochukhan V.D., Goryunova N.A. "Obtaining of macrocrystalline boron phosphide from the solution in the Cu_3P melt" [in Russian]. *Izv. AN SSSR Neorgan. Mater.*, 3(9), 1691–1692 (1967).

Barsoum M.W. "The $M_{N+1}AX_N$ phases: A new class of solids; thermodynamically stable nanolaminates." *Progr. Solid State Chem.*, 28(1–4), 201–281 (2000).

Barsoum M.W., Ali M., El-Raghy T. "Processing and characterization of Ti_2AlC, Ti_2AlN and $Ti_2AlC_{0.5}N_{0.5}$." *Metall. Trans. A*, 31(7), 1857–1865 (2000a).

Barsoum M.W., Farber L., Levin I., Procopio A., El-Raghy T., Berner A. "High-resolution transmission electron microscopy of Ti_4AlN_3, or $Ti_3Al_2N_2$ revisited." *J. Am. Ceram. Soc.*, 82(9), 2545–2547 (1999).

Barsoum M.W., Rawn C.J., El-Raghy T., Procopio A.T., Porter W.D., Wang H., Hubbard C.R. "Thermal properties of Ti_4AlN_3." *J. Appl. Phys.*, 87(12), 8407–8414 (2000b).

Barsoum M.W., Schuster J.C. "Comment on 'New ternary nitride in the Ti–Al–N system.'" *J. Am. Ceram. Soc.*, 81(3), 785–786 (1998).

Bartnitskaya T.S., Butylenko A.K., Lugovskaya E.S., Timofeeva I.I. "Study of the quasi-binary cross section of AN–BN at high pressures" [in Russian]. In *Vysok. Davleniya Svoistva Mater*. Kiev: Nauk. Dumka Publish., 90–94 (1980).

Bartram S.F., Slack G.A. "$Al_{10}N_8O_3$ and $Al_9N_7O_3$, two new repeated-layer structures in the $AlN–Al_2O_3$ system." *Acta Crystallogr.*, B35(9), 2281–2283 (1979).

Baruch P., Desse M. "Étude des alliages étain–antimoniure d'indium." *C. R. Acad. Sci.*, 241(16), 1040–1043 (1955).

Basalaev Yu.M. "Sublattice effect on the formation of the band structure of crystals with the chalcopyrite lattice: B_2CN, BC_2N, BCN_2." *J. Struct. Chem.*, 57(1), 8–13 (2016).

Bassoul P., Lefebvre A., Gilles J.C. "Phases ε obtenues par decomposition des spinelles non-stoechiometriques dans les systemes Ga_2O_3–MgO, Al_2O_3–NiO, Al_2O_3–AlN et Al_2O_3–Li_2O." *Mater. Res. Bull.*, 11(1), 11–14 (1976).

Basu S., Adhikari T. "Bulk growth, composition and morphology of gallium manganese antimonide—A new ternary alloy system." *J. Alloys Compd.*, 205(1–2), 81–85 (1994).

Basu S., Adhikari T. "Variation of band gap with Mn concentration in $Ga_{1-x}Mn_xSb$—A new III-V diluted magnetic semiconductor." *Solid State Commun.*, 95(1), 53–55 (1995).

Batyrev N.I., Selin A.A., Shumilin V.P. "Analysis of the interaction in the In–Ga–P ternary system" [in Russian]. *Izv. AN SSSR Neorgan. Mater.*, 13(10), 1733–1735 (1977).

Baumgartner O., Behmer M., Preisinger A. "Die Kristallstruktur von $AlAsO_4$ bei 20°C, 500°C und 750°C." *Z. Kristallogr.*, 187(1–2), 125–131 (1989).

Baumgartner O., Preisinger A., Krempl P.W., Mang H. "Die Kristallstruktur von $GaPO_4$ bei 20°C, 500°C und 750°C." *Z. Kristallogr.*, Kiev: Nauk. Dumka Publish 168(1–4), 83–91 (1984).

Bayer G. "$InSbO_4$ and $ScSbO_4$, new compounds with the rutile structure." *Z. Kristallogr.*, 118(1–2), 158–160 (1963).

Baykal A., Kizilyalli M., Torpak M., Kniep R. "Hydrothermal and microwave synthesis of boron phosphate, BPO_4." *Turk. J. Chem.*, 25(4), 425–432 (2001).

Beck W.R. "Crystallographic inversions of the aluminum orthophosphate polymorphs and their relation to those of silica." *J. Am. Ceram. Soc.*, 32(4), 147–151 (1949).

Beckmann O., Boiler H., Nowotny H., Benesovsky F. "Einige Komplexcarbide und -nitride in den Systemen Ti-{Zn, Cd, Hg}-{C, N} und Cr–Ga–N." *Monatsch. Chem.*, 100(5), 1465–1470 (1969).

Bedair S.M. "Growth and characterization of $Al_xGa_{1-x}Sb$." *J. Electrochem. Soc.*, 122(8), 1150–1152 (1975).

Beensh-Marchwicka G., Król-Stępniewska L., Posadowski W. "Structure of thin films prepared by the cosputtering of titanium and aluminium or titanium and silicon." *Thin Solid Films*, 82(4), 313–320 (1981).

Belaya A.D., Zemskov V.S. "Distribution coefficient of tellurium in InSb" [in Russian]. *Izv. AN SSSR Neorgan. Mater.*, 9(6), 897–899 (1973).

Belotskiy D.P., Dundich M.S., Kotsyumakha M.P., Makhova M.K., Lesina N.V., Noval'kovskiy N.P. "Cd-Sb-Al system" [in Russian]. *Izv. AN SSSR Neorgan. Mater.*, 21(7), 1093–1096 (1985).

Belotskiy D.P., Kotsumakha M.P. "Physico-chemical investigation of the Cd_3Sb_2–InSb system" [in Russian]. *Dop. AN URSR*, (2), 222–226 (1965).

Belotskiy D.P., Kotsumakha M.P., Dundich M.S., Makhova M.K., Franchuk V.I., Lesina N.V. "In–Sb–Se system" [in Russian]. *Izv. AN SSSR Neorgan. Mater.*, *14*(5), 838–842 (1978).

Belousov A., Karpinski J., Batlogg B. "Thermodynamics of the Al–Ga–N_2 system." *J. Cryst. Growth*, *312*(18), 2579–2584 (2010a).

Belousov A., Katrych S., Hametner K., Günther R., Karpinski J., Batlogg B. "$Al_xGa_{1-x}N$ bulk crystal growth: Crystallographic properties and *p-T* phase diagram." *J. Cryst. Growth*, *312*(18), 2585–2592 (2010b).

Belousov A., Katrych S., Jun J., Günther R., Sobolewski J., Karpinski J., Batlogg B. "Bulk single-crystal growth of ternary $Al_xGa_{1-x}N$ from solution in gallium under high pressure." *J. Cryst. Growth*, *311*(16), 3971–3974 (2009).

Benaissa H., Zaoui A., Ferhat M. "Strong composition-dependent disorder in $InAs_{1-x}N_x$ alloys." *J. Alloys Compd.*, *475*(1–2), 592–594 (2009).

Benayad N., Rached D., Khenata R., Litimein F., Reshak A.H., Rabah M., Baltache H. "First principles study of the structural, elastic and electronic properties of Ti_2InC and Ti_2InN." *Mod. Phys. Lett. B*, *25*(10), 747–761 (2011).

Benko E., Barr T.L., Bernasik A., Hardcastle S., Hoppe E., Bielańska E., Klimczyk P. "Experimental and calculated phase equilibria in the cubic BN–Ta–C system." *Ceram. Intern.*, *30*(1), 31–40 (2004).

Benkraouda M., Amrane N. "Band gap calculations of the semiconductor BN_xP_{1-x} using modified Becke-Johnson approximation." *J. Alloys Compd.*, *546*, 151–156 (2013).

Bennarndt W., Boehm G., Amann M.-C. "Domains of molecular beam epitaxial growth of Ga(In)AsBi on GaAs and InP substrates." *J. Cryst. Growth*, *436*, 56–61 (2016).

Bentama J., Durand J., Cot L. "Structure cristalline du trimétaphosphate d'indium(III)." *Z. Anorg. Allg. Chem.*, *556*(1), 227–232 (1988).

Bessolov V.N., Konnikov S.G., Umanskiy V.Ye., Yakovlev Yu.P. "Lattice parameter of $Al_xGa_{1-x}P$" [in Russian]. *Fiz. Tv. Tela*, *24*(5), 1528–1531 (1982).

Beyers R., Sinclair R., Thomas M.E. "Phase equilibria in thin-film metallizations." *J. Vac. Sci. Technol.*, *B2*(4), 781–784 (1984).

Beznosikov B.V. "Predicted nitrides with an antiperovskite structure." *J. Struct. Chem.*, *44*(5), 885–888 (2003).

Bhansali A.S., Sinclair R., Morgan A.E. "A thermodynamic approach for interpreting metallization layer stability and thin-film reactions involving four elements: Application to integrated circuit contact metallurgy." *J. Appl. Phys.*, *68*(3), 1043–1049 (1990).

Bi W.G., Tu C.W. "N incorporation in GaN_xP_{1-x} and InN_xP_{1-x} using a RF N plasma source." *J. Cryst. Growth*, *175–176*, 145–149 (1997).

Bi W.G., Tu C.W. "N incorporation in InP and band gap bowing of InN_xP_{1-x}." *J. Appl. Phys.*, *80*(3), 1934–1936 (1996).

Birdwhistell T.L.T., Klein C.L., Jeffries T., Stevens E.D., O'Connor C.J. "Synthesis and crystal structures of the layered I–III–V Zintl phases, $K_4In_4X_6$, where X = As, Sb." *J. Mater. Chem.*, *1*(4), 555–558 (1991).

Birdwhistell T.L.T., Stevens E.D., O'Connor C.J. "Synthesis and crystal structure of a novel layered Zintl phase: $K_3Ga_3As_4$." *Inorg. Chem.*, *29*(19), 3892–3894 (1990).

Blanc J., Weisberg L.R. "Electrical activity of copper in GaAs." *J. Phys. Chem. Solids*, *25*(2), 221–223 (1964).

Blase W., Cordier G., Ludwig M., Kniep R. "$Sr_3[Al_2N_4]$: Ein Nitridoaluminat mit gewellten Tetraederketten $^1_\infty[AlN_{4/2}^{3-}]$." *Z. Naturforsch.*, *49B*(4), 501–505 (1994).

Blase W., Cordier G., Peters K., Somer M., von Schnering H.G. "$[InAs_3]^{6-}$ and $[AlSb_3]^{6-}$, trigonal planar anions in Cs_6InAs_3 and Cs_6AlSb_3." *Angew. Chem. Int. Ed. Engl.*, *30*(3), 326–328 (1991a).

Blase W., Cordier G., Peters K., Somer M., von Schnering H.G. "$[InAs_3]^{6-}$ und $[AlSb_3]^{6-}$, trigonal-planare Anionen in Cs_6InAs_3 und Cs_6AlSb_3." *Angew. Chem.*, *103*(3), 335–336 (1991b).

Blase W., Cordier G., Poth L., Weil K.G. "Crystal structure of dicaesium phyllo-triantimonidoindate, $Cs_2In_2Sb_3$." *Z. Kristallogr.*, *210*(1), 60 (1995).

Blase W., Cordier G., Somer M. "Crystal structure of caesium catena-di-μ-phosphidoindate di-μ-phosphido bis(phosphidoindate), $Cs_{12}(InP_2)_2(In_2P_4)$." *Z. Kristallogr.*, *195*(1–2), 123–124 (1991c).

Blase W., Cordier G., Somer M. "Crystal structure of dipotassium catena-diphosphidogallate, K_2GaP_2." *Z. Kristallogr.*, *195*(1–2), 115–116 (1991d).

Blase W., Cordier G., Somer M. "Crystal structure of hexacaesium triarsenidoindate, Cs_6InAs_3." *Z. Kristallogr.*, *195*(1–2), 117–118 (1991e).

Blase W., Cordier G., Somer M. "Crystal structure of hexapotassium triphosphidoindate, K_6InP_3." *Z. Kristallogr.*, *195*(1–2), 121–122 (1991f).

Blase W., Cordier G., Somer M. "Crystal structure of tripotassium catena-di-μ-phosphido-indate, K_3InP_2." *Z. Kristallogr.*, *195*(1–2), 109–110 (1991g).

Blase W., Cordier G., Somer M. "Crystal structure of trisodium tectodiphosphidoindate, Na_3InP_2." *Z. Kristallogr.*, *195*(1–2), 119–120 (1991h).

Blase W., Cordier G., Somer M. "Crystal structure of hexacaesium triantimonidogallate, Cs_6GaSb_3." *Z. Kristallogr.*, *199*(3–4), 277–278 (1992a).

Blase W., Cordier G., Somer M. "Crystal structure of hexarubidium triantimonidoaluminate, Rb_6AlSb_3." *Z. Kristallogr.*, *199*(3–4), 279–280 (1992b).

Blase W., Cordier G., Somer M. "Crystal structure of hexapotassium triarsenidoindate, K_6InAs_3." *Z. Kristallogr.*, *206*(1–2), 141–142 (1993a).

Blase W., Cordier G., Somer M. "Crystal structure of hexasodium triphosphidogallate, Na_6GaP_3." *Z. Kristallogr.*, *206*(1–2), 143–144 (1993b).

Blom G.M. "The In–Ga–P ternary phase diagram and its application to liquid state epitaxial growth." *J. Electrochem. Soc.*, *118*(11), 1834–1836 (1971).

Blom G.M., Plaskett T.S. "The In–Ga–Sb ternary phase diagram." *J. Electrochem. Soc.*, 118(11), 1831–1834 (1971).

Bobev S., Fritsch V., Thompson J.D., Sarrao J.L., Eck B., Dronskowski R., Kauzlarich S.M. "Synthesis, structure and properties of the new rare-earth Zintl phase $Yb_{11}GaSb_9$." *J. Solid State Chem.*, 178(4), 1071–1079 (2005).

Bobev S., Hullmann J., Harmening T., Pöttgen R. "Novel ternary alkaline-earth and rare-earth metal antimonides from gallium or indium flux. Synthesis, structural characterization and [121]Sb and [151]Eu Mössbauer spectroscopy of the series $A_7Ga_8Sb_8$ (A = Sr, Ba, Eu) and $Ba_7In_8Sb_8$." *Dalton Trans.*, 39(30), 6049–6053 (2010).

Bobrov V.I., Gorelova N.N., Lange V.N., Molodyan I.P., Matkus M.M., Radautsan S.I. "On the phase equilibria in the indium–antimony–tellurium system" [in Russian]. In *Poluprovodn. Soedin. i Ikh Tverdye Rastvory*. Kishinev: Publish. House of AN MSSR, 167–174 (1970).

Bobrov V.I., Kotrubenko B.P., Lange V.N. "On the InAs–Te phase diagram" [in Russian]. *Izv. AN SSSR Neorgan. Mater.*, 3(7), 1273–1275 (1967).

Bobrov V.I., Novik F.S., Radautsan S.I. "Construction of the liquidus surface of the InSb–Sb–InTe system" [in Russian]. *Izv. AN SSSR Neorgan. Mater.*, 7(12), 2156–2161 (1971).

Bochkareva V.A., Maksimov S.K., Sokolov E.B. "Precipitation of supersaturated chromium solutions in gallium arsenide." *Cryst. Res. Technol.*, 17(2), 237–243 (1982).

Böddeker K.W., Shore S.G., Bunting R.K. "Boron–nitrogen chemistry. I. Syntheses and properties of new cycloborazanes, $(BH_2NH_2)_n$." *J. Am. Chem. Soc.*, 88(19), 4396–4401 (1961).

Bodnar I.V. "On the phase diagram of the solid solutions in the InAs–GaAs system" [in Russian]. *Izv. AN BSSR Ser. Khim. N.*, (5), 49–52 (1975).

Bodnar I.V., Makovetskaya L.A., Shishonok N.A., Yaroshevich G.P. "Investigation of the $Ga_xIn_{1-x}P$ solid solutions" [in Russian]. In *Khimiya i Tekhnol. Fosfidov i Fosforsoderzh. Splavov*. Kiev: Nauk. Dumka Publish., 138–140 (1979).

Bodnar I.V., Makovetskaya L.A., Smirnova G.F. "Single crystals growth and investigation of the properties of the $Ga_xIn_{1-x}P$ solid solutions" [in Russian]. *Izv. AN SSSR Neorgan. Mater.*, 14(8), 1370–1373 (1978).

Bodnar I.V., Matyas E., Makovetskaya L.A. "Phase diagram of $Ga_xIn_{1-x}P$ solid solution." *Phys. Status Solidi (A)*, 36(2), K141–K144 (1976).

Bodnar I.V., Matyas E.E. "Phase diagram of the InAs–InP system" [in Russian]. *Zhurn. Neorgan. Khimii*, 22(3), 796–798 (1977).

Boese R., Maulitz A.H., Stellberg P. "Solid-state borazine: Does it deserve to be entitled 'inorganic benzene'?" *Chem. Ber.*, 127(10), 1887–1889 (1994).

Bogorodskiy O.V., Nashel'skiy A.Ya., Ostrovskaya V.Z. "X-ray study of the InAs–InP solid solutions" [in Russian]. *Kristallografiya*, 6(1), 119–121 (1961).

Boller H. "Kristallchemische Untersuchungen an Komplexcarbiden und -nitriden mit aufgefülltem Re_3B-Typ." *Monatsch. Chem.*, 102(2), 431–437 (1971).

Boller H., Parthé E. "On the possibility of forming 'pseudosilicides'." *Acta Crystallogr.*, 16(8), 830–833 (1963).

Boltaks B.I., Kulikov G.S., Nikulitsa I.N., Shishiyanu F.S. "Diffusion and solubility of Fe in GaAs" [in Russian]. *Izv. AN SSSR Neorgan. Mater.*, 11(2), 348–350 (1975).

Bonapace C.R., Kahn A. "An AES-ELEED study of the Al/GaP (110) interface." *J. Vac. Sci. Technol.*, A1(2), Pt. 1, 588–591 (1983).

Bondarenko V.P., Baranovskiy A.M., Pugach E.A. "Investigation of cobalt interaction with cubic boron nitride" [in Russian]. In *Vysok. Davleniya i Svoistva Materialov. Materialy III Resp. Nauch. Seminara*. Kiev: Nauk. Dumka Publish., 101–105 (1980).

Bondarenko V.P., Khalepa A.P. "Thermodynamic investigation of boron nitride interaction with transition metals" [in Russian]. *Sintetich. Almazy*, (2), 13–18 (1977).

Bondarenko V.P., Khalepa A.P., Cherepinina E.S. "Investigation of cubic boron nitride interaction with transition metals and their carbides" [in Russian]. *Sintetich. Almazy*, (4), 22–25 (1978).

Borisov I.I., Manasijević D.M., Živković D.T. "Calculation of the phase equilibria in the Ag–In–Sb system by the *CALPHAD* method" [in Croatian]. *Hem. Ind.*, 61(3), 152–156 (2007).

Borisova L.A., Akkerman Z.L., Dorokhov A.N. "Solubility of oxygen and its chemical state in the gallium arsenide lattice" [in Russian]. In *Svoystva Legirovan. Poluprovodn*, Moscow: Nauka Publish., 215–221 (1977a).

Borisova L.A., Akkerman Z.L., Dorokhov A.N. "Solubility of oxygen in GaAs" [in Russian]. *Izv. AN SSSR Neorgan. Mater.*, 13(5), 908–909 (1977b).

Borisova L.A., Artyukhin P.I., Akkerman Z.L. "Solubility of carbon in the GaAs layers grown from the gallium agitated solutions" [in Russian]. *Izv. AN SSSR Neorgan. Mater.*, 14(10), 1790–1792 (1978).

Borisova L.A., Valisheva N.A. "Phase diagram of the GaAs–Si system" [in Russian]. Izv. AN SSSR Neorgan. Mater., 10(7), 1263–1267 (1974).

Boudin S., Lii K.-H. "Ga_3PO_7." *Acta Crystallogr.*, C54(1), 5–7 (1998).

Bouhemadou A. "Structural, electronic and elastic properties of MAX phases M_2GaN (M = Ti, V and Cr)." *Solid State Sci.*, 11(11), 1875–1881 (2009).

Bowden M., Heldebrant D.J., Karkamkar A., Proffen T., Schenter G.K., Autrey T. "The diammoniate of diborane: Crystal structure and hydrogen release." *Chem. Commun.*, 46(45), 8564–8566 (2010).

Brebrick R.F., Panlener R.J. "A systematic investigation of quantitative fits to III-V pseudobinaries using the quasiregular model and its special cases." *J. Electrochem. Soc.*, 1974, 121(7), 932–942 (1974).

Brock S.L., Weston L.J., Olmstead M.M., Kauzlarich S.M. "Synthesis, structure, and properties of $A_{14}AlSb_{11}$ (A = Ca, Sr, Ba)." *J. Solid State Chem.*, 107(2), 513–523 (1993).

Brousseau L.C., Williams D., Kouvetakis J., O'Keeffe M. "Synthetic routes to $Ga(CN)_3$ and $MGa(CN)_4$ (M = Li, Cu) framework structures." *J. Am. Chem. Soc.*, *119*(27), 6292–6296 (1997).

Bubenzer A., Nitsche R., Raufer A. "Vapour growth and piezoelectric effect of indium thiophosphate, $InPS_4$." *J. Cryst. Growth*, *29*(3), 237–240 (1975).

Bublik V.T., Gorelik S.S., Shumskiy M.G., Fomin V.G. "Investigation of the lattice parameter concentration dependence of the GaAs–GaP and GaAs–AlAs solid solutions" [in Russian]. *Nauch. Tr. Mosk. In-T Stali Splavov*, (83), 61–69 (1974).

Bublik V.T., Karman M.B., Kleptsin V.F., Leikin V.N. "Determination of the GaP–GaAs solid solution interaction parameter." *Phys. Stat. Sol. (A)*, *32*(2), 631–638 (1975).

Bublik V.T., Leikin V.N. "Calculation of the pseudobinary alloy semiconductor phase diagrams." *Phys. Stat. Sol. (A)*, *46*(1), 365–372 (1978).

Bucher G., Ellner M., Sommer F., Predel B. "Zur Gültigkeit des Zenschen Gesetzes und des Le Chatelierschen Prinzips in einigen Phasen mit Grimm-Sommerfeld-Binding." *Monatsch. Chem.*, *117*(12), 1367–1378 (1986).

Buchtová V., Živković D., Vřešt'ál J., Manasijević D., Kroupa A. "Comparison of prediction of phase equilibria in the Ag–In–Sb system at 200°C with experimental data." *Monatsh. Chem.*, *136*(11), 1939–1945 (2005).

Buck P., Carpentier C.D. "The crystal structure of gallium thiophosphate, $GaPS_4$." *Acta Crystallogr.*, *29*(9), 1864–1868 (1973).

Buck P., Nitsche R. "$GaPS_4$—Eine neue Verbindung im System Ga–P–S." *Z. Naturforsch.*, *26B*(7), 731 (1971).

Budzyński V., Mitsiuk V.I., Ryzhkovskii V.M., Surowiec Z., Tkachenka T.M. "Preparation and properties of $Mn_{1.1}Sb_{1-y}Al_y$ and $Mn_{1.1}Sb_{1-y}Si_y$ solid solutions." *Inorg. Mater.*, *49*(2), 115–119 (2013).

Buehler E., Bachmann K.J. "Solubilities of InP and CdS in Cd, Sn, In, Bi and Pb." *J. Cryst. Growth*, *35*(1), 60–64 (1976).

Buehler E., Bachmann K.J., Kammlott G.W., Shay J.L. "Phase relations, crystal growth and heteroepitaxy in the quaternary system Cd, Sn, In and P." *J. Cryst. Growth*, *24–25*, 260–265 (1974).

Bühl M., Steinke T., von Ragué Schleyer P., Boese R. "Einfluß des Lösungsmittels auf Geometrie und chemische Verschiebung; Auflösung scheinbarer experimenteller Widersprüche für $H_3B•NH_3$ mit ab-initio/IGLO-Rechnungen." *Angew. Chem.*, *103*(9), 1179–1181 (1991a).

Bühl M., Steinke T., von Ragué Schleyer P., Boese R. "Solvation effects on geometry and chemical shifts. An ab initio/IGLO reconciliation of apparent experimental inconsistencies on $H_3B•NH_3$." *Angew. Chem. Int. Ed. Engl.*, *30*(9), 1160–1161 (1991b).

Burghaus J., Sougrati M.T., Möchel A., Houben A., Hermann R.P., Dronskowski R. "Local ordering and magnetism in $Ga_{0.9}Fe_{3.1}N$." *J. Solid State Chem.*, *184*(9), 2315–2321 (2011).

Burghaus J., Wessel M., Houben A., Dronskowski R. "Ternary nitride $GaFe_3N$: An experimental and quantum-theoretical study." *Inorg. Chem.*, *49*(21), 10148–10155 (2010).

Burkhardt W., Schubert K. "Über messingartige Phasen mit A3-verwandter Struktur." *Z. Metallkde*, *50*(8), 442–452 (1959).

Cadien K.C., Zilko J.L., Eltoukhy A.H., Greene J.E. "Growth of single crystal metastable $InSb_{1-x}Bi_x$ and $(GaSb)_{1-x}Ge_x$ semiconducting films." *J. Vac. Sci. Technol.*, *17*(1), 441–444 (1980).

Caetano C., Teles L.K., Marques M., Dal Pino Jr. A., Ferreira L.G. "Phase stability, chemical bonds, and gap bowing of $In_xGa_{1-x}N$ alloys: Comparison between cubic and wurtzite structures." *Phys. Rev. B*, *74*(20), 045215_1–045215_8 (2006).

Campbell J.P., Hwang J.-W., Young Jr. V.G., Von Dreele R.B., Cramer C.J., Gladfelter W.L. "Crystal engineering using the unconventional hydrogen bond. Synthesis, structure, and theoretical investigation of cyclotrigallazane." *J. Am. Chem. Soc.*, *120*(3), 521–531 (1998).

Cannard P.J., Ekström T., Tilley R.J.D. "The formation of phases in the AlN-rich corner of the Si–Al–O–N system." *J. Eur. Ceram. Soc.*, *8*(6), 375–382 (1991).

Cao Z., Xie W., Wang K., Du G., Qiao Z. "Thermodynamic assessment of the In–Ni–Sb system and predictions for thermally stable contacts to InSb." *J. Electron. Mater.*, *42*(8), 2615–2629 (2013).

Caron M., Gagnon G., Fortin V., Currie J.F., Ouellet L., Tremblay Y., Biberger M., Reynolds R. "Calculation of a Al–Ti–O–N quaternary isotherm diagram for the prediction of stable phases in TiN/Al alloy contact metallization." *J. Appl. Phys.*, *79*(8), 4468–4470 (1996).

Caron-Popowich R., Washburn J., Sands T., Kaplan A.S. "Phase formation in the Pd–InP system." *J. Appl. Phys.*, *64*(10), 4909–4913 (1988).

Carpentier C.D., Diehl R., Nitsche R. "Die Kristallstruktur InPS." *Naturwissenschaften*, *57*(8), 393 (1970).

Carrillo-Cabrera W., Somer M., Peters K., von Schnering H.G. "Crystal structure of trieuropium bis(dinitridoborate), $Eu_3[BN_2]_2$." *Z. Kristallogr. New Cryst. Struct.*, *216*(1), 43–44 (2001).

Carrillo-Cabrera W., Somer M., Peters K., von Schnering H.G. "Synthesis, structure and vibrational spectra of $Ba_{14}InP_{11}$." *Chem. Ber.*, *129*(9), 1015–1023 (1996).

Casey Jr. H.C., Pearson G.L. "Rare earths in covalent semiconductors: The thullium–gallium arsenide system." *J. Appl. Phys.*, *35*(11), 3401–3407 (1964).

Chambers S.A. "Fermi-level movement and atomic geometry at the Al/GaAs (001) interface." *Phys. Rev. B*, *39*(17), 12664–12671 (1989).

Chan J.Y., Wang M.E., Rehr A., Kauzlarich S.M., Webb D.J. "Synthesis, structure, and magnetic properties of the rare-earth Zintl compounds $Eu_{14}MnPn_{11}$ and $Eu_{14}InPn_{11}$ (Pn = Sb, Bi)." *Chem. Mater.*, *9*(10), 2131–2138 (1997).

Chang K.M., Coughanowr C.A., Anderson T.J. "Determination of the gallium activity in the gallium–indium–antimony pseudobinary liquid mixture by a solid state electrochemical method." *Chem. Eng. Commun.*, *38*(3–6), 275–297 (1985).

Chang L.L., Pearson G.L. "Diffusion and solubility of zinc in gallium phosphide single crystals." *J. Appl. Phys.*, 35(2), 374–378 (1964a).

Chang L.L., Pearson G.L. "The solubilities and distribution coefficients of Zn in GaAs and GaP." *J. Phys. Chem. Solids*, 25(1), 23–30 (1964b).

Chang Y.H.R., Yoon T.L., Lim T.L., Rakitin M. "Thorough investigations of the structural and electronic properties of $Al_xIn_{1-x}N$ ternary compound via ab initio computations." *J. Alloys Compd.*, 682, 338–344 (2016).

Chen G., Sundman B. "Thermodynamic assessment of the Ti-Al-N system." *J. Phase Equilibria*, 19(2), 146–160 (1998).

Chen S.-W., Chien Y.-C., Chen W.-A., Chang J.-S., Tseng S.-M., Snyder G.J. "Liquidus projection and isothermal section at 650°C of ternary Co–Sb–Ga system." *J. Alloys Compd.*, 637, 98–105 (2015).

Chen Y., Feng Q., Zeng L., He J., He W. "Phase relationships of the La–Al–Sb system at 773K from 0 to 50.0 at.% Sb." *J. Alloys Compd.*, 492(1–2), 208–212 (2010).

Cheng K.Y., Pearson G.L. "The Al–Ga–Sb ternary phase diagram and its application to liquid phase epitaxial growth." *J. Electrochem. Soc.*, 124(5), 753–757 (1977).

Cheviré F., DiSalvo F.J. "A new ternary nitride La_2GaN_3: Synthesis and crystal structure." *J. Alloys Compd.*, 457(1–2), 372–375 (2008).

Cho S.-A. "Thermodynamic characteristics of the III-V semiconductor pseudobinary alloys." *Z. Metallkde*, 67(7), 479–487 (1976).

Christie D.M., Chelikowsky J.R. "Structural properties of α-berlinite ($AlPO_4$)." *Phys. Chem. Miner.*, 25(3), 222–226 (1998).

Chudinova N.N., Grunze I., Guzeeva L.S. "Binary ammonium–gallium phosphates" [in Russian]. *Izv. AN SSSR Neorgan. Mater.*, 23(4), 616–621 (1987).

Chudinova N.N., Tananaev I.V., Avaliani M.A. "Synthesis of binary gallium–potassium polyphosphates in polyphosphoric acid melts" [in Russian]. *Izv. AN SSSR Neorgan. Mater.*, 13(12), 2234–2235 (1977).

Clarke S.J., DiSalvo F.J. "Synthesis and structure of one-, two-, and three-dimensional alkaline earth metal gallium nitrides: $Sr_3Ga_2N_4$, $Ca_3Ga_2N_4$, and $Sr_3Ga_3N_5$." *Inorg. Chem.*, 36(6), 1143–1148 (1997).

Clarke S.J., DiSalvo F.J. "Synthesis and structure of β-$Ca_3Ga_2N_4$, a ternary nitride with two interpenetrating three dimensional nets." *J. Alloys Compd.*, 274(1–2), 118–121 (1998).

Cohen L.H., Klement Jr. W. "Determination of the high-low inversion in berlinite ($AlPO_4$) to 6 kbar." *Am. Mineral.*, 58(7–8), 796–798 (1973).

Colinet C., Tedenac J.-C. "Structural stability of ternary $D8_m$–Ti_5Sb_2X (X = Al, Ga, In, Si, Ge, Sn) compounds." *CALPHAD*, 49, 8–18 (2015).

Collongues R., Gilles J.C., Lejus A.M., Perez y Jorba M., Michel D. "Recherches sur les oxynitrures metalliques." *Mater. Res. Bull.*, 2(9), 837–848 (1967).

Compeán García V.D., Orozco Hinostroza I.E., Escobosa Echavarría A., López Luna E., Rodríguez A.G., Vidal M.A. "Bulk lattice parameter and band gap of cubic $In_xGa_{1-x}N$ (001) alloys on MgO (100) substrates." *J. Cryst. Growth*, 418, 120–125 (2015).

Corbin N.D. "Aluminum oxynitride spinel: A review." *J. Eur. Ceram. Soc.*, 5(3), 143–154 (1989).

Cordier G. "Darstellung und Kristallstruktur von $Ca_5Ga_2N_4$." *Z. Naturforsch.*, 43B(10), 1253–1255 (1988).

Cordier G., Höhn P., Kniep R., Rabenau A. "Ca_6GaN_5 und Ca_6FeN_5. Verbindungen mit $[CO_3]^{2-}$isosteren Anionen $[GaN_3]^{6-}$ und $[FeN_3]^{6-}$." *Z. Anorg. Allg. Chem.*, 591(1), 58–66 (1990).

Cordier G., Ludwig M., Stahl D., Schmidt P.C., Kniep R. "$(Sr_6N)[Ga_5]$ and $(Ba_6N)[Ga_5]$: Compounds with discrete (M_6N) octahedra and $[Ga_5]$ clusters." *Angew. Chem. Int. Ed. Engl.*, 34(16), 1761–1763 (1995a).

Cordier G., Ludwig M., Stahl D., Schmidt P.C., Kniep R. "$(Sr_6N)[Ga_5]$ und $(Ba_6N)[Ga_5]$: Verbindungen mit diskreten (M_6N)-Oktaedern und $[Ga_5]$-Clustern." *Angew. Chem.*, 107(16), 1879–1881 (1995b).

Cordier G., Ochmann H. "Na_3AlAs_2, eine Zintlphase mit SiS_2-isosterer Anionenteilstruktur." *Z. Naturforsch.*, 43B(12), 1538–1540 (1988).

Cordier G., Ochmann H. "$K_{20}Ga_6Sb_{12.66}$ und $K_{20}Ga_6As_{12.66}$, Verbindungen mit zu $B_3O_6^{3-}$ isostrukturelen Anionen $Ga_3Sb_6^{9-}$ und $Ga_3As_6^{9-}$." *Z. Naturforsch.*, 45B(3), 277–282 (1990).

Cordier G., Ochmann H. "Crystal structure of dicaesium catena-diantimonidogallate, Cs_2GaSb_2." *Z. Kristallogr.*, 197(3–4), 310–311 (1991a).

Cordier G., Ochmann H. "Crystal structure of dipotassium catenadiarsenidogallate, K_2GaAs_2." *Z. Kristallogr.*, 195(1–2), 111–112 (1991b).

Cordier G., Ochmann H. "Crystal structure of dipotassium phyllo-triantimonidodigallate, $K_2Ga_2Sb_3$." *Z. Kristallogr.*, 197(3–4), 289–290 (1991c).

Cordier G., Ochmann H. "Crystal structure of dipotassium phyllo-triantimonidodiindate, $K_2In_2Sb_3$." *Z. Kristallogr.*, 197(3–4), 291–292 (1991d).

Cordier G., Ochmann H. "Crystal structure of dipotassium phyllo-triarsenidodigallate, $K_2Ga_2As_3$." *Z. Kristallogr.*, 197(3–4), 287–288 (1991e).

Cordier G., Ochmann H. "Crystal structure of dipotassium phyllo-triarsenidodiindate, $K_3In_2As_3$." *Z. Kristallogr.*, 197(3–4), 295–296 (1991f).

Cordier G., Ochmann H. "Crystal structure of dirubidium catenadiantimonidogallate, Rb_2GaSb_2." *Z. Kristallogr.*, 195(1–2), 125–126 (1991g).

Cordier G., Ochmann H. "Crystal structure of dirubidium catenadiarsenidogallate, Rb_2GaAs_2." *Z. Kristallogr.*, 195(1–2), 113–114 (1991h).

Cordier G., Ochmann H. "Crystal structure of disodium phyllotriantimonidodiindate, $Na_2In_2Sb_3$." *Z. Kristallogr.*, 197(3–4), 281–282 (1991i).

Cordier G., Ochmann H. "Crystal structure of disodium phyllotriarsenidodialuminate, $Na_2Al_2As_3$." *Z. Kristallogr.*, *197*(3–4), 283–284 (1991j).

Cordier G., Ochmann H. "Crystal structure of disodium phyllotriarsenidodigallate, $Na_2Ga_2As_3$." *Z. Kristallogr.*, *197*(3–4), 285–286 (1991k).

Cordier G., Ochmann H. "Crystal structure of potassium phyllodiantimonidogallate, $KGaSb_2$." *Z. Kristallogr.*, *197*(3–4), 297–298 (1991l).

Cordier G., Ochmann H. "Crystal structure of potassium tecto-tetraantimonidoaluminate, $KAlSb_4$." *Z. Kristallogr.*, *197*(3–4), 308–309 (1991m).

Cordier G., Ochmann H. "Crystal structure of potassium tecto-tetraantimonidogallate, $KGaSb_4$." *Z. Kristallogr.*, *195*(3–4), 306–307 (1991n).

Cordier G., Ochmann H. "Crystal structure of tripotassium phyllo-triarsenidodiindate, $K_2In_2As_3$." *Z. Kristallogr.*, *197*(3–4), 293–294 (1991o).

Cordier G., Ochmann H. "Crystal structure of trisodium tectoantimonidoindate, Na_3InSb_2." *Z. Kristallogr.*, *195*(1–2), 107–108 (1991p).

Cordier G., Ochmann H. "Crystal structure of trisodium tectodiarsenidoindate, Na_3InAs_2." *Z. Kristallogr.*, *195*(1–2), 105–106 (1991q).

Cordier G., Ochmann H., Schäfer H. "Neueartige Polyanionen in Zintl-phasen. Zur Kenntnis von $Na_7Al_2Sb_5$." *Z. Anorg. Allg. Chem.*, *517*(10), 118–124 (1984a).

Cordier G., Ochmann H., Schäfer H. "$Na_2Ga_3Sb_3$, eine neue Zintlphase mit Schichstruktur." *Mater. Res. Bull.*, *21*(3), 331–336 (1986a).

Cordier G., Ochmann H., Schäfer H. "Zur Isosterie bei Halbmetallanionen: Dem $(BS_2)\infty$ isostere $GaSb_2^{2-}$-Anionen im K_2GaSb_2." *J. Less Common Metals*, *119*(2), 291–296 (1986b).

Cordier G., Rönninger S. "Darstellung und Kristallstrukturen von Ca_4In_2N und Sr_4In_2N." *Z. Naturforsch.*, *42*B(7), 825–827 (1987).

Cordier G., Savelsberg G., Schäfer H. "Zintlephasen mit komplexen Anionen: Zur Kenntnis von Ca_3AlAs_3 und Ba_3AlSb_3." *Z. Naturforsch.*, *37*B(8), 975–980 (1982).

Cordier G., Schäfer H. "Ca_3AlAs_3—An intermetallic analogue of the chain silicates." *Angew. Chem. Int. Ed. Engl.*, *20*(5), 466 (1981a).

Cordier G., Schäfer H. "Ca_3AlAs_3—Ein intermetallisches Analogon zu den Kettensilicaten." *Angew. Chem.*, *93*(5), 474 (1981b).

Cordier G., Schäfer H., Stelter M. "$Ca_3Al_3Sb_3$ und $Ca_5Al_2Bi_6$, zwei neue Zintlphasen mit kettenförmigen Anionen." *Z. Naturforsch.*, *39*B(6), 727–732 (1984b).

Cordier G., Schäfer H., Stelter M. "Darstellung und Struktur der Verbundung $Ca_{14}AlSb_{11}$." *Z. Anorg. Allg. Chem.*, *519*(12), 183–188 (1984c).

Cordier G., Schäfer H., Stelter M. "$Ca_{11}InSb_9$, eine Zintlphase mit diskreten $InSb_4^{9-}$-Anionen." *Z. Naturforsch.*, *40*B(7), 868–871 (1985a).

Cordier G., Schäfer H., Stelter M. "Neue Zintlphasen: Ba_3GaSb_3, Ca_3GaAs_3 und Ca_3InP_3." *Z. Naturforsch.*, *40*B(8), 1100–1104 (1985b).

Cordier G., Schäfer H., Stelter M. "Perantimonidogallate und -indate: Zur Kenntnis von $Ca_5Ga_2Sb_6$, $Ca_5In_2Sb_6$ und $Sr_5In_2Sb_6$." *Z. Naturforsch.*, *40*B(1), 5–8 (1985c).

Cordier G., Schäfer H., Stelter M. "Grenzfälle des Zintl-Konzepts von $Ba_7Ga_4Sb_9$." *Z. Anorg. Allg. Chem.*, *534*(3), 137–142 (1986c).

Cordier G., Schäfer H., Stelter M. "$Sr_3In_2P_4$ und $Ca_3In_2As_4$, Zintlphasen mit Bänderanionen aus kanten- und eckenverknüpften InP_4- bzw. InA_4-Tetraedren." *Z. Naturforsch.*, *41*B(11), 1416–1419 (1986d).

Cordier G., Schäfer H., Stelter M. "Sr_3GaSb_3 und Sr_3InP_3, zwei neue Zintlphasen mit komplexen Anionen." *Z. Naturforsch.*, *42*B(10), 1268–1272 (1987).

Cordier G., Stelter M. "$Sr_5Al_2Sb_6$ und $Ba_5In_2Sb_6$: Zwei neue Zintlphasen mit unterschiedlichen Bänderanionen." *Z. Naturforsch.*, *43*B(4), 463–466 (1988).

Cordier G., Stelter M., Schäfer H. "Zintlephasen mit komplexen Anionen: Zur Kenntnis von $Sr_6Al_2Sb_6$." *J. Less Common Metals*, *98*(2), 285–290 (1984d).

Corfield P.W.R., Shore S.G. "Crystal and molecular structure of cyclotriborazane, the inorganic analog of cyclohexane." *J. Am. Chem. Soc.*, *95*(5), 1480–1487 (1973).

Cui Y., Ishihara S., Liu X.J., Ohnuma I., Kainuma R., Ohtani H., Ishida K. "Thermodynamic calculation of phase diagram in the Bi–In–Sb ternary system." *Mater. Trans.*, *43*(8), 1879–1886 (2002).

Dachille F., Glasser L.S.D. "High pressure forms of BPO_4 and $BAsO_4$; quartz analogues." *Acta Crystallogr.*, *12*(10), 820–821 (1959).

Dachille F., Roy R. "High-pressure region of the silica isotypes." *Z. Kristallogr.*, *111*(1–6), 451–461 (1959).

Dahle G.H., Schaeffer R. "Studies of boron–nitrogen compounds. III. Preparation and properties of hexahydroborazole, $B_3N_3H_{12}$." *J. Am. Chem. Soc.*, *83*(14), 3032–3034 (1961).

Dantas N.S., Almeida de J.S., Ahuja R., Persson C., Ferreira da Silva A. "Novel semiconducting materials for optoelectronic applications: $Al_{1-x}Tl_xN$ alloys." *Appl. Phys. Lett.*, *92*(12), 121914_1–121914_3 (2008).

Dashevskiy M.Ya., Kolobrodov L.N., Yakovleva V.G. "Investigation of the Ga–As–Te and Ga–As–Se phase diagrams in the region of the formation of the solid solution based on gallium arsenide" [in Russian]. In *Arsenid Galliya*. Vyp. 4. Tomsk: Publish. House of Tomsk Un-T, 129–132 (1974a).

Dashevskiy M.Ya., Kolobrodov L.N., Yakovleva V.G., Pinchuk P.A., Belousov V.I., Bykanova V.N., Stepanova T.N., Sukhova N.V. "Solid solutions based on InSb and GaAs in the In–Sb–Te and Ga–As–Te systems" [in Russian]. *Nauch. Tr. Mosk. In-T Stali Splavov*, (83), 80–96 (1974b).

Dashevskiy M.Ya., Kolobrodov L.N., Yakovleva V.G., Pinchuk P.A., Stepanova T.N. "Investigation of the solid solution region based on InSb in the In–Sb–Te system" [in Russian]. *Izv. AN SSSR Neorgan. Mater.*, *11*(9), 1564–1569 (1975).

Dashevskiy M.Ya., Kukuladze G.V., Chikhladze G.G. "Investigation of the solid solution region based on GaSb in the Ga–Sb–Te system" [in Russian]. *Izv. AN SSSR Neorgan. Mater.*, *16*(8), 1374–1379 (1980).

Deal M.D., Gasser R.A., Stevenson D.A. "Ternary phase equilibria in the Ga–As–Cr system and its relevance to diffusion and solubility studies." *J. Phys. Chem. Solids*, 46(7), 859–867 (1985).

Deal M.D., Stevenson D.A. "The solubility of chromium in gallium arsenide." *J. Electrochem. Soc.*, 131(10), 2343–2347 (1984).

Dehnicke K., Krüger N. "Dijodo- und Dibromometallazide X_2MN_3 von Aluminium und Gallium." *Z. Anorg. Allg. Chem.*, 444(1), 71–76 (1978).

Demchenko L.E., Chaus I.S., Kompanichenko N.M., Sukhenko V.D. "In_2S_3–As_2S_3 system" [in Russian]. *Zhurn. Neorgan. Khimii*, 20(2), 493–496 (1975).

Dement'ev Yu.S., Fedorov V.A., Bletskan N.I., Okunev Yu.A., Severtsev V.N. "Solubilities and distribution coefficients of transition metals impurities in gallium phosphide" [in Russian]. *Izv. AN SSSR Neorgan. Mater.*, 16(7), 1164–1167 (1980).

Deneke K., Rabenau A. "Über die Natur der Phase In_3SbTe_2 mit Kochsalzstruktur." *Z. Anorg. Allg. Chem.*, 333(4–6), 201–208 (1964).

Denisov V.M., Dubovikov G.S. "Kinetic of dissolution of iron group metals in the InSb, GaSb, Bi_2Te_3, Sb_2Te_3, SnTe melts" [in Russian]. *Izv. AN SSSR Neorgan. Mater.*, 23(11), 1792–1795 (1987).

Députier S., Barrier N., Guérin R., Guivarc'h A. "Solid state phase equilibria in the Fe–Ga–Sb system at 600°C." *J. Alloys Compd.*, 340(1–2), 132–140 (2002).

Députier S., Guérin R., Ballini Y., Guivarc'h A. "Solid state phase equilibria in the Er–Ga–As system." *J. Alloys Compd.*, 202(1–2), 95–100 (1993).

Députier S., Guérin R., Ballini Y., Guivarc'h A. "Solid state phase equilibria in the Ni–Al–As system." *J. Alloys Compd.*, 217(1), 13–21 (1995a).

Députier S., Guérin R., Guivarc'h A. "Solid state phase equilibria in the Fe–Ga–As system." *Eur. Phys. J. Appl. Phys.*, 2(2), 127–133 (1998).

Députier S., Guérin R., Lépine B., Guivarc'h A., Jézéquel G. "The ternary compound $Fe_3Ga_{2-x}As_x$: A promising candidate for epitaxial and thermodynamically stable contacts on GaAs." *J. Alloys Compd.*, 262–263, 416–422 (1997).

Députier S., Guivarc'h A., Caulet J., Minier M., Guérin R. "Étude des interdiffusions en phase solide dans le contact Er/GaAs." *J. Phys. III France*, 4(5), 867–880 (1994).

Députier S., Guivarc'h A., Caulet J., Poudoulec A., Guenais B., Monier M., Guérin R. "Etude des interdiffusions en phase solide dans le contact Ni/AlAs." *J. Phys. III France*, 5(4), 373–388 (1995b).

Deputier S., Pivan J.Y., Guerin R. "The crystal structures of two ternary $M_x(Ga,As)_y$ phases (M ≡ Rh, Pd) with Rh_5Ge_3- and $Cr_{12}P_7$-type derivative structures." *J. Less Common Metals*, 171(2), 357–368 (1991).

Derviševič I., Todorović A., Talijan N., Djokic J. "Experimental investigation and thermodynamic calculation of the Ga–Sb–Zn phase diagram." *J. Mater. Sci.*, 45(10), 2725–2731 (2010).

DeVries R.C., Fleischer J.F. "The system Li_3BN_2 at high pressures and temperatures." *Mater. Res. Bull.*, 4(7), 433–441 (1969).

DeVries R.C., Fleischer J.F. "Phase equilibria pertinent to the growth of cubic boron nitride." *J. Cryst. Growth*, 13–14, 88–92 (1972).

Diehl R., Carpentier C.-D. "The structural chemistry of indium phosphorus chalcogenides." *Acta Crystallogr.*, B34(4), 1097–1105 (1978).

Diehl R., Carpentier C.-D., Nitsche R. "The crystal structure of γ-In_2S_3 stabilized by As or Sb." *Acta Crystallogr.*, B32(4), 1257–1261 (1976).

Diehl R., Nitsche R. "Vapour and flux growth of γ-In_2S_3, a new modification of indium sesquisulphid." *J. Cryst. Growth*, 20(1), 38–46 (1973).

Diehl R., Nitsche R. "Vapour growth of three In_2S_3 modifications by iodine transport." *J. Cryst. Growth*, 28(3), 306–310 (1975).

Diehl R., Nitsche R., Ottemann J. "Eine neue arsenhaltige Indiumtrisulfid-Phase." *Naturwissenschaften*, 57(12), 670 (1970).

Ding X., Wang W., Han Q. "Thermodynamic calculation of Fe–P–j system melt" [in Chinese]. *Acta Metall. Sin. (China)*, 29(12), B527–B532 (1993).

Dinsdale A., Watson A., Kroupa A., Vřešt'ál J., Zemanová A., Vízdal J. "In–Sb–Sn system." *COST 531 Lead-Free Solders: Atlas of Phase Diagrams for Lead-Free Soldering*, 1, 265–271 (2008).

Djoudi L., Lachebi A., Merabet B., Abid H. "First-principles investigation of structural and electronic properties of the $B_xGa_{1-x}N$, $B_xAl_{1-x}N$, $Al_xGa_{1-x}N$ and $B_xAl_yGa_{1-x-y}N$ compounds." *Acta Phys. Polon. A*, 122(4), 748–753 (2012).

Dörner P., Gauckler L.J., Krieg H., Lukas H.L., Petzow G., Weiss J. "Calculation of heterogeneous phase equilibria in the SiAlON system." *J. Mater. Sci.*, 16(4), 935–943 (1981).

Dravid V.P., Sutliff J.A., Westwood A.D., Notis M.R., Lyman C.E. "On the space group of aluminium oxynitride spinel." *Phil. Mag.*, 61(3), 417–434 (1990).

Dridi Z., Bouhafs B., Ruterana P. "First-principles investigation of lattice constants and bowing parameters in wurtzite $Al_xGa_{1-x}N$, $In_xGa_{1-x}N$ and $In_xAl_{1-x}N$ alloys." *Semicond. Sci. Technol.*, 18(9), 850–856 (2003).

Dronskowski R. "$In_{2.24}(NCN)_3$ and $NaIn(NCN)_2$: synthesis and crystal structures of new main group metal cyanamides", *Z. Naturforsch.*, 50B(7), 1245–1251 (1995).

Dronkowski R., Simon A., Miller G.J. "Comment on the existence of phases $A_xMo_5As_4$ with A_x = Al_2, Ca_2, Cu_2 or Cu_4." *J. Chem. Soc. Dalton Trans.*, (9), 2483–2484 (1991).

Du Y., Wenzel R., Schmid-Fetzer R. "Thermodynamic analysis of reactions in the Al–N–Ta and Al–N–V systems." *CALPHAD*, 22(1), 43–58 (1998).

Dumitrescu L., Sundman B. "A thermodynamic reassessment of the Si–Al–O–N system." *J. Eur. Ceram. Soc.*, 15(3), 239–247 (1995).

Dumont H., Dazord J., Monteil Y., Alexandre F., Goldstein L. "Growth and characterization of high quality $B_xGa_{1-x}As$/GaAs(001) epilayers." *J. Cryst. Growth*, 248, 463–467 (2003).

Dumont H., Monteil Y. "Some aspects on thermodynamic properties, phase diagram and alloy formation in the ternary system BAs–GaAs. Part II. BGaAs alloy formation." *J. Cryst. Growth*, 290(2), 419–425 (2006).

Durand J., Lopez M., Cot L., Retout O., Saint-Gregoire P. "Evidence for the incommensurate phase in $AlPO_4$ near α-β transition: A differential scanning calorimetry study." *J. Phys. C Solid State Phys.*, 16(11), L311–L315 (1983).

Durlu N., Gruber U., Pietzka M.A., Schmidt H., Schuster J.C. "Phases and phase equilibria in the quaternary system Ti–Cu–Al–N at 850°C." *Z. Metallkde*, 88(5), 390–400 (1997).

Eilertsen J., Li J., Rouvimov S., Subramanian M.A. "Thermoelectric properties of indium-filled of $In_xRh_4Sb_{12}$ skutterudites." *J. Alloys Compd.*, 509(21), 6289–6295 (2011).

Eilertsen J., Rouvimov S., Plachinda P., Hendricks III T.J., Subramanian M. "Thermoelectric and structural characterization of $In_xRh_4Sb_{12}$ (0 < x < 0.2) skutterudites." *Microsc. Microanal.*, 16(Suppl. S2), 1504–1505 (2010).

El-Boragy M., Schubert K. "Über eine verzerrte dichteste Kugelpackung mit Leerstellen." *Z. Metallkde*, 61(8), 579–584 (1970).

El-Boragy M., Schubert K. "Kristallstrukturen einiger ternärer Phasen in T–B–B′-Systemen." *Z. Metallkde*, 62(9), 667–675 (1971).

El-Boragy M., Schubert K. "On the mixtures $PdZn_MGe_N$, $PdAl_MSi_N$, $PdGa_MSi_N$, $PdAl_MGe_N$ and $PdGa_MAs_N$." *Z. Metallkde*, 72(4), 279–282 (1981).

El Haj Hassan F., Breidi A., Ghemid S., Amrani B., Meradji H., Pagès O. "First-principles study of the ternary semiconductor alloys (Ga,Al)(As,Sb)." *J. Alloys Compd.*, 499(1), 80–89 (2010).

Ellner M., El-Boragy M. "Einfluss der Valenzelektronenkonzentration auf die Bildung der Phasen vom Ni_2In-Strukturtyp am Beispiel der ternären Verbindungen $Mn_2Ga_{0.5}As_{0.5}(m)$, Fe_3GaAs und $Co_2Ga_{0.5}As_{0.5}$." *J. Appl. Crystallogr.*, 19(2), 80–85 (1986).

Ellner M., El-Boragy M. "Über die Stabilität einiger Isotypen des $Pd_{25}Ge_9$." *J. Less Common Metals*, 161(1), 147–158 (1990).

Ellner M., El-Boragy M., Predel B. "Zur Struktur der ternären Phosphide $Pd_2Zn_{0.3}P_{0.7}$, $Pd_2Al_{0.5}P_{0.5}$ und $Pd_2Ga_{0.4}P_{0.6}$." *J. Less Common Metals*, 114(1), 175–182 (1985).

El-Masry N.A., Piner E.L., Liu S.X., Bedair S.M. "Phase separation in InGaN grown by metalorganic chemical vapor deposition." *Appl. Phys. Lett.*, 72(1), 40–42 (1998).

Elyutin V.P., Polushin N.I., Burdina K.P., Polyakov V.P., Kalashnikov Ya.A., Semenenko K.N., Pavlov Yu.A. "About interaction in the Mg_3N_2–BN system" [in Russian]. *Dokl. AN SSSR*, 259(1), 112–116 (1981).

Emi T., Shimoji M. "Thermodynamic properties of indium–tin–antimony liquid alloys." *Ber. Bunsenges. Phys. Chem.*, 71(4), 367–371 (1967).

Endo T., Fukunaga O., Iwata M. "Precipitation mechanism of BN in the ternary system of B–Mg–N." *J. Mater. Sci.*, 14(7), 1676–1680 (1979).

Endo T., Fukunga O., Iwata M. "The synthesis of cBN using $Ca_3B_2N_4$." *J. Mater. Sci.*, 16(8), 2227–2232 (1981).

Ensslin F., Dreyer H., Lessman O. "Zur Chemie des Indiums. XI Verbindungen des lndiums mit den Sauerstoffsäuren des Phosphors." *Z. Anorg. Allg. Chem.*, 254(5–6), 315–318 (1947).

Evans D.B., Pehlke R.D. "The boron–nitrogen equilibrium in liquid iron." *Trans. Metallurg. Soc. AIME*, 230(12) 1657–1662 (1964).

Evers J., Münsterkötter M., Oehlinger G., Polborn K., Sendlinger B. "Natriumdinitridoborate mit dem linear gebauten, symmetrischen $BN_2{}^{3-}$-Anionen." *J. Less Common Metals*, 162(1), L17–L22 (1990).

Farber L., Barsoum M.W. "Isothermal sections in the Cr–Ga–N system in the 650–1000°C temperature." *J. Mater. Res.*, 14(6), 2560–2566 (1999).

Farber L., Levin I., Barsoum M.W., El-Raghy T., Tzenov T. "High-resolution transmission electron microscopy of some $Ti_{n+1}AX_n$ compounds (n = 1, 2; A = Al or Si; X = C or N)." *J. Appl. Phys.*, 86(5), 2540–2543 (1999).

Fedorov P.I., Smarina E.I., Roshchina A.V. "Sections of the Ga–Sb–Se system" [in Russian]. *Izv. AN SSSR Neorgan. Mater.*, 6(10), 1877–1878 (1970).

Fedorov P.I., Smarina E.I., Roshchina A.V., Sergeeva L.V. "Immiscibility in the Ga–Sb–Se system" [in Russian]. *Izv. AN SSSR Neorgan. Mater.*, 8(2), 372–373 (1972).

Fedotova J.A., Stanek J., Shishonok N.A., Uglov V.V. "Mössbauer study of iron impurities in cubic boron nitride." *J. Alloys Compd.*, 352(1–2), 296–303 (2003).

Feng Q., Wu X., Xu Z., Yang G., Zeng L., He W. "Investigation on phase equilibria of the Al–Sb–Tb ternary system at 773 K utilizing Rietveld analysis." *J. Phase Equilibria Diffus.*, 37(5), 564–573 (2016).

Ferguson M.J., Ellenwood R.E., Mar A. "A new rare-earth indium antimonide, $(RE)In_{1-x}Sb_2$ (RE = La-Nd), featuring in zigzag chains and Sb square nets." *Inorg. Chem.*, 38(20), 4503–4509 (1999).

Ferhat M., Bechstedt F. "First-principles calculations of gap bowing in $In_xGa_{1-x}N$ and $In_xAl_{1-x}N$ alloys: Relation to structural and thermodynamic properties." *Phys. Rev. B*, 65(7), 075213_1–075213_7 (2002).

Filonenko V.P., Khabashesku V.N., Davydov V.A., Zibrov I.P., Agafonov V.N. "Synthesis of a new cubic phase in the B–C–N system." *Inorg. Mater.*, 44(4), 395–400 (2008).

Filonenko V.P., Zibrov I.P., Sidorov V.A., Davydov V.A., Trenikhin M.V. "Heterographene BCN phase prepared at high pressures and temperatures: Formation kinetics, structure, and properties." *Inorg. Mater.*, 50(4), 349–357 (2014).

Finkel P., Barsoum M.W., El-Raghy T. "Low temperature dependencies of the elastic properties of Ti_4AlN_3, $Ti_3Al_{1.1}C_{1.8}$, and Ti_3SiC_2." *J. Appl. Phys.*, 87(4), 1701–1703 (2000).

Fistul' V.I., Ufimtsev V.B., Arbenina V.V., Kudasova M.M., Ukharskaya T.A. "Solid solutions in the Cd–Ga–Sb system" [in Russian]. *Izv. AN SSSR Neorgan. Mater.*, *18*(2), 197–202 (1982).

Flandorfer H., Richter K.W., Giester G., Ipser H. "The ternary compounds $Pd_{13}In_{5.25}Sb_{3.75}$ and $PdIn_{1.26}Sb_{0.74}$: Crystal structure and electronic structure calculations." *J. Solid State Chem.*, *164*(1), 110–118 (2002).

Flörke O.W. "Kristallisation und Polymorphie von $AlPO_4$ und $AlPO_4$–SiO_2-Mischkristallen." *Z. Kristallogr.*, *125*(1–6), 134–146 (1967).

Folberth O.G. "Mischkristallbildung bei $A^{III}B^V$-Verbindungen." *Z. Naturforsch.*, *10A*(6), 502–503 (1955).

Fornasini M.L., Manfrinetti P. "Crystal structure of ytterbium aluminium antimonide, $Yb_2Al_5Sb_6$." *Z. Kristallogr. New Cryst. Struct.*, *224*(3), 345–346 (2009).

Foster L.M., Scardefield J.E. "The solidus boundary in the GaP–InP pseudobinary system." *J. Electrochem. Soc.*, *117*(4), 534–536 (1970).

Foster L.M., Scardefield J.E. "The solidus boundary in the InAs–AlAs pseudobinary system." *J. Electrochem. Soc.*, *118*(3), 495–496 (1971).

Foster L.M., Scardefield J.E., Woods J.F. "Thermodynamic analysis of the III-V alloy semiconductor phase diagrams. III. The solidus boundary in the $Ga_{1-x}Al_xAs$ pseudobinary system." *J. Electrochem. Soc.*, *119*(6), 765–766 (1972a).

Foster L.M., Scardefield J.E., Woods J.F. "The solidus boundary in the GaAs–GaP pseudobinary phase diagram." *J. Electrochem. Soc.*, *119*(10), 1426–1427 (1972b).

Foster L.M., Woods J.F. "Thermodynamic analysis of the III-V alloy semiconductor phase diagrams. I. InSb–GaSb, InAs–GaAs, and InP–GaP." *J. Electrochem. Soc.*, *118*(7), 1175–1183 (1971).

Foster L.M., Woods J.F. "Thermodynamic analysis of the III-V alloy semiconductor phase diagrams. II. The GaSb–GaAs system." *J. Electrochem. Soc.*, *119*(4), 504–507 (1972).

Fountain R.W., Chipman J. "Solubility and precipitation of boron nitride in iron–boron alloys." *Trans. Metallurg. Soc. AIME*, *224*(6) 599–606 (1962).

Foxon C.T., Cheng T.S., Novikov S.V., Lacklison D.E., Jenkins L.C., Johnston D., Orton J.W., Hooper S.E., Baba-Ali N., Tansley T.L., Tret'yakov V.V. "The growth and properties of group III nitrides." *J. Cryst. Growth*, *150*, Pt. 2, 892–896 (1995).

Francoeur S., Seong M.-J., Mascarenhas A., Tixier S., Adamcyk M., Tiedje T. "Band gap of $GaAs_{1-x}Bi_x$, $0 < x < 3.6\%$." *Appl. Phys. Lett.*, *82*(22), 3874–3876 (2003).

Fujimoto M. "Solubility and distribution coefficient of cadmium in gallium arsenide." *Rev. Phys. Chem. Jpn.*, *40*(1), 34–38 (1970).

Fukuhara M. "Phase relations in the Si_3N_4-rich portion of the Si_3N_4–AlN–Al_2O_3–Y_2O_3 system." *J. Am. Ceram. Soc.*, *71*(7), C-359–C-361 (1988).

Fuller C.S., Whelan J.M. "Diffusion, solubility, and electrical behavior of copper in gallium arsenide." *J. Phys. Chem. Solids*, *6*(2–3), 173–177 (1958).

Furukawa Y., Thurmond C.D. "Effect of arsenic pressure on solubility of copper in GaAs." *J. Phys. Chem. Solids*, *26*(9), 1535–1542 (1965).

Gago R., Jiménez I., Agulló-Rueda F., Albella J.M., Czigány Zs., Hultman L. "Transition from amorphous boron carbide to hexagonal boron carbon nitride thin films induced by nitrogen ion assistance." *J. Appl. Phys.*, *92*(9), 5177–5182 (2002).

Galagovets I.V., Peresh E.Yu., Lazarev V.B., Barchiy I.E., D'ordyay V.S. "Phase equilibria and some properties of the compounds in the $In_2(Ga_2)S_3$–P_2S_5 systems" [in Russian]. *Neorgan. Mater.*, *31*(2), 202–205 (1995).

Gallagher M.J., Gerards J.F. "Berlinite from Rwanda." *Mineral. Mag.*, *33*(262), 613–615 (1963).

Gamarnik M.Y., Barsoum M.W., El-Raghy T. "Improved x-ray powder diffraction data for Ti_2AlN." *Powder Diffr.*, *15*(4), 241–242 (2000).

Gao J., Li C., Wang N., Du Z. "Thermodynamic analysis of the Ti–Al–N system." *J. Univ. Sci. Technol. Beijing*, *15*(4), 420–424 (2008).

Gascoin F., Sevov S.C. "Synthesis and characterization of $Cs_5In_3As_4$ with a structure of two coexisting 'polymorphic' forms of different dimensionality." *Inorg. Chem.*, *40*(24), 6254–6257 (2001).

Gather B., Schröter P., Blachnik R. "Mischungsenthalpien in ternären Systemen. IV. Das System Indium–Antimon–Zinn." *Z. Metallkde.*, *76*(8), 523–527 (1985).

Gather B., Schröter P., Blachnik R. "Mischungsenthalpien der ternären Systeme Ag–In–Sn, Ag–Sn–Sb, Ag–In–Sb und In–Pb–Sb." *Z. Metallkde.*, *78*(4), 280–285 (1987).

Gaudé J. "$LnBN_2$ (Ln = Nd, Sm): Premier exemple de nitrure double associant le bore et un élément lanthanides." *C. R. Acad. Sci. Sér. II*, *297*(9), 717–719 (1983).

Gaudé J., L'Haridon P., Guyader J., Lang J. "Étude structurale d'un nouveau nitrure double $Ce_{15}B_8N_{25}$." *J. Solid State Chem.*, *59*(2), 143–148 (1985).

Gebesh V.Yu., Potoriy M.V., Voroshilov Yu.V. "Phase equilibria in the In–P–S system" [in Russian]. *Ukr. Khim. Zhurn.*, *57*(8), 803–806 (1991).

Geis D.R., Peretti E.A. "Constitution of the lead-rich and antimony-rich regions of the indium–antimony–lead ternary system." *J. Less Common Metals*, *4*(6), 523–532 (1962a).

Geis D.R., Peretti E.A. "The system indium antimonide-lead." *J. Chem. Eng. Data*, *7*(3), 392 (1962b).

Geis D.R., Peretti E.A. "The ternary system indium–lead–indium antimonide." *J. Chem. Eng. Data*, *8*(3), 470–472 (1963).

Geiser P., Jun J., Kazakov S.M., Wägli P., Karpinski J., Batlogg B., Klemm L. "$Al_xGa_{1-x}N$ bulk single crystals." *Appl. Phys. Lett.*, *86*(8), 081908_1–081908_3 (2005).

Geisz J.F., Friedman D.J., Kurtz S., Olson J.M., Swartzlander A.B., Reedy R.C., Norman A.G. "Epitaxial growth of BGaAs and BGaInAs by MOCVD." *J. Cryst. Growth*, *225*(2–4), 372–376 (2001a).

Geisz J.F., Friedman D.J., Kurtz S., Reedy R.C., Barber G. "Alternative boron precursors for BGaAs epitaxy." *J. Electron. Mater.*, 30(11), 1387–1391 (2001b).

Geisz J.F., Friedman D.J., Olson J.M., Kurtz S.R., Reedy R.C., Swartzlander A.B., Keyes B.M., Norman A.G. "BGaInAs alloys lattice matched to GaAs." *Appl. Phys. Lett.*, 76(11), 1443–1445 (2000).

Gendry M., Durand J., Villeneuve J.M., Cot L. "Étude du diagramme des phases condensées du système In–P–S." *Thin Solid Films*, 139(1), 53–59 (1986).

Gerdes F., Predel B. "Thermodynamische Eigenschaften flüssiger Legierungen der Systeme InSb–Bi, GaSb–Bi und AlSb–Bi." *J. Less Common Metals*, 65(1), 41–49 (1979a).

Gerdes F., Predel B. "Thermodynamische Untersuchung der Systeme GaSb–InSb, AlSb–GaSb und AlSb–InSb." *J. Less Common Metals*, 64(2), 285–294 (1979b).

Gerdes F., Predel B. "Phasengleichgewichte der pseudobinären Systeme GaSb–Sn, InSb–Sn und AlSb–Sn." *J. Less Common Metals*, 79(2), 281–288 (1981a).

Gerdes F., Predel B. "Thermodynamische Eigenschaften flüssiger Legierungen der Systeme AlSb–Pb, GaSb–Pb und InSb–Pb." *J. Less Common Metals*, 79(2), 315–321 (1981b).

Gerdes F., Predel B. "Thermodynamische Eigenschaften flüssiger Legierungen der Systeme InSb–Sn, GaSb–Sn und AlSb–Sn." *J. Less Common Metals*, 78(1), 19–28 (1981c).

Germogenov V.P., Otman Ya.I., Chaldyshev V.V., Shmartsev Yu.V. "Band gap of the GaSb$_{1-x}$Bi$_x$ solid solution" [in Russian]. *Fiz. Tekhn. Poluprovodn.*, 23(8), 1517–1518 (1989).

Ghasemi M., Johansson J. "Thermodynamic assessment of the As–Zn and As–Ga–Zn systems." *J. Alloys Compd.*, 638, 95–102 (2015).

Gierlotka W. "Thermodynamic description of the binary Au–Sb and ternary Au–In–Sb systems." *J. Alloys Compd.*, 579, 533–539 (2013).

Gillet P., Badro J., Varrel B., McMillan P.F. "High-pressure behavior in α-AlPO$_4$: Amorphization and the memory-glass effect." *Phys. Rev. B*, 51(17), 11262–11269 (1995).

Gimel'farb F.A., Layner B.D., Mil'vidskiy M.G., Pelevin O.V., Fistul' V.I. "GaAs–Te polythermal section in the Ga–As–Te system" [in Russian]. *Izv. AN SSSR Neorgan. Mater.*, 8(6), 1055–1058 (1972).

Girard C., Miane J.M., Riou J., Baret R., Bros J.P. "Enthalpy of formation of Al–Sb and Al–Ga–Sb liquid alloys." *J. Less Common Metals*, 128(1–2), 101–115 (1987).

Girard C., Miane J.M., Riou J., Baret R., Bros J.P. "Potentiometrie investigations of the liquid Al + Ga + Sb alloy between 623 K and 1300 K." *Ber. Bunsenges. Phys. Chem.*, 92(2), 132–139 (1988).

Gladkaya I.S., Kremkova G.N., Bendeliani N.A., Lorenz H., Kuehne U. "The binary system of BN–Mg$_3$N$_2$ under high pressures and temperatures." *J. Mater. Sci.*, 29(24), 6616–6619 (1994).

Glazov V.M. "On the interaction of germanium with gallium antimonide in the melt" [in Russian]. *Zhurn. Neorgan. Khimii*, 6(4), 933–936 (1961).

Glazov V.M., Evdokimov A.V., Pavlova L.M. "About diffusionless crystallization in the Ge–AIIIBV systems" [in Russian]. *Dokl. AN SSSR*, 232(2), 371–374 (1977a).

Glazov V.M., Kiselev A.N., Shvedkov E.I. "Solubility and donor-acceptor interaction in InAs, doped with S, Se, and Zn" [in Russian]. *Izv. AN SSSR Neorgan. Mater.*, 15(3), 390–394 (1979).

Glazov V.M., Krestovnikov A.N., Malyutina G.L. "About interaction of germanium with indium antimonide in the liquid" [in Russian]. *Izv. AN SSSR Neorgan. Mater.*, 2(9), 1692–1694 (1966).

Glazov V.M., Lyu Chzhen'-yuan'. "Investigation of the phase equilibria in the Ga–AlSb and Si–AlSb systems" [in Russian]. *Zhurn. Neorgan. Khimii*, 7(3), 582–589 (1962).

Glazov V.M., Malyutina G.L. "Investigation of germanium interaction with gallium and indium arsenides" [in Russian]. *Zhurn. Neorgan. Khimii*, 8(8), 1921–1927 (1963a).

Glazov V.M., Malyutina G.L. "Solubility of gallium and indium phosphide in germanium" [in Russian]. *Zhurn. Neorgan. Khimii*, 8(10), 2372–2375 (1963b).

Glazov V.M., Pavlova L.M. "Phase diagram GaSb–Ag" [in Russian]. *Izv. AN SSSR Neorgan. Mater.*, 13(1), 15–17 (1977).

Glazov V.M., Pavlova L.M., Perederiy L.I. "Investigation of the Ge interaction with InP and GaP" [in Russian]. *Izv. AN SSSR Neorgan. Mater.*, 13(2), 209–213 (1977b).

Glazov V.M., Pavlova L.M., Perederiy L.I. "Calculation of the liquidus in the Ge–In(Ga)P quasi-binary systems based on the components atomic volumes and electronegativity" [in Russian]. *Izv. AN SSSR Neorgan. Mater.*, 16(4), 595–598 (1980).

Glazov V.M., Pavlova L.M., Perederiy L.I. "Experimental and thermodynamic investigation of retrograde solidus in the InP–Ge quasi-binary system" [in Russian]. *Zhurn. Fiz. Khimii*, 57(4), 880–884 (1983).

Glazov V.M., Pavlova L.M., Sokurenko S.N. "Obtaining of metastable Ge–GaP and Ge–InP solid solutions at ultrafast cooling" [in Russian]. *Dokl. AN SSSR*, 279(1), 125–129 (1984).

Glazov V.M., Petrov D.A., Chizhevskaya S.N. "On the joint solubility of the elements of III and V groups in germanium" [in Russian]. *Izv. AN SSSR OTN Metallurgiya Toplivo*, (4), 153–155 (1959).

Glazov V.M., Poyarkov K.B. "Crystallization of InSb–InAs alloys at very high cooling rates" [in Russian]. *Dokl. RAN*, 364(5), 630–635 (1999).

Glazov V.M., Poyarkov K.B. "Crystallization of InSb–InAs alloys at very high cooling rates (10^6–10^8 K/s)" [in Russian]. *Neorgan. Mater.*, 36(10), 1181–1187 (2000).

Glazov V.M., Tszyan' T. Lyu C. "Investigation of separated and joint solubility of aluminium and antimony in germanium" [in Russian]. *Zhurn. Neorgan. Khimii*, 7(3), 576–581 (1962).

Goel N.C., Cahoon J.R. "The Al–Li–X systems (X = Ag, As, P, B, Cd, Fe, Ga, H, In, N, Pb, S, Sb and Sb." *Bull. Alloy Phase Diagr.*, 10(5), 546–548 (1989).

Goforth A.M., Hope H., Condron C.L., Kauzlarich S.M., Jensen N., Klavins P., MaQuilon S., Fisk Z. "Magnetism and negative magnetoresistance of two magnetically ordering, rare-earth-containing Zintl phases with a new structure type: $EuGa_2Pn_2$ (Pn = P, As)." *Chem. Mater.*, 21(19), 4480–4489 (2009).

Goforth A.M., Klavins K., Fettinger J.C., Kauzlarich S.M. "Magnetic properties and negative colossal magnetoresistance of the rare earth Zintl phase $EuIn_2As_2$." *Inorg. Chem.*, 47(23), 11048–11056 (2008).

Goggin P.L., McColm I.J., Shore R. "Indium tricyanide and indium trithiocyanate." *J. Chem. Soc. A Inorg. Phys. Theor.*, 1314–1317 (1966).

Goiffon A., Jumas J.-C., Maurin M., Philippot E. "Etude comparée à diverses températures (173, 293 et 373°K) des structures de type quartz α des phases $M^{III}X^VO_4$ (M^{III} = Al, Ga et X^V = P, As)." *J. Solid State Chem.*, 61(3), 384–396 (1986).

Goliusov V.A., Voronin V.A., Chechmarev S.K., Plakhotnaya L.S. "Investigation of the phase equilibria in the In–Ga–P system" [in Russian]. Deposited in ONIITEKhim, N. 237 khp-D81 (1980).

Golubev L.V., Khachaturyan O.A., Shmartsev Yu.V. "Solubility of GaSb in bismuth and based on it eutectic alloys" [in Russian]. *Elektronnaya Tekhnika. Ser. 6 Materialy*, (5), 156–158 (1972).

Golubev L.V., Novikov S.V., Chaldyshev V.V., Yastrebov S.G. "Analysis of the In–Sb–P phase diagram" [in Russian]. *Izv. AN SSSR Neorgan. Mater.*, 26(6), 1321–1323 (1990).

Gomidželović L., Živković D., Manasijević D., Minić D. "Comparative analysis of thermodynamic characteristics for ternary Me–In–Sb (Me = Sn, Ga) systems." *J. Univ. Chem. Technol. Metall.*, 45(3), 327–334 (2010).

Gomidzelovic L., Zivkovic D., Mihajlovic I., Trujic V. "Predicting of thermodynamic properties of ternary Au–In–Sb system." *Arch. Metall. Mater.*, 51(3), 355–363 (2006).

Gomidzelović L., Živković D., Štrbac N., Živković Ž. "Calculation of mixing enthalpies for ternary Au–In–Sb alloys." *J. Univ. Chem. Technol. Metall.*, 42(2), 207–210 (2007a).

Gomidzelović L., Živković D., Talijan N., Manasiević D., Ćosović V., Grujić A. "Phase equilibria investigation and characterization of the Au–In–Sb system." *Metalurgija J. Metallurg.*, 13(4), 269–276 (2007b).

Gomidzelović L., Živković D., Talijan N., Manasiević D., Ćosović V., Grujić A. "Phase equilibria investigation and characterization of the Au–In–Sb system." *J. Optoelectron. Adv. Mater.*, 10(2), 455–460 (2008).

Goreshnik E., Mazej Z. "Single crystal structures of $InPF_6$, $InAsF_6$, $TlPF_6$ and $TlAsF_6$." *Solid State Sci.*, 10(3), 303–306 (2008).

Gorshkov I.E., Goryunova N.A. "GaSb–InSb quasi-binary section of the gallium–indium–antimony ternary system" [in Russian]. *Zhurn. Neorgan. Khimii*, 3(3), 668–672 (1958).

Goryacheva V.I., Pashin S.F., Geyderikh V.A. "Thermodynamic properties and liquidus of the In–Bi–Sb system" [in Russain]. *Vestn. MGU. Khimiya*, 26(2), 142–150 (1985).

Goryunova N.A. "Substitutional solid solutions in the compounds with the structure of sphalerite" [in Russian]. In *Voprosy Teorii i Issled. Poluprovodn. i Protsesov Poluprovodn. Metalutghii*. Moscow Publish. House of AN SSSR, 29–37 (1955).

Goryunova N.A., Baranov B.V., Valov Yu.A., Prochukhan V.D. "About solubility of boron phosphide in various melts" [in Russian]. In *Fizika*. Leningrad: Publish. House of Lening. Inst. Civil Eng., 15–19 (1964).

Goryunova N.A., Burdiyan I.I. "Solid solutions in the InAs–In_2Te_3 system" [in Russian]. *Dokl. AN SSSR*, 121(5), 848–849 (1958).

Goryunova N.A., Fedorova N.N. "On the isomorphism of III-V compounds" [in Russian]. *Zhurn. Tekhnich. Fiziki*, 25(7), 1339–1341 (1955).

Goryunova N.A., Grigor'eva V.S. "On the gallium arsenoselenide" [in Russian]. *Zhurn. Tekhnich. Fiziki*, 26(10), 2157–2161 (1956).

Goryunova N.A., Radautsan S.I. "Solid solutions in the AlSb–GaSb system" [in Russian]. *Dokl. AN SSSR*, 120(5), 1031–1034 (1958a).

Goryunova N.A., Radautsan S.I. "Solid solutions in the InAs–In_2Se_3 system" [in Russian]. *Zhurn. Tekhnich. Fiziki*, 28(9), 1917–1921 (1958b).

Goryunova N.A., Radautsan S.I. "On the defect diamond-like semiconductors" [in Russian]. In *Issled. Po Poluprovodn*. Kishinev: Kartya Moldovenyake Publish., 3–43 (1964).

Goryunova N.A., Radautsan S.I., Deryabina V.I. "Homogenization of the InAs–In_2Se_3 alloys by the annealing under pressure" [in Russian]. *Fiz. Tv. Tela*, 1(3), 512–514 (1959a).

Goryunova N.A., Radautsan S.I., Kiosse G.A. "On the new semiconductor compound in the In–Sb–Te system" [in Russian]. *Fiz. Tv. Tela*, 1(12), 1858–1860 (1959b).

Goryunova N.A., Sokolova V.I. "About complex phosphide" [in Russian]. *Izv. Mold. Fil. AN SSSR*, 3(69), 31–35 (1960a).

Goryunova N.A., Sokolova V.I. "Solid solutions in the InP–GaP system" [in Russian]. *Izv. Mold. Fil. AN SSSR*, 3(69), 97–98 (1960b).

Goryunova N.A., Takhtareva N.K. "Formation of the solid solutions between indium antimonide and indium arsenide" [in Russian]. *Izv. AN MSSR*, 10(88), 89–90 (1961).

Gottschalch V., Leibiger G., Benndorf G. "MOVPE growth of $B_xGa_{1-x}As$, $B_xGa_{1-x-y}In_yAs$, and $B_xAl_{1-x}As$ alloys on (001) GaAs." *J. Cryst. Growth*, 248, 468–473 (2003).

Goubeau J., Anselment W. "Über ternäre Metal–Bornitride." *Z. Anorg. Allg. Chem.*, 310(4–6), 248–260 (1961).

Gourdon O., Boucher F., Gareh J., Evain M., O'Connor C., Jin-Seung J. "Rubidium indium antimonide, $Rb_2In_2Sb_3$." *Acta Crystallogr.*, C52(12), 2963–2964 (1996).

Goursat P., Goeuriot P., Billy M. "Contribution a l etude du systeme Al/O/N. I. Reactivite de l oxynitrure d aluminium γ." *Mater. Chem.*, *1*(2), 131–149 (1976).

Graetsch H. "Two forms of aluminium phosphate tridymite from x-ray powder data." *Acta Crystallogr.*, *C56*(4), 401–403 (2000).

Graetsch H.A. "Hexagonal high-temperature form of aluminium phosphate tridymite from x-ray powder data." *Acta Crystallogr.*, *C57*(6), 665–667 (2001).

Graetsch H.A. "Monoclinic AlPO₄ tridymite at 473 and 463 K from x-ray powder data." *Acta Crystallogr.*, *C58*(1), i18–i20 (2002).

Graetsch H.A. "Thermal expansion and thermally induced variations of the crystal structure of AlPO₄ low cristobalite." *Neues Jb. Miner. Mh.*, (7), 289–301 (2003).

Gratton M.F., Goodchild R.G., Juravel L.Y., Woolley J.C. "Miscibility gap in the GaAs$_y$Sb$_{1-y}$ system." *J. Electron. Mater.*, *8*(1), 25–29 (1979).

Gratton M.F., Woolley J.C. "Phase diagram and lattice parameter data for the GaAs$_y$Sb$_{1-y}$ system." *J. Electron. Mater.*, *2*(3), 455–464 (1973).

Gratton M.F., Woolley J.C. "Solidus isotherms and isoconcentration lines in the system Ga–In–Sb." *J. Electrochem. Soc.*, *125*(4), 657–661 (1978).

Gratton M.F., Woolley J.C. "Investigation of two- and three-phase fields in the Ga–As–Sb system." *J. Electrochem. Soc.*, *127*(1), 55–62 (1980).

Greaves C., Devlin E.J., Smith N.A., Harris I.R., Cockayne B., MacEwan W.R. "Structural identification of the new magnetic phases Fe$_3$Ga$_{2-x}$As$_x$." *J. Less Common Metals*, *157*(2), 315–325 (1990).

Grekov F.F., Demidov D.M., Zykov A.M. "Obtaining of gallium oxynitride from the vapor phase" [in Russian]. *Zhurn. Prikl. Khimii*, *52*(6), 1394–1396 (1979).

Grigor'ev O.N., Lyashenko V.I., Timofeeva I.I., Rogozinskaya A.A., Tomila T.V., Dubovik T.V., Panashenko V.M. "Study of the synthesis of the ternary compound B–N–C." *Powder Metall. Met. Ceram.*, *44*(9–10), 415–419 (2005).

Grigor'eva V.S., Kradinova L.V., Prochukhan V.D. "Solid solutions in the BP–Si system" [in Russian]. *Tr. Kishinevsk. Politekhn. In-T*, (12), 67–69 (1968).

Grinberg Ya.Kh., Luzhnaya N.P., Medvedeva Z.S. "Equlibrium investigation in the boron phosphide–iodine system" [in Russian]. *Izv. AN SSSR Neorgan. Mater.*, *2*(12), 2130–2133 (1966).

Gröbner J., Wenzel R., Fischer G.G., Schmid-Fetzer R. "Thermodynamic calculation of the binary systems M–Ga and investigation of ternary M–Ga–N phase equilibria (M = Ni, Co, Pd, Cr)." *J. Phase Equilibria*, *20*(6), 615–625 (1999).

Groenert M.E., Averbeck R., Hösler W., Schuster M., Riechert H. "Optimized growth of BGaAs by molecular beam epitaxy." *J. Cryst. Growth*, *264*(1–3), 123–127 (2004).

Gromova T.I., Yevgen'ev S.B., Borisov S.R. "Phase equilibria in the In–Sb–As system" [in Russian]. *Izv. AN SSSR Neorgan. Mater.*, *19*(3), 341–343 (1983).

Gromyko S.N., Zelyavskiy V.B., Kurdyumov A.V., Pilyankevich A.N., Dobryanskiy V.M., Kosyarev O.M. "X-ray investigation of the BN–Si and BN–Si$_3$N$_4$ solid solutions" [in Russian]. *Poroshk. Metallurgiya*, (7), 84–88 (1990).

Grytsiv A., Rogl P., Michor H., Bauer E., Giester G. "In$_y$Co$_4$Sb$_{12}$ skutterudite: Phase equilibria and crystal structure." *J. Electron. Mater.*, *42*(10), 2940–2952 (2013).

Gu Y., Wang K., Zhou H., Li Y., Cao C., Zhang L., Zhang Y., Gong Q., Wang S. "Structural and optical characterizations of InPBi thin films grown by molecular beam epitaxy." *Nanoscale Res. Lett.*, *9*(1), 24 (5 pp.) (2014).

Gubskaya G.F., Van Bin-nan', Luzhnaya N.P., Kudryavtsev D.L. "About interaction in the Ag–III–V ternary systems" [in Russian]. *Izv. AN SSSR Neorgan. Mater.*, *1*(2), 188–192 (1965).

Guechi N., Bouhemadou A., Guechi A., Reffas M., Louail L., Bourzami A., Chegaar M., Bin-Omran S. "First-principles prediction of the structural, elastic, electronic and optical properties of the Zintl phases MIn$_2$P$_2$ (M = Ca, Sr)." *J. Alloys Compd.*, *577*, 587–599 (2013).

Guérin R., Guivarc'h A. "Metallurgical study of Ni/GaAs contacts. I. Experimental determination of the solid portion of the Ni–Ga–As ternary-phase diagram." *J. Appl. Phys.*, *66*(5), 2122–2128 (1989).

Guérin R., Guivarc'h A., Ballini Y., Secoué M. "Métallurgie du système Rh–Ga–As: Détermination du diagramme ternaire et interdiffusion en phase solide dans le contact Rh/GaAs." *Rev. Phys. Appl.*, *25*(5), 411–422 (1990).

Guertler W., Bergmann A. "Studien am Dreistoffsystem Aluminium–Antimon–Magnesium." *Z. Metallkde.*, *25*(4), 111–116 (1933).

Guliev A.N. "On the formation of the Mn$_x$Ga$_{1-x}$Sb and Sm$_x$Ga$_{1-x}$Sb solid solutions" [in Russian]. In *Materialy Nauchn. Konf. Aspirantov*. Baku: Publish. House of Azerb. Gos. Ped. In-T, 218–222 (1973).

Guliev T.N. "Growth of the single crystals of the semiconductor compounds A^{II}B^{VI}, A^{III}$_2$B^{VI}, A^{III}B^{VI}$_3$, A^{II}A^{III}$_2$B^{VI}$_4$, A^{III}A^{V}B^{VI}$_3$, A^{II}A^V$_2$B^{VI}$_4$ (where A^{II}–Ca, Sr, Ba; A^{III}–Ga, In; A^V–As, Sb, Bi; B^{VI}–S, Se, Te) from the gas phase and mechanism of their transport" [in Russian]. *Izv. Vuzov. Khim. Khim. Tekhnol.*, *40*(5), 3–12 (1997).

Guliev T.N., Magerramov E.V., Rustamov P.G. "Investigation of the In$_2$Se$_3$–Sb$_2$Se$_3$ system" [in Russian]. *Izv. AN SSSR Neorgan. Mater.*, *13*(4), 627–629 (1977a).

Guliev T.N., Rustamov P.G., Sinechek V., Magerramov E.V. "Synthesis, single crystals preparation and properties investigation of the InSbS$_3$ and InSbSe$_3$ compounds" [in Russian]. *Izv. AN SSSR Neorgan. Mater.*, *13*(4), 630–632 (1977b).

Guo C., Li C., Du Z. "Thermodynamic modeling of the In–Pt–Sb system." *Int. J. Mater. Res.*, *105*(6), 525–536 (2014).

Guo C., Li C., Du Z. "Thermodynamic modeling of the Ga–Pt–Sb system." *CALPHAD*, *52*, 169–179 (2016).

Guo Q., Ogawa H., Yoshida A. "Growth of $Al_xIn_{1-x}N$ single crystal films by microwave-excited metalorganic vapor phase epitaxy." *J. Cryst. Growth*, *146*(1–4), 462–466 (1995).

Guo W., Shi Y., Yang P., Ma H.A., Jia X., Wang S. "Characterization of boron nitride phase transformations in the Li–B–N system under high pressure and high temperature." *J. Alloys Compd.*, *644*, 888–892 (2015).

Gupta V.K., Koch M.W., Watkins N.J., Gao Y., Wicks G.W. "Molecular beam epitaxial growth of BGaAs ternary compounds." *J. Electron. Mater.*, *29*(12), 1387–1391 (2000).

Gupta V.K., Wamsley C.C., Koch M.W., Wicks G.W. "Molecular beam epitaxy growth of boron-containing nitrides." *J. Vac. Sci. Technol.*, *B17*(3), 1246–1248 (1999).

Guseinova M., Markus M.M., Bobrov V.I. "Investigation of the (GaSb)3–(Ga2Se3) section of the Ga–Sb–Se ternary system" [in Russian]. Tr. Kishinevsk. Politekhn. In-T, (*12*), 19–22 (1968).

Guseinova M.G., Rustamov P.G. "On the interaction of III2V3 gallium chalcogenides with gallium antimonide" [in Russian]. In Issled. v Obl. Neorgan. i Fiz. Khimii i Ikh Rol' v Khim. Prom-Sti. Ser. Khimiya i Neftepererab. Baku: Publish. House of AzINTI, 11–14 (1969).

Guseinova M.A., Rustamov P.G. "Investigation of the interaction in the GaSb–Se system" [in Russian]. In Issled. v Obl. Neorgan. i Fiz. Khimii. Baku: Elm Publish., 97–102 (1970).

Häberlen M., Glaser J., Meyer H.-J. "$Ca_3(BN_2)N$—Eine fehlende Verbindung im quasi-binären System Ca_3N_2–BN." *Z. Anorg. Allg. Chem.*, *628*(9–10), 1959–1962 (2002a).

Häberlen M., Glaser J., Meyer H.-J. "Zwei neue Nitridoborate des Calciums." *Z. Anorg. Allg. Chem.*, *628*(9–10), 2169 (2002b).

Häberlen M., Glaser J., Meyer H.-J. "Phase transitions in $Ca_3(BN_2)_2$ and $Sr_3(BN_2)_2$." *J. Solid State Chem.*, *178*(5), 1478–1487 (2005).

Hagenmayer R.M., Müller U., Benmore C., Neuefeind J., Jansen M. "Structural studies on amorphous silicon boron nitride $Si_3B_3N_7$: Neutron contrast technique on nitrogen and high energy x-ray diffraction." *J. Mater. Chem.*, *9*(11), 2865–2870 (1999).

Hahn H. "Über die Beeinflussung der Struktur ternärer und quaternärer Phasen bzw. Verbindungen durch Wahl geeigneter Komponenten. Über die Struktur von Mischkristallen des InAs mit InSe, In_2Se_3 und InTe." *Naturwissenschaften*, *44*(20), 534 (1957).

Hahn H., Thiele D. "Über die Beeinflussung der Struktur ternärer und quaternärer Phasen durch die Wahl geeigneter Komponenten. III. Über die Systeme In_2Se_3/InP, In_2Se_3/InAs, In_2Te_3/InAs, InSe/InAs und InTe/InAs." *Z. Anorg. Allg. Chem.*, *303*(3–4), 147–154 (1960).

Hall R.N., Racette J.H. "Diffusion and solubility of copper in extrinsinc and intrinsic germanium, silicon and gallium arsenide." *J. Appl. Phys.*, *35*(2), 379–397 (1964).

Han Q., Schmid-Fetzer R. "Experimental investigation of polythermal ternary Mo–Ga–As phase equilibria." *J. Phase Equilibria*, *13*(6), 588–600 (1992).

Han Q., Schmid-Fetzer R. "Phase equilibria in ternary Ga-As–M systems (M = W, Re, Si)." *J. Mater. Sci. Mater. Electron.*, *4*(2), 113–119 (1993a).

Han Q., Schmid-Fetzer R. "Reaction diffusion in Ta/GaAs contacts." *Mater. Sci. Eng. B*, *17*(1–3), 147–151 (1993b).

Han Y.S., Kalmykov K.B., Dunaev S.F., Zaitsev A.I. "Solid state phase equilibria in the titanium–aluminum–nitrogen system." *J. Phase Equilibria Diffus.*, *25*(5), 427–436 (2004).

Hannemann A., Schön J.C., Jansen M. "Thermodynamic stability of solid and fluid phases in the $Si_3B_3N_7$ system." *Phil. Mag.*, *88*(7), 1037–1057 (2008).

Hansen S.C., Loper Jr. C.R. "Effect of antimony on the phase equilibrium of binary Al–Si alloys." *CALPHAD*, *24*(3), 339–352 (2000).

Haris M., Hayakawa Y., Chou F.C., Veeramani P., Babu S.M. "Structural, constitutional and optical analysis of $InAs_xSb_{1-x}$ crystals grown by vertical directional solidification method." *J. Alloys Compd.*, *548*, 23–26 (2013).

Harris I.R., Smith N.A., Cockayne B., MacEwan W.R. "Phase identification in Fe-doped GaAs single crystals." *J. Cryst. Growth*, *82*(3), 450–458 (1987).

Harris I.R., Smith N.A., Devlin E., Cockayne B., Macewan W.R., Longworth G. "Structural, magnetic and constitutional studies of a new family of ternary phases based on the compound Fe_3GaAs." *J. Less Common Metals*, *146*(1–2), 103–119 (1989).

Haschke H., Nowotny H., Benesovsky F. "Neodym-Perowskitcarbide und -nitride." *Monatsh. Chem.*, *98*(6), 2157–2163 (1967).

Hashizume H., Hiraguchi H. "Determination of the Mg_3BN_3 structure according to the powder x-ray diffraction" [in Japanese]. *Repts. Asahi Glass Found*, *58*, 1–7 (1991).

Hassam S., Boa D., Rogez J. "Calorimetric investigations of Au–In, In–Sb and Au–In–Sb systems at 973 K." *J. Alloys Compd.*, *520*, 65–71 (2012).

Hassan El Haj F., Akbarzadeh H. "First-principles investigation of BN_xP_{1-x}, BN_xAs_{1-x} and BP_xAs_{1-x} ternary alloys." *Mater. Sci. Eng. B*, *121*(1–2), 170–177 (2005).

Hatano K., Nakamura K., Akiyama T., Ito T. "Theoretical study of alloy phase stability in zincblende $Ga_{1-x}Mn_xAs$." *J. Cryst. Growth*, *301–302*, 631–633 (2007).

Hatch D.M., Ghose S., Bjorkstam J.L. "The α-β phase transition in $AlPO_4$ cristobalite: Symmetry analysis, domain structure and transition dynamics." *Phys. Chem. Miner.*, *21*(1–2), 67–77 (1994).

He H., Stearrett R., Nowak E.R., Bobev S. "$BaGa_2Pn_2$ (Pn = P, As): New semiconducting phosphides and arsenides with layered structures." *Inorg. Chem.*, *49*(17), 7935–7940 (2010).

He H., Stearrett R., Nowak E.R., Bobev S. "Gallium pnictides of the alkaline earth metals, synthesized by means of the flux method: Crystal structures and properties of $CaGa_2Pn_2$, $SrGa_2As_2$, $Ba_2Ga_5As_5$, and $Ba_4Ga_5Pn_8$ (Pn = P or As)." *Eur. J. Inorg. Chem.*, (26), 4025–4036 (2011).

He H., Stoyko S., Bobev S. "New insights into the application of the valence rules in Zintl phases. Crystal and electronic structures of $Ba_7Ga_4P_9$, $Ba_7Ga_4As_9$, $Ba_7Al_4Sb_9$, $Ba_6CaAl_4Sb_9$, and $Ba_6CaGa_4Sb_9$." *J. Solid State Chem.*, 236, 116–122 (2016).

He H., Tyson C., Saito M., Bobev S. "Synthesis and structural characterization of the ternary Zintl phases $AE_3Al_2Pn_4$ and $AE_3Ga_2Pn_4$ (AE = Ca, Sr, Ba, Eu; Pn = P, As)." *J. Solid State Chem.*, 188, 59–65 (2012).

He H., Tyson C., Saito M., Bobev S. "Synthesis, crystal and electronic structures of the new Zintl phases $Ba_3Al_3Pn_5$ (Pn = P, As) and $Ba_3Ga_3P_5$." *Inorg. Chem.*, 52(1), 499–505 (2013).

He J.L., Guo L.C., Wu E., Luo X.G., Tian Y.J. "First-principles study of B_2CN crystals deduced from the diamond structure." *J. Phys. Condens. Matter*, 16(46), 8131–8138 (2004).

He J.L., Tian Y.J., Yu D.L., Guo L.C., Wang T.S., Xiao F.R., Liu S.M., et al. "B_2CN compounds prepared by high pressure and temperature." *Polzunov Bull.*, (2), 48–52 (2002).

He J.L., Tian Y.L., Yu D.L., Wang T.S., Liu S.M., Guo L.C., Li D.C., Jia X.P., Chen L.X., Zou G.T., Yanagisawa O. "Orthorhombic B_2CN crystal synthesized by high pressure and temperature." *Chem. Phys. Lett.*, 340(5–6), 431–436 (2001).

He T., Chen J., Rosenfeld H.D., Subramanian M.A. "Thermoelectric properties of indium-filled skutterudites." *Chem. Mater.*, 18(3), 759–762 (2006).

Hendricks S.B., Wyckoff R.W.G. "Space group of aluminum metaphosphate." *Am. J. Sci.*, 13(78), 491–496 (1927).

Hermann M., Furtmayr F., Bergmaier A., Dollinger G., Stutzmann M., Eickhoff M. "Highly Si-doped AlN grown by plasma-assisted molecular-beam epitaxy." *Appl. Phys. Lett.*, 86(19), 192108_1–192108_3 (2005).

Hernández-Zarazúa R., Hernández-Sustaita M., Anda de F., Mishurnyi V.A., Gorbatchev A.Yu., Asomoza R., Kudriavtsev Yu., Godines J.A. "Investigation of the phase diagram of the Pb–Ga–Sb system." *Thin Solid Films*, 461(2), 233–236 (2004).

Hillert M., Jonsson S. "An assessment of the Al–Fe–N system." *Metall. Trans. A*, 23(11), 3141–3149 (1992a).

Hillert M., Jonsson S. "Prediction of the Al–Si–N system." *CALPHAD*, 16(2), 199–205 (1992b).

Hillert M., Jonsson S. "Thermodynamic calculation of the Al–N–O system." *Z. Metallkde.*, 83(10), 714–719 (1992c).

Hintze F., Hummel F., Schmidt P.J., Wiechert D., Schnick W. "$Ba_3Ga_3N_5$—A novel host lattice for Eu^{2+}-doped luminescent materials with unexpected nitridogallate substructure." *Chem. Mater.*, 24(2), 402–407 (2012).

Hintze H., Johnson N.W., Seibald M., Muir M., Moewes A., Schnick W. "Magnesium double nitride Mg_3GaN_3 as new host lattice for Eu^{2+}-doping—Synthesis, structural studies, luminescence and band-gap determination." *Chem. Mater.*, 25(20), 4044–4052 (2013).

Hiraguchi H., Hashizume H., Fukunaga O., Takenaka A., Sakata M. "Structure determination of magnesium boron nitride, Mg_3BN_3, from x-ray powder diffraction data." *J. Appl. Crystallogr.*, 24(4), 286–292 (1991).

Hiraguchi H., Hashizume H., Sasaki S., Nakano S., Fukunaga O. "Structure of a high-presure polymorph of Mg_3BN_3 determined from x-ray powder data." *Acta Crystallogr.*, B49(3), 478–483 (1993).

Hirano S.-I., Kim P.-C. "Hydrothermal synthesis of gallium orthophospha/te crystals." *Bull. Chem. Soc. Jpn.*, 62(1), 275–278 (1989).

Hirano S.-I., Miwa K., Naka S. "Hydrothermal synthesis of gallium orthophosphate crystals." *J. Cryst. Growth*, 79(1–3), 215–218 (1986).

Hitova L., Yakimova R., Kishmar I.N. "Evaluation of InAs solubility in the In–As–Sn system." *Thin Solid Films*, 198(1–2), 387–392 (1991).

Ho I.H., Stringfellow G.B. "Incomplete solubility in nitride alloys." *Mater. Res. Soc. Symp. Proc.*, 449, 871–880 (1996a).

Ho I.-H., Stringfellow G.B. "Solid phase immiscibility in GaInN." *Appl. Phys. Lett.*, 69(18), 2701–2703 (1996b).

Ho J.C., Hamdeh H.H., Barsoum M.W., El-Raghy T. "Low temperature heat capacities of $Ti_3Al_{1.1}C_{1.8}$, Ti_4AlN_3, and Ti_3SiC_2." *J. Appl. Phys.*, 86(7), 3609–3611 (1999).

Hockings E.F., Kudman I., Seidel T.E., Schmelz C.M., Steigmeier E.F. "Thermal and electrical transport in InAs–GaAs alloys." *J. Appl. Phys.*, 37(7), 2879–2887 (1966).

Höglund C., Birch J., Beckers M., Alling B., Czigány Z., Mücklich A., Hultman L. "Sc_3AlN—A new perovskite." *Eur. J. Inorg. Chem.*, (8), 1193–1195 (2008).

Hohlfeld C. "Pecularities of the cubic boron nitride formation in the system $BN–Mg_3N_2$." *J. Mater. Sci. Lett.*, 8(9), 1082–1084 (1989).

Höhn P., Auffermann G., Ramlau R., Rosner H., Schnelle W., Kniep R. "$(Ca_7N_4)[M_x]$ (M = Ag, Ga, In, Tl): Linear metal chains as guests in a subnitride host." *Angew. Chem. Int. Ed.*, 45(40), 6681–6685 (2006).

Höhn P., Ramlau R., Rosner H., Schnelle W., Kniep R. "$(Ca_7N_4)M_x$ (M = Ag, Ga, In, Tl): Subnitride und Metallketten $^1_\infty M_x$." *Z. Anorg. Allg. Chem.*, 630(11), 1704 (2004).

Holec D., Rovere F., Mayrhofer P.H., Barna P.B. "Pressure-dependent stability of cubic and wurtzite phases within the TiN–AlN and CrN–AlN systems." *Scr. Mater.*, 62(6), 349–352 (2010).

Holleck H. "Metastable coatings—Prediction of composition and structure." *Surf. Coat. Technol.*, 36(1–2), 151–159 (1988).

Holleck H. "Neue Entwicklungen bei PVD-Hartstoffbeschichtungen." *Metall*, 43(7), 614–624 (1989).

Hoon C.F., Reynhardt E.C. "Molecular dynamics and structures of amine boranes of the type $R_3N\,BH_3$. I. X-ray investigation of $H_3N\,BH_3$ at 295 K and 110 K." *J. Phys. C*, 16(20), 6129–6136 (1983).

Hou T., Zhu Z., Zhao Y., Yin F., Li Z., Liu Y. "450°C isothermal section of the Fe–Al–Sb ternary phase diagram." *J. Phase Equilibria Diffus.*, 34(3), 188–195 (2013).

Hou T., Zhu Z., Zhao Y., Yin F., Zhao M. "The 680°C and 800°C isothermal sections of the Fe–Al–Sb system." *J. Alloys Compd.*, 586, 295–301 (2014).

Houben A., Burghaus J., Dronskowski R. "The ternary nitrides $GaFe_3N$ and $AlFe_3N$: Improved synthesis and magnetic properties." *Chem. Mater.*, 21(18), 4332–4338 (2009).

Huang J., Zhu Y.T., Mori H. "Structure and phase characteristics of amorphous boron–carbon–nitrogen under high pressure and high temperature." *J. Mater. Res.*, 16(4), 1178–1184 (2001).

Huang Z.-K., Greil P., Petzow G. "Formation of α-Si_3N_4 solid solutions in the system Si_3N_4–AlN–Y_2O_3." *J. Am. Ceram. Soc.*, 66(6), C-96–C-97 (1983).

Huang Z.-K., Tien T.-Y., Yen T.-S. "Subsolidus phase relationships in Si_3N_4–AlN–rare-earth oxide systems." *J. Am. Ceram. Soc.*, 69(10), C-241–C-242 (1986).

Hug G., Jaouen M., Barsoum M.W. "X-ray absorption spectroscopy, EELS, and full-potential augmented plane wave study of the electronic structure of Ti_2AlC, Ti_2AlN, Nb_2AlC, and $(Ti_{0.5}Nb_{0.5})_2AlC$." *Phys. Rev. B*, 71(2), 024105_1–024105_12 (2005).

Hughes E.W. "The crystal structure of ammonia-borane, H_3NBH_3." *J. Am. Chem. Soc.*, 78(2), 502–503 (1956).

Hullmann J., Xia S., Bobev S. "$Sr_{11}InSb_9$ grown from molten In." *Acta Crystallogr.*, E63(9), i178 (2007).

Hummel F.A., Kupinski T.A. "Thermal properties of the compound BPO_4." *J. Am. Chem. Soc.*, 72(11), 5318–319 (1950).

Huttenlocher H.F. "Kristallstruktur des Aluminiumorthophosphates $AlPO_4$." *Z. Kristallogr.*, 90(1), 508–16 (1935).

Igarashi O., Okada Y. "Epitaxial growth of $GaN_{1-x}P_x$ ($x \le$ 0.04) on sapphire substrates." *Jpn. J. Appl. Phys.*, 24(10), Pt. 2, L792–L794 (1985).

Ikeda T., Satoh H. "Phase formation and characterization of hard coatings in the Ti–Al–N system prepared by the cathodic arc ion platting method." *Thin Solid Films*, 195(1–2), 99–110 (1991).

Ilegems M., Panish M.B. "Phase diagram of the system Al–Ga–P." *J. Cryst. Growth*, 20(2), 77–81 (1973).

Il'yasov T.M., Rustamov P.G. "Chemical interaction and glass formation in the As_2X_3–GaX chalcogenide systems" [in Russian]. *Zhurn. Neorgan. Khimii*, 27(10), 2651–2654 (1982).

Il'yin Yu.L., Ovchinnikov S.Yu., Sorokin V.S. "Crystallization of the InP–GaP system solid solutions from the stoichiometric melts" [in Russian]. *Izv. AN SSSR Neorgan. Mater.*, 12(12), 2131–2133 (1976).

Il'yin Yu.L., Saenko I.V. "Solubility of nitrogen in GaP" [in Russian]. *Izv. AN SSSR Neorgan. Mater.*, 16(3), 376–379 (1980).

Il'yin Yu.L., Saenko I.V., Yas'kov D.A. "Preparation and properties of GaP crystals doping with nitrogen" [in Russian]. *Izv. AN SSSR Neorgan. Mater.*, 16(1), 5–8 (1980).

Inamura S., Nubugai K., Kanamaru F. "The preparation of NaCl-type $Ti_{1-x}Al_xN$ solid solution." *J. Solid State Chem.*, 68(1), 124–127 (1987).

Ingerly D.B., Swenson D., Jan C.-H., Chang Y.A. "Phase equilibria of the Ga–Ni–As ternary system." *J. Appl. Phys.*, 80(1), 543–550 (1996).

Inoue A., Kitamura A., Masumoto T. "The effect of aluminium on mechanical properties and thermal stability of (Fe,Ni)–Al–P ternary amorphous alloys." *J. Mater. Sci.*, 18(3) 753–758 (1983).

Inoue J., Osamura K. "Study of the phase diagram of the GaAs–GaSb quasi-binary system." *Trans. Jpn. Inst. Met.*, 12(1), 13–16 (1971).

Ipser H., Richter K.W. "Ni, Pd, or Pt as contact materials for GaSb and InSb semiconductors: Phase diagrams." *J. Electron. Mater.*, 32(11), 1136–1140 (2003).

Isayev V.F., Morozov A.N. "Nitrogen solubility and nitrides formation in the iron–boron melts" [in Russian]. *Izv. AN SSSR Metallurg. Gorn. Delo*, (2), 13–16 (1964).

Ishida K., Nomura T., Tokunaga H., Ohtani H., Nishizawa T. "Miscibility gaps in the GaP–InP, GaP–GaSb, InP–InSb and InAs–InSb systems." *J. Less Common Metals*, 155(2), 193–206 (1989a).

Ishida K., Shumiya T., Nomura T., Ohtani H., Nishizawa T. "Phase diagram of the Ga–As–Sb system." *J. Less Common Metals*, 142, 135–144 (1988a).

Ishida K., Shumiya T., Ohtani H., Hasebe M., Nishizawa T. "Phase diagram of the Al–In–Sb system." *J. Less Common Metals*, 143(1–2), 279–289 (1988b).

Ishida K., Tokunaga H., Ohtani H., Nishizawa T. "Data base for calculating phase diagrams of III-V alloy semiconductors." *J. Cryst. Growth*, 98(1–2), 140–147 (1989b).

Ishihara S., Ohtani H., Saito T., Ishida K. "Thermodynamic analysis of the Sn–In–Sb ternary phase diagram" [in Japanese]. *J. Jpn. Inst. Met.*, 63(6), 695–701 (1999).

Islamov S.A., Evgen'ev S.B., Sorokina O.V., Ufimtsev V.B., Mikhalov A.V. "Study of the phase equilibria in the In–As–Bi and In–P–Bi systems" [in Russian]. *Zhurn. Neorgan. Khimii*, 29(9), 2369–2372 (1984).

Itabashi S., Yamaguchi K., Kameda K. "Activity measurement of indium in liquid In–Sb–Ag and In–Sb–Te alloys by EMF method with zirconia solid electrolyte" [in Japanese]. *J. Jpn. Inst. Met.*, 65(9), 888–892 (2001).

Ivanovskii A.L., Sabiryanov R.F., Skazkin A.N., Zhukovskii V.M., Shveikin G.P. "Electronic structure and bonding configuration of the H-phases Ti_2MC and Ti_2MN (M = Al, Ga, In)." *Inorg. Mater.*, 36(1), 28–31 (2000).

Ivashchenko A.I., Kopanskaya F.Ya., Tarchenko V.P. "Calculation of homogeneous region in $Ga_{1-x}Al_xAs$ and $GaAs_{1-x}P_x$ ternary solid solutions." *Cryst. Res. Technol.*, 25(6), 661–666 (1990).

Ivchenko V.I., Kosolapova T.Y. "Study of preparation conditions and some properties of ternary compounds in the Ti–Al–C and Ti–Al–N systems" [in Russian]. *Nauch. Tr. Mosk. In-T Stali Splavov*, (99), 86–90 (1977).

Ivchenko V.I., Kosolapova T.Ya. "Investigation of abrasive properties of ternary compounds in the systems Ti–Al–C and Ti–Al–N" [in Russian]. *Poroshk. Metallurgia*, (8), 56–59 (1976).

Ivey D.G., Zhang L., Jian P. "Interfacial reactions between palladium thin films and InP." *J. Mater. Sci. Mater. Electron.*, 2(1), 21–27 (1991).

Ivin G.F. "Interaction of InSb with Ga" [in Russian]. *Izv. AN SSSR Neorgan. Mater.*, 13(3), 522–523 (1977).

Jambor J.L., Puziewicz J., Roberts A.C. "New mineral names." *Am. Mineral.*, 80(11–12), 1328–1333 (1995).

Jan C.-H., Chang Y.A. "The existence of Ni_3MSb phases in ternary nickel–M–antimony systems (where 'M' represents aluminium, gallium or indium)." *J. Mater. Res.*, 6(12), 2660–2665 (1991).

Jeffrey G.A., Wu V.Y. "The structures of the aluminum carbonitrides." *Acta Crystallogr.*, 16(6), 559–566 (1963).

Jeffrey G.A., Wu V.Y. "The structures of the aluminum carbonitrides. II." *Acta Crystallogr.*, 20(4), 538–547 (1966).

Jehn H.A., Hofmann S., Rückborn V.-E., Münz W.-D. "Morphology and properties of magnetron-sputtered (Ti,Al)N layers on high speed steel substrates as a function of deposition temperatures and sputtering atmosphere." *J. Vac. Sci. Technol.*, A4(6), 2701–2704 (1986).

Jeitschko W., Nowotny H., Benesovsky F. "Ti_2AlN, eine stickstoffhaltige H-Phase." *Monatsch. Chem.*, 94(6), 1198–1200 (1963).

Jeitschko W., Nowotny H., Benesovsky F. "Die H-Phasen: Ti_2CdC, Ti_2GaC, Ti_2GaN, Ti_2InN, Zr_2InN und Nb_2GaC." *Monatsch. Chem.*, 95(1), 178–179 (1964a).

Jeitschko W., Nowotny H., Benesovsky F. "Die Kristallstruktur von Ti_3InC, Ti_3InN, Ti_3TlC und Ti_3TlN." *Monatsch. Chem.*, 95(3), 436–438 (1964b).

Jeitschko W., Nowotny H., Benesovsky F. "Verbindungen vom Typ T_5M_3X. " *Monatsch. Chem.*, 95(4–5), 1242–1246 (1964c).

Jeitschko W., Nowotny H., Benesovsky F. "Ternäre Carbide und Nitride in Systemen: Übergangsmetall–Metametall–Kohlenstoff (Stickstoff)." *Monatsch. Chem.*, 95(1), 156–157 (1964d).

Jeitschko W., Nowotny H., Benesovsky F. "Phasen mit aufgefüllter β-Manganstruktur." *Monatsch. Chem.*, 95(4–5), 1212–1218 (1965).

Jenny D.A., Braunstein R. "Some properties of gallium arsenide–germanium mixtures." *J. Appl. Phys.*, 29(3), 596–597 (1958).

Jiang J., Kauzlarich S.M. "Colossal magnetoresistance in a rare earth Zintl compound with a new structure type: $EuIn_2P_2$." *Chem. Mater.*, 18(2), 435–441 (2006).

Jiang J., Olmstead M.M., Kauzlarich S.M., Lee H.-O., Klavins P., Fisk Z. "Negative magnetoresistance in a magnetic semiconducting Zintl phase $Eu_3In_2P_4$." *Inorg. Chem.*, 44(15), 5322–5327 (2005a).

Jiang J., Payne A.C., Olmstead M.M., Lee H.-O., Klavins P., Fisk Z., Kauzlarich S.M., Hermann R.P., Grandjean F., Long G.L. "Complex magnetic ordering in Eu_3InP_3: A new rare earth metal Zintl compound." *Inorg. Chem.*, 44(7), 2189–2197 (2005b).

Jianrong Y., Watson A. "An assessment of phase diagram and thermodynamic properties of the gallium–indium–antimony system." *CALPHAD*, 18(2), 165–175 (1994).

Jing H., Blaschkowski B., Meyer H.-J. "-Eine Festkörper-Metathese-Route zur Synthese von Nitridoboraten des Lanthans." *Z. Anorg. Allg. Chem.*, 628(9–10), 1955–1958 (2002).

Jing H., Meyer H.-J. "Über das metallreiche Lanthannitridoboratnitrid $La_5(B_2N_4)N_2$." *Z. Anorg. Allg. Chem.*, 626(2), 514–517 (2000).

Jing H., Pickardt J., Meyer H.-J. "Anionische Fragmente aus h-BN in der Struktur $La_6B_4N_{10}$." *Z. Anorg. Allg. Chem.*, 627(9), 2070–2074 (2001a).

Jing H., Reckeweg O., Blaschkowski B., Meyer H.-J. "Synthese und Struktur der Nitridoborat–Nitride $Ln_4(B_2N_4)N$ (Ln = La, Ce) des Formeltyps $Ln_{3+x}(B_2N_4)N_x$ (x = 0, 1, 2)." *Z. Anorg. Allg. Chem.*, 627(4), 774–778 (2001b).

Jordan A.S. "The liquidus surfaces of ternary systems involving compound semiconductors. II. Calculation of the liquidus isotherms and component partial pressures in the Ga–As–Zn and Ga–P–Zn systems." *Met. Trans.*, 2(7), 1965–1970 (1971a).

Jordan A.S. "The solid solubility isotherms of Zn in GaP and GaAs." *J. Electrochem. Soc.*, 118(5), 781–787 (1971b).

Jordan A.S. "Calculation of phase equilibria in the Ga–Bi and Ga–P–Bi systems based on a theory of regular associated solutions." *Met. Trans. B*, 7(2), 191–202 (1976).

Jordan A.S., Trumbore F.A., Nash D.L., Kowalchik M. "The solid solubility of Bi in GaP." *J. Electrochem. Soc.*, 123(7), 1083–1088 (1976).

Jordan A.S., Weiner M.E. "Calculation of the liquidus isotherms and component activities in the Ga–As–Si and Ga–P–Si ternary systems." *J. Electrochem. Soc.*, 121(12), 1634–1641 (1974).

Joukoff B., Jean-Louis A.M. "Growth of $InSb_{1-x}Bi_x$ single crystals by Czochralski method." *J. Cryst. Growth*, 12(2), 169–172 (1972).

Joullié A., Aulombard R., Bougnot G. "Growth of $Ga_xIn_{1-x}Sb$ liquid phase epitaxial layers from a knowledge of the ternary phase diagram." *J. Cryst. Growth*, 24–25, 276–281 (1974).

Joullié A., Gautier P., Monteil E. "The Al–Ga–Sb ternary phase diagram and its application to solution growth." *J. Cryst. Growth*, 47(1), 100–108 (1979).

Juza R., Hund F. "Die Kristallstrukturen LiMgN, LiZnN, Li_3AlN_2 und Li_3GaN_2." *Naturwissenschaften*, 33(4), 121–122 (1946).

Juza R., Hund F. "Die ternären Nitride Li_3AlN_2 und Li_3GaN_2. 17. Mitteilung über Metallamide und Metallnitride." *Z. Anorg. Chem.*, 257(1–3), 13–25 (1948).

Juza R., Schulz W. "Herstellung und Eigenschaften der Verbindungen Li$_3$AlP$_2$ und Li$_3$AlAs$_2$." *Z. Anorg. Allg. Chem.*, 269(1–2), 1–12 (1952).

Kabutov K., Korobov O.E., Maslov V.N., Nechaev V.V. "Obtaining of the Ga$_{1-x}$Al$_x$As solid solutions from the vapor phase" [in Russian]. *Izv. AN SSSR Neorgan. Mater.*, 16(8), 1366–1369 (1980).

Kajikawa Y., Kubota H., Asahina S., Kanayama N. "Growth of TlGaAs by low-temperature molecular-beam epitaxy." *J. Cryst. Growth*, 237–239, Pt. 2, 1495–1498 (2002).

Kajiyama K. "The In–Ga–P ternary phase diagram." *Jpn. J. Appl. Phys.*, 10(5), 561–565 (1971).

Kameda K., Yamaguchi K., Kon T. "Activity of indium in molten In–Pb–Ag and In–Bi–Sb alloys measured by EMF method using zirconia electrolyte" [in Japanese]. *J. Jpn. Inst. Met.*, 61(5), 444–448 (1997).

Kamilla S.K., Samantaray B.K., Basu S. "Effect of Ni concentrations on the microhardness of GaNiSb ternary alloys." *J. Alloys Compd.*, 414(1–2), 235–239 (2006).

Kamińska E., Piotrowska A., Barcz A., Ilka L., Guziewicz M., Kasjaniuk S., Dynowska E., Kwiatkowski S., Bremser M.D., Davis R.F. "Ohmic contacts to GaN by solid-phase regrowth." *Acta Phys. Pol. A*, 92(4), 819–823 (1997).

Kanchana V. "Mechanical properties of Ti$_3$AlX (X = C, N): Ab initio study." *Europhys. Lett.*, 87(2), 26006_p1–26006_p6 (2009).

Kaneko H., Nishizawa T., Tamaki K. "Phosphide-phases in ternary alloys of iron, phosphorus and other elements" [in Japanese]. *Nippon Kinzoku Gakkai-shi*, 29(2), 159–165 (1965a).

Kaneko H., Nishizawa T., Tamaki K., Tanifuji A. "Solubility of phosphorus in α- and γ-iron" [in Japanese]. *Nippon Kinzoku Gakkai-shi*, 29(2), 166–170 (1965b).

Kaner R.B., Kouvetakis J., Warble C.E., Sattler M.L., Bartlett N. "Boron–carbon–nitrogen materials of graphite-like structure." *Mater. Res. Bull.*, 22(3), 399–404 (1987).

Kao Le Din', Yevgen'ev S.B., Il'chenko A.V. "Investigation of the vapor pressure in the Ga–As–Bi system" [in Russian]. *Elektron. Tekhnika. Ser. 6. Materialy*, 12(197), 18–22 (1984).

Karavaev V.A. "On the phase transformation in the system indium antimonide–indium bismuthide" [in Russian]. *Sb. Nauch. Tr. Po Probl. Mikroelektron. Mosk. In-T Electron. Tekhn.*, (19), 83–86 (1974).

Karpiński J., Jun J., Grzegory I., Bugajski M. "Crystal growth of GaP doped with nitrogen under high nitrogen pressure." *J. Cryst. Growth*, 72(3), 711–716 (1985).

Kasper B. "Phasengleichgewichte im System B–C–N–Si." Thesis, Max-Planck-Institut, Stuttgart, 1–225 (1996).

Kasu M., Taniyasu Y., Kobayashi N. "Formation of solid solution of Al$_{1-x}$Si$_x$N (0 < x ≤ 12%) ternary alloy." *Jpn. J. Appl. Phys.*, 40(10A), Pt. 2, L1048–L1050 (2001).

Katayama I., Fukuda Y., Hattori Y. "Measurement of activity of gallium in liquid Ga–Sb–Ge alloys by EMF method with zirconia as solid electrolyte." *Ber. Bunsenges. Phys. Chem.*, 102(9), 1235–1239 (1998a).

Katayama I., Fukuda Y., Hattori Y., Maruyama T. "Measurement of activity of gallium in liquid Ga–Sb–Sn alloys by emf method with zirconia as solid electrolyte." *Thermochim. Acta*, 314(1–2), 175–181 (1998b).

Katayama I., Ikura T., Maki K., Iida T. "Activity of gallium in Ga–Sb–Te alloys determined by EMF method with zirconia as solid electrolyte." *Mater. Trans.*, 36(1), 41–43 (1995).

Katayama I., Nakai T., Inomoto T., Kozuka Z. "Activity measurements of Ga in GaAs–InAs solid solutions by the EMF method." *Mater. Trans.*, 30(5), 354–359 (1989).

Katayama I., Nakayama J.-I., Ikura T., Kozuka Z., Iida T. "Activity measurements of gallium in Ga–Bi and Ga–Sb–Bi alloys by EMF method using zirconia as solid electrolyte." *Mater. Trans.*, 34(9), 792–795 (1993a).

Katayama I., Nakayama J.-I., Ikura T., Tanaka T., Kozuka Z., Iida T. "Measurement of activity of gallium in Ga–Sb–In alloys by emf method using zirconia as solid electrolyte." *J. Non-Cryst. Solids*, 156–158, Pt. 1, 393–397 (1993b).

Katty A., Castro C.A., Odile J.P., Soled S., Wold A. "Crystal growth and characterization of In$_{1.9}$As$_{0.1}$Se$_3$." *J. Solid State Chem.*, 24(1), 107–111 (1978).

Katty A., Soled S., Wold A. "Crystal growth and characterization of In$_{2/3}$PSe$_3$." *Mater. Res. Bull.*, 12(6), 663–666 (1977).

Kaufman L. "Calculation of quasibinary and quasiternary oxynitride systems. III." *CALPHAD*, 3(4), 275–291 (1979).

Kaufman L., Ditchek B. "Calculation of ternary and quaternary phase diagrams containing semiconductor/metallic compound pseudo-binary eutectics." *J. Less Common Metals*, 168(1), 115–126 (1991).

Kaufman L., Nell J., Taylor K., Hayes F. "Calculation of ternary systems containing III-V and II-VI compound phases." *CALPHAD*, 5(3), 185–215 (1981).

Kaufman M.J., Konitzer D.G., Shull R.D., Fraser H.L. "An analytical electron microscopy study of the recently reported "Ti$_2$Al phase" in γ-TiAl alloys." *Scr. Metall.*, 20(1), 103–108 (1986).

Kauzlarich S.M., Kuromoto T.Y. "Exploring the structure and bonding of the Zintl compounds: A$_{14}$GaAs$_{11}$ (A = Ca, Sr)." *Croat. Chem. Acta*, 64(3), 343–352 (1991).

Kauzlarich S.M., Thomas M.M., Odink D.A., Olmstead M.M. "Ca$_{14}$GaAs$_{11}$: A new compound containing discrete GaAs$_4$ tetrahedra and a hypervalent As$_3$ polyatomic unit." *J. Am. Chem. Soc.*, 113(19), 7205–7208 (1991).

Kawaguchi M., Kawashima T. "Synthesis of a new graphite-like layered material of composition BC$_3$N." *Chem. Commun.*, (13), 1133–1134 (1993).

Keimes V., Blume H.-M., Mewis A. "Darstellung und Kristallstruktur von ANi$_{10}$P$_3$ (A: Zn, Ga, Sn, Sb)." *Z. Anorg. Allg. Chem.*, 625(2), 207–210 (1999).

Kevorkov M.N., Nagibin O.V., Popkov A.N., Yurova E.S. "Investigation of InSb doped with gallium" [in Russian]. *Izv. AN SSSR Neorgan. Mater.*, 24(9), 1443–1445 (1988).

Khan Yu.S., Kalmykov K.B., Dunaev S.F., Zaitsev A.I. "Phase equilibria in the Zr–Al–N system at 1273 K" [in Russian]. *Metally*, (5), 54–63 (2004a).

Khan Yu.S., Kalmykov K.B., Dunaev S.F., Zaitsev A.I. "Phase equilibria in the Ti–Al–N system at 1273 K" [in Russian]. *Dokl. RAN*, 396(6), 788–792 (2004b).

Kharif Ya.L., Kovtunenko P.V., Mayer A.A., Avetisov I.Kh. "Calculation of the phase equilibria in the Ga–Zn–As ternary system" [in Russian]. *Izv. Vuzov. Khimiya Khim. Tekhnol.*, 27(12), 1434–1436 (1984).

Khlystovskaya M.D., Chuveleva N.P. "Investigation of iodide transport of the InAs–GaAs solid solutions" [in Russian]. *Izv. AN SSSR Neorgan. Mater.*, 9(7), 1256–1257 (1973).

Khlystovskaya M.D., Zotova G.N., Elanskaya L.G., Rashevskaya E.P. "Properties of InAs–FeAs semiconductor eutectic" [in Russian]. *Izv. AN SSSR Neorgan. Mater.*, 11(3), 413–417 (1975).

Khoruzhaya V.G., Martsenyuk P.S., Meleshevich K.A., Velikanova T.Ya. "Physicochemical investigation of the reaction of components of the Si–Al–O–N system in the Si_3N_4–Al_2O_3–AlN–SiO_2–TiN–TiO_2 region. II. Subsystems Si_3N–AlN, TiN–AlN, Al_2O_3–AlN." *Powder Metall. Met. Ceram.*, 41(5–6), 278–283 (2005).

Khuber D.V. "Theory of *n*-component alloys with participation of III-V compounds. Calculation of liquidus and solidus curves of mixed semiconductors. Calculation of Al–Ga–As, Al–Ga–P, Si–Ge–Au and Si–Ge–Al ternary phase diagrams" [in Russian]. In *Protsesy Rosta i Sinteza Poluprovodn. Kristallov i Plenok. Part 2.* Novosibirsk: Nauka Publish., 212–218 (1975).

Kiessling R., Liu Y.H. "Thermal stability of the chromium, iron, and tungsten borides in streaming ammonia and the existence of a new tungsten nitride." *Trans. Metallurg. Soc. AIME*, 191(8), 639–642 (1951).

Kim S.-J., Hu S., Uher C., Kanatzidis M.G. "$Ba_4In_8Sb_{16}$: Thermoelectric properties of a new layered Zintl phase with infinite zigzag Sb chains and pentagonal tubes." *Chem. Mater.*, 11(11), 3154–3159 (1999).

Kim S.-J., Ireland J.R., Kannewurf C.R., Kanatzidis M.G. "$Yb_5In_2Sb_6$: A new rare earth Zintl phase with a narrow band gap." *J. Solid State Chem.*, 155(1), 55–61 (2000).

Kim S.-J., Kanatzidis M.G. "A unique framework in $BaGa_2Sb_2$: A new Zintl phase with large tunnels." *Inorg. Chem.*, 40(15), 3781–3785 (2001).

Kimura T., Doi H., Hashimoto K., Abe E., Isoda Y. "Phase equilibria in the TiAl-rich portion of Ti–Al–Sb system at 1373 and 1573 K" [in Japanese]. *Nippon Kinzoku Gakkaishi*, 61(5), 385–390 (1997).

Kinski I., Maurer F., Winkler H., Riedel R. "Synthesis of $In_xGa_{1-x}N$ solid solutions." *Z. Kristallogr.*, 220(2–3), 196–200 (2005a).

Kinski I., Miehe G., Heymann G., Theissmann R., Riedel R., Huppertz H. "High-pressure synthesis of a gallium oxonitride with a spinel-type structure." *Z. Naturforsch.*, 60B(8), 831–836 (2005b).

Kiosse G.A., Malinovskiy T.I., Radautsan S.I. "X-ray investigation of the ingots of the In–Sb–Te system" [in Russian]. *Izv. Mold. fil. AN SSSR*, 3(69), 3–9 (1960).

Kirchner M., Schnelle W., Wagner F.R., Kniep R., Niewa R. "$(A_{19}N_7)[In_4]_2$ (A = Ca, Sr) and $(Ca_4N)[In_2]$: Synthesis, crystal structures, physical properties, and chemical bonding." *Z. Anorg. Allg. Chem.*, 631(8), 1477–1486 (2005).

Kirchner M., Schnelle W., Wagner F.R., Niewa R. "Preparation, crystal structure and physical properties of ternary compounds $(R_3N)In$, R = rare-earth metal." *Solid State Sci.*, 5(9), 1247–1257 (2003).

Kirchner M., Wagner F.R., Schnelle W., Niewa R. "Beiträge zur Verbindungsbildung im System Ca–In–N." *Z. Anorg. Allg. Chem.*, 630(11), 1735 (2004).

Kiriy V.G., Marenkin S.F., Kiriy A.V., Cherkashina Yu.G., Marenkin D.S. "Synthesis of the alloys and construction of the phase diagram of the CdSb–InSb system" [in Russian]. *Zhurn. Neorgan. Khimii*, 41(10), 1720–1724 (1996).

Klančnik G., Medved J. "Ternary invariant point at 374°C in the three phase region AlSb–Al–Zn inside the Al–Sb–Zn ternary system." *J. Min. Metall. Sect. B Metall.*, 47(2), 179–192 (2011a).

Klančnik G., Medved J. "Thermodynamic investigation of the Al–Sb–Zn system." *Mater. Technol.*, 45(4), 317–323 (2011b).

Klančnik G., Medved J. "The isothermal section at 800°C and an AlSb–Zn quasi-binary cut in the Al–Sb–Zn system." *Comput. Mater. Sci.*, 66, 14–19 (2013).

Klesnar H., Rogl P. "The ternary system uranium–boron–nitrogen." *Boron-Rich Solids AIP Conference Proc.*, 231, 414–422 (1991).

Klesnar H., Rogl P. "Phase relations in the ternary systems Nd–B–N, Sm–B–N and Gd–B–N." *J. Am. Ceram. Soc.*, 75(10), 2825–2827 (1992).

Klesnar H.P., Rogl P. "Phase relations in the ternary systems rare-earth metal (RE)–boron–nitrogen, where RE = Tb, Dy, Ho, Er, Tm, Lu, Sc, and Y." *High Temp. High Pressure*, 22(4), 453–457 (1990).

Klimova L.N., Sokolov L.I., Goncharov E.G., Bityutskaya L.A., Ugai A.Ya. "Ternary semiconductor systems based on silicon" [in Russian]. *Tr. Voronezh. Un-T*, 76, 169–173 (1969).

Klingbeil J., Schmid-Fetzer R. "Interaction of metals with AlAs and InAs: Estimation of ternary Al–As–M and In–As–M phase diagrams." *CALPHAD*, 13(4), 367–388 (1989).

Klingbeil J., Schmid-Fetzer R. "Phase equilibria in some In–As–metal and Al–As–metal systems." *CALPHAD*, 18(4), 429–440 (1994).

Klooster W.T., Koetzle T.F., Siegbahn P.E.M., Richardson T.B., Crabtree R.H. "Study of the N–H●●H–B dihydrogen bond including the crystal structure of BH_3NH_3 by neutron diffraction." *J. Am. Chem. Soc.*, 121(27), 6337–6343 (1999).

Knittle E., Kaner R.B., Jeanloz R., Cohen M.L. "High-pressure synthesis, characterization, and equation of state of cubic C–BN solid solutions." *Phys. Rev. B*, 51(18), 12149–12155 (1995).

Knotek O., Böhmer M., Leyendecker T. "On structure and properties of sputtered Ti and Al based hard compound film." *J. Vac. Sci. Technol.*, A4(6), 2695–2700 (1986).

Kodentsov A.A., Markovski S.L., van Loo F.J.J. "Phase relations and interfacial reactions in the Ga–P–Co system at 500°C." *J. Phase Equilibria*, 22(3), 227–231 (2001).

Kolm C., Kulin S.A., Averbach B.L. "Studies on Group III–V intermetallic compounds." *Phys. Rev.*, 108(4), 965–971 (1957).

Komatsu T. "Bulk synthesis and characterization of graphite-like B–C–N and B–C–N heterodiamond compounds." *J. Mater. Chem.*, 14(2), 221–227 (2004).

Komatsu T., Nomura M., Kakudate Y., Fujiwara S. "Synthesis and characterization of a shock-synthesized cubic B–C–N solid solution of composition $BC_{2.5}N$." *J. Mater. Chem.*, 6(11), 1799–1803 (1996).

Kompanichenko N.M., Chaus I.S., Sukhenko V.D., Sheka I.A., Lugin V.N. "In_2S_3–Sb_2S_3 system" [in Russian]. *Zhurn. Neorgan. Khimii*, 18(4), 1080–1083 (1973).

Kompanichenko N.M., Omel'chuk A.A., Kozin V.F. "Interaction of indium arsenide and gallium antiminide with sulfur" [in Russian]. *Neorgan. Mater.*, 39(3), 276–281 (2003).

Kondow M., Uomi K., Hosomi K., Mozume T. "Gas-source molecular beam epitaxy of GaN_xAs_{1-x} using a N radical as the N source." *Jpn. J. Appl. Phys.*, 33(8A), Pt. 2, L1056–L1058 (1994).

Koppel Kh.D., Luzhnaya N.P., Medvedeva Z.S. "Cd–In–As system" [in Russian]. *Zhurn. Neorgan. Khimii*, 10(10), 2315–2319 (1965).

Koppel Kh.D., Luzhnaya N.P., Medvedeva Z.S. "Interaction of indium arsenide with some salts" [in Russian]. *Izv. AN SSSR Neorgan. Mater.*, 3(8), 1354–1359 (1967a).

Koppel Kh.D., Medvedeva Z.S., Luzhnaya N.P. "Interaction of indium arsenide with some metals" [in Russian]. *Izv. AN SSSR Neorgan. Mater.*, 3(2), 300–310 (1967b).

Korzun B.V., Ignatenko O.V., Lebedev S.A. "Interaction between BN_c and titanium in vacuum." *Inorg. Mater.*, 45(6), 626–630 (2009).

Kosten K., Arnold H. "Die III-V-Analoga des SiO_2." *Z. Kristallogr.*, 152(1–2), 119–133 (1980).

Köster W. "Das System Zink–Aluminium–Antimon." *Z. Metallkde*, 34(11), 257–259 (1942).

Köster W., Thoma B. "Aufbau ternärer Systeme von Metallen der dritten und fünften Gruppe des periodischen Systems." *Z. Metallkde*, 46(4), 293–297 (1955).

Köster W., Ulrich W. "Das Dreistoffsystem Gallium–Arsen–Zink." *Z. Metallkde*, 49(7), 361–364 (1958a).

Köster W., Ulrich W. "Zur Isomorphie der Verbindungen desa Typs $A^{III}B^V$." *Z. Metallkde*, 49(7), 365–367 (1958b).

Kostov A., Živković D., Živković Ž. "Comparative thermodynamic analysis of Ga–$GeSb_{0.855}$ section in the ternary system Ga–Ge–Sb." *J. Therm. Anal. Calorim.*, 60(2), 473–487 (2000).

Kostov A., Živković D., Živković Ž. "Thermodynamic analysis and characterization of Ga–$GeSb_{0.855}$ section in Ga–Ge–Sb ternary system." *J. Therm. Anal. Calorim.*, 65(3), 955–964 (2001).

Kotrubenko B.P. "On the solubility of the elements of VIb subgroup of the periodic system in indium arsenide" [in Russian]. In *Materialy IV Konf. Molodykh Uchenykh Moldavii*, 1964. Sekts. Fiz.-Mat. Kishinev: Stiintsa Publish., 18–19 (1965).

Kotrubenko B.P., Lange V.N., Lange T.I. "Physico-chemical properties of the alloys of the indium arsenide–tellurium section" [in Russian]. *Izv. AN SSSR Ser. Fiz.*, 28(6), 1007–1009 (1964).

Kotrubenko B.P., Lange V.N., Markus M.M. "Chemical bond and structure of the ternary compound InAsTe." *Phys. Status Solidi*, 22(1), K15–K17 (1967).

Kotrubenko B.P., Lange V.N., Radautsan S.I. "Homogeneous solid solutions based on InAs in the In–As–Te ternary system" [in Russian]. *Izv. AN MoldSSR Ser. Fiz.-Tekhn. Matem. N.*, (8), 60–61 (1966).

Kouvetakis J., McMurran J., Matsunaga P., O'Keeffe M., Hubbard J.L. "Synthesis and structure of a novel Lewis acid-base adduct, $(H_3C)_3SiN_3•GaCl_3$, en route to Cl_2GaN_3 and its derivatives: Inorganic precursors to heteroepitaxial GaN." *Inorg. Chem.*, 36(9), 1792–1797 (1997).

Kouvetakis J., O'Keefe M., Brouseau L., McMurran J., Williams D., Smith D.J. "New pathways to heteroepitaxial GaN by inorganic CVD synthesis and characterization of related Ga–C–N novel systems." *Mater. Res. Soc. Symp. Proc.*, 449, 313–318 (1996).

Kouvetakis J., Sasaki T., Shen C., Hagiwara R., Lerner M., Krishnan K.M., Bartlett N. "Novel aspects of graphite intercalation by fluorine and fluorides and new B/C, C/N and B/C/N materials based on the graphite network." *Synth. Met.*, 34(1–3), 1–7 (1989).

Kovaleva I.S., Luzhnaya N.P., Martikyan S.B. "Investigation of the interaction of the components in the Ga–In–As system" [in Russian]. *Zhurn. Neorgan. Khimii*, 14(10), 2860–2863 (1969).

Kozhina I.I., Tolkachev S.S., Borshchevskiy A.S., Goryunova N.A. "Investigation of the $GaAs$–Ga_2S_3 system" [in Russian]. *Vestn. Lenigr. Un-Ta*, (4), 122–127 (1962).

Kozlov A.Yu., Pavlyuk V.V. "New ternary antimonides Ti_5XSb_2 with W_5Si_3 structure type." *Intermetallics*, 11(3), 237–239 (2003).

Kroll P. "Spinel-type gallium oxynitrides attainable at high pressure and high temperature." *Phys. Rev. B*, 72(14), 144407_1–144407_4 (2005).

Kroll P., Dronskowski R., Martin M. "Formation of spinel-type gallium oxynitrides: A density-functional study of binary and ternary phases in the system Ga–O–N." *J. Mater. Chem.*, 15(32), 3296–3302 (2005).

Kroll P., Hoffmann R. "Silicon boron nitrides: Hypothetical polymorph of $Si_3B_3N_7$." *Angew. Chem. Int. Ed.*, 37(18), 2527–2530 (1998).

Kruger M.B., Jeanloz R. "Memory glass: An amorphous material formed from $AlPO_4$." *Science*, 249(4969), 647–649 (1990).

Ku S.M. "Preparation and properties of boron arsenides and boron arsenide–gallium arsenide mixed crystals." *J. Electrochem. Soc.*, 113(8), 813–816 (1966).

Kuan T.S., Freeouf J.L., Batson P.E., Wilkie E.L. "Reactions of Pd on (100) and (110) GaAs surfaces." *J. Appl. Phys.*, 58(4), 1519–1526 (1985).

Kubiak R., Schubert K. "Über das Legierungssystem Au–In–Sb." *Z. Metallkde*, 71(10), 635–637 (1980).

Kudaka K., Konno H., Matoba T. "Parameters for the crystal growth of cubic boron nitride" [in Japanese]. *Kogyo Kagaku Zasshi*, 69, 365–369 (1966).

Kuhn A., Eger R., Ganter P., Duppel V., Nuss J., Lotsch B.V. "In search of aluminum hexathiohypodiphosphate: Synthesis and structures of ht-AlPS$_4$, lt-AlPS$_4$, and Al$_4$(P$_2$S$_6$)$_3$." *Z. Anorg. Allg. Chem.*, 640(14), 2663–2668 (2014).

Kühn G., Leonhardt A. "Zur Ermittlung der thermodynamischen Exzeßgrößen von A^{III}–B^{V}-Mischkritallen." *Krist. Tech.*, 7(5), K57–K62 (1972).

Kuhnen C.A., Figueiredo de R.S., Santos dos A.V. "Mössbauer spectroscopy, crystallographic, magnetic and electronic structure of ZnFe$_3$N and InFe$_3$N." *J. Magn. Magn. Mater.*, 219(1), 58–68 (2000).

Kuliev A.N., Safaraliev G.I., Guseynov G.A. "Sm$_x$Ga$_{1-x}$Sb solid solutions" [in Russian]. *Izv. AN SSSR Neorgan. Mater.*, 26(3), 500–503 (1990).

Kulinich S.A., Sevast'yanova L.G., Leonova M.E., Burdina K.P., Semenenko K.N. "Formation and properties of new compound in the Ca$_3$N$_2$–BN system" [in Russian]. *Zhurn. Obshch. Khimii*, 65(7), 1065–1067 (1995).

Kulinich S.A., Sevast'yanova L.G., Zhukov A.N., Burdina K.P. "Boronitrides of alkali and alkaline earth metals containing N^{3-} anion" [in Russian]. *Zhurn. Obshch. Khimii*, 70(2), 190–196 (2000).

Kumar S., Joshi S., Joshi B., Auluck S. "Thermodynamical and electronic properties of B$_x$Al$_{1-x}$N alloys: A first principle study." *J. Phys. Chem. Solids*, 86, 101–107 (2015).

Kumar N., Raidongia K., Mishra A.K., Waghmare U.V., Sundaresan A., Rao C.N.R. "Synthetic approaches to borocarbonitrides, BC$_x$N (x = 1, 2)." *J. Solid State Chem.*, 184(11), 2902–2908 (2011).

Kurata K., Hirai T. "Semiconducting properties of several III$_B$–V$_B$–VI$_B$ ternary materials and their metallurgical aspects." *Solid State Electron.*, 9(6), 633–640 (1966).

Kuriyama K., Anzawa J., Kushida K. "Growth and band gap of the filled tetrahedral semiconductor Li$_3$AlP$_2$." *J. Cryst. Growth*, 310(7–9), 2298–2300 (2008).

Kuriyama K., Ishikawa T., Kushida K. "Optical band gap and bonding character of Li$_3$GaN$_2$." *AIP Conf. Proc.*, 893(1), 1479–1480 (2007).

Kuriyama K., Kaneko Y., Kushida K. "Synthesis and characterization of AlN-like Li$_3$AlN$_2$." *J. Cryst. Growth*, 275(1–2), e395–e399 (2005).

Kushida K., Kaneko Y., Kuriyama K. "Filled tetrahedral semiconductor Li$_3$AlN$_2$ studied with optical absorption: Application of the interstitial insertion rule." *Phys. Rev. B*, 70(23), 233303_1–233303_4 (2004).

Kuwatsuka H., Tanahashi T., Anayama C., Nishiyama S., Mikawa T., Nakajima K. "Liquid phase epitaxial growth of Al$_x$Ga$_{1-x}$Sb from Sb-rich solution." *J. Cryst. Growth*, 94(4), 923–928 (1989).

Kuzenkova M.A., Kislyi P.S., Pshenichnaya O.V. "The structure and properties of composite materials based on the nitrides of Ti, Zr and Al" [in Russian]. *Izv. AN SSSR Neorgan. Mater.*, 12(3), 430–434 (1976).

Kuznetsov G.M., Barsukov A.D., Pelevin O.V., Olenin V.V., Berdavtseva N.N. "Investigation of Se solid solutions in GaAs" [in Russian]. *Izv. AN SSSR Neorgan. Mater.*, 9(6), 1053–1054 (1973).

Kuznetsov G.M., Bobrov A.P. "Investigation of the InSb interaction with metals" [in Russian]. *Izv. AN SSSR Neorgan. Mater.*, 7(5), 766–768 (1971).

Kuznetsov G.M., Kuznetsova S.K. "Solidus calculations in the binary system based on germanium and silicon and in the III–V–metal sections" [in Russian]. *Izv. AN SSSR Neorgan. Mater.*, 2(4), 643–649 (1966).

Kuznetsov G.M., Kuznetsova S.K. "Investigation of the alloys of the GaAs–Sn section" [in Russian]. *Izv. AN SSSR Neorgan. Mater.*, 3(6), 981–985 (1967).

Kuznetsov G.M., Kuznetsova S.K. "Investigation of the alloys of the Pb–GaAs and Ag–GaAs sections" [in Russian]. *Izv. AN SSSR Neorgan. Mater.*, 4(8), 1243–1247 (1968).

Kuznetsov G.M., Kuznetsova S.K., Bulannikova N.D. "Intereaction of GaAs with Ga–As–In melts" [in Russian]. *Arsenide Galliya*. Vyp. 2. Tomsk: Tomsk Un-T Publish., 198–200 (1969).

Kuznetsov G.M., Kuznetsova S.K., Sokolkova E.G. "Investigation of the GaAs–Cu section" [in Russian]. *Izv. AN SSSR Neorgan. Mater.*, 4(11), 1934–1937 (1968).

Kuznetsov G.M., Rotenberg V.A. "Calculation of the liquidus surface of the Al–Si–P system" [in Russian]. *Izv. AN SSSR Neorgan. Mater.*, 7(6), 943–946 (1971).

Kuznetsov G.M., Tsurgan L.S., Kraynyuchenko I.A. "Investigation of the alloys of the GaP–Cd, GaP–Sn and GaP–Pb sections" [in Russian]. *Izv. AN SSSR Neorgan. Mater.*, 12(7), 1168–1170 (1976a).

Kuznetsov G.M., Tsurgan L.S., Kraynyuchenko I.A. "Investigation of the In interaction with GaP" [in Russian]. *Izv. AN SSSR Neorgan. Mater.*, 12(8), 1352–1356 (1976b).

Kuznetsov V.V., Sorokin V.S. "Solubility of GaP in antimony and bismuth melts" [in Russian]. *Izv. AN SSSR Neorgan. Mater.*, 13(5), 780–784 (1977).

Kuznetsov V.V., Sorokin V.S. "The solidus in the GaP–Sb quasi-binary system in the region of GaP primary crystallization" [in Russian]. *Izv. AN SSSR Neorgan. Mater.*, 14(7), 1212–1216 (1978).

Kuznetsova S.K. "Calculated of the solidus in the systems based on GaAs" [in Russian]. *Izv. AN SSSR Neorgan. Mater.*, 11(5), 950–951 (1975).

Kyffin W.J., Rainforth W.M., Jones H. "The formation of aluminum phosphide in aluminum melt treated with an Al–Fe–P inoculant addition." *Z. Metallkde*, 92(4), 396–398 (2001).

Lambrecht W.R.L., Segall B. "Anomalous band-gap behavior and phase stability of c-BN–diamond alloys." *Phys. Rev. B*, 47(15), 9289–9296 (1993).

Labbe J.C., Jeanne A., Roult G. "Synthesis of the gamma phase of aluminium oxynitride by reaction of the boron nitride/alumina system." *Ceram. Int.*, 18(2), 81–84 (1992).

Lahav A., Eizenberg M. "Interfacial reactions of Ni–Ta thin films on GaAs." *Appl. Phys. Lett.*, 45(3), 256–258 (1984).

Lahav A., Eizenberg M., Komem Y. "Epitaxial phases formation due to interaction between Ni thin films and GaAs." *Mater. Res. Soc. Symp. Proc.*, 37, 641–646 (1985).

Lahav A., Eizenberg M., Komem Y. "Interfacial reactions between Ni films and GaAs." *J. Appl. Phys.*, 60(3), 991–1001 (1986).

Lakeenkov V.M., Morgulis L.M., Mil'vidskiy M.G., Pelevin O.V. "Phase diagram of the Ge–As–Se condensed system and liquid-epitaxy of the GaAs:Se layers" [in Russian]. *Izv. AN SSSR Neorgan. Mater.*, 13(8), 1373–1376 (1977).

Land P.L., Wimmer J.M., Barns R.W., Choudhury N.S. "Compounds and properties of the system Si–Al–O–N." *J. Am. Ceram. Soc.*, 61(1–2), 56–60 (1978).

Lane N.J., Vogel S.C., Barsoum M.W. "Temperature-dependent crystal structures of Ti_2AlN and Cr_2GeC as determined from high temperature neutron diffraction." *J. Am. Ceram. Soc.*, 94(10), 3473–3479 (2011).

Langenhorst F., Solozhenko V.L. "ATEM-EELS study of new diamond-like phases in the B–C–N system." *Phys. Chem. Chem. Phys.*, 4(20), 5183–5188 (2002).

Lange S., Bawohl M., Weihrich R., Nilges T. "Mineralisator-Konzept für Polyphosphide: Cu_2P_{20} and Cu_5InP_{16}." *Angew. Chem.*, 120(30), 5736–5739 (2008a).

Lange S., Bawohl M., Weihrich R., Nilges T. "Mineralization routes to polyphosphides: Cu_2P_{20} and Cu_5InP_{16}." *Angew. Chem. Int. Ed.*, 47(30), 5654–5657 (2008b).

Lapkina I.A., Sorokina O.V., Voloshin A.E., Shcherbovskiy E.Ya., Viryasova T.B. "Solubility of zinc in indium antimonides at various sections of the In–Sb–Zn ternary system" [in Russian]. *Izv. AN SSSR Neorgan. Mater.*, 23(4), 533–536 (1987).

Lapkina I.A., Voloshin A.E., Sorokina O.V., Shcherbovskiy E.Ya. "Solubility of cadmium in the indium antimonide at various sections of the In–Sb–Cd system" [in Russian]. *Izv. AN SSSR Neorgan. Mater.*, 25(3), 360–362 (1989).

Latypov Z.M., Savel'ev V.P., Aver'yanov I.S., Fayzullina N.R., Prikhodtsev M.N. "Investigation of the solid solution region based on InSb in the In–Sb–Bi system" [in Russian]. *Izv. AN SSSR Neorgan. Mater.*, 13(5), 910–911 (1977).

Lavrishchev T.T., Khludkov S.S. "Influence of arsenic vapor pressure on the germanium diffusion in gallium arsenide" [in Russian]. *Izv. AN SSSR Neorgan. Mater.*, 7(2), 310–311 (1971).

Leadbetter A.J., Wright A.F. "The α-β-transition in the cristobalite phases of SiO_2 and $AlPO_4$. I. X-ray studies." *Phil. Mag.*, 33(1), 105–112 (1976).

Le Clanche M.C., Députier S., Jégaden J.C., Guérin R., Ballini Y., Guivarc'h A. "Solid state phase equilibria in the Ni–Ga–Sb system: Experimental and calculated determination." *J. Alloys Compd.*, 206(1), 21–29 (1994).

Lee H.D., Petuskey W.T. "New ternary nitride in the Ti–Al–N system." *J. Am. Ceram. Soc.*, 80(3), 604–608 (1997).

Lee H.D., Petuskey W.T. "Reply to 'Comment on "New ternary nitride in the Ti–Al–N system."'" *J. Am. Ceram. Soc.*, 81(3), 787–788 (1998).

Legendre B., Dichi E., Vassiliev V. "The phase diagram of the In–Sb–Sn System." *Z. Metallkde*, 92(4), 328–335 (2001).

Lejus A. "Formation at high temperature of nonstoichiometric spinels and of derived phases in several oxide systems based on alumina and in the system alumina–aluminum nitride." *Rev. Int. Hautes Temp. Refract.*, 1(1), 53–95 (1964).

Lejus A.-M. "Préparation par reaction à l'état solid et principales propriétés des oxynitrures d'aluminium." *Bull. Soc. Chim. Fr.*, (11–12), 2123–2126 (1962).

Lendvay E. "Ternary $A^{III}B^V$ antimonides." *Prog. Cryst. Growth Charact.*, 8(4), 371–425 (1984).

Leonhardt A., Buchcheiser K., Kühn G. "Phase diagram of semiconducting compounds (III). Thermodynamic calculation of the ternary melting diagram of Ga–Al–As." *Krist. Tech.*, 9(3), 197–203 (1974).

Leonhardt A., Kühn G. "Die Löslichkeiten von GaP und GaAs in Sn und Bi." *Krist. Tech.*, 9(1), 77–85 (1974a).

Leonhardt A., Kühn G. "Phasendiagramme von Verbindungenhalbleiten. IV. Löslichkeiten von $A^{III}B^V$-Verbindungen in Metallen." *J. Less Common Metals*, 37(2), 310–312 (1974b).

Leonhardt A., Kühn G. "Phasendiagramme von halbleitenden Verbindungen. V. Löslichkeiten von $A^{III}B^V$-Verbindungen in metallischen Schmelzen." *J. Less Common Metals*, 39(2), 247–264 (1975).

Lesunova R.P., Burmakin E.I. "Electroconductivity of the solid solutions in the AlN–Mg_3N_2 system" [in Russian]. *Neorgan. Mater.*, 41(7), 819–822 (2005).

Lewin E., Parlinska-Wojtan M., Patscheider J. "Nanocomposite Al–Ge–N thin films and their mechanical and optical properties." *J. Mater. Chem.*, 22(33), 16761–16773 (2012).

Lewin E., Patscheider J. "Structure and properties of sputter-deposited Al–Sn–N thin films." *J. Alloys Compd.*, 682, 42–51 (2016).

Lewis R.B., Masnadi-Shirazi M., Tiedje T. "Growth of high Bi concentration $GaAs_{1-x}Bi_x$ by molecular beam epitaxy." *Appl. Phys. Lett.*, 101(8), 082112_1–082112_4 (2012).

Li Ch., Li J.-B., Du Z., Zhang W. "A thermodynamic assessment of the Ga–In–P system." *J. Phase Equilibria*, 21(4), 357–363 (2000).

Li Ch., Li J.-B., Lu L., Zhang W. "A thermodynamic reassessment of the Al–As–Ga system." *J. Phase Equilibria*, 22(1), 26–33 (2001).

Li D., Yu D., He J., Xu B., Liu Z., Tian Y. "First-principle calculation on structures and properties of diamond-like $B_3C_{10}N_3$ compound." *J. Alloys Compd.*, 481(1–2), 855–857 (2009).

Li J.-B., Zhang W., Li Ch., Du Zh. "A thermodynamic assessment of the Ga–As–Sb system." *J. Phase Equilibria*, *19*(5), 466–472 (1998a).

Li J.-B., Zhang W., Li Ch., Du Zh. "A thermodynamic assessment of the In–As–Sb system." *J. Phase Equilibria*, *19*(5), 473–478 (1998b).

Li J.-B., Zhang W., Li Ch., Du Zh. "Assessment of phase diagram and thermodynamic properties of the Al–Ga–Sb system." *J. Phase Equilibria*, *20*(3), 316–323 (1999).

Li N.Y., Wong W.S., Tomich D.H., Dong H.K., Solomon J.S., Grant J.T., Tu C.W. "Growth study of chemical beam epitaxy of GaN_xP_{1-x} using NH_3 and tertiarybutylphosphine." *J. Cryst. Growth*, *164*(1–4), 180–184 (1996).

Li Q., Zhou D., Wang H., Chen W., Wu B., Wu Z., Zheng W. "Crystal and electronic structures of superhard B_2CN: An *ab initio* study." *Solid State Commun.*, *152*(2), 71–75 (2012).

Liao P.-K., Su C.-H., Tung Tse, Brebrick R.F. "Quantitative simultaneous fit to the liquidus surface and thermodynamic data for the Ga–In–Sb system using an associated solution model for the liquid." *CALPHAD*, *6*(2), 141–169 (1982).

Lidov A., Gibbons J.F., Deline V.R., Evans C.A. "Solid solubility of selenium in GaAs as measured by secondary ion mass spectrometry." *Appl. Phys. Lett.*, *32*(9), 572–573 (1978).

Lieth R.M.A., Heyligers H.J.M. "Phase diagram of the pseudo-binary system Ge–GaAs." *J. Electrochem. Soc.*, *113*(1), 96 (1966).

Lightowlers E.C. "Boron and nitrogen in gallium phosphide grown by the liquid encapsulated Czochralski process." *J. Electron. Mater.*, *1*(1), 38–52 (1972).

Lin C.F., Mohney S.E., Chang Y.A. "Phase equilibria in the Pt–In–P system." *J. Appl. Phys.*, *74*(7), 4398–4402 (1993).

Lin J.-C., Hsieh K.-C., Schulz K.J., Chang Y.A. "Reactions between palladium and gallium arsenide: Bulk versus thin-film studies." *J. Mater. Res.*, *3*(1), 148–163 (1988).

Lin Z., Wu L., Yang X., Liu X., Li G. "A study of a $GaP_{1-x}N_x$ alloy system." *J. Lumin.*, *40–41*, 909–910 (1988).

Lindeberg I., Andersson Y. "The ternary system Co–Ga–As." *J. Less Common Metals*, *175*(1), 155–162 (1991).

Lingam H.K., Chen X., Zhao J.-C., Shore S.G. "A convenient synthesis and a NMR study of the diammoniate of diborane." *Chem. Eur. J.*, *18*(12), 3490–3492 (2011).

Lingam H.K., Wang C., Gallucci J.C., Chen X., Shore S.G. "New syntheses and structural characterization of NH_3BH_2Cl and $(BH_2NH_2)_3$ and thermal decomposition behavior of NH_3BH_2Cl." *Inorg. Chem.*, *51*(24), 13430–13436 (2012).

Liou B.-T., Liu C.-W. "Electronic and structural properties of zincblende $Al_xIn_{1-x}N$." *Optics Commun.*, *274*(2), 361–365 (2007).

Liou B.-T., Yen S.-H., Kuo Y.-K. "Vegard's law deviation in band gap and bowing parameter of $Al_xIn_{1-x}N$." *Appl. Phys. A*, *81*(3), 651–655 (2005).

Lippert E.L., Lipscomb W.N. "The structure of H_3NBH_3." *J. Am. Chem. Soc.*, *78*(2), 503–504 (1956).

Litvak A.M. "Melting diagrams of multinary systems with neutral solvent (on the example of the Pb–In–Ga–As–Sb system)" [in Russian]. *Zhurn. Neorgan. Khimii*, *37*(2), 470–478 (1992).

Litvak A.M., Charykov N.A. "New thermodynamic method for calculation of binary and ternary phase diagram containing In, Ga, As and Sb" [in Russian]. *Izv. AN SSSR Neorgan. Mater.*, *27*(2), 225–230 (1991).

Litvin B.N., Popolitov V.I., Simonov M.A., Yakubovich O.V., Yaroslavskiy I.M. "Solubility, preparation and properties of gallium orthophosphate crystals (α-$GaPO_4$)" [in Russian]. *Kristallografiya*, *32*(2), 486–489 (1987).

Liu H.S., Liu C.L., Wang C., Jin Z.P., Ishida K. "Thermodynamic modelling of the Au–In–Sb ternary system." *J. Electron. Mater.*, *32*(2), 81–88 (2003).

Liu W.E., Mohney S.E. "Condensed phase equilibria in transition metal–Ga–Sb systems and predictions for thermally stable contacts to GaSb." *J. Electron. Mater.*, *32*(10), 1090–1099 (2003a).

Liu W.E., Mohney S.E. "Condensed phase equilibria in transition metal–In–Sb systems and predictions for thermally stable contacts to InSb." *Mater. Sci. Eng. B*, *103*(2), 189–201 (2003b).

Liu X.J., Yamaki T., Ohnuma I., Kainuma R., Ishida K. "Thermodynamic calculations of phase equilibria, surface tension and viscosity in the In–Ag–X (X = Bi, Sb) system." *Mater. Trans.*, *45*(3), 637–645 (2004).

Löhberg K. "Über das System Zinn–Antimon–Indium." *Metall*, *22*(8), 777–784 (1968).

Lomnitskaya Ya.F., Doskoch A.M. "Interaction in the Zr(Nb)–Al–P systems" [in Russian]. *Neorgan. Mater.*, *29*(6), 870–872 (1993).

Lomnitskaya Ya.F., Kuz'ma Yu.B. "Interaction of aluminum and phosphorus with molybdenum" [in Russian]. *Poroshk. Metallurgiya*, (5–6), 70–74 (1996).

Long G., Foster L.M. "Crystal phases in the system Al_2O_3–AlN." *J. Am. Ceram. Soc.*, *44*(6), 255–258 (1961).

López-Pérez W., Simon-Olivera N., Molina-Coronell J., González-García A., González-Hernández R. "Structural parameters, band-gap bowings and phase diagrams of zinc-blende $Sc_{1-x}In_xP$ ternary alloys: A FP-LAPW study." *J. Alloys Compd.*, *574*, 124–130 (2013).

Lorenz H., Orgzall I. "Formation of cubic boron nitride in the system Mg_3N_2–BN: A new contribution to the phase diagram." *Diamond Relat. Mater.*, *4*(8), 1046–1049 (1995).

Lorenz H., Orgzall I., Hinze E. "Rapid formation of cubic boron nitride in the system Mg_3N_2–hBN." *Diamond Relat. Mater.*, *4*(8), 1050–1055 (1995).

Lorenz H., Orgzall I., Hinze E., Kremmler J. "Structural phase transitions of Mg_3BN_3 under high pressures and temperatures." *Scr. Metal. Mater.*, *27*(8), 993–997 (1992).

Lorenz H., Orgzall I., Hinze E., Kremmler J. "Formation of Mg_3BN_3 under high pressures and temperatures in the system Mg_3N_2–hBN." *Scr. Metal. Mater.*, *28*(8), 925–930 (1993).

Lowther J.E., Wagner T., Kinski I., Riedel R. "Potential gallium oxynitrides with a derived spinel structure." *J. Alloys Compd.*, *376*(1–2), 1–4 (2004).

Lu J., Bi J.F., Wang W.Z., Gan H.D., Meng H.J., Deng J.J., Zheng H.Z., Zhao J.H. "Growth parameter dependence of structural characterizations of diluted magnetic semiconductor (Ga,Cr)As." *IEEE Trans. Magn.*, *44*(11), 2692–2695 (2008).

Ludwig M., Jäger J., Niewa R., Kniep R. "Crystal structure of two polymorphs of $Ca_3[Al_2N_4]$." *Inorg. Chem.*, *39*(26), 5909–5911 (2000).

Ludwig M., Niewa R., Kniep R. "Dimers $[Al_2N_6]^{12-}$ and chains $^1_\infty[AlN_{4/2}^{3-}]$ in the crystal structures of $Ca_6[Al_2N_6]$ and $Ba_3[Al_2N_4]$." *Z. Naturforsch.*, *54B*(4), 461–465 (1999).

Luedecke C.M., Doench J.P., Huggins R.A. "Thermodynamic studies of the lithium–aluminum–antimony system." *Proc. Electrochem. Soc.*, *83*(7), 105–114 (1983).

Luef C., Flandorfer H., Richter K.W., Ipser H. "Palladium as a contact material for InSb semiconductors. The In–Pd–Sb phase diagram." *J. Electron. Mater.*, *32*(2), 43–51 (2003).

Lundström T., Andreev Y.G. "Superhard boron-rich borides and studies of the B–C–N system." *Mater. Sci. Eng. A*, *209*(1–2), 16–22 (1996).

Lushnaja N.P., Medwedjewa Z.S., Koppel H.D. "Löslichkeit einiger III-V-Verbindungen in Schmelzen." *Z. Anorg. Allg. Chem.*, *344*(5–6), 323–328 (1966).

Luzhnaya N.P., Nikol'skaya G.F., Van Bin-nan'. "About $A^I_2B^{III}C^V_2$ semoconductor compounds" [in Russian]. *Izv. AN SSSR Neorgan. Mater.*, *1*(8), 1328–1334 (1965).

Luzhnaya N.P., Slavnova G.K., Medvedeva Z.S., Eliseev A.A. "In–As–Se system" [in Russian]. *Zhurn. Neorgan. Khimii*, *9*(5), 1174–1181 (1964).

Lyalikov Yu.S., Radautsan S.I., Kopanskaya L.S., Molodyan I.P. "Synthesis of complex semiconductors and their chemical analysis" [in Russian]. *Vestn. AN SSSR*, (9), 75–78 (1964).

Lyutaya M.D., Bartnitskaya T.S. "Formation pecularities and chemical stability of complex in the Ga–B–N system" [in Russian]. *Izv. AN SSSR Neorgan. Mater.*, *9*(8), 1372–1376 (1973a).

Lyutaya M.D., Bartnitskaya T.S. "Formation pecularities and chemical stability of complex nitrides of III-B subgroup elements (Al–B–N system)" [in Russian]. *Izv. AN SSSR Neorgan. Mater.*, *9*(8), 1367–1371 (1973b).

Ma K.Y., Fang Z.M., Cohen R.M., Stringfellow G.B. "Organometallic vapor-phase epitaxy growth and characterization of Bi-containing III/V alloys." *J. Appl. Phys.*, *68*(9), 4586–4591 (1990).

Ma K.Y., Fang Z.M., Cohen R.M., Stringfellow G.B. "OMVPE growth and characterization of Bi-containing III-V alloys." *J. Cryst. Growth*, *107*(1–4), 416–421 (1991).

Ma K.Y., Fang Z.M., Cohen R.M., Stringfellow G.B. "Ultra-low temperature OMVPE of InAs and InAsBi." *J. Electron. Mater.*, *21*(2), 143–148 (1992).

Ma K.Y., Fang Z.M., Jaw D.H., Cohen R.M., Stringfellow G.B., Kosar W.P., Brown D.W. "Organometallic vapor phase epitaxial growth and characterization of InAsBi and InAsSbBi." *Appl. Phys. Lett.*, *55*(23), 2420–2422 (1989).

Mabbitt A.W. "Calculation of the In/Ga/P ternary phase diagram and its relation to liquid phase epitaxy." *J. Mater. Sci.*, *5*(12), 1043–1046 (1970).

Machatschki F. "Die Kristallstruktur des Aluminiumarsenates $AlAsO_4$. Ein weiterer Beitrag zur Frage der Isomorphie von Kristallarten des Siliciums und fünfwertigen Arsens." *Z. Kristallogr.*, *90*(1), 314–321 (1935).

Machatschki F. "Die Kristallstruktur von Tiefquarz SiO_2 und Aluminiumorthoarsenat $AlAsO_4$." *Z. Kristallogr.*, *94*(1), 222–230 (1936).

Mackenzie J.D., Roth W.L., Wentorf R.H. "New high pressure modifications of BPO_4 and $BAsO_4$." *Acta Crystallogr.*, *12*(1), 79 (1959).

Magerramov E.V., Rustamov P.G., Guliev T.N. "Investigation of the interaction in the $In_2S_3–Sb_2S_3$ system" [in Russian]. *Uch. Zap. Azerb. Un-T Ser. Khim. N.*, (2), 12–17 (1975).

Magnavita E.T., Rettori C., Osorio-Guillén J.M., Ferreira F.F., Mendonça-Ferreira L., Avila M.A., Ribeiro R.A. "Low temperature transport and thermodynamic properties of the Zintl compound $Yb_{11}AlSb_9$: A new Kondo lattice semiconductor." *J. Alloys Compd.*, *669*, 60–65 (2016).

Mallik R.C., Stiewe C., Karpinski G., Hassdorf R., Müller E. "Thermoelectric properties of Co_4Sb_{12} skutterudite materials with partial In filling and excess In additions." *J. Electron. Mater.*, *38*(7), 1337–1343 (2009).

Mamedov A.N., Bagirov Z.B., Kerimova R.A. "Thermodynamic functions of mixing of liquid alloys of the Cd–In–Sb system" [in Russian]. *Zhurn. Fiz. Khimii*, *61*(3), 597–601 (1987).

Manasijević D., Minić D., Živković D., Vřešt'ál J., Aljilji A., Talijan N., Stajić-Trošić J., Marjanović S., Todorović R. "Experimental investigation and thermodynamic calculation of the Cu–In–Sb phase diagram." *CALPHAD*, *33*(1), 221–226 (2009).

Manasijević D., Vřešt'ál J., Minić D., Kroupa A., Živković D., Živković Ž. "Experimental investigation and thermodynamic description of the In–Sb–Sn ternary system." *J. Alloys Compd.*, *450*(1–2), 193–199 (2008).

Manasijević D., Živković D., Cocić M., Janjić D., Živković Ž. "Phase equilibria in the quasibinary GaSb–Pb system." *Thermochim. Acta*, *419*(1–2), 295–297 (2004).

Manasijević D., Živković D., Talijan N., Ćosović V., Gomidželović L., Todorović R., Minić D. "Thermal analysis and thermodynamic prediction of phase equilibria in the ternary Au–Ga–Sb system." *J. Phys. Chem. Solids*, *74*(2), 280–285 (2013).

Manasijević D., Živković D., Živković Ž. "Prediction of the thermodynamic properties for the Ga–Sb–Pb ternary system." *CALPHAD*, *27*(4), 361–366 (2003).

Mani H., Joullie A., Karouta F., Schiller C. "Low-temperature phase diagram of the Ga–As–Sb system and liquid-phase-epitaxial growth of lattice-matched GaAsSb on (100) InAs substrates." *J. Appl. Phys.*, 59(8), 2728–2734 (1986).

Manoun B., Kulkarni S., Pathak N., Saxena S.K., Amini S., Barsoum M.W. "Bulk moduli of Cr_2GaC and Ti_2GaN up to 50 GPa." *J. Alloys Compd.*, 505(1), 328–331 (2010).

Manoun B., Saxena S.K., Barsoum M.W. "High pressure study of Ti_4AlN_3 to 55 GPa." *Appl. Phys. Lett.*, 86(10), 101906_1–101906_3 (2005).

Manoun B., Zhang F.X., Saxena S.K., El-Raghy T., Barsoum M.W. "X-ray high-pressure study of Ti_2AlN and Ti_2AlC." *J. Phys. Chem. Solids*, 67(9–10), 2091–2094 (2006).

Mao H., Selleby M. "Thermodynamic reassessment of the Si_3N_4–AlN–Al_2O_3–SiO_2 system—Modeling of the SiAlON and liquid phase." *CALPHAD*, 31(2), 269–280 (2007).

Marenkin S.F. "Phase diagrams of the Zn(Cd)–III-V ternary systems" [in Russian]. *Zhurn. Neorgan. Khimii*, 38(11), 1890–1901 (1993).

Marenkin S.F., Lazarev V.B., Babievskaya I.Z., Pashkova O.N. "Phase equilibria in the Cd–Ga–As ternary system" [in Russian]. *Izv. AN SSSR Neorgan. Mater.*, 23(8), 1241–1246 (1987a).

Marenkin S.F., Pashkova O.N., Babievskaya I.Z., Lazarev V.B. "Phase equilibria in the Cd–In–P system" [in Russian]. *Izv. AN SSSR Neorgan. Mater.*, 24(7), 1075–1079 (1988a).

Marenkin S.F., Pashkova O.N., Lazarev V.B. "Cd–GaAs polythermal section of the Cd–Ga–As system" [in Russian]. *Zhurn. Neorgan. Khimii*, 31(6), 1606–1608 (1986a).

Marenkin S.F., Pashkova O.N., Lazarev V.B. "Interaction in the Cd_3As_2–GaAs system" [in Russian]. *Zhurn. Neorgan. Khimii*, 31(5), 1271–1273 (1986b).

Marenkin S.F., Pashkova O.N., Sanygin V.P., Kvardakov A.M. "Cd–InP vertical section of the Cd–In–P system" [in Russian]. *Izv. AN SSSR Neorgan. Mater.*, 23(9), 1558–1560 (1987b).

Marenkin S.F., Pashkova O.N., Sanygin V.P., Semenova E.Yu., Lazarev V.B. "$CdAs_2$–GaAs polythermal section of the Cd–Ga–As system" [in Russian]. *Zhurn. Neorgan. Khimii*, 31(7), 1821–1824 (1986c).

Marenkin S.F., Pishchikov D.I., Lazarev V.B. "Interaction in the system $ZnAs_2$–InAs" [in Russian]. *Zhurn. Neorgan. Khimii*, 33(4), 988–990 (1988b).

Marenkin S.F., Raukhman A.M., Lazarev V.B. "Interaction in the $CdAs_2$–InAs system" [in Russian]. *Izv. AN SSSR Neorgan. Mater.*, 24(9), 1440–1442 (1988c).

Marenkin S.F., Trukhan V.M., Trukhanov S.V., Fedorchenko I.V., Novotortsev V.M. "Phase equilibria and electrical and magnetic properties of a eutectic in the GaSb–MnSb system." *Russ. J. Inorg. Chem.*, 58(11), 1324–1329 (2013).

Marian C.M., Gastreich M., Gale J.D. "Empirical two-body potential for solid silicon nitride, boron nitride, and borosilazane modifications." *Phys. Rev. B*, 62(5), 3117–3124 (2000).

Markovski S.L., Micke K., Richter K.W., van Loo F.J.J., Ipser H. "Phase relationships in the ternary Ga–Ni–Sb system." *J. Alloys Compd.*, 302(1–2), 128–136 (2000).

Markovski S.L., Pleumeekers M.C.L.P., Kodentsov A.A., van Loo F.J.J. "Phase relations in the Pt–Ga–Sb system at 500°C." *J. Alloys Compd.*, 268(1–2), 188–192 (1998).

Martin M., Dronskowski R., Janek J., Becker K.-D., Roehrens D., Brendt J., Lumey M.W., Nagarajan L., Valov I., Börger A. "Thermodynamics, structure and kinetics in the system Ga–O–N." *Progr. Solid State Chem.*, 37(2–3), 132–152 (2009).

Matar S., Lelogeais M., Michau D., Demazeau G. "Investigations on the high-pressure varieties of $GaAsO_4$." *Mater. Lett.*, 10(1–2), 45–48 (1990).

Mathieu J., Achey R., Park J.-H., Purcell K.M., Tozer S.W., Latturner S.E. "Flux growth and electronic properties of $Ba_2In_5Pn_5$ (Pn = P, As): Zintl phases exhibiting metallic behavior." *Chem. Mater.*, 20(17), 5675–5681 (2008).

Matselko O., Drukhardt U., Grin Yu., Gladyshevskii R. "The ternary system Pd–Ga–Sb at 500°C." In *XIII Intern. Conf. on Crystal Chem. of Intermetallic Compounds*. L viv: Publish. House of L viv Univ., 28 (2016).

Matsuoka T. "Unstable mixing region in wurtzite $In_{1-x-y}Ga_xAl_yN$." *J. Cryst. Growth*, 189–190, 19–23 (1998).

Mattesini M., Matar S.F. "Search for ultra-hard materials: Theoretical characterisation of novel orthorhombic BC_2N crystals." *Int. J. Inorg. Mater.*, 3(7), 943–957 (2001).

Matyas E.E. "Phase diagram of the InAs–AlAs pseudobinary system." *Phys. Status Solidi (A)*, 42(2), K129–K131 (1977).

Mavlonov Sh., Makhmatkulov M. "Study of separate and joint solubility of thallium and selenium in indium arsenide" [in Russian]. *Izv. AN TadzhSSR Otd-Nie Fiz.-Mat. Khim. Geol. N.*, (4), 86–90 (1983).

Mayrhofer P.H., Hultman L., Schneider J.M., Staron P., Clemens H. "Spinodal decomposition of cubic $Ti_{1-x}Al_xN$: Comparison between experiments and modeling." *Int. J. Mat. Res.*, 98(11), 1054–1059 (2007).

Mayrhofer P.H., Music D., Schneider J.M. "*Ab initio* calculated binodal and spinodal of cubic $Ti_{1-x}Al_xN$." *Appl. Phys. Lett.*, 88(7), 071922_1–071922_3 (2006).

Mazej Z. "Indium(I) hexafluoropnictates ($InPnF_6$; Pn = P, As, Sb)." *Eur. J. Inorg. Chem.*, (19), 3983–3987 (2005).

Mbarki M., Rebey A. "First principles calculations of structural and electronic properties of $GaN_{1-x}Bi_x$ alloys." *J. Alloys Compd.*, 530, 36–39 (2012).

McCaldin J.O. "Solubility of zinc in gallium arsenide." *J. Appl. Phys.*, 34(6), 1748–1753 (1963).

McCauley J.W., Corbin N.D. "Phase relations and reaction sintering of transparent cubic aluminum oxynitride spinel (ALON)." *J. Am. Ceram. Soc.*, 62(9–10), 476–479 (1979).

McCauley J.W., Patel P., Chen M., Gilde G., Strassburger E., Paliwal B., Ramesh K.T., Dandekar D.P. "ALON: A brief history of its emergence and evolution." *J. Eur. Ceram. Soc.*, 29(2), 223–236 (2009).

McCluskey M.D., Van de Walle C.G., Master C.P., Romano L.T., Johnson N.M. "Large band gap bowing of In$_x$Ga$_{1-x}$N alloys." *Appl. Phys. Lett.*, *72*(21), 2725–2726 (1998).

McCluskey M.D., Van de Walle C.G., Romano L.T., Krusor B.S., Johnson N.M. "Effect of composition on the band gap of strained In$_x$Ga$_{1-x}$N." *J. Appl. Phys.*, *93*(7), 4340–4342 (2003).

McHale J.M., Navrotsky A., DiSalvo F.J. "Energetics of ternary nitride formation in the (Li,Ca)–(B,Al)–N system." *Chem. Mater.*, *11*(4), 1148–1152 (1999).

Medraj M., Hammond R., Thompson W.T., Drew R.A.L. "Phase equilibria, thermodynamic modeling and neutron diffraction of the AlN–Al$_2$O$_3$–Y$_2$O$_3$." *Can. Metallurg. Q.*, *42*(4), 495–507 (2003).

Micke K., Markovski S.L., Ipser H., van Loo F.J.J. "The nickel-rich part of the Ga–Ni phase diagram and the corresponding relations in the ternary Ga–Ni–Sb system." *Ber. Bunsenges. Phys. Chem.*, *102*(9), 1240–1244 (1998).

Micke K., Richter K.W., Ipser H. "The ternary Ga–Ni–Sb system: Invariant reactions and liquidus surface." *Z. Metallkde*, *92*(1) 14–21 (2001).

Miettinen J., Louhenkilpi S., Vassilev G. "Thermodynamic description of ternary Fe–X–P systems. Part 9. Fe–Al–P." *J. Phase Equilibria Diffus.*, *36*(4), 317–326 (2015).

Mikhalenko S.I., Chernogorenko V.B., Kuz'ma Yu.B., Muchnik S.V., Lavrov Yu.V. "Phase equilibria in the Ni–B–P system" [in Russian]. *Izv. AN SSSR Neorgan. Mater.*, *20*(10), 1746–1748 (1984).

Miki H., Segawa K., Otsubo M., Shirahata K., Fujibayashi K. "The Ga–In–Sb ternary phase diagram at low growth temperature." *Jpn. J. Appl. Phys.*, *17*(12), 2079–2084 (1978).

Miller J.F., Goering H.L., Himes R.C. "Preparation and properties of AlSb–GaSb solid solution alloys." *J. Electrochem. Soc.*, *107*(6), 527–533 (1960).

Mills A.M., Mar A. "Rare-earth gallium antimonides La$_{13}$Ga$_8$Sb$_{21}$ and RE_{12}Ga$_4$Sb$_{23}$ (RE = La–Nd, Sm): Linking Sb ribbons by Ga$_6$-rings or Ga$_2$-pairs." *Inorg. Chem.*, *39*(20), 4599–4607 (2000).

Mills A.M., Mar A. "Layered rare-earth gallium antimonide REGaSb$_2$ (RE = La–Nd, Sm)." *J. Am. Chem. Soc.*, *123*(5), 1151–1158 (2001).

Mil'vidskiy M.G., Pelevin O.V. "About determination of distribution coefficients of volatile impurities at the growing of the gallium arsenide single crystals by the direction crystallization" [in Russian]. *Izv. AN SSSR Neorgan. Mater.*, *1*(9), 1454–1458 (1965).

Mil'vidskiy M.G., Pelevin O.V., Belyaev A.I. "About concentration dependence of zinc distribution coefficient in gallium arsenide" [in Russian]. *Izv. AN SSSR Neorgan. Mater.*, *2*(2), 249–250 (1966).

Minaev V.S. "New glasses and peculiarities of glass formation in the ternary telluride systems" [in Russian]. *Fiz. Khimiya Stekla*, *9*(4), 432–436 (1983).

Minić D., Manasijević D., Živković D., Stajić-Trošić J., Đokić J., Petković D. "Experimental investigation and thermodynamic calculation of Bi–Ga–Sb phase diagram." *Mater. Sci. Technol.*, *27*(5), 884–889 (2011).

Minić D., Manasijević D., Živković D., Štrbac N., Stanković Z. "Prediction of phase equilibria in the In–Sb–Pb system." *J. Serb. Chem. Soc.*, *73*(3), 377–384 (2008).

Minić D., Manasijević D., Živković D., Živković Ž. "Phase equilibria in the In–Sb–Bi system at 300°C." *J. Serb. Chem. Soc.*, *71*(7), 843–847 (2006).

Minić D., Premović M., Ćosović V., Manasijević D., Živković D. "Experimental investigation and thermodynamic calculations of the Al–Cu–Sb phase diagram." *J. Alloys Compd.*, *555*, 347–356 (2013).

Minic D., Zivkovic D., Zivkovic Z. "Thermodynamic and structure analysis of the Pb–InSb system." *Thermochim. Acta*, *400*(1–2), 143–152 (2003).

Mirgalovskaya M.S., Alekseeva E.M. "Ga–Sb–Cd system" [in Russian]. *Izv. AN SSSR Neorgan. Mater.*, *1*(2), 193–200 (1965).

Mirgalovskaya M.S., Komova E.M. "Phase interaction in the Ga–Sb–Zn system" [in Russian]. *Izv. AN SSSR Neorgan. Mater.*, *11*(1), 45–50 (1975).

Mirgalovskaya M.S., Komova E.M., Il'chenko L.N., Karpinskiy O.G. "Properties of the solid solution based on GaSb in the Ga–Sb–Te ternary system" [in Russian]. *Izv. AN SSSR Neorgan. Mater.*, *13*(9), 1574–1577 (1977).

Mirgalovskaya M.S., Skudnova E.V. "Investigation of the alloys of the AlSb–Al$_2$Te$_3$ system" [in Russian]. *Zhurn. Neorgan. Khimii*, *4*(5), 1113–1120 (1959).

Mironov A., Kazakov S., Jun J., Karpinski J. "MgNB$_9$, a new magnesium nitridoboride." *Acta Crystallogr.*, *C58*(7), i95–i97 (2002).

Mirtskhulava A.A., Rakov V.V., Lainer B.D., Mil'vidskiy M.G., Sakvarelidze L.G. "Investigation of the phase equilibrium in the gallium arsenide–aluminum arsenide system" [in Russian]. Deposited in VINITI, N. 3039–71Dep (1971).

Mirtskhulava A.A., Sakvarelidze L.G. "Investigation of the phase equilibria in the GaAs–InAs and GaAs–AlAs systems" [in Russian]. *Zhurn. Fiz. Khimii*, *51*(2), 513–516 (1977).

Miyake M., Nakamura K., Akiyama T., Ito T. "Structural stability of Mn-doped GaInAs and GaInN alloys." *J. Cryst. Growth*, *362*, 324–326 (2013).

Miyoshi S., Yaguchi H., Onabe K., Ito R., Shiraki Y. "Metalorganic vapor phase epitaxy of GaP$_{1-x}$N$_x$ alloys on GaP." *Appl. Phys. Lett.*, *63*(25), 3506–3508 (1993).

Mohammad R., Katircioğlu Ş. "The structural and electronic properties of BAs and BP compounds and BP$_x$As$_{1-x}$ alloys." *J. Alloys Compd.*, *485*(1–2), 687–694 (2009a).

Mohammad R., Katircioğlu Ş. "The structural and electronic properties of BN and BP compounds and BN$_x$P$_{1-x}$ alloys." *J. Alloys Compd.*, *478*(1–2), 531–537 (2009b).

Mohney S.E. "Condensed phase equilibria in the transition metal–In–P systems." *J. Electron. Mater.*, *27*(1), 26–31 (1998).

Mohney S.E., Chang Y.A. "Phase equilibria and ternary phase formation in the In–Ni–P system." *J. Mater. Res.*, *7*(4), 955–960 (1992).

Mohney S.E., Chang Y.A. "Solid phase equilibria in the In–P–Pd system." *Mater. Sci. Eng. B*, 18(1), 94–99 (1993).

Mohney S.E., Lin X. "Estimated phase equilibria in the transition metal–Ga–N systems: Consequences for electrical contacts to GaN." *J. Electron. Mater.*, 25(5), 811–818 (1996).

Mohney S.E., MacMahon D.J., Whitmire K.A. "Condensed phase equilibria in the Cr–Ga–N system." *Mater. Sci. Eng. B*, 49(2), 152–154 (1997).

Moldovyan N.A. "Photoconductivity and optical absorption of $InSbS_3$ crystals" [in Russian]. *Izv. AN SSSR Neorgan. Mater.*, 26(6), 1325–1327 (1990).

Molodyan I.P. "Electrical properties of the alloys of the $(InSb)_x$–$(InTe)_{1-x}$ section" [in Russian]. In *Tr. 3 Konf. Molodykh Uchenykh Moldavii. Yestestv.-Tekhn. N. Vyp. 1*. Kishinev: Kartya Moldovenyaske Publish., 33 (1964).

Molodyan I.P., Radautsan S.I. "Solid solutions based on indium antimonide in the indium–antimony tellurium system" [in Russian]. In *Issled. Po Poluprovodn*. Kishinev: Kartya Moldovenyaske Publish., 143–152 (1964a).

Molodyan I.P., Radautsan S.I. "Some homogeneous phases of indium antimonide-telluride" [in Russian]. *Izv. AN SSSR Ser. Fiz.*, 28(6) 1017–1022 (1964b).

Molodyan I.P., Radautsan S.I., Madan I.A. "Some structural and thermic investigations of the In_4SbTe_3 compound" [in Russian]. *Izv. AN MSSR*, 10(88), 91–94 (1961).

Mooney R.C.L. Crystal structure of anhydrous indium phosphate and thallic phosphate by x-ray diffraction." *Acta Crystallogr.*, 9(2), 113–117 (1956a).

Mooney R.C.L. "The crystal structure of aluminium phosphate and gallium phosphate, low-cristobalite type." *Acta Crystallogr.*, 9(9), 728–734 (1956b).

Mooney R.C.L., Kissinger H., Perloff A. "X-ray analysis of orthophosphate of some trivalent elements." *Acta Crystallogr.*, 7(10), 642–643 (1954).

Morjan I., Conde O., Oliveira M., Vasiliu F. "Structural characterization of $C_xB_yN_z$ (x = 0.1 to 0.2) layers obtained by laser-driven synthesis." *Thin Solid Films*, 340(1–2), 95–105 (1999).

Morozov A.N., Bublik V.T., Grigor'eva T.P., Karataev V.V., Mil'vidskiy M.G. "Calculation of the homogeneity region of indium arsenide doped with tin" [in Russian]. *Kristallografiya*, 30(3), 548–554 (1985).

Morozov A.N., Bublik V.T., Morozova O.Yu. "Calculation of the homogeneity region of tin doped gallium arsenide" [in Russian]. *Izv. AN SSSR Neorgan. Mater.*, 23(9), 1429–1433 (1987).

Morrison C.B., Bedair S.M. "Phase diagram for In–Ga–P ternary system." *J. Appl. Phys.*, 53(12), 9058–9062 (1982).

Moze O., Paoluzi A., Pareti L., Turilli G., Bouree-Vigneron F., Cockayne B., Macewan W.R., Greaves C., Smith N.A., Harris I.R. "Magnetic and crystal structure and magnetocrystalline anisotropy of hexagonal $Fe_{60}Ga_{40-x}As_x$ (x = 6, 23) and $Fe_{60}Ga_{23}Sb_{17}$ compounds." *IEEE Trans. Magn.*, 30(2), 1039–1041 (1994).

Müller A., Wilhelm M. "Über den gerichteten einbau von Schwermetallphasen in $A^{III}B^V$-Verbindungen: Die Eutektika GaSb–CrSb, GaSb–$FeGa_{1.3}$, GaSb–$CoGa_{1.3}$, InAs–CrAs und InAs–FeAs." *J. Phys. Chem. Solids*, 26(12), 2029–2035 (1965a).

Müller A., Wilhelm M. "Über den gerichteten einbau von Schwermetallphasen in $A^{III}B^V$-Verbindungen: Die Eutektika InSb–NiSb, InSb–FeSb, InSb–MnSb und InSb–CrSb." *J. Phys. Chem. Solids*, 26(12), 2021–2028 (1965b).

Müller A., Wilhelm M. "Der quasibinäre Schnitt InSb–NiSb in Dreistoffsystem In–Sb–Ni." *Mater. Res. Bull.*, 2(7), 531–534 (1967a).

Müller A., Wilhelm M. "Gefügefermen im Eutektikum InSb–NiSb." *Z. Naturforsch.*, 22A(2), 264–269 (1967b).

Müller A., Wilhelm M. "Die binaren Eutektika GaSb–GaV_3Sb_5 und GaSb–V_2Sb_5." *J. Phys. Chem. Solids*, 28(2), 219–224 (1967c).

Müller C.J., Lidin S. "On squaring triangles—Structural motifs in Cu–In–Sb compounds." *J. Solid State Chem.*, 231, 25–35 (2015).

Müller C.J., Lidin S., Ramos de Debiaggi S., Deluque Toro C.E., Guillermet A.F. "Synthesis, structural characterization, and ab initio study of $Cu_{5+}\delta In_{2+x}Sb_{2-x}$: A new B8-related structure type." *Inorg. Chem.*, 51(20), 10787–10792 (2012).

Müller U. "Die Kristall- und Molekularstruktur von Bordichloridazid $(BCl_2N_3)_3$." *Z. Anorg. Allg. Chem.*, 382(2), 110–122 (1971).

Munekata H., Ohno H., Ruf R.R., Gambino R.J., Chang L.L. "p-Type diluted magnetic III-V semiconductors." *J. Cryst. Growth*, 111(1–4), 1011–1015 (1991).

Munekata H., Ohno H., von Molnar S., Segmüller A., Chang L.L., Esaki L. "Diluted magnetic III-V semiconductors." *Phys. Rev. Lett.*, 63(17), 1849–1852 (1989).

Muraoka Y., Kihara K. "The temperature dependence of the crystal structure of berlinite, a quartz-type form of $AlPO_4$." *Phys. Chem. Miner.*, 24(4), 243–253 (1997).

Muravieva A.A., Zarechnyuk O.S., Gladyshevskiy E.I. "The Y–Al–Si(Ge, Sb) systems within the 0-33.3 at.% Y region" [in Russian]. *Izv. AN SSSR Neorgan. Mater.*, 7(1), 38–40 (1971).

Musil J., Šašek M., Zeman P., Čerstvý R., Heřman D., Han J.B., Šatava V. "Properties of magnetron sputtered Al–Si–N thin films with a low and high Si content." *Surf. Coat. Technol.*, 202(15), 3485–3493 (2008).

Muszyński Z., Ryabcev N. "The determination of liquidus surface in ternary phase diagrams of Ga-As–P, In-As–P, Ga-In-As, Ga-Al-As and Ga-Al-Sb by simplex lattice method." *Electron. Technol.*, 8(3–4), 119–130 (1975).

Muszyński Z., Ryabcev N. "A new method for the determination of liquidus surfaces in ternary phase diagrams of Ga-As–P, In-As–P, Ga-In-As, Ga-Al-As and Ga-Al-Sb." *J. Cryst. Growth*, 36(2), 335–341 (1976).

Nahory R.E., Pollack M.A., DeWinter J.C., Williams K.M. "Growth and properties of liquid-phase epitaxial $GaAs_{1-x}Sb_x$." *J. Appl. Phys.*, 48(4), 1607–1614 (1977).

Nakae H., Kihara K., Okuno M., Hirano S. "The crystal structure of the quartz-type form of $GaPO_4$ and its temperature dependence." *Z. Kristallogr.*, 210(10), 746–753 (1995).

Nakahara S., Gallagher P.K., Felder E.C., Lawry R.B. "Interaction between zinc metallization and indium phosphide." *Solid State Electron.*, 27(6), 557–564 (1984).

Nakajima K., Okazaki J. "Substrate orientation dependence of the In–Ga–As phase diagram for liquid phase epitaxial growth of $In_{0.53}Ga_{0.47}As$ on InP." *J. Electrochem. Soc.*, 132(6), 1424–1432 (1985).

Nakajima K., Osamura K., Murakami Y. "Study of the InP–InSb quasi-binary phase diagram" [in Japanese]. *J. Jpn. Inst. Met.*, 37(12), 1276–1278 (1973).

Nakajima K., Tanahashi T., Akita K., Yamaoka T. "Determination of In–Ga–As phase diagram at 650°C and LPE growth of lattice-matched $In_{0.53}Ga_{0.47}As$ on InP." *J. Appl. Phys.*, 50(7), 4975–4981 (1979).

Nakamura K., Hatano K., Akiyama T., Ito T., Freeman A.J. "Lattice expansion, stability, and Mn solubility in substitutionally Mn-doped GaAs." *Phys. Rev. B*, 75(20), 205205_1–205205_6 (2007).

Nakano S., Akaishi M., Sasaki T., Yamaoka S. "Segregative crystallization of several diamond-like phases from the graphitic BC_2N without an additive at 7.7 GPa." *Chem. Mater.*, 6(12), 2246–2251 (1994a).

Nakano S., Ikawa H., Fukunaga O. "Synthesis of cubic boron nitride by decomposition of magnesium boron nitride." *J. Am. Ceram. Soc.*, 75(1), 240–243 (1992).

Nakano S., Ikawa H., Fukunaga O. "High pressure reactions and formation mechanism of cubic BN in the system $BN–Mg_3N_2$." *Diamond Relat. Mater.*, 2(8), 1168–1174 (1993).

Nakano S., Ikawa H., Fukunaga O. "Synthesis of cubic boron nitride using Li_3BN_2, $Sr_3B_2N_4$ and $Ca_3B_2N_4$ as solvent-catalysts." *Diamond Relat. Mater.*, 3(1–2), 75–82 (1994b).

Nanjundaswamy K.S., Gopalakrishnan J. "Synthesis and characterization of Ti_5Te_4-type molybdenum cluster compounds, $A_xMo_5As_4$ (A = Cu, Al, or Ga)." *J. Chem. Soc. Dalton Trans.*, (1), 1–5 (1988).

Nashelskiy A.Ya., Yakobson S.V., Minakova N.Yu. "Effective distribution coefficients of dopants in crystals of indium phosphide" [in Russian]. In *Legir. Poluprovodn.*, Moscow: Nauka Publish., 75–77 (1982).

Negreskul V.V. "Investigation of the semiconductor alloys of the $GaP–Ga_2S_3$ system" [in Russian]. In *Tr. 3 Konf. Molodykh Uchenykh Moldavii. Estestv.-Tekhn. N.* Vyp. 1. Kishinev: Kartia Moldoveniaske Publish., 35–36 (1964).

Negreskul V.V., Radautsan S.I. "Some properties of the solid solutions based on gallium phosphide" [in Russian]. *Izv. AN SSSR Ser. Fiz.*, 28(6), 1002–1006 (1964).

Ng H.N., Calvo C. "X-ray study of the twinning and phase transformation of phosphocristobalite ($AlPO_4$)." *Can. J. Phys.*, 55(7–8), 677–683 (1977).

Nicolich J.P., Hofer F., Brey G., Riedel R. "Synthesis and structure of three-dimensionally ordered graphitelike BC_2N ternary crystals." *J. Am. Ceram. Soc.*, 84(2), 279–282 (2001).

Nikitina V.K., Babitsyna A.A., Lobanova Yu.K. "On the interaction of indium antimonide with $AuIn_2$," [in Russian]. *Izv. AN SSSR Neorgan. Mater.*, 3(2), 311–314 (1967).

Nikitina V.K., Lobanova Yu.K. "Phase diagram of the AuIn–Sb system" [in Russian]. *Izv. AN SSSR Neorgan. Mater.*, 10(9), 1596–1599 (1974).

Nitsche R., Wild P. "Crystal growth of metal–phosphorus–sulfur compounds by vapor transport." *Mater. Res. Bull.*, 5(6), 419–423 (1970).

Nordman C.E., Reimann C. "The molecular and crystal structures of ammonia–triborane." *J. Am. Chem. Soc.*, 81(14), 3538–3543 (1959).

Novikov M.V., Bezhenar M.P., Bozhko S.A. "Evolution of the crystal structure of sphaleritic boron nitride at the sintering of BNcϕ–AlN and BNcϕ–TiC components and its influence on the hardness" [in Ukrainian]. *Dop. Nats. AN Ukrainy*, (6), 118–122 (1997).

Novotortsev V.M., Kochura A.V., Marenkin S.F., Fedorchenko I.V., Drogunov S.V., Lashkul A., Lähderanta E. "Synthesis and magnetic properties of the InSb–MnSb eutectic" [in Russian]. *Zhurn. Neorgan. Khimii*, 56(12), 2038–2044 (2011).

Nowotny H., Jeitschko W., Benesovsky F. "Neue Komplexkarbide und -nitride, ihre Beziehung zu Hartstoffphasen." *Planseeber. Pulvermet.*, 12, 31–43 (1964).

Nowotny H., Rudy E., Benesovsky F. "Untersuchungen in den Systemen: Zirkonium–Bor–Kohlenstoff und Zirkonium–Bor–Stickstoff." *Monatsch. Chem.*, 91(5), 963–974 (1960).

Nozaki H., Itoh S. "Structural stability of BC_2N." *J. Phys. Chem. Solids*, 57(1), 41–49 (1996).

Nye S.D., Banus M.D., Gatos H.C. "Superconductivity in the high-pressure InSb–beta-Sn system." *J. Appl. Phys.*, 35(4), 1361–1362 (1964).

Oe K., Okamoto H. "New semiconductor alloy $GaAs_{1-x}Bi_x$ grown by metal organic vapor phase epitaxy." *Jpn. J. Appl. Phys.*, 37(11A), Pt. 2, L1283–L1285 (1998).

Ogawa M. "Alloying reaction in thin nickel films deposited on GaAs." *Thin Solid Films*, 70(1), 181–189 (1980).

Ohnuma I., Cui Y., Liu X.J., Inohana Y., Ishihara S., Ohtani H., Kainuma R., Ishida K. "Phase equilibria of Sn–In based micro-soldering alloys." *J. Electron. Mater.*, 29(10), 1113–1121 (2000).

Ohse L., Somer M., Blase W., Cordier G. "Verbindungen mit SiS_2-isosteren Anionen $^1_\infty[AlX_{4/2}{}^{3-}]$ und $^1_\infty[InP_{4/2}{}^{3-}]$: Synthesen, Kristallstrukturen und Schwingungsspektren von $Na_3[AlX_2]$. $K_2Na[AlX_2]$ und $K_3[InP_2]$ (X = P, As)." *Z. Naturforsch.*, 48B(8), 1027–1034 (1993).

Ohshima H., Tanaka A., Sukegawa T. "Calculation of the Ga–Al–Sb phase diagram using the Redlich-Kister expression." *Phys. Stat. Sol. (A)*, 87(2), K131–K134 (1985).

Ohtani H., Kobayashi K., Ishida K. "Thermodynamic study of phase equilibria in strained III-V alloy semiconductors." *J. Phase Equilibria*, 22(3), 276–286 (2001).

O'Kane D.F., Stemple N.R. "Phase diagram for the pseudo-binary system InSb–InSe." *J. Electrochem. Soc.*, 113(3), 289–290 (1966).

Okeke O.U., Lowthe J.E. "Theoretical electronic structures and relative stabilities of the spinel oxynitrides M_3NO_3 (M = B, Al, Ga, In)." *Phys. Rev. B*, 77(9), 094192_1– 094192_9 (2008).

Olowolafe J.O., Ho P.S., Hovel H.J., Lewis J.E., Woodall J.M. "Contact reactions in Pd/GaAs junctions." *J. Appl. Phys.*, 50(2), 955–962 (1979).

Onabe K. "Liquid-solid equilibrium phase diagrams for III-V ternary solid solutions with miscibility gap." *Jpn. J. Appl. Phys.*, 22(1), Pt. 1, 201 (1983).

Onac B.P., Effenberger H.S. "Re-examination of berlinite ($AlPO_4$) from the Cioclovina Cave, Romania." *Am. Mineral.*, 92(11–12), 1998–2001 (2007).

Onac B.P., White W.B. "First reported sedimentary occurrence of berlinite ($AlPO_4$) in phosphate-bearing sediments from Cioclovina Cave, Romania." *Am. Mineral.*, 88(8–9), 1395–1397 (2003).

Onodera A., Matsumoto K., Hirai T., Goto T., Motoyama M., Yamada K., Kohzuki H. "Synthesis of dense forms of B–C–N system using chemical-vapor-deposition/high-pressure process." *J. Mater. Sci.*, 36(3), 679–684 (2001).

Orth M., Schnick W. "Zur Kenntnis von Tripraseodym-hexanitridotriborat $Pr_3B_3N_6$: Neue Synthese und Verfeinerung der Kristallstruktur." *Z. Anorg. Allg. Chem.*, 625(4), 551–554 (1999).

Osamura K., Inoue J., Murakami Y. "Experiments and calculation of the Ga–GaAs–GaP ternary phase diagram." *J. Electrochem. Soc.*, 119(1), 103–108 (1972a).

Osamura K., Murakami Y. "Phase diagram of GaAs–GaP quasi-binary system." *Jpn. J. Appl. Phys.*, 8(7), 967 (1969).

Osamura K., Naka S., Murakami Y. "Preparation and optical properties of $Ga_{1-x}In_xN$ thin films." *J. Appl. Phys.*, 46(8), 3432–3437 (1975).

Osamura K., Nakajima K., Murakami Y. "Calculation of III-V quasi-binary phase diagram and theoretical analysis of the excess free energies for their solid solutions" [in Japanese]. *J. Jpn. Inst. Metals*, 36(8), 744–750 (1972b).

Osamura K., Nakajima K., Murakami Y. "Experiments and calculation of the Al–Ga–Sb ternary phase diagram." *J. Electrochem. Soc.*, 126(11), 1992–1997 (1979).

Ovchinnikov S.Yu., Sorokin V.S., Yas'kov D.A. "Phase equilibria in the GaAs–GaP system" [in Russian]. *Izv. AN SSSR Neorgan. Mater.*, 10(2), 234–239 (1974).

Paetzold P.I. "Beiträge zur Chemie der Bor-Azide. I. Zur Kenntnis von Dichlorborazid." *Z. Anorg. Allg. Chem.*, 326(1–2), 47–52 (1963).

Paetzold P.I., Gayoso M., Dehnicke K. "Darstellung, Eigenschaften und Schwingungsspektren der trimeren Bordihalogenidazide ($BCl_2N_3)_3$ und ($BBr_2N_3)_3$." *Chem. Ber.*, 98(4), 1173–1180 (1965).

Paliwal M., Jung I.-H. "Thermodynamic modeling of the Al–Bi, Al–Sb, Mg–Al–Bi and Mg–Al–Sb systems." *CALPHAD*, 34(1), 51–63 (2010).

Palkina K.K., Maksimova S.I., Chibiskova N.T., Chudinova N.N., Karmanovskaya N.B. "Crystal structure of indium polyphosphate $In(PO_3)_3$-C " [in Russian]. *Zhurn. Neorgan. Khimii*, 38(8), 1270–1272 (1993).

Pan W., Wang P., Wu X., Wang K., Cui J., Yue L., Zhang L., Gong Q., Wang S. "Growth and material properties of InPBi thin films using gas source molecular beam epitaxy." *J. Alloys Compd.*, 656, 777–783 (2016).

Pan Z., Sun H., Chen C. "*Ab initio* structural identification of high density cubic BC_2N." *Phys. Rev. B*, 73(21), 214111_1–214111_4 (2006).

Pang W.K., Low I.M., Kennedy S.J., Smith R.I. "*In situ* diffraction study on decomposition of Ti_2AlN at 1500– 1800°C in vacuum." *Mater. Sci. Eng. A*, 528(1), 137–142 (2010).

Panish M.B. "The arsenic-rich region of the Ga–As–Zn ternary phase system—Modification of the ternary diagram." *J. Electrochem. Soc.*, 113(8), 861 (1966a).

Panish M.B. "The Ga–As–Si ternary phase system." *J. Electrochem. Soc.*, 113(11), 1226–1228 (1966b).

Panish M.B. "The gallium–arsenic–tin and gallium–arsenic–germanium ternary systems." *J. Less Common Metals*, 10(6), 416–424 (1966c).

Panish M.B. "The gallium–arsenic–zinc system." *J. Phys. Chem. Solids*, 27(2), 291–298 (1966d).

Panish M.B. "The gallium–phosphorus–zinc ternary phase diagram." *J. Electrochem. Soc.*, 113(3), 224–226 (1966e).

Panish M.B. "The psedobinary system Ge-GaAs." *J. Electrochem. Soc.*, 113(6), 626 (1966f).

Panish M.B. "Ternary condensed phase systems of gallium and arsenic with Group IB elements." *J. Electrochem. Soc.*, 114(5), 516–521 (1967a).

Panish M.B. "The Ga–P–Te system." *J. Electrochem. Soc.*, 114(11), 1161–1164 (1967b).

Panish M.B. "The ternary condensed phase diagram of the Ga–As–Te system." *J. Electrochem. Soc.*, 114(1), 91–95 (1967c).

Panish M.B. "The Ga–GaAs–GaP system: Phase chemistry and solution growth of $GaAs_xP_{1-x}$." *J. Phys. Chem. Solids*, 30(5), 1083–1090 (1969).

Panish M.B. "Liquidus isotherms in the Ga–In–As system." *J. Electrochem. Soc.*, 117(9), 1202–1203 (1970a).

Panish M.B. "The Ga + In + P system." *J. Chem. Thermodyn.*, 2(3), 319–331 (1970b).

Panish M.B. "The system Ga–As–Sn: Incorporation of Sn into GaAs." *J. Appl. Phys.*, 44(6), 2659–2666 (1973).

Panish M.B., Ilegems M. "Phase equilibria in ternary III-V systems." *Progr. Solid State Chem.*, 7, 39–83 (1972).

Panish M.B., Sumski S. "Ga–Al–As: Phase, thermodynamic and optical properties." *J. Phys. Chem. Solids*, 30(1), 129–137 (1969).

Panish M.B., Sumski S. "Ga–As–Si: Phase studies and electrical properties of solution-grown Si-doped GaAs." *J. Appl. Phys.*, 41(7), 3195–3196 (1970).

Papadakis A.C., Heasell E.L., Barber H.D. "The solubility of indium antimonide in tin." *Solid State Electron.*, 8(10), 825–827 (1965).

Park D.G., Gál Z.A., DiSalvo F.J. "Synthesis and structure of Sr_3GaN_3 and Sr_6GaN_5: Strontium gallium nitrides with isolated planar $[GaN_3]^{6-}$ anions." *Inorg. Chem.*, 42(5), 1779–1785 (2003a).

Park S.-M., Choi E.S., Kang W., Kim S.-J. "$Eu_5In_2Sb_6$, $Eu_5In_{22-x}Zn_xSb_6$: Rare earth Zintl phases with narrow band gaps." *J. Mater. Chem.*, 12(6), 1839–1843 (2002).

Park S.-M., Kim S.-J., Kanatzidis M.G. "Ga–Ga bonding and tunnel framework in the new Zintl phase $Ba_3Ga_4Sb_5$." *J. Solid State Chem.*, 175(2), 310–315 (2003b).

Park S.-M., Kim S.-J., Kanatzidis M.G. "Eu7Ga6Sb8: A Zintl phase with Ga–Ga bonds and polymeric gallium anti-monide chains." *J. Solid State Chem.*, 177(8), 2867–2874 (2004).

Parry R.W., Schultz D.R., Girardot P.R. "The preparation and properties of hexamminecobalt(III) borohydride, hexamminechromium(III) borohydride and ammonium borohydride." *J. Am. Chem. Soc.*, 80(1), 1–3 (1958).

Pashkova O.N., Sanygin V.P., Ivanov V.A., Padalko A.G., Novotortsev V.M. "Synthesis and properties of the $In_{1-x}Mn_xSb$ solid solutions" [in Russian]. *Neorgan. Mater.*, 42(5), 519–522 (2006).

Patrone M., Richter K.W., Borzone G., Ipser H. "The In–Pt–Sb phase diagram." *Int. J. Mater. Res.*, 97(5), 533–538 (2006).

Pavlova L.M., Antonov V.A., Vanyukova N.V., Karamov A.G., Perederiy L.I., Selin V.V. "Doping of indium phosphide by germanium" [in Russian]. *Izv. AN SSSR Neorgan. Mater.*, 18(9), 1444–1447 (1982).

Pavlyuchkov D., Fabrichnaya O., Herrmann M., Seifert H.J. "Thermodynamic assessments of the $Al_2O_3–Al_4C_3–AlN$ and $Al_4C_3–AlN–SiC$ systems." *J. Phase Equilibria Diffus.*, 33(5), 357–368 (2012).

Pauling L., Sherman J. "The crystal structure of aluminum metaphosphate, $Al(PO_3)_3$." *Z. Kristallogr.*, 96(1), 481–487 (1937).

Pelevin O.V. "Phase diagrams of the Ga–GaAs–Si, Ga–GaAs–Ge and Ga–GaAs–Sn" [in Russian]. *Izv. AN SSSR Neorgan. Mater.*, 11(1), 51–55 (1975).

Pelevin O.V., Chupakhina V.M. "Phase diagram of the Ga–AlSb–GaSb system and liquid phase epitaxy of $(AlSb)_x(GaSb)_{1-x}$ solid solution" [in Russian]. *Izv. AN SSSR Neorgan. Mater.*, 11(9), 1698–1699 (1975).

Pelevin O.V., Gimel'farb F.A., Mil'vidskiy M.G., Ukhorskaya T.A. "Express method of triangulations of the Ga–As–impurity ternary systems" [in Russian]. *Izv. AN SSSR Neorgan. Mater.*, 8(10), 1744–1750 (1972a).

Pelevin O.V., Gimel'farb F.A., Mil'vidskiy M.G., Zherdev B.P. "GaAs–Zn polythermal section" [in Russian]. *Izv. AN SSSR Neorgan. Mater.*, 8(6), 1049–1054 (1972b).

Pelevin O.V., Girich B.G., Mil'vidskiy M.G. "The behavior of iron at the crystallization of gallium arsenide" [in Russian]. *Izv. AN SSSR Neorgan. Mater.*, 5(7), 1200–1202 (1969).

Pelevin O.V., Mil'vidskiy M.G. "Some pecularities of zinc behavior at the growth of doping gallium arsenide single crystals" [in Russian]. *Izv. AN SSSR Neorgan. Mater.*, 4(11), 1864–1868 (1968).

Pelevin O.V., Mil'vidskiy M.G., Zherdev B.P. "Phase diagram of the GaAs–Sn quasi-binary system" [in Russian]. *Izv. AN SSSR Neorgan. Mater.*, 8(3), 446–450 (1972c).

Pelevin O.V., Shershakova I.N., Gimel'farb F.A., Mil'vidskiy M.G., Ukhorskaya T.A. "Crystallization of gallium arsenide doped with the elements of Group IVA" [in Russian]. *Kristallografiya*, 16(3), 622–627 (1971).

Pelevin O.V., Shershakova I.N., Mil'vidskiy M.G. "GaAs–Si and GaAs–Ge quasi-binary system" [in Russian]. *Nauch. Tr. N.-I Proekt. In-T Redkomet. Prom-Sti*, 46, 69–74 (1973).

Pelevin O.V., Ufimtseva L.V. "Solubility of zinc in solid gallium arsenide in connection with Ga–As–Zn phase diagram" [in Russian]. *Nauch. Tr. N.-I Proekt. In-T Redkomet. Prom-Sti*, 55, 123–130 (1974).

Pélisson A., Parlinska-Wojtan M., Hug H.J., Patscheider J. "Microstructure and mechanical properties of Al–Si–N transparent hard coatings deposited by magnetron sputtering." *Surf. Coat. Technol.*, 202(4–7), 884–889 (2007).

Pélisson-Schecker A., Hug H.J., Patscheider J. "Morphology, microstructure evolution and optical properties of Al–Si–N nanocomposite coatings." *Surf. Coat. Technol.*, 257, 114–120 (2014).

Peltier V., L'Haridon P., Marchand R., Laurent Y. "Synthèse et caractérisation structurale d'un phosphate d'indium $In_3P_2O_8$ présentant des paires In–In." *Acta Crystallogr.*, B52(6), 905–908 (1996).

Peretti E.A. "The system InSb–InBi." *Trans. Metallurg. Soc. AIME*, 212(1), 79 (1958).

Peretti E.A. "Constitution studies of the antimony- and bismuth-rich portions of the Sb–Bi–In system." *Trans. ASM*, 52, 1046–1058 (1960).

Peretti E.A. "Constitution studies of the indium-rich portion of the system antimony–bismuth–indium." *Trans. ASM*, 54, 12–19 (1961).

Peretti E.A.J., Jelinek T.M., Peretti E.A. "Contribution to the system indium–arsenic–tellurium." *J. Less Common Metals*, 35(2), 293–297 (1974).

Pérez W.L., Rodríguez J.A.M., Mancera L. "In-composition effect on band gap width of $Sc_{1-x}In_xN$ alloys." *J. Alloys Compd.*, 481(1–2) 697–703 (2009).

Pérez W.L., Rodríguez J.A.M., Moreno-Armenta M.G. "First-principles calculations of structural properties of $Sc_{1-x}In_xN$ compound." *Physica B*, 398(2), 385–388 (2007).

Perloff A. "Temperature inversions of anhydrous gallium orthophosphate." *J. Am. Ceram. Soc.*, 39(3), 83–88 (1956).

Pessetto J.R., Stringfellow G.B. "$Al_xGa_{1-x}As_ySb_{1-y}$ phase diagram." *J. Cryst. Growth*, 62(1), 1–6 (1983).

Peters K., Carrillo-Cabrera W., Somer M., von Schnering H.G. "Crystal structure of hexabarium di-μ-phosphido-bis(diphosphidogallate(III)), $Ba_6[Ga_2P_6]$." *Z. Kristallogr.*, 211(1), 53 (1996).

Peuschel G.-P., Knobloch G., Butter E., Apelt R. "Thermodynamic studies on ternary mixed III-V systems." *Cryst. Res. Technol.*, *16*(1), 13–18 (1981).

Pietzka M.A., Schuster J.C. "Phase equilibria in the quaternary system Ti–Al–C–N." *J. Am. Ceram. Soc.*, *79*(9), 2321–2330 (1996).

Pikalov S.N., Tarasenko N.V., Sobokar' O.A., Bondarevich K.A. "Composition of magnesium boronitride synthesized from boron nitride and magnesium in nitrogen atmosphere at 1400 K" [in Russian]. *Poroshk. Metallurgia*, (3), 71–74 (1989).

Piotrowska A., Auvray P., Guivarc'h A., Pelous G., Henoc P. "On the formation of binary compounds in Au/InP system." *J. Appl. Phys.*, *52*(8), 5112–5117 (1981).

Pizzarello F.A. "Preparation of solid solutions of GaP and GaAs by a gas phase reaction." *J. Electrochem. Soc.*, *109*(3), 226–229 (1962).

Pohl S. "$Al_2Sb_2I_{12}$: Vierkern-Einheiten in einem 1:1-Addukt aus AlI_3 und SbI_3." *Z. Naturforsch.*, *38B*(12), 1539–1542 (1983).

Pollak M.A., Nahory R.E., Deas L.V., Wonsidler D.R. "Liquidus-solidus isotherms in the In–Ga–As system." *J. Electrochem. Soc.*, *122*(11), 1550–1552 (1975).

Polyakov A.Y., Shin M., Skowronski M., Greve D.W., Wilson R.G., Govorkov A.V., Desrosiers R.M. "Growth of GaBN ternary solutions by organometallic vapor phase epitaxy." *J. Electron. Mater.*, *26*(3), 237–242 (1997a).

Polyakov A.Y., Shin M., Qian W., Skowronski M., Greve D.W., Wilson R.G. "Growth of AlBN solid solutions by organometallic vapor-phase epitaxy." *J. Appl. Phys.*, *81*(4), 1715–1719 (1997b).

Popkov A.N., Braginskaya A.G., Vekshina V.S., Kolchina G.P., Grabarskiy D.P. "Investigation of the properties of the GaSb single crystals doped with tin" [in Russian]. *Izv. AN SSSR Neorgan. Mater.*, *16*(7), 1288–1289 (1980).

Popov A.S., Koinova A.M., Tzeneva S.L. "The In–As–Sb phase diagram and LPE growth of InAsSb layers on InAs at extremely low temperatures." *J. Cryst. Growth*, *186*(3), 338–343 (1998).

Popova S.V., Sazanova O.A., Brazhkin V.V., Kalyaeva N.V., Kondrin M.B., Lyapin A.G. "High-pressure phases in the GaSb–Mn system." *Phys. Solid State*, *48*(11), 2177–2182 (2006).

Potoriy M.V., Milyan P.M. "Pecularities and features of the component interaction in the Me–P–S(Se) systems where M – Cu, Ag, Zn, Cd, In, Tl, Sn, Pb, Sb, Bi" [in Russian]. *Ukr. Khim. Zhurn.*, *82*(2), 71–78 (2016).

Predel B., Gerdes F. "Thermodynamische Eigenschaften flüssiger Legierungen der Systeme InSb–Pb und InSb–Bi." *J. Less Common Metals*, *59*(2), 153–164 (1978).

Procopio A.T., Barsoum M.W., El-Raghy T. "Characterization of Ti_4AlN_3." *Metal. Mater. Trans.*, *A31*(2), 333–337 (2000a).

Procopio A.T., El-Raghy T., Barsoum M.W. "Synthesis of Ti_4AlN_3 and phase equilibria in the Ti–Al–N system." *Metal. Mater. Trans.*, *A31*(2), 373–378 (2000b).

Prudaev I.A., Khludkov S.S., Gutakovskiy A.K., Novikov V.A., Tolbanov O.P., Ivonin I.V. "Decomposition of the supersaturated solid solution of iron in GaAs" [in Russian]. *Neorgan Mater.*, *48*(2), 133–135 (2012).

Pshenichnaya O.V., Verkhovodov P.A., Kislyi P.S., Kuzenkova M.A., Goncharuk A.B. "Test methods and properties of powder metallurgical materials. Nitriding of the intermetallic compound $TiAl_3$." *Soviet Powder Metall. Met. Ceram.*, (10), 851–855 (1983).

Puchinger M., Kisailus D.J., Lange F.F., Wagner T. "Microstructural evolution of precursor-derived gallium nitride thin films." *J. Cryst. Growth*, *245*(3–4), 219–227 (2002).

Pyshkin S.L., Negreskul V.V. "Formation of the solid solutions and some electrical properties of gallium phosphidotelluride" [in Russian]. In *Materialy IV Konf. Molodykh Uchenykh Moldavii*, 1964. Sekts. Fiz.-Mat. Kishinev: Publish. House of AN MSSR, 2932 (1965).

Qiu C., Metselaar R. "Phase relations in the aluminum carbide–aluminum nitride–aluminum oxide system." *J. Am. Ceram. Soc.*, *80*(8), 2013–2020 (1997).

Qiu Y., Xi L., Shi X., Qiu P., Zhang W., Chen L., Salvador J.R., et al. "Charge-compensated compound defects in Ga-containing thermoelectric skutterudites." *Adv. Funct. Mater.*, *23*(25), 3194–3203 (2013).

Radautsan S.I. "Investigation of the $InAs–In_2Se_3$ section of the In–As–Se system" [in Russian]. *Zhurn. Neorgan. Khimii*, *4*(5), 1121–1124 (1959).

Radautsan S.I. "Investigation of some complex semiconductor solid solutions and compounds based on indium" [in Russian]. *Chekhosl. Fiz. Zhurn.*, *12*(5), 382–391 (1962).

Radautsan S.I., Madan I.A. "Solid solutions of the indium phosphides–selenides" [in Russian]. *Izv. AN MSSR*, (5), 92–98 (1962).

Radautsan S.I., Madan I.A. "Obtaining and investigation of physico-chemical properties of some alloys in the In–P–Se, In–P–S ternary system" [in Russian]. In *Nekot. Voprosy Khimii i Fiziki Poluprovodn. Slozhn. Sostava*. Uzhgorod: Publish. House of Uzhgorod Un-Ta, 186–189 (1970).

Radautsan S.I., Madan I.A., Molodyan I.P., Ivanova R.A. "Formation of the solid solutions in the $InP–In_2Se_3$ system" [in Russian]. *Izv. Mold. Fil. AN SSSR*, *3*(69), 107–109 (1960).

Radautsan S.I., Madan I.A., Molodyan I.P., Ivanova R.A. "Solid solutions of gallium phosphide and selenide" [in Russian]. *Izv. AN MSSR*, *10*(88), 98–101 (1961a).

Radautsan S.I., Molodyan I.P. "Homogenization of the alloys of the $InSb–In_2Te_3$ section of the indium–antimone–tellurium ternary system" [in Russian]. *Izv. Mold. Fil. AN SSSR*, *3*(69), 37–47 (1960).

Radautsan S.I., Negreskul V.V. "Solid solutions of the gallium phosphidosulphide" [in Russian]. In *Issled. Po Poluprovodn.* Kishinev: Kartia Moldoveniaske Publish., 158–163 (1964).

Radautsan S.I., Negreskul V.V. Madan I.A. "Some solid solutions based on a new In$_4$SbTe$_3$ compound" [in Russian]. *Izv. AN MSSR*, 10(88), 57–63 (1961b).

Raghavan V. "The B–Fe–N (boron–iron–nitrogen) system." *Phase Diagrams Ternary Iron Alloys*, 1, 149–151 (1987).

Raghavan V. "The Al–Fe–P system (aluminium–iron–phosphorus)." *J. Alloy Phase Diagrams*, 5(1), 32–39 (1989).

Raghavan V. "Al–Fe–N (aluminium–iron–titanium)." *J. Phase Equilibria*, 14(5), 617–618 (1993a).

Raghavan V. "B–Fe–N (boron–iron–nitrogen)." *J. Phase Equilibria Diffus.*, 14(5), 619–620 (1993b).

Raghavan V. "Fe–In–P (iron–indium–phosphorus)." *J. Phase Equilibria Diffus.*, 19(3), 284 (1998).

Raghavan V. "As–Fe–Ga (arsenic–iron–gallium)." *J. Phase Equilibria Diffus.*, 25(1), 77–78 (2004a).

Raghavan V. "Fe–Ga–Sb (iron–gallium–antimony)." *J. Phase Equilibria Diffus.*, 25(1), 85–86 (2004b).

Raghavan V. "Al–Sb–Ti (aluminum–antimony–titanium)." *J. Phase Equilibria Diffus.* 26(2), 190 (2005).

Raghavan V. "Al–As–Ni (aluminum–arsenic–nickel)." *J. Phase Equilibria Diffus.*, 27(5), 484–485 (2006a).

Raghavan V. "Al–C–N (aluminum–carbon–nitrogen)." *J. Phase Equilibria Diffus.*, 27(2), 147 (2006b).

Raghavan V. "Al–N–Ti (aluminum–nitrogen–titanium)." *J. Phase Equilibria Diffus.*, 27(2), 159–162 (2006c).

Raghavan V. Al–Mg–Sb (aluminum–magnesium–antimony)." *J. Phase Equilibria Diffus.*, 28(6), 553–554 (2007).

Raghavan V. "Al–Sb–Y (aluminum–antimony–yttrium)." *J. Phase Equilibria Diffus.*, 29(2), 190–191 (2008).

Raghavan V. "Al–Dy–Sb (aluminum–dysprosium–antimony)." *J. Phase Equilibria Diffus.*, 31(1), 43 (2010a).

Raghavan V. "Al–La–Sb (aluminum–lanthanum–antimony)." *J. Phase Equilibria Diffus.*, 31(6), 560 (2010b).

Raghavan V. "Al–P–Zn (aluminum–phosphorus–zinc)." *J. Phase Equilibria Diffus.*, 31(6), 562 (2010c).

Raghavan V. "Al–Sb–Si (aluminum–antimony–silicon)." *J. Phase Equilibria Diffus.*, 31(3), 295–296 (2010d).

Raghavan V. "Al–Sb–Y (aluminum–antimony–yttrium)." *J. Phase Equilibria Diffus.*, 31(1), 60 (2010e).

Raghavan V. Al–Mg–Sb (aluminum–magnesium–antimony)." *J. Phase Equilibria Diffus.*, 32(5), 462–464 (2011).

Raghavan V. "Al–Sb–Zn (aluminum–antimony–zinc)." *J. Phase Equilibria Diffus.*, 33(2), 149–150 (2012).

Raghavan V. "Phase diagram updates and evaluations of the Al–Fe–P, B–Fe–U, Bi–Fe–Zn, Cu–Fe–Zn, Fe–Si–Zn and Fe–Ti–V systems." *J. Phase Equilibria Diffus.*, 34(3), 230–243 (2013).

Ragimova V.M., Alieva Z.G., Movsum-zade A.A., Sadykhova S.A. "Interaction of InSb with InSe, In$_2$Se$_3$ and Sb$_2$Se$_3$" [in Russian]. *Zhurn. Neorgan. Khimii*, 29(11), 2888–2890 (1984).

Ragimova V.M., Alieva Z.G., Rustamov P.G., Sadykhova S.A. "In–Sb–Se system" [in Russian]. *Zhurn. Neorgan. Khimii*, 30(5), 1274–1278 (1985).

Ragimova V.M., Alieva Z.G., Sadykhova S.A., Allazov M.R., Alieva R.A. "Investigation of the InSb–Se section" [in Russian]. *Azerb. Khim. Zhurn.*, (2), 120–123 (1983).

Rajpalke M.K., Linhart W.M., Birkett M., Yu K.M., Alaria J., Kopaczek J., Kudrawiec R., Jones T.S., Ashwin M.J., Veal T.D. "High Bi content GaSbBi alloys." *J. Appl. Phys.*, 116(4), 043511_1–043511_5 (2014).

Rajpalke M.K., Linhart W.M., Birkett M., Yu K.M., Scanlon D.O., Buckeridge J., Jones T.S., Ashwin M.J., Veal T.D. "Growth and properties of GaSbBi alloys." *Appl. Phys. Lett.*, 103(14), 142106_1–142106_4 (2013).

Rakov V.V., Ufimtsev V.B. "Phase equilibrium in the gallium arsenide–indium arsenide system" [in Russian]. *Zhurn. Fiz. Khimii*, 43(2), 493–495 (1969).

Ramachandran P.V., Gagare P.D. "Preparation of ammonia borane in high yield and purity, methanolysis, and regeneration." *Inorg. Chem.*, 46(19), 7810–7817 (2007).

Rasmussen S.E., Hazell R.G. "Preparation of single phases and single crystals in the vanadium–gallium–antimony system. Crystal structure of V$_6$GaSb." *Acta Chem. Scand.*, A32(9), 785–788 (1978).

Raukhman M.R., Zemskov V.S., Kharakhorin F.F., Boyarintsev P.K., Laptev A.V., Yankauskas S.M. "On the bismuth solubility in indium antimonide" [in Russian]. In *Svoystva Legir. Poluprovodn.* Moscow: Nauka Publish., 62–68 (1977).

Rauscher J.F., Condron C.L., Beault T., Kauzlarich S.M., Jensen N., Klavins P., MaQuilon S., Fisk Z., Olmstead M.M. "Flux growth and structure of two compounds with the EuIn$_2$P$_2$ structure type, AIn$_2$P$_2$ (A = Ca and Sr), and a new structure type, BaIn$_2$P$_2$." *Acta Crystallogr.*, C65(10), i69–i73 (2009).

Rawn C.J., Barsoum M.W., El-Raghy T., Procipio A., Hoffmann C.M., Hubbard C.R. "Structure of Ti$_4$AlN$_3$—A layered M$_{n+1}$AX$_n$ nitride." *Mater. Res. Bull.*, 35(11), 1785–1796 (2000).

Rayson H.W., Goulding C.W., Raynor G.V. "Binary and ternary alloys of tin with the elements of Groups IIB, IIIB and VB of the periodic table." *Metallurgia*, 59, 125–130 (1959).

Reckeweg O., DiSalvo F.J., Somer M. "Orthorhombic Ba$_3$[BN$_2$]$_2$: A new structure type for an alkaline earth metal nitrido borate." *J. Alloys Compd.*, 361(1–2), 102–107 (2003).

Reckeweg O., Meyer H.-J. "Lanthanide nitrido borates with six-membered B$_3$N$_6$ rings: Ln$_3$B$_3$N$_6$." *Angew. Chem. Int. Ed.*, 38(11), 1607–1609 (1999a).

Reckeweg O., Meyer H.-J. "Lanthanoidnitridoborate mit sechsgliedrigen B$_3$N$_6$-Ringen: Ln$_3$B$_3$N$_6$." *Angew. Chem.*, 111(11), 1714–1716 (1999b).

Reckeweg O., Meyer H.-J. "The nitrido borates Ln$_3$B$_2$N$_4$ (Ln = La–Nd) and La$_5$B$_4$N$_9$: Syntheses, structures, and properties." *Z. Anorg. Allg. Chem.*, 625(6), 866–874 (1999c).

Reiss B., Renner Th. "Der gerichtete Einbau von Schwermetallphasen in GaAs." *Z. Naturforsch.*, 22A(5), 546–548 (1966).

Reiter A.E., Derflinger V.H., Hanselmann B., Bachmann T., Sartory B. "Investigation of the properties of $Al_{1-x}Cr_xN$ coatings prepared by cathodic arc evaporation." *Surf. Coat. Technol.*, *200*(7), 2114–2122 (2005).

Rekalova G.I., Kebe U., Mezrina L.A., Gurevich M.K., Marchenko T.E. "Solubility and distribution coefficient of tellurium in InSb" [in Russian]. *Izv. AN SSSR Neorgan. Mater.*, *7*(8), 1314–1316 (1971).

Riane R., Boussahla Z., Matar S.F., Zaoui A. "Structural and electronic properties of zinc blende-type nitrides $B_xAl_{1-x}N$." *Z. Naturforsch.*, *63B*(9), 1069–1076 (2008).

Richter K.W. "The ternary In–Ni–Sb phase diagram in the vicinity of the binary In–Ni system." *J. Phase Equilibria*, *19*(5), 455–465 (1998).

Richter K.W., Ipser H. "Phase equilibria in the ternary Ga–Pt–Sb system." *J. Alloys Compd.*, *281*(2), 241–248 (1998a).

Richter K.W., Ipser H. "The ternary Ga–Pd–Sb phase diagram: A system relevant to contact materials for GaSb." *Ber. Bunsenges. Phys. Chem.*, *102*(9), 1245–1251 (1998b).

Richter K.W., Ipser H. "Contact materials for GaSb and InSb: A phase diagram approach." *Acta Metallurg. Sin.*, *15*(2), 143–148 (2002).

Richter K.W., Micke K., Ipser H. "Antimony activities in the ternary NiAs-phase of the In–Ni–Sb system." *Termochim. Acta*, *314*(1–2), 137–144 (1998a).

Richter K.W., Micke K., Ipser H. "Contact materials for III–V semiconductors: Phase equilibria of InSb in the ternary system In–Ni–Sb." *Mater. Sci. Eng. B*, *55*(1–2), 44–52 (1998b).

Rizzol C., Salamakha P.S., Sologub O.L., Bocell G. "X-ray investigation of the Al–B–N ternary system: Isothermal section at 1500°C: Crystal structure of the $Al_{0.185}B_6CN_{0.256}$ compound." *J. Alloys Compd.*, *343*(1–2), 135–141 (2002).

Robbins M., Lambrecht Jr. V.G. "Preparation and structural investigation of compounds of the type $In_{3-a}Ga_aX_3Y$ where X = S, Se, Y = P, As and a = 1, 2." *Mater. Res. Bull.*, *10*(5), 331–334 (1975).

Rogl P. "Materials science of ternary metal boron nitrides." *Int. J. Inorg. Mater.*, *3*(3), 201–209 (2001).

Rogl P., Klesnar H. "The crystal structure of $PrBN_2$ and isotypic compounds $REBN_2$ (RE = Nd, Sm, Gd)." *J. Solid State Chem.*, *98*(1), 99–104 (1992).

Rogl P., Klesnar H., Fischer P. "Neutron powder diffraction of Nb_2BN_{1-x}." *J. Am. Ceram. Soc.*, *71*(10), C450–C452 (1988a).

Rogl P., Klesnar H., Fischer P. "Neutron powder diffraction studies of $Ce_3B_2N_4$ and isotypic $RE_3B_2N_4$ compounds (RE = La, Pr, Nd, MM)." *J. Am. Ceram. Soc.*, *73*(9), 2634–2639 (1990).

Rogl P., Klesnar H., Fischer P., Chevalier B., Buffat B., Demazeau G., Etourneau J. "Structural chemistry and phase equilibria in the ternary system niobium–boron–nitrogen." *J. Mater. Sci. Lett.*, *7*(11), 1229–1230 (1988b).

Rogl P., Schuster J.C. "Al–B–N (aluminum–boron–nitrogen)." In *Phase Diagrams of Ternary Boron Nitride and Silicon Nitride Systems*. Monogr. Ser. of Alloy Phase Diag. Materials Park, OH: American Society for Metals International, 3–5 (1992).

Romanenko V.N., Kheyfets V.S. "Distribution coefficients and solubility of some rare earth elements in GaAs" [in Russian]. *Izv. AN SSSR Neorgan. Mater.*, *9*(2), 190–196 (1973).

Rosenberg A.J., Strauss A.J. "Solid solutions of In_2Te_3 in Sb_2Te_3 and Bi_2Te_3." *J. Phys. Chem. Solids*, *19*(1–2), 105–116 (1961).

Rubinstein M. "Solubilities of GaAs in metallic solvents." *J. Electrochem. Soc.*, *113*(7), 752–753 (1966).

Rudy E., Benesovsky F. "Untersuchungen in den Systemen: Hafnium–Bor–Stickstoff und Zirkonium–Bor–Stickstoff." *Monatsch. Chem.*, *92*(2), 415–441 (1961).

Rugg B.C., Silk N.J., Bryant A.W., Argent B.B. "Calorimetric measurements of the enthalpies of formation and of mixing of II/VI and III/V compounds." *CALPHAD*, *19*(3), 389–398 (1995).

Ruh R., Kearns M., Zangvil A., Xu Y. "Phase and property studies of boron carbide–boron nitride composites." *J. Am. Ceram. Soc.*, *75*(4), 864–872 (1992).

Rundqvist S. "X-ray investigations of the ternary system Fe–P–B. Some features of the systems Cr–P–B, Mn–P–B, Co–P–B and Ni–P–B." *Acta Chem. Scand.*, *16*(1), 1–19 (1962).

Rustamov P.G., Alieva Z.G., Alidzhanov M.A. Cherstvova V.B. "Investigation of the $GaSb–Ga_2Se_3$ section of the Ga–Sb–Se ternary system" [in Russian]. In *Issled. v Obl. Neorgan. i Fiz. Khimii*. Part 1. Baku: Elm Publish., 273–278 (1971).

Rustamov P.G., Alieva Z.G., Guseynova G.A., Cherstvova V.B. "Interaction in the Ga–Sb–Se semiconductor system" [in Russian]. In *Nekot. Voprosy Khimii i Fiziki Poluprovodn. Slozhn. Sostava*. Uzhgorod: Publish. House of Uzhgorod Un-T, 205–210 (1970).

Rustamov P.G., Azhdarova D.S., Dzhalil-zade T.A. "Investigation of the Sb_2S_3 interaction with gallium antimonide and sulfide" [in Russian]. *Azerb. Khim. Zhurn.*, (2), 106–109 (1974a).

Rustamov P.G., Babaeva B.K., Cherstvova V.B. "Investigation of the interaction in the Ga–As–Te system and obtaining of the indium chalcogenoantimonides" [in Russian]. In *Khalkogenidy*. Vyp. 3. Kiev: Nauk. Dumka Publish., 106–111 (1974b).

Rustamov P.G., Cherstvova V.B., Guseynova M.A. "Phase diagram of the GaSb–GaSe system" [in Russian]. *Uch. Zap. Azerb. Un-T Ser. Khim. N.*, (2), 28–31 (1969).

Rustamov P.G., Geydarova E.A. "Investigation of the Ga–Sb–Te ternary system" [in Russian]. *Azerb. Khim. Zhurn.*, (3), 108–112 (1977).

Rustamov P.G., Il'yasov T.M., Safarov M.G. "Nature of the interaction in the As_2S_3–GaS system" [in Russian]. *Zhurn. Neorgan. Khimii*, *27*(3), 758–760 (1982).

Rustamov P.G., Il'yasov T.M., Safarov M.G., Sadykhova S.A. "Projection of the liquidus of the Ga,As ‖ Se,Te system" [in Russian]. *Zhurn. Neorgan. Khimii*, *24*(2), 476–480 (1979).

Ruys A.J., Sorrell C.C. "Ternary phase equilibria in the system Zr–N–B–C at 1500°C." *Key Eng. Mater.*, 53–55, 92–100 (1991).

Sadowski J., Domagała J.Z., Bąk-Misiuk J., Koleśnik S., Świątek K., Kanski J., Ilver L. "Structural properties of MBE grown GaMnAs layers." *Thin Solid Films*, 367(1–2), 165–167 (2000).

Saenko I.V. "Temperature dependence of bismuth solubility in gallium phosphide" [in Russian]. *Izv. Leningr. Elektrotekhn. In-Ta*, (211), 28–32 (1977).

Safaraliev G.I., Dadashev I.Sh., Vagabova L.K., Azimov K.A. "Artificially-anisotropic ferromagnetic alloys in the GaSb–MnSb system" [in Russian]. *Izv. AN SSSR Neorgan. Mater.*, 27(10), 2196–2198 (1991).

Safaraliev G.I., Vagabova L.K. "Phase diagram of the Mn₂Sb–GaSb system" [in Russian]. *Izv. AN SSSR Neorgan. Mater.*, 24(4), 550–554 (1988).

Safarov M.G. "Physico-chemical investigation of the As_2Te_3–InTe system" [in Russian]. *Zhurn. Neorgan. Khimii*, 33(2), 537–539 (1988).

Safarov M.G., Gamidov R.S. "As_2Se_3–InSe system" [in Russian]. *Zhurn. Neorgan. Khimii*, 35(2), 495–498 (1990).

Safarov M.G., Gamidov R.S., Poladov P.M., Bagirova E.M. "InS–As_2S_3–As_2Te_3 ternary system" [in Russian]. *Zhurn. Neorgan. Khimii*, 37(6), 1384–1391 (1992).

Safarov M.G., Poladov P.M., Gamidov R.S. "$InAs$–$InSe$–As_2Se_3 system" [in Russian]. *Zhurn. Neorgan. Khimii*, 35(2), 556–558 (1990).

Safarov M.G., Poladov P.M., Safarov R.M. "InSe–InAs system" [in Russian]. *Zhurn. Neorgan. Khimii*, 33(2), 524–526 (1988).

Saidi-Houat N., Zaoui A., Belabbes A., Ferhat M. "Ab initio study of the fundamental properties of novel III–V nitride alloys $Ga_{1-x}Tl_xN$." *Mater. Sci. Eng. B*, 162(1), 26–31 (2009).

Saito T., Arakawa Y. "Atomic structure and phase stability of $In_xGa_{1-x}N$ random alloys calculated using a valence-force-field method." *Phys. Rev. B*, 60(3), 1701–1706 (1999).

Sakai T. "Hot-pressing of the AlN–Al_2O_3 system." *Yogyo-Kyokai-Shi*, 86(991), 125–130 (1978).

Sakakibara W., Hayashi Y., Takizawa H. "$MnGa_2Sb_2$, a new ferromagnetic compound synthesized under high pressure." *J. Ceram. Soc. Jpn.*, 117(1), 72–75 (2009).

Sakakibara W., Hayashi Y., Takizawa H. "High-pressure synthesis of a new ferromagnetic compound, $CrGa_2Sb_2$." *J. Alloys Compd.*, 496(1–2), L14–L17 (2010).

Sands T., Chang C.C., Kaplan A.S., Keramidas V.G., Krishnan K.M., Washburn J. "Ni–InP reaction: Formation of amorphous and crystalline ternary phases." *Appl. Phys. Lett.*, 50(19), 1346–1348 (1987).

Sands T., Keramidas V.G., Gronsky R., Washburn J. "Ternary phases in the Pd–GaAs system: Implications for shallow contacts to GaAs." *Mater. Lett.*, 3(9–10), 409–413 (1985).

Sands T., Keramidas V.G., Gronsky R., Washburn J. "Initial stages of the Pd–GaAs reaction: Formation and decomposition of ternary phases." *Thin Solid Films*, 136(1), 105–122 (1986a).

Sands T., Keramidas V.G., Washburn J., Gronsky R. "Structure and composition of Ni_xGaAs." *Appl. Phys. Lett.*, 48(6), 402–404 (1986b).

Sandulova A.V., Berezhkova G.V., Dronyuk M.I., Petrushko I.A. "Investigation of the gas transport processes in the $InSb$–I_2 system" [in Russian]. *Izv. AN SSSR Neorgan. Mater.*, 8(7), 1224–1228 (1972).

Sandulova A.V., Ostrovskiy P.I. "The study of equilibrium and mechanism of transport in the GaP–I and GaP–Ga–I systems" [in Russian]. *Izv. AN SSSR Neorgan. Mater.*, 7(1), 10–12 (1971).

Sanygin V.P., Pashkova O.N., Filatov A.V., Novotortsev V.M. "Manganese solubility in quenched InSb" [in Russian]. *Neorgan. Mater.*, 46(8), 901–906 (2010).

Sasaki T., Akaishi M., Yamaoka S., Fujiki Y., Oikawa T. "Simultaneous crystallization of diamond and cubic boron nitride from the graphite relative BC_2N under high pressure/high temperature conditions." *Chem. Mater.*, 5(5), 695–699 (1993).

Schenck H., Steinmetz E. "Chemische Aktivitäten von Bor und Stickstoff im flüssigen Bereich des Systems Eisen–Bor–Stickstoff bei 1600°C." *Arch. Eisenhuettenwes.*, 39(4), 255–257 (1968).

Schlechte A., Prots Yu., Niewa R. "Crystal structure of hexabarium mononitride pentaindide, $(Ba_6N)[In_5]$." *Z. Kristallogr. New Cryst. Struct.*, 219(4), 349–350 (2004).

Schlesinger H.I., Ritter D.M., Burg A.B. "Hydrides of boron. X. The preparation and preliminary study of the new compound B_2H_7N." *J. Am. Chem. Soc.*, 60(10), 2297–2300 (1938).

Schmid-Fetzer R. "Stability of metal/GaAs-interfaces: A phase diagram survey." *J. Electron. Mater.*, 17(2), 193–200 (1988).

Schmid-Fetzer R. "The Al–As–P system (aluminum–arsenic–phosphorus)." *Bull. Alloy Phase Diagr.*, 10(5), 530–532 (1989a).

Schmid-Fetzer R. "The Al–As–Sb system (aluminum–arsenic–antimony)." *Bull. Alloy Phase Diagr.*, 10(5), 532–533 (1989b).

Schmid-Fetzer R., Zeng K. "Nitridation of Ti–Al alloys: A thermodynamic approach." *Metal. Mater. Trans. A*, 28(9), 1949–1951 (1997).

Schmidt M., Ewald B., Prots Yu., Cardoso-Gil R., Armbrüster M., Loa I., Zhang L., Huang Y.-X., Schwarz U., Kniep R. "Growth and characterization of BPO_4 single crystals." *Z. Anorg. Allg. Chem.*, 630(5), 655–662 (2004).

Schmitz G., Eumann M., Stapel D., Franchy R. "Combined EELS/STM study of the adsorption of nitric oxide and the formation of GaO_xN_y on CoGa(001)." *Surf. Sci.*, 427–428, 91–96 (1999).

Schneider G., Gauckler L.J., Petzow G. "Phase equilibria in the system AlN–Si_3N_4–Be_3N_2." *J. Am. Ceram. Soc.*, 63(1–2), 32–35 (1980).

Schulz K.J., Musbah O.A., Chang Y.A. "An investigation of the Ir–Ga–As system." *Bull. Alloy Phase Diagr.*, 11(3), 211–215 (1990).

Schulz K.J., Musbah O.A., Chang Y.A. "A phase investigation of the Rh–Ga–As system." *J. Phase Equilibria*, *12*(1), 10–14 (1991).

Schulz K.J., Zheng X.-Y., Chang Y.A. "Phase equilibria in the Nb–Ga–As system below 1000°C." *Bull. Alloy Phase Diagr.*, *10*(4), 314–318 (1989).

Schulze G.E.R. "Die Kristallstruktur von BPO_4 und $BAsO_4$." *Naturwissenschaften*, *21*(30), 562 (1933).

Schuster J.C. "Über die Verbreitung ternärer Komplexnitride mit Perowskitstruktur in SE–Al–N–Systemen (SE ≡ La, Ce, Pr, Nd, Sm, Gd, Tb, Dy, Ho, Er, Tm, Yb, Lu)." *J. Less Common Metals*, *105*(2), 327–332 (1985).

Schuster J.C. "The crystal structure of Zr_3AlN." *Z. Kristallogr.*, *175*(3–4), 211–215 (1986).

Schuster J.C., Bauer J. "Investigation of phase equilibria related to fusion reactor materials. II. The ternary system Hf–Al–N." *J. Nucl. Mater.*, *120*(2–3), 133–136 (1984a).

Schuster J.C., Bauer J. "The ternary system titanium–aluminium–nitrogen." *J. Solid State Chem.*, *53*(2), 260–265 (1984b).

Schuster J.C., Bauer J. "The ternary systems Sc–Al–N and Y–Al–N." *J. Less Common Metals*, *109*(2), 345–350 (1985).

Schuster J.C., Bauer J., Debuigne J. "Investigation of phase equilibria related to fusion reactor materials. I. The ternary system Zr–Al–N." *J. Nucl. Mater.*, *116*(2–3), 131–135 (1983).

Schuster J.C. Bauer J., Nowotny H. "Applications to materials science of phase diagrams and crystal structures in the ternary systems transition metal–aluminium–nitrogen." *Rev. Chim. Miner.*, *22*, 546–554 (1985).

Schuster J.C., Nowotny H. "Phase equilibria in the ternary systems Nb–Al–N and Ta–Al–N." *Z. Metallkde*, *76*(11), 728–729 (1985a).

Schuster J.C., Nowotny H. "Phase relationships in the ternary systems (V, Cr, Mo, W, Mn, Re)–Al–N." *J. Mater. Sci.*, *20*(8), 2787–2793 (1985b).

Schwartz G.P. "Analysis of native oxide films and oxide-substrate reactions on III-V semiconductors using thermochemical phase." *Thin Solid Films*, *103*(1–3), 3–16 (1983).

Schwartz G.P., Gualtieri G.J., Griffiths J.E., Thurmond C.D., Schwartz B. "Oxide-substrate and oxide-oxide chemical reactions in thermally annealed anodic films on GaSb, GaAs and GaP." *J. Electrochem. Soc.*, *127*(11), 2488–2499 (1980).

Schwartz G.P., Sunder W.A., Griffiths J.E. "The In–P–O phase diagram: Construction and applications." *J. Electrochem. Soc.*, *129*(6), 1361–1367 (1982a).

Schwartz G.P., Sunder W.A., Griffiths J.E., Gualtieri G.J. "Condensed phase diagram for the In–As–O system." *Thin Solid Films*, *94*(3), 205–212 (1982b).

Schwarzenbach D. "Verfeinerung der Struktur der Tiefquarz-Modifikation von $AlPO_4$." *Z. Kristallogr.*, *123*(3–4), 161–185 (1966).

Schweitz K.O., Mohney S.E. "Phase equilibria in transition metal Al–Ga–N systems and thermal stability of contacts to AlGaN." *J. Electron. Mater.*, *30*(3), 175–182 (2001).

Scolfaro L.M.R. "Phase separation in cubic Group-III nitride alloys." *Phys. Stat. Sol. (A)*, *190*(1), 15–22 (2002).

Sedmidubský D., Leitner J., Sofer Z. "Phase relations in the Ga–Mn–N system." *J. Alloys Compd.*, *452*(1), 105–109 (2008).

Sellinschegg H., Stuckmeyer S.L., Hornbostel M.D., Johnson D.C. "Synthesis of metastable post-transition-metal iron antimony skutterudites using the multilayer precursor method." *Chem. Mater.*, *10*(4), 1096–1101 (1998).

Semenova G.V., Grekova I.I., Kalyuzhnaya M.I., Goncharov E.G. "Composition of the vapor phase in the In–As–P ternary system" [in Russian]. *Zhurn. Neorgan. Khimii*, *37*(7), 1635–1637 (1992).

Semenova G.V., Sidorov A.V., Goncharov E.G. "Determination of nonstoichiometry of $In_{1-\delta}(As_{1-x}P_x)_{1+}\delta$ solid solution" [in Russian]. *Zhurn. Neorgan. Khimii*, *44*(4), 544–546 (1999).

Semenova G.V., Sushkova T.P., Goncharov E.G. "Thermodynamic calculation of deviation from stoichiometry in the $InAs_{1-x}P_x$ solid solutions" [in Russian]. *Zhurn. Neorgan. Khimii*, *38*(10), 1717–1720 (1993).

Semenova G.V., Sushkova T.P., Shumskaya O.N. "Phase diagram of the In–Sb–P ternary system" [in Russian]. *Zhurn. Neorgan. Khimii*, *50*(4), 710–713 (2005).

Semenova G.V., Sushkova T.P., Shumskaya O.N. "Stability and thermodynamic mixing functions of the solid solutions in the InP–InAs–InSb system" [in Russian]. *Kondens. Sredy Mezhfaz. Granitsy*, *10*(2), 149–155 (2008).

Semenova G.V., Ugay Ya.A., Goncharov E.G., Grekova I.I. "Phase diagram of the In–As–P system" [in Russian]. *Zhurn. Neorgan. Khimii*, *33*(4), 1000–1003 (1988).

Semenova G.V., Ugay Ya.A., Grekova I.I., Kalyuzhnaya M.I. "p-T-x-y phase diagram of the In–As–P system" [in Russian]. *Zhurn. Neorgan. Khimii*, *35*(8), 2130–2133 (1990).

Semenova T.F., Vergasova L.P., Filatov S.K., Ananev V.V. "Alarsite $AlAsO_4$—A new mineral from volcanic exhalations" [in Russian]. *Dokl. AN SSSR*, *338*(4), 501–505 (1994).

Šesták J., Leitner J., Yokokawa H., Štěpánek B. "Thermodynamics and phase equilibria data in the S–Ga–Sb system auxiliary to the growth of doped GaSb single crystals." *Thermochim. Acta*, *245*, 189–206 (1994).

Šesták J., Štěpánek B., Yokokawa H., Šestáková V. "Copper solubility and distribution in doped GaSb single crystals." *J. Therm. Anal.*, *43*(2), 389–397 (1995).

Shafer E.C., Roy R. "Studies of silica-structure phases: I, $GaPO_4$, $GaAsO_4$, and $GaSbO_4$." *J. Am. Ceram. Soc.*, *39*(10), 330–336 (1956).

Shafer E.C., Roy R. "Studies of silica structure phases, III: New data on the system $AlPO_4$." *Z. Phys. Chem.*, *11*(1–2), 30–40 (1957).

Shafer E.C., Shafer M.W., Roy R. "Studies of silica structure phases II: Data on $FePO_4$, $FeAsO_4$, $MnPO_4$, BPO_4, $AlVO_4$ and others." *Z. Kristallogr.*, *108*(3–4), 263–275 (1956).

Sharan B. "A new modification of aluminium ortho-arsenate." *Acta Crystallogr.*, *12*(11), 948–949 (1959).

Sharma R.C., Mukerjee I. "Thermodynamic analysis and phase equilibria calculations of the Ga–In–Sb system." *J. Phase Equilibria*, 13(1), 5–15 (1992).

Sharma R.C., Srivastava M. "Phase equilibria calculations of Al–In and Al–In–Sb systems." *CALPHAD*, 16(4), 409–426 (1992a).

Sharma R.C., Srivastava M. "Phase equilibria calculations of Al–Sb, Al–Ga and Al–Ga–Sb systems." *CALPHAD*, 16(4), 387–408 (1992b).

Sharma S.M., Garg N., Sikka S.K. "High-pressure x-ray-diffraction study of α-AlPO$_4$." *Phys. Rev. B*, 62(13), 8824–8827 (2000).

Shchegoleva V.F., Marina L.I., Nashel'skiy A.Ya. "Behavior of tellurium in GaP at the horizontal zone melting" [in Russian]. *Izv. AN SSSR Neorgan. Mater.*, 10(8), 1406–1408 (1974).

Shchegoleva V.F., Marina L.I., Nashel'skiy A.Ya., Marunina N.I. "Investigation of distribution coefficient of Zn in GaP" [in Russian]. *Izv. AN SSSR Neorgan. Mater.*, 7(8), 1449–1450 (1971).

Shen A., Ohno H., Matsukura F., Sugawara Y., Akiba N., Kuroiwa T., Oiwa A., Endo A., Katsumoto S., Iye Y. "Epitaxy of (Ga,Mn)As, a new diluted magnetic semiconductor based on GaAs." *J. Cryst. Growth*, 175–176, Pt. 2, 1069–1074 (1997).

Shen J.-Y., Chatillon C. "Thermodynamic calculations of congruent vaporization in III-V systems; applications to the In–As, Ga–As and Ga–In–As systems." *J. Cryst. Growth*, 106(4), 543–552 (1990).

Shen J.-Y., Chatillon C., Ansara I., Watson A., Rugg B., Chart T. "Optimisation of the thermodynamic and phase diagram data in the ternary As–Ga–In system." *CALPHAD*, 19(2), 215–226 (1995).

Shi L., Duan Y., Yang X., Tang G., Qin L., Qiu L. "Structural, electronic and elastic properties of wurtzite-structured Tl$_x$Al$_{1-x}$N alloys from first principles." *Mater. Sci. Semicond. Process.*, 15(5), 499–504 (2012).

Shiau F.-Y., Zuo Y., Lin J.-C., Zheng X.-Y., Chang Y.A. "Solid state equilibria in the Ga–Co–As system at 600°C." *Z. Metallkde*, 80(8), 544–548 (1989).

Shiau F.-Y., Zuo Y., Zheng X.-Y., Lin J.-C., Chang Y.A. "Interfacial reactions between Co and GaAs." *Mater. Res. Soc. Symp. Proc.*, 119, 171–176 (1988).

Shih C., Peretti E.A. "The system InAs–InSb." *J. Am. Chem. Soc.*, 75(3), 608–609 (1953).

Shih C., Peretti E.A. "The phase diagram of the system InAs–Sb." *Trans. Am. Soc. Metals*, 46, 389–396 (1954).

Shih C.H., Peretti E.A. "The constitution of indium–arsenic–antimony alloys." *Trans. Am. Soc. Metals*, 48, 706–725 (1956).

Shih K.K., Allen J.W., Pearson G.L. "Gas phase equilibria in the system Ga–As–Zn." *J. Phys. Chem. Solids*, 29(2), 367–377 (1968).

Shishiyanu F.S., Gheorghiu V.Gh., Palazov S.K. "Diffusion, solubility and electrical activity of Co and Fe in InP." *Phys. Status Solidi (A)*, 40(1), 29–35 (1977).

Shumilin V.P., Batyrev N.I., Selin A.A. "Investigation of the liquidus surface of the In–InP–GaP system" [in Russian]. *Izv. AN SSSR Neorgan. Mater.*, 13(3), 520–521 (1977).

Shumilin V.P., Ufimtsev V.B., Vigdorovich V.N. "Components distribution coefficients and phase diagram of the InP–GaP system" [in Russian]. *Izv. AN SSSR Neorgan. Mater.*, 8(4), 693–695 (1972).

Sirota N.N., Matyas E.Ye. "On the phase diagram of the gallium antimonide and arsenide quasi-binary system" [in Russian]. *Dokl. AN BSSR*, 17(8), 699–701 (1973).

Sirota N.N., Matyas E.Ye., Golodushko V.Z. "Determination of a vapor pressure of the GaAs–GaSb solid solutions" [in Russian]. *Dokl. AN BSSR*, 22(3), 217–219 (1978).

Sirota N.N., Shipilo V.B., Zakhrabova F.A. "Obtaining of homogeneous solid solutions of indium and gallium antimonides" [in Russian]. *Izv. AN SSSR Neorgan. Mater.*, 4(6), 858–862 (1968).

Sirota N.N., Zhuk M.M., Mazurenko A.M., Olekhnovich A.I. "About solubility of boron carbide in cubic boron nitride" [in Russian]. *Vestsi AN BSSR Ser. Fiz.-Mat. Navuk*, (2), 111–112 (1977).

Sitte W., Weppner W. "Thermodynamics and phase stabilities of the ternary system Li–In–Sb." *Appl. Phys. A*, 38(1), 31–36 (1985).

Skudnova E.V., Karaseva T.P., Mirgalovskaya M.S. "Investigation of the In–Sb–Zn system" [in Russian]. *Zhurn. Neorgan. Khimii*, 9(2), 367–371 (1964).

Skudnova E.V., Mirgalovskaya M.S. "On the distribution coefficient of sulfur in indium antimonide" [in Russian]. *Izv. AN SSSR Neorgan. Mater.*, 1(2), 184–187 (1965).

Smid I., Rogl P. "Phase equilibria and structure chemistry in ternary systems: Transition metal–boron–nitrogen." *Inst. Phys. Conf. Ser.*, (75), 249–259 (1986).

Smirnova T.P., Golubenko A.N., Zacharchuk N.F., Belyi V.I., Kokovin G.A., Valisheva N.A. "Phase composition of thin oxide films on InSb." *Thin Solid Films*, 76(1), 11–21 (1981).

Smith N.A., Hill P.J., Devlin E., Forsyth H., Harris I.R., Cockayne B., MacEwan W.R. "Structural and magnetic studies of B8$_2$-type alloys represented by Fe$_3$Ga$_{2-x}$Sb." *J. Alloys Compd.*, 179(1–2), 111–124 (1992).

Soignard E., Machon D., McMillan P.F., Dong J., Xu B., Leinenweber K. "Spinel-structured gallium oxynitride (Ga$_3$O$_3$N) synthesis and characterization: An experimental and theoretical study." *Chem. Mater.*, 17(22), 5465–5472 (2005).

Sokolov E.B., Karamov A.G., Afanas'ev S.P., Prokof'eva V.K., Okunev Yu.A. "Macrosegregetion of the iron transition group metals in gallium arsenide and phosphide" [in Russian]. In *Legir. Poluprovodn.* Moscow: Nauka Publish., 178–182 (1982).

Sokolov L.I., Goncharov E.G. "Investigation of the phase diagram of the Si–InP system" [in Russian]. *Tr. Voronezh. Un-Ta*, 74, 26–28 (1971).

Soled S., Wold A. "Crystal growth and characterization of In$_{2/3}$PS$_3$." *Mater. Res. Bull.*, 11(6), 657–662 (1976).

Solozhenko V.L., Andrault D., Fiquet G., Mezouar M., Rubie D.C. "Synthesis of superhard cubic BC$_2$N." *Appl. Phys. Lett.*, 78(10), 1385–1387 (2001).

Solozhenko V.L., Novikov N.V. "Cubic boron carbonitride—A new superhard phase" [in Russian]. *Dop. Nats. AN Ukr.*, (11), 81–86 (2001).

Solozhenko V.L., Turkevich V.Z., Holzapfel W.B. "On nucleation of cubic boron nitride in the BN–MgB$_2$ system." *J. Phys. Chem. B*, 103(38), 8137–8140 (1999).

Solozhenko V.L., Turkevich V.Z., Sato T. "Phase stability of graphitelike BC$_4$N up to 2100 K and 7 GPa." *J. Am. Ceram. Soc.*, 80(12), 3229–3232 (1997).

Somer M., Carrillo-Cabrera W., Peters E.-M., Peters K., von Schnering H.G. "Crystal structure of tribarium tetraphosphidodiindate, Ba$_3$In$_2$P$_4$." *Z. Kristallogr. New Cryst. Struct.*, 213(1), 4 (1998a).

Somer M., Carrillo-Cabrera W., Peters E.-M., Peters K., von Schnering H.G. "Crystal structures of trirubidium diarsenidoborate, Rb$_3$BAs$_2$, and trirubidium diphosphidoborate, Rb$_3$BP$_2$." *Z. Kristallogr.*, 210(10), 779–780 (1995a).

Somer M., Carillo-Cabrera W., Peters E.-M., Peters K., von Schnering H.G. "Crystal structure of trisodium catena-di-μ-phosphidoaluminate, Na$_3$AlP$_2$." *Z. Kristallogr.*, 210(10), 777 (1995b).

Somer M., Carrillo-Cabrera W., Peters K., von Schnering H.G. "Crystal structure of rubidium *catena*-di-μ-phosphidoindate di-μ-phosphido-bis(phosphidoindate), Rb$_{12}$^{1}∞[InP$_2$]$_2$[In$_2$P$_4$]." *Z. Kristallogr. New Cryst. Struct.*, 213(1), 5–6 (1998b).

Somer M., Carrillo-Cabrera W., Peters K., von Schnering H.G. "Crystal structure of tetradecaeuropium tetraphosphide tetraphosphidoaluminate polytriphosphide, Eu$_{14}$(P$_4$)(AlP$_4$)(P$_3$)." *Z. Kristallogr. New Cryst. Struct.*, 213(1), 7 (1998c).

Somer M., Carrillo-Cabrera W., Peters K., von Schnering H.G. "Crystal structure of tristrontium tetraphosphidodialuminate, Sr$_3$Al$_2$P$_4$." *Z. Kristallogr. New Cryst. Struct.*, 213(2), 230 (1998d).

Somer M., Carrillo-Cabrera W., Peters K., von Schnering H.G. "Darstellung, Kristallstrukturen und Svhwingungsspertren von Verbindungen mit den linearen Dipnictidoborat(3–)-Anionen [P–B–P]$^{3-}$, [As–B–As]$^{3-}$ und [P–B–As]$^{3-}$." *Z. Anorg. Allg. Chem.*, 626(4), 897–904 (2000b).

Somer M., Carrillo-Cabrera W., Peters K., von Schnering H.G., Cordier G. "Crystal structure of hexaeuropium di-μ-phosphido-bis(diphosphidogallate(III)), Eu$_6$Ga$_2$P$_6$." *Z. Kristallogr.*, 211(4), 257 (1996a).

Somer M., Carrillo-Cabrera W., Peters K., von Schnering H.G., Cordier G. "Crystal structure of tribarium triarsenidoindate, Ba$_3$In$_3$As$_3$." *Z. Kristallogr.*, 211(9), 632 (1996b).

Somer M., Gül C., Müllmann R., Mosel B.D., Kremer R.K., Pöttgen R. "Vibrational spectra and magnetic properties of Eu$_3$[BN$_2$]$_2$ and LiEu$_4$[BN$_2$]$_3$." *Z. Anorg. Allg. Chem.*, 630(3), 389–393 (2004).

Somer M., Hartweg M., Peters K., von Schnering H.G. "Crystal structure of potassium diarsenidoborate, K$_3$BAs$_2$." *Z. Kristallogr.*, 191(3–4), 313–314 (1990a).

Somer M., Hartweg M., Peters K., von Schnering H.G. "Crystal structure of potassium diphosphidoborate, K$_3$BP$_2$." *Z. Kristallogr.*, 191(3–4), 311–312 (1990b).

Somer M., Peters K., Popp T., von Schnering H.G. "Crystal structure of tricaesium diarsenidogallate, Cs$_3$GaAs$_2$." *Z. Kristallogr.*, 192(3–4), 273–274 (1990c).

Somer M., Peters K., Thiery D., von Schnering H.G. "Crystal structure of trirubidium diphosphidogallate, Rb$_3$GaP$_2$." *Z. Kristallogr.*, 192(3–4), 271–272 (1990d).

Somer M., Popp T., Peters K., von Schnering H.G. "Crystal structure of caesium diarsenidoborate, Cs$_3$(BAs$_2$)." *Z. Kristallogr.*, 193(3–4), 295–296 (1990e).

Somer M., Popp T., Peters K., von Schnering H.G. "Crystal structure of caesium diphosphidoborate, Cs$_3$(BP$_2$)." *Z. Kristallogr.*, 193(3–4), 297–298 (1990f).

Somer M., Popp T., Peters K., von Schnering H.G. "Crystal structure of sodium diphosphidoborate, Na$_3$BP$_2$." *Z. Kristallogr.*, 193(3–4), 281–282 (1990g).

Somer M., Thiery D., Hartweg M., Walz L., Peters K., von Schnering H.G. "Crystal structure of caesium di-μ-phosphido-bis(phosphidogallate), Cs$_6$(Ga$_2$P$_4$)." *Z. Kristallogr.*, 193(3–4), 287–288 (1990h).

Somer M., Thiery D., Hartweg M., Walz L., Popp T., Peters K., von Schnering H.G. "Crystal structure of tricaesium diarsenidoaluminate, Cs$_3$AlAs$_2$." *Z. Kristallogr.*, 192(3–4), 269–270 (1990i).

Somer M., Walz L., Peters K., von Schnering H.G. "Crystal structure of potassium catena-di-μ-phosphidoaluminate di-μ-phosphido-bis(phosphidoaluminate), K$_{12}$(AlP$_2$)$_2$(Al$_2$P$_4$)." *Z. Kristallogr.*, 192(3–4), 301–302 (1990j).

Somer M., Walz L., Thiery D., von Schnering H.G. "Crystal structure of caesium di-μ-phosphido-bis(phosphidoaluminate), Cs$_6$(Al$_2$P$_4$)." *Z. Kristallogr.*, 193(3–4), 303–304 (1990k).

Sorokin V.P., Vesnina B.I., Klimova N.S. "New method of ammine-borane obtaining and its properties" [in Russian]. *Zhurn. Neorgan. Khimii*, 8(1), 66–68 (1963).

Sorokina O.V., Morgun A.I. "Investigation of gallium antimonide crystallization conditions in the Ga–Sb–Bi system" [in Russian]. *Zhurn. Neorgan. Khimii*, 30(12), 3174–3176 (1985).

Sorokina O.V., Yevgen'ev S.B., Mashchenko I.V. "Heterogeneous equilibria in the InSb–Tl system" [in Russian]. *Zhurn. Neorgan. Khimii*, 28(6), 1621–1623 (1983).

Sorokina O.V., Yevgen'ev S.B., Zhukov A.B. "Vapor pressure in the In–Sb–As system" [in Russian]. *Izv. AN SSSR Neorgan. Mater.*, 20(2), 200–203 (1984).

Sorokina O.V., Yevgen'ev S.B., Zhukov A.B. "In–Sb–Tl system" [in Russian]. *Zhurn. Neorgan. Khimii*, 30(10), 2642–2644 (1985).

Sowa H. "The crystal structure of AlAsO$_4$ at high pressure." *Z. Kristallogr.*, 194(3–4), 291–304 (1991).

Sowa H. "The crystal structure of GaPO$_4$ at high pressure." *Z. Kristallogr.*, 209(12), 954–960 (1994).

Sowa H., Macavei J., Schultz H. "The crystal structure of berlinite $AlPO_4$ at high pressure." *Z. Kristallogr.*, 192(1–4), 119–136 (1990).

Spencer P.J. "Computational thermochemistry: From its early CALPHAD days to a cost-effective role in materials development and processing." *CALPHAD*, 25(2), 163–174 (2001).

Spiesser M., Gruska R.P., Subbarao S.N., Castro C.A., Wold A. "Preparation and properties of two indium antimony selenides." *J. Solid State Chem.*, 26(2), 111–114 (1978).

Stadelmaier H.H., Fraker A.S. "Stickstofflegirungen der T-Metalle Mangan, Eisen, Kobalt and Nickel mit Gallium, Germanium, Indium und Zinn." *Z. Metallkde*, 53(1), 48–51 (1962).

Stadelmaier H.H., Yun T.S. "Stickstofflegirungen der T-Metalle Mangan, Eisen, Kobalt and Nickel mit Magnesium, Aluminium, Zink und Kadmium." *Z. Metallkde*, 52(7), 477–480 (1961).

Stanley J.M. "Hydrothermal growth of aluminum arsenate crystals." *Am. Mineral.*, 41(11–12), 947–952 (1956).

Stearns R.I., Greene P.E. "Phase stability regions of boron phosphides in the Ni–B–P system." *J. Electrochem. Soc.*, 112(12), 1239–1240 (1965).

Stegman R.L., Peretti E.A. "An investigation of the indium telluride–indium antimonide quasibinary phase diagram." *J. Inorg. Nucl. Chem.*, 28(8), 1589–1592 (1966a).

Stegman R.L., Peretti E.A. "The ternary system indium telluride–indium antimonide–antimony." *J. Chem. Eng. Data*, 11(4), 496–500 (1966b).

Steininger J. "Thermodynamics and calculation of the liquidus-solidus gap in homogeneous, monotonic alloy systems." *J. Appl. Phys.*, 41(6), 2713–2724 (1970).

Stone P.E., Egan Jr. E.P., Lehr J.R. "Phase relationships in the system $CaO–Al_2O_3–P_2O_5$." *J. Am. Ceram. Soc.*, 39(3), 89–98 (1956).

Storonkin A.V., Krivousov V.Yu., Vasil'kova I.V. "Thermodynamic investigation and calculation of some metallic systems" [in Russian]. *Zhurn. Fiz. Khimii*, 60(12), 2927–2931 (1986).

Stoyko S.S., Voss L.H., He H., Bobev S. "Synthesis, crystal and electronic structures of the pnictides AE_3TrPn_3 (AE = Sr, Ba; Tr = Al, Ga; Pn = P, As), *Crystals*, 5(4), 433–446 (2015).

Straumanis M.E., Kim C.D. "Solid solubility in the system GaSb–GaAs." *J. Electrochem. Soc.*, 112(1), 112–113 (1965).

Straumanis M.E., Krumme J.-P., Rubenstein M. "Thermal expansion coefficients and lattice parameters between 10 and 65°C in the system GaP–GaAs." *J. Electrochem. Soc.*, 114(6), 640–641 (1967).

Strel'nikova I.A., Mirgalovskaya M.S. "On the coefficient of zinc distribution in aluminum antimonide" [in Russian]. *Izv. AN SSSR Neorgan. Mater.*, 1(1), 91–95 (1965a).

Strel'nikova I.A., Mirgalovskaya M.S. "Study of the phase equilibria in the Al–Sb–S system" [in Russian]. *Izv. AN SSSR Neorgan. Mater.*, 1(1), 96–99 (1965b).

Stringfellow G.B. "Calculation of the Ga–In–P ternary phase diagram using the quasi-chemical equilibrium model." *J. Electrochem. Soc.*, 117(10), 1301–1305 (1970).

Stringfellow G.B. "Calculation of ternary phase diagrams of III-V systems." *J. Phys. Chem. Solids*, 33(3), 665–677 (1972a).

Stringfellow G.B. "Calculation of the solubility and solid-gas distribution coefficient of N in GaP." *J. Electrochem. Soc.*, 119(12), 1780–1783 (1972b).

Stringfellow G.B. "Miscibility gaps in quaternary Ill/V alloys." *J. Cryst. Growth*, 58(1), 194–202 (1982a).

Stringfellow G.B. "Spinodal decomposition and clustering in III/V alloys." *J. Electron. Mater.*, 11(5), 903–918 (1982b).

Stringfellow G.B. "Miscibility gaps and spinodal decomposition in III/V quaternary alloys of the type $A_xB_yC_{1-x-y}D$." *J. Appl. Phys.*, 54(1), 404–409 (1983).

Stringfellow G.B., Greene P.E. "Calculation of III-V ternary phase diagrams: In–Ga–As and In–As–Sb." *J. Phys. Chem. Solids*, 30(7), 1779–1791 (1969).

Stringfellow G.B., Greene P.E. "Liquid phase epitaxial growth of $InAs_{1-x}Sb_x$." *J. Electrochem. Soc.*, 118(5), 805–810 (1971).

Strunz H. "Isotypie von Berlinit mit Quarz." *Z. Kristallogr.*, 103(1), 228–229 (1941).

Su X., Zhu Z., Wu C., Yin F., Wang J., Li Z. "Experimental investigation and thermodynamic calculation of the Zn–Al–Sb system." *Int. J. Mater. Res.*, 102(3), 241–247 (2011).

Subbarao U., Sarkar S., Gudelli V.K., Kanchana V., Vaitheeswaran G., Peter S.C. "$Yb_5Ga_2Sb_6$: A mixed valent and narrow-band gap material in the $RE_5M_2X_6$ family." *Inorg. Chem.*, 52(23), 13631–13638 (2013).

Sugiura T., Sugiura H., Tanaka A., Sukegawa T. "Low temperature phase diagram of In–Ga–P ternary system." *J. Cryst. Growth*, 49(3), 559–562 (1980).

Sugiyama K., Oe K. "Phase diagram of the $GaAs_{1-x}Sb_x$ system." *Jpn. J. Appl. Phys.*, 16(1), 197–198 (1977).

Sun H., Jhi S.-H., Roundy D., Cohen M.L., Louie S.G. "Structural forms of cubic BC_2N." *Phys. Rev. B*, 64(9), 094108_1–094108_6 (2001).

Sun J., Zhou X.-F., Qian G.-R., Chen J., Fan Y.-X., Wang H.-T., Guo X., He J., Liu Z., Tian Y. "Chalcopyrite polymorph for superhard BC_2N." *Appl. Phys. Lett.*, 89(15), 151911_1–151911_3 (2006).

Sun W.-Y., Huang Z.K., Tien T.-Y., Yen T.-S. "Phase relationships in the system Y–Al–O–N." *Mater. Lett.*, 11(3–4), 67–69 (1991).

Suriwong T., Kurosaki K., Thongtem S., Harnwunggmoung A., Sugahara T., Plirdpring T., Ohishi Y., Muta H., Yamanaka S. "Synthesis and high-temperature thermoelectric properties of Ni_3GaSb and Ni_3InSb." *J. Alloys Compd.*, 511(9), 4014–4017 (2011).

Sutopo. "Phase equilibria in the system Ga–Ni–Sb at 500°C." *Int. J. Eng. Technol.*, 11(2), 102–110 (2011a).

Sutopo. "Phase equilibria in the system In–Ni–Sb at 450°C." *J. Mater. Sci. Eng. B*, 1, 106–112 (2011b).

Swenson D., Chang Y.A. "Phase equilibria in the system In–Co–P at 450°C." *Mater. Sci. Eng. B*, 8(3), 225–230 (1991).

Swenson D., Chang Y.A. "Phase equilibria in the In–Pt–As system at 600°C." *Mater. Sci. Eng. B*, 22(2–3), 267–273 (1994).

Swenson D., Chang Y.A. "On the constitution of some Ga–M–P systems (where M represents Co, Rh, Ir, Ni or Pt)." *Mater. Sci. Eng. B*, 39(1), 52–61 (1996a).

Swenson D., Chang Y.A. "Phase equilibria in the In–Ni–As system at 600°C." *Mater. Sci. Eng. B*, 39(3), 232–240 (1996b).

Swenson D.J., Sutopo, Chang Y.A. "Phase equilibria in the system In–Ir–As at 600°C." *Z. Metallkde*, 85(4), 228–231 (1994a).

Swenson D., Sutopo, Chang Y.A. "Phase equilibria in the In–Rh–As systems at 600°C." *J. Alloys Compd.*, 216(1), 67–73 (1994b).

Szapiro S. "Calculation of the phase diagram of the Ga–In–Sb system in the (Ga + In)-rich region." *J. Phys. Chem. Solids*, 41(3), 279–290 (1980).

Tabary P., Servant C. "Thermodynamic reassessment of the AlN–Al$_2$O$_3$ system." *CALPHAD*, 22(2), 179–201 (1998).

Tabary P., Servant C. "Crystalline and microstructure study of the AlN–Al$_2$O$_3$ section in the Al–N–O system. I. Polytypes and γ-AlON spinel phase." *J. Appl. Crystallogr.*, 32(2), 241–252 (1999).

Tabary P., Servant C., Alary J.A. "Effects of a low amount of C on the phase transformations in the AlN–Al$_2$O$_3$ pseudo-binary system." *J. Eur. Ceram. Soc.*, 20(12), 1915–1921 (2000a).

Tabary P., Servant C., Alary J.A. "Microstructure and phase transformations in the AlN–Al$_2$O$_3$ pseudo-binary system." *J. Eur. Ceram. Soc.*, 20(7), 913–926 (2000b).

Takayama T., Yuri M., Itoh K., Baba T., Harris Jr. J.S. "Analysis of unstable two-phase region in wurtzite Group III nitride ternary alloy using modified valence force field model." *Jpn. J. Appl. Phys.*, 39(9A), 5057–5062 (2000).

Takayama T., Yuri M., Itoh K., Baba T., Harris Jr. J.S. "Analysis of phase-separation region in wurtzite group III nitride quaternary material system using modifed valence force field model." *J. Cryst. Growth*, 222(1–2), 29–37 (2001).

Takeda Y., Hirai T., Hirao M. "Phase diagram for the pseudo-binary system germanium and gallium arsenide." *J. Electrochem. Soc.*, 112(3), 363–364 (1965).

Talis L.D., Karataev V.V., Mil'vidskiy M.G. "Calculation of the homogeneity region of indium arsenide doped with sulfur" [in Russian]. *Izv. AN SSSR Neorgan. Mater.*, 25(10), 1609–1613 (1989).

Tanaka A., Sugiura T., Sukegawa T. "Low temperature phase diagram of Ga–Al–P ternary system." *J. Cryst. Growth*, 60(1), 120–122 (1982).

Tanaka Y., Gür T.M., Kelly M., Hagstrom S.B., Ikeda T., Wakihira K., Satoh H. "Properties of (Ti$_{1-x}$Al$_x$)N coating tools prepared by the cathodic arc ion plating method." *J. Vac. Sci. Technol.*, A10(4), 1749–1756 (1992).

Taniyasu Y., Kasu M., Kobayashi N. "Lattice parameters of wurtzite Al$_{1-x}$Si$_x$N ternary alloys." *Appl. Phys. Lett.*, 79(26), 4351–4353 (2001).

Tao J.G., Yao B., Yang J.H., Zhang S.J., Zhang K., Bai S.Z., Ding Z.H., Wang W.R. "Mechanism of formation of Fe–N alloy in the solid-state reaction process between iron and boron nitride." *J. Alloys Compd.*, 384(1–2), 268–273 (2004).

Teles L.K., Scolfaro L.M.R., Leite J.R., Furthmüller J., Bechstedt F. "Phase diagram, chemical bonds and gap bowing of cubic In$_x$Al$_{1-x}$N alloys: *Ab initio* calculations." *Appl. Phys. Lett.*, 92(12), 7109–7113 (2002a).

Teles L.K., Scolfaro L.M.R., Leite J.R., Furthmüller J., Bechstedt F. "Spinodal decomposition in B$_x$Ga$_{1-x}$N and B$_x$Al$_{1-x}$N alloys." *Appl. Phys. Lett.*, 80(7), 1177–1179 (2002b).

Testova N.A., Golubenko A.N., Kokovin G.A., Sysoev S.V. "Prediction of the phase composition of the transition layers formed at the interface between the gallium arsenide and nickel" [in Russian]. *Izv. AN SSSR Neorgan. Mater.*, 22(11), 1781–1785 (1986).

Thompson A.G., Wagner J.W. "Preparation and properties of InAs$_{1-x}$P$_x$ alloys." *J. Phys. Chem. Solids*, 32(11), 2613–2619 (1971).

Thong N., Schwarzenbach D. "The use of electric field gradient calculations in charge density refinements. II. Charge density refinement of the low-quartz structure of aluminum phosphate." *Acta Crystallogr.*, A35(4), 658–664 (1979).

Thurmond C.D., Schwartz G.P., Kammlott G.W., Schwartz B. "GaAs oxidation and the Ga–As–O equilibrium phase diagram." *J. Electrochem. Soc.*, 127(6), 1366–1371 (1980).

Tisch U., Finkman E., Salzman J. "The anomalous bandgap bowing in GaAsN." *Appl. Phys. Lett.*, 81(3), 463–465 (2002).

Tixier S., Adamchyk M., Tiedje T., Francoeur S., Mascarenhas A., Wei P., Schiettekatte F. "Molecular beam epitaxy growth of GaAs$_{1-x}$Bi$_x$." *Appl. Phys. Lett.*, 82(14), 2245–2247 (2003).

Tkachuk A.V., Tam T., Mar A. "Ce$_6$ZnBi$_{14}$ and Pr$_6$InSb$_{15}$: Ternary rare-earth intermetallics with extended pnicogen ribbons." *Chem. Met. Alloys*, 1(1), 76–83 (2008).

Toberer E.S., Zevalkink A., Crisosto N., Snyder G.J. "The Zintl compound Ca$_5$Al$_2$Sb$_6$ for low-cost thermoelectric power generation." *Adv. Funct. Mater.*, 20(24), 4375–4380 (2010).

Todorov I., Chung D.Y., Ye L., Freeman A.J., Kanatzidis M.G. "Synthesis, structure and charge transport properties of Yb$_5$Al$_2$Sb$_6$: A Zintl phase with incomplete electron transfer." *Inorg. Chem.*, 48(11), 4768–4776 (2009).

Tokaychuk I., Tokaychuk Y., Gladyshevskii R. "The ternary system Hf–Ga–Sb at 600°C." *Chem. Met. Alloys*, 6(1–2), 75–80 (2013).

Tokaychuk I., Tokaychuk Y., Gladyshevskii R.E. "Crystal structure of Hf$_2$GaSb$_3$." *Solid State Phenomena*, 194, 1–4 (2012).

Tombs N.C., Fitzgerald J.F., Croft W.J. "Solid solution in the gallium arsenide–indium arsenide system." *Inorg. Chem*, 2(5), 1073–1074 (1963).

Trumbore F.A., Freeland P.E., Mills A.D. "Extent of solid solution in the GaSb–InSb system from crystal pulling experiments." *J. Electrochem. Soc.*, 109(7), 645–647 (1962).

Tsai C.T., Williams R.S. "Solid phase equilibria in the Au–Ga–As, Au–Ga–Sb, Au–In–As, and Au–In–Sb ternaries." *J. Mater. Res.*, 1(2), 352–360 (1986).

Tuck B. "The GaAs/Zn solidus at 1000°C." *J. Phys. D Appl. Phys.*, 9(14), 2061–2073 (1976).

Tuck B., Jay P.K. "Application of the phase diagram to the diffusion of Zn in GaP." *J. Phys. D Appl. Phys.*, 9(15), 2225–2238 (1976).

Tu H., Yin F., Su X., Liu Y., Wang X. "Experimental investigation and thermodynamic modeling of the Al–P–Zn ternary system." *CALPHAD*, 33(4), 755–760 (2009).

Turkevich V.Z. "Phase diagram and synthesis of cubic boron nitride." *J. Phys. Condens. Matter*, 14(44), 10963–10968 (2002).

Tyurin A.V., Gribchenkova N.A., Guskov V.N., Gavrichev K.S. "Thermodynamic properties of aluminum oxynitride from 0 to 340 K." *Inorg. Mater.*, 51(4), 340–344 (2015).

Ufimtsev V.B., Gimel'farb F.A. "Investigation of chemical interaction in the heavy doped III-V crystals" [in Russian]. *Izv. AN SSSR Neorgan. Mater.*, 9(12), 2073–2078 (1973).

Ufimtsev V.B., Shumilin V.P., Krestovnikov A.N., Uglichina G.N. "p-T-X phase diagram and the thermodynamics of the indium phosphide–gallium phosphide system" [in Russian]. *Dokl. AN SSSR*, 193(3), 602–604 (1970).

Ufimtsev V.B., Timoshin A.S., Kostin G.V. "Intermolecular interaction in the InSb–GaSb system" [in Russian]. *Izv. AN SSSR Neorgan. Mater.*, 7(11), 2090–2091 (1971).

Ufimtsev V.B., Zinov'ev V.G., Raukhman M.R. "Heterogeneous equilibria in the In–Sb–Bi system and liquid-phase epitaxy of the solid solutions based on InSb" [in Russian]. *Izv. AN SSSR Neorgan. Mater.*, 15(10), 1740–1743 (1979).

Ugay Ya.A., Bityutskaya L.A., Goncharov E.G., Belousova G.P. "Investigation of p-T-x-phase diagram In–InP and InP–InAs" [in Russian]. *Izv. AN SSSR Neorgan. Mater.*, 6(6), 1179–1181 (1970a).

Ugay Ya.A., Goncharov E.G., Kitina Z.V., Bityutskaya L.A. "Investigation of p-T-x-phase diagram of the In–InAs–InP system" [in Russian]. In *Khimiya Fosfidov s Poluprovodn. Svoystvami*. Novosibirsk: Nauka Publish., 42–46 (1970b).

Ugay Ya.A., Goncharov E.G., Kitina Z.V., Shvyreva T.N. "On the InAs–InP phase diagram" [in Russian]. *Izv. AN SSSR Neorgan. Mater.*, 4(3), 348–351 (1968).

Ugay Ya.A., Grekova I.I., Semenova G.V., Goncharov E.G., Kalyuzhnaya M.I. "Obtaining of the solid solution crystals in the In–As–P system" [in Russian]. *Izv. AN SSSR Neorgan. Mater.*, 27(8), 1560–1562 (1991).

Ugay Ya.A., Murav'eva S.N., Goncharov E.G., Afinogenov Yu.P. "Interaction in the InP + As = InAs + P system" [in Russian]. *Tr. Voronezh. Un-Ta*, 74, 29–33 (1971).

Ugay Ya.A., Samoylov A.M., Semenova G.V., Goncharov E.G., Abramova A.G. "InAs–InP system" [in Russian]. *Zhurn. Neorgan. Khimii*, 30(8), 2112–2115 (1985).

Ugay Ya.A., Semenova G.V., Goncharov E.G., Grekova I.I., Kalyuzhnaya M.I. "Vapor composition and thermodynamic analysis of the InAs–InP system" [in Russian]. *Zhurn. Neorgan. Khimii*, 31(3), 775–778 (1986).

Vakiv M.M., Krukovskyi S.I., Krukovskyi R.S. "Investigation of the phase equilibria in the Bi–InSb system" [in Ukrainian]. *Visnyk Nats. Un-Tu "L'vivs'ka Politekhnika" Elektronika*, (708), 105–109 (2011).

van Hook H.J., Lenker E.S. "The system InAs–GaAs." *Trans. Metallurg. Soc. AIME*, 227(1), 220–222 (1963).

van Nguyen Mau A., Ance C., Bougnot G. "Croissance par épitaxie en phase liquide et diagramme de phase du système Ga$_{1-x}$Al$_x$Sb." *J. Cryst. Growth*, 36(2), 273–277 (1976).

van Schilfgaarde M., Chen A.-B., Krishnamurthy S., Sher A. "InTlP—A proposed infrared detector material." *Appl. Phys. Lett.*, 65(21), 2714–2716 (1994).

van Schilfgaarde M., Sher A., Chen A.-B. "InTlSb: An infrared detector material?" *Appl. Phys. Lett.*, 62(16), 1857–1859 (1993).

van Thyne R.J., Kessler H.D. "Influence of oxygen, nitrogen and carbon on the phase relationships in the Ti–Al System." *Trans. AIME, J. Met.*, (2), 193–199 (1954).

Varfolomeev M.B., Sotnikova M.N., Chidirova F.Ch., Shpinel V.S. "Some structural differencies of double oxides formed by indium and gallium with Sb^V" [in Russian]. *Zhurn. Neorgan. Khimii*, 20(5) 1163–1166 (1975a).

Varfolomeev M.B., Sotnikova M.N., Plyushchev V.E., Strizhkov B.V. "Reactions of indium oxide with antimony oxides" [in Russian]. *Izv. AN SSSR Neorgan. Mater.*, 11(8), 1416–1419 (1975b).

Vasil'yev A.P., Vyatkin A.P. "Phase diagram of the gallium arsenide–tin system" [in Russian]. *Izv. Vyssh. Uchebn. Zaved. Fizika*, (3), 152–153 (1965).

Vasil'yev E.A., Leusenko A.A., Mazurenko A.M., Rakitskiy E.B., Shablovskiy A.V. "Investigation of the cBN polycrystals synthesized from α-BN with iron addition using NMR method" [in Russian]. *Sverkhtverdyye Materialy*, (6), 5–7 (1982).

Vasil'yev M.G., Vigdorovich V.N., Selin A.A., Khanin V.A. "Solibility of III-V compounds in tin" [in Russian]. *Izv. AN SSSR Neorgan. Mater.*, 19(11), 1933–1934 (1983).

Vasil'yev M.G., Vigdorovich V.N., Selin A.A., Khanin V.A. "Phase equilibria in the Sn–In–P, Sn–Ga–In–As and Sn–Ga–In–As–P systems" [in Russian]. In *Legir. Poluprovodn. Materialy*. Moscow: Nauka Publish., 61–65 (1985).

Vasil'yev V.P., Legendre B. "Thermodynamic properties of alloys and phase equilibria in the In–Sn–Sb system." *Inorg. Mater.*, 43(8), 803–815 (2007).

Vassilev V., Aljihmani L., Milanova V., Hristova-Vasileva T. "Phase equilibria in the Sb$_2$Te$_3$–InSb system." *J. Phase Equilibria Diffus.*, 37(5), 524–531 (2016).

Vassiliev V., Borzone G., Gambino M., Bros J.P. "Thermodynamic properties of the ternary system InSb–NiSb–Sb in the temperature range 640–860 K." *Intermetallics*, *11*(11–12), 1211–1215 (2003).

Vassiliev V., Feutelais Y., Sghaier M., Legendre B. "Thermodynamic investigation in In–Sb, Sb–Sn and In–Sb–Sn liquid systems." *J. Alloys Compd.*, *314*(1–2), 198–205 (2001).

Vaughey J.T., Corbett J.D. "Synthesis and structure of $Ca_{14}GaP_{11}$ with the new hypervalent P_3^{7-} anion. Matrix effects within the family of isostructural alkaline-earth metal pnictides." *Chem. Mater.*, *8*(3), 671–675 (1996).

Vaypolin A.A., Sinitsyn M.A., Yakovenko A.A. "Ga_4InAs_3 superstructure" [in Russian]. *Fiz. Tv. Tela*, *43*(4), 594–597 (2001).

Vecher A.A., Mechkovskii L.A., Poluyan A.F., Skoropanov A.S. "Enthalpies of formation of indium–tin–antimony melts." *Russ. J. Phys. Chem.*, *51*(3), 413–414 (1977).

Vecher A.A., Voronova E.I., Mechkovskiy L.A., Skoropanov A.S. "Determination of mixing enthalpy in the Ga–In–Sb and Bi–Sn–Sb systems through quantitative differential thermal analysis" [in Russian]. *Zhurn. Fiz. Khimii*, *48*(4), 1007–1009 (1974).

Venugopalan H.S., Mohney S.E. "Condensed phase equilibria in the Mo–Ga–N system at 800°C." *Z. Metallkde.*, *89*(3), 184–186 (1998).

Verdier P., L'Haridon P., Maunaye M., Laurent Y. "Etude structurale de $Ca_5Ga_2As_6$." *Acta Crystallogr.*, *B32*(3), 726–728 (1976).

Verdier P., L'Haridon P., Maunaye M., Marchand R. "Etude structurale de CaGaN." *Acta Crystallogr.*, *B30*(1), 226–228 (1974).

Verdier P., Marchand R., Lang J. "Un nouveau nitrure double contenant du gallium: Mg_3GaN_3." *C. R. Acad. Sci.*, *C271*(16), 1002–1004 (1970).

Verdier P., Marchand R., Lang J. "Préparation et étude d'un nitrure de formule CaGaN." *C. R. Acad. Sci.*, *C276*(7), 607–609 (1973).

Verdier P., Maunaye M., Marchand R. "Préparation et caractérisation d'arséniures ternaires dans le système calcium–gallium–arsenic." *Rev. Chim. Minér.*, *12*(5), 433–439 (1975a).

Verdier P., Maunaye M., Marchand R., Lang J. "Préparation de deux composés dans le système Ca–Ga–As: $Ca_5Ga_2As_6$ et $Ca_4Ga_3As_5$." *C. R. Acad. Sci.*, *C281*(12), 457–459 (1975b).

Vezin V., Yatagai S., Shiraki H., Uda S. "Growth of $Ga_{1-x}B_xN$ by molecular beam epitaxy." *Jpn. J. Appl. Phys.*, *36*(11), Pt. 2, L1483–L1485 (1997).

Vishnevskiy A.S., Delevi V.G., Mukovoz Yu.A., Ositinskaya T.D., Chapalyuk V.P. "Interaction of the boron nitride with steels and titanium" [in Russian]. *Sintetich. Almazy*, (4), 17–22 (1978).

Vogel R., Klose H. "Das Zustandsschaubild Eisen–Eisenphosphid–Aluminiumphosphid–Aluminium." *Arch. Eisenhuettenwes.*, *23*(7–8), 287–291 (1952).

von Schnering H.G., Somer M., Hartweg M., Peters K. "$[BP_2]^{3-}$ and $[BAs_2]^{3-}$, Zintl-anions with propadiene structure." *Angew. Chem. Int. Ed. Engl.*, *29*(1), 65–67 (1990a).

von Schnering H.G., Somer M., Hartweg M., Peters K. "$[BP_2]^{3-}$ und $[BAs_2]^{3-}$, Zintl-Anionen mit Propadienstruktur." *Angew. Chem.*, *102*(1), 63–64 (1990b).

von Schnering H.G., Somer M., Peters K., Blase W., Cordier G. "Crystal structure of caesium triantimonidoaluminate, Cs_6AlSb_3." *Z. Kristallogr.*, *193*(3–4), 283–284 (1990c).

von Schnering H.G., Somer M., Walz L., Peters K., Cordier G., Blase W. "Crystal structure of potassium catena-di-μ-arsenidoaluminate di-μ-arsenido-bis(arsenidoaluminate), $K_{12}(AlAs_2)_2(Al_2As_4)$." *Z. Kristallogr.*, *193*(3–4), 299–300 (1990d).

Voroshilov Yu.V., Gebesh V.Yu., Potoriy M.V. "Phase equilibria in the In–P–Se system and crystal structure of β-$In_4(P_2Se_6)_3$" [in Russian]. *Izv. AN SSSR Neorgan. Mater.*, *27*(12), 2495–2498 (1991).

Voroshilov Yu.V., Potoriy M.V., Milyan P.M., Milyan Zh.I. "Crystal-chemistry analysis of ternary phases in the Me–P–S(Se) systems, where Me – Ag, Cu, Zn, Cd, In, Tl, Sn, Pb" [in Ukrainian]. *Fizyka Khimiya Tv. Tila*, *17*(3), 412–420 (2016).

Voznyak I., Tokaychuk Ya., Gladyshevskii R. "Crystal structure of the $Hf_5Ga_{1.5}Sb_{1.5}$ compound" [in Ukrainian]. *Visnyk L'viv Un-Tu Ser. Khim.*, (53), 109–114 (2012).

Wang F., Li S.F., Sun Q., Jia Y. "First-principles study of structural and electronic properties of zincblende $Al_xIn_{1-x}N$." *Solid State Sci.*, *12*(9), 1641–1644 (2010).

Wang G., Yuan W.X., Jian J.K., Bao H.Q., Wang J.F., Chen X.L., Liang J.K. "Growth of GaN single crystals by Ca_3N_2 flux." *Scr. Mater.*, *58*(4), 319–322 (2008).

Wang J., Chen Z.-Q., Li C.-M., Li F., Nie C.-Y. "Electronic structures, elastic properties, and minimum thermal conductivities of cermet M_3AlN." *J. Solid State Chem.*, *216*, 1–8 (2014).

Wang J., Yuan W., Li M. "Thermodynamic modeling of the Ga–N–Na system." *J. Cryst. Growth*, *307*(1), 59–65 (2007).

Wang J., Zhou Y., Lin Z., Liao T. "Pressure-induced polymorphism in Al_3BC_3: A first-principles study." *J. Solid State Chem.*, *179*(8), 2703–2707 (2006).

Wang K., Gu Y., Zhou H.F., Zhang L.Y., Kang C.Z., Wu M.J., Pan W.W., Lu P.F., Gong Q., Wang S.M. "InPBi single crystals grown by molecular beam epitaxy." *Sci. Rep.*, *4*, 5449 (6 pp.) (2014).

Wang W.J., Chen X.L., Song Y.T., Yuan W.X., Cao Y.G., Wu X. "Assessment of Li–Ga–N ternary system and GaN single crystal growth." *J. Cryst. Growth*, *264*(1–3), 13–16 (2004).

Wei C.H., Edgar J.H. "Unstable composition region in the wurtzite $B_{1-x-y}Ga_xAl_yN$ system." *J. Cryst. Growth*, *208*(1–4), 179–182 (2000).

Wei S.-H., Ferreira L.G., Zunger A. "First-principles calculation of temperature-composition phase diagrams of semiconductor alloys." *Phys. Rev. B*, *41*(12), 8240–8269 (1990).

Wei X., Liu H., Zeng L. "Phase equilibria in the Al–Nd–Sb system at 500°C." *J. Rare Earths*, *25*(Suppl. 2), 218–220 (2007).

Weiss A., Schäfer H. "Zur Kenntnis von Aluminiumthiophosphat AlPS$_4$." *Naturwissenschaften*, *47*(21), 495 (1960).

Weiss A., Schäfer H. "Zur Kenntnis von Bortetrathiophosphat BPS$_4$." *Z. Naturforsch.*, *18B*(1), 81–82 (1963).

Weiss H. "Über die elektrischen Eigenschaften von Mischkristallen der Form In(As$_y$P$_{1-y}$)." *Z. Naturforsch.*, *11A*(6), 430–434 (1956).

Weitzer F., Remschnig K., Schuster J.C., Rogl P. "Phase equilibria and structural chemistry in the ternary systems M–Si–N and M–B–N (M = Al, Cu, Zn, Ag, Cd, In, Sn, Sb, Au, Tl, Pb, Bi)." *J. Mater. Res.*, *5*(10), 2152–2159 (1990).

Wen H. "Thermodynamische Berechungen der Konstitution des Systemes Al–B–C–N–Si–Ti." Thesis, Univ. Stuttgart, 1–183 (1993).

Westrum Jr. E.F., Levitin N.E. "Ammonia–triborane. The heat capacity, heat of transition and thermodynamic properties from 5 to 317°K." *J. Am. Chem. Soc.*, *81*(14), 3544–3547 (1959).

Weyers M., Sato M. "Growth of GaAsN alloys by low-pressure metalorganic chemical vapor deposition using plasma-cracked NH$_3$." *Appl. Phys. Lett.*, *62*(12), 1396–1398 (1993).

Wiberg E., Michaud H. "Zur Kenntnis eines ätherlöslichen Lithiumborazids LiB(N$_3$)$_4$." *Z. Naturforsch.*, *9B*(7), 499 (1954).

Wiberg N., Joo W.-Ch., Schmid K.H. "Uber einige Azide des Berylliums, Magnesiums, Bors und Aluminiums (Zur Reaktion von Silylaziden mit Elementhalogeniden)." *Z. Anorg. Allg. Chem.*, *394*(1–2), 197–208 (1972).

Widany J., Verwoerd W.S., Frauenheim Th. "Density-functional based tight-binding calculations on zinc-blende type BC$_2$N-crystals." *Diamond Relat. Mater.*, *7*(11–12), 1633–1638 (1998).

Willems H.X., de With G., Metselaar R., Helmholdt R.B., Petersen K.K. "Neutron diffraction of γ-aluminium oxynitride." *J. Mater. Sci. Lett.*, *12*(18), 1470–1472 (1993).

Willems H.X., Hendrix M.M.R.M., de With G., Metselaar R. "Thermodynamics of Alon. II: Phase relations." *J. Eur. Ceram. Soc.*, *10*(4), 339–346 (1992a).

Willems H.X., Hendrix M.M.R.M., Metselaar R., de With G. "Thermodynamics of Alon. I: Stability at lower temperatures." *J. Eur. Ceram. Soc.*, *10*(4), 327–337 (1992b).

Williams D., Kouvetakis J., O'Keeffe M. "Synthesis of nanoporous cubic In(CN)$_3$ and In$_{1-x}$Ga$_x$(CN)$_3$ and corresponding inclusion compounds." *Inorg. Chem.*, *37*(18), 4617–4620 (1998).

Williams D., Pleune B., Kouvetakis J., Williams M.D., Andersen R.A. "Synthesis of LiBC$_4$N$_4$, BC$_3$N$_3$ and related C–N compounds of boron: New precursors to light element ceramics." *J. Am. Chem. Soc.*, *122*(32), 7735–7741 (2000).

Williams D.J., Partin D.E., Lincoln F.J., J. Kouvetakis, O'Keeffe M. "The disordered crystal structures of Zn(CN)$_2$ and Ga(CN)$_3$." *J. Solid State Chem.*, *134*(1), 164–169 (1997).

Winiarski M.J., Polak M., Scharoch P. "Anomalous band-gap bowing of AlN$_{1-x}$P$_x$ alloy." *J. Alloys Compd.*, *575*, 158–161 (2013).

Winiarski M.J., Scharoch P., Polak M.P. "First principles prediction of structural and electronic properties of Tl$_x$In$_{1-x}$N alloy." *J. Alloys Compd.*, *613*, 33–36 (2014).

Wipf S., Coles B.R. "A note on the tin–indium–antimony system." *J. Inst. Met.*, *90*, 216 (1961–1962).

Wise W.S., Loh S.E. "Equilibria and origin of minerals in the system Al$_2$O$_3$–AlPO$_4$–H$_2$O." *Am. Mineral.*, *61*(5–6), 409–413 (1976).

Wittig G., Bille H. "Über Aluminium-tricyanid und Lithium-aluminiumtetracyanid." *Z. Naturforsch.*, *6B*(4), 226 (1951).

Womelsdorf H., Meyer H.-J. "Zur Kenntnis der Struktur von Sr$_3$(BN$_2$)$_2$." *Z. Anorg. Allg. Chem.*, *620*(2), 262–265 (1994).

Woolley J.C. "Alloys of InSb with InTe, In$_2$Te$_3$ and tellurium." *J. Electrochem. Soc.*, *113*(5), 465–466 (1966).

Woolley J.C., Blazey K.W. "Some properties of GaSb–Ga$_2$Se$_3$ and GaSb–Ga$_2$Te$_3$ alloys." *J. Electrochem. Soc.*, *111*(8), 951–954 (1964).

Woolley J.C., Gillett C.M., Evans J.A. "Electrical and optical properties of GaAs–InAs alloys." *Proc. Phys. Soc.*, *77*(3), 700–704 (1961).

Woolley J.C., Keating P.N. "Solid solubility of In$_2$Se$_3$ in some compounds of zinc blende structure." *J. Less Common Metals*, *3*(3), 194–201 (1961a).

Woolley J.C., Keating P.N. "Some electrical and optical properties of InAs–In$_2$Se$_3$ and InSb–In$_2$Se$_3$ alloys." *Proc. Phys. Soc.*, *78*(5), 1009–1016 (1961b).

Woolley J.C., Lees D.G. "Equilibrium diagrams with InSb as one component." *J. Less Common Metals*, *1*(3), 192–198 (1959).

Woolley J.C., Pamplin B.R., Evans J.A. "Some electrical and optical properties of InSb–In$_2$Te$_3$ alloys." *J. Phys. Chem. Solids*, *16*(1–2), 138–143 (1960).

Woolley J.C., Smith B.A. "Solid solution in the GaAs–InAs system." *Proc. Phys. Soc.*, *B70*(1), 153–154 (1957).

Woolley J.C., Smith B.A. "Solid solution in A^{III}B^V compounds." *Proc. Phys. Soc.*, *72*(2), 214–223 (1958a).

Woolley J.C., Smith B.A. "Solid solutions in zinc blende type A$^{III}_2$B$^{VI}_3$ compounds." *Proc. Phys. Soc.*, *72*(5), 867–873 (1958b).

Woolley J.C., Smith B.A., Lees D.G. "Solid solution in the GaSb–InSb system." *Proc. Phys. Soc.*, *B69*(12), 1339–1343 (1956).

Woolley J.C., Warner J. "Preparation of InAs–InSb." *J. Electrochem. Soc.*, *111*(10), 1142–1145 (1964).

Wörle M., Meyer zu Altenschildesche H., Nesper R. "Synthesis, properties and crystal structures of α-Ca(BN$_2$)$_2$ and Ca$_{9+x}$(BN$_2$, CBN)$_6$—Two compounds with BN$_2^{3-}$ and CBN^{4-} anions." *J. Alloys Compd.*, 264(1–2), 107–114 (1998).

Wright A.F., Leadbetter A.J. "The structures of the β-cristobalite phases of SiO$_2$ and AlPO$_4$." *Phil. Mag.*, 31(6), 1391–1401 (1975).

Wu B., Birdwhistel L.T., Jun M.-J., O'Connor C.J. "Semiconducting behavior in the polymeric Zintl phase material K$_2$Ga$_2$Sb$_4$." *Bull. Korean Chem. Soc.*, 11(5), 464–466 (1990).

Wu C., Huang W., Su X., Peng H., Wang J., Liu Y. "Experimental investigation and thermodynamic calculation of the Al–Fe–P system at low phosphorus contents." *CALPHAD*, 38(1), 1–6 (2012).

Wu T.Y., Pearson G.L. "Phase diagram, crystal growth and band structure of In$_x$Ga$_{1-x}$As." *J. Phys. Chem. Solids*, 33(2), 409–415 (1972).

Wu Q.-H., Hu Q.-K., Luo X.-G., Yu D.-L., Li D.-C., He J.-L. "First-principles study of structural and electronic properties of layered B$_2$CN crystals." *Chin. Phys. Lett.*, 24(1), 180–183 (2007).

Wu Z.L., Pope D.P., Vitek V. "Ti$_2$NAl in L1$_2$ Al$_3$Ti-base alloys." *Metal. Mater. Trans.*, A26(3), 521–524 (1995).

Xia S.-Q., Hullmann J., Bobev S. "Gallium substitutions as a means to stabilize alkaline-earth and rare-earth metal pnictides with the cubic Th$_3$P$_4$ type: Synthesis and structure of A_7Ga$_2$Sb$_6$ (A = Sr, Ba, Eu)." *J. Solid State Chem.*, 181(8), 1909–1914 (2008).

Xia S.-Q., Hullmann J., Bobev S., Ozbay A., Nowak E.R., Fritsch V. "Synthesis, crystal structures, magnetic and electric transport properties of Eu$_{11}$InSb$_9$ and Yb$_{11}$InSb$_9$." *J. Solid State Chem.*, 180(7), 2088–2094 (2007).

Xu X.-W., Hu L., Yu X., Lu Z.-M., Fan Y., Li Y.-X., Tang C.-C. "*Ab initio* study of the electronic structure and elastic properties of Al$_5$C$_3$N." *Chin. Phys. B.*, 20(12), 126201_1–126201_6 (2011).

Yaghmaee M.S., Kaptay G. "The solubility of nitrogen and nitrides in ternary liquid iron alloys. The limits of the 'solubility product' concept." *Mater. Sci. Forum*, 414–415, 491–501 (2003).

Yaghmaee M.S., Kaptay G., Jánosfy G. "Equilibria in the ternary Fe–Al–N system." *Mater. Sci. Forum*, 329–330, 519–524 (2000).

Yakimova R., Hitova L. "Influence of tin on InAs dissolution and epitaxy in In–Sn melts." *Thin Solid Films*, 209(1), 26–31 (1992).

Yakusheva N.A., Chikichev S.I. "Solubility of gallium arsenide in bismuth-gallim melts" [in Russian]. *Izv. AN SSSR Neorgan. Mater.*, 23(10), 1607–1609 (1987).

Yamada K., Kato E. "Mass spectrometric determination of activities of phosphorus in liquid Fe–P–Si, Al, Ti, V, Cr, Co, Ni, Nb and Mo alloys" [in Japanese]. *Tetsu-to-Hagane (J. Iron Steel Inst. Jpn.)*, 65(2), 273–280 (1979).

Yamada K., Kato E. "Effect of dilute concentrations of Si, Al, Ti, V, Cr, Co, Ni, Nb and Mo on the activity coefficient of P in liquid iron." *Trans. Iron Steel Inst. Jpn.*, 23(1), 51–55 (1983).

Yamaguchi G., Yanagida H. "Study on the reductive spinel—A new spinel formula AlN–Al$_2$O$_3$ instead of the previous one Al$_3$O$_4$." *Bull. Chem. Soc. Jpn.*, 32(11), 1264–1265 (1959).

Yamamoto A., Hasan Md.T., Kodama K., Shigekawa N., Kuzuhara M. "Growth temperature dependent critical thickness for phase separation in thick (~1 μm) In$_x$Ga$_{1-x}$N (x = 0.2–0.4)." *J. Cryst. Growth*, 419, 64–68 (2015).

Yamamoto A., Hasan Md.T., Mihara A., Narita N., Shigekawa N., Kuzuhara M. "Phase separation of thick (~1 μm) In$_x$Ga$_{1-x}$N (x ~ 0.3) grown on AlN/Si(111): Simultaneous emergence of metallic In–Ga and GaN-rich InGaN." *Appl. Phys. Express*, 7(3), 035502_1–035502_4 (2014).

Yamane H., DiSalvo F.J. "Ba$_3$Ga$_2$N$_4$." *Acta Crystallogr.*, C52(4), 760–761 (1996).

Yamane H., Kikkawa S., Horiuchi H., Koizumi M. "Structure of a new polymorph of lithium boron nitride, Li$_3$BN$_2$." *J. Solid State Chem.*, 65(1), 6–12 (1986).

Yamane H., Kikkawa S., Koizumi M. "Lithium aluminium nitride, Li$_3$AlN$_2$ as a lithium solid electrolyte." *Solid State Ionics*, 15(1), 51–54 (1985).

Yamane H., Kikkawa S., Koizumi M. "High- and low-temperature phases of lithium boron nitride, Li$_3$BN$_2$: Preparation, phase relation, crystal structure and ionic conductivity." *J. Solid State Chem.*, 71(1), 1–11 (1987).

Yamane H., Sasaki S., Kajiwara T., Yamada T., Shimada M. "Ba$_{19}$In$_9$N$_9$, a subnitride containing isolated [In$_5$]$^{5-}$ and [In$_8$]$^{12-}$ Zintl anions." *Acta Crystallogr.*, E60(10), i120–i123 (2004).

Yan Y.L., Wang Y.X. "Crystal structure, electronic structure, and thermoelectric properties of Ca$_5$Al$_2$Sb$_6$." *J. Mater. Chem.*, 21(33), 12497–12502 (2011).

Yan Y.L., Wang Y.X., Zhang G.B. "A key factor improving the thermoelectric properties of Zintl compounds A$_5$M$_2$Pn$_6$ (A = Ca, Sr, Ba; M = Ga, Al, In; Pn = As, Sb)." *Comput. Mater. Sci.*, 85, 88–93 (2014).

Yang X., Lin Z., Li Z., Wu L., Mao C. "GaP$_{1-x}$N$_x$ alloys formed by ion implantation." *J. Appl. Phys.*, 77(11), 5553–5557 (1995).

Yang Z.J., Li R., Linghu R.F., Cheng X.L., Yang X.D. "First-principle investigations on the structural dynamics of Ti$_2$GaN." *J. Alloys Compd.*, 574, 573–579 (2013).

Ye L., Wang Y.X., Yang J., Yan Y., Zhang J., Guo L., Feng Z. "Electronic structure and thermoelectric properties of the Zintl compounds Sr$_5$Al$_2$Sb$_6$ and Ca$_5$Al$_2$Sb$_6$: First-principles study." *RSC Adv.*, 5(63), 50720–50728 (2015).

Yegorov L.A., Doronin V.N., Medvedeva Z.S. "On the equilibrium vapor pressure in the InAs–InP–I$_2$ system" [in Russian]. *Izv. AN SSSR Neorgan. Mater.*, 6(11), 1944–1946 (1970).

Yegorov L.A., Kochnev V.N. "Equilibrium in the system indium arsenide–iodine" [in Russian]. *Izv. AN SSSR Neorgan. Mater.*, 5(8), 1373–1376 (1969).

Yemel'yanenko O.V., Ivkin E.B., Kotrubenko B.P., Lange V.N. "Electrical properties of InAsTe" [in Russian]. *Tr. Kishinevsk. Politekhn. In-T*, (12), 69–72 (1968).

Yevgen'ev S.B. "Investigation of the liquid–vapor phase equilibria in the In–P–Bi system" [in Russian]. *Zhurn. Neorgan. Khimii*, 32(11), 2781–2784 (1987).

Yevgen'ev S.B., Ganina N.V. "Phase equilibria in the Ga–Bi–GaAs system" [in Russian]. *Izv. AN SSSR Neorgan. Mater.*, 20(4), 561–562 (1984).

Yevgen'ev S.B., Kuz'michova G.M., Kholopova L.V. "Investigation of the phase equilibria liquid–vapor in the In–As–Bi system" [in Russian]. *Zhurn. Neorgan. Khimii*, 33(11), 2913–2917 (1988).

Yevgen'ev S.B., Le Van Khuan, Kharkevich S.I. "Investigation of the phase equilibria in the GaSb–Yb system" [in Russian]. *Izv. AN SSSR Neorgan. Mater.*, 26(6), 1323–1325 (1990).

Yevgen'ev S.B., Sorokina O.V., Mashchenko I.V. "Solubility of thallium in indium antimonide" [in Russian]. *Zhurn. Neorgan. Khimii*, 29(12), 3128–3130 (1984).

Yevgen'ev S.B., Sorokina O.V., Zinov'yev V.G. "Physicochemical behavior of thallium in indium antimonide" [in Russian]. *Izv. AN SSSR Neorgan. Mater.*, 21(12), 2000–2002 (1985).

Yi T., Cox C.A., Toberer E.S., Snyder G.F., Kauzlarich S.M. "High-temperature transport properties of the Zintl phases $Yb_{11}GaSb_9$ and $Yb_{11}InSb_9$." *Chem. Mater.*, 22(3), 935–941 (2010).

Yordanova I., Tret'yakov D.N. "Phase equilibrium of the Ga–Al–Sb system." *Cryst. Res. Technol.*, 17(12), 1469–1475 (1982).

Young A.B., Sclar C.B., Schwartz C.M. "Synthesis and properties of a high-pressure rutile-type $AlAsO_4$." *Z. Kristallogr.*, 118(3–4), 223–232 (1963).

Young D.M., Kauzlarich S.M. "Preparation, structure, and electronic properties of $Ca_{11}MSb_9$ (M = Al, Ga, In)." *Chem. Mater.*, 7(1), 206–209 (1995).

Yuge K. "Phase stability of boron carbon nitride in a heterographene structure: A first-principles study." *Phys. Rev. B*, 79(14), 144109_1–144109_6 (2009).

Yuge K., Seko A., Koyama Y., Oba F., Tanaka I. "First-principles-based phase diagram of the cubic BNC ternary system." *Phys. Rev. B*, 77(9), 094121_1–094121_6 (2008).

Yunusov M.S., Abdurakhmanov Yu.Yu., Mukhtarova N.N., Ob'edkov E.V. "Phase diagram of the Ga–In–Sb system in a region with a high content of antimony" [in Russian]. *Izv. AN SSSR Neorgan. Mater.*, 25(9), 1426–1429 (1989).

Zakharenkov L.F., Zykov L.F., Romanov V.V., Samorukov B.E. "Manganese behavior in indium phosphide" [in Russian]. *Izv. AN SSSR Neorgan. Mater.*, 19(8), 1245–1249 (1983).

Zakharenkov L.F., Zykov L.F., Samorukov B.E. "Doping of indium phosphide by the elements of IV group" [in Russian]. *Izv. AN SSSR Neorgan. Mater.*, 21(1), 10–13 (1985).

Zarechnyuk O.S., Tsygulya E.A. "Investigation of the phase equilibria in the Al-rich alloys of the Ce–Al–Sb system" [in Russian]. *Izv. AN SSSR Neorgan. Mater.*, 2(1), 210–211 (1966).

Zemskov V.S., Suchkova A.D., Zhurkin B.G. "Investigation of the heterogeneous equilibrium in the Ge–In–Sb system" [in Russian]. *Zhurn. Fiz. Khimii*, 36(9), 1914–1918 (1962).

Zeng L., He J., Yan J., He W. "The phase equilibria of the Dy–Al–Sb ternary system at 500°C." *J. Alloys Compd.*, 479(1–2), 173–179 (2009).

Zeng L., Liao J., Qin P., Qin H., Nong L. "Phase relations in the Al–Pr–Sb system at 773 K." *J. Alloys Compd.*, 450(1–2), 252–254 (2008).

Zeng L., Wang S. "The 800 K isothermal section of the Y–Al–Sb phase diagram." *J. Alloys Compd.*, 351, 176–179 (2003).

Zeng X.-F., Chung D.D.L. "Structural characterization of the interfacial reactions between palladium and gallium arsenide." *J. Vac. Sci. Technol.*, 21(2), 611–614 (1982).

Zevalkink A., Pomrehn G.S., Johnson S., Swallow J., Gibbs Z.M., Snyder G.J. "Influence of the triel elements (M = Al, Ga, In) on the transport properties of $Ca_5M_2Sb_6$ Zintl compounds." *Chem. Mater.*, 24(11), 2091–2098 (2012).

Zhang H., Yao S., Widom M. "Predicted phase diagram of boron–carbon–nitrogen." *Phys. Rev. B*, 93(14), 144107_1–144107_9 (2016).

Zhang L.G., Chen H.M., Dong H.Q., Huang G.X., Liu L.B., Jin Z.P. "Thermodynamic description of the Al–Sb–Y system." *J. Alloys Compd.*, 475(1–2), 233–237 (2009).

Zhang R.F., Veprek S. "Metastable phases and spinodal decomposition in $Ti_{1-x}Al_xN$ system studied by ab initio and thermodynamic modeling, a comparison with the $TiN–Si_3N_4$ system." *Mater. Sci. Eng. A*, 448(1–2), 111–114 (2007).

Zhang Y., Li J.-B., Liang J.K., Zhang Q., Sun B.J., Xiao Y.G., Rao G.H. "Subsolidus phase relations of the Cu–Ga–N system." *J. Alloys Compd.*, 438(1–2), 158–164 (2007a).

Zhang Y., Liang J.K., Li J.-B., Liu Q.L., Xiao Y.G., Zhang Q., Sun B.J., Rao G.H. "Phase relations of the Ag–Ga–N system." *J. Alloys Compd.*, 429(1–2), 184–191 (2007b).

Zhao J.C., Wu Z.Q. "Reactions in Pd/GaAs (001) contacts at 7×10^8 Pa pressure." *Appl. Phys. Lett.*, 58(4), 349–351 (1991).

Zhao J.T., Dong Z.C., Vaughey J.T., Ostenson J.E., Corbett J.D. "Synthesis, structures and properties of cubic R_3In and R_3InZ phases (R = Y, La; Z = B, C, N, O): The effect of interstitial Z on the superconductivity of La_3In." *J. Alloys Compd.*, 230(1), 1–12 (1995).

Zhao Y., He D.W., Daemen L.L., Shen T.D., Schwarz R.B., Zhu Y., Bish D.L., et al. "Superhard B–C–N materials synthesized in nanostructured bulks." *J. Mater. Res.*, 17(12), 3139–3145 (2002).

Zheng J.-C., Wang H.-Q., Wee A.T.S., Huan C.H.A. "Possible complete miscibility of $(BN)_x(C_2)_{1-x}$ alloys." *Phys. Rev. B*, 66(9), 092104_1–092104_4 (2002).

Zheng X.-Y., Lin J.-C., Hsieh K.-C., Chang Y.A. "Phase equilibria of Ga–Ni–As at 600°C and the structural relationship between γ-Ni$_3$Ga$_2$, γ′-Ni$_{13}$Ga$_9$ and T–Ni$_3$GaAs." *Mater. Sci. Eng. B*, 5(1), 63–72 (1989a).

Zheng X.-Y., Schulz K.J., Lin J.-C., Chang Y.A. "Solid state phase equilibria in the Pt–Ga–As system." *J. Less Common Metals*, 146(1–2), 233–239 (1989b).

Zhitar' V.F. "Interaction of indium arsenide-sulfide" [in Russian]. In *Materialy IV Konf. Molodykh Uchenykh Moldavii*, 1964. Sekts. Fiz.-Mat. Kishinev: Stiintsa Publish., 7–9 (1965).

Zhou P., Hu C., Liu Z., Wang F., Zhou M., Hu C., Zheng Z., Ji Y., Wang X. "First-principle study on thermodynamic property of superhard BC$_2$N under extreme conditions." *J. Mater. Res.*, 29(12), 1326–1333 (2014).

Zhou X.-F., Sun J., Fan Y.-X., Chen J., Wang H.-T. "Most likely phase of superhard BC$_2$N *ab initio* calculations." *Phys. Rev. B*, 76(10), 100101_1–100101_4 (2007).

Zhu Y., Wang J., Wang X., Peng H., Li X., Li Z., Su X. "Redetermination of the 450°C isothermal section of the Al–P–Zn ternary phase diagram." *J. Alloys Compd.*, 515, 161–165 (2012).

Zhu Z., Su X., Yin F., Wang J., Wu C. "Experimental investigation of the Zn–Al–Sb system at 450°C." *J. Phase Equilibria Diffus.*, 30(6), 595–601 (2009).

Zhukov A.N., Burdina K.P., Semenenko K.N. "Composition and polymorphism of magnesium boron nitride" [in Russian]. *Zhurn. Obshch. Khimii*, 64(3), 357–360 (1994).

Zhukov A.N., Burdina K.P., Semenenko K.N. "Investigation of the interaction in the AlN–Mg$_3$N$_2$ system at atmospheric and high pressures" [in Russian]. *Zhurn. Obshch. Khimii*, 66(7), 1073–1077 (1996).

Zhukov A.N., Burdina K.P., Semenenko K.N. "Interaction in the Mg$_3$N$_2$–BN system at high and atmospheric pressure" [in Russian]. *Zhurn. Obshch. Khimii*, 68(2), 180–184 (1998).

Zilko J.L., Greene J.E. "Growth of metastable InSb$_{1-x}$Bi$_x$ thin films by multitarget sputtering." *Appl. Phys. Lett.*, 33(3), 254–256 (1978).

Zinov'ev V.G., Morgun A.I., Ufimtsev V.B. "Behavior of bismuth in the GaSb<Bi> epitaxial layers" [in Russian]. *Neorgan. Mater.*, 29(2), 177–180 (1993).

Zitter R.N. "InSb–Sn system." *Trans. Metallurg. Soc. AIME*, 212(1), 31–32 (1958).

Živković D., Katayama I., Kostov A., Živković Ž. "Comparative thermodynamic investigation of the GaSb–Sn system." *J. Therm. Anal. Calorim.*, 71(2), 567–582 (2003).

Živković D., Minić D., Manasijević D., Talijan N., Balanović L.J., Mitovski A., Ćosović V., Rangelov I. "Phase diagram investigation and characterization of alloys in Bi–Ga$_{10}$Sb$_{90}$ section of Ga–Bi–Sb system." *J. Optoelectr. Adv. Mater.*, 12(6), 1262–1267 (2010).

Živković D., Živković Ž., Šestak J. "Predicting of the thermodynamic properties for the ternary system Ga–Sb–Bi." *CALPHAD*, 23(1), 113–131 (1999).

Živković D., Živković Ž, Stuparević L., Rančić S. "Comparative thermodynamic investigation of the Bi–GaSb system." *J. Therm. Anal. Calorim.*, 65(3), 805–819 (2000).

Živković D., Živković Ž., Vučinic B. "Comparative thermodynamic analysis of the Bi–Ga$_{0.1}$Sb$_{0.9}$ section in the Bi–Ga–Sb system." *J. Therm. Anal. Calorim.*, 61(1), 263–271 (2001).

Index

Printed and bound by CPI Group (UK) Ltd, Croydon, CR0 4YY

01/11/2024

01782600-0016